Lectures on Quantum Mechanics

Quantum mechanics is one of the principal pillars of modern physics. It also remains a topic of great interest to mathematicians. Since its discovery it has inspired and been inspired by many topics within modern mathematics, including functional analysis and operator algebras, Lie groups, Lie algebras and their representations, principal bundles, distribution theory, and much more.

Written with beginning mathematics graduate students in mind, this book provides a thorough treatment of nonrelativistic quantum mechanics in a style that is leisurely and without the usual theorem–proof grammar of pure mathematics, while remaining mathematically honest. The author takes the time to develop fully the required mathematics and employs a consistent mathematical presentation to clarify the often-confusing notation of physics texts. Along the way the reader encounters several topics requiring more advanced mathematics than found in many discussions of quantum mechanics. This all makes for a fascinating course in how mathematics and physics interact.

PHILIP L. BOWERS is the Dwight B. Goodner Professor of Mathematics at the Florida State University.

Lectures on Quantum Mechanics

Lectures on Quantum Mechanics

A Primer for Mathematicians

PHILIP L. BOWERS
Florida State University

CAMBRIDGE
UNIVERSITY PRESS

University Printing House, Cambridge CB2 8BS, United Kingdom

One Liberty Plaza, 20th Floor, New York, NY 10006, USA

477 Williamstown Road, Port Melbourne, VIC 3207, Australia

314–321, 3rd Floor, Plot 3, Splendor Forum, Jasola District Centre,
New Delhi – 110025, India

79 Anson Road, #06–04/06, Singapore 079906

Cambridge University Press is part of the University of Cambridge.

It furthers the University's mission by disseminating knowledge in the pursuit of education, learning, and research at the highest international levels of excellence.

www.cambridge.org
Information on this title: www.cambridge.org/9781108429764
DOI: 10.1017/9781108555241

First published 2020

A catalogue record for this publication is available from the British Library.

Library of Congress Cataloging-in-Publication Data
Names: Bowers, Philip L., 1956– author.
Title: Lectures on quantum mechanics : a primer for mathematicians.
Philip L. Bowers, Florida State University.
Description: New York : Cambridge University Press, [2020] |
Includes bibliographical references and index.
Identifiers: LCCN 2020007818 (print) | LCCN 2020007819 (ebook) |
ISBN 9781108429764 (hardback) | ISBN 9781108555241 (epub)
Subjects: LCSH: Nonrelativistic quantum mechanics. | Quantum theory.
Classification: LCC QC174.24.N64 B69 2020 (print) | LCC QC174.24.N64
(ebook) | DDC 530.12–dc23
LC record available at https://lccn.loc.gov/2020007818
LC ebook record available at https://lccn.loc.gov/2020007819

ISBN 978-1-108-42976-4 Hardback

Contents

Preface *page* xi
Prolegomenon xix

1 The Harmonic Oscillator: Classical versus Quantum 1
1.1 The Classical Harmonic Oscillator 3
1.2 The Quantum Mechanical Treatment 4
1.3 What Does It All Mean? 7
1.4 Foundational Issues 12
1.5 End Notes 20

2 The Mathematical Structure of Quantum Mechanics 24
2.1 The Minimalist Rules 25
2.2 Wave Mechanics 26
2.3 Adjoints and Self-Adjoint Operators 29
2.4 The Position and Momentum Operators 32
2.5 End Notes 35

3 Observables and Expectation Values 38
3.1 Elementary Properties of Expectation Values 39
3.2 Can a Quantum Observable have a Precise Value? 43
3.3 What Happens Upon Measurement? 44
3.4 The Measurement Problem 46
3.5 End Notes 46

4 The Projection Postulate Examined 48
4.1 The Physicist's Approach 49
4.2 The Mathematician's Rigor 52
4.3 An Important Class of Operators 56
4.4 The Spectrum of a Self-Adjoint Operator 58
4.5 End Notes 63

5 Rigged Hilbert Space and the Dirac Calculus 65

5.1 Gelfand Triples and the Rigging of $\mathcal{H}$ 67
5.2 The Position and Momentum Operators 69
5.3 Products of Bras and Kets 73
5.4 Spectral Decomposition and the Dirac Calculus 75
5.5 End Notes 78

6 A Review of Classical Mechanics 80

6.1 Newtonian Mechanics 81
6.2 Lagrangian Mechanics 84
6.3 Hamiltonian Mechanics and Poisson Brackets 88
6.4 Noether's Theorem 91
6.5 End Notes 93

7 Hamilton–Jacobi Theory ⋆ 95

7.1 Generalized Coordinates Reexamined 96
7.2 Canonical Transformations 99
7.3 The Hamilton–Jacobi Equation 104
7.4 Some Sample Applications 106
7.5 End Notes 109

8 Classical Mechanics Regain'd 111

8.1 The Quantum Evolution Equation 112
8.2 Commutation Relations and the Ehrenfest Theorems 114
8.3 Commuting Self-Adjoint Operators 118
8.4 The Baker–Hausdorff Formula 121
8.5 End Notes 123

9 Wave Mechanics I: Heisenberg Uncertainty 124

9.1 Statement of the Principle 125
9.2 Interpretation 127
9.3 Minimal-Uncertainty States 130
9.4 The Fourier Transform and Uncertainty 131
9.5 End Notes 133

10 Wave Mechanics II: The Fourier Transform 135

10.1 The Fourier Transform 136
10.2 Eigenvalues and Eigenfunctions: Hermite Functions 142
10.3 The Position and Momentum Representations 144
10.4 Wave Packets and Superposition 147
10.5 End Notes 152

11 Wave Mechanics III: The Quantum Oscillator 154

11.1 Ladder Operators and the Ground State 155

11.2	Higher Energy States	156
11.3	Generating Function and Completeness	159
11.4	Coherent States of the Oscillator	162
11.5	End Notes	169
12	**Angular Momentum I: Basics**	170
12.1	Angular Momentum Operators	171
12.2	Eigenvectors and Eigenvalues	173
12.3	Derivation	174
12.4	Further Remarks on Angular Momentum	177
12.5	End Notes	178
13	**Angular Momentum II: Representations of** $\mathfrak{su}(2)$	179
13.1	The Pauli Spin Matrices and the Lie Algebra $\mathfrak{su}(2)$	180
13.2	Angular Momentum and $\mathfrak{su}(2)$-Representations	183
13.3	The Quaternions $\mathbb{H}$, $\mathbf{S}^3$, SU(2), SO(3), and SO(4)	185
13.4	Representations of Lie Groups SU(2) and SO(3)	191
13.5	End Notes	195
14	**Angular Momentum III: The Central Force Problem**	197
14.1	Orbital Angular Momentum	198
14.2	Legendre Functions and Spherical Harmonics	200
14.3	Radial Symmetry and Representations	205
14.4	A Single Particle in a General Central Potential	210
14.5	End Notes	212
15	**Wave Mechanics IV: The Hydrogenic Potential**	214
15.1	An Algebraic Approach to the Radial Equation	216
15.2	Power Series and Hypergeometric Functions	220
15.3	The Full Solution for Bound State Electrons	226
15.4	The Unbound Electron in the Coulomb Potential	229
15.5	End Notes	231
16	**Wave Mechanics V: Hidden Symmetry Revealed**	233
16.1	Quantum Numbers, Degeneracy, and Fine Structure	234
16.2	The Laplace–Runge–Lenz Vector	239
16.3	Hidden Symmetry	243
16.4	Momentum Representation and SO(4) Symmetry	246
16.5	End Notes	251
17	**Wave Mechanics VI: Hidden Symmetry Solved**	252
17.1	Fock's Treatment of the Momentum Space Equation	253
17.2	Group Characters and Representations	257
17.3	The Momentum Space Equation and Characters	260

17.4 The Infinitesimal Generators of SO(4) ⋆ 264
17.5 End Notes 269

18 Angular Momentum IV: Addition Rules and Spin 270
18.1 Coupled Angular Momenta 271
18.2 The Selection Rules 273
18.3 Spin-$\frac{1}{2}$ Systems 275
18.4 Rotations of Wave Functions and Spin-$\frac{1}{2}$ Particles 279
18.5 End Notes 283

19 Wave Mechanics VII: Pauli's Spinor Theory 285
19.1 Tensor Products and Internal Degrees of Freedom 288
19.2 Action of the Double Cover of SO(3) on $L^2(\mathbb{R}^3)\otimes V^{1/2}$ 291
19.3 The Spin and Magnetic Moment of the Electron 295
19.4 The Hydrogenic Potential with Spin 297
19.5 End Notes 300

20 Clifford Algebras and Spin Representations ⋆ 301
20.1 Clifford Algebras 302
20.2 Low-Dimensional Algebras 309
20.3 The Groups Pin and Spin 314
20.4 Spin Representations and Spinors 320
20.5 End Notes 324

21 Many-Particle Quantum Systems 325
21.1 Multi-Particle States and Tensor Products 326
21.2 The Axiom for Multi-Component Systems 329
21.3 Coupled Angular Momenta, Again 331
21.4 A Mathematical Interlude: Bases for Tensor Products 333
21.5 End Notes 335

22 The EPR Argument and Bell's Inequalities 336
22.1 The EPR Criticism of Quantum Mechanics 339
22.2 The Coupled Spin-$\frac{1}{2}$ System in Quantum Mechanics 344
22.3 Bell Inequalities, Realism, and Nonlocality 348
22.4 The GHZ Scheme for Spin Triplets 355
22.5 End Notes 359

23 Ensembles and Density Operators 361
23.1 The Spin-$\frac{1}{2}$ System Revisited 361
23.2 Density Operators I: Finite-Dimensional Setting 364
23.3 Matrix Calculations in the Spin-$\frac{1}{2}$ System 370
23.4 Density Operators II: Infinite-Dimensional Setting 375
23.5 End Notes 378

24 Bosons and Fermions 379
24.1 Bosons and Fermions 381
24.2 N Indistinguishable Quanta 386
24.3 Algebraic Structure of the Tensor Algebra 392
24.4 Analytic Structure of the Tensor Algebra 393
24.5 End Notes 395

25 The Fock Space for Indistinguishable Quanta 397
25.1 Notation for Fock Space 399
25.2 Annihilation and Creation Operators 401
25.3 Bosonic Fock Space 402
25.4 Fermionic Fock Space 404
25.5 End Notes 406

26 An Introduction to Quantum Statistical Mechanics 407
26.1 Statistical Mechanics 408
26.2 The Most Probable Configuration 411
26.3 The Mechanics of Maximizing $Q(\mathbf{n})$ 413
26.4 The Fundamental Law and the Canonical Ensemble 417
26.5 End Notes 420

27 Quantum Dynamics 421
27.1 The Schrödinger Picture 422
27.2 Deriving the Schrödinger Equation: Stone's Theorem 426
27.3 The Heisenberg Picture and the Heisenberg Equation 428
27.4 Synthesis: The Dirac, or Interaction, Picture 431
27.5 End Notes 434

28 Unitary Representations and Conservation Laws 435
28.1 Euclidean Symmetries 436
28.2 Conservation Laws 440
28.3 Phase Transformations 442
28.4 Local Phase Symmetry Requires a Mediating Field 444
28.5 End Notes 450

29 The Feynman Formulation of Quantum Mechanics 451
29.1 A New Look at Quantum Mechanics 453
29.2 Feynman's Propagator Calculus 457
29.3 Evaluating Path Integrals: Quadratic Lagrangians 461
29.4 The Free Particle Propagator 464
29.5 End Notes 466

30 A Mathematical Interlude: Gaussian Integrals 467
30.1 Fresnel and Gaussian Integrals 468

30.2 The Complex Gaussian Integral in N Dimensions 471
30.3 Generalizing $F_N(a, c, d)$ 474
30.4 Hilbert Matrices and Cauchy Determinants 476
30.5 End Notes 478

31 Evaluating Path Integrals I 479
31.1 The Piecewise Linear Family 481
31.2 The Constant Force Propagator 484
31.3 The Filtration Measures for the Polynomial Family 485
31.4 From the Dirac Calculus to the Path Integral 489
31.5 End Notes 493

32 Evaluating Path Integrals II 495
32.1 The Proposal 496
32.2 The Harmonic Oscillator and Fourier Sums 499
32.3 The Harmonic Oscillator and Polynomial Sums ⋆ 505
32.4 The Forced Harmonic Oscillator 510
32.5 End Notes 513

Epilogue 515
Resources for Individual Exploration 517
Bibliography 537
Index 547

Preface

> ...I have written this work in order to learn the subject myself, in a form I find comprehensible. And readers familiar with some of my previous books probably realize that this has pretty much been the reasons for those works also ... Perhaps this travelogue of an innocent abroad in a very different field will also turn out to be a book that mathematicians will enjoy (though physicists probably will not).
>
> Michael Spivak
> in the Preface to *Physics for Mathematicians: Mechanics I*
> (2010)

Apologia

I imagine many authors write first and foremost for themselves. I expect that, often, writing a textbook on a technical subject is the author's way of mastering the subject matter, of coming to grips with technical difficulties, and of arranging topics of keen interest to himself or herself, certainly with the reader in mind but motivated even more so by his or her own desire for ordering and framing a difficult subject. I must confess guilt in these matters. This book ultimately is written to myself of 40 years ago when I was a graduate student in mathematics with several core graduate courses in pure mathematics under my belt.

I was originally a physics major as an undergraduate. Like many of my colleagues who ended up in pure mathematics, I ran from the messy mathematical imprecision of physics to the elegant, precise, and rigorous garden of pure mathematics. By the time I had acquired the mathematical tools to understand, with the rigor my psychology demanded, what the physicists were doing, I was captured by the beauty and elegance of pure mathematics, and any professional aspirations I had held for a career in physics had morphed into aspirations for one in mathematics. But that keen interest in physics that had been fueled and fanned into flame upon reading George Gamow's *One, Two, Three, Infinity* and the Isaac Asimov books on physics, while in high school, remained. Over the past four decades I have, sometimes with less, sometimes with greater intensity, continually filled in the gaps in my understanding of quantum mechanics, and this book is my report to my 20-year-old self. It is a roadmap that would have made my attempt to gain a precise understanding of the mathematical difficulties of quantum mechanics a bit easier had it sat on my desk four decades ago. It is my hope that this book might be of use to the student of mathematics who is in the same predicament.

The reader I target is that student of mathematics who has perhaps a year of graduate training, who has seen Hilbert spaces or measure, who knows something of manifolds or Lie groups, who has been introduced to tensor products, though much of this might still seem quite mysterious and exotic. I am not looking for familiarity with functional analysis at the level of the spectral theorem or Stone's theorem, nor familiarity of group theory at the level of representation theory – these will be explained when needed – but I am looking for a desire to understand quantum mechanics coupled with just the right amount of mathematical sophistication and preparation that allows for a mathematically honest and thorough, if not quite rigorous, treatment of quantum mechanics. The level of sophistication necessary is at least, but really no more than that of the graduate student of mathematics at the end of the first year of study at a US university.

Quite a bit of modern mathematics that is absent from most physicists' expositions is used, reviewed, or developed in this book. I hope that the student of mathematics will find in it just the right amount of mathematical explanation to satisfy the need for rigor yet a presentation that is not overly pedantic.

The exposition and choice of topics bear the strong imprint of the author, and so the reader will find herein some rather unorthodox topics as well as some rather unorthodox approaches to standard topics, all of which are flavored by the author's personal mathematical prejudices as well as his personal pedagogical perspective. This, I imagine, is true of most works of most authors, and some of the particularly distinctive features of this presentation will be discussed in more detail next.

Some Distinctive Features

This is not a textbook. There are no exercises. It fails to present many of the practical tools – the various approximation methods for example – that physicists use to do calculations, a deficiency for use as a textbook for the student of physics. It repeats material found in a variety of proper mathematics courses, but without the degree of completeness and generality one expects of a good mathematics text, and often describes mathematical results without the benefit of rigorous justification – the mathematics is always honest but not always rigorously justified! Thus, I do not consider this work an exposition of mathematics but rather an explanation of quantum mechanics in as mathematically thorough a way as I know, and I have chosen to present the mathematics as needed as part of the physics story. In particular, I let the physics take center stage and drive the mathematics. For this, prose seems more appropriate than technical mathematical syntax for the telling of this story, and so I have dispensed with the familiar and precise mathematical syntax of axiom–definition–lemma–proposition–theorem–corollary–QED so dear to pure mathematicians. In fact, I try to present much of the material in a style that physics students appreciate but all the while taking great care with the mathematics. Of course this requires some extensive pauses to consider the relevant mathematics, some of which is developed precisely and in detail. This approach already differentiates this work from many that serve first as textbooks of mathematics, with the physics simply the motivation for the mathematics.

As an example of how this book differs from those, we employ the spectral theorem for unbounded operators to make sense of the so-called projection postulate of physicists, but we omit a proof of the spectral theorem even though we do not expect the reader to be familiar with this result; rather, we carefully explain the spectral theorem and how it may be used to encode the physical content of the theory. This should have the effect of reassuring students that the quantum mechanical formalism rests on a sound, rigorous footing and may inspire the more curious and motivated ones to learn the proof on their own, either by taking an appropriate course in functional analysis or by consulting the list of references provided at the end of the book.

Aside from expectations of the reader that are more mathematically advanced than those for most books on quantum mechanics, there are some rather distinctive features of this book in terms of both the choice of topics and their development. These features are detailed in the prolegomenon that follows this preface, but I should like to point out some of them now. Herein the reader may find the following: a rigorous treatment of the measurement axiom and the projection postulate that employs the spectral theorem for unbounded operators; a development of the Dirac calculus, the famous bra-ket notation, in the context of rigged Hilbert spaces; a careful discussion of commuting operators; a detailed review of the Fourier transform for use in the momentum representation; angular momentum operators as $\mathfrak{su}(2)$ representations; the employment of quaternions in both the development of SU(2) representations and in transforming the central force problem in three-dimensional space to the 3-sphere; a detailed discussion of the hidden SO(4) symmetry of the hydrogenic potential using stereographic projection and unit quaternions; Fock's treatment of the momentum space equation as well as a modern treatment using group characters; the use of tensor products to incorporate internal variables, with Pauli's theory of the electron as an example; a development of the Clifford algebra approach to spin; a careful discussion of the Einstein–Podolsky–Rosen paradox and nonlocality, including the Greenberger–Horne–Zeilinger scheme; a careful development of Fock space; the use of Stone's theorem on unitary representations to "derive" the Schrödinger equation and to understand conservation laws; a somewhat novel interpretation of the Feynman approach via path integrals; and a careful exposition of Gaussian integrals.

My hope for the student studying these topics is the same excitement and satisfaction that I found when discovering for myself the great work of the physicists and mathematicians who constructed quantum mechanics during the first half of the last century.

Homage

It is a pleasure to acknowledge three treatises on quantum mechanics that have been instrumental in shaping the general tenor and tone of the present work. These made a great impression on me, not only in my understanding of some key features of the subject but, more importantly, on how to present the subject with one's own personal elegance and élan. The present work may be considered an extended homage to these three books and my fond hope is that it may be seen as a worthy complement to them, though emphasizing different aspects of the subject and from a perspective that differs in some significant ways from theirs.

Eugene Merzbacher published the first edition of his book *Quantum*

Mechanics in 1961. I embarked on my first systematic study of quantum mechanics by a rather sporadic reading of the second edition, published in 1970, a copy of which I had acquired in October 1977. I was delighted when Professor Merzbacher produced a substantially updated third edition in 1998, which I consider to be as definitive an exposition of quantum mechanics from the perspective of a practitioner of the discipline as one can find. The publishing house Wiley has done a great service by keeping this work in print for so many years, a work that I think deserves the term *magisterial*. It is the case even today that I can turn randomly to its pages and find some bit of the subject that has eluded my notice, and I can learn from an expositor so well versed and at home with the physical reality behind the mathematics that his explications seem effortless and inevitable. Merzbacher has a genius for presenting the physics front and center but with enough mathematical honesty to satisfy even this most unrepentant mathematician.

In January 1999 at the Annual American Mathematical Society meeting in San Antonio I had the good fortune of stumbling upon Keith Hannabuss's beautiful text, *An Introduction to Quantum Theory*. This had been published two years earlier as the first work in the series *Oxford Graduate Texts in Mathematics*. If Merzbacher was my introduction to quantum theory, Hannabuss was my guide on how to present the theory with great clarity and elegance by adopting rigorous modern mathematics for its telling. I had read von Neumann's historically important and mathematically sound rendition of the subject in his *Mathematical Foundations of Quantum Mechanics* and was familiar with Chris Isham's beautiful little gem of 1995, *Lectures on Quantum Theory*, a mathematically robust "second course" in the theory written by a famed theoretical physicist with a strong training in pure mathematics, but it was the text by Hannabuss that captured my imagination in its choice of material and style of presentation. I absorbed this book over the following semester and based many lectures on Hannabuss's exposition in a seminar on quantum mechanics that I organized over the next two years. Among the scores of books and other sources from which I have learnt the subject over the past four decades, this book owes its greatest debt to Hannabuss. I consider the present work as a companion to Hannabuss's and would hope that the reader who appreciates his book would find mine quite pleasing to his or her temperament.

Finally I must express my admiration of Shlomo Sternberg's 1994 treatise *Group Theory and Physics*, certainly one of the more unique contributions to the literature of physics and mathematics and one of my favorite books. The ease with which Sternberg mixes the facts of the physical world with sophisticated mathematics and uses the mathematics to provide rigorous explanations and perceptive insights is impressive. He flows seamlessly from a physical description of a problem to a development of the rigorous mathematics needed to attack the problem, and then back to the problem with a beautiful application. He does this while dispensing, like this book, with formal mathematical syntax, and so his explication of the subject unfolds like a good story being told. I am rather captured by Sternberg's style of laying out the story he tells, and the present work in style and temperament is most akin to his. Included as an appendix to his book is a rather wonderful essay on the history of nineteeth century spectroscopy. It is amazing that the seemingly pedestrian act of measuring the characteristic spectral lines of nature's elements leads invariably to an explanation in which group representation theory takes a central role.

Gratitude

I would be remiss were I to fail to acknowledge and thank the folks and institutions that have made it possible for me to write this book. It has been a journey of 45 years. First I must acknowledge the University of North Carolina at Asheville, whose dedicated professors first nurtured the wide-eyed enthusiasm for physics and mathematics they found in a rather naïve freshman in 1974, as well as the University of Tennessee, which brought me to maturity in mathematics, and my doctoral mentor, John J. Walsh, who taught me how to think like a mathematician. To The Florida State University in Tallahassee I owe a great debt of gratitude for providing a stimulating environment over the past 36 years in which to build a career in mathematics, which gave me time to think and learn and do research wherever my interests turned and provided stimulating students and classes to teach. I should mention Cambridge University, venue of my 1996 sabbatical, where my interest in quantum mechanics was renewed after a hiatus of several years. The town of Cambridge provided exceptional bookstores to browse, where I found stimulating books on aspects of quantum mechanics that had escaped my off-and-on study of the subject since my graduate school days.

The two individuals who have had the greatest impact on my mathematical work are Jim Cannon and Ken Stephenson. Jim's work has influenced mine significantly, and I greatly admire his mathematical tastes and contributions. To collaborate with Ken has been a joy over the past two decades. We have been co-authors on a number of research articles and his down-to-earth approach to the understanding of mathematics has been a constant check on my tendency toward flights of fancy. I have learned from him how to tell a good story about a mathematical topic.

I must extend thanks to all those who attended my lectures over the years, but especially to Ettore Aldrovandi, with whom I often discussed topics in depth, to Paolo Aluffi, one of my chief cheerleaders in this effort and another discussion participant, and to Matilde Marcolli, who listened to some of my far-fetched notions on the Feynman path integral. I must also thank the students in two graduate classes I taught – Quantum Mechanics for Mathematicians, I and II – in the spring and fall semesters of 2011, whose questions helped fine tune much of the presentation of this book. I mention with fondness my former office neighbor, Jack Quine, who shares my enthusiasm for both quantum mechanics and our other guilty pleasure, number theory. Washington Mio and I had a brief flirtation with quantum computing, and he graciously tolerated my intrusions and prodding to speak in seminar about the topic.

It has been a pleasure to work with the wonderful folks at Cambridge University Press in preparing this book for publication. Over a period of several years at annual AMS Joint Mathematics Meetings I spoke with mathematics editor Roger Ashley about a book of this type and he was always a great encouragement. When finally I was able to submit a manuscript, I was placed in the capable hands of mathematics editor Tom Harris and his able assistant Anna Scriven. Tom and Anna have been a source of support and great patience as I juggled my professional responsibilities as a professor at a research university with the technical editing of a book of this length. It has been a joy to work with senior content manager Esther Migueliz Obanos and copy editor Susan Parkinson in this process. Esther was always understanding when I failed to meet suggested deadlines and with gentle prodding directed me towards manageable goals. Susan is a copy editor par excellence whose edits, suggestions, and questions were always spot-on and insightful. Susan has a PhD in spectroscopy and has taught quantum mechanics at the Open University in the UK. Her work greatly improved the original manuscript, not only from her incredible attention to the fine points of the English language, but also because her expert knowledge of quantum mechanics more than once pointed me toward a clearer way to frame some technical aspect of physics.

Finally, I must thank my family – my parents, Harry and Mary, for providing as great an environment in which to come of age as I can imagine, my children, John, Thomas, and Maddy, who are a constant source of joy and surprise, and especially my wife, Kris, steadfast companion and best friend whose support and encouragement throughout these past 36 years have made my mathematical career possible.

Tallahassee, Florida

Prolegomenon

> Never in the history of science has there been a theory which has had such a profound impact on human thinking as quantum mechanics; nor has there been a theory which scored such spectacular successes in the prediction of such an enormous variety of phenomena (atomic physics, solid state physics, chemistry, etc.). Furthermore, for all that is known today, quantum mechanics is the only consistent theory of elementary processes.
>
> Thus although quantum mechanics calls for a drastic revision of the very foundations of traditional physics and epistemology, its mathematical apparatus or, more generally, its abstract *formalism* seems to be firmly established. In fact, no other formalism of a radically different structure has ever been generally accepted as an alternative. The *interpretation* of this formalism, however, is today, almost half a century after the advent of the theory, still an issue of unprecedented dissension. In fact, it is by far the most controversial problem of current research in the foundations of physics and divides the community of physicists and philosophers of science into numerous opposing "schools of thought."
>
> Max Jammer
> in the Preface to *The Philosophy of Quantum Mechanics*
> (1974)

A Wealth of Mathematics

One of the unexpected windfalls for the mathematician who studies quantum mechanics is that here, in this one topic, the umbrella theory of the whole of physics, one finds a host of modern, sophisticated mathematics that is pertinent to its sound telling. This serves to clothe the bones of the theoretical mathematics that the student learns in his or her graduate studies with the concrete flesh of the physical world. Here the abstraction and generality of graduate pure mathematics take on a specificity of meaning in its application to the understanding of the workings of the micro-world. This helps to solidify in the student the effectiveness of his or her craft and serves to showcase the interconnectedness of many topics that have been presented in isolation from one another. There is no other theory within the natural sciences that surveys in its sound telling so much of the landscape of modern pure mathematics. Here I present a brief listing of the disciplines and subdisciplines within both classical and modern mathematics that make an appearance in a mathematically honest study of quantum mechanics. All three primary divisions of pure mathematics – analysis, algebra, and geometry/topology – are suitably represented in this listing. The student of quantum mechanics cannot be expected to have mastered all the topics in the list, not even the majority of them, but he or she should be willing to work through just enough to apply to the discussion at hand.

Analysis

Of the three broad divisions, perhaps analysis has the greatest claim on quantum mechanics. The very playing field of Schrödinger's wave mechanics is the Hilbert space of square-integrable Lebesgue measurable functions. In the axiomatic treatment of the theory, other Hilbert spaces make their appearance as important instantiations of the axioms. It is not surprising that the isometry of $L^2(\mathbb{R}^n)$, known as the Fourier transform, makes a central appearance as the transformation from the position to the momentum representation of the theory. As much of the language of the elementary applications of quantum mechanics is expressed in the form of partial differential equations, there is an abundance of classical special function theory in the mix – the polynomials of Hermite, Laguerre, and Legendre and their associated functions; spherical harmonics, general hypergeometric functions, and the gamma function. Contour integration makes its appearance as do stereographic projection, Green's formula, the general Stokes theorem, Gaussian integrals, and the differential operators of Laplace, d'Alembert, and Dirac.

It is not an overstatement to say that functional analysis is at the very core of a rigorous treatment of quantum mechanics. Self-adjoint operators in all their subtlety are pivotal for the development of the theory of measurement, and unitary operators for a description of quantum evolution and for a clarification of the role of symmetry in determining conservation laws in the theory. Not surprisingly, the two great theorems of the subject, the spectral theorem for self-adjoint operators and Stone's theorem on generators for one-parameter unitary groups, are crucial for providing rigor to quantum measurement, unitary evolution, and quantum conservation laws. Distribution theory and generalized functions, with the machinery of Schwartz spaces and test functions and tempered distributions, appear both in the development of the theory of the Fourier transform and as a way to explain the *lingua franca* of the subject, the Dirac calculus with its bra-ket notation for state vectors.

Finally, the calculus of variations and the Euler equation are important in the development of the Lagrangian approach, and canonical transformations for the introduction of the Hamiltonian approach, and Noether's beautiful theorem is important for understanding how conservation laws arise from symmetries.

Algebra

Groups, vector spaces, and algebras all have their place in informing and building quantum theory. Infinite-dimensional vector spaces are really the stuff of functional analysis, but their finite-dimensional cousins show up as state spaces of finite-dimensional quantum systems and as representation spaces of groups and algebras. Lie groups, Lie algebras, and their real, complex, and quaternionic representations are used to develop a proper theory of spin and angular momentum, and unitary representations appear in discussions of symmetry and conservation laws. Of primary importance are the-lower dimensional classical matrix groups and algebras, particularly $\mathrm{SU}(2)$, $\mathrm{SO}(3)$, $\mathrm{SO}(4)$, $\mathrm{Sp}(1)$, $\mathrm{SL}(2,\mathbb{R})$, $\mathrm{SL}(2,\mathbb{C})$, the quaternions $\mathbb{H}$, the Lie algebra $\mathfrak{su}(2)$, the matrix algebras $\mathbf{M}_2(\mathbb{K})$ for $\mathbb{K} = \mathbb{C}, \mathbb{H}$, and $\mathbf{M}_4(\mathbb{C})$. Clifford algebras as well as spin groups and spin representations play important roles in the Pauli and Dirac theories of fermions. Group characters are used in solving the momentum space equation for the hydrogenic potential. Representations of the Lorentz and Poincaré groups uncover some intricacies of special relativity, and tensor products are used to construct state spaces for multi-part quantum systems and as a tool for incorporating novel aspects into a known quantum system as, for instance, in the Pauli theory of spin. Of particular interest are the exterior and symmetric graded subalgebras of the tensor algebra of a Hilbert space and their completions, used as multi-particle quantum state spaces for fermions and bosons.

Geometry/Topology

Differentiable and Riemannian manifolds make their appearance along with some of their associated machinery – tangent and cotangent bundles, tensor fields and differential forms, covariant derivatives and connections. These are used in placing some traditional topics of physics into a global, modern setting. This occurs in discussions of Lagrangian and Hamiltonian mechanics, of gauge symmetry of the electromagnetic field, and, to a lesser extent, in some comments on special relativity. Covering spaces arise in the context of examining the topology of SU(2) and SO(3), as well as in the completely unrelated topics of the Aharonov–Bohm effect and Dirac monopoles. The quantum version of electromagnetic theory may be placed in the sophisticated setting of complex line bundles and Koszul connections, where all the gauge equivalent theories are seen as trivializations of a single theory. This latter version using complex line bundles is rather esoteric, but the payoff in mastering it is the unity this sophisticated approach brings to the theory.

Starred Sections

Lest this listing of modern mathematics should frighten the reader, many of these topics, particularly the ones from topology and geometry, do not appear in this book. They are needed for building the mathematical edifice of relativistic quantum mechanics, which is avoided here. Some lectures and sections in this book that employ more advanced topics are marked with a star, $\star$, and may be skipped without affecting the flow of the book.

The Lie of the Land

This is a large book with an abundance of topics. I think it may be as well to present something of a guided tour for the prospective reader. Before that, though, allow a bit of advice from an old hand at learning mathematics and physics. The strict training of many students of mathematics imprints a certain psychological pressure on their practice of learning a new topic. Since the rallying cry of pure mathematics, especially in the foundational courses at the advanced undergraduate and beginning graduate levels, is rigorous argument – nothing accepted without airtight proof argued from axioms or theorems – the student often enters the second year of graduate work with a psychological need to learn any new technical subject linearly, accepting nothing until its proof is understood. Now that the student is conversant with the craft of constructing, reading, and understanding proofs, this psychological need for linear learning and complete rigor should be broken. Most research in mathematics takes place at a frontier so far removed from the foundational courses that one cannot afford to dot every i and cross every t in order to get to the frontier. The student must learn to understand the statements of well-established results and how to use them, hoping to place them in his or her mathematical toolbox, and use the fact that these tools are very useful to motivate him or her to go back later, when time affords the luxury, to dot the i's and cross the t's. Learning how a result can be used to derive new results and to bring understanding and organization to a technical subject, even without understanding the rigors of the proof of the result, can be a great motivator for delving deeply into that topic later. So, my advice to the reader is to allow himself or herself an indulgence, namely, to browse this book and pick and choose parts that interest him or her. Read a chapter without necessarily having read the background material needed for that chapter; skip the more technical discussions on some of the minutiae in order to get to the heart of the physical theory; do not read this book linearly! I think you will learn more and have a more satisfying experience with the book if you heed this advice. With this in mind, I begin the guided tour.

A Guide to this Book

The book you are holding comprises 32 chapters, which from now on are called lectures. It presents the basic results of the nonrelativistic theory of quantum mechanics. While all the important theoretical features of quantum mechanics make their appearance, including some nontraditional topics, the treatment largely neglects some issues found in standard texts. For example, I make little mention of the experimental work that motivated the development of the theory. I omit the one-dimensional scattering discussion of wave packets hitting potential barriers and the several standard perturbation techniques used to approximate solutions to the Schrödinger equation. And, finally, I largely avoid some standard metaphysical musings on the subject that I find opaque.

On this last point, I do not mean that philosophical and interpretative issues are abandoned altogether. In fact, I have a healthy respect for these issues and encourage the student to explore them here and on his or her own. I only mean that some claims found in some of the standard texts that champion the standard (Copenhagen) interpretation of the theory are either, on the one hand, rather abstruse or unintelligible, or, on the other hand, outright dogmatic and lend nothing to an understanding of the theory.

Lecture 1 presents a comparison of the classical treatment of a physical system by Newtonian mechanics and the mathematical treatment using the rules of quantum mechanics. This is designed to illustrate the striking differences between the two approaches, both in methodology and in results. The system chosen for presentation is that of a particle in a central quadratic potential – the simple harmonic oscillator – whose classical analysis is familiar from elementary mathematics courses. The quantum treatment should seem rather odd and its meaning will not be apparent from the mathematical analysis applied. It will not even be clear from the quantum analysis that the results say anything about the motion of the particle in this central potential. This state of affairs is remedied by interpreting the quantum results using an understanding that developed along with the theory and imprints meaning on the quantum calculations in terms of measurement probabilities. There follows a more philosophical discussion on foundational issues that lays out for the reader some of the very interesting and deep questions that arise concerning what quantum mechanics says about the nature of reality at the micro-level. My hope is that this initial lecture will give that reader who is a novice to the subject an intellectual jolt that will inspire him or her to delve deeply into the mathematical and philosophical issues of this intriguing theory.

The book's treatment of quantum mechanics proper begins with *Lecture 2*. Here the basic mathematical structure of the theory is encased in a set of axioms that are rather minimal with respect to assumptions. Over the next three lectures these axioms are unpacked and expanded, and the mathematics necessary to bring rigor to the subject is explained. This entails an excursion into functional analysis with emphases on self-adjoint operators and the subtleties of their definition, the spectral theory of these operators, and rigged Hilbert spaces. As the book moves through these mathematical issues, it concurrently develops the physical interpretation necessary to give meaning to the quantum calculations. In particular, the probability interpretation of the state vector is presented, quantum measurement is discussed along with the knotty problem this leaves unresolved, and the Dirac bra-ket calculus is presented and placed on a sound mathematical footing. This takes us through *Lecture 5.*

Lecture 6 presents a review of classical mechanics, including the Lagrangian and Hamiltonian formulations, as well as Noether's theorem on conserved quantities. For the student who is unfamiliar with these topics, this lecture should serve as an accessible introduction with enough detail sufficient for their use in the development of quantum mechanics. *Lecture 7* on Jacobi's contributions to classical mechanics is a bonus lecture that completes the classical picture but may be skipped without affecting the flow of the book. *Lecture 8* relates the quantum development to the classical, suggesting how the classical treatment of the macro-world of everyday objects emerges from the quantum picture.

Schrödinger's wave mechanics is introduced briefly in Lecture 2 as the most important instantiation of the minimal axioms for quantum mechanics, but is developed rather sparsely over the following few lectures. It is developed more fully in *Lectures 9, 10, 11, 15, 16, 17,* and *19. Lecture 9* attempts to clear away a bit of the fog that has accumulated over popular accounts of the Heisenberg uncertainty principle, and it introduces minimal uncertainty states as well as the Fourier transform. The Fourier transform is reviewed quite carefully in *Lecture 10* and used to introduce the momentum representation of wave mechanics. The first fully quantum mechanical calculation is performed in *Lecture 11*, giving details of the quantum analysis of the simple harmonic oscillator reported in the first lecture. A thorough treatment of the hydrogenic potential, begun in *Lecture 14*, is presented in *Lectures 15, 16,* and *17.* In particular, the usual quantum mechanical treatment appears in *Lectures 14* and *15*; an unusual treatment is presented in *Lecture 16*, where the hidden symmetry of the hydrogenic potential is revealed from an analysis of the momentum space equation. This is followed with two solutions to the momentum space equation in *Lecture 17*, the first derived from Fock's original treatment of the equation and the second from a modern treatment based on group characters. Finally, Pauli's ad hoc addition to the quantum formalism to include spin, his *spinor theory*, is introduced in *Lecture 19* as an example of how to use the formalism of tensor products to include novel phenomena in a quantum theory.

Angular momentum operators are treated in *Lectures 12, 13, 14,* and *18*, with bonus material in *Lecture 20.* The orbital angular momentum operators are defined in *Lecture 12* by quantizing the classical angular momentum formulae of classical mechanics, after which the eigenvalues and eigenvectors of these operators are determined. Realizing that the representation theory of the Lie algebra $\mathfrak{su}(2)$ is central to a mathematical understanding of these operators, *Lecture 13* introduces the Pauli spin matrices in the context of the algebra $\mathfrak{su}(2)$ and reinterprets the results of the preceding lecture in light of representation theory. This lecture becomes a short course on the low-dimensional Lie groups and their Lie algebras, which have much currency in modern physics. In particular, detailed developments of the quaternion algebra $\mathbb{H}$, the 3-sphere group $\mathbf{S}^3$, and the Lie groups SU(2), SO(3), and SO(4), as well as their corresponding Lie algebras, are presented. The representation theory of SU(2) and SO(3) is worked out in detail and the topological structure of these groups is uncovered. This mathematical work pays off in *Lecture 14*, where the central force problem is analyzed. Here the representation theory of SO(3) is found to be crucial for understanding the finer points of the space of spherical harmonics, which themselves arise in the analysis of the central force problem. This material is necessary for a complete understanding of the hydrogenic potential analyzed in the lecture following. After a foray into the analysis of the hydrogenic potential, *Lecture 18* returns to the subject of angular momentum operators.

Up to this point the full integral representations of $\mathfrak{su}(2)$ have been used to build a quantum theory of orbital angular momentum, but this begs the question of whether the half-integral representations have anything to add to the quantum mix. After presenting the addition rules for angular momentum operators, *Lecture 18* continues by analyzing the simplest half-integral representation, the spin one-half system, that turns out to be crucial for an understanding of spin in quantum theory. Spin is a purely quantum mechanical property of particles that has no classical counterpart, and is integrated into the quantum development in the next lecture by incorporating the spin one-half representations via Pauli's spinor theory. With this introduction of spin into the quantum formalism, it seems natural to pause for a moment and include a mathematical lecture, *Lecture 20*, on Clifford algebras and spin representations. This material is not needed for the remainder of the development and may be skipped, but I have decided to place it here to position the Pauli theory in a context that is modern and one that anticipates the more natural Dirac theory, a critical ingredient of relativistic quantum mechanics.

Thus far in the development of the theory I will have dealt only with quantum systems that are indivisible or systems with a single quantum particle. *Lectures 21* through *26* explain how to incorporate multi-part quantum systems, or multi-particle systems, into the theory. This begins in *Lecture 21* where the tensor product machinery is used to construct state spaces for multi-particle quantum systems. As an important example, I work out a theory of coupled angular momenta for a pair of particles. This is put to work in *Lecture 22* where the most important nonclassical feature of quantum mechanics is uncovered in an articulation of the central role of inseparability in the theory. Here the nonlocality of quantum theory is elucidated and the famous Einstein–Podolsky–Rosen criticism of the completeness of quantum theory is analyzed. The equally famous Bell inequalities are derived, and it is shown how quantum mechanics easily accommodates nonlocality and the violation of the Bell inequalities in its formalism. This lecture closes with a presentation of the GHZ scheme for spin triplets, which even more forcefully illustrates the quantum weirdness of inseparability.

Lecture 23 develops von Neumann's approach to incorporating statistical ensembles of particles into the quantum formalism, using density operators to represent quantum states and traces to calculate expectation values. An example of this new formalism in action is presented in the context of spin one-half systems, where matrix calculations are illustrated and the topology of spin states is uncovered. In *Lecture 24* bosons and fermions are introduced and their indistinguishable multi-particle state spaces are constructed. This is done first algebraically, by identifying the bosonic state space as the completion of the symmetric algebra generated by the state Hilbert space of a single boson and, similarly, the fermionic state space as the completion of the antisymmetric (or exterior) algebra generated by the state Hilbert space of a single fermion. These state spaces for bosons and fermions are reintroduced in *Lecture 25* where they are seen to be isomorphic to state spaces defined by Fock in a more natural, physical way than is afforded by the algebraic description of the preceding lecture. Finally, the statistical mechanics of an ensemble of quantum systems is explored in *Lecture 26*.

The primary concerns of the book up to this point have been to articulate the mathematical structure of the theory, to develop some of the requisite mathematics, to apply the theory to an analysis of the stationary states of important quantum systems, to construct a theory of spin and angular momentum, and to build appropriate state spaces for the great variety of quantum systems of interest. The emphasis has been on the time-independent parts of the theory and on deriving the eigenstates of Hamiltonians and angular momentum operators whose eigenvalues provide possible dial readings on measuring instruments. Though there have been brief excursions into the time evolution of quantum systems, a clear and general development of quantum dynamics awaits *Lecture 27.* Stone's theorem on unitary representations and infinitesimal generators is used to "derive" the Schrödinger equation. Both the Schrödinger and the Heisenberg pictures of quantum evolution are presented, and Dirac's synthesis of the two approaches is explicated.

Stone's theorem of Lecture 27 is put to great effect in *Lecture 28* in illuminating the role of symmetry in uncovering quantum conservation laws. This leads naturally to a discussion of the distinction between global symmetries and symmetries that are local in nature. Here one sees the first hints of gauge theory in the argument that a local symmetry requires the existence of a mediating field. The reader is left in the somewhat unsatisfactory position of having seen the connection between local symmetry and a gauge field but without a full explanation. This is meant as a motivation for the reader to advance to the next step in the study of quantum mechanics, the study of the relativistic theory, where gauge symmetries are given their due after the machinery of special relativity is introduced.

Lectures 29 through *32* introduce Feynman's idiosyncratic approach to quantum mechanics, employing his highly original development of path integrals. This approach is ripe with both deep physical insight and confusing mathematical chicanery. Even seven decades after Feynman introduced his very powerful method for calculating quantum amplitudes, the Feynman path integral remains on an unsound footing with no rigorous development of the integral on the horizon. In *Lecture 29*, I present an overview of Feynman's approach, giving a physically motivated plausibility argument for the method. It is hoped that what the discussion lacks in rigor is compensated by a psychologically attractive, heuristic, approach to the calculation of quantum amplitudes. At the heart of the actual calculations using the Feynman integral are various complex Gaussian definite integrals. As most pure mathematics students will never have seen these creatures, *Lecture 30* presents a mathematical interlude where the main Gaussian integrals of interest are determined in a rigorous manner. Though there is no rigorous treatment of the path integral available, I present an interpretation of the integral based on Feynman's original approach using piecewise linear families of paths, and I abstract from this the notion of filtration measures for various dense families of paths. This method of bringing order to the madness via filtration measures is explored in *Lecture 31*, where the method is applied to derive the constant force propagator using polynomial sums.

The lecture closes with a derivation via the Dirac calculus of the Feynman formula for calculating path integrals, which along the way demonstrates its equivalence to the more standard Schrödinger approach of wave mechanics. *Lecture 32* then presents the derivation of the propagator for the simple harmonic oscillator using Fourier sums, and also presents a very difficult calculation using polynomial sums as a test case for the filtration method. The lecture, and indeed the book, closes with an analysis of the forced harmonic oscillator using path integrals, this being the bare beginning of the quantum treatment of fields.

Resources for Concurrent Study

There are a great many books on all things quantum. At the end of this book following the Epilogue the student will find Resources for Individual Exploration, a list of 99 references with commentary for the further study of quantum mechanics. I have divided the listed references into categories and given detailed recommendations for using the references. I would hope that a student would go ahead and read through these recommendations and use them as a resource to supplement his or her study of this book.

The author desires for the student the same excitement and pleasure that he has experienced over the years in learning the workings of nature at the micro-level. May the reader be inspired to learn more deeply the mathematics needed to make rigorous this quantum theory, and may he or she more profoundly appreciate the interconnectedness of pure mathematics and theoretical physics as his or her studies progress.

1
The Harmonic Oscillator: Classical versus Quantum

> It seems to me that there are deep philosophical lessons to be learned in the way in which the practicing theoretical physicist thinks about the foundations of the subject, i.e., the manner in which he approaches the problems, the general criteria he brings to bear on what is a reasonable solution. So, the important thing then is to display the general world view, the world picture that the theoretical physicist has.
>
> This is particularly significant in connection with the philosophical implications of quantum physics, because quantum physics or quantum mechanics – by which I think we mean finally the rational mode of understanding of microscopic or atomic phenomena – has perhaps had the greatest impact of any of the developments of physics upon the mode of thinking or the world picture of the physicist and thereby, indirectly, of the general citizen.
>
> Julian Schwinger
> in the Prologue to *Quantum Mechanics: Symbolism of Atomic Measurement*, 2001

"Quantum mechanics – by which I think we mean finally the rational mode of understanding of microscopic or atomic phenomena," this, in a nutshell, is the topic of this book. We seek to present a calculus that encodes this rational understanding. This is not an easy task. It is in no way straightforward, and a cursory reading of the history of the development of the subject illustrates nothing if not how very difficult it was to discover the right calculus. It does not help that the right calculus is written in the language of some rather esoteric and sophisticated mathematics, and necessarily so, it seems, and that there are competing ways to interpret the resulting abstract formalism. Rather than repeat the history of the founding of the subject, which covers a period from roughly 1900 until the mid-1930s and involves the work of scores of physicists and even a few mathematicians, which involves many experiments and observations as well as many false starts and stops, we present the calculus as a *fait accompli* and then back track to fill in our understanding. The reader should understand at the outset, though, that there is a wealth of experimental justification for this calculus.

In this opening lecture, we present a contrast between classical mechanics and quantum mechanics through the lens of a single example. This example illustrates in stark relief the differences between the classical view of the world as articulated by Newton's laws and the modern view as articulated by the rules of quantum mechanics. The harmonic oscillator is the quintessential physical system and, as such, its analysis, whether classical or quantum, serves as an archetype of the discipline. In this lecture we review the classical treatment of the harmonic oscillator and outline the quantum treatment. The quantum treatment will seem ad hoc and unmotivated and should elicit some unease and, perhaps, even perplexity. The reader will see that the utter simplicity of both the methodology and results of the classical treatment stands in sharp contrast to the sophistication of the quantum treatment. Indeed, whereas both the application and meaning of the classical treatment are apparent from the mathematics itself, the methods and results of the quantum treatment cry out for explanation and interpretation. We present a standard interpretation of the quantum treatment here, but the reader should find our explanations, though internally coherent, nonetheless unmotivated. This interpretation was developed over a period of years concurrent with the development of the machinery of quantum mechanics itself, but the reader is warned that other interpretations are possible.

At the end of this lecture, we delve into some foundational issues surrounding the interpretation of quantum mechanics. This is something of a departure from our practice throughout the remainder of the book, where the development of the formalism takes precedence over philosophical issues.[1] Nonetheless, we want the reader to be aware at the outset that the quantum mechanical world view is a drastic departure from that of the classical and leaves open many deep philosophical issues.

Welcome to the world of the quantum!

[1] The exception is Lecture 22, where a precise discussion of nonlocality in real systems is presented, along with its accommodation by the quantum formalism and its implications for hidden variables theories.

1.1 The Classical Harmonic Oscillator

Consider a particle moving in a one-dimensional quadratic potential well $V(x) = \frac{1}{2}kx^2$, where the **spring constant** k is positive and x represents position along a line. The force experienced by the particle is $F = -\nabla V = -kx$, and serves to restore the particle to its equilibrium position $x = 0$. The force on the particle diminishes as it moves toward equilibrium but, since the particle has positive speed as it reaches the equilibrium point, it overshoots past equilibrium and begins to feel the restorative force again. Eventually the force slows the particle to rest at some maximum displacement from equilibrium, at which point the particle begins its journey back toward the equilibrium position, overshooting again, and reaching a maximum displacement opposite to that of its previous maximum displacement. And so it goes ad infinitum. This is the classical harmonic oscillator, the quintessential physical system of classical mechanics.

The newtonian analysis of the harmonic oscillator is straightforward. From Newton's second law, the equation of motion of the harmonic oscillator is

$$m\ddot{x} + kx = 0,$$

where m is the positive mass of the particle and $x = x(t)$ is position at time t. Here we are using Newton's notation, where the derivative of a function of the time variable is indicated by a dot above the function. The solution to the equation of motion, equally transparent, is

$$x(t) = a\cos\omega t + b\sin\omega t,$$

where the **natural** or **characteristic frequency** is

$$\omega = \sqrt{\frac{k}{m}},$$

and the initial conditions are $x(0) = a$ and $\dot{x}(0) = \omega b$. For simplicity, take $x(0) = X > 0$ and $\dot{x}(0) = 0$, so that $x(t) = X\cos\omega t$. Of course X is the magnitude of the maximum displacement. The kinetic energy of the system is $K = \frac{1}{2}m\dot{x}^2 = \frac{1}{2}m\omega^2X^2\sin^2\omega t$ and the potential energy is $V = \frac{1}{2}kx^2 = \frac{1}{2}kX^2\cos^2\omega t$, so that, using $k = m\omega^2$, the total energy $E = K + V$ is a constant of the motion:

$$E = \frac{1}{2}kX^2 = \frac{1}{2}m\omega^2X^2.$$

The picture presented by classical mechanics is clear and compelling.

The equation of motion involves the particle position $x = x(t)$ as a function of time and its solution provides a complete description of the location of the particle at all times. The motion, described precisely by sinusoidal terms, is periodic, oscillating between extremes $-X$ and X with a natural frequency determined by the mass of the particle and the spring constant, a characteristic of the potential well. The total energy is a constant of the motion with well-defined contributions at each instant in time, the kinetic energy from the particle's speed and the potential energy from the particle's position in the potential well. We now examine how the quantum mechanical rules treat the harmonic oscillator.

1.2 The Quantum Mechanical Treatment

The quantum mechanical analysis of the harmonic oscillator begins with the classical Hamiltonian $H(x, p)$ for the system, the total energy written in terms of the position x and its conjugate momentum $p = m\dot{x}$. The momentum conjugate to the position variable x is by definition $p = \partial L/\partial \dot{x}$, where the Lagrangian is $L(x, \dot{x}) = K - V = \frac{1}{2}m\dot{x}^2 - \frac{1}{2}kx^2$. The Lagrangian and Hamiltonian formulations of classical mechanics are reviewed in Lectures 6 and 7. The energy E of the system is given in Hamiltonian form by

$$H(x, p) = \frac{p^2}{2m} + \frac{1}{2}m\omega^2 x^2 = E.$$

We now perform a surprising procedure that will yield the quantum mechanical equation of motion for the harmonic oscillator. To **quantize** the Hamiltonian equation, we replace the three dynamical variables, or **observables** – the position x, the momentum p, and the energy E – by differential operators according to the following prescription:

$$\begin{aligned} x &\rightsquigarrow \mathsf{X} \\ p &\rightsquigarrow \mathsf{P} = -\mathbf{i}\hbar\nabla = -\mathbf{i}\hbar\frac{\partial}{\partial x} \\ E &\rightsquigarrow \mathsf{E} = \mathbf{i}\hbar\frac{\partial}{\partial t}. \end{aligned}$$

Here X is the operator that multiplies by x, i.e., for a function f, $\mathsf{X}f(x) = xf(x)$ and $\hbar$ is the **reduced Planck constant**, the proportionality constant relating the energy of a photon to its angular frequency. These operators act linearly on a suitable class of functions. Upon replacing x,

p, and E in the Hamiltonian equation by these operators, we obtain the operator equation

$$H(\mathsf{X},\mathsf{P}) = -\frac{\hbar^2}{2m}\frac{\partial^2}{\partial x^2} + \frac{1}{2}m\omega^2\mathsf{X}^2 = \mathbf{i}\hbar\frac{\partial}{\partial t} = \mathsf{E},$$

which operates on a suitable function $\Psi(x,t)$ to give the quantum mechanical equation of motion, the **Schrödinger equation**, for the harmonic oscillator:

$$-\frac{\hbar^2}{2m}\frac{\partial^2\Psi}{\partial x^2} + \frac{1}{2}m\omega^2x^2\Psi = \mathbf{i}\hbar\frac{\partial\Psi}{\partial t}.$$

At this point the novice might ask why we are transforming a classical Hamiltonian equation into an operator equation that then produces the Schrödinger equation. What does this have to do with the simple system of a particle in a quadratic potential well? It turns out to be quite a lot, but the way in which the Schrödinger equation encodes information about the physical oscillator is quite sophisticated. Some textbooks give plausibility arguments to "derive" the Schrödinger equation that, ultimately, in our opinion, offer little insight or comfort. Others proceed along the historical path, easing us into the quantum world view gently, and keeping the uncomfortable "why" questions at bay by recording the struggle to understand experimental results that, ultimately, led to the creation of the machinery. Still others report such results and then show how simple matrix manipulations can be used to predict the observed outcomes. This then is developed and generalized rapidly to arrive at the Schrödinger equation and the rest of the machinery. The real answer as to why we are performing this procedure is **that it works to give accurate statistical predictions and does so in the face of the failure of classical mechanics**. It is our view that there is no way to make quantum mechanics palatable to minds informed at every level of experience by the classical world. It is something we must get used to, and we must get used to it because it works. Our tactic then is to present the formalism for this simple physical system in this first lecture, without candy-coating, and then to jump into the formal mathematical structure immediately in the next lecture. So, read on and see this quantum mechanical calculation to its end, and then we will ask what happened and try to understand how to become comfortable with what we have done.

To solve the Schrödinger equation we separate variables by writing

$\Psi(x,t) = \psi(x)f(t)$, which gives the eigenvalue equations

$$\mathrm{i}\hbar\dot{f}(t) = Ef(t),$$

$$-\frac{\hbar^2}{2m}\psi''(x) + \frac{1}{2}m\omega^2x^2\psi(x) = E\psi(x),$$

where E is an unknown constant. Our use of the symbol E for the constant arising from the separation of variables anticipates its eventual interpretation as something to do with the energy of the system. The first equation has the solution

$$f(t) = C\exp\left(-\mathrm{i}\frac{E}{\hbar}t\right)$$

so that

$$\Psi(x,t) = Ce^{-\mathrm{i}Et/\hbar}\psi(x).$$

Here C may be taken to be an arbitrary complex constant, and the unimodular term $e^{-\mathrm{i}Et/\hbar}$ is called a **phase factor** of Ψ.

We are left with the second eigenvalue equation, known as the **time-independent Schrödinger equation**:

$$-\frac{\hbar^2}{2m}\psi''(x) + \frac{1}{2}m\omega^2x^2\psi(x) = E\psi(x).$$

This may be written succinctly as $\mathsf{H}\psi = E\psi$, where H is the linear differential operator given by

$$\mathsf{H} = -\frac{\hbar^2}{2m}\frac{d^2}{dx^2} + \frac{1}{2}m\omega^2\mathsf{X}^2.$$

The boundary condition imposed on this equation is that ψ and its derivatives must decrease to 0 sufficiently rapidly as $x \to \pm\infty$, for example, by requiring that all the functional terms of the equation – ψ, $x^2\psi$, ψ'' – be functions in $L^2(\mathbb{R})$, the Hilbert space of square-integrable complex-valued functions on the real line. This requirement forces the eigenvalue E to take on only certain restricted values, namely, values in the discrete sequence

$$E_N = \left(N + \frac{1}{2}\right)\hbar\omega, \quad N = 0, 1, 2, \ldots$$

The eigenfunction ψ_N corresponding to the eigenvalue E_N, normalized to the unit L^2-norm, is

$$\psi_N(x) = \left(2^N N!\right)^{-1/2}\left(\frac{m\omega}{\pi\hbar}\right)^{1/4}\exp\left(-\frac{m\omega x^2}{2\hbar}\right)H_N\left(\sqrt{\frac{m\omega}{\hbar}}x\right),$$

where H_N is the Nth Hermite polynomial. The family $\{\psi_N : N = 0, 1, 2, \dots\}$ forms a complete orthonormal (Schauder) basis of the Hilbert space $L^2(\mathbb{R})$. One can solve the time-independent Schrödinger equation and arrive at E_N and ψ_N via infinite series methods or via a beautifully elegant algebraic method using **annihilation** and **creation operators**. We present the details of the algebraic method in Lecture 11.

At this point we have applied the initial rules of quantum mechanics to the harmonic oscillator and have solved its quantum mechanical equation of motion, but we do not seem to have any information about the usual objects of interest for an understanding of the motion of the oscillator. We do not have a position function, we cannot calculate the momentum or the energy; we cannot even tell that the oscillator oscillates from the mathematics of this section. What we have done is to use rather unintuitive rules to quantize a classical Hamiltonian equation. This leads to the Schrödinger equation and, upon solving this resulting partial differential equation, a discrete sequence of numbers E_N with corresponding eigenfunctions ψ_N. The question before us serves as the title of the next section, where we present a standard, orthodox answer to the titled question.

1.3 What Does It All Mean?

The meaning given here of the calculations of the previous section is the result of a quantum world view that developed concurrently with the development of the machinery of quantum mechanics and was in place as the standard interpretation of quantum mechanics emerged by the nineteen thirties. The very fact that the machinery needed an interpretation, and that there are even today various competing interpretations, is part of what makes quantum mechanics a truly revolutionary break with the past. Indeed, the meaning or interpretation of classical mechanics is transparent in the mathematics of the discipline. After all, the primary physical attributes of a system are the particle positions. If appropriate initial or boundary conditions and the forces are known, then Newton's laws provide equations of motion from which the positions as functions of time can be derived, at least in principle. From these functions, various subsidiary physical attributes, like momentum and energy, can be calculated in a straightforward way. The classical world view provides a strongly intuitive mathematical model of the physical world where the correspondence between the mathematical objects and the physical attributes they represent is thought to accurately reflect the reality of the physical world. Nonetheless, as successful as classical mechanics is in its domain, it breaks down at the micro-level of the atom. The quantum world view offers a sophisticated calculus that works to describe the outcomes of experiments at the micro-level accurately. Indeed, it is primarily in the predictions of the outcomes of experiments that quantum mechanics derives its standard meaning.

We begin with an explanation of the eigenvalues E_N of the **energy equation** $\mathsf{H}\psi = E\psi$. These eigenvalues are, in fact, the possible values of the energy of the oscillator that are obtained upon measurement. This means that, if one performs an experiment to measure the energy of the oscillator, the value measured will, with certainty, be E_N for some value of $N = 0, 1, 2, \ldots$ Quantum mechanics thus predicts two surprising, nonclassical results: first, the energy of the system, rather than the continuously variable quantity $E = \frac{1}{2}m\omega^2 X^2$, $X \geq 0$, as derived in the classical analysis of the oscillator, is **quantized** into discrete units E_0, E_1, E_2, $\ldots$; second, the minimum possible energy of the oscillator is not zero, as in the classical case when the oscillator is at rest at equilibrium, but $E_0 = \frac{1}{2}\hbar\omega$, a positive quantity. This behavior is typical of micro-systems.

In the standard interpretation, the **energy eigenfunctions**, ψ_N, contain all the information that can be mined concerning the usual attributes we assign to physical systems, attributes such as the positions and momenta of particles and the energy of the system. We use these eigenfunctions to predict the outcomes of experimental measurements of these classical attributes by interpreting a general solution Ψ of the Schrödinger equation, expanded in terms of the eigenfunctions, as a **probability amplitude** that can be used to give probabilities for the various possible numerical values of these experimental measurements. A description of how these probabilities are determined from solutions to the Schrödinger equation will occupy the remainder of this section.

The full Schrödinger equation,

$$H(\mathsf{X}, \mathsf{P})\,\Psi = \mathsf{E}\Psi,$$

is a linear homogeneous equation, so that the complex linear combination $a\Psi + b\Phi$ is a solution whenever Ψ and Φ are solutions. The most general solution to the Schrödinger equation for the quadratic potential well is the **wave function**

$$\Psi(x,t) = \sum_{N=0}^{\infty} c_N \psi_N(x) e^{-\mathrm{i}E_N t/\hbar},$$

for complex constants c_N restricted only to ensure L^2-convergence of the infinite sum. The initial-value problem, solving the Schrödinger equation subject to $\Psi(x,0) = \psi(x)$, where $\psi \in L^2(\mathbb{R})$, has the solution $\Psi(x,t)$ given above when the coefficients are chosen so that $\psi = \sum c_N \psi_N$, which is possible since the family $\{\psi_N\}$ forms an orthonormal basis for $L^2(\mathbb{R})$. The immediate implication is that c_N is given by the Fourier coefficient

$$c_N = \int_{\mathbb{R}} \overline{\psi}_N \psi \, dx.$$

As the eigenfunctions ψ_N form an orthonormal collection, we may guarantee that the **state vector** $\psi_t = \Psi(-,t)$ is normalized in that, for each value of t,

$$\|\psi_t\|_2^2 = \int_{\mathbb{R}} |\psi_t(x)|^2 \, dx = 1,$$

provided that the coefficients satisfy $\sum |c_N|^2 = 1$. Assuming this, the normalized state vector ψ_t is interpreted as a **probability amplitude** for the position of the oscillator at time t. This means precisely that, when the oscillator has initial state vector $\psi_{t=0}$, at time t its state vector is ψ_t and the squared modulus $|\psi_t|^2 = \overline{\psi}_t \psi_t$ is the probability density for position measurements in the sense that, upon a position measurement of the oscillator at time t, the probability that the measured result is in the numerical range $[\alpha, \beta]$ is

$$\int_{\alpha}^{\beta} |\psi_t(x)|^2 \, dx.$$

Another nonclassical result appears here. It turns out that no matter what values $\alpha < \beta$ are chosen, the probability expressed by this integral is nonzero and positive when the state vector is one of the energy eigenfunctions ψ_N, as these eigenfunctions are nonconstant on every nontrivial interval. In particular, when the state vector of the oscillator is ψ_N with energy E_N, there is a positive probability of measuring the position of the oscillator outside its classical range of motion, the interval $[-X_N, X_N]$ where $E_N = \frac{1}{2} m\omega^2 X_N^2$. Of course, the expected value of the measured position is

$$\langle \mathsf{X} \rangle = \int_{\mathbb{R}} \overline{\psi_t(x)} \mathsf{X} \psi_t(x) \, dx = \int_{\mathbb{R}} \overline{\psi_t(x)} x \psi_t(x) \, dx$$

whenever the oscillator is in state ψ_t.

It is in this sense – that the state vector at time t gives probabilities for experimental measurements of position at that time – that the eigenfunctions contain information about the position of the oscillator at time t. But there is more, much more. The wave function contains not only information about the position, but, indeed, about any observable attribute of the system. For example, the expected value of a momentum measurement of the oscillator with state vector ψ_t is

$$\langle \mathsf{P} \rangle = \int_{\mathbb{R}} \overline{\psi_t(x)} \mathsf{P} \psi_t(x)\, dx = -\mathbf{i}\hbar \int_{\mathbb{R}} \overline{\psi_t(x)} \frac{\partial}{\partial x} \psi_t(x)\, dx,$$

and, it turns out, the probability amplitude for the momentum of the oscillator is the slightly modified Fourier transform

$$\widehat{\psi}_t(p) = \frac{1}{\sqrt{2\pi\hbar}} \int_{\mathbb{R}} \psi_t(x) e^{-\mathbf{i}px/\hbar}\, dx.$$

In particular, the probability that the result of a momentum measurement at time t is in the numerical range $[\alpha, \beta]$ is

$$\int_\alpha^\beta |\widehat{\psi}_t(p)|^2\, dp.$$

Why the interpretation articulated thus far should hold is still not in any way clear. Intrepidly, we continue. It is particularly enlightening to consider, in a bit more detail, energy measurements on the harmonic oscillator. The expected value of an energy measurement on the system with state vector ψ_t is

$$\langle \mathsf{E} \rangle = \int_{\mathbb{R}} \overline{\psi_t(x)} \mathsf{E} \psi_t(x)\, dx = \mathbf{i}\hbar \int_{\mathbb{R}} \overline{\psi_t(x)} \frac{\partial}{\partial t} \psi_t(x)\, dx.$$

A straightforward calculation using the orthonormality of the sequence of eigenfunctions ψ_N gives

$$\langle \mathsf{E} \rangle = \sum_{N=0}^{\infty} \bar{c}_N E_N c_N = \sum_{N=0}^{\infty} E_N |c_N|^2.$$

Now recall that an energy measurement of the system yields one of the eigenvalues E_N with certainty. In fact, if the state vector is the energy eigenfunction ψ_N, then the energy measurement returns the corresponding eigenvalue E_N with certainty. What happens in the general case with an arbitrary state vector ψ_t? Comparing the formula for the expected value $\langle \mathsf{E} \rangle$ with the usual formula for the expected value of a discrete random variable in probability theory suggests the answer. The coefficient c_N of ψ_N that arises when the state vector ψ_t is expanded in energy eigenfunctions is the probability amplitude, and its square modulus $|c_N|^2$ the probability, that the value E_N will be observed upon an energy measurement. Moreover, the orthodox interpretation demands that when an energy measurement of the system with state vector ψ_t yields the eigenvalue E_N, then, immediately after the measurement, the state vector of the system is the energy eigenfunction ψ_N. This is often described quite graphically by saying that, upon measurement with resulting value E_N, the wave function ψ_t **collapses** or is **projected** onto the state vector ψ_N.

The connection with classical mechanics is provided by Ehrenfest's theorems, described in Lecture 8, a glimpse of which is appropriate here. We shall see how the classical equation of motion arises out of the quantum treatment. The expected values of position and momentum measurements as functions of time are

$$\langle \mathsf{X} \rangle(t) = \int_{\mathbb{R}} \overline{\psi_t(x)} \mathsf{X} \psi_t(x)\, dx = \int_{\mathbb{R}} \overline{\psi_t(x)} x \psi_t(x)\, dx,$$

$$\langle \mathsf{P} \rangle(t) = \int_{\mathbb{R}} \overline{\psi_t(x)} \mathsf{P} \psi_t(x)\, dx = -\mathrm{i}\hbar \int_{\mathbb{R}} \overline{\psi_t(x)} \frac{\partial}{\partial x} \psi_t(x)\, dx.$$

A calculation using the full Schrödinger equation for $\psi_t(x) = \Psi(x, t)$ and a couple of integrations by part, using also the fact that ψ_t and its derivatives up to second order vanish at infinity, provides that

$$\frac{d\langle \mathsf{P} \rangle}{dt} = -m\omega^2 \langle \mathsf{X} \rangle = -k \langle \mathsf{X} \rangle,$$

recovering the classical equation of motion for the harmonic oscillator when the position x and momentum p are interpreted as the expected values of those attributes.

The answer we have given to the question of the title of this section is termed by Redhead (1987) the **minimal instrumentalist interpretation** of quantum mechanics. It posits that the meaning of quantum mechanics is found in the calculation of probabilities for various possible instrument readings when experiments are performed to measure system attributes. It tells how the formalism of the mathematics is related to possible experimental outcomes, but it does not attempt to construct a correspondence between mathematical objects like the wave function and various physical attributes that in any way models or reflects the "real status" of those attributes. Instead, the correspondence constructed is between fairly abstract mathematical expressions and very concrete instrument readings. In this view, quantum mechanics is a calculus for determining probabilities that give accurate statistical predictions of physical measurements, not a theory that provides a mental image of the micro-world that to some degree comports with reality.

1.4 Foundational Issues

I am convinced of the reciprocal usefulness of a dialogue between physics and philosophy. This dialogue has played a major role during other periods in which science faced foundational problems. In my opinion, most physicists underestimate the effect of their own epistemological prejudices on their research, and many philosophers underestimate the influence – either positive or negative – they have on fundamental research. On the other hand, a more accurate philosophical awareness would greatly help physicists engaged in fundamental research: Newton, Heisenberg, and Einstein could not have done what they did if they were not nurtured by (good or bad) philosophy ... I wish that contemporary philosophers concerned with science would be more interested in foundational problems that science is facing today. It is here, I believe, that stimulating and vital issues lie.

Carlo Rovelli[2]

There are several natural concerns that arise when the quantum world view is first encountered. Anyone who wishes for a deeper understanding of these issues is forced to face profound questions concerning both the nature of reality and what can be known about that reality. As the deepest questions are, ultimately, both ontological and epistemological in nature, a serious engagement with metaphysics is warranted for those who wish more than an instrumentalist answer to the question titling the previous section. Many scientists do not bother with these foundational issues and are content with the standard interpretation of quantum mechanics, known as the **Copenhagen interpretation**. This interpretation was worked out by Born, Heisenberg, and, especially, Bohr, in the decade following the introduction of wave mechanics by Schrödinger and matrix mechanics by Heisenberg, and it has become the accepted dogma of most quantum mechanics texts. It goes beyond the minimal instrumentalist interpretation in certain key ways, purporting to explain something of the gap between the quantum description of the micro-world and the very successful classical newtonian description

[2] From "Quantum spacetime: What do we know?" in *Physics Meets Philosophy at the Planck Scale*, 2001.

of the macro-world. It dogmatically asserts that the quantum formalism gives a complete description of micro-processes, in that there is no more fundamental theory of these processes from which quantum mechanics emerges. The physics community has settled upon the Copenhagen interpretation as its orthodox interpretation and tends to dismiss any philosophical questions that arise from students as having been already examined and answered as far as is possible by the orthodox interpretation. This of course is quite understandable, as the training of the professional practitioner must be aimed at mastering the practical technicalities of quantum mechanics and field theory; this is significantly difficult and perhaps leaves little time for philosophical reflection, especially as the job of a professional physicist is to advance the science with new discoveries and fresh explanations.

It seems to this writer, however, that even those whose natural inclination is toward a broader or deeper understanding than that given by the orthodox interpretation are discouraged from any sort of further investigation of the foundations. Indeed, even asking questions beyond the minimal interpretation is frowned upon, as an engagement in metaphysical speculation that has no place in modern science. This view, though unfortunately widespread, suffers from an impoverished modern misunderstanding of what constitutes a metaphysical position, caricaturing it as mere opinion embellishing the facts with mental baggage that has no scientific worth precisely because it is thought to have no falsifiable physical implications.

This view is at odds with the ancient, mature view of Aristotle, for example, that understands the role of metaphysics as the analysis of the generic parts of what exists, analysis informed by experience and the facts of the matter and that frames reality in an ordered mental landscape. Physics provides a calculus for understanding laboratory readings, and metaphysics, in its conversation with physics, provides a mental landscape full of images that are meant to comport with what really exists underneath and behind the calculus. This view also is at odds with the history of the generation that discovered the machinery of quantum mechanics. That generation was conversant with the metaphysical milieu of the intellectual world of their day, which informed the framework in which they developed their science. Indeed, most leading scientists of that day had exposure to both classical and modern philosophical ideas in their general education, something, unfortunately, modern scientific education has jettisoned as so much bombast. Einstein was influenced greatly in his thinking about physics by the philosopher Mach and Bohr by the existentialism of Kirkegaard, and the influence of positivism and the currency of the Vienna Circle on the physicists of that generation is unmistakable. Not that philosophy guided the development of quantum mechanics, for the history of that development shows how contingent was the development on the desire to get a theory that works, and the metaphysical assumptions of the principle quantum protagonists were surprisingly malleable toward that goal. Although it may be the case that philosophical considerations fail to lead to discoveries, it certainly is also the case that constructing a meaningful understanding of the ontological status of those discoveries is impossible without such considerations. Interpretation does not take place in a philosophical vacuum, despite that popular, and fictional, account of science as the unprejudiced, purest form of knowledge that accepts only that which is forced upon it by the hard reality of the laboratory. The Copenhagen interpretation seems content to give up on the possibility of even discussing the ontological status of the micro-world, prematurely cutting off routes of exploration that might prove fruitful in our understanding of the world.

One of the author's goals is to challenge the reader to struggle with the foundational questions if for no other reason than to become conversant with some of the most interesting issues concerning the natural world. At this point, we list and discuss some concerns that might have arisen for the novice upon an introduction to the machinery of quantum mechanics.

1.4.1 Dynamics and Collapse: The Problem with Measurement

The astute reader might have noted a rather bizarre feature of orthodox quantum mechanics, namely, that there seemingly are two different ways in which a state vector $\psi = \psi_{t=0}$ may evolve in time past its initial value. On the one hand, the Schrödinger equation implies continuous time evolution via $\psi_t = \Psi(-,t)$ while, on the other, an energy measurement induces an abrupt, discontinuous collapse $\psi_t \rightsquigarrow \psi_N$ to an eigenfunction; this collapse is certainly not described by the Schrödinger equation.

A closer look at the continuous time evolution of the Schrödinger equation reveals that this evolution is unitary. Indeed, setting $\psi = \psi_{t=0}$, we may write the wave function as

$$\Psi(-,t) = \psi_t = \mathsf{U}_t\psi,$$

where U_t is the unitary operator that acts on the basis eigenfunctions via

$$\mathsf{U}_t\psi_N = e^{-\mathrm{i}E_N t/\hbar}\psi_N.$$

The picture this paints is that a system described by a state vector ψ at time $t = 0$ will evolve deterministically, linearly, continuously, and via unitary evolution, thus preserving the L^2-norm of the state vector and ensuring its use to calculate probabilities. However, when, for example, an energy measurement is made with result E_N, out of a number of possible results, each with a nonzero probability of occurrence, the state vector immediately collapses to the energy eigenfunction ψ_N, a discontinuous jump not governed by the Schrödinger equation. Are there really two different sorts of time evolution demarcated by the presence of a measurement? Going even further, exactly what is it about a measurement that makes it a measurement, as opposed to merely the traversal of the atomic particle through a potential energy environment governed by the Schrödinger equation? More directly, how does the quantum system under investigation know that it has entered the realm of measurement, as opposed to unitary evolution, and therefore should collapse rather than continue its unitary evolution? The orthodox interpretation is silent on these issues.

One possible response is that we have incomplete knowledge of the procedure that attempts to prepare the system in its initial state described by ψ as well as the microphysics of the measuring device itself. If we were able to include a more precise procedure for identifying both the initial state vector and the potential field of the oscillator during measurement, and were able to quantize a Hamiltonian of the form

$$\frac{p^2}{2m} + \frac{1}{2}m\omega^2 x^2 + W(x, p, t),$$

where W describes the contributions to the potential energy environment from the measuring device, would the evolution, including the supposed collapse, be seen as continuous and unitary and described by the Schrödinger equation? Under this view then, collapse would be seen as a bookkeeping device that arises from incomplete knowledge of both the initial state vector and the potential energy environment. A critical, even fatal, objection to this view of collapse asks how is it that several systems described initially by a very small variation from a fixed state vector ψ, when subjected to what seems to be exactly the same measurement environment, yield a disparate range of results, each with positive probability. In our harmonic oscillator example, consider a system prepared in an initial state $\psi = \frac{1}{\sqrt{2}}\psi_0 + \frac{1}{\sqrt{2}}\psi_N$, where N is so large that the two possible results of an energy measurement – E_0 and $E_N = E_0 + N\omega\hbar$ – are vastly different. In this example there is very little room to allow for variations from ψ – the coefficients may be slightly different from $\frac{1}{\sqrt{2}}$ and some other eigenfunctions may have coefficients not exactly equal to zero – yet, upon energy measurement, these two vastly different possible values appear with essentially equal probability. Is it really the case that micro-differences in system preparation and in W are amplified into incredibly vast differences in measured values? General mathematical considerations on the continuous dependence of solutions on parameters, in this case parameters that would appear in W, suggest that this is implausible. The struggle to resolve these issues is known as the **measurement problem**; it asks deep questions about the nature of reality, what constitutes a measurement, and even the status of the observer in the evolution of a quantum system.

1.4.2 Is Quantum Mechanics Complete?

A natural question to ask about the quantum mechanical formalism is whether it is a theory of aggregates, something akin to statistical mechanics. Whereas statistical mechanics describes aggregates of particles in space, quantum mechanics, perhaps, describes aggregates of similarly prepared systems in time. Conceivably there is a deeper theory that is not inherently probabilistic and contains exact positions and momenta and energy in its formalism, and which might predict experimental outcomes without the uncertainties entailed in the quantum formalism. Quantum mechanics would be understood as the statistical mechanics of this deeper theory and would describe the average behavior of systems prepared in the same state. This deeper theory would, of course, have to reproduce the statistical predictions of quantum mechanics as those predictions have a century of unbroken success. Perhaps even this deeper theory could predict nothing more than can the quantum formalism, in the way that the behavior of large aggregates in classical mechanics are inaccessible to particle by particle analysis so that the "correct" theory, Newtonian mechanics, offers no advantage over the emergent theory of statistical mechanics. What would be the advantage then of a deeper theory? Perhaps it would offer a lens into the true nature of the microworld, the understanding of which might yield insights of use in the further development of fundamental physics.

The orthodox view rejects the possibility of quantum mechanics as a theory emergent from a deeper one and posits that quantum mechanics is a **complete** theory underlying reality. In particular, the orthodox view holds that the wave function $\Psi(x, t)$ of a system contains all the information that can be obtained about the system, so all predictions must be based on its analysis and manipulation. There is nothing more basic or deeper to understand about the system. This paradigm has held sway as the dominant one since the 1930s, and, unfortunately, has been successful at cutting off debate prematurely on foundational issues. Only since the work of John Bell in the 1960s, which began to have an influence a couple of decades ago, has the orthodox hegemony been challenged by mainstream physicists. Though orthodoxy with its completeness hypothesis is still pervasive among professional practitioners, students of the philosophy of science and a minority of professional physicists are not so sanguine about its triumphalist status. What must be acknowledged is that the completeness of quantum theory is ultimately, rather than a view dictated by science, a dogmatic philosophical position, a metaphysical view. That there is in fact at least one proposed deeper theory – Bohmian mechanics, often dismissed by professional physicists – that provably reproduces the statistical results of quantum theory, and from which quantum mechanics is emergent, falsifies the view that science demands the completeness hypothesis.

The most devastating critique of completeness is still that of Einstein, Podolsky, and Rosen (1935), the so-called **EPR paradox**, extended by Bohm (1951) and J.S. Bell (1964). Obscure metaphysical arguments won the day in the 1930s when the majority of physicists considered that Bohr had answered this critique adequately, though his explanations and arguments are unclear and doubtful. Bohr's answer reinforced von Neumann's earlier proof in 1932 of the impossibility of a deeper theory based on so-called "hidden variables," which was presented in his influential work *The Mathematical Foundations of Quantum Mechanics.* With Bohr declared the victor and the foundational issues settled, professional physicists went about their calling of technically refining quantum mechanics with incredible success over the next four decades. Though physics proper flourished, a side effect of this was a three decade delay in understanding the importance of the most profound difference between physical reality and the classical world view, viz., the presence of nonlocality in the physical universe. In 1964, John Stewart Bell gave a cogent analysis of EPR that articulated precisely the role of inseparability and nonlocality in quantum mechanics, and two years later his analysis of von Neumann's impossibility argument, in Bell (1966), showed that it applied only to rather naïve hidden variables theories. He further elucidated Bohm's hidden variables theory of 1952 and demonstrated that it is exactly of the type not ruled out by von Neumann. Physicists by and large paid little attention to Bell's metaphysical musings, but philosophers of science and a few renegade physicists took notice, and there began a modest, but fruitful, research program in the foundations of the subject that has produced highly refined results which place strict constraints on how we are to understand the micro-world. It has also sharpened our understanding of the incompatibilities between orthodox quantum theory and special relativity that have lain hidden for decades, incompatibilities as yet unresolved.

We will present Bell's analysis of the EPR paradox in Lecture 22, which leads to an understanding of this most profound difference between reality and classical mechanics in its articulation of inseparability and nonlocality. Quantum theory is nonlocal, and any proposed deeper theory from which quantum mechanics emerges must itself be nonlocal. Not only that, irrespective of whatever theory is advanced to describe nature, whether quantum or any other, deterministic or stochastic, the Aspect experiments of 1982 confirmed that nonlocality is in fact a property of the real world. Whatever theory one advances must be nonlocal, as of course is quantum mechanics as well as Bohmian mechanics.

1.4.3 Determinism and Causality: Is Reality Irreducibly Probabilistic?

If quantum mechanics is complete, then reality is irreducibly and inherently probabilistic. This is a profound insight into the working of nature and has significant implications for determinism and causality. The Schrödinger equation is entirely deterministic. It provides for the deterministic evolution of the wave function, as long as measurements are not applied to the system. A measurement produces a non-deterministic, statistically allowable collapse of the wave function. This feature of quantum mechanics is part of the orthodox understanding, but what is often not articulated is that the implication of the possibility of collapse into different mutually orthogonal states from a single initial wave function, when completeness is assumed, is that nature, at its basic micro-level, is acausal, at least under the traditional understanding of causation. If the measurement of energy produces E_0 rather than the equally likely E_N when the system starts in state $\psi = \frac{1}{\sqrt{2}}\psi_0 + \frac{1}{\sqrt{2}}\psi_N$, it is not because of some difference in the system immediately before measurement that somehow led to one energy value being chosen over the other. One might say that the energy value E_0 resulted from the fact that the system was prepared initially in a state ψ where, in the state vector's eigenfunction expansion, the eigenfunction ψ_0 has a nonzero coefficient. But at the micro-level, at the juncture of decision between E_0 and E_N, there is no cause that explains the one measured value over the other. All that can be said is that there was a possibility of E_0 upon measurement, with a well-defined probability, and the micro-mechanics of the measurement allowed, but did not cause, the realization of that measurement and the immediate collapse of the system into the state ψ_0. With exactly identical antecedents, though, the outcome could equally well have been E_N. This point cannot be over-emphasized and bears careful consideration by the reader. Reiterating, in the micro-world, *if orthodox quantum mechanics holds sway then with exactly identical antecedents, two mutually exclusive outcomes are possible, neither one specified nor individually caused, but both allowed, by the antecedents.*

1.4.4 What about Relativity?

Of course we should not expect the Schrödinger equation for the harmonic oscillator to describe an oscillator where relativistic speeds are allowed. Indeed, the Schrödinger equation is obviously not symmetric in the time and space derivatives, as would be expected for a relativistically correct equation, and its derivation from a classical nonrelativistic Hamiltonian makes it suspect. Perhaps the reader already has anticipated a way forward: instead of quantizing the nonrelativistic Hamiltonian, why not begin with the relativistic expression for the energy and quantize it? In fact, such a procedure produces the Klein–Gordon equation, which does describe some relativistic particles but not all. The surprising thing about constructing a fully relativistic quantum theory of particles is that the resulting theory automatically includes much of what we observe in nature but which the quantum mechanics of these lectures does not describe. This includes the two basic kinds of particles, bosons and fermions, as well as antiparticles, which fall out of the mathematics of the Dirac equation. Further, when the fields of classical physics are quantized, particle pair production and annihilation can be described, and various relativistic field corrections to the calculations of the Schrödinger and Dirac equations can be computed. These achievements are some of the most sublime and profound of twentieth century physics.

Unfortunately, not all is well in the marriage of quantum mechanics with relativity. Achieving a quantum version of gravity is still well beyond current understanding, and even when restricted to the special theory of relativity, problems arise. The deepest difficulties are in harmonizing the nonlocality in quantum theory with Lorentz invariance. This is a problem even if one subscribes to the orthodox interpretation, as the collapse of the wave function seems to require preferred reference frames along which such collapse occurs. The problem remains for most heterodox interpretations, including Bohmian mechanics.

In the course of these lectures, even more unexpected and counter-intuitive implications of the quantum world view than so far elucidated will be encountered. This introductory lecture will have served its intended purpose if the reader has been inspired to learn more deeply the machinery of quantum mechanics and to engage with its foundational questions at a more than cursory level.

1.5 End Notes

In the End Notes to each lecture, I will not only mention the most significant references used for the lecture but also will point the reader to resources that I have found helpful in expanding my understanding of the issues.[3] The End Notes for Lecture 1 concentrate on resources for exploring the philosophical issues that arise in the study of quantum mechanics and the historical development of various approaches to resolving these philosophical issues. Since the emphasis in the remaining lectures is on the mathematical theory rather than on the philosophical and historical issues, the opportunity is taken here for a rather more extensive end note discussion than appears in subsequent chapters.

Jammer (1974) *The Philosophy of Quantum Mechanics: The Interpretations of Quantum Mechanics in Historical Perspective* is the standard

[3] The observant reader may have noticed that I have reverted to the first person singular voice in this section. Throughout the main body of this book I am following the fairly common practice in the mathematical literature of using the first person plural rather than the singular, this signifying "the reader and the author," that we are on a journey of discovery and understanding together. Nonetheless, in the End Notes (as well as in the Prolegomenon), I revert to the singular as the discussion concerns personal opinion and practice and advice to the reader or the references that I, personally, have found valuable in either developing the ideas of the lecture or in gaining deeper understanding of the topics.

work on the interpretations of quantum mechanics, their philosophical implications, and their interrelations that were current up to the year of its publication. The point of view of that work is historical and its goal is a critical analysis of all the major interpretive proposals. The dominant interpretation of quantum mechanics is the Copenhagen interpretation of Bohr, Born, Heisenberg, and others, and it is used routinely in physics textbooks to provide meaning to the quantum mathematical machinery. It is criticized today more than ever, and physicists generally are less committed to it than were those of preceding generations. John Bell's analyses of the EPR paradox, the von Neumann impossibility proof of hidden variables, and the mechanics of Bohm and de Broglie have done more to overthrow the dominance of the Copenhagen interpretation than anything else. Beller (1999) *Quantum Dialogue: The Making of a Revolution* is a critical-historical study of the evolution and acceptance of the Copenhagen interpretation that challenges the coherence of the interpretation. This scholarly work demonstrates how contingent was the acceptance of the dominant interpretation on historical accident, rhetorical skill, forceful personalities, scientific ambition, and philosophical beliefs. Wick (1995) *The Infamous Boundary: Seven Decades of Controversy in Quantum Physics* recounts the major controversial interpretive features of quantum mechanics in an engaging historical retelling of the development of the theory. Wick's use of biographical vignettes of the principle players is especially effective in capturing the interest of the reader. Whitaker (2006) *Einstein, Bohr, and the Quantum Dilemma* thrashes out the arguments for and against the status of quantum mechanics as a complete theory through the historical lens of the famous Bohr–Einstein debates of the 1930s. This book reviews the physical arguments on both sides, updating them with the insights of Bohm, Bell, and others, and in doing so penetrates deeply into the conceptual problems of quantum theory. Bub (1997) *Interpreting the Quantum World* concentrates on the measurement problem introduced by the orthodox interpretation and succeeds in showing how to construct all possible no-collapse interpretations, subject to certain constraints imposed by nature. Laloë (2012) *Do We Really Understand Quantum Mechanics?* is a recent entry into this genre, which surveys the current understanding of the meaning of quantum mechanics and compares and contrasts the various interpretations. It includes not only those interpretations covered by Jammer but also more recent contributions to the field such as various modal interpretations, continuous spontaneous localizations (the GRW model), the transactional interpretation, the consistent histories approach, as well as the impact that quantum computing is having on the conceptual basis of the theory.

It is not an exaggeration to say that the work of John Bell in the 1960s has had a great impact on critical analysis of the conceptual foundations of quantum mechanics. Though Einstein and Schrödinger had pointed out the nonlocality of quantum mechanics and Bohr had responded with the idea of the inseparability of quantum systems, it wasn't until Bell's work that physicists really began to appreciate the radical nature of these ideas, and how nonlocality and inseparability are quantum mechanics's most profound break with classical mechanics. Bell's assertion that locality is subject to experimental falsification came as a surprise to the physics community, and the Aspect experiments of the 1980s, which ruled out locality and affirmed that nature is nonlocal, led to a reexamination of the conceptual basis of the theory by many researchers. Bell (1987) *Speakable and Unspeakable in Quantum Mechanics* is a compilation of all Bell's published and unpublished papers on the conceptual basis of the subject. Cushing and McMullin (1989) *Philosophical Consequences of Quantum Theory: Reflections on Bell's Theorem* is a collection of papers by philosophers and physicists who wrestled with the meaning and implications of Bell's insights and the experimental verification of nonlocality. Baggott (1992) *The Meaning of Quantum Theory* presents a historically based, detailed, and lucid account at an advanced undergraduate level of the development of quantum theory, the Copenhagen interpretation, the Bohr–Einstein debates, the EPR paradox, Bell's theorem, locality and the Aspect experiments, and various interpretations.

The following two references are sophisticated philosophical treatises that concentrate attention on incompleteness, nonlocality, and measurement and are written by philosophers with a clear and deep understanding of quantum mechanics, including its mathematical machinery. Both are very careful and deep studies that are very precise in their explications and claims. After carefully developing the formalism of quantum mechanics, Redhead (1987) *Incompleteness, Nonlocality, and Realism: A Prolegomenon to the Philosophy of Quantum Mechanics* explores the question of its completeness in the context of the EPR paradox, Bell's theorem, and the Kochen–Specker theorem. Dickson (1998) *Quantum Chance and Non-locality: Probability and Non-locality in the Interpretations of Quantum Mechanics* gives a very careful examination of determinism and nonlocality, concentrating on the measurement problem. The nonlocality of the physical world poses problems for physicists, since how to harmonize nonlocality with relativity is in no way obvious. The work of yet another philosopher merits attention in the context of Bell's theorem, namely that of Tim Maudlin, who examined the incompatibility of nonlocality with relativity in Maudlin (2011) *Quantum Non-Locality and Relativity: Metaphysical Intimations of Modern Physics*, recognized as "the premier philosophical study of Bell's theorem and its implication for the relativistic account of space and time."[4] Finally, the book M. Bell and S. Gao (2016) is a compilation of scholarly articles centered on the theme of its title, *Quantum Nonlocality and Reality: 50 Years of Bell's Theorem.*

Aharonov and Rohrlich (2005) *Quantum Paradoxes: Quantum Theory for the Perplexed* takes the novel approach of beginning each chapter with a paradox that motivates the rest of the chapter. Quantum mechanics is developed as a remedy to paradoxes, whose resolutions help the reader understand the rather unintuitive quantum calculus. This leaves open many philosophical issues as the cost of resolving paradoxes.

To close this section I mention a valuable nonphilosophical work that presents a purely historical account of the development of the idea of spin in quantum mechanics. Detailed in its mathematical explanations, Tomonaga (1997) *The Story of Spin*[5] presents an enlightening historical study of the main protagonists and their attempts to understand this nonclassical, purely quantum mechanical, attribute of elementary particles.

[4] From the back cover of the 2011 third edition.
[5] A translation by Tekeshi Oka of the 1974 Japanese original.

2

The Mathematical Structure of Quantum Mechanics

By 1929, however, [Marshall] Stone had moved into the abstract theory of unbounded self-adjoint operators in Hilbert space. This new work culminated in 1932 with the publication by the American Mathematical Society of his massive treatise, *Linear Transformations in Hilbert Space and Their Applications to Analysis*, a book that has been deemed "one of the great classics of twentieth-century mathematics" [by George Mackey]. In it, Stone succeeded in extending David Hilbert's spectral theorem from bounded to unbounded operators. As George Mackey put it "... his extension was made necessary by the problem of making mathematically coherent sense of the newly discovered refinement of classical mechanics known as quantum mechanics. Here an important part of the problem was discovering the 'correct' definition of self-adjointness for unbounded operators. This correct definition is rather delicate and the extension of the older theory of Hilbert and others was a major task."

Karen Hunger Parshall
Marshall Stone and the Internationalization of the American Mathematical Research Community, 2009

In this lecture we present an abstract, minimalist, list of the quantum mechanical rules for calculating statistical predictions of the outcomes of measurements on pure-state single-particle quantum systems. In the next lecture we will see how these rules are used to calculate expected outcomes of experimental measurements and what they imply about the status of the physical attributes that a system possesses. The rules presented here are deficient in that they make no mention of collapse at the point of measurement. This deficiency will be removed at the end of the next lecture when the **projection postulate** is presented, which in turn is dissected in Lecture 4. We also introduce **wave mechanics** as an instantiation of these minimalist axioms.

Self-adjoint operators on Hilbert spaces are of primary importance in understanding the mathematical structure of quantum mechanics. However, the theory of self-adjoint operators is vast, intricate, subtle, and complicated. Our aim in these lectures is not to give a maximally complete and rigorous treatment of the mathematics that arises from quantum mechanics, but to present a mathematically honest exploration of quantum mechanics that acknowledges the points of difficulty in making the subject rigorous but accepts some incompleteness in order to cover a significant number of topics in modern quantum physics. We will flesh out some intricacies of the theory of adjoint operators by presenting a quick tour of the basics and applying the theory to analyze the position and momentum operators of wave mechanics. However, in subsequent lectures we will take a practical approach to these matters and not worry too much about technical mathematical details whenever an elucidation of those details offers little insight in understanding quantum mechanics. The End Notes to the lecture will provide references to the literature to fill in any gaps.

2.1 The Minimalist Rules

We condense in a series of four apparently uninformative axioms the overarching methodology of quantum mechanics. These will be expanded upon, explained, and added to, over the course of the book. We take at this point the terms **state, observable, measurement**, and **quantum system** as primitive terms that will be given substance and weight as the discussion develops. The formalism of much of quantum mechanics may be given as follows.

Axiom 1: State space. The states of a quantum system are represented by nonzero vectors in a separable complex Hilbert space $\mathcal{H}$ that forms the **state space** of the system. Two vectors represent the same state $\mathfrak{S}$ whenever they are nonzero complex multiples of one another. For **state vectors** φ and ψ in $\mathcal{H}$, their Hermitian inner product is denoted by $\langle\varphi|\psi\rangle$, which is conjugate linear in φ and linear in ψ. The state vector ψ is **normalized** if $\langle\psi|\psi\rangle = \|\psi\|^2 = 1$.

Axiom 2: Observables. The observables of the system are represented by self-adjoint linear operators defined on dense linear subspaces of $\mathcal{H}$.

Axiom 3: Expectation values. If an observable a is represented by the operator A, then the **expected** or **expectation value** for measuring a on a system represented by state vector ψ is

$$\langle \mathsf{A} \rangle_\psi = \text{Exp}_\psi(\mathsf{A}) = \frac{\langle \psi | \mathsf{A}\psi \rangle}{\|\psi\|^2},$$

or, if ψ is normalized,

$$\langle \mathsf{A} \rangle_\psi = \langle \psi | \mathsf{A}\psi \rangle.$$

Axiom 4: Time evolution. Let ψ_t represent the state of the system at time t. The time evolution of the state vector ψ_t is governed by the **Schrödinger equation**:

$$\mathrm{i}\hbar \frac{d\psi_t}{dt} = \mathsf{H}\psi_t,$$

where H is a self-adjoint operator on a dense linear subspace of $\mathcal{H}$ and is known as the **Hamiltonian operator** for the system.

The next few lectures will unpack these axioms and examine in some detail their content and implications. Before one can use these rules to perform quantum mechanical calculations on a physical system, one must first answer the following questions for the system under investigation.

Question 1. What Hilbert space $\mathcal{H}$ serves as the state space for the system?

Question 2. Which self-adjoint operators A represent which observables a of interest?

Question 3. What is the Hamiltonian operator H for the system?

One answer to these questions, of sufficient generality to include the statistical predictions of most measurements obtained from the position and momentum attributes of a single particle, is known as **wave mechanics**.

2.2 Wave Mechanics

A wave mechanical analysis of the simple harmonic oscillator system, i.e., that of a single particle in a quadratic potential well, was presented in the previous lecture. The formalism is due primarily to Schrödinger and is, perhaps, the model interpretation of the abstract rules of quantum mechanics. In this interpretation, which governs one-particle quantum systems, the axioms are interpreted as follows.

Axiom 1. The state space is $\mathcal{H} = L^2(\mathbb{R}^3)$, the Hilbert space of complex-valued, square-integrable Lebesque measurable functions on $\mathbb{R}^3$. The inner product is

$$\langle \varphi | \psi \rangle = \int_{\mathbb{R}^3} \overline{\varphi}\psi.$$

For many applications, the state vector of a system may be taken to be piecewise continuous and the integral of the inner product may be taken to be the Riemann integral.

Axiom 2. The basic observables are the three components of position $\mathbf{x} = (x_1, x_2, x_3)$ and the canonically conjugate momentum $\mathbf{p} = (p_1, p_2, p_3)$ of the particle. The operators representing position and momentum are given by the components of the triples $\mathsf{X} = (\mathsf{X}_1, \mathsf{X}_2, \mathsf{X}_3)$ and $\mathsf{P} = (\mathsf{P}_1, \mathsf{P}_2, \mathsf{P}_3)$, where for $j = 1, 2, 3$,

$$(\mathsf{X}_j\psi)(\mathbf{x}) = x_j\psi(\mathbf{x}),$$
$$(\mathsf{P}_j\psi)(\mathbf{x}) = -\mathrm{i}\hbar\frac{\partial\psi}{\partial x_j}(\mathbf{x}).$$

We will examine the domains of these operators as well as their self-adjointness when we discuss operators in greater detail. For now, notice that $\mathsf{X}^2 = \mathsf{X}\cdot\mathsf{X} = \mathsf{X}_1^2 + \mathsf{X}_2^2 + \mathsf{X}_3^2$ is the operator $(\mathsf{X}^2\psi)(\mathbf{x}) = \|\mathbf{x}\|^2\psi(\mathbf{x})$, and $\mathsf{P}^2 = \mathsf{P}\cdot\mathsf{P} = \mathsf{P}_1^2 + \mathsf{P}_2^2 + \mathsf{P}_3^2$ is the Laplacian[1]

$$\mathsf{P}^2 = -\hbar^2\Delta = -\hbar^2\left(\frac{\partial^2}{\partial x_1^2} + \frac{\partial^2}{\partial x_2^2} + \frac{\partial^2}{\partial x_3^2}\right).$$

Other operators, such as angular momentum operators, are defined in terms of these, the **position operator triad X** and the **momentum operator triad P**.

Axiom 3. The expectation value for the operator A when the system is represented by ψ is

$$\langle \mathsf{A}\rangle_\psi = \langle \psi | \mathsf{A}\psi \rangle = \int_{\mathbb{R}^3} \overline{\psi}\mathsf{A}\psi,$$

[1] Mathematicians use the symbol Δ for the Laplacian operator $\partial_1^2 + \partial_2^2 + \partial_3^2$ while physicists use $\nabla^2 = \nabla \cdot \nabla$. We follow the mathematicians' notation in these lectures. The reader will have noticed that we are using the physicists' convention that the square of a vector quantity is to be interpreted as the self-dot product.

when ψ is normalized.

Axiom 4. The Hamiltonian operator is obtained by **quantizing** the classical Hamiltonian of the system, i.e., by replacing the canonically conjugate dynamical variables x_j and p_j by the respective operators X_j and P_j. This takes some explanation. In the simplest case, the classical Hamiltonian with time-independent potential V is

$$H(\mathbf{x},\mathbf{p}) = \frac{\mathbf{p}\cdot\mathbf{p}}{2m} + V(\mathbf{x}).$$

Quantization then proceeds on replacing x_j by X_j and p_j by P_j, from which comes the prescription

$$V(\mathbf{x}) \rightsquigarrow V(\mathbf{X}),$$
$$\mathbf{p}\cdot\mathbf{p} \rightsquigarrow \mathbf{P}\cdot\mathbf{P} = -\hbar^2\Delta.$$

The operator $V(\mathbf{X})$ multiplies a function of the variable $\mathbf{x}$ by the potential function $V(\mathbf{x})$. Setting $\Psi(\mathbf{x},t) = \psi_t(\mathbf{x})$ gives the Schrödinger equation as

$$\mathrm{i}\hbar\frac{\partial\Psi}{\partial t} = -\frac{\hbar^2}{2m}\Delta\Psi + V(\mathbf{x})\Psi.$$

More complicated Hamiltonians, such as those for a charged particle in an electromagnetic field and those that are time-dependent, are handled in a similar fashion and will be discussed in the course of these lectures.

Furthermore, this Hamiltonian operator,

$$\mathsf{H} = \frac{\mathbf{P}^2}{2m} + V(\mathbf{X}) = -\frac{\hbar^2}{2m}\Delta + V(\mathbf{x}),$$

is self-adjoint[2] and, see Axiom 2, is the operator that corresponds to energy measurements of the system.

It will take several lectures to develop wave mechanics fully. A preview has been given in Lecture 1, where an overview of the wave mechanical analysis of the harmonic oscillator was presented, details of which are given in Lecture 11 where the Schrödinger equation is solved by an elegant algebraic analysis. Lecture 14 presents a wave mechanical analysis of a particle in a central Coulomb potential, which gives the first quantum mechanical treatment of the hydrogen atom. The success of the quantum mechanical analysis of the central potential in describing the gross features of the spectrum of the hydrogen atom is one of the

[2] This is surprisingly difficult to prove in most applications.

crowning achievements of nonrelativistic quantum mechanics. For the remainder of the present lecture, we review the basic mathematics of self-adjoint operators on Hilbert spaces and examine the position and momentum operators of wave mechanics in greater detail.

2.3 Adjoints and Self-Adjoint Operators

What is a self-adjoint operator? The quick and easy, but incomplete, answer is that the adjoint A^* of the operator A is defined by the **adjoint equation**

$$\langle \mathsf{A}^*\varphi|\psi\rangle = \langle\varphi|\mathsf{A}\psi\rangle,$$

for all vectors φ and ψ, and A is self-adjoint if $\mathsf{A} = \mathsf{A}^*$. The details are a bit more involved. This description is accurate and complete if the Hilbert space is finite-dimensional, but the situation is more intricate and in general full of subtlety. In fact it was von Neumann's and Stone's development of a rigorous mathematical treatment of the then fledgling quantum mechanics that led to an understanding of the complexities of self-adjointness in infinite dimensions. We begin with a discussion of the dual correspondence $\mathcal{H} \leftrightarrow \mathcal{H}^*$ between the Hilbert space $\mathcal{H}$ and its dual $\mathcal{H}^*$, followed by an illumination of self-adjointness in the case where $\mathcal{H}$ is finite-dimensional and, finally, an explanation of the general, unrestricted case.

2.3.1 The Dual Correspondence

The **dual space** $\mathcal{H}^*$ to the Hilbert space $\mathcal{H}$ is the linear space whose members are the continuous linear functionals defined on $\mathcal{H}$. The adjective "continuous" is unnecessary if $\mathcal{H}$ is finite-dimensional as every linear functional is then continuous, but it is necessary in general. The **dual correspondence** between $\mathcal{H}$ and its dual space $\mathcal{H}^*$ is the mapping $* : \mathcal{H} \to \mathcal{H}^*$ defined by $*(\varphi) = \varphi^*$, where φ^* denotes the continuous linear functional $\langle\varphi|-\rangle$, i.e., $\varphi^*(\psi) = \langle\varphi|\psi\rangle$ for any $\psi \in \mathcal{H}$. It is easy to see that this correspondence is conjugate linear (recall that the inner product is conjugate linear in its first argument) and injective, but the interesting fact is that it also is surjective, and hence a conjugate linear isomorphism. To see this, we need to prove that every continuous linear functional on $\mathcal{H}$ arises from the inner product in the following sense: given the continuous linear functional $f : \mathcal{H} \to \mathbb{C}$, there exists a (unique) vector $\varphi = \varphi_f \in \mathcal{H}$ such that $f = \varphi^*$, i.e., such that $f(\psi) = \langle\varphi|\psi\rangle$ for all $\psi \in \mathcal{H}$. If such a vector φ exists then it is unique, which follows easily from the injectivity of $*$. For the existence, note that $0^* = 0$ so we may assume that $f \neq 0$. It follows that $\ker f$ is not the whole of $\mathcal{H}$. Set $\varphi_f = \overline{f(u)}u$, where u is any normalized vector in $(\ker f)^\perp$, the orthogonal complement of the kernel of f. Now, since f is continuous, $\ker f = f^{-1}(0)$ is a closed linear subspace of $\mathcal{H}$ and, as such, $\mathcal{H}$ decomposes as the orthogonal direct sum $(\ker f) \perp \mathbb{C}u$. This is precisely where the continuity of f is used, for this orthogonal decomposition of $\mathcal{H}$ fails if $\ker f$ is not closed. It follows that any vector ψ may be written as $v + \lambda u$ for unique $v \in \ker f$ and $\lambda \in \mathbb{C}$. We have

$$f(\psi) = f(v + \lambda u) = \lambda f(u) = \langle\overline{f(u)}u|\lambda u\rangle = \langle\overline{f(u)}u|v + \lambda u\rangle = \langle\varphi_f|\psi\rangle.$$

Thus $f = (\varphi_f)^*$ and the dual correspondence is shown to be surjective with inverse conjugate linear isomorphism given by $f \mapsto \varphi_f$.

2.3.2 The Finite-Dimensional Case

Assume now that $\mathcal{H}$ is finite-dimensional and let A be a linear operator on $\mathcal{H}$, i.e., a linear transformation $\mathsf{A} : \mathcal{H} \to \mathcal{H}$. The **adjoint** of A is the operator $\mathsf{A}^* : \mathcal{H} \to \mathcal{H}$ defined as follows: for any vector $\varphi \in \mathcal{H}$, the composition $\varphi^*\mathsf{A}$ is a linear functional, so, by the dual correspondence, there exists a unique vector $\phi \in \mathcal{H}$ such that $\varphi^*\mathsf{A} = \phi^*$. Define $\mathsf{A}^*\varphi = \phi$. Thus, $\mathsf{A}^*\varphi$ is the pullback under the dual correspondence of the functional $\varphi^*\mathsf{A}$ or, equivalently, $\mathsf{A}^*\varphi$ is uniquely defined by the equation $\varphi^*\mathsf{A} = (\mathsf{A}^*\varphi)^*$. The operator A is said to be **self-adjoint** if $\mathsf{A} = \mathsf{A}^*$. Here is a list of elementary properties of adjoints, each of which is proved quite easily.

- The adjoint operator $\mathsf{A}^* : \mathcal{H} \to \mathcal{H}$ is linear.
- $\langle\varphi|\mathsf{A}\psi\rangle = \langle\mathsf{A}^*\varphi|\psi\rangle$ for all $\varphi, \psi \in \mathcal{H}$.
- $\langle\varphi|\mathsf{A}\psi\rangle = \langle\mathsf{A}\varphi|\psi\rangle$ for all $\varphi, \psi \in \mathcal{H}$ if A is self-adjoint.
- For linear operators A and B, $(\mathsf{AB})^* = \mathsf{B}^*\mathsf{A}^*$ and $(\mathsf{A}+\mathsf{B})^* = \mathsf{A}^* + \mathsf{B}^*$, and, for $c \in \mathbb{C}$, $(c\mathsf{A})^* = \overline{c}\mathsf{A}^*$.
- $\mathsf{A}^{**} = \mathsf{A}$.

As an example, fix two nonzero vectors α and β in $\mathcal{H}$ and consider the operator $\mathsf{A} = \alpha\beta^*$ defined by $\mathsf{A}\varphi = \langle\beta|\varphi\rangle\alpha$. Then $\mathsf{A}^* = \beta\alpha^*$ since, for all $\varphi, \psi \in \mathcal{H}$,

$$\begin{aligned}\langle\mathsf{A}^*\varphi|\psi\rangle = \langle\varphi|\mathsf{A}\psi\rangle &= \langle\varphi|\langle\beta|\psi\rangle\alpha\rangle = \langle\beta|\psi\rangle\langle\varphi|\alpha\rangle \\ &= \langle\overline{\langle\varphi|\alpha\rangle}\beta|\psi\rangle = \langle\langle\alpha|\varphi\rangle\beta|\psi\rangle = \langle\beta\alpha^*(\varphi)|\psi\rangle.\end{aligned}$$

2.3.3 The General Case

When the Hilbert space $\mathcal{H}$ is infinite-dimensional, more care is needed to uncover the correct notion of self-adjointness. This was accomplished by the American mathematician Marshall Stone in response to the mathematical needs of the infant quantum theory of 1929. The adjoint of a bounded, or continuous, linear operator A is defined exactly as in the finite-dimensional case. Indeed, the important point is that, regardless of the dimensionality of $\mathcal{H}$, the composition $\varphi^*\mathsf{A}$ is then a continuous linear functional, hence a member of the dual space $\mathcal{H}^*$, and thus the dual correspondence $\mathcal{H} \leftrightarrow \mathcal{H}^*$ may be exploited to pull $\varphi^*\mathsf{A}$ back to $\mathcal{H}$ to define $\mathsf{A}^*\varphi$, exactly as before. Unfortunately this is inadequate for quantum mechanics as many linear operators of interest, including the position and momentum operators of wave mechanics, are not bounded. To include all the linear operators appropriate for representing quantum mechanical observables, we must relax our definition of a linear operator on a Hilbert space to allow for operators whose domain is not the whole space.

John von Neumann, in his magisterial 1932 classic *The Mathematical Foundations of Quantum Mechanics*, and Marshall Stone, in his equally famous 1932 classic *Linear Transformations in Hilbert Space and Their Applications to Analysis*, articulate the following concepts, which are now the mainstay of modern functional analysis. An **operator** on a Hilbert space $\mathcal{H}$ is a linear map $\mathsf{A} : \mathcal{D}(\mathsf{A}) \to \mathcal{H}$ defined on a linear subspace $\mathcal{D}(\mathsf{A})$ of $\mathcal{H}$. Almost always the operators we deal with are **densely defined**, meaning that the domain of the operator is dense in the Hilbert space. The adjoint A^* of the densely defined operator A is the mapping defined by

$$\mathcal{D}(\mathsf{A}^*) = \{\varphi \in \mathcal{H} : (\exists\phi \in \mathcal{H})(\forall\psi \in \mathcal{D}(\mathsf{A})), \langle\varphi|\mathsf{A}\psi\rangle = \langle\phi|\psi\rangle\}$$

and

$$\mathsf{A}^*\varphi = \phi.$$

The adjoint is well-defined, i.e., ϕ is uniquely determined, since the domain of A is dense. The operator A is said to be **self-adjoint** if $\mathcal{D}(\mathsf{A}) = \mathcal{D}(\mathsf{A}^*)$, and on this common domain $\mathsf{A} = \mathsf{A}^*$.

If A and B are operators on $\mathcal{H}$ with $\mathcal{D}(\mathsf{A}) \subset \mathcal{D}(\mathsf{B})$ and if $\mathsf{A} = \mathsf{B}$ on $\mathcal{D}(\mathsf{A})$ then B **extends** A, or A **restricts** B, and we write $\mathsf{A} \subset \mathsf{B}$. It can be very difficult to verify that a given operator is self-adjoint. Often the adjoint equation $\langle\varphi|\mathsf{A}\psi\rangle = \langle\mathsf{A}\varphi|\psi\rangle$ is easy to verify for φ and ψ from a suitable dense domain $\mathcal{D}(\mathsf{A})$, which shows that $\mathsf{A} \subset \mathsf{A}^*$, but verifying that $\mathcal{D}(\mathsf{A}^*) \subset \mathcal{D}(\mathsf{A})$ can be quite challenging. A densely defined operator A that satisfies $\mathsf{A} \subset \mathsf{A}^*$ is said to be **symmetric**. Obviously a self-adjoint operator is precisely a symmetric operator such that $\mathcal{D}(\mathsf{A}^*) \subset \mathcal{D}(\mathsf{A})$. It is easy to verify that $\mathsf{B}^* \subset \mathsf{A}^*$ whenever $\mathsf{A} \subset \mathsf{B}$ thus extending A by enlarging its domain restricts the adjoint and decreases its domain. This suggests a strategy for building self-adjoint operators, viz., to symmetrically extend a symmetric operator A by enlarging its domain. Since $\mathsf{A} \subset \mathsf{A}^*$, as the domain of A enlarges, that of its adjoint reduces in size, and the goal is to force the two domains to coincide. We will see an example of this strategy when we analyze the momentum operator.

Here is a list of elementary properties of adjoints, each of which is proved quite easily.

- The domain $\mathcal{D}(\mathsf{A}^*)$ is a linear subspace of $\mathcal{H}$.
- The adjoint mapping $\mathsf{A}^* : \mathcal{D}(\mathsf{A}^*) \to \mathcal{H}$ is linear.
- $\langle \varphi | \mathsf{A}\psi \rangle = \langle \mathsf{A}^*\varphi | \psi \rangle$ for all $\varphi \in \mathcal{D}(\mathsf{A}^*)$ and $\psi \in \mathcal{D}(\mathsf{A})$.
- $\langle \varphi | \mathsf{A}\psi \rangle = \langle \mathsf{A}\varphi | \psi \rangle$ for all $\varphi, \psi \in \mathcal{D}(\mathsf{A})$ if A is symmetric.
- For operators A and B, $(\mathsf{AB})^* = \mathsf{B}^*\mathsf{A}^*$ and $(\mathsf{A} + \mathsf{B})^* = \mathsf{A}^* + \mathsf{B}^*$ on suitably restricted domains, and, for $c \in \mathbb{C}$, $(c\mathsf{A})^* = \overline{c}\mathsf{A}^*$.
- If $\mathcal{D}(\mathsf{A}^*)$ is dense in $\mathcal{H}$ then A^{**} is defined and $\mathsf{A} \subset \mathsf{A}^{**}$.

This relaxation of the notion of a linear operator to include those whose domain is not the whole of the Hilbert space turns out to be related intimately to continuity. For example, a self-adjoint operator A with $\mathcal{D}(\mathsf{A}) = \mathcal{H}$ is necessarily continuous; this known as the Hellinger–Toeplitz Theorem.

2.4 The Position and Momentum Operators

In this section, we verify that appropriately defined one-dimensional position and momentum operators are self-adjoint. In the Hilbert space $\mathcal{H} = L^2(\mathbb{R})$, let

$$\mathcal{D}(\mathsf{X}) = \{\varphi \in L^2(\mathbb{R}) : x\varphi(x) \in L^2(\mathbb{R})\}.$$

Notice that $\mathcal{D}(\mathsf{X})$ is a linear subspace of $L^2(\mathbb{R})$ that contains $C_c^\infty(\mathbb{R})$, the set of smooth complex-valued functions with compact support, a dense subspace of $L^2(\mathbb{R})$. Define the operator $\mathsf{X} : \mathcal{D}(\mathsf{X}) \to L^2(\mathbb{R})$ by $(\mathsf{X}\varphi)(x) = x\varphi(x)$. For any $\varphi, \psi \in \mathcal{D}(\mathsf{X})$, we have

$$\langle \varphi | \mathsf{X}\psi \rangle = \int_{\mathbb{R}} \overline{\varphi} x \psi = \int_{\mathbb{R}} \overline{x\varphi} \psi = \langle \mathsf{X}\varphi | \psi \rangle.$$

It follows that $\mathcal{D}(\mathsf{X}) \subset \mathcal{D}(\mathsf{X}^*)$ and $\mathsf{X} = \mathsf{X}^*$ on $\mathcal{D}(\mathsf{X})$, i.e., $\mathsf{X} \subset \mathsf{X}^*$. To verify that X is self-adjoint it remains only to prove that $\mathcal{D}(\mathsf{X}^*) \subset \mathcal{D}(\mathsf{X})$. Let $\varphi \in \mathcal{D}(\mathsf{X}^*)$. Then, for all $\psi \in \mathcal{D}(\mathsf{X})$, we have $\langle \mathsf{X}^*\varphi|\psi\rangle = \langle\varphi|\mathsf{X}\psi\rangle$, which implies that

$$\int_{\mathbb{R}} \overline{(\mathsf{X}^*\varphi)}\psi = \int_{\mathbb{R}} \overline{\varphi}\mathsf{X}\psi = \int_{\mathbb{R}} \overline{\varphi}x\psi = \int_{\mathbb{R}} \overline{x\varphi}\psi.$$

As this integral equation holds for all $\psi \in \mathcal{D}(\mathsf{X})$, which is dense in $L^2(\mathbb{R})$, we conclude that $(\mathsf{X}^*\varphi)(x) = x\varphi(x)$ almost everywhere and, since $\mathsf{X}^*\varphi \in L^2(\mathbb{R})$, we get $x\varphi \in L^2(\mathbb{R})$. Therefore both φ and $x\varphi$ are square-integrable and, hence, $\varphi \in \mathcal{D}(\mathsf{X})$. Thus, $\mathcal{D}(\mathsf{X}^*) \subset \mathcal{D}(\mathsf{X})$, and X is self-adjoint.

The analysis of the momentum operator is both more delicate and more complicated than that of the position operator, primarily because it involves subtleties in understanding the correct definition of the domain $\mathcal{D}(\mathsf{P})$. We shall ease into the subtleties by examining first an operator Q, with domain $\mathcal{D}(\mathsf{Q}) = C_c^1(\mathbb{R})$, the dense linear subspace of $L^2(\mathbb{R})$ of continuously differentiable functions with compact support defined by

$$\mathsf{Q}\varphi = -\mathbf{i}\hbar\frac{d\varphi}{dx}$$

for φ in its domain. Notice that $d\varphi/dx$ is continuous with compact support, hence $\mathsf{Q}\varphi$ is square-integrable so that $\mathsf{Q} : \mathcal{D}(\mathsf{Q}) \to L^2(\mathbb{R})$. Integration by parts implies that, for all $\varphi, \psi \in \mathcal{D}(\mathsf{Q})$,

$$\langle\varphi|\mathsf{Q}\psi\rangle = -\mathbf{i}\hbar\int_{\mathbb{R}} \overline{\varphi}\frac{d\psi}{dx} = \mathbf{i}\hbar\int_{\mathbb{R}} \overline{\frac{d\varphi}{dx}}\psi = \langle\mathsf{Q}\varphi|\psi\rangle.$$

It follows that $\mathsf{Q} \subset \mathsf{Q}^*$, i.e., Q is symmetric. Is Q self-adjoint? We will show that the answer is "no" by exhibiting an element of the domain of the adjoint that is not continuously differentiable. To see that $\mathcal{D}(\mathsf{Q}^*)$ is strictly larger than $\mathcal{D}(\mathsf{Q})$, let $\mu(x) = 1 - x^2$ if $-1 \le x \le 1$ and $\mu(x) = 0$ otherwise. Of course $\mu \in L^2(\mathbb{R})$, but μ is not differentiable at $x = -1$ and $x = 1$, so is not in the domain of Q. Let $\nu(x) = 2x\hbar\mathbf{i}$ if $-1 \le x \le 1$ and $\nu(x) = 0$ otherwise. Note that $\nu \in L^2(\mathbb{R})$ and is precisely equal to $-\mathbf{i}\hbar(d\mu/dx)$ except at the two points of nondifferentiability of μ. An easy calculation using integration by parts confirms that $\langle\mu|\mathsf{Q}\psi\rangle = \langle\nu|\psi\rangle$ for every $\psi \in \mathcal{D}(\mathsf{Q})$. By the definition of an adjoint, $\mu \in \mathcal{D}(\mathsf{Q}^*)$ and $\mathsf{Q}^*\mu = \nu$. As $\mu \notin \mathcal{D}(\mathsf{Q})$, Q is not self-adjoint.

We wish to extend Q to a self-adjoint operator P. This is not always possible for arbitrary symmetric operators but may be shown to work in this case. We need to enlarge the domain of Q to include functions such as the μ of the preceding paragraph that are almost differentiable. Perhaps the cleanest way to accomplish our goal is to generalize the usual notion of a derivative to include functions such as μ. Here is one way forward. The function $w : \mathbb{R} \to \mathbb{C}$ is said to be a **generalized derivative** of the element $\varphi \in L^2(\mathbb{R})$ if

$$\int_{\mathbb{R}} \varphi \frac{d\psi}{dx} = -\int_{\mathbb{R}} w\psi$$

for all $\psi \in C_c^\infty(\mathbb{R})$. The function w is trying to play the role of $d\varphi/dx$ in integration by parts, but whereas the actual derivative of φ may fail to exist, such a function w may exist. We denote the generalized derivative using a prime, writing $w = \varphi'$. Four things to note: first, if w and $\widetilde{w}$ are both generalized derivatives of φ then $w = \widetilde{w}$ almost everywhere. Second, if $\varphi \in C_c^1(\mathbb{R})$ then $\varphi' = d\varphi/dx$, i.e., the usual derivative of φ, almost everywhere and $\varphi' \in L^2(\mathbb{R})$. Third, if φ' exists then so does $(\overline{\varphi})'$, and $(\overline{\varphi})' = \overline{\varphi'}$. Fourth, the operation of taking generalized derivatives is linear, i.e., $(a\varphi + b\psi)' = a\varphi' + b\psi'$ whenever all the generalized derivatives exist. In the example of the previous paragraph, $\mu' = \mathbf{i}\nu/\hbar$.

We are now in a position to extend the operator Q to a self-adjoint operator P, called the **momentum operator**. Let

$$\mathcal{D}(\mathsf{P}) = \{\varphi \in L^2(\mathbb{R}) : \varphi' \text{ exists and } \varphi' \in L^2(\mathbb{R})\}.$$

Since the linear space $\mathcal{D}(\mathsf{P})$ contains $C_c^1(\mathbb{R})$ as a dense subset, the domain $\mathcal{D}(\mathsf{P})$ is dense in $L^2(\mathbb{R})$. Define $\mathsf{P} : \mathcal{D}(\mathsf{P}) \to L^2(\mathbb{R})$ by

$$\mathsf{P}\varphi = -\mathbf{i}\hbar\varphi',$$

and observe that $\mathsf{Q} \subset \mathsf{P}$, i.e., P extends Q. To verify that P is self-adjoint, we need one more fact about generalized derivatives, namely, that the appropriate version of integration by parts holds for generalized derivatives. Precisely, if $\varphi, \psi \in \mathcal{D}(\mathsf{P})$ then

$$\int_{\mathbb{R}} \varphi\psi' = -\int_{\mathbb{R}} \varphi'\psi.$$

This is just the usual integration by parts formula if the functions φ and ψ happen to be in $C_c^1(\mathbb{R})$, and the general formula follows by approximating elements of $\mathcal{D}(\mathsf{P})$ and their generalized derivatives by appropriate elements of $C_c^1(\mathbb{R})$ and taking limits.[3]

Armed with integration by parts, the proof of self-adjointness proceeds

[3] The details of the proof may be found in Zeidler (1995).

quickly. Indeed, for any $\varphi, \psi \in \mathcal{D}(\mathsf{P})$, we have

$$\langle \varphi | \mathsf{P}\psi \rangle = -\mathbf{i}\hbar \int_{\mathbb{R}} \overline{\varphi}\psi' = \mathbf{i}\hbar \int_{\mathbb{R}} \overline{\varphi}'\psi = \langle \mathsf{P}\varphi | \psi \rangle.$$

Therefore $\mathsf{P} \subset \mathsf{P}^*$ and P is symmetric. We now show that $\mathcal{D}(\mathsf{P}^*) \subset \mathcal{D}(\mathsf{P})$, confirming that P is self-adjoint. Let $\varphi \in \mathcal{D}(\mathsf{P}^*)$. Then, for all $\psi \in \mathcal{D}(\mathsf{P})$, we have $\langle \varphi | \mathsf{P}\psi \rangle = \langle \mathsf{P}^*\varphi | \psi \rangle$, which implies that

$$\int_{\mathbb{R}} \overline{\varphi} \frac{d\psi}{dx} = \frac{\mathbf{i}}{\hbar} \int_{\mathbb{R}} \overline{\varphi} \mathsf{P}\psi = \frac{\mathbf{i}}{\hbar} \int_{\mathbb{R}} \overline{(\mathsf{P}^*\varphi)}\psi$$

whenever ψ is in $C_c^\infty(\mathbb{R})$. The definition of a generalized derivative now implies that $\overline{\varphi}'$, and therefore φ', exists and, in fact,

$$\overline{\varphi}' = -\frac{\mathbf{i}}{\hbar} \overline{(\mathsf{P}^*\varphi)}, \qquad \text{so} \qquad \varphi' = \frac{\mathbf{i}}{\hbar} \mathsf{P}^*\varphi.$$

Since $\mathsf{P}^*\varphi$ is square-integrable, so too is φ' and $\varphi' \in L^2(\mathbb{R})$. This confirms that $\varphi \in \mathcal{D}(\mathsf{P})$ and, therefore, P is self-adjoint.

2.5 End Notes

Jammer (1974) greatly influenced the approach I have taken, of encapsulating the theory of quantum mechanics in a system of axioms.

The example of the momentum operator suggests that, generally, it can be difficult to identify the domain of a self-adjoint operator. Usually the operator arises as a symmetric differential operator on a class of differentiable functions, and often it is neither possible nor useful to find the precise conditions that describe the domain of the adjoint that extends the given operator. This is where a little abstraction pays off. The theory provides for certain abstract techniques that have the possibility of producing self-adjoint extensions without the need to explicitly identify the appropriate domain. The reader is referred to Chapter VIII, especially VIII.1 and VIII.2, of Reed and Simon (1980) and Chapter 33 of Lax (2002) for a detailed discussion of self-adjoint extensions of symmetric operators. In what follows, I will take a fairly cavalier attitude regarding the appropriate domains for operators, unless important physical or mathematical considerations warrant more care.

The discipline of mathematics that studies self-adjoint operators is functional analysis, which developed out of a theoretical approach to the study of partial differential equations at the end of the nineteenth century. Perhaps the culmination of the subject before the days of quantum mechanics was David Hilbert's spectral theorem for bounded operators. It was the need to place the theory of quantum mechanics, with its unbounded, discontinuous operators, on a sound mathematical footing that inspired John von Neumann and Marshall Stone to define self-adjoint operators in the general setting and prove the spectral theorem for self-adjoint operators, which is a sort of a generalized diagonalization of the operator in the spirit of the diagonalization of matrices. The spectral theorem is at the heart of a mathematically sound treatment of measurement in quantum mechanics, the details of which are presented in Lecture 4. It is not an exaggeration to say that in the rich soil of quantum mechanics germinated the most fruitful ideas in functional analysis, leading to the most enlightening and useful tools of the subject. The tools that have their origin in quantum mechanics include the right understanding of self-adjoint operators, various spectral theorems, Stone's theorem on infinitesimal generators, generalized functions and distributions, rigged Hilbert spaces, and the link between unitary operators and symmetries. The reader will see most of these tools applied to the understanding of quantum mechanics in the course of these lectures.

The parts of functional analysis needed for quantum mechanics center around the various spectral theorems for operators and unitary representations. These are beautifully covered in the classic text Riesz and Sz.-Nagy (1955). I have a weakness for the classics, but it is good to have a variety that includes expositions at a more modest level as well as treatments from a modern vantage and from a variety of viewpoints. Fano (1971) presents a gentler introduction to the functional analysis useful in quantum mechanics, prefaced by a nicely designed course in linear spaces, measure theory, and Hilbert spaces. It also includes as a capstone a final brief chapter on axiomatic quantum mechanics. Lax (2002) is a modern treatment from the viewpoint of one of the foremost applied mathematicians of our age, and Rudin (1991) is a treatment from the viewpoint of one of the finest expositors of pure mathematics of the last half century. Zeidler (1995) is a rather unique presentation of the functional analysis of quantum mechanics. Applications-oriented in its presentation of rigorous mathematics, Zeidler's book offers a valuable bilingual dictionary that translates between the language of mathematicians and that of physicists. The present author, a mathematician, has found this very helpful in getting a handle on the mathematical liberties taken by physicists. Strichartz (2003) is a highly accessible account of the modern theory of distributions and the Fourier transform. The book is remarkable for its clarity and sufficient for the needs of the student of quantum mechanics. Blanchard and Brüning (2003) gives a detailed account of the subjects of its subtitle and is one of the more complete accounts of the modern analytic methods needed for a sound treatment of quantum mechanics. Finally, Teschl (2009) presents a careful development of the spectral theory of unbounded operators, as its central objective in Part 1, and applies this to understand the technical mathematics of Schrödinger operators in Part 2.

Special mention belongs to the four-volume set entitled *Methods of Modern Mathematical Physics* by the mathematical physicists Michael Reed and Barry Simon. The first volume, Reed and Simon (1980) *Functional Analysis*, is one of the best available expositions of functional analysis as used in modern physics. The subject matter of the remaining volumes is indicated by their titles: (1975) *Volume II: Fourier Analysis, Self-Adjointness*; (1979) *Volume III: Scattering Theory*; and (1978) *Volume IV: Analysis of Operators.* Taken together, these represent a definitive exposition and thoroughgoing analysis of the operators of quantum mechanics, the Schrödinger equation, and its spectral analysis. They are not for the faint-hearted as the level of detail and mathematical sophistication is high. A final entry in this guide to the subject of functional analysis for quantum mechanics is Schechter (1981). This nicely written work presents operator theory in the context of a single quantum particle moving along a straight line. Its style is to ask physical questions of this one-particle system, and then to develop the powerful mathematical methods that help answer the questions. The student can learn a lot of modern hard-core analysis by a careful study of this text.

3
Observables and Expectation Values

> Quantum mechanics, as far as I know, is the first theory in the history of Western science to predict randomness but to deny that *either* measurement errors *or* "real fluctuations" accounts for it.
>
> David Wick
> *The Infamous Boundary: Seven Decades of Controversy in Quantum Physics*, 1995

In the preceding lecture we spent some time developing an understanding of the self-adjoint operators that, according to Axiom 2 of our minimalist list (see Section 2.1), represent the observables of a quantum system. The connection of this rather abstract assertion – that observables that encode system attributes such as position and momentum are somehow connected to rather esoteric mathematical objects such as self-adjoint operators – with the hard-nosed, concrete world of laboratory instrument readings is given by Axiom 3. This axiom asserts that if an observable a is represented by the self-adjoint operator A then the expected value on measuring the observable a in the laboratory is $\langle\psi|\mathsf{A}\psi\rangle$ if the system is represented by the normalized state vector ψ. This means that if many systems are prepared in the state ψ and a-measurements are made on them, the statistical distribution of the resulting instrument readings will give experimental mean values for a that approach $\langle\psi|\mathsf{A}\psi\rangle$ in the limit of infinitely many measurements. Notice that the axioms afford no mechanism for predicting exact instrument readings on one system prepared in the state ψ. This rather unfortunate feature of the axioms will be ameliorated somewhat by unpacking the implications of Axiom 3 in the course of the present lecture, arriving finally, in the next lecture, at a refinement of the axiom that describes in more detail the act of measurement. We will find, for instance, that lurking in the shadows of Axiom 3 is information about the spectrum of possible a-measurements as well as their respective individual statistical occurrences upon measurement. It is the accurate prediction, correct without exception over the past 80 years, of the statistical distribution of measurements in the micro-world of atoms that has given quantum mechanics the mantle of the physical theory *par excellence.*

This lecture begins with an illumination of some elementary properties of expectation values in quantum mechanics, and it then examines the question whether a quantum mechanical observable can have a precise value. Answering this, we then ask how one should understand what happens in a quantum system upon measurement. This culminates in the **von Neumann–Lüders projection postulate**, which describes the (in)famous collapse of the state vector upon measurement. The next lecture, Lecture 4, then examines a proposed refinement of Axiom 3 that is found wanting in its mathematical details, and a mathematically satisfactory replacement is presented and analyzed in the remainder of the lecture.

3.1 Elementary Properties of Expectation Values

Throughout this lecture, ψ denotes a normalized state vector in the domain $\mathcal{D}(\mathsf{A})$ of a self-adjoint operator A on a Hilbert space $\mathcal{H}$. The operator A represents the observable a. Notice that

$$\langle \mathsf{A} \rangle_\psi = \mathrm{Exp}_\psi(\mathsf{A}) = \langle \psi | \mathsf{A}\psi \rangle = \langle \mathsf{A}\psi | \psi \rangle = \overline{\langle \psi | \mathsf{A}\psi \rangle}$$

implies that the expectation value $\langle \mathsf{A} \rangle_\psi$ is real, which is required if self-adjoint operators are to represent values for physically observable quantities measured in the laboratory. Notice also that $\langle \mathsf{A} \rangle_\psi$ depends linearly on A. Let $1 = 1_{\mathcal{H}}$ be the identity operator on $\mathcal{H}$ and let c be a complex number. Then $c1$ is the operator that multiplies every state vector in $\mathcal{D}(c1) = \mathcal{H}$ by c and its adjoint is the operator $(c1)^* = \overline{c}1$. Hence, $c1$ is self-adjoint if and only c is real. By a slight abuse of notation, we will use c to denote the operator $c1$, in which case $c^* = \overline{c}$.

3.1.1 Interlude: The Probability Interpretation of the State Vector in Wave Mechanics

In classical probability theory, the expected value, or the mean, of a function $g(X)$ of a continuous random variable X whose probability density function is $f(x)$ is given by

$$\mathbf{E}_f(g) = \operatorname{Exp}_f(g(X)) = \int_{-\infty}^{\infty} g(x)f(x)\,dx.$$

In one-dimensional wave mechanics, the expectation value of the position operator X is

$$\langle \mathsf{X} \rangle_\psi = \int_{\mathbb{R}} \overline{\psi}\mathsf{X}\psi = \int_{\mathbb{R}} x|\psi|^2.$$

If g is a polynomial function then the operator $g(\mathsf{X})$ makes sense,[1] and its expectation value is

$$\langle g(\mathsf{X}) \rangle_\psi = \int_{\mathbb{R}} \overline{\psi}g(\mathsf{X})\psi = \int_{\mathbb{R}} g(x)|\psi|^2.$$

Comparing this with the formula for the mean of a function of a random variable suggests that, in wave mechanics, $|\psi|^2 = \overline{\psi}\psi$ plays the role of the probability density function and X the role of the random variable. It is natural then to interpret the square modulus of the state vector ψ as the probability density function for position measurements on our quantum system. This was Max Born's fundamental insight into the meaning of the wave function, which he proposed in 1926 within a year of his and Heisenberg's formulation of matrix mechanics and Schrödinger's formulation of wave mechanics. The state vector itself is called the **probability amplitude (density)** for position measurements. The analogy remains valid for the momentum operator P if the Fourier transform $\widehat{\psi}$ is interpreted as the probability amplitude density for momentum measurements. Indeed, using the (slightly modified)[2] Fourier transform of ψ given for appropriately integrable functions by

$$\widehat{\psi}(p) = \frac{1}{\sqrt{2\pi\hbar}} \int_{\mathbb{R}} \psi(x)e^{-\mathrm{i}px/\hbar}\,dx,$$

[1] Even if g is not a polynomial but merely continuous, $g(\mathsf{X})$ makes sense as will be seen in subsequent lectures.

[2] The modification is the presence of the $\hbar$'s in the formula for $\widehat{\psi}$, which conveniently normalizes the transform for use with the momentum operator.

we obtain for the expected value of a momentum measurement the expression

$$\langle \mathsf{P} \rangle_\psi = \int_{\mathbb{R}} \overline{\psi} \mathsf{P} \psi = \frac{\hbar}{\mathbf{i}} \int_{\mathbb{R}} \overline{\psi} \psi' = \frac{\hbar}{\mathbf{i}} \int_{\mathbb{R}} \overline{\widehat{\psi}} \widehat{\psi'} = \frac{\hbar}{\mathbf{i}} \int_{\mathbb{R}} \overline{\widehat{\psi}} \frac{\mathbf{i}}{\hbar} p \widehat{\psi} = \int_{\mathbb{R}} p |\widehat{\psi}|^2.$$

The third equality of this calculation uses the Plancherel theorem from the theory of Fourier transforms. This says that the usual Fourier transform is not only a linear isomorphism of $L^2(\mathbb{R})$ but also an isometry that preserves the Hermitian inner product. A straightforward calculation verifies that the particular placement of $\hbar$ in the definition of our modified Fourier transform preserves this feature.[3] The fourth equality uses the easily verified fact that

$$\widehat{\psi'}(p) = \widehat{\frac{d\psi}{dx}}(p) = \frac{\mathbf{i}}{\hbar} p \, \widehat{\psi}(p).$$

For the general operator A representing an observable in wave mechanics, the expectation value takes the form

$$\langle \mathsf{A} \rangle_\psi = \int_{\mathbb{R}} \overline{\psi} \mathsf{A} \psi.$$

If A is not a simple polynomial function of one of the operators X or P, the analogy with classical probability theory is not quite so strong, as ψ is operated upon by A before multiplication with $\overline{\psi}$.

In general, when an instantiation of the quantum rules other than wave mechanics is used, there may not be such a straightforward connection with classical probability theory, but, as per Axiom 3, the formula $\langle \mathsf{A} \rangle_\psi = \langle \psi | \mathsf{A} \psi \rangle$ still gives the expectation value for the observable a when the system is prepared in the state ψ, which has always been found to agree with experimental observations.

3.1.2 The Variance and Dispersion of a Quantum Observable

Recall that the variance σ^2 of a continuous random variable X with probability distribution function $f(x)$ is in classical probability theory just the second moment about the mean $\mu = \mathbf{E}_f(X)$, i.e.,

$$\sigma^2 = \mathbf{E}_f((X - \mu)^2) = \int_{-\infty}^{\infty} (x - \mu)^2 f(x) \, dx,$$

[3] Wave mechanics and the Fourier transform are treated in more detail in Lecture 10, where the extension of the transform to $L^2(\mathbb{R})$ and the Plancherel theorem are reviewed.

and the positive square root σ of the variance is the standard deviation. This motivates the following definitions. The **variance** $\nu_\psi(\mathsf{A})$ of the observable represented by A and the state vector ψ is the expectation value of the operator $(\mathsf{A} - \langle\mathsf{A}\rangle_\psi)^2$, and the **dispersion** $\Delta_\psi\mathsf{A}$, serving in the role of standard deviation, is the positive square root of the variance.[4] In classical probability theory, an easy calculation shows that $\sigma^2 = \mu_2 - \mu^2$, where μ_2 is the second moment about the origin, given by $\mu_2 = \mathbf{E}_f(X^2)$. A similarly easy calculation, using the linearity of $\langle\mathsf{A}\rangle_\psi$ in A and the immediate fact that $\langle c\rangle_\psi = c$ for any complex number c, gives an analogous formula for the variance of an observable:

$$\begin{aligned}\nu_\psi(\mathsf{A}) = \mathrm{Exp}_\psi\left[(\mathsf{A} - \langle\mathsf{A}\rangle_\psi)^2\right] &= \langle\mathsf{A}^2 - 2\langle\mathsf{A}\rangle_\psi\mathsf{A} + \langle\mathsf{A}\rangle_\psi^2\rangle_\psi \\ &= \langle\mathsf{A}^2\rangle_\psi - 2\langle\mathsf{A}\rangle_\psi\langle\mathsf{A}\rangle_\psi + \langle\mathsf{A}\rangle_\psi^2 \\ &= \langle\mathsf{A}^2\rangle_\psi - \langle\mathsf{A}\rangle_\psi^2.\end{aligned}$$

This then gives a useful characterization of the dispersion as

$$\Delta_\psi\mathsf{A} = \left(\langle\mathsf{A}^2\rangle_\psi - \langle\mathsf{A}\rangle_\psi^2\right)^{1/2} = \left(\|\mathsf{A}\psi\|^2 - \langle\mathsf{A}\rangle_\psi^2\right)^{1/2}.$$

Notice that as ψ is normalized, $\langle\mathsf{A}\rangle_\psi = \langle\psi|\mathsf{A}\psi\rangle$ is just the magnitude of the projection of the vector $\mathsf{A}\psi$ in the direction of ψ, and so the dispersion attains its maximum value $\|\mathsf{A}\psi\|$ when $\mathsf{A}\psi$ is orthogonal to ψ, and attains its minimum value 0 when $\mathsf{A}\psi$ is aligned with ψ. A closer look reveals the following important result: the dispersion $\Delta_\psi\mathsf{A} = 0$ if and only if ψ is an eigenvector of A with associated eigenvalue $\langle\mathsf{A}\rangle_\psi$. To see this, first notice that if c is real then $\mathsf{A} - c$ is self-adjoint with $\mathcal{D}(\mathsf{A} - c) = \mathcal{D}(\mathsf{A})$. Applying this with $c = \langle\mathsf{A}\rangle_\psi$ gives

$$\begin{aligned}(\Delta_\psi\mathsf{A})^2 &= \langle\psi|(\mathsf{A} - \langle\mathsf{A}\rangle_\psi)^2\psi\rangle \\ &= \langle(\mathsf{A} - \langle\mathsf{A}\rangle_\psi)\psi|(\mathsf{A} - \langle\mathsf{A}\rangle_\psi)\psi\rangle \\ &= \|(\mathsf{A} - \langle\mathsf{A}\rangle_\psi)\psi\|^2.\end{aligned}$$

It follows that the dispersion $\Delta_\psi\mathsf{A}$ is equal to 0 if and only if $\mathsf{A}\psi = \langle\mathsf{A}\rangle_\psi\psi$, verifying our assertion. It also is easy to see that if ψ is any eigenvector of A with eigenvalue λ then $\lambda = \langle\mathsf{A}\rangle_\psi$, from which it follows that every eigenvalue λ of A is real.

One more observation before tackling the question posed in the title of

[4] As promised in the preface to the previous lecture and more explicitly stated at the end of the lecture, we do not concern ourselves with the proper domain of definition of operators such as A^2 unless this is warranted by physical or mathematical considerations. Under the usual circumstances, we make the standing assumption that the relevant state vector is in the domain of the relevant operator.

the next section: if φ and ψ are eigenvectors of the self-adjoint operator A with distinct respective eigenvalues κ and λ, then

$$0 = \langle\varphi|\mathsf{A}\psi\rangle - \langle\varphi|\mathsf{A}\psi\rangle = \langle\mathsf{A}\varphi|\psi\rangle - \langle\varphi|\mathsf{A}\psi\rangle = (\overline{\kappa} - \lambda)\langle\varphi|\psi\rangle.$$

Since κ is real, $\overline{\kappa} = \kappa \neq \lambda$, and hence $\langle\varphi|\psi\rangle = 0$. Thus the eigenvectors of a self-adjoint operator with distinct eigenvalues are orthogonal.

3.2 Can a Quantum Observable have a Precise Value?

We now consider the special case of a nondegenerate self-adjoint operator A with a **complete set of eigenstates**. This means that the Hilbert space $\mathcal{H}$ has a Schauder basis of normalized eigenvectors $\{\psi_i\}$ for A with pairwise distinct eigenvalues λ_i associated with distinct basis vectors. Assuming that $\mathcal{H}$ is infinite-dimensional, since it is separable the index i ranges over the natural numbers. As the eigenvalues are distinct for distinct basis vectors, the basis is orthogonal as well as normalized. Let ψ be any normalized state vector and expand ψ in the basis so that $\psi = \sum_i c_i\psi_i$, where $c_i = \langle\psi_i|\psi\rangle$. Calculating the expectation value gives

$$\begin{aligned}\langle\mathsf{A}\rangle_\psi = \langle\psi|\mathsf{A}\psi\rangle &= \Big\langle \sum_i c_i\psi_i \,\Big|\, \mathsf{A}\sum_j c_j\psi_j \Big\rangle \\ &= \sum_i\sum_j \overline{c}_i c_j \lambda_j \langle\psi_i|\psi_j\rangle = \sum_i \lambda_i |c_i|^2.\end{aligned}$$

As ψ is normalized, $\sum_i |c_i|^2 = 1$. Comparing these facts with the classical expected value of a discrete random variable strongly suggests that we interpret the coefficient $c_i = \langle\psi_i|\psi\rangle$ as an amplitude, and its square modulus $|c_i|^2$ as the probability that an experimental instrument records the value λ_i when an a-measurement is performed on a system represented by the state vector ψ. Notice that the total probability of obtaining some result upon measurement is given by $\sum_i |c_i|^2$, which, as required, is unity. This interpretation has over 90 years of successful experimental verification and will be incorporated into a modified Axiom 3 in the next lecture.

With this interpretation in hand, that $\langle\psi_i|\psi\rangle$ represents the amplitude for measuring the experimental value λ_i upon an a-measurement, we may answer the question posed in the section title. Indeed, when a system is represented by a state vector ψ, the observable a will have the exact value λ precisely when an a-measurement yields λ with certainty, i.e., with unit probability. This can happen only when each amplitude $\langle\psi_i|\psi\rangle$ is zero except for one, say the one with index $i = N$; moreover, that amplitude squared, $|\langle\psi_N|\psi\rangle|^2$, must be unity and the eigenvalue necessarily satisfies $\lambda = \lambda_N$. In this case, then, the expansion of ψ in terms of the basis of eigenvectors becomes $\psi = u\psi_N$ for the unimodular constant $u = \langle\psi_N|\psi\rangle$. In conclusion, the observable a has the exact value λ if and only if the state vector ψ is an eigenvector of A with eigenvalue $\lambda = \langle\mathsf{A}\rangle_\psi$. In terms of dispersion, the results of the previous section imply that the observable a has an exact value if and only if the dispersion $\Delta_\psi\mathsf{A} = 0$, and in this case the exact value, measured with certainty, is the corresponding eigenvalue $\langle\mathsf{A}\rangle_\psi$.

3.3 What Happens Upon Measurement?

In a-measurements of a system for which the representative operator A satisfies the constraints of the previous section, in which there is a complete orthonormal basis of eigenvectors of A with distinct eigenvalues, the only experimental values obtained are, within experimental error, elements of the **spectrum** $\mathrm{Spec}(\mathsf{A})$, the set of eigenvalues $\{\lambda_i\}_i$ of A. Moreover, in measurements of certain quantities, such as the energy and the angular momentum of a bound quantum system, once an eigenvalue λ is registered, any subsequent a-measurement, if the system is otherwise isolated and undisturbed, yields precisely λ with certainty. The experimental dispersion of later a-measurements is zero. For other observables represented by more general operators with, for example, a continuous spectrum,[5] near-in-time consecutive measurements produce essentially the same value. The discussion of the previous section suggests that the natural interpretation of these facts is that when an a-measurement produces the value λ, an eigenvalue of A, the state vector ψ undergoes an abrupt change and **collapses** to an eigenvector of A with associated eigenvalue λ. This collapse is sometimes referred to as a **reduction** or **projection** of ψ onto the λ-eigenspace.

Examining this in a bit more detail requires a look at Axiom 4 of the minimalist rules in Section 2.1. Recall that this axiom asserts that the time evolution of a quantum system is governed by the Schrödinger equation, the primary component of which is the self-adjoint Hamiltonian operator H for the system. We will examine the Hamiltonian and

[5] The relevant definitions and mathematics of these operators are presented in the next lecture.

time evolution of a quantum system in some detail in Lecture 27, but, for the present, we will borrow from that discussion some facts to help clarify the measurement process. For the present discussion, we make the simplifying assumption that A commutes with H. We will see in Lecture 27 that the solution to the Schrödinger equation may be written as $\psi_t = \mathsf{U}_t\psi$, where, for each time t, U_t is a unitary operator determined by H that commutes with A, since H commutes with A. This means that U_t preserves the Hermitian product, i.e., $\langle \mathsf{U}_t\varphi|\mathsf{U}_t\psi\rangle = \langle\varphi|\psi\rangle$ for all vectors φ and ψ, and that $\mathsf{U}_t\mathsf{A} = \mathsf{A}\mathsf{U}_t$ on appropriate domains. Whenever an a-measurement yields the eigenvalue λ, the state vector ψ collapses to an eigenvector ψ_λ with that eigenvalue. At that point, unitary evolution takes over and, at time $t > 0$ after the measurement, the system is represented by state vector $\mathsf{U}_t\psi_\lambda$. What happens upon another a-measurement at time t? Since U_t commutes with A, $\mathsf{A}\mathsf{U}_t\psi_\lambda = \mathsf{U}_t\mathsf{A}\psi_\lambda = \lambda(\mathsf{U}_t\psi_\lambda)$, so $\mathsf{U}_t\psi_\lambda$ is an eigenvector of A with eigenvalue λ. Therefore, an a-measurement at time t will yield λ with certainty and with zero dispersion. This shows that once an a-measurement yields a value, subsequent a-measurements will always yield the same value if the operator representing a commutes with the Hamiltonian and, between measurements, there are no other influences. Thus, between measurements, the state vector evolves via unitary evolution by tracing out a path in the λ-eigenspace of A. If the operator A does not commute with the Hamiltonian then the state vector may evolve out of the λ-eigenspace.

Considerations such as these motivate the following **projection postulate**.

von Neumann–Lüders projection postulate. Let Proj be the orthogonal projection of $\mathcal{H}$ onto the closed linear subspace spanned by the eigenvectors of A that are consistent with the outcome of the measurement. Then, after measurement, the state vector has changed (been reduced or projected) to $\mathsf{Proj}\psi$.

In particular, if an a-measurement yields the precise value λ the state vector ψ reduces to $\langle\psi_\lambda|\psi\rangle\psi_\lambda$, if λ is **nondegenerate**, meaning that the λ-eigenspace is one-dimensional. More generally, if A has degeneracy at λ, let Proj_λ be the orthogonal projection onto the λ-eigenspace of A. Then ψ reduces to $\mathsf{Proj}_\lambda\psi$ if the measurement result is λ.

3.4 The Measurement Problem

What are we to make of this story? How is it that a theory that purports to be the fundamental theory of nature gives such a prominent place to the pedestrian act of measurement? Indeed, exactly what act constitutes a measurement? In classical physics there is no room for measurement in the theoretical edifice of the science. More worrisome still, in quantum mechanics there seem to be two different, mutually exclusive modes of evolution of the state vector. On the one hand, 'normal" time evolution is governed by the Schrödinger equation and proceeds via a deterministic, linear, unitary, continuous development. On the other hand, intervals of unitary evolution are interrupted by discontinuous, nonunitary collapse. The collapse is nondeterministic if the dispersion of the observable is nonzero so that there are multiple possible measurement values available, each with positive probability of occurrence. Asking again a question from Lecture 1, are there really two different sorts of time evolution demarcated by the presence of a measurement? And again, just what constitutes a measurement?

Almost a century after its recognition, the **measurement problem** has no entirely satisfactory resolution. Several proposals that work to varying degrees, all of which have supporters and detractors, and all of which have unconventional, even exotic, features, have been advanced. These are the *consistent histories interpretation*, the *many-worlds interpretation*, the *many-minds interpretation*, various *modal interpretations*, *quantum logic*, and various *hidden variables* proposals including *Bohmian mechanics*. We encourage the reader to explore any of these interpretations in both the popular and technical literature.

3.5 End Notes

The general tenor of this lecture owes much to the development of the same topics in Hannabuss (1997), a book I cannot praise enough for study by students of mathematics. The introduction of the probability interpretation of the wave function in this lecture begins this book's treatment of wave mechanics, which continues in full force in Lectures 9–11 and 15–19. Three of my favorite treatments of wave mechanics are those of Peebles (1992), Thaller (2000), and Merzbacher (1998). Dickson (1998) provides deep analyses and cogent criticisms of the measurement problem, the projection postulate, and the orthodox interpretation of the quantum formalism.

4
The Projection Postulate Examined

> ...the projection postulate is extremely suspect, if not grossly ad hoc. It is explicitly designed to solve the measurement problem. Moreover, supposing that quantum mechanics is to be considered a fundamental theory – the theory of quantum phenomena, from which we are to derive our theories of everyday phenomena – it is extremely unsatisfactory to include "measurement" as a primitive notion in the theory. At the least we should require that the notion of "measurement" be unambiguously reducible to fundamental elements (in particular, processes and interactions) of the theory. Thus far, no fully satisfactory reduction has been proposed.
>
> W. Michael Dickson
> *Quantum Chance and Non-Locality*, 1998

The insights of the previous lecture crystallize here into a refinement of Axiom 3. Our first attempt proposes as an appropriate refinement a version of the axiom that conforms to common usage among physicists. Many physicists see no particular problem with the refined axiom, but its exact statement upon close scrutiny offends the sensibilities of the mathematician. Of course, physicists are right to gloss over fine points of rigor as they use the axiom to guide calculations that, ultimately, yield correct predictions. Like many uses of mathematics beyond the sanctions of the pure and precise, useful and convenient fictions are invaluable for summarizing succinctly a calculus whose rigorous development and justification would offer no practical benefit to the professional. The trick is to produce a fiction that approximates the rigorous picture closely enough that the practitioner is neither in danger of gross miscalculation nor misled by the intuitive picture that the approximation affords. This "physicist's fiction," expressed perhaps most effectively and elegantly in Dirac (1930), succeeds admirably on these terms.

Regardless of the success of the refined axiom to the physicist, the mathematician is compelled to ask foundational questions any time a mathematical scheme is proposed. It is in this way that mathematics is enriched by science, and one would be hard pressed to find a more fruitful example of this than the development of the theory of self-adjoint and unitary operators on Hilbert spaces, due principally to von Neumann and Stone as a response to the incredible success of quantum mechanics. In fact, it is hard to overestimate the importance and extent of the influence of quantum mechanics on the development of functional analysis in the last century. Though some might criticize the lack of mathematical rigor in a scientific theory, the proper attitude of the mathematician when faced with a practical calculus that works is to take its success at prediction as a sign of hidden truths, possibly of great depth and sublimity, that warrant mathematical investigation, exposure, and exploitation.

After articulating the physicist's revision of Axiom 3, we present two examples that falsify the precise statement of the physicist. We then present von Neumann's approach for laying a rigorous mathematical foundation for quantum mechanics. This leads us to consider one of the most beautiful and powerful results of modern mathematics, viz., the **spectral theorem for self-adjoint operators**, which lies at the heart of a rigorous version of the physicist's revision. Next, we derive the physicist's revision from von Neumann's in the context of an important class of operators, namely, those with a complete orthonormal set of eigenvectors. Then we refine the notion of "eigenvalue" for an operator in a way that, eventually, will help to bridge the gap between the physicist's and mathematician's versions of Axiom 3. In the next lecture, we give a brief description of Gelfand's alternative formulation in terms of **rigged Hilbert spaces**, which also supplies the desired mathematical rigor to the subject, and we present Dirac's elegant calculus for quantum mechanics.

4.1 The Physicist's Approach

The projection postulate of the previous lecture is quite reasonable in the context of an observable represented by a self-adjoint operator with a complete set of eigenstates. What about a general observable? What happens during measurement? What is the state immediately after measurement? Many physics texts on quantum mechanics list axioms essentially like our four, except that Axiom 3 is replaced by the following refinement, which incorporates the von Neumann–Lüders projection postulate as well as the discussion of the previous lecture on the probability interpretation of the Fourier coefficients of the eigenstate expansion of the state vector.

Physicist's Axiom 3. If a system is represented by a normalized state vector ψ then a measurement of the observable a represented by the self-adjoint operator A will record one of the eigenvalues λ of A with probability $\langle\psi|\mathsf{Proj}_\lambda\psi\rangle = \|\mathsf{Proj}_\lambda\psi\|^2$, where Proj_λ is the projection onto the λ-eigenspace. The state of the system immediately after measurement is $\mathsf{Proj}_\lambda\psi$.

This is consistent with the content of the previous lecture, agreeing with that content for an operator with a complete set of eigenvectors; however, a closer examination uncovers some mathematical problems. Though it remains a useful fiction accurate for certain observables, a consideration of the position and momentum operators of wave mechanics serves to reveal the problems.

The first problem is that the position operator X of one-dimensional wave mechanics has no eigenvalues. Indeed, assume otherwise; say that $\psi \in \mathcal{D}(\mathsf{X})$ is an eigenstate with associated eigenvalue λ. Then $x\psi(x) = \mathsf{X}\psi(x) = \lambda\psi(x)$ and, therefore, $(x-\lambda)\psi(x) = 0$ for almost all real numbers x. It follows that $\psi = 0$ almost everywhere, so that $\psi = 0$ in $L^2(\mathbb{R})$ and, hence, ψ is not a state vector. This contradiction shows that no quantum state vector is an eigenvector of the position operator.

Now a physicist will resolve this seemingly ironclad criticism as follows. One can measure the position of a particle in principle as accurately as desired. By Physicist's Axiom 3, were a position measurement to yield the exact result $\lambda \in \mathbb{R}$ then λ would be an eigenvalue of the position operator X with, say, associated state vector ψ_λ. This implies that $x\psi_\lambda(x) = \lambda\psi_\lambda(x)$ for all real numbers x. Substituting $x = \lambda$ gives $\lambda\psi_\lambda(\lambda) = \lambda\psi_\lambda(\lambda)$ – nothing wrong there. Now substitute $x = \mu \neq \lambda$ to get $\mu\psi_\lambda(\mu) = \lambda\psi_\lambda(\mu)$. Since $\mu \neq \lambda$, this equation implies that $\psi_\lambda(\mu) = 0$. Since ψ_λ is a state vector, we may assume it to be normalized so that its square-integral is unity, but this can happen only if its value at $x = \lambda$ is infinite in such a way that the square-integral is unity. But we have seen this before – the Dirac delta function concentrated at $x = \lambda$ works, modulo a small matter about the square-integral rather than the integral being unity. Thus we can conclude that $\psi_\lambda(x) = \delta(x-\lambda)$, where δ is the Dirac delta function that has value 0 everywhere except at 0, where it takes on the value ∞. The "small matter" mentioned about the square -integral rather than the integral turns out to be a real problem, since we cannot seem to make sense of $\int_{\mathbb{R}} \delta^2$.[1] We resolve this by saying that ψ_λ is a **nonnormalizable** state vector, which serves as an eigenvector of the position operator with eigenvalue λ.

What is a mathematician to say to this? One reaction is to remember that Dirac's fictitious delta function, maligned by generations of mathematicians, does have a completely satisfactory and rigorous manifestation in Sobolev's theory of generalized functions and Schwartz's theory of distributions. Perhaps Physicist's Axiom 3 can be interpreted in terms of generalized eigenstates using the perfectly rigorous theory of distributions. This reaction is, in fact, appropriate, for such a program does work to bring rigor to Physicist's Axiom 3 using rigged Hilbert spaces and generalized eigenfunctions. The next lecture presents this approach for laying a rigorous foundation for quantum mechanics.

Before Sobolev's 1935 work and Schwartz's 1940 work on generalized functions, von Neumann, in 1932, had offered a completely rigorous treatment of quantum mechanics in his classic work entitled *The Mathematical Foundations of Quantum Mechanics.* von Neumann called the treatment of quantum mechanics by the physicists, presented perhaps most elegantly in the magnificently arid prose of Dirac in his highly influential 1930 masterpiece *The Principles of Quantum Mechanics,*[2] "fictitious mathematics." From the preface to von Neumann's *Mathematical Foundations,*

The method of Dirac ... in no way satisfies the requirements of mathematical rigor – not even if these are reduced in a natural and proper fashion to the extent common elsewhere in theoretical physics. For example, the method

[1] Laurent Schwartz incorporates the delta function in his theory of distributions but proved that the theory of distributions is purely linear in that the product of two distributions has no consistent formulation.

[2] It was in this work that the 28-year-old Dirac introduced his delta function and the bra-ket notation that we present in Lecture 5, synthesized the two competing approaches to quantum mechanics – Heisenberg's matrix mechanics and Schrödinger's wave mechanics – into a single formalism using Hilbert spaces and operators, offered an insightful comment that inspired Feynman in his path integral formulation of quantum mechanics that we introduce in Lecture 29, gave the first hints at quantizing fields in his treatment of the interaction of electrons with radiation, gave the first relativistic treatment of quantum mechanics in his derivation of the Dirac equation, and used the Dirac equation to derive the fine structure of the hydrogen spectrum. It would be difficult to classify this work as anything but a monumental watershed in the history of scientific publishing, and its close study today is profitable to both mathematician and physicist.

adheres to the fiction that each self-adjoint operator can be put in diagonal form. In the case of those operators for which this is not actually the case, this requires the introduction of "improper" functions with self-contradictory properties. The insertion of such a mathematical "fiction" is frequently necessary in Dirac's approach, even though the problem at hand is merely one of calculating numerically the result of a clearly defined experiment.

There is also a problem with the momentum operator of wave mechanics. The solution to the eigenvalue equation $-\mathbf{i}\hbar\psi' = \mathsf{P}\psi = \lambda\psi$ is $\psi(x) = C\exp(\mathbf{i}\lambda x/\hbar)$, which is not an element of $L^2(\mathbb{R})$ since it has infinite L^2-norm. Physicists interpret such solutions ψ as representing a stream of particles of momentum λ, yet such functions are not square-integrable and so do not fit into the standard axiomatic scheme.

4.2 The Mathematician's Rigor

Continuing von Neumann's quote from *Mathematical Foundations*,

> There would be no objection here if these concepts, which cannot be incorporated into the present day framework of analysis, were intrinsically necessary for the physical theory. Thus, as Newtonian mechanics first brought about the development of the infinitesimal calculus, which, in its original form, was undoubtedly not self-consistent, so quantum mechanics might suggest a new structure for our "analysis of infinitely many variables" – i.e., the mathematical techniques would have to be changed, but not the physical theory. But this is by no means the case. It should rather be pointed out that the quantum mechanical "Transformation Theory" can be established in a manner which is just as clear and unified, but which is also without mathematical objections. It should be emphasized that the correct structure need not consist in a mathematical refinement and explanation of the Dirac method, but rather that it requires a procedure differing from the beginning, namely, the reliance on the Hilbert space theory of operators.

von Neumann's approach makes rigorous and precise Physicist's Axiom 3 in the following refinement.

von Neumann's Axiom 3. If a system is represented by the normalized state vector ψ then the probability that the result of measurement of the observable a represented by the self-adjoint operator A will record a number between the two values $\lambda_1 < \lambda_2$ is

$$\langle\psi|(\mathsf{E}_{\lambda_2} - \mathsf{E}_{\lambda_1})\psi\rangle = \|(\mathsf{E}_{\lambda_2} - \mathsf{E}_{\lambda_1})\psi\|^2,$$

where $\{\mathsf{E}_\lambda : \lambda \in \mathbb{R}\}$ is the **resolution of the identity** belonging to A. When the measurement yields a result between λ_1 and λ_2, the state of the system immediately after measurement is an eigenvector of the operator $\mathsf{E}_{\lambda_2} - \mathsf{E}_{\lambda_1}$.

To understand what this refinement of Axiom 3 says, we need to discuss the spectral theorem for self-adjoint operators on a Hilbert space $\mathcal{H}$. Motivated by the mathematical needs of the then blossoming quantum mechanics, von Neumann and, independently, Stone, proved this in 1929, generalizing Hilbert's treatment of bounded and compact operators of a generation earlier. After discussion of the spectral theorem in this section, we will show in the next section that von Neumann's Axiom 3 reduces to Physicist's Axiom 3 when the relevant operator has a complete set of eigenstates.

We begin with some definitions and observations. Let $\mathcal{L}$ be a closed linear subspace of the Hilbert space $\mathcal{H}$. Since $\mathcal{L}$ is closed, each vector $\psi \in \mathcal{H}$ decomposes uniquely as an orthogonal sum $\psi = \psi_{\mathcal{L}} + \psi_{\mathcal{L}}^{\perp}$ where $\psi_{\mathcal{L}} \in \mathcal{L}$ and $\psi_{\mathcal{L}}^{\perp} \in \mathcal{L}^{\perp}$. Here $\mathcal{L}^{\perp}$ is the orthogonal complement of $\mathcal{L}$ and $\mathcal{H}$ thus decomposes as the orthogonal direct sum $\mathcal{H} = \mathcal{L} \oplus \mathcal{L}^{\perp}$. The mapping $\psi \mapsto \psi_{\mathcal{L}}$ is the **projection operator** $\mathsf{Proj}_{\mathcal{L}} : \mathcal{H} \to \mathcal{L}$, a bounded, self-adjoint, idempotent operator that maps onto the closed subspace $\mathcal{L}$. Conversely, a straightforward argument shows that every bounded, self-adjoint, idempotent operator on $\mathcal{H}$ is the projection onto its range.

Notice that if A is a projection operator then $\langle\psi|\mathsf{A}\psi\rangle = \|\mathsf{A}\psi\|^2 \geq 0$ for any vector $\psi \in \mathcal{H}$. Two projection operators, A_1 and A_2, satisfy $\mathsf{A}_1 \leq \mathsf{A}_2$ if, for all $\psi \in \mathcal{H}$, $\langle\psi|\mathsf{A}_1\psi\rangle \leq \langle\psi|\mathsf{A}_2\psi\rangle$.

Spectral theorem for self-adjoint operators *Let* A *be a self-adjoint operator on the separable Hilbert space* $\mathcal{H}$. *Then* A *has a unique* ***resolution of the identity*** $\{\mathsf{E}_\lambda : \lambda \in \mathbb{R}\}$, *where each* E_λ *is a projection operator. The resolution of the identity satisfies the following properties.*

(i) $\mathsf{E}_{\lambda_1} \leq \mathsf{E}_{\lambda_2}$ *if* $\lambda_1 \leq \lambda_2$;

(ii) $\mathsf{E}_{-\infty} = \lim_{\lambda \to -\infty} \mathsf{E}_\lambda = 0$;

(iii) $\mathsf{E}_{\infty} = \lim_{\lambda \to \infty} \mathsf{E}_\lambda = 1_{\mathcal{H}}$, *the identity map on* $\mathcal{H}$;

(iv) *for each* $\lambda \in \mathbb{R}$, $\mathsf{E}_{\lambda+0} = \lim_{\mu \to \lambda+} \mathsf{E}_\mu = \mathsf{E}_\lambda$;

(v) $1_{\mathcal{H}} = \int_{\mathbb{R}} d\mathsf{E}_\lambda$;

(vi) $\mathsf{A} = \int_{\mathbb{R}} \lambda \, d\mathsf{E}_\lambda$;

(vii) A *commutes with the operator* B *iff, for each* $\lambda \in \mathbb{R}$, E_λ *commutes with* B; *each* E_λ *commutes with* A, *and thus the elements of the resolution commute pairwise. In fact, for any* $\mu, \nu \in \mathbb{R}$, $\mathsf{E}_\mu \mathsf{E}_\nu = \mathsf{E}_{\min\{\mu,\nu\}}$.

For those unfamiliar with spectral theorems of operators and the language of functional analysis, this is quite a lot to digest. We will comment on each item, providing detailed explanations as well as exposing pertinent implications of each. We also give references to the proof in the End Notes to this lecture.

Item (i) guarantees that for any $\psi \in \mathcal{H}$, the function $\lambda \mapsto \langle \psi | \mathsf{E}_\lambda \psi \rangle$ is a monotonically increasing function on $\mathbb{R}$. In items (ii) through (iv), the limits are taken in the strong sense. This means that when $\lim \mathsf{E}_\lambda = \mathsf{E}$, where the limit is any of the limits in (ii) through (iv), then, for any $\psi \in \mathcal{H}$, it holds that $\lim \mathsf{E}_\lambda \psi = \mathsf{E}\psi$ in the Hilbert space norm of $\mathcal{H}$. This implies, upon taking the Hermitian product with any $\varphi \in \mathcal{H}$, that $\lim \langle \varphi | \mathsf{E}_\lambda \psi \rangle = \langle \varphi | \mathsf{E}\psi \rangle$. Item (iv) then guarantees that the function $\lambda \mapsto \langle \psi | \mathsf{E}_\lambda \psi \rangle$ is continuous from the right.

The integrals in items (v) and (vi) are Riemann–Stieltjes integrals with respect to a monotone family of projections. Instead of giving the definition, it perhaps is more useful to give the implication that captures the primary use of these integrals. In fact, the implication is far-reaching, and allows us to define a function of a self-adjoint operator. Let $f : \mathbb{R} \to \mathbb{C}$ be any complex-valued function of a real variable. Our aim is to define a linear operator $f(\mathsf{A}) : \mathcal{D}(f(\mathsf{A})) \to \mathcal{H}$, where the domain is given by

$$\mathcal{D}(f(\mathsf{A})) = \left\{ \psi \in \mathcal{H} : \int_{-\infty}^{\infty} |f(\lambda)|^2 \, d\langle \psi | \mathsf{E}_\lambda \psi \rangle < \infty \right\}.$$

The integral is the usual Riemann–Stieltjes integral with respect to the right-continuous monotone function $\lambda \mapsto \langle \psi | \mathsf{E}_\lambda \psi \rangle$. Notice that if the function f is not sufficiently well-behaved to be Riemann–Stieltjes integrable with respect to a sufficiently large class of functions, the domain may be nothing more than the trivial subspace of $\mathcal{H}$. Fix a vector $\psi \in \mathcal{D}(f(\mathsf{A}))$ and define the function $\ell_\psi : \mathcal{H} \to \mathbb{C}$ by

$$\ell_\psi(\varphi) = \overline{\int_{-\infty}^{\infty} f(\lambda) \, d\langle \varphi | \mathsf{E}_\lambda \psi \rangle},$$

where the overline means complex conjugation. The integral in this case is the usual Riemann–Stieltjes integral with respect to the function $\lambda \mapsto \langle \varphi | \mathsf{E}_\lambda \psi \rangle$, which is of bounded variation, and the fact that ψ is in $\mathcal{D}(f(\mathsf{A}))$ guarantees convergence of the integral. It is easy to see that ℓ_ψ is a continuous linear functional on $\mathcal{H}$ and, as such, the dual correspondence provides a unique vector ψ such that $\ell_\psi = \psi^* = \langle\psi|-\rangle$. We define the action of the operator $f(\mathsf{A})$ by $f(\mathsf{A})\psi = \psi$. The domain $\mathcal{D}(f(\mathsf{A}))$ is a linear subspace of $\mathcal{H}$ and $f(\mathsf{A})$ is a linear operator on that domain. It now follows that, for all $\varphi \in \mathcal{H}$ and $\psi \in \mathcal{D}(f(\mathsf{A}))$,

$$\langle\varphi|f(\mathsf{A})\psi\rangle = \int_{-\infty}^{\infty} f(\lambda)\, d\langle\varphi|\mathsf{E}_\lambda\psi\rangle.$$

Symbolically we write $f(\mathsf{A}) = \int_{\mathbb{R}} f(\lambda)\, d\mathsf{E}_\lambda$.[3] With this machinery in hand, we now restate items (v) and (vi).

(v) Let $\mathfrak{c}_1$ be the constant mapping on $\mathbb{R}$ with value 1. Then $1_{\mathcal{H}} = \mathfrak{c}_1(\mathsf{A})$. In particular, for all $\varphi, \psi \in \mathcal{H}$, $\langle\varphi|\psi\rangle = \int_{-\infty}^{\infty} d\langle\varphi|\mathsf{E}_\lambda\psi\rangle$.

(vi) Let $1_{\mathbb{R}}$ be the identity map on $\mathbb{R}$. Then $\mathsf{A} = 1_{\mathbb{R}}(\mathsf{A})$. In particular, for all $\varphi \in \mathcal{H}$ and $\psi \in \mathcal{D}(\mathsf{A})$, $\langle\varphi|\mathsf{A}\psi\rangle = \int_{-\infty}^{\infty} \lambda\, d\langle\varphi|\mathsf{E}_\lambda\psi\rangle$.

It is interesting to note a couple of facts about functions of operators. First, it can be shown that $\mathsf{E}_\lambda = \chi_{(-\infty,\lambda]}(\mathsf{A})$, where χ denotes a characteristic function such that $\chi_{(-\infty,\lambda]}$ has the value 1 on the interval $(-\infty, \lambda]$ and 0 otherwise. Second, this **functional calculus** allows us to define the exponential of a self-adjoint operator, which generalizes the exponential of a bounded operator defined in terms of the usual exponential series; this series is easily seen to converge for bounded operators but not necessarily for the general self-adjoint operator. This will be important when we examine Axiom 4 and quantum dynamics and evolution in Lecture 27.

Finally, item (vii) states a nice compatibility condition. In fact, it allows us to see that, for any $\lambda_1 < \lambda_2$, the operator $\mathsf{E}_{\lambda_2} - \mathsf{E}_{\lambda_1}$ is itself a projection operator. We leave the verification to the reader. It follows that, as for all projection operators, $\langle\psi|(\mathsf{E}_{\lambda_2} - \mathsf{E}_{\lambda_1})\psi\rangle = \|(\mathsf{E}_{\lambda_2} - \mathsf{E}_{\lambda_1})\psi\|^2$, as in the statement of von Neumann's Axiom 3.

It is instructive to examine the spectral theorem in the context of a specific example. We invite the reader to do so for the position operator X of wave mechanics. In this case, the resolution of the identity of X is given by $\mathsf{E}_\lambda\psi = \chi_{(-\infty,\lambda]}\psi$; thus, for $\psi \in L^2(\mathbb{R})$,

$$\mathsf{E}_\lambda\psi(x) = \begin{cases} \psi(x) & \text{if } x \le \lambda, \\ 0 & \text{if } x > \lambda. \end{cases}$$

[3] Alternatively, we may define this integral in terms of appropriate limits of Riemann–Stieltjes sums of operators and then calculate $\langle\varphi|f(\mathsf{A})\psi\rangle$.

With this, the von Neumann Axiom 3 implies that the probability of measuring the position to be in the interval (μ, ν) is

$$\langle \psi | (\mathsf{E}_\nu - \mathsf{E}_\mu) \psi \rangle = \int_{\mathbb{R}} \overline{\psi} \chi_{(\mu,\nu]} \psi = \int_\mu^\nu |\psi|^2,$$

agreeing with the Born probability interpretation of the wave function as the probability amplitude for position measurements, which was articulated in the previous lecture.

4.3 An Important Class of Operators

The operators used in Lecture 3 to expose the quantum mechanical understanding of measurement and collapse, though not general enough to account for many physical phenomena, do nonetheless form an important class of operators. These are the self-adjoint operators that are totally nondegenerate, meaning that each eigenspace is one-dimensional, and for which there is an orthonormal basis of eigenvectors. Throughout this section, let A be such an operator and $\{\psi_i : i \in \mathbb{Z}\}$ an orthonormal basis for $\mathcal{H}$ of eigenvectors of A. Let λ_i be the eigenvalue associated with ψ_i and assume that $\lambda_i < \lambda_{i+1}$ for each i and, further, that the spectrum $\{\lambda_i\}_i$ is a discrete subset of $\mathbb{R}$.

For each integer i, let $\mathsf{P}_i = \mathsf{Proj}_{\lambda_i}$ be the projection to the λ_i-eigenspace so that $\mathsf{P}_i(\psi) = \langle \psi_i | \psi \rangle \psi_i$. As $\{\psi_i : i \in \mathbb{Z}\}$ is a complete orthonormal basis, each $\psi \in \mathcal{H}$ may be expanded as $\psi = \sum_i \langle \psi_i | \psi \rangle \psi_i = \sum_i \mathsf{P}_i \psi$. These sums are doubly infinite, and convergence as i approaches $-\infty$ or ∞ holds separately. For each real number λ, define the operator E_λ by $\mathsf{E}_\lambda = \sum \{\mathsf{P}_i : \lambda_i \leq \lambda\}$, so that $\mathsf{E}_\lambda \psi = \sum \{\langle \psi_i | \psi \rangle \psi_i : \lambda_i \leq \lambda\}$. The family $\{\mathsf{E}_\lambda : \lambda \in \mathbb{R}\}$ is the resolution of the identity belonging to A. Indeed, items (i) through (iv) of the Spectral Theorem are verified easily. For items (v) and (vi), let f be a complex-valued function of a real variable and let us derive a useful formula for $f(\mathsf{A})$. We first derive a condition that guarantees that a vector in $\mathcal{H}$ is an element of the domain $\mathcal{D}(f(\mathsf{A}))$ of $f(\mathsf{A})$. For a vector $\psi \in \mathcal{H}$, let $c_i = \langle \psi_i | \psi \rangle$, the ith coefficient in the expansion of ψ with respect to the basis $\{\psi_i : i \in \mathbb{Z}\}$. We have $\sum_i |c_i|^2 = \|\psi\|^2$. Let $\alpha_\psi : \mathbb{R} \to \mathbb{C}$ be the monotone function $\alpha_\psi(\lambda) = \langle \psi | \mathsf{E}_\lambda \psi \rangle = \sum \{|c_i|^2 : \lambda_i \leq \lambda\}$. Thus α_ψ is a step function that limits to 0 as $\lambda \to -\infty$, limits to $\|\psi\|^2$ as $\lambda \to \infty$, is constant on $[\lambda_{i-1}, \lambda_i)$, and has a jump discontinuity at λ_i of height $|c_i|^2$. The condition that ψ be an element of the domain $\mathcal{D}(f(\mathsf{A}))$ translates to

$$\int_{-\infty}^{\infty} |f(\lambda)|^2 \, d\langle\psi|\mathsf{E}_\lambda\psi\rangle = \int_{-\infty}^{\infty} |f(\lambda)|^2 \, d\alpha_\psi = \sum_{-\infty}^{\infty} |f(\lambda_i)|^2 |c_i|^2 < \infty.$$

The function $f(\mathsf{A})$ is then defined on such a ψ by

$$f(\mathsf{A})\psi = \sum_{-\infty}^{\infty} f(\lambda_i) c_i \psi_i = \sum_{-\infty}^{\infty} f(\lambda_i) \langle\psi_i|\psi\rangle\psi_i.$$

When $f = \mathfrak{c}_1$, the constant function with value 1, then $\mathcal{D}(f(\mathsf{A})) = \mathcal{D}(\mathfrak{c}_1(\mathsf{A})) = \mathcal{H}$. Item (v) becomes $\psi = \mathfrak{c}_1(\mathsf{A})\psi = \sum_i \mathfrak{c}_1(\lambda_i)\langle\psi_i|\psi\rangle\psi_i = \sum_i \langle\psi_i|\psi\rangle\psi_i$, which just expresses the expansion of ψ in the basis of eigenvectors of A. If $f = 1_{\mathbb{R}}$, the identity function on $\mathbb{R}$, then $\mathcal{D}(f(\mathsf{A})) = \mathcal{D}(1_{\mathbb{R}}(\mathsf{A})) = \{\psi : \sum_i \lambda_i^2 |c_i|^2 < \infty\} = \mathcal{D}(\mathsf{A})$. Item (vi) becomes $\mathsf{A}\psi = 1_{\mathbb{R}}(\mathsf{A})\psi = \sum_i \lambda_i \langle\psi_i|\psi\rangle\psi_i$, which expresses A as a diagonalized operator.

We now unpack von Neumann's Axiom 3 and see that it leads to Physicist's Axiom 3 in the present context. Suppose that a quantum system is represented by the normalized state vector ψ and the observable a is represented by A. Let $\mu < \nu$ be real numbers. If the interval $(\mu, \nu]$ does not contain any of the eigenvalues λ_i then $\mathsf{E}_\mu = \mathsf{E}_\nu$, so $\mathsf{E}_\nu - \mathsf{E}_\mu = 0$, the zero operator. From von Neumann's Axiom 3, the probability that a a-measurement will yield a result in the interval $(\mu, \nu]$ is then zero. On the other hand, assume that $\mu < \lambda_i \leq \nu$ and that the interval $(\mu, \nu]$ does not contain any eigenvalues of A other than λ_i. Then $\mathsf{E}_\nu - \mathsf{E}_\mu = \mathsf{P}_i = \mathsf{Proj}_{\lambda_i}$, the projection onto the λ_i-eigenspace. From von Neumann's Axiom 3, the probability that an a-measurement will yield a result in the interval $(\mu, \nu]$ is now $\langle\psi|\mathsf{P}_i\psi\rangle = \langle\psi_i|\psi\rangle\langle\psi|\psi_i\rangle = |\langle\psi_i|\psi\rangle|^2$, and, as the probability that it will yield a result in any interval not containing an eigenvalue of A is zero, the only possible result of the measurement from the interval $(\mu, \nu]$ is the eigenvalue λ_i. Hence, the first sentence of von Neumann's Axiom 3 reduces to the first of Physicist's Axiom 3 in the context of a nondegenerate operator with a basis of eigenstates. What about collapse?

Assume that the value λ_i is obtained upon an a-measurement. Again let $(\mu, \nu]$ be an interval that contains only λ_i among the eigenvalues of A. Immediately after measurement, von Neumann's Axiom 3 implies that the state has collapsed to φ, an eigenvector of the operator P_i. The eigenvalue equation $\langle\psi_i|\varphi\rangle\psi_i = \mathsf{P}_i\varphi = \lambda\varphi$ implies, upon application of the functional $\langle\psi_j|-\rangle$, that $\langle\psi_i|\varphi\rangle\delta_{ij} = \lambda\langle\psi_j|\varphi\rangle$, where δ_{ij} is the Kronecker delta. In particular, the Fourier coefficient $\langle\psi_j|\varphi\rangle$ of φ with respect to the basis $\{\psi_j : j \in \mathbb{Z}\}$ is zero if $j \neq i$. When $j = i$, we find that the Fourier coefficient $\langle\psi_i|\varphi\rangle$ cannot be zero since $\varphi \neq 0$ and all the other coefficients are zero, and therefore $\lambda = 1$. It follows that $\varphi = \langle\psi_i|\varphi\rangle\psi_i$. This differs from the requirement of Physicist's Axiom 3, which asserts collapse to $\mathsf{P}_i\psi = \langle\psi_i|\psi\rangle\psi_i$, with at most a nonzero constant multiple. By Axiom 1, these two state vectors, φ and $\mathsf{P}_i\psi$, represent the same quantum state.

Now, the nondegeneracy of A is not in any way crucial to this development; we chose that restriction merely for convenience. The important feature that makes the arguments of this section work is not that the eigenspaces are one-dimensional but rather that there is a complete orthonormal family of eigenvectors of the operator. Mild adjustments to the arguments allow a generalization to the case where the λ_i-eigenspace has dimension greater than one, and still the von Neumann axiom reduces to the physicist's axiom.

4.4 The Spectrum of a Self-Adjoint Operator

We have seen how important the eigenvalues of the self-adjoint operators representing observables are for quantum mechanics. We also have seen that operators central to the development of quantum mechanics, such as the position and momentum operators of wave mechanics, have no eigenvalues yet they represent observables. What are the quantities that take the place of eigenvalues for such operators? We already have used the term "spectrum" for the collection of eigenvalues of an operator. In this section we expand the definition of this term in a natural way and, in so doing, appropriately generalize the notion of an eigenvalue. We will find that the possible measured values of an observable a are precisely those numbers in the spectrum of the operator representing a.

Let A be an operator defined on the dense domain $\mathcal{D}(\mathsf{A})$ of the Hilbert space $\mathcal{H}$. A complex number λ is in $\rho(\mathsf{A})$, the **resolvent set** of A, if the operator $\mathsf{A} - \lambda 1_{\mathcal{H}}$ is a bijection of $\mathcal{D}(\mathsf{A})$ onto $\mathcal{H}$ with bounded inverse $(\mathsf{A} - \lambda 1_{\mathcal{H}})^{-1}$. The **spectrum** of A is the complement of the resolvent set in $\mathbb{C}$ and is denoted as $\operatorname{Spec}(\mathsf{A})$ or $\sigma(\mathsf{A})$. Note that every eigenvalue λ of A is in $\operatorname{Spec}(\mathsf{A})$ since then $\mathsf{A} - \lambda 1_{\mathcal{H}}$ is not injective, because any nonzero eigenvector associated with λ is in the kernel of the operator $\mathsf{A} - \lambda 1_{\mathcal{H}}$. However, the spectrum may contain other elements, elements for which the operator $\mathsf{A} - \lambda 1_{\mathcal{H}}$ is indeed injective, so that λ is not an eigenvalue, but for which this operator is not surjective. The spectrum is always a closed subset of $\mathbb{C}$. The **point spectrum** is the subset $\sigma_p(\mathsf{A})$ of the spectrum that consists of the eigenvalues of A and the **discrete spectrum** is the subset $\sigma_d(\mathsf{A})$ of the point spectrum that consists of those eigenvalues that are isolated points of the spectrum and have finite **multiplicity**, meaning that such an eigenvalue's corresponding eigenspace is finite-dimensional. The **continuous spectrum** $\sigma_c(\mathsf{A})$ is the complement of the point spectrum in $\operatorname{Spec}(\mathsf{A})$. We say that the operator A has a **discrete spectrum** if $\operatorname{Spec}(\mathsf{A}) = \sigma_d(\mathsf{A})$ and a **continuous spectrum** if $\operatorname{Spec}(\mathsf{A}) = \sigma_c(\mathsf{A})$.

4.4.1 Examples

Some examples will help acclimate the reader to new features that arise in infinite dimensions when generalizing from eigenvalues to spectra. As the examples at this point are merely intended to be illustrative, we do not now give detailed derivations of the spectra; in the course of subsequent lectures, more details will emerge.[4]

Finite-dimensional case. We should mention first the easily verified fact that the spectrum of a self-adjoint operator coincides with its point spectrum, which itself is finite with at most $\dim \mathcal{H}$ values, whenever the Hilbert space $\mathcal{H}$ is finite-dimensional. Indeed, in this case the spectrum is precisely the set of points λ for which $\mathsf{A} - \lambda 1_{\mathcal{H}}$ is not injective and, by elementary linear algebra, this is precisely the set of roots of the characteristic polynomial $\det(\mathsf{A} - \lambda 1_{\mathcal{H}})$, a polynomial of degree $\dim \mathcal{H}$.

Fourier transform. The Fourier transform $\mathcal{F} : L^2(\mathbb{R}) \to L^2(\mathbb{R})$ was defined in the previous lecture for appropriately integrable functions and extends uniquely to a unitary mapping of $L^2(\mathbb{R})$. Its spectrum has precisely four points, viz., $\operatorname{Spec}(\mathcal{F}) = \sigma_p(\mathcal{F}) = \{\pm 1, \pm \mathbf{i}\}$. This is not an operator with a discrete spectrum since each eigenvalue has infinite multiplicity. We review the Fourier transform and give a detailed description of the eigenspaces in Lecture 10.

Unitary Operator. Generalizing from the previous example, the spectrum of any unitary operator on the Hilbert space $\mathcal{H}$ is a closed subset of the unit circle $\{z \in \mathbb{C} : |z| = 1\}$.

[4] These examples are taken primarily from Blanchard and Brüning (2003), pp. 322–324, where details abound for the reader who desires instant "proof gratification."

Position operator. The position operator X of one-dimensional wave mechanics has an empty point spectrum; its spectrum is in fact $\mathrm{Spec}(\mathsf{X}) = \mathbb{R}$. This is an example of an operator with a continuous spectrum.

Multiplication by a function. Generalizing the example of the position operator, let $\xi : \mathbb{R} \to \mathbb{C}$ be a continuous function and define the **multiplication operator** M_ξ on $L^2(\mathbb{R})$ by $\mathsf{M}_\xi \psi = \xi\psi$ for all $\psi \in \mathcal{D}(\mathsf{M}_\xi) = \{\psi \in L^2(\mathbb{R}) : \xi\psi \in L^2(\mathbb{R})\}$, a dense subset of $L^2(\mathbb{R})$ since it contains $C_c(\mathbb{R})$. The adjoint of M_ξ is $\mathsf{M}_{\overline{\xi}}$, multiplication by the complex conjugate of ξ. The spectrum of M_ξ is the closure of the range of ξ in $\mathbb{C}$: $\mathrm{Spec}(\mathsf{M}_\xi) = \mathrm{Cl}_{\mathbb{C}}(\xi(\mathbb{R}))$. Notice that $\mathsf{X} = \mathsf{M}_\xi$ for $\xi = 1_{\mathbb{R}}$.

Sequence. Let $\{\psi_n : n \in \mathbb{N}\}$ be an orthonormal basis in the Hilbert space $\mathcal{H}$. Let $\{a_n : n \in \mathbb{N}\}$ be an arbitrary sequence of complex numbers. Define

$$\mathcal{D}(\mathsf{A}) = \left\{ \psi \in \mathcal{H} : \sum_{n=1}^{\infty} |a_n|^2 |\langle \psi_n | \psi \rangle|^2 < \infty \right\},$$

and, for $\psi \in \mathcal{D}(\mathsf{A})$, define

$$\mathsf{A}\psi = \sum_{n=1}^{\infty} a_n \langle \psi_n | \psi \rangle \psi_n.$$

The domain $\mathcal{D}(\mathsf{A})$ is dense since it contains the dense subset of $\mathcal{H}$ of all finite linear combinations of the basis vectors. The adjoint A^* has the same domain as A and satisfies

$$\mathsf{A}^*\psi = \sum_{n=1}^{\infty} \overline{a}_n \langle \psi_n | \psi \rangle \psi_n.$$

The point spectrum $\sigma_p(\mathsf{A})$ is just the sequence $\{a_n : n \in \mathbb{N}\}$ and the spectrum is its closure in $\mathbb{C}$. This shows that every closed subset of the complex plane serves as the spectrum of some operator. Applying this with $a_n = 1/n$ gives an example where the continuous spectrum is the singleton $\sigma_c(\mathsf{A}) = \{0\}$.

4.4.2 Spectral Characteristics for Self-Adjoint Operators

The resolvent set of a self-adjoint operator A is an open subset of the complex plane that contains all nonreal complex numbers. Thus, the spectrum of A is a closed subset of the real axis $\mathbb{R}$. Further, if A is unbounded, then the spectrum is an unbounded subset of $\mathbb{R}$. We will show how to use the spectral theorem to verify that the spectrum of a self-adjoint operator is a subset of $\mathbb{R}$ and will give references in the End Notes for the remaining claims. Before we present our argument, however, we ask what has this discussion to do with quantum mechanics and Axiom 3? The answer is that Physicist's Axiom 3 is correct for operators with a discrete spectrum but needs to be modified in general to say that the result of a measurement, rather than an eigenvalue, is an element of the spectrum of A. Herein lies the importance of the fact that an element of the spectrum is always real. The way this works is as follows. Recall from von Neumann's Axiom 3 that the probability of obtaining a value in the interval (μ, ν) is $\langle\psi|(\mathsf{E}_\nu - \mathsf{E}_\mu)\psi\rangle$. It turns out that the spectral projections E_λ are constant on any open interval that does not meet the spectrum of A, so $\mathsf{E}_\nu - \mathsf{E}_\mu = 0$ and Axiom 3 implies that, upon measurement, there is a zero possibility of obtaining a value not in the spectrum.

We now use the spectral theorem to prove that $\mathrm{Spec}(\mathsf{A}) \subset \mathbb{R}$. For this, we need a couple more facts from the functional calculus that arises from the spectral theorem.

First, if f and g are continuous complex-valued functions defined on $\mathbb{R}$ then, for any $\varphi \in \mathcal{D}(f(\mathsf{A}))$ and $\psi \in \mathcal{D}(g(\mathsf{A}))$,

$$\langle f(\mathsf{A})\varphi|g(\mathsf{A})\psi\rangle = \int_{-\infty}^{\infty} \overline{f(\lambda)}g(\lambda)\, d\langle\varphi|\mathsf{E}_\lambda\psi\rangle.$$

Using this, we observe that if the continuous function f is bounded, say $|f(x)| \leq M$ for all $x \in \mathbb{R}$, then $\mathcal{D}(f(\mathsf{A})) = \mathcal{H}$ and $f(\mathsf{A})$ is a bounded operator. Indeed, for any vector $\psi \in \mathcal{H}$, we have

$$\begin{aligned}\int_{-\infty}^{\infty} |f(\lambda)|^2\, d\langle\psi|\mathsf{E}_\lambda\psi\rangle &\leq M^2 \int_{-\infty}^{\infty} d\langle\psi|\mathsf{E}_\lambda\psi\rangle \\ &= M^2\langle\psi|\psi\rangle = M^2\|\psi\|^2 < \infty,\end{aligned}$$

verifying by the criterion of Section 4.2 that $\psi \in \mathcal{D}(f(\mathsf{A}))$; further,

$$\begin{aligned}\|f(\mathsf{A})\psi\|^2 = \langle f(\mathsf{A})\psi|f(\mathsf{A})\psi\rangle &= \int_{-\infty}^{\infty} \overline{f(\lambda)}f(\lambda)\, d\langle\psi|\mathsf{E}_\lambda\psi\rangle \\ &\leq M^2 \int_{-\infty}^{\infty} d\langle\psi|\mathsf{E}_\lambda\psi\rangle = M^2\|\psi\|^2,\end{aligned}$$

verifying that $f(\mathsf{A})$ is bounded with norm $\|f(\mathsf{A})\| \leq M$.

Second, a straightforward calculation shows that the adjoint of $f(\mathsf{A})$

is $\overline{f}(\mathsf{A})$ and that it has the same domain as $f(\mathsf{A})$; in particular, we have that $f(\mathsf{A})^* = \int_{\mathbb{R}} \overline{f(\lambda)}\, d\mathsf{E}_\lambda$.

With these facts in hand, for any nonreal complex number z the function $f(\lambda) = (\lambda - z)^{-1}$ is a bounded continuous function of the real variable λ into the complex plane, bounded by $M = |\mathrm{Im}\, z|^{-1}$. Thus $f(\mathsf{A})$ is a bounded operator with domain $\mathcal{H}$. The reciprocal of f is the function $g(\lambda) = \lambda - z$, and it is easy to see that $g(\mathsf{A}) = \mathsf{A} - z1_{\mathcal{H}}$ on the domain $\mathcal{D}(g(\mathsf{A})) = \mathcal{D}(\mathsf{A})$. It now follows from the first fact presented above that, for all $\varphi \in \mathcal{H}$ and $\psi \in \mathcal{D}(\mathsf{A})$,

$$\langle \varphi | f(\mathsf{A}) g(\mathsf{A}) \psi \rangle = \langle f(\mathsf{A})^* \varphi | g(\mathsf{A}) \psi \rangle = \langle \varphi | \psi \rangle$$

and

$$\langle \psi | g(\mathsf{A}) f(\mathsf{A}) \varphi \rangle = \langle g(\mathsf{A})^* \psi | f(\mathsf{A}) \varphi \rangle = \langle \psi | \varphi \rangle.^5$$

Since $\mathcal{D}(\mathsf{A})$ is dense in $\mathcal{H}$, these equations imply that $f(\mathsf{A})g(\mathsf{A}) = 1_{\mathcal{D}(\mathsf{A})}$ and $g(\mathsf{A})f(\mathsf{A}) = 1_{\mathcal{H}}$. It follows that $z \in \rho(\mathsf{A})$.

4.4.3 Almost an Eigenvalue: Weyl Sequences

We finish this lecture with a discussion of how close to an eigenvalue is any element of the spectrum of a self-adjoint operator. Weyl argued that the elements of the spectrum may be characterized as those real numbers that are almost an eigenvalue in a certain rigorous sense. Indeed, he proved that a real number λ is in the spectrum of the self-adjoint operator A if and only if there is a sequence $\{\psi_n\}$ in $\mathcal{D}(\mathsf{A})$ of normalized vectors such that $\|(\mathsf{A} - \lambda 1_{\mathcal{H}})\psi_n\| \to 0$ as $n \to \infty$. Of course, if ψ is a normalized eigenvector with eigenvalue λ then taking $\psi_n = \psi$ for all n demonstrates that each eigenvalue is in the spectrum. In general, though, this **Weyl sequence** $\{\psi_n\}$ may be nonconvergent and may even contain no convergent subsequence. In this extreme case, an orthonormal Weyl sequence always exists for λ.

As an example, consider the position operator X of wave mechanics. If $\psi_n = \sqrt{n}\chi_{[\lambda,\lambda+1/n]}$, then $\|\psi_n\| = 1$ and $\|(\mathsf{X} - \lambda 1_{\mathcal{H}})\psi_n\| = 1/\sqrt{3n} \to 0$ as $n \to \infty$. This shows that $\sigma(\mathsf{X}) = \mathbb{R}$. Note that the Weyl sequence $\{\psi_n\}$ contains no convergent subsequence in $L^2(\mathbb{R})$ but, in defense of the physicists, does converge pointwise to the Dirac delta function $\delta(x - \lambda)$. We leave it to the reader to adjust this example to describe a Weyl sequence with these properties that is also orthonormal.

[5] Note that the validity of this depends on the fact that $f(\mathsf{A})\varphi$ is a vector in $\mathcal{D}(\mathsf{A})$. The slightly involved proof of this fact is left to the interested reader.

4.5 End Notes

The rigorous approach to quantum mechanics using the spectral theorem was first laid out in 1932 in the now classic text von Neumann (1953). It has aged well. Stone's (1932) massive work on operators in Hilbert space is still a great reference to the mathematics that appears in this lecture. These two works provide a synthesis of the research into self-adjoint operators carried out from 1927 until 1932, principally by von Neumann and Stone, somewhat independently but each acknowledging the influence of the other.

A good modern treatment that concentrates on a careful development of the spectral theorem and its applications to Schrödinger operators appears in Teschl (2009). Takhtajan (2008) presents a mathematical formulation of quantum mechanics using the spectral theorem that is at roughly the same level of difficulty as this book but which adopts the more traditional theorem–proof presentation of mathematics. My favorite developments of the spectral theorem are those of Reed and Simon (1980) and Zeidler (1995) and the classic text of Riesz and Sz.-Nagy (1955). Especially well-written and well-adapted to the needs of quantum theory, the treatment by Reed and Simon in the first volume of their four-volume methods in mathematical physics course deserves a look. I like the fact that Reed and Simon present differing versions of the spectral theorem and am particularly fond of their treatment of the functional calculus form of the theorem. Zeidler (1995), entitled *Applied Functional Analysis*, is a rigorous development of the functional analysis of the spectral theorem from the point of view of an applied mathematician and mathematical physicist who works on quantum mechanics and quantum field theory. His goal is to present the mathematical theory in the language of physicists, and on this point the book succeeds admirably.

Schechter (1981) develops a spectral theorem approach to the study of a single quantum particle restricted to a line. He has a rather extensive discussion, presented at an advanced undergraduate level, of the spectrum of an operator in Chapters 2–5 that provides an in-depth look at many of the intricacies and idiosyncrasies of spectral theory. Found on p. 28 of Schechter is a proof of Weyl's result stated in Section 4.4.3. An invaluable detailed mathematical account of the spectral theory is given in Akheizer and Glazman (1993), especially in their Volume 2, which concentrates attention on the spectral theory of self-adjoint and unitary operators and on extensions of symmetric operators. Section 43 of Volume 1 of Akheizer and Glazman contains a proof that the spectrum of a self-adjoint operator is closed. Finally, I must mention Chapter 24, entitled "Elements of Spectral Theory," in Blanchard and Brüning (2003) as the source of several examples in this lecture and worthy of further study.

5

Rigged Hilbert Space and the Dirac Calculus

> The centrepiece of the notation was the symbol $\langle q$ for a quantum state labelled a and the complementary $q\rangle$: together they can be combined to form mathematical constructions such as $\langle q|q\rangle$, a bracket. With his rectilinear logic, Dirac named each part of the "bracket" after its first and last three letters, *bra* and *ket*, new words that took several years to reach the dictionaries, leaving thousands of non-English-speaking physicists wondering why a mathematical symbol in quantum mechanics had been named for an item of lingerie. They were not the only ones to be flummoxed. A decade later, after an evening meal in St John's, Dirac was listening to dons reflecting on the pleasures of coining a new word, and, during a lull in the conversation, piped up with four words: "I invented the bra." There was not a flicker of a smile on his face. The dons looked at one another anxiously, only just managing to suppress a fit of giggling, and one of them asked him to elaborate. But he shook his head and returned to his habitual silence, leaving his colleagues mystified.
>
> Graham Farmelo
> *The Strangest Man: The Hidden Life of Paul Dirac, Mystic of the Atom*, 2009

Over a quarter century after von Neumann and Stone constructed a rigorous foundation for quantum mechanics by developing the spectral theory of self-adjoint operators, Gelfand developed an alternative approach based on distribution theory that has a close affinity with the approach of the physicists. In fact, Gelfand's approach provides a rigorous foundation for the Dirac calculus, the lingua franca of the professional physicist, which is articulated eloquently in Dirac's *The Principles of Quantum Mechanics*. This approach "rigs" the state Hilbert space $\mathcal{H}$

with useful auxiliary spaces that are rich enough to contain generalized eigenvectors for self-adjoint operators that have continuous spectra, such as the position and momentum operators of wave mechanics.

A nuisance of von Neumann's approach via the spectral theorem and self-adjoint operators is that the operators are not defined on the whole of $\mathcal{H}$. This causes two problems – the one philosophical, the other technical. The first is that when a-measurements are discussed for a quantum system, only those quantum states represented by state vectors in the domain of the corresponding operator A can be considered. Are we to assume that the elements of $\mathcal{H}$ not in $\mathcal{D}(\mathsf{A})$ are irrelevant physically? If so, what makes this so? What is the physical mechanism for ruling out such states? What about those state vectors that are in the domain of A and therefore physically relevant, but are not in the domain of another operator that represents an observable of interest? Why does the theory contain irrelevant data? The second is that the composition of two operators requires great care, as either the range of one must be contained in the domain of the other, in order to define the composition, or a suitable restriction of the domain of the first must be specified. This causes technical problems when forming appropriate combinations of operators. Even the self-composition of an operator becomes problematical, as the domain may not be invariant under application of the operator. For example, to define X^2, the self-composition of the position operator of wave mechanics, we must restrict the domain of X from those $\psi \in L^2(\mathbb{R})$ with $x\psi(x)$ square-integrable to those with $x^2\psi(x)$ square-integrable. These issues can cause technical mathematical problems when calculating expectation values, uncertainty relations, and commutation relations for typical operators of physical interest.

Gelfand's approach is to think of the Hilbert space $\mathcal{H}$ as providing a rich source of possible state vectors equipped with a calculus for calculating probabilities via the Hermitian product, a stage upon which operators representing observables have their natural existence and the stage upon which unitary evolution proceeds. However, the physically relevant state vectors, those that actually represent physical states, are corralled into a smaller linear subset of $\mathcal{H}$, the **physical state space** $\mathcal{D}$, determined by the operators representing the relevant observables and dense in $\mathcal{H}$. These operators representing the relevant observables are defined on all of $\mathcal{D}$, which also is invariant under the action of each such operator. The possible results of measurement are recorded by the generalized eigenvalues of those operators associated with generalized eigenvectors that live, not necessarily in $\mathcal{H}$, but in a linear extension of $\mathcal{H}$, the dual $\mathcal{D}'$, or anti-dual $\overline{\mathcal{D}'}$, to the physical state space. The use of these **Gelfand triples** $(\mathcal{D}, \mathcal{H}, \mathcal{D}')$ and $(\mathcal{D}, \mathcal{H}, \overline{\mathcal{D}'})$ addresses the two problems of the previous paragraph by identifying at the start, and separating out, the physically relevant states and allowing for arbitrary compositions.

In this lecture, we give a brief taste of Gelfand's theory of **rigged Hilbert spaces** by illustrating how the position and momentum operators of wave mechanics may be thought of as having a complete set of (generalized) eigenvectors. We introduce the **Dirac calculus** and examine in some detail two formulae in the Dirac calculus that express the identity and the position operators of wave mechanics that cause some discomfort for the mathematician. These are dissected in the light of Gelfand's theory and found to be adequate expressions of the functional calculus of the spectral theorem, as discussed in the previous lecture.

5.1 Gelfand Triples and the Rigging of $\mathcal{H}$

Let $\mathcal{H}$ be a Hilbert space and let $\mathcal{D}$ be a dense, linear subspace of $\mathcal{H}$. Recall that the **dual space** to $\mathcal{D}$ is the linear space of bounded, $\mathbb{C}$-valued, linear functionals on the domain $\mathcal{D}$ and is denoted as $\mathcal{D}'$. The **anti-dual space** is the linear space of bounded, $\mathbb{C}$-valued, conjugate-linear functionals on the domain $\mathcal{D}$ and is denoted as $\overline{\mathcal{D}'}$.[1] There is an obvious conjugate-linear isomorphism $\mathcal{D}' \to \overline{\mathcal{D}'}$ given by conjugation: the correspondence is $\ell \mapsto \overline{\ell}$ where $\overline{\ell}(\varphi) = \overline{\ell(\varphi)}$. The triples $(\mathcal{D}, \mathcal{H}, \mathcal{D}')$ and $(\mathcal{D}, \mathcal{H}, \overline{\mathcal{D}'})$ are **associated Gelfand triples**.

The dual correspondence of $\mathcal{H}$ to $\mathcal{H}^*$ induces a conjugate-linear embedding of $\mathcal{H}$ into $\mathcal{D}'$ via $\psi \mapsto \psi^* = \langle\psi|-\rangle$, where it is understood that ψ^* is restricted to the domain $\mathcal{D}$. That this correspondence is injective follows from the fact that $\mathcal{D}$ is dense in $\mathcal{H}$ but, in general, the correspondence is not surjective. The representative example of this non-surjectivity occurs when, for example, $\mathcal{D} = C_c(\mathbb{R}) \subset L^2(\mathbb{R}) = \mathcal{H}$. The Dirac delta distribution δ_x concentrated at x and defined by $\delta_x(\varphi) = \varphi(x)$ when $\varphi \in C_c(\mathbb{R})$ is an element of the dual space $\mathcal{D}'$ but is not equal to ψ^* for any $\psi \in \mathcal{H}$.

[1] By a **conjugate-linear mapping** we mean a mapping ℓ, defined on $\mathcal{D}$, into a complex linear space that satisfies, for all $\varphi, \psi \in \mathcal{D}$ and $a, b \in \mathbb{C}$,

$$\ell(a\varphi + b\psi) = \overline{a}\ell(\varphi) + \overline{b}\ell(\psi).$$

The Dirac **bra-ket notation** is particularly apt in the present development. Following Dirac, we denote the functional ψ^* as $\langle\psi|$ and call it a **bra-vector**. The **bra-space** $\mathcal{B}_\mathcal{D}$, the collection of all $\langle\psi|$ for $\psi \in \mathcal{H}$, is the image of $\mathcal{H}$ under this conjugate-linear embedding into $\mathcal{D}'$. The image of the bra-space under the isomorphism $\mathcal{D}' \to \overline{\mathcal{D}'}$ is the **ket-space** $\mathcal{K}_\mathcal{D}$. For the bra-vector $\langle\psi|$, the corresponding element of $\mathcal{K}_\mathcal{D}$ is $\overline{\langle\psi|}$ and its action on $\varphi \in \mathcal{D}$ is

$$\overline{\langle\psi|}\varphi = \overline{\langle\psi|\varphi\rangle} = \langle\varphi|\psi\rangle.$$

Thus, $\overline{\langle\psi|} = \langle -|\psi\rangle$ on the domain $\mathcal{D}$. In the Dirac calculus, $\overline{\langle\psi|}$ is denoted by $|\psi\rangle$ and called a **ket-vector**; the Hilbert space $\mathcal{H}$ is identified with the ket-space $\mathcal{K}_\mathcal{D}$ and its dual $\mathcal{H}^*$ with the bra-space $\mathcal{B}_\mathcal{D}$. There is the apparent conjugate bilinear pairing $\mathcal{B}_\mathcal{D} \times \mathcal{K}_\mathcal{D} \to \mathbb{C}$ given by $(\langle\varphi|, |\psi\rangle) \mapsto \langle\varphi|\psi\rangle$. In the uninteresting case where $\mathcal{D} = \mathcal{H}$, the Dirac calculus identifies $\mathcal{H}$ with $\mathcal{K}_\mathcal{H} = \overline{\mathcal{H}'}$ and the dual correspondence $\overline{\mathcal{H}'} = \mathcal{H} \to \mathcal{H}^* = \mathcal{H}'$ is given by $|\psi\rangle \mapsto \overline{|\psi\rangle} = \langle\psi|$.

With these definitions we can refine the picture a bit and build the following commutative diagram,

$$\begin{array}{ccccc} \mathcal{H} & \cong & \mathcal{K}_\mathcal{D} & \subset & \overline{\mathcal{D}'} \\ \updownarrow & & \updownarrow & & \updownarrow \\ \mathcal{H}^* & \cong & \mathcal{B}_\mathcal{D} & \subset & \mathcal{D}' \end{array}$$

given on elements by

$$\begin{array}{ccccc} \psi & \mapsto & |\psi\rangle & \in & \overline{\mathcal{D}'} \\ \updownarrow & & \updownarrow & & \updownarrow \\ \psi^* & \mapsto & \langle\psi| & \in & \mathcal{D}'. \end{array}$$

The horizontal mappings are linear injections while the vertical mappings are conjugate-linear isomorphisms, viz., the dual correspondences.

For the use of these Gelfand triples in quantum mechanics, the idea is to choose $\mathcal{D}$ in such a way that all the relevant operators and their compositions are defined on $\mathcal{D}$ and $\mathcal{D}$ is invariant under the action of those operators. The particular potential that appears in the Schrödinger equation of wave mechanics is pertinent to the choice. The most important choices occur when $\mathcal{D}$ is chosen to be a suitable space of smooth functions. When $\mathcal{D} = C_c^\infty(\mathbb{R}^3)$, the smooth functions with compact support, the dual $\mathcal{D}'$ is the space of distributions and when $\mathcal{D} = \mathcal{S}$, the Schwartz space of rapidly decreasing functions, the dual $\mathcal{S}' = \mathcal{T}$ is the space of tempered distributions.[2]

[2] See Lecture 10 for a review of distribution theory with detailed definitions.

Gelfand used this marriage of Hilbert space with distribution theory to generalize the notion of an eigenvector of a self-adjoint operator. To see how this is done, we first reinterpret the usual notion of an eigenvector. Let A be a self-adjoint operator on $\mathcal{H}$ whose domain contains $\mathcal{D}$, and let $\psi \in \mathcal{H}$ be an eigenvector of A with associated eigenvalue λ. Let $\ell_\psi = |\psi\rangle$. Then, for all $\varphi \in \mathcal{D}$, we have

$$\ell_\psi(\mathsf{A}\varphi) = \langle \mathsf{A}\varphi|\psi\rangle = \langle \varphi|\mathsf{A}\psi\rangle = \langle \varphi|\lambda\psi\rangle = \lambda\ell_\psi\varphi.$$

This equation makes sense for any ℓ in the anti-dual space; thus we say that $\ell \in \overline{\mathcal{D}'}$ is a **generalized eigenvector** of the self-adjoint operator A, with associated **generalized eigenvalue** λ, if $\ell(\mathsf{A}\varphi) = \lambda\ell\varphi$ for every $\varphi \in \mathcal{D}$. In terms of functionals, the eigenvalue equation may be written as $\ell\mathsf{A} = \lambda\ell$. The discussion above shows that if $\psi \in \mathcal{H}$ is an eigenvector of A with eigenvalue λ then the ket $|\psi\rangle$ is a generalized eigenvector of A with generalized eigenvalue λ. In the Dirac notation, $|\psi\rangle$ often is denoted by its eigenvalue as $|\lambda\rangle$ and is called an **eigenket** of the operator A. This notation is somewhat ambiguous and highly context-dependent and care should be taken when there are several operators under consideration, but it makes for elegant formulae as we shall see subsequently.

5.2 The Position and Momentum Operators in the Dirac Calculus

Two examples from one-dimensional wave mechanics are instructive. In both, we choose $\mathcal{D}$ to be the Schwartz space $\mathcal{S}$ of test functions, i.e., the smooth functions on $\mathbb{R}$ that, along with all their derivatives, decrease to zero as $x \to \pm\infty$ faster than any negative power of x. The domains of both the position and momentum operators contain $\mathcal{S}$, which is invariant under both. Let δ_x be the Dirac distribution concentrated at the real number x and let $\varphi \in \mathcal{S}$ be a test function. Then $\overline{\delta_x}(\mathsf{X}\varphi) = x\overline{\delta_x}\varphi$, which shows that $\overline{\delta_x} \in \overline{\mathcal{S}'}$ is a generalized eigenvector of the position operator with generalized eigenvalue x. Consider now the functional $\varepsilon_p : \mathcal{S} \to \mathbb{C}$, where $p \in \mathbb{R}$, given by

$$\varepsilon_p(\varphi) = \frac{1}{\sqrt{2\pi\hbar}} \int_{\mathbb{R}} \overline{\varphi(x)} e^{\mathrm{i}px/\hbar}\, dx, \tag{5.1}$$

which converges since $\varphi \in \mathcal{S}$. Now $\varepsilon_p \in \overline{\mathcal{S}'}$ and a straightforward calculation using integration by parts verifies that $\varepsilon_p(\mathsf{P}\varphi) = p\varepsilon_p(\varphi)$ and, therefore, that ε_p is a generalized eigenvector of the momentum operator with generalized eigenvalue p. Extending the Dirac notation where an eigenket is denoted by its eigenvalue, we write $|x\rangle = \overline{\delta_x}$ and $|p\rangle = \varepsilon_p$, and their corresponding elements of $\mathcal{S}'$ as $\langle x| = \delta_x$ and $\langle p| = \overline{\varepsilon_p}$. The action of these functionals on a test function $\varphi \in \mathcal{S}$ is usually written as an action on the corresponding bra-vector $\langle\varphi|$ on the right or ket-vector $|\varphi\rangle$ on the left as

$$\varphi|x\rangle = \langle\varphi|x\rangle = \overline{\varphi(x)} \quad \text{and} \quad \langle x|\varphi = \langle x|\varphi\rangle = \varphi(x), \tag{5.2}$$

$$\varphi|p\rangle = \langle\varphi|p\rangle = \frac{1}{\sqrt{2\pi\hbar}}\int_{\mathbb{R}} \overline{\varphi(x)}e^{\mathrm{i}px/\hbar}\,dx = \overline{\widehat{\varphi}}(p),$$

and

$$\langle p|\varphi = \langle p|\varphi\rangle = \frac{1}{\sqrt{2\pi\hbar}}\int_{\mathbb{R}} \varphi(x)e^{-\mathrm{i}px/\hbar}\,dx = \widehat{\varphi}(p),$$

where $\widehat{\varphi}$ is the Fourier transform. Notice that the functional $\langle p|$ is precisely $\delta_p\mathcal{F}$, the composition of the delta distribution concentrated at p with the Fourier transform. This turns out to reflect the fact that the Fourier transform is a precise symmetry of the Schrödinger wave mechanical formalism that exchanges the roles of position and momentum. This is explored carefully in Sections 10.3 and 10.4. Note that we use the bra-notation for the functional $\langle p|$ and the Dirac delta notation for $\langle x| = \delta_x$. The reason is that there is an obvious ambiguity in the notation: does $\langle 2.7|$ denote the Dirac distribution concentrated at $x = 2.7$, or the evaluation of the Fourier transform at $p = 2.7$? Both the context and the labeling of the variables as x and p help one keep track of the meaning, but, as this example shows, care is still warranted. After all, the equation $\langle p| = \delta_p\mathcal{F}$ is unambiguous while $\langle p| = \langle p|\mathcal{F}$ causes one to pause. In wave mechanics, $\langle x|\varphi\rangle = \varphi(x)$ is called the **position representation**, and $\langle p|\varphi\rangle = \widehat{\varphi}(p)$ the **momentum representation**, of the wave function φ.

We have seen that $|x\rangle = \overline{\delta_x}$ is a generalized eigenvector of the position operator X with generalized eigenvalue x. The eigenvalue equation is $\overline{\delta_x}(\mathsf{X}\varphi) = x\overline{\delta_x}\varphi$, which may be written now using the first of formulae 5.2 as $\langle\mathsf{X}\varphi|x\rangle = x\langle\varphi|x\rangle$. Using this, we may rewrite the eigenvalue equation so that not only the functional $|x\rangle$, but also the operator X, acts on a test function on the right instead of the left to get

$$\varphi\cdot\mathsf{X}|x\rangle = \langle\mathsf{X}\varphi|x\rangle = x\langle\varphi|x\rangle = \varphi\cdot|x\rangle x.$$

The centered dot is just a reminder that the functions are operating on φ from the right.[3] This justifies the physicists' formula $\mathsf{X}|x\rangle = x|x\rangle$.

Another clever invention of the physicists is

$$\langle x|p\rangle = \frac{1}{\sqrt{2\pi\hbar}} e^{\mathrm{i}px/\hbar}. \tag{5.3}$$

To explain this, let $\psi_p(x) = \frac{1}{\sqrt{2\pi\hbar}} e^{\mathrm{i}px/\hbar}$ and notice that $\psi_p \notin L^2(\mathbb{R})$, so that, technically, $|\psi_p\rangle$ is not a ket-vector. However, the notation works formally in the sense that $\varepsilon_p(\varphi) = \langle\varphi|p\rangle = \langle\varphi|\psi_p\rangle$, so that the function ψ_p represents $|p\rangle$ in the same way that ψ represents the functional $|\psi\rangle$ when $\psi \in L^2(\mathbb{R})$. Writing $|p\rangle$ as $|\psi_p\rangle$, formula 5.3 now follows from formula 5.2.

One must take care not to think of these uses of the Hermitian product notation for $|x\rangle$ and $|p\rangle$ as an indication that the Hermitian product on $L^2(\mathbb{R})$ somehow extends to a conjugate bilinear pairing of the dual and the anti-dual spaces $\mathcal{S}'$ and $\overline{\mathcal{S}'}$. Indeed, suppose that a conjugate bilinear pairing extends to $\mathcal{S}' \times \overline{\mathcal{S}'}$ with $\langle x|\varphi\rangle = \varphi(x)$ for all $x \in \mathbb{R}$ and $\varphi \in \mathcal{S}$ and suppose, for example, that $\langle 0|0\rangle = 1$. The Schwarz inequality then implies that $|\langle 0|\varphi\rangle|^2 \leq \langle 0|0\rangle\langle\varphi|\varphi\rangle$, or $|\varphi(0)|^2 \leq \|\varphi\|^2 = \int_{\mathbb{R}} |\varphi|^2$. However, it is easy to construct a smooth function of compact support φ with, say, $\varphi(0) = 2$ while $\|\varphi\|^2 = 1$. The only way to assign a value to $\langle x|x\rangle$ consistent with the Schwarz inequality is to assign the value ∞. Physicists say that $|x\rangle$ is a **nonnormalizable state vector**, indicating that the product $\langle x|x\rangle$ cannot be defined as a complex number, but they also say that the kets $|x\rangle$ and $|x'\rangle$ are orthogonal when $x \neq x'$. This information is usually symbolized by

$$\langle x|x'\rangle = \delta(x - x'), \tag{5.4}$$

where δ is the original Dirac delta function with $\delta(y) = 0$ when $y \neq 0$, and $\delta(0) = \infty$. How are we to make sense of the orthogonality of $|x\rangle$ and $|x'\rangle$ when $x \neq x'$? We have observed that the generalized eigenvalue equation for the position operator that recognizes $|x\rangle$ as an generalized eigenvector of X with generalized eigenvalue x may be written as

$$\mathsf{X}|x\rangle = x|x\rangle$$

in $\overline{\mathcal{S}'}$, meaning that $\langle\mathsf{X}\varphi|x\rangle = x\langle\varphi|x\rangle$ for all test functions φ. This equa-

[3] In the section following, we will define the action of a linear operator A on a ket-vector $|\psi\rangle \in \mathcal{K}_{\mathcal{D}}$ as $\mathsf{A}|\psi\rangle = |\mathsf{A}\psi\rangle$. The action of this functional on a test function φ is $\langle\varphi|\mathsf{A}\psi\rangle = \langle\mathsf{A}^*\varphi|\psi\rangle$, so that $\mathsf{A}|\psi\rangle = |\psi\rangle \circ \mathsf{A}^*$ when written in the usual order of composition. This definition suggests that $\mathsf{X}|x\rangle$ should be defined by $\varphi \cdot \mathsf{X}|x\rangle = \langle\mathsf{X}^*\varphi|x\rangle$ but, since X is self-adjoint, these agree.

tion may also be expressed as $\langle x|\mathsf{X} = x\langle x|$ in $\mathcal{S}'$, meaning that $\langle x|\mathsf{X}\varphi\rangle = x\langle x|\varphi\rangle$ for all test functions φ. If we are to give a meaning to $\langle x|x'\rangle$, the symbol $\langle x|\mathsf{X}|x'\rangle$ now has two interpretations: $\langle x|\mathsf{X}|x'\rangle = \langle x|(x'|x'\rangle)$ or $\langle x|\mathsf{X}|x'\rangle = (x\langle x|)|x'\rangle$, depending on whether X acts on $|x'\rangle$ or on $\langle x|$. If linearity is to be preserved, we must have $x\langle x|x'\rangle = x'\langle x|x'\rangle$, which implies that $\langle x|x'\rangle = 0$ as $x \neq x'$. Though a mathematician might cringe at this explanation, we will see, nonetheless, that it is a useful fiction and, in fact, can be given a precise derivation in terms of distributions. This will be worked out in the next section.

For the record, just as the physicist's eigenvalue equation $\mathsf{X}|x\rangle = x|x\rangle$ has now been given a precise meaning in wave mechanics, similarly, the eigenvalue equation

$$\mathsf{P}|p\rangle = p|p\rangle \tag{5.5}$$

now makes sense. These expressions as well as the development of the results of the Dirac calculus presented in the remainder of this lecture form the mainstay of much of the technical calculational toolbox of the working physicist.

Returning to the general case of associated Gelfand triples determined by a dense linear subspace $\mathcal{D}$ of $\mathcal{H}$, we expand the terminology by calling any element of the dual space $\mathcal{D}'$ a bra-vector, or a **bra** for short, and any element of the anti-dual space $\overline{\mathcal{D}'}$ a ket-vector, or a **ket** for short. Though Dirac's route to bras and kets was quite different from ours, he observed that *The space of bra and ket vectors when the vectors are restricted to be of finite length and to have finite scalar products is called by mathematicians a* Hilbert space. *The bra and ket vectors that we now use form a more general space than a Hilbert space,*[4] which, of course, is an expression of the fact that the linear embeddings of $\mathcal{H}$ into the dual and anti-dual spaces are not surjective. A brief warning is in order. As indicated for wave mechanics, the pairing $\mathcal{B}_{\mathcal{D}} \times \mathcal{K}_{\mathcal{D}} \to \mathbb{C}$ in general does not extend to a bilinear pairing of the dual and anti-dual spaces, and, in general, $\mathcal{D}'$ and $\overline{\mathcal{D}'}$ are not Hilbert spaces.

Recall that a collection $\{\psi_i\}_i$ of vectors in a Hilbert space $\mathcal{H}$ is complete if, for all $\psi \in \mathcal{H}$, $\psi = 0$ if and only if $\langle\psi_i|\psi\rangle = 0$ for every i. If a complete collection is also orthogonal, meaning that $\langle\psi_i|\psi_j\rangle = 0$ when $i \neq j$, then it forms a basis and every vector in $\mathcal{H}$ may be expanded uniquely in terms of that basis. Generalizing this, we say that a collection of kets, $\{\zeta_\lambda\}_\lambda$, is **complete** for the Gelfand triples determined by $\mathcal{D}$ if, for all $\varphi \in \mathcal{D}$, $\varphi = 0$ if and only if $\zeta_\lambda(\varphi) = 0$ for all λ. Notice that we

[4] Dirac (1930), p. 40.

have left unspecified the range of the index λ, indicating that it may be a continuous index. It is obvious that the collection $\{|x\rangle : x \in \mathbb{R}\}$ is an orthogonal complete collection of kets for the Gelfand triples determined by the Schwartz space $\mathcal{S}$ in one-dimensional wave mechanics. We will address in what sense kets may be expanded in terms of $\{|x\rangle : x \in \mathbb{R}\}$ in Section 5.4.

5.3 Products of Bras and Kets

With the basics of the Dirac calculus in place, we expand its role in the next section where we see how the position and momentum operators of one-dimensional wave mechanics are tamed even further by the calculus. There, these two primary operators of wave mechanics are seen to possess a complete family of generalized eigenvectors that allow for eigenvector expansions of wave functions and expressions for the functional calculus of an operator that mimic those of Section 4.3.

First though, we distill from the previous section the essentials of Dirac's elegant bra-ket notation for elements of $\mathcal{H}$ and its dual $\mathcal{H}^*$, a notation designed cleverly to take advantage of the dual correspondence. In identifying the Hilbert space $\mathcal{H}$ with the ket-space $\mathcal{K}_{\mathcal{D}}$, the linear subspace of $\overline{\mathcal{D}'}$ of ket-vectors, each element ψ of $\mathcal{H}$ is now denoted as a ket $|\psi\rangle$. The dual correspondence is $\mathcal{H} = \mathcal{K}_{\mathcal{D}} \to \mathcal{B}_{\mathcal{D}} = \mathcal{H}^*$, given by $|\psi\rangle \mapsto \overline{|\psi\rangle} = \langle\psi|$. The conjugate bilinear pairing realizes the Hermitian product and induces a right-action of $\mathcal{H}$ on $\mathcal{H}^*$ via $\langle\varphi| \mapsto \langle\varphi|\psi\rangle$ and a left-action of $\mathcal{H}^*$ on $\mathcal{H}$ via $|\psi\rangle \mapsto \langle\varphi|\psi\rangle$. The dual correspondence is conjugate-linear, and the notation handles this naturally: for $\varphi, \psi \in \mathcal{H}$ and $a, b \in \mathbb{C}$,

$$a|\varphi\rangle + b|\psi\rangle = |a\varphi + b\psi\rangle \mapsto \langle a\varphi + b\psi| = \overline{a}\langle\varphi| + \overline{b}\langle\psi|.$$

Let A be a linear operator on $\mathcal{H}$. The adjoint equation may be written as $\langle\psi|\mathsf{A}^*\varphi\rangle = \langle\mathsf{A}\psi|\varphi\rangle$ for appropriate φ and ψ in $\mathcal{H}$. This implies that $\mathsf{A}|\psi\rangle = |\mathsf{A}\psi\rangle \mapsto \langle\mathsf{A}\psi| = \langle\psi|\mathsf{A}^*$ under the dual correspondence, where $\langle\psi|\mathsf{A}^*$ is understood as composition of operators. This is an example of one kind of multiplication that makes sense in the Dirac calculus. Expanding on this, consider the following products that may be formed by bras $\langle\varphi|$, kets $|\psi\rangle$, and operators A: bra×ket, ket×bra, operator×ket, and bra×operator.

bra×ket=scalar: This is the **inner product** of $\langle\varphi|$ and $|\psi\rangle$, defined as $\langle\varphi|\cdot|\psi\rangle = \langle\varphi|\psi\rangle$, which is just the conjugate bilinear pairing, the Hermitian product of φ and ψ.

ket×bra=operator: This is the **outer product** of $|\psi\rangle$ and $\langle\varphi|$, defined as $|\psi\rangle\cdot\langle\varphi| = |\psi\rangle\langle\varphi|$, an operator that acts on the ket-space on the left via $|\xi\rangle \mapsto (|\psi\rangle\langle\varphi|)|\xi\rangle = |\psi\rangle\langle\varphi|\xi\rangle = \langle\varphi|\xi\rangle|\psi\rangle$, and on the bra-space on the right via $\langle\xi| \mapsto \langle\xi|(|\psi\rangle\langle\varphi|) = \langle\xi|\psi\rangle\langle\varphi|$.

operator×ket=ket: $\mathsf{A}\cdot|\psi\rangle = \mathsf{A}|\psi\rangle = |\mathsf{A}\psi\rangle$.

bra×operator=bra: $\langle\varphi|\cdot\mathsf{A} = \langle\varphi|\mathsf{A}$, the composition of the functional $\langle\varphi|$ with A, and $\langle\varphi|\mathsf{A} = \langle\mathsf{A}^*\varphi|$.

This notation is both skillfully crafted and effective, and often makes calculations simple and somewhat automatic. Notice that the expression $\langle\varphi|\mathsf{A}|\psi\rangle$ is defined unambiguously as either $\langle\varphi|(\mathsf{A}|\psi\rangle)$ or $(\langle\varphi|\mathsf{A})|\psi\rangle$, and the adjoint equation becomes

$$\langle\varphi|\mathsf{A}|\psi\rangle = \overline{\langle\psi|\mathsf{A}^*|\varphi\rangle}.$$

Here are a couple of sample calculations that condense almost automatically from the use of the Dirac calculus. First, we prove that $(|\alpha\rangle\langle\beta|)^* = |\beta\rangle\langle\alpha|$. Let $\mathsf{A} = |\alpha\rangle\langle\beta|$. From the adjoint equation, for $\varphi, \psi \in \mathcal{H}$,

$$\langle\varphi|\mathsf{A}^*|\psi\rangle = \overline{\langle\psi|\mathsf{A}|\varphi\rangle} = \overline{\langle\psi|\alpha\rangle\langle\beta|\varphi\rangle} = \langle\alpha|\psi\rangle\langle\varphi|\beta\rangle = \langle\varphi|\beta\rangle\langle\alpha|\psi\rangle.$$

Therefore, $\mathsf{A}^* = |\beta\rangle\langle\alpha|$. Our second easy computation begins with a ket $|\alpha\rangle$ of unit length. Then $\mathsf{Proj}_\alpha = |\alpha\rangle\langle\alpha|$ is the projection operator in the direction of $|\alpha\rangle$. Let W be any operator on $\mathcal{H}$. Then

$$\mathsf{Proj}_\alpha \mathsf{W}\,\mathsf{Proj}_\alpha = \langle\alpha|\mathsf{W}|\alpha\rangle\mathsf{Proj}_\alpha,$$

since $\mathsf{Proj}_\alpha \mathsf{W}\,\mathsf{Proj}_\alpha = |\alpha\rangle\langle\alpha|\mathsf{W}|\alpha\rangle\langle\alpha| = \langle\alpha|\mathsf{W}|\alpha\rangle|\alpha\rangle\langle\alpha| = \langle\alpha|\mathsf{W}|\alpha\rangle\mathsf{Proj}_\alpha$. These calculations rely on the associativity of these products in various combinations whose identification and confirmation we leave to the reader.

One final observation is in order. If $\{\xi_i : i \in \mathbb{N}\}$ is any complete orthonormal basis of the Hilbert space $\mathcal{H}$ then the identity operator $1_{\mathcal{H}}$ may be written in the Dirac calculus as $1_{\mathcal{H}} = \sum_{i=1}^{\infty} |\xi_i\rangle\langle\xi_i|$. This nifty little formula allows one to quickly expand a vector $\psi \in \mathcal{H}$ in its ket realization as

$$|\psi\rangle = 1_{\mathcal{H}}|\psi\rangle = \sum_{i=1}^{\infty} |\xi_i\rangle\langle\xi_i|\psi\rangle = \sum_{i=1}^{\infty} c_i|\xi_i\rangle,$$

where $c_i = \langle\xi_i|\psi\rangle$ is the Fourier coefficient.

5.4 Spectral Decomposition and the Dirac Calculus

We return briefly to the setting of Section 4.3. Let A be a self-adjoint operator on $\mathcal{H}$, identified with the ket space $\mathcal{K}_{\mathcal{H}}$, for which there is an orthonormal basis $\{|\lambda_i\rangle : i \in \mathbb{N}\}$ of eigenkets labeled by their eigenvalues. The projection operator in the direction of $|\lambda_i\rangle$ is $|\lambda_i\rangle\langle\lambda_i|$ and the identity operator $1_{\mathcal{H}}$, as just noted, may be written as

$$1_{\mathcal{H}} = \sum_{i=1}^{\infty} |\lambda_i\rangle\langle\lambda_i|.$$

This expression for the identity operator is called by physicists the **resolution of the identity** determined by A. Note that the eigenvalue equation implies that $\mathsf{A}|\lambda_i\rangle = \lambda_i|\lambda_i\rangle$ for each i. The **spectral decomposition** of A is

$$\mathsf{A} = \mathsf{A}1_{\mathcal{H}} = \mathsf{A}\sum_{i=1}^{\infty} |\lambda_i\rangle\langle\lambda_i| = \sum_{i=1}^{\infty} \mathsf{A}|\lambda_i\rangle\langle\lambda_i| = \sum_{i=1}^{\infty} \lambda_i|\lambda_i\rangle\langle\lambda_i|.$$

Immediately, since $\mathsf{A}^n|\lambda_i\rangle = \lambda_i^n|\lambda_i\rangle$ for each positive integer n, the operator $f(\mathsf{A})$ is given by

$$f(\mathsf{A}) = \sum_{i=1}^{\infty} f(\lambda_i)|\lambda_i\rangle\langle\lambda_i|$$

whenever f is a polynomial function. Even if f is not a polynomial, this equation may be used to define $f(\mathsf{A})$, with $\mathcal{D}(f(\mathsf{A}))$ the collection of vectors ψ in $\mathcal{H}$ for which the sum $\sum |f(\lambda_i)|^2|\langle\lambda_i|\psi\rangle|^2$converges.

We now turn our attention to the position and momentum operators of one-dimensional wave mechanics. These are self-adjoint operators with continuous spectra and are handled in a rigorous manner by the spectral theorem. First we express the spectral decomposition of the position operator in terms of the Dirac calculus and generalized eigenvectors. The resulting expressions play a central role in the physicists' development, but, at least as presented in most physics texts, cause some distress to the mathematician's sensibilities and quest for precision. Our route here is first to present the development from the perspective of the physicists, and then to reinterpret their expressions in terms of the spectral theorem, hopefully easing the mathematician's distress.

Not all products of bras and kets are allowed. For example, as mentioned earlier the generalized eigenkets $|x\rangle$ of the position operator X of wave mechanics are not normalizable as the inner product $\langle x|x\rangle$ cannot be defined in a consistent manner. Despite this, the outer product

$|x\rangle\langle x|$ does make sense as an operator on the Schwartz space $\mathcal{S}$, acting on both the left and the right. For $\varphi, \psi \in \mathcal{S}$, we have $\varphi|x\rangle\langle x|\psi = \langle\varphi|x\rangle\langle x|\psi\rangle = \overline{\varphi(x)}\psi(x)$. Recall that, for $x \neq x'$, $\langle x|x'\rangle = 0$ and the collection $\{|x\rangle : x \in \mathbb{R}\}$ is a complete collection of kets. Under the principle often employed in physics textbooks that *properties valid for summations are valid for integrations as well*,[5] the physicist might formulate the resolution of the identity and the spectral decomposition of the position operator X by analogy with the expressions of the opening paragraph of this section. This would lead to the following expressions. First, the resolution of the identity for X becomes

$$1_{\mathcal{S}} = \int_{-\infty}^{\infty} |x\rangle\langle x|\, dx, \tag{5.6}$$

its spectral decomposition becomes

$$\mathsf{X} = \int_{-\infty}^{\infty} x|x\rangle\langle x|\, dx,$$

and, for any continuous function f,

$$f(\mathsf{X}) = \int_{-\infty}^{\infty} f(x)|x\rangle\langle x|\, dx.$$

If $\psi \in \mathcal{S}$ is a test function, then the expression

$$|\psi\rangle = 1_{\mathcal{S}}|\psi\rangle = \int_{-\infty}^{\infty} |x\rangle\langle x|\, dx \cdot |\psi\rangle = \int_{-\infty}^{\infty} |x\rangle\langle x|\psi\rangle\, dx = \int_{-\infty}^{\infty} \psi(x)|x\rangle\, dx$$

is supposed to express the ket $|\psi\rangle$ in the position basis.

How do we make precise mathematical sense of these physicists' expressions? The answer is found by observing that their primary use is in formulae using the Hermitian product. For example, formally, this last expression leads to

$$\langle\varphi|\psi\rangle = \langle\varphi| \cdot \int_{-\infty}^{\infty} \psi(x)|x\rangle\, dx = \int_{-\infty}^{\infty} \psi(x)\langle\varphi|x\rangle\, dx = \int_{-\infty}^{\infty} \overline{\varphi(x)}\psi(x)\, dx.$$

Similarly,

$$\begin{aligned}\langle\varphi|\mathsf{X}|\psi\rangle &= \langle\varphi|\mathsf{X} \cdot \int_{-\infty}^{\infty} \psi(x)|x\rangle\, dx \\ &= \int_{-\infty}^{\infty} \psi(x)\langle\varphi|\mathsf{X}|x\rangle\, dx = \int_{-\infty}^{\infty} \overline{\varphi(x)}x\psi(x)\, dx.\end{aligned}$$

[5] This is comes from the preface of Fano's (1971) beautiful book *Mathematical Methods of Quantum Mechanics*. Therein he gives three principles, used by physicists to manipulate and derive expressions, at which mathematicians would blush.

These equations yield correct expressions for both $\langle\varphi|\psi\rangle$ and $\langle\varphi|\mathsf{X}|\psi\rangle$. To ease the distress of the mathematician, we compare the expression here for $\langle\varphi|\psi\rangle$ with that in Section 4.2. We have, assuming that $\varphi, \psi \in \mathcal{S}$,

$$\langle\varphi|\psi\rangle = \int_{-\infty}^{\infty} d\langle\varphi|\mathsf{E}_x\psi\rangle,$$

where, it may be recalled, $\{\mathsf{E}_x : x \in \mathbb{R}\}$ is the resolution of the identity promised by the spectral theorem, with E_x the projection operator defined by $\mathsf{E}_x\psi = \chi_{(-\infty,x]}\psi$. This integral is a Riemann–Stieltjes integral with respect to the function $\alpha(x) = \langle\varphi|\mathsf{E}_x\psi\rangle$, a function with bounded variation. Notice that the expression $\langle\varphi|\mathsf{E}_x\psi\rangle = \int_{-\infty}^{x} \overline{\varphi}\psi$, so that $d\alpha = \overline{\varphi(x)}\psi(x)\,dx = \langle\varphi|x\rangle\langle x|\psi\rangle\,dx$. This may be written symbolically as $d\mathsf{E}_x = d\langle -|\mathsf{E}_x-\rangle = |x\rangle\langle x|\,dx$. We find then that

$$1_{\mathcal{S}} = \int_{-\infty}^{\infty} d\mathsf{E}_x = \int_{-\infty}^{\infty} |x\rangle\langle x|\,dx,$$

and more generally,

$$f(\mathsf{X}) = \int_{-\infty}^{\infty} f(x) d\mathsf{E}_x = \int_{-\infty}^{\infty} f(x)|x\rangle\langle x|\,dx,$$

defined on an appropriately restricted domain of $\mathcal{S}$; notice, though, that $f(\mathsf{X})$ is defined on all of $\mathcal{S}$ whenever f is a polynomial. In this way we see that the physicist's expression $|x\rangle\langle x|\,dx$ is an appropriate notation for the mathematician's expression $d\mathsf{E}_x$, at least when our attention is restricted to the physical state space $\mathcal{S}$.

We now derive equation 5.4, the orthogonality relation for the generalized position eigenvectors $|x\rangle$. Indeed if $\varphi \in \mathcal{S}$ then inserting the resolution of the identity, equation 5.6, gives

$$\begin{aligned}\varphi(x) = \langle x|\varphi\rangle = \langle x|1_{\mathcal{S}}|\varphi\rangle &= \langle x|\int_{-\infty}^{\infty} |x'\rangle\langle x'|\,dx'|\varphi\rangle \\ &= \int_{\infty}^{\infty} \langle x|x'\rangle\langle x'|\varphi\rangle\,dx' = \int_{\infty}^{\infty} \langle x|x'\rangle\varphi(x')\,dx'. \qquad (5.7)\end{aligned}$$

Since this holds for all elements φ of $\mathcal{S}$, we see that $\langle x|x'\rangle$ under an integral sign acts as a delta distribution, the defining property of Dirac's function $\delta(x - x')$. This, of course, justifies the orthogonality relation $\langle x|x'\rangle = \delta(x - x')$.

Here is another sample calculation using equations 5.2, 5.3, and 5.6.

We have

$$\begin{aligned}\widehat{\varphi}(p) = \langle p|\varphi\rangle = \langle p|1_{\mathcal{S}}|\varphi\rangle = \langle p| \int_{-\infty}^{\infty} |x\rangle\langle x|\, dx\, |\varphi\rangle \\ = \int_{-\infty}^{\infty} \langle p|x\rangle\langle x|\varphi\rangle\, dx = \frac{1}{\sqrt{2\pi\hbar}} \int_{-\infty}^{\infty} \varphi(x) e^{-\mathbf{i}px/\hbar}\, dx, \qquad (5.8)\end{aligned}$$

a perfectly legitimate formula for a test function φ.

The development for the momentum operator P parallels that of the position operator. Indeed, the collection $\{|p\rangle : p \in \mathbb{R}\}$ is a complete orthogonal family of kets, as a test function $\varphi \in \mathcal{S}$ satisfies $\varphi = 0$ if and only if $\langle \varphi|p\rangle = \overline{\widehat{\varphi}}(p) = 0$ for all $p \in \mathbb{R}$. Thus, the resolution of the identity for P becomes

$$1_{\mathcal{S}} = \int_{-\infty}^{\infty} |p\rangle\langle p|\, dp, \qquad (5.9)$$

and its spectral decomposition becomes

$$\mathsf{P} = \int_{-\infty}^{\infty} p|p\rangle\langle p|\, dp.$$

Recall from Lecture 3 that the Fourier transform of the derivative of $\psi \in \mathcal{S}$ satisfies

$$\widehat{\psi'}(p) = \widehat{\frac{d\psi}{dx}}(p) = \frac{\mathbf{i}}{\hbar}\, p\, \widehat{\psi}(p).$$

Using this and the Plancherel theorem, that $\int_{\mathbb{R}} \overline{\widehat{\varphi}}\widehat{\psi} = \int_{\mathbb{R}} \overline{\varphi}\psi$ for all $\varphi, \psi \in \mathcal{S}$, we have

$$\begin{aligned}\langle \varphi|\mathsf{P}|\psi\rangle = \langle \varphi| \int_{-\infty}^{\infty} p|p\rangle\langle p|\, dp\, |\psi\rangle = \int_{-\infty}^{\infty} \overline{\widehat{\varphi}}(p) p \widehat{\psi}(p)\, dp \\ = \int_{-\infty}^{\infty} \overline{\widehat{\varphi}}(p) \frac{\hbar}{\mathbf{i}} \widehat{\frac{d\psi}{dx}}(p)\, dp = \int_{-\infty}^{\infty} \overline{\varphi}(x) \frac{\hbar}{\mathbf{i}} \frac{d}{dx} \psi(x)\, dx = \int_{\mathbb{R}} \varphi \mathsf{P} \psi,\end{aligned}$$

as expected.

5.5 End Notes

Zeidler (1995) employs the Dirac calculus in his development of functional analysis and provides a bilingual dictionary for translating between the notation and language of the physicists and that of the mathematicians. I am very fond of his treatment, having found it a great help in sorting out all kinds of confusion that arose for me when trying to understand the physicists' explanations. The development of the Dirac calculus using rigged Hilbert spaces as presented in this lecture developed from my reading of Madrid (2005). Of course, Dirac (1930) originated his calculus in his classic work on the principles of quantum mechanics, which nearly a century after publication is still one of the finest presentations of the subject. Rigged Hilbert spaces using Gelfand triples are really a marriage of Hilbert space theory with distribution theory which makes Dirac's calculus rigorous. Strichartz (2003) is my favorite introduction to distribution theory; it presents as simply as possible all the main developments needed to understand the use of distribution theory in quantum mechanics.

6

A Review of Classical Mechanics

> ...the idea of enlarging reality by including "tentative" possibilities and then selecting one of these by the condition that it minimizes a certain quantity, seems to bring a *purpose* to the flow of natural events.
>
> Cornelius Lanczos
> *The Variational Principles of Mechanics*, 1949

Any theory that purports to refine, correct, or supersede an already marvelously successful theory like classical mechanics must revert to that theory in its realm of successful application. Thus, if classical mechanics is to be the large–scale, or large–quantum–number, or macroscopic limit of quantum mechanics, the equations of motion of classical mechanics must be recoverable, at least in principle, from the quantum mechanical formalism. The results that demonstrate this are known collectively as **Ehrenfest's theorems**, after Paul Ehrenfest, who first articulated precisely the idea that classical equations should hold for average values of quantum observables.

Lecture 8 will be devoted to a discussion of the Ehrenfest theorems. In preparation for the discussion in that lecture, in this lecture we give a review of classical mechanics, the large-scale limit of quantum mechanics. For simplicity, initially we restrict our discussion to the one-dimensional motion of a particle of constant mass moving along a straight line under a conservative force. Our starting point is the Newtonian treatment, from which we derive, for this special case, the Euler–Lagrange equation of motion in Cartesian coordinates. This leads to a discussion of generalized coordinates and Hamilton's action principle, which is used to derive the general Euler–Lagrange equation of motion, the central ingredient of Lagrangian mechanics. The action principle is used throughout the development of quantum mechanics, especially in deriving the equations that govern quantum fields. For this reason it is essential that the student of quantum mechanics masters the content of this lecture.

After our presentation of Lagrangian mechanics and an examination of the Euler–Lagrange equations, we move on to Hamiltonian mechanics. We derive the Hamilton equations of motion and cast them into a particularly elegant form using Poisson brackets. Finally, Noether's theorem, which shows how conservation laws arise from symmetries of the Lagrangian, is proved and several examples are presented. The quantum version of Noether's theorem will be used in Lecture 28 to derive several conservation laws in quantum mechanics. Our review of classical mechanics is continued in Lecture 7, where we forge the whole structure of classical mechanics into the mold of the Hamilton–Jacobi equation.

This review of classical mechanics will serve us throughout the remainder of these lectures as, time and again, quantum mechanics draws from classical mechanics for both inspiration and substantive discourse.

6.1 Newtonian Mechanics and the Euler–Lagrange Equation in Cartesian Coordinates

Consider a particle of mass m moving in a one-dimensional potential well determined by the potential $V(x)$, where x is a Cartesian coordinate along the line of motion. The force experienced by the particle is $F = -\nabla V = -dV/dx$. Newton's law for this conservative force acting in the system gives the equation of motion of the particle as

$$m\ddot{x} + \frac{dV}{dx} = 0.$$

The momentum of the particle is $p = m\dot{x}$, so that

$$\dot{p} = m\ddot{x} = \frac{d}{dt}\frac{d}{d\dot{x}}\left(\frac{m\dot{x}^2}{2}\right) = \frac{d}{dt}\frac{dT}{d\dot{x}},$$

where $T = T(\dot{x}) = \frac{1}{2}m\dot{x}^2$ is the kinetic energy of the particle. Define the **Lagrangian** L for this system as the difference of the kinetic and potential energy as a function of the position x and velocity $\dot{x}$: $L = L(x, \dot{x}) = T(\dot{x}) - V(x)$. The equation of motion now can be written as an **Euler–Lagrange equation** in cartesian coordinates as

$$\frac{d}{dt}\frac{\partial L}{\partial \dot{x}} - \frac{\partial L}{\partial x} = 0.$$

The reader may well ask what advantage is gained from the Euler–Lagrange formalism. There are two advantages that we will articulate, which will lead us to a consideration of Lagrangian mechanics, a beautifully elegant formalism for classical Newtonian mechanics. The first advantage of the Euler–Lagrange formalism involves a change in perspective that moves one from the extremely localized view of the system in terms of instantaneous positions and velocities of the particle to a global viewpoint that examines at once all the possible paths of the particle as it moves from point to point. It entails a principle under which nature seems to act that has scarcely been surpassed in the history of science in its elegance and simplicity. To explain this in the present context, suppose the particle of our system begins at point x_0 at time t_0 and finds itself at point x_1 at time t_1. Consider all possible continuously differentiable paths that the particle may traverse from x_0 to x_1 during the time interval $[t_0, t_1]$. The collection of such paths is a subset of $C^1([t_0, t_1])$, the space of real-valued continuously differentiable functions defined on the interval $[t_0, t_1]$, and may be described as

$$\mathcal{D} = \{x \in C^1([t_0, t_1]) : x(t_0) = x_0 \text{ and } x(t_1) = x_1\}.$$

Define a functional[1] $S : \mathcal{D} \to \mathbb{R}$ by

$$S[x] = \int_{t_0}^{t_1} L(x(t), \dot{x}(t))\, dt$$

called the **action functional**. Students of the calculus of variations will have noticed that the Euler–Lagrange equation of motion is precisely the Euler condition on the path in $\mathcal{D}$ that gives a stationary value for the action functional. Thus, the Euler–Lagrange equation of motion may be understood, rather than as a consequence of Newton's Law, as a consequence of the fundamental

Hamilton's principle of stationary action. The actual path $x(t)$ traversed by the particle in its movement from x_0 to x_1 during the time interval $[t_0, t_1]$ gives a stationary value for the action functional S.[2]

[1] A function whose domain is itself a space of functions often is called a *functional.*

[2] Often called the **principle of least action** and named after the Irish mathematician and physicist, Sir William Rowan Hamilton, who first applied a variational principle in 1834 to derive the equations of motion from the Lagrangian, the principle itself has a long history involving such luminaries as Leibnitz, Fermat, Maupertuis, Euler, and Lagrange before Hamilton, and Jacobi and Morse after him.

In the next section we develop enough of the calculus of variations to derive the Euler–Lagrange equation from Hamilton's principle but in a more general context that frees us from the constraint of Cartesian coordinates and provides a wide latitude in the choice of coordinates employed. The importance and centrality of Hamilton's principle cannot be overemphasized. Indeed, exactly the same variational principle, when applied to the appropriate Lagrangian densities in the context of fields rather than particles, produces the equations of motion for both classical electromagnetism and general relativity. Importantly for our study, an extension of the principle reproduces the quantum mechanical formalism and is indispensable for the quantum theory of fields.

The second advantage of the Euler–Lagrange formalism over the Newtonian formalism is the freedom it allows in choosing the coordinates with which to specify the motion of a particle. This observation is very powerful in both practical and theoretical considerations in mechanics. To understand this point, notice that the action $S[x]$ is a physical quantity dependent on the energy characteristics of the particle in its motion from point x_0 to point x_1. It is the time-integral of the difference in the kinetic and the potential energy at time t, integrated from t_0 to t_1. The value of this difference between the kinetic and the potential energy at any given time t is a number that is independent of the coordinates one may use to describe the location of the particle. It is clear, then, that we are free to choose any **generalized coordinate** $q = q(t)$ to specify the location of the particle at time t, and so we may express the Lagrangian $L = T - V$ at time t in terms of the generalized coordinate q and the **generalized velocity** $\dot{q} = \dot{q}(t)$. Technically, this means that there is a function $\mathcal{L}$ such that $L(x(t), \dot{x}(t)) = \mathcal{L}(q(t), \dot{q}(t))$ for every t in the interval $[t_0, t_1]$, where $q(t)$ uniquely determines $x(t)$. By Hamilton's principle, the actual path traversed by the particle gives a stationary value for the action, expressed as

$$S[q] = \int_{t_0}^{t_1} \mathcal{L}(q(t), \dot{q}(t))\, dt,$$

where each path in $\mathcal{D}$ is expressed in the generalized coordinate q rather than the Cartesian coordinate x. This freedom to choose generalized coordinates whose only constraint is that $(q, \dot{q})$ must determine $(x, \dot{x})$ uniquely is one of the qualities that make the Euler–Lagrange formulation of mechanics so powerful. We will take the point of view that the Lagrangian L is a scalar quantity with a value at each point of the two-dimensional manifold of possible position and velocity values available to the particle, and that we may express L in different coordinate charts $(q, \dot{q})$ on this manifold and, thus, write $L(q, \dot{q})$ instead of $\mathcal{L}(q, \dot{q})$.

Our goal now is to develop Lagrangian mechanics from Hamilton's principle, the task that next occupies our efforts.

6.2 Lagrangian Mechanics

For the moment, we continue to consider the one-dimensional system of a single particle moving along a curve subject to a conservative force that arises from the potential V, which may vary with time. We choose a coordinate q that uniquely specifies the location of the particle so that the potential energy may be expressed as $V = V(q,t)$ and the endpoints of the path have coordinates $q_0 = q(t_0)$ and $q_1 = q(t_1)$. The Lagrangian, then, is expressed as $L = L(q, \dot{q}, t)$ and the action functional evaluated at the path $q = q(t)$ is

$$S[q] = \int_{t_0}^{t_1} L(q(t), \dot{q}(t), t)\, dt.$$

A stationary value for the action functional occurs at paths in $\mathfrak{D} = \{q \in C^1([t_0, t_1]) : q_0 = q(t_0) \text{ and } q_1 = q(t_1)\}$ at which the action functional takes on a relative extremum. To unpack the meaning here, define a norm $\| - \|_1$ on $C^1([t_0, t_1])$ by

$$\|q\|_1 = \sup\{|q(t)| : t_0 \le t \le t_1\} + \sup\{|\dot{q}(t)| : t_0 \le t \le t_1\}, \tag{6.1}$$

for all $q \in C^1([t_0, t_1])$. A path $\tilde{q} \in \mathfrak{D}$ is a **relative extremum** for the action functional S if $S[q] - S[\tilde{q}]$ does not change sign in a neighborhood of $\tilde{q}$ in $\mathfrak{D}$, meaning that there exists some $\varepsilon > 0$ such that $S[q] - S[\tilde{q}]$ does not change sign for all $q \in \mathfrak{D}$ for which $\|q - \tilde{q}\|_1 < \varepsilon$. To define a stationary value, we must first define the **variational derivative**, or the **differential**, of the action functional and for this it is convenient to introduce the normed linear space of **increments** defined as

$$\mathfrak{D}_0 = \{h \in C^1([t_0, t_1]) : h(t_0) = h(t_1) = 0\},$$

with norm $\| - \|_1$. Note that, indeed, $\mathfrak{D}_0$ is a real vector space under the usual point-wise addition and scalar multiplication since the boundary constraints are preserved under these operations. Like the derivative of a function f at a point $x \in \mathbb{R}^n$, which is the best linear approximation to f at x, the variational derivative of S at the path $q \in \mathfrak{D}$ is the best linear approximation to S at q. Precisely, the variational derivative, or **variation** for short, of S at q is the continuous linear functional $\delta S_q : \mathfrak{D}_0 \to \mathbb{R}$ such that, for all $h \in \mathfrak{D}_0$,

$$S[q+h] = S[q] + \delta S_q(h) + \varepsilon(h)\|h\|_1, \tag{6.2}$$

where $\varepsilon(h) \to 0$ as $\|h\|_1 \to 0$, provided that such a functional exists. If δS_q exists then it is unique and is in fact the only linear functional that approximates $S[q+h] - S[q]$ with an error that is $o(\|h\|_1)$. We say that S is a **differentiable action** provided that δS_q exists for all $q \in \mathfrak{D}$.

The number $S[q]$ is a **stationary value** and the path $q = q(t)$ is a **stationary path** for the differentiable action S if the variation δS_q vanishes, i.e., if $\delta S_q(h) = 0$ for all $h \in \mathfrak{D}_0$. The connection to extrema is stated now more precisely, viz., that a necessary condition for the differentiable action S to have a relative extremum at q is that $S[q]$ be a stationary value. To prove this, assume that S takes on a relative minimum value at the path $q \in \mathfrak{D}$, say that $S[q+h] - S[q] \geq 0$ for all $h \in \mathfrak{D}_0$ with $\|h\|_1 < \zeta$. If $\delta S_q \neq 0$ then there exists $h_0 \in \mathfrak{D}$ with $\delta S_q(h_0) > 0$ and, as δS_q is linear, we may assume that $\|h_0\|_1 < \zeta$. Then, for $|t| < 1$,

$$\delta S_q(th_0) + \varepsilon(th_0)|t|\|h_0\|_1 = S[q+th_0] - S[q] \geq 0.$$

In particular, for all $-1 < t < 0$, we have that

$$t(\delta S_q(h_0) - \varepsilon(th_0)\|h_0\|_1) \geq 0.$$

Since $t < 0$, it follows that $\delta S_q(h_0) \leq \varepsilon(th_0)\|h_0\|$ and taking the limit as $t \to 0-$ gives $\delta S_q(h_0) \leq 0$, a contradiction. It follows that $\delta S_q = 0$ and that $S[q]$ is a stationary value.

Hamilton's principle was stated originally as a principle of least action, viz., that the actual path taken minimizes the action among all paths in $\mathfrak{D}$. Of course, having a stationary value at q is merely a necessary condition for an extremum and not a sufficient one. It turns out that verifying that a path that gives a stationary value is actually an extremum, much less a minimum, can be daunting, and normally we are content with finding conditions on a path that guarantee that it gives a stationary value for the action. Finding such conditions that are useful and practical is our next task, and leads to the canonical equations of Lagrangian mechanics.

Using Taylor's theorem and assuming that the Lagrangian is of class

C^2 in the independent variables q and $\dot{q}$, we expand the Lagrangian as

$$L(q+h,\dot{q}+\dot{h},t) = L(q,\dot{q},t) + \frac{\partial L}{\partial q}(q,\dot{q},t)h + \frac{\partial L}{\partial \dot{q}}(q,\dot{q},t)\dot{h} + \text{ord}(2),$$ [3]

where ord(2), which represents terms of order greater than 1 relative to h and $\dot{h}$, has the form $Ah^2 + Bh\dot{h} + C\dot{h}^2$. From Taylor's theorem, the coefficients A, B, and C are appropriate second partial derivatives of L evaluated at time t at a point in the line segment connecting the points $(q,\dot{q})$ and $(q+h,\dot{q}+\dot{h})$ in the $q\dot{q}$–plane. Applying this at each point $q = q(t)$ of a path and for an increment $h = h(t)$ and integrating from t_0 to t_1 gives

$$\begin{aligned} S[q+h] = S[q] &+ \int_{t_0}^{t_1} \left(\frac{\partial L}{\partial q}(q(t),\dot{q}(t),t)h(t) + \frac{\partial L}{\partial \dot{q}}(q(t),\dot{q}(t),t)\dot{h}(t) \right) dt \\ &+ \int_{t_0}^{t_1} \left(A(t)h(t)^2 + B(t)h(t)\dot{h}(t) + C(t)\dot{h}(t)^2 \right) dt. \end{aligned}$$

Since q and $\dot{q}$ are continuous, the subset $\{(q(t),\dot{q}(t)) : t_0 \le t \le t_1\}$ of the $q\dot{q}$–plane is bounded and, hence, so too is $\{(q(t)+h(t),\dot{q}(t)+\dot{h}(t)) : t_0 \le t \le t_1 \text{ and } \|h\|_1 \le 1\}$; therefore this collection is contained in a convex compact subset D. As L is of class C^2, the second partial derivatives of L are bounded on $D \times [t_0,t_1]$ and, therefore, there exists $M > 0$ such that $|A(t)+B(t)+C(t)| < M$ for all $t_0 \le t \le t_1$, provided also that $\|h\|_1 \le 1$. Observing that $|h(t)| \le \|h\|_1$ and $|\dot{h}(t)| \le \|h\|_1$ for all $t_0 \le t \le t_1$, it follows that

$$\left| \int_{t_0}^{t_1} \left(A(t)h(t)^2 + B(t)h(t)\dot{h}(t) + C(t)\dot{h}(t)^2 \right) dt \right| \le M(t_1 - t_0)\|h\|_1^2$$

for all $\|h\|_1 \le 1$, confirming this integral term to be $o(\|h\|_1)$. Define the functional $\ell_q : \mathfrak{D}_0 \to \mathbb{R}$ via

$$\ell_q(h) = \int_{t_0}^{t_1} \left(\frac{\partial L}{\partial q}(q(t),\dot{q}(t),t)h(t) + \frac{\partial L}{\partial \dot{q}}(q(t),\dot{q}(t),t)\dot{h}(t) \right) dt.$$

It can easily be seen that ℓ_q is a continuous linear functional on $\mathfrak{D}_0$ and satisfies

$$S[q+h] = S[q] + \ell_q(h) + o(\|h\|_1),$$

verifying that $\delta S_q = \ell_q$. The condition that gives a stationary value of

[3] This sometimes causes confusion the first time it is seen. Here, q and $\dot{q}$ are independent variables forming a $q\dot{q}$–plane rather than the position and velocity coordinates associated with a path.

the action functional becomes

$$\delta S_q(h) = \int_{t_0}^{t_1} \left(\frac{\partial L}{\partial q}(q, \dot{q}, t)h + \frac{\partial L}{\partial \dot{q}}(q, \dot{q}, t)\dot{h} \right) dt = 0$$

for all $h \in \mathfrak{D}_0$. Integrating the second term in the integrand by parts and using that $h(t_0) = h(t_1) = 0$ gives the condition for a stationary value as

$$\delta S_q(h) = \int_{t_0}^{t_1} \left(\frac{\partial L}{\partial q}(q, \dot{q}, t) - \frac{d}{dt}\frac{\partial L}{\partial \dot{q}}(q, \dot{q}, t) \right) h\, dt = 0$$

for all $h \in \mathfrak{D}_0$. Clearly this holds for every increment h if and only if

$$\frac{\partial L}{\partial q}(q, \dot{q}, t) - \frac{d}{dt}\frac{\partial L}{\partial \dot{q}}(q, \dot{q}, t) = 0. \tag{6.3}$$

This is the famous **Euler–Lagrange equation** of Lagrangian mechanics and is the equation of motion whose solution gives the path $q = q(t)$ that the particle traverses from q_0 to q_1 in time $t_1 - t_0$.

The general case offers no further difficulties. Let the instantaneous state of a system of possibly many particles under the influence of a conservative field of force derived from a potential V be determined uniquely by the N generalized position coordinates $q_1, \ldots, q_N$ with generalized velocities $\dot{q}_1, \ldots, \dot{q}_N$. The Lagrangian L is the difference in the total kinetic and potential energy of the system and is a function of the $2N$ generalized coordinates $q_1, \ldots, q_n, \dot{q}_1, \ldots, \dot{q}_N$, abbreviated as $q = (q_1, \ldots, q_N)$ and $\dot{q} = (\dot{q}_1, \ldots, \dot{q}_N)$. The previous discussion when extended to N dimensions implies that the actual dynamical development of the system, given by a function $q(t) = (q_1(t), \ldots, q_N(t))$, must give a stationary value for the action

$$S[q] = S[q_1, \ldots, q_N] = \int_{t_0}^{t_1} L(q, \dot{q}, t)\, dt.$$

The necessary conditions for $q = q(t)$ to give a stationary value are then given by the N Euler–Lagrange equations

$$\frac{\partial L}{\partial q_i}(q, \dot{q}, t) - \frac{d}{dt}\frac{\partial L}{\partial \dot{q}_i}(q, \dot{q}, t) = 0 \quad \text{for} \quad i = 1, \ldots, N.$$

A path $q = q(t)$ in the N-dimensional position space that gives a stationary value for the action functional is called a **stationary trajectory** for the system.

6.3 Hamiltonian Mechanics and Poisson Brackets

A version of mechanics due to Hamilton replaces the N second-order Euler–Lagrange equations of motion with $2N$ coupled first-order equations, known as **Hamilton's equations**. As for our development of Lagrangian mechanics, we first develop Hamiltonian mechanics for the case of a single particle moving along a one-dimensional curve under the influence of a conservative force $F = -\nabla V$ and whose position is uniquely determined by the generalized position coordinate q.

First, consider the quantity

$$\widetilde{H}(q, \dot{q}, t) = \dot{q}\frac{\partial L}{\partial \dot{q}} - L(q, \dot{q}, t),$$

and observe that, along a path $q = q(t)$ of actual motion, i.e., one that satisfies the Euler–Lagrange equation, we have

$$\begin{aligned}\frac{d\widetilde{H}}{dt} &= \ddot{q}\frac{\partial L}{\partial \dot{q}} + \dot{q}\frac{d}{dt}\frac{\partial L}{\partial \dot{q}} - \frac{\partial L}{\partial q}\dot{q} - \frac{\partial L}{\partial \dot{q}}\ddot{q} - \frac{\partial L}{\partial t} \\ &= \dot{q}\left(\frac{d}{dt}\frac{\partial L}{\partial \dot{q}} - \frac{\partial L}{\partial q}\right) - \frac{\partial L}{\partial t} = -\frac{\partial L}{\partial t}.\end{aligned}$$

It follows that, when the potential V does not depend explicitly on t, $\partial L/\partial t = 0$ and the function $\widetilde{H}$ is constant along a stationary path q. In this case, in Cartesian coordinates, an easy exercise shows that $\widetilde{H}$ is the energy $T + V$. By considering the transformation equations connecting coordinates $x, \dot{x}$ and $q, \dot{q}$, it follows that $\widetilde{H}(q, \dot{q}) = T(\dot{q}) + V(q)$, and hence energy is conserved along a stationary path as long as the transformation equation from q to x does not explicitly depend on time.

Consider now the momentum in Cartesian coordinates. It satisfies

$$p = m\dot{x} = \frac{d}{d\dot{x}}\left(\frac{1}{2}m\dot{x}^2\right) = \frac{\partial L}{\partial \dot{x}}.$$

This motivates the definition of the **generalized momentum p conjugate to q** as

$$p = p(q, \dot{q}) = \frac{\partial L}{\partial \dot{q}},$$

indicating that generally p is a function of both q and $\dot{q}$. Under appropriate conditions, the equation $p = p(q, \dot{q})$ may be inverted to solve for $\dot{q}$ in terms of q and p, and we may write $\dot{q} = \dot{q}(q, p)$. For example, in Cartesian coordinates where $q = x$, we have $p = p(x, \dot{x}) = m\dot{x}$ and, inverting, $\dot{x} = \dot{x}(x, p) = p/m$. When this inversion is allowed we may write the total energy $\widetilde{H}$ as a function of the generalized position q and its conjugate momentum p as

$$H(q,p,t) = \widetilde{H}(q,\dot{q}(q,p),t) = \dot{q}(q,p)\,p - L(q,\dot{q}(q,p),t).$$

The expression $H(q,p,t)$ in terms of the generalized position q and its conjugate momentum p is called the **Hamiltonian** of the system. For example, in Cartesian coordinates we have

$$\widetilde{H}(x,\dot{x},t) = \tfrac{1}{2}m\dot{x}^2 + V(x,t),$$

while the Hamiltonian is given as

$$H(x,p,t) = \frac{p^2}{2m} + V(x,t).$$

Assuming now that $q = q(t)$ is a stationary path for the action functional so that it is a solution of the Euler–Lagrange equation, take the differentials of both sides of the expression $H(q,p,t) = \dot{q}p - L(q,\dot{q},t)$, where it is understood that $\dot{q} = \dot{q}(q,p)$; this gives

$$\begin{aligned}\frac{\partial H}{\partial q}dq + \frac{\partial H}{\partial p}dp + \frac{\partial H}{\partial t}dt &= p\,d\dot{q} + \dot{q}\,dp - \frac{\partial L}{\partial q}dq - \frac{\partial L}{\partial \dot{q}}d\dot{q} - \frac{\partial L}{\partial t}dt \\ &= -\dot{p}\,dq + \dot{q}\,dp - \frac{\partial L}{\partial t}dt,\end{aligned}$$

where we have used that $p = \partial L/\partial \dot{q}$ and, from the Euler–Lagrange equation,

$$\frac{\partial L}{\partial q} = \frac{d}{dt}\frac{\partial L}{\partial \dot{q}} = \dot{p}.$$

Equating coefficients of the coordinate differentials dq and dp gives **Hamilton's equations of motion** as

$$\dot{q} = \frac{\partial H}{\partial p} \quad \text{and} \quad \dot{p} = -\frac{\partial H}{\partial q}.$$

For a system of several particles with generalized position coordinates $q = (q_1, \ldots, q_N)$ and conjugate momenta $p = (p_1, \ldots, p_N)$, where $p_i = \partial L/\partial q_i$, the Hamiltonian is given by a so-called **Legendre transformation** as

$$H(q,p,t) = \sum_{i=1}^{N} \dot{q}_i p_i - L(q,\dot{q},t),$$

where $\dot{q} = \dot{q}(q,p)$. When the potential V does not depend explicitly on time, $H(q,p)$ is a constant of the motion that gives the fixed total energy along a stationary trajectory. In general, a stationary trajectory $q = q(t)$ satisfies the $2N$ Hamiltonian equations

$$\dot{q}_i = \frac{\partial H}{\partial p_i} \quad \text{and} \quad \dot{p}_i = -\frac{\partial H}{\partial q_i} \quad \text{for} \quad i = 1, \ldots, N.$$

We have derived Hamilton's equations from the Euler–Lagrange equations. It is easy to perform the reverse, deriving the Euler–Lagrange from Hamilton's equations, showing these two flavors of mechanics to be fully equivalent.

Hamilton's equations can be cast in a particularly elegant form using Poisson brackets. Aesthetics aside, their importance in quantum mechanics flows from a suggested correspondence between the Poisson bracket of two classical mechanical observables and the commutator of the self-adjoint operators representing those observables. This correspondence, suggested by Dirac, is not exact and requires interpretation but has been an accurate guide for certain developments in quantum mechanics. We will consider this correspondence in the two lectures following.

For two differentiable functions A and B of the generalized position and momentum coordinates q_i and p_i, the **Poisson bracket** of A and B is

$$\{A, B\} = \sum_{i=1}^{N} \left(\frac{\partial A}{\partial q_i} \frac{\partial B}{\partial p_i} - \frac{\partial A}{\partial p_i} \frac{\partial B}{\partial q_i} \right).$$

Quick and easy examples are $\{q_i, p_j\} = \delta_{ij}$ and $\{q_i, q_j\} = \{p_i, p_j\} = 0$ for $1 \leq i, j \leq N$.

If A has an explicit dependence on time then, along a stationary trajectory $q = q(t)$ of the system, A has the form $A(q(t), p(t), t)$. Applying Hamilton's equations, its total time derivative is easily seen to be given by the classical **evolution equation**

$$\frac{dA}{dt} = \frac{\partial A}{\partial t} + \{A, H\}, \tag{6.4}$$

which expresses the time rate of change along the trajectory of a mechanical quantity evaluated from the system. Applying this with $A = q_i$ and $A = p_i$ reproduces Hamilton's equations aesthetically as

$$\dot{q}_i = \{q_i, H\} \quad \text{and} \quad \dot{p}_i = \{p_i, H\} \quad \text{for} \quad i = 1, \ldots, N.$$

The evolution equation gives a useful condition for determining when a physical quantity is constant along a stationary solution. If A does not depend explicitly on time, so that $\partial A/\partial t = 0$, then A is constant along a stationary trajectory $q = q(t)$ precisely when the Poisson bracket

$\{A, H\} = 0$. For example, as $\{H, H\} = 0$, the Hamiltonian is conserved in the system exactly when it does not depend explicitly on time.

Easy calculations show that the Poisson bracket satisfies $\{A, A\} = 0$, is antisymmetric in that $\{A, B\} = -\{B, A\}$, and linear in both A and B. A somewhat involved calculation, whose verification is left to the reader, shows that the **Jacobi identity** holds for any three C^2 functions A, B, and C, viz., that

$$\{A, \{B, C\}\} + \{B, \{C, A\}\} + \{C, \{A, B\}\} = 0,$$

demonstrating that the Poisson bracket makes the vector space of real-valued C^2 functions of q and p into a real Lie algebra.

A final comment on the Hamiltonian formalism is in order. Let $\mathcal{C} = \{p \in C^1([t_0, t_1]) : p_0 = p(t_0) \text{ and } p_1 = p(t_1)\}$, where p_0 and p_1 are the initial and terminal momenta of the solutions to Hamilton's equations. Define the functional $S : \mathcal{D} \times \mathcal{C} \to \mathbb{R}$ via $S[q, p] = \int_{t_0}^{t_1} \sum \dot{q}_i p_i - H(q, p)\, dt$. Then the Hamilton equations of motion are exactly the Euler–Lagrange equations for a stationary trajectory $q = q(t)$, $p = p(t)$ of the functional $S[q, p]$.

6.4 Noether's Theorem

The eminent German mathematician Emmy Noether proved an important result in 1918 that shows how conservation laws arise from symmetries of the Lagrangian. Indeed, the conservation of energy, linear momentum, and angular momentum in physical systems in classical mechanics follow from various symmetries – temporal, spacial, rotational – of the Lagrangian of the system. Ultimately we will see how the various quantum equations of motion may be derived from the action principle applied to appropriate Lagrangians, and Noether's theorem will be used in identifying some quantum mechanical conservation laws that have no classical counterparts.

The case of interest is that of a Lagrangian that is invariant under the action of a suitable one-parameter family of transformations. For this discussion, let the Lagrangian L describe an **autonomous** physical system, meaning that $L = L(q, \dot{q})$ has no explicit time dependence. Let $h : (-\varepsilon, \varepsilon) \to \mathrm{Diff}_0(U)$ be a differentiable one-parameter family of diffeomorphisms of an open, connected, domain U of $\mathbb{R}^N$. We use the notation h_s for the diffeomorphism at parameter value $s \in (-\varepsilon, \varepsilon)$ and $Dh_s(q)$ for its derivative at the point $q \in U$. We assume that $h_0 = 1_U$ and that the stationary trajectory $q = q(t)$ has its trace in U, meaning that $q(t) \in U$ for all $t_0 \leq t \leq t_1$. When we say that h is a differentiable family, we mean that the function $\varphi(s, q) = h_s(q) = (\varphi_1(s, q), \ldots, \varphi_N(s, q))$ has continuous partial derivatives for all $(s, q) \in (-\varepsilon, \varepsilon) \times U$.

Now suppose that the Lagrangian is invariant under the one-parameter family h_s. This means that, for all $q \in U$, $L(q, \dot{q}) = L(h_s(q), Dh_s(q)\dot{q})$. Our claim is that the function

$$J(q, \dot{q}) = \sum_{i=1}^{N} \frac{\partial L}{\partial \dot{q}_i}(q, \dot{q}) \frac{\partial \varphi_i}{\partial s}(0, q)$$

is constant along the stationary trajectory $q = q(t)$. Indeed, let $\Phi(s, t) = \varphi(s, q(t))$ with $\dot{\Phi} = d\Phi/dt$. The invariance of L under h_s then implies that, for each s, the trajectory $\Phi(s, -)$ is stationary and therefore solves the Euler–Lagrange equation, so that

$$\frac{d}{dt}\frac{\partial L}{\partial \dot{q}}(\Phi, \dot{\Phi}) = \frac{\partial L}{\partial q}(\Phi, \dot{\Phi}). \tag{6.5}$$

The invariance of L also implies that the expression $L(\Phi(s, t), \dot{\Phi}(s, t))$ is independent of s, so that the total derivative $dL(\Phi, \dot{\Phi})/ds = 0$. Calculating, we have

$$0 = \frac{d}{ds} L(\Phi, \dot{\Phi}) = \sum_{i=1}^{N} \left(\frac{\partial L}{\partial q_i}\frac{d\Phi_i}{ds} + \frac{\partial L}{\partial \dot{q}_i}\frac{d\dot{\Phi}_i}{ds} \right).$$

Using equation 6.5 and exchanging the order of differentiation in the second summand under the summation sign, we have when evaluating along the trajectory $\Phi(s, -)$,

$$0 = \sum_{i=1}^{N} \left[\left(\frac{d}{dt}\frac{\partial L}{\partial \dot{q}} \right) \frac{d\Phi_i}{ds} + \frac{\partial L}{\partial \dot{q}_i} \left(\frac{d}{dt}\frac{d\Phi_i}{ds} \right) \right] = \frac{d}{dt} \left(\sum_{i=1}^{N} \frac{\partial L}{\partial \dot{q}_i}\frac{d\Phi_i}{ds} \right).$$

It follows that the expression in parentheses on the right-hand side of this equation is independent of t. Evaluating at $s = 0$ gives $J(q(t), \dot{q}(t))$, which, being independent of t, must be constant. This is known as Noether's theorem, and we now apply it to derive some standard conservation laws.

Consider the simple case of a single particle with generalized coordinates that are just the Cartesian coordinates $\mathbf{x} = (x_1, x_2, x_3)$, and assume the Lagrangian $L(\mathbf{x}, \dot{\mathbf{x}}) = \frac{1}{2}m\dot{x}^2 - V(\mathbf{x})$ is invariant under translations along the direction of the unit vector $\mathbf{v} = (v_1, v_2, v_3)$. This means that L is invariant under the action of the family $h_s(\mathbf{x}) = \mathbf{x} + s\mathbf{v} =$

$\varphi(s,\mathbf{x})$ of translations. Then $\partial L/\partial\dot{x}_i = m\dot{x}_i$, $\partial\varphi_i/\partial s = v_i$, and $J(\mathbf{x},\dot{\mathbf{x}}) = \sum_{i=1}^{3} m\dot{x}_i v_i = \mathbf{p}\cdot\mathbf{v}$, where $\mathbf{p} = m\dot{\mathbf{x}}$ is the linear momentum. Hence, J is the component of linear momentum in the direction of $\mathbf{v}$ and is a constant of the motion. It follows that if L is invariant under all spatial translations then the linear momentum $\mathbf{p}$ is conserved. Of course, for L to be invariant under translations is the same as saying that the potential function V is constant. This means that no force acts and Newton's first law is recovered.

As a second example, assume that this Lagrangian for a single particle is invariant under rotations about the axis $\mathbf{v}$. Then L is invariant under the action of the family $h_s(\mathbf{x}) = R(s,\mathbf{v})\mathbf{x} = \varphi(s,\mathbf{x})$, where $R(s,\mathbf{v})$ is a positive rotation through an angle s about the axis $\mathbf{v}$. To ease the calculations, let $\mathbf{v} = (0,0,1) = \mathbf{k}$, the unit basis vector in the x_3-direction. The rotation $R(s,\mathbf{k})$ is then represented by the standard rotation matrix:

$$R(s,\mathbf{k}) = \begin{pmatrix} \cos s & -\sin s & 0 \\ \sin s & \cos s & 0 \\ 0 & 0 & 1 \end{pmatrix}.$$

Now $(\partial\varphi/\partial s)(0,\mathbf{x}) = (-x_2, x_1, 0)$ and, hence, $J = m(x_1\dot{x}_2 - \dot{x}_1 x_2) = \mathbf{l}\cdot\mathbf{k} = l_3$, where $\mathbf{l} = m\mathbf{x}\times\dot{\mathbf{x}}$ is the angular momentum vector. Therefore J is the component l_3 of angular momentum in the direction of $\mathbf{k}$ and is a constant of the motion. In the general case where v is arbitrary, slightly more complicated calculations show that $J = \mathbf{l}\cdot\mathbf{v}$, the component of the angular momentum in the direction of $\mathbf{v}$, which is conserved. If the Lagrangian is invariant under rotations about all axes, which is the same as asserting that the potential V is centrally symmetric, then the angular momentum $\mathbf{l}$ itself is conserved.

We will see more sophisticated applications of Noether's theorem subsequently.

6.5 End Notes

Goldstein (1980) *Classical Mechanics* sets the gold standard of expositions of classical mechanics. It was updated in Goldstein, Poole, and Safko (2002), which includes new material while maintaining Goldstein's conciseness of language and standards for well-written prose. I have found the presentations of Lagrangian, Hamiltonian, and Jacobian mechanics particularly useful. Landau and Liftshitz (1972) is another classic that concisely tells the tale of mechanics in the flavor of Lagrange,

Hamilton, and Jacobi. A very quick and skillfully clear introduction to generalized coordinates and to the mechanics of Lagrange and Hamilton appears in the very short book Byerly (1916). A scholarly study of the variational principles of mechanics, written in a lovely old-fashioned and elegant style of prose and which does not neglect the philosophical issues surrounding the principles, is found in Lanczos (1970). This book deserves more study than it has engendered, though it might be somewhat difficult to approach by students trained as they are today bereft of a general philosophical education. Two further recommendations are Greiner (2010), which offers clear explanations and a wealth of examples and exercises, many of which are worked out in the text, and Scheck (2010), which emphasizes geometry and includes a very nice introduction to mechanics on symplectic manifolds.

The beautifully written book Gelfand and Fomin (1963) *Calculus of Variations* presents a thorough, rigorous, study of the topic of its title. It gives a rather complete development of various Euler–Lagrange equations and their uses in geometry and physics. The rigorous mathematical development of my Section 6.2 is taken from Sections 2 through 4 and 9 of Gelfand and Fomin. A more recent addition to the literature that the student will find useful for a rigorous treatment of the Euler–Lagrange equations and variational principles is van Brunt (2004), which includes chapters on Lagrangian, Hamiltonian, and Jacobian mechanics as well as Noether's theorem.

7
Hamilton–Jacobi Theory

> The philosophical value of these methods, the entirely new understanding they furnished for the deeper problems of mechanics, remained unnoticed except by a few scientists who were impressed by the extraordinary beauty of the Hamiltonian developments ... Since the advent of Schroedinger's wave mechanics, which is based essentially on Hamilton's researches, the leading ideas of Hamiltonian mechanics have found their way into textbooks on theoretical physics. Yet even so, the *technical* side of the theory is primarily stressed, at the cost of the philosophical. From the point of view adopted here, the purely technical side of the subject is of minor importance.
>
> Cornelius Lanczos
> *The Variational Principles of Mechanics*, 1949

Hamilton–Jacobi theory is the most esoteric flavor of classical mechanics that we will consider. In this lecture[1] we will outline the beautiful theory of Jacobi, which is useful classically for finding conserved quantities of mechanical systems. The Hamilton–Jacobi equation is a classical equation in which the motion of a particle can be represented as a wave, and the resulting wave equation has affinity with the Schrödinger equation. For this reason it is considered as the closest approach of classical mechanics to quantum mechanics and, in the hands of David Bohm,

[1] The material in this lecture is not needed for any of the remaining development of quantum mechanics found in this book and so may be skipped safely by the reader, hence the star underneath the title. It is included, however, both for completeness of the classical picture and because a basic knowledge of the Hamilton–Jacobi theory is needed for some of the developments that the student will find in the literature on quantum mechanics. For example, it is crucial to understanding the semi-classical approximation to quantum mechanics, which the author has omitted from this book.

becomes the basis of his hidden variables theory from which quantum mechanics emerges. Historically, it also inspired some developments in quantum theory.

Our primary interest in this lecture is in deriving the Hamilton–Jacobi equation. Hamilton–Jacobi theory seeks to find generalized position and momentum coordinates that are even more general than those discussed in the previous lecture. The payment we must remit for this generality is that Lagrangians and Hamiltonians are not preserved under the new coordinates, and those coordinates themselves are likely to be quite complicated, but the payoff is that appropriately chosen new coordinates allow for simple solutions of the dynamical equations in terms of these complicated coordinates. To explain this properly, we will reexamine in detail the type of coordinate transformations allowed thus far in the development of Lagrangian and Hamiltonian mechanics, and will derive from scratch the fact that the Euler–Lagrange equation and Hamilton's equations are independent of the generalized position coordinates chosen. This will place in sharp relief the differences between coordinates allowed in the Hamilton–Jacobi theory and those of Lagrangian and Hamiltonian theory.

After a thorough discussion of generalized coordinates in the Lagrangian and Hamiltonian formalism, we introduce the notion of **canonical coordinates**, which generalizes further the already general generalized coordinates. These are related to the generalized Hamiltonian position and momentum coordinates by **canonical transformations**. We develop a method for recognizing and constructing canonical coordinates that leads to the Hamilton–Jacobi differential equation. Its solutions are seen to provide an elegant, if complicated, solution to mechanical problems. Finally, we illustrate the Hamilton–Jacobi formalism by working out several enlightening examples, including a careful examination of the simple harmonic oscillator.

7.1 Generalized Coordinates Reexamined

The discussion at the end of the first section of Lecture 6 that articulates a second advantage of the Euler–Lagrange formalism over the Newtonian formalism explains why the Lagrangian may be written with respect to a generalized position coordinate q and generalized velocity coordinate $\dot{q}$; then the motion of the particle is given by solutions to the Euler–Lagrange equations, whose form does not change under changes in the generalized coordinates. In this section, we examine in detail these coordinate changes allowed in the Lagrangian formalism.

For simplicity, for the time being we consider the one-dimensional setting where a single generalized position coordinate q determines uniquely the location of a particle. We consider the change to a new generalized coordinate Q related to the old by a diffeomorphism f. This just means that $Q = f(q)$ on an appropriate open connected domain in the q-coordinate line, where f is differentiable and injective and $f'(q) \neq 0$ for all q. The new generalized velocity coordinate is given by $\dot{Q} = f'(q)\dot{q}$, and the coordinate transformation $(Q, \dot{Q}) = F(q, \dot{q}) = (f(q), f'(q)\dot{q})$ is a diffeomorphism from an appropriate open connected domain of the $q\dot{q}$-plane to the $Q\dot{Q}$-plane.

Using $L(q, \dot{q}, t)$ and $\mathcal{L}(Q, \dot{Q}, t)$ for the Lagrangians expressed in the two sets of coordinates, we have

$$L(q, \dot{q}, t) = \mathcal{L}(Q, \dot{Q}, t) = \mathcal{L}(F(q, \dot{q}), t) = \mathcal{L}(f(q), f'(q)\dot{q}, t),$$

and, calculating, we get

$$\frac{\partial L}{\partial q} = \frac{\partial \mathcal{L}}{\partial Q}\frac{\partial Q}{\partial q} + \frac{\partial \mathcal{L}}{\partial \dot{Q}}\frac{\partial \dot{Q}}{\partial q} = \frac{\partial \mathcal{L}}{\partial Q} f'(q) + \frac{\partial \mathcal{L}}{\partial \dot{Q}} f''(q)\dot{q}.$$

Along a curve $q = q(t)$, since $\partial f(q)/\partial \dot{q} = 0$, we have

$$\frac{d}{dt}\frac{\partial L}{\partial \dot{q}} = \frac{d}{dt}\left(\frac{\partial \mathcal{L}}{\partial Q}\frac{\partial f(q)}{\partial \dot{q}} + \frac{\partial \mathcal{L}}{\partial \dot{Q}} f'(q)\right) = \left(\frac{d}{dt}\frac{\partial \mathcal{L}}{\partial \dot{Q}}\right) f'(q) + \frac{\partial \mathcal{L}}{\partial \dot{Q}} f''(q)\dot{q}.$$

Subtracting, we find that

$$\frac{\partial L}{\partial q} - \frac{d}{dt}\frac{\partial L}{\partial \dot{q}} = f'(q)\left(\frac{\partial \mathcal{L}}{\partial Q} - \frac{d}{dt}\frac{\partial \mathcal{L}}{\partial \dot{Q}}\right).$$

Since $f'(q) \neq 0$, a curve $q = q(t)$ is a stationary path for the action functional satisfying the Euler–Lagrange equation for $L(q, \dot{q}, t)$ if and only if the curve $Q = Q(t) = f(q(t))$ is a stationary path for the action functional satisfying the Euler–Lagrange equation for $\mathcal{L}(Q, \dot{Q}, t)$.

The conjugate momenta p and P are related by

$$p = \frac{\partial L}{\partial \dot{q}} = \frac{\partial \mathcal{L}}{\partial \dot{Q}} f'(q) = f'(q)P,$$

which gives, for the Hamiltonians,

$$H(q, p, t) = \dot{q}p - L(q, \dot{q}, t) = \dot{Q}P - \mathcal{L}(Q, \dot{Q}, t) = \mathcal{H}(Q, P, t).$$

Direct calculation gives

$$\frac{\partial H}{\partial p} = \frac{1}{f'(q)}\frac{\partial \mathcal{H}}{\partial P} \quad \text{and} \quad \frac{\partial H}{\partial q} = \frac{\partial \mathcal{H}}{\partial Q} f'(q) - \frac{\partial \mathcal{H}}{\partial P}\frac{f''(q)}{(f'(q))^2} p.$$

Letting $q = q(t)$ be a stationary path, so that Hamilton's equations hold for $\dot{q}$ and $\dot{p}$, we find for $Q = Q(t) = f(q(t))$ and $P = P(t) = p(t)/f'(q(t))$ that

$$\dot{Q} = f'(q)\dot{q} = f'(q)\frac{\partial H}{\partial p} = \frac{\partial \mathcal{H}}{\partial P}$$

and

$$\dot{P} = \frac{\dot{p}}{f'(q)} - \frac{f''(q)}{(f'(q))^2}\dot{q}p = -\frac{1}{f'(q)}\frac{\partial H}{\partial q} - \frac{f''(q)}{(f'(q))^2}\dot{q}p$$

$$= -\frac{1}{f'(q)}\left(\frac{\partial \mathcal{H}}{\partial Q} f'(q) - \frac{\partial \mathcal{H}}{\partial P}\frac{f''(q)}{(f'(q))^2} p\right) - \frac{f''(q)}{(f'(q))^2}\dot{q}p$$

$$= -\frac{\partial \mathcal{H}}{\partial Q} + \frac{\partial H}{\partial p}\frac{f''(q)}{(f'(q))^2} p - \frac{f''(q)}{(f'(q))^2}\dot{q}p = -\frac{\partial \mathcal{H}}{\partial Q}.$$

Thus if $q = q(t)$ satisfies Hamilton's equations in the coordinates (q, p) then $Q = Q(t) = f(q(t))$ satisfies Hamilton's equations in the coordinates (Q, P). The argument is reversible, so that the converse also holds.

Of course we knew that the calculations would have to bear out the argument of Section 6.1, but the details of these calculations show how the form $(Q, P) = (f(q), f'(q)\dot{q})$ of the transformation of coordinates leads to the preservation of the form of both the Euler–Lagrange equations and Hamilton's equations of motion for the stationary path under this coordinate transformation.

For the general N-dimensional case, suppose that a system is described uniquely by the N-vector $q = (q_1, \ldots, q_N)$ of generalized position coordinates with generalized velocities $\dot{q} = (\dot{q}_1, \ldots, \dot{q}_N)$ and conjugate momenta $p = (p_1, \ldots, p_N)$. Consider a change to new position coordinates $Q = (Q_1, \ldots, Q_N)$, related to the old by the diffeomorphism f defined on an open connected region of $\mathbb{R}^N$. With $Q = f(q)$, the new velocity coordinates are given by $\dot{Q} = Df_q(\dot{q})$ and the coordinate transformation $(Q, \dot{Q}) = F(q, \dot{q}) = (f(q), Df_q(\dot{q}))$ is a diffeomorphism of an appropriate open connected domain of $\mathbb{R}^{2N}$ into $\mathbb{R}^{2N}$. A straightforward calculation, the N-dimensional version of the previous one-dimensional calculation, shows that, along a trajectory $q = q(t)$ in the domain of f, with $Q = Q(t) = f(q(t))$, we have

$$\omega(q(t), \dot{q}(t)) = (Df_{q(t)})^{\mathsf{Tr}}(\Omega(Q(t), \dot{Q}(t))),$$

where ω is the N-dimensional vector function whose ith coordinate is $\omega_i = \omega_i(q, \dot{q}) = (\partial L/\partial q_i) - d(\partial L/\partial \dot{q}_i)/dt$ and, similarly, Ω is the N-dimensional vector function whose ith coordinate is $\Omega_i = \Omega_i(Q, \dot{Q}) = \Omega_i(F(q, \dot{q})) = (\partial \mathcal{L}/\partial Q_i) - d(\partial \mathcal{L}/\partial \dot{Q}_i)/dt$. Since f is a diffeomorphism, the derivative Df_q is invertible at each $q = q(t)$, $t_0 \leq t \leq t_1$, and therefore the trajectory $q = q(t)$ satisfies the Euler–Lagrange equation $\omega(q, \dot{q}) = \mathbf{0}$ if and only if the trajectory $Q = Q(t) = f(q(t))$ satisfies the Euler–Lagrange equation $\Omega(Q, \dot{Q}) = \mathbf{0}$.

Another straightforward calculation gives the relationship between the conjugate momentum vectors p and P as $p = (Df_q)^{\mathsf{Tr}}(P)$. Interpreting vectors as column vectors, we may write the Hamiltonians as

$$\begin{aligned} H(q, p, t) &= p^{\mathsf{Tr}}\dot{q} - L(q, \dot{q}, t) \\ &= [(Df_q)^{\mathsf{Tr}}(P)]^{\mathsf{Tr}}\,(Df_q)^{-1}(\dot{Q}) - \mathcal{L}(Q, \dot{Q}, t) \\ &= P^{\mathsf{Tr}}\dot{Q} - \mathcal{L}(Q, \dot{Q}, t) = \mathcal{H}(Q, P, t). \end{aligned}$$

Since the Euler–Lagrange equations are fully equivalent to Hamilton's equations, and we have already seen the equivalence of the two Euler–Lagrange equations $\omega(q, \dot{q}) = \mathbf{0}$ and $\Omega(Q, \dot{Q}) = \mathbf{0}$, we conclude that Hamilton's equations hold for both sets of coordinates whenever they hold for one set.

7.2 Canonical Transformations

We now recast mechanics into a formalism due to Jacobi that is of more theoretical than practical interest but provides a starting point for some developments in quantum mechanics. Herbert Goldstein in the introduction to canonical transformations in his venerable and timeless text *Classical Mechanics* says

> The advantages in the Hamiltonian formulation lie not in its use as a calculational tool, but rather in the deeper insights it affords into the formal structure of mechanics ... As a result we are led to newer, more abstract ways of presenting the physical content of mechanics ... these more abstract formulations are primarily of interest to us today because of their essential role in constructing the more modern theories of matter. Thus, one or another of

these formulations of classical mechanics serves as a point of departure for both statistical mechanics and quantum theory.[2]

The coordinate transformations allowed in Lagrangian and Hamiltonian mechanics are of a very specialized form. The transformation $(Q, \dot{Q}) = F(q, \dot{q})$ is generated by a transformation of the position coordinates only, $Q = f(q)$, with the velocity transformation derived from the derivative of f via $\dot{Q} = Df_q(\dot{q})$. We have seen, both on physical grounds in Section 6.1 and by calculation in the preceding section, that these specialized coordinate transformations preserve the Lagrangian, the Hamiltonian, and the forms of the equations of motion in both Lagrangian and Hamiltonian mechanics. We now expand our attention to more general coordinate transformations. These are allowed to change the Lagrangian and Hamiltonian but only in restricted ways that are determined by our requirement that the equations of motion in the transformed coordinates be derivable from an action principle, and so they have the same form as the Euler–Lagrange and Hamilton equations of motion.

Let (q, p) be N-dimensional generalized position and momentum coordinates with Hamiltonian $H(q, p, t)$ and Lagrangian $L(q, \dot{q}, t)$. The coordinate transformation $(Q, P) = F(q, p, t) = (F_1(q, p, t), F_2(q, p, t))$ of interest to us is restricted only by the requirement that there exists a function $\mathcal{H}(Q, P, t)$ such that Hamilton's equations $\dot{Q}_i = \partial\mathcal{H}/\partial P_i$ and $\dot{P}_i = -\partial\mathcal{H}/\partial Q_i$ hold along any trajectory $Q = Q(t) = F_1(q(t), p(t), t)$ for which $q = q(t)$ is a stationary trajectory of the action functional $S[q]$. The transformation F is called a **canonical transformation**, and the coordinates (Q, P) are called **canonical coordinates**. Though the function $\mathcal{H}$ generally is not an expression of the total energy of the system, we nonetheless call $\mathcal{H}$ the **Hamiltonian** for the system in the canonical coordinates (Q, P). The extent to which $\mathcal{H}$ resembles the Hamiltonian of the system in generalized coordinates (q, p) will be examined in a moment, but first we complete the portrait offered by canonical coordinates.

Defining the canonical velocity coordinates $\dot{Q}_i$ as $\partial\mathcal{H}/\partial P_i$, we may then define the **Lagrangian** for the system in canonical coordinates as $\mathcal{L}(Q, \dot{Q}, t) = P^{\mathrm{Tr}}\dot{Q} - \mathcal{H}(Q, P, t)$, where, of course, P is written as $P = P(Q, \dot{Q})$. The equivalence of the Hamiltonian and Lagrangian formalisms then implies that, since stationary trajectories are solutions of Hamilton's equations in the canonical variables with Hamiltonian $\mathcal{H}$, the stationary trajectories are also solutions of the Euler–Lagrange

[2] Goldstein (1980), p. 378.

equations $\partial\mathcal{L}/\partial Q_i - d(\partial\mathcal{L}/\partial\dot{Q}_i)/dt = 0$. Of course, this means that the trajectories of the system give stationary values for the action $\mathcal{S}[Q] = \int_{t_0}^{t_1} \mathcal{L}(Q, \dot{Q}, t)\, dt$ so that, under a canonical change of coordinates, the whole structure of the Lagrangian and Hamiltonian formalisms is preserved.

Notice that all transformations generated by a transformation of the position coordinates only, $Q = f(q)$, with the velocity and momentum transformations derived from the derivative of f via $\dot{Q} = Df_q(\dot{q})$ and $p = (Df_q)^{\mathsf{Tr}}(P)$, are canonical transformations. These have the particular form $(Q, P) = F(q, p) = (f(q), T_q p)$, where T_q is a q-indexed collection of linear maps. The general canonical transformation allows for the dependence of Q on not only q but also both p and t, and allows for the dependence of P to be nonlinear in p as well as dependent on t, subject only to the requirement that the stationary-action formalism provides the equations of motion from appropriate Lagrangian and Hamiltonian functions. The question arises naturally whether there are any canonical transformations other than those generated by a transformation of position coordinates only, and, if so, how one goes about discovering them. Our answer to this question is quite interesting and leads to the concept of a generating function for a set of canonical transformations. This in turn will be used to derive the Hamilton–Jacobi equation, which encodes in one differential equation the complete mechanics of the system. Had we more time to spend on this topic, we could elucidate the general form of a canonical transformation and list some standard types; instead, we mention Chapter 5 of Pollard (1976) for an elegant presentation of the general form of a canonical transformation when it does not depend explicitly on the time variable, and Chapter 5 of Dittrich and Reuter (2001) for a detailed explication of general canonical transformations, with several useful examples of generating functions.

Recall from the end of Section 6.3 that Hamilton's equations are the Euler–Lagrange equations for the functional

$$S[q, p] = \int_{t_0}^{t_1} \left(p^{\mathsf{Tr}} \dot{q} - H(q, p) \right) dt$$

defined on the product $\mathcal{D} \times \mathcal{C}$. It follows then that an equivalent definition of canonical transformations is that $(Q, P) = F(q, p, t)$ is canonical provided that there exists a Hamiltonian function $\mathcal{H}$ such that the trajectory $(Q, P) = (Q(t), P(t)) = F(q(t), p(t), t)$ is stationary for the functional $\mathcal{S}[Q, P] = \int_{t_0}^{t_1} [P^{\mathsf{Tr}} \dot{Q} - \mathcal{H}(Q, P)] dt$ precisely when $(q, p) = (q(t), p(t))$ is stationary for the functional $S[q, p]$. To search for sufficient conditions to give canonical transformations, then, we first concentrate on a sufficient condition to ensure that this property holds for a function $\mathcal{H}$, viz., that a trajectory $(q, p) = (q(t), p(t))$ is stationary for $S[q, p]$ when and only when the trajectory $(Q, P) = (Q(t), P(t)) = F(q(t), p(t), t)$ is stationary for $\mathcal{S}[Q, P]$.

Assume that the transformation $(Q, P) = F(q, p, t)$ is given. We ask whether F is canonical. To answer this, first assume that the function $\mathcal{H}(Q, P, t)$ is proposed as an appropriate Hamiltonian for F. Let Σ be an arbitrary differentiable function of the coordinates q, p, and time t and define a function Σ' of five independent vector variables $q, p, \dot{q}, \dot{p}, t$ as

$$\Sigma'(q, p, \dot{q}, \dot{p}, t) = \sum_{i=1}^{N} \frac{\partial \Sigma}{\partial q_i} \dot{q}_i + \sum_{i=1}^{N} \frac{\partial \Sigma}{\partial p_i} \dot{p}_i + \frac{\partial \Sigma}{\partial t}.$$

Note that if $(q, p) = (q(t), p(t))$ is a trajectory in qp-space then

$$\frac{d\Sigma}{dt}(q(t), p(t), t) = \Sigma'(q(t), p(t), \dot{q}(t), \dot{p}(t), t).$$

If it holds that

$$P^{\mathsf{Tr}} \dot{Q} - \mathcal{H}(Q, P, t) = p^{\mathsf{Tr}} \dot{q} - H(q, p, t) - \Sigma'(q, p, \dot{q}, \dot{p}, t) \tag{7.1}$$

then, integrating along a trajectory $(q, p) = (q(t), p(t))$ gives $\mathcal{S}[Q, P] = S[q, p] + (\Sigma(q(t_1), p(t_1), t_1) - \Sigma(q(t_0), p(t_0), t_0)) = S[q, p] + C$, where C is a constant. Therefore, at $(q, p) = (q(t), p(t))$ and $(Q, P) = (Q(t), P(t)) = F(q(t), p(t), t)$, the variational derivatives agree: $\delta\mathcal{S}_{(Q,P)} = \delta S_{(q,p)}$. It follows that a sufficient condition to ensure that the transformation F is canonical with Hamiltonian $\mathcal{H}(Q, P, t)$ is that there exists a function $\Sigma(q, p, t)$ such that condition 7.1 holds.

As it stands, condition 7.1 is not very useful for identifying canonical transformations. To derive a more useful form, first write Condition 7.1 in differential form as

$$P^{\mathsf{Tr}} dQ - \mathcal{H}(Q, P, t)dt = p^{\mathsf{Tr}} dq - H(q, p, t)dt - d\Sigma(q, p, t), \tag{7.2}$$

where $dQ = (dQ_1, \ldots, dQ_N)$ is a column vector and similarly for dq. Among the $4N$ variables q, p, Q, P, only $2N$ are independent. Indeed, the transformation F displays the dependence of the $2N$ variables Q, P on the $2N$ variables q, p at each time t. Under mild conditions, whose precise nature need not concern us, at any time t any subset of $2N$ variables among the $4N$ may be solved in terms of the remaining $2N$, so we may choose any subset of $2N$ variables from the list q, p, Q, P as independent variables. It turns out to be fruitful for the development we are pursuing to choose the $2N$ variables q, P as independent variables with the dependence of Q, p on q, P given by a coordinate transformation $(Q, p) = \mathcal{F}(q, P, t)$.

Consider now the differential condition 7.2 as a relation among independent variables q, P and dependent variables Q, p. Using $P^{\mathrm{Tr}} dQ = d(P^{\mathrm{Tr}} Q) - Q^{\mathrm{Tr}} dP$, condition 7.2 may be written as

$$d(\Sigma + P^{\mathrm{Tr}} Q) = p^{\mathrm{Tr}} dq + Q^{\mathrm{Tr}} dP + (\mathcal{H} - H)dt. \tag{7.3}$$

The expression $S = \Sigma + P^{\mathrm{Tr}} Q$ is a function of the independent variables q, P when the transformation $(Q, p) = \mathcal{F}(q, P, t)$ is used to write Q and p in terms of q and P, and we write $S = S(q, P, t)$ to denote this expression. Expanding the differential dS and equating its coefficients to the corresponding coefficients of the coordinate differentials on the right-hand side of equation 7.3 yields

$$p_i = \frac{\partial S}{\partial q_i}, \quad Q_i = \frac{\partial S}{\partial P_i}, \quad \mathcal{H} = H + \frac{\partial S}{\partial t}. \tag{7.4}$$

The function S is called a **generating function** for the canonical transformation F, and the reason for using the same notation for the generating function as used for the action functional will be clear by the end of the next section.

By reversing this discussion, we now can answer the question, posed earlier, whether there are any canonical transformations other than those generated by a transformation of position coordinates only, and, if so, how one goes about discovering them. Indeed, let $S = S(q, P, t)$ be any twice continuously differentiable function of the $2N + 1$ independent variables q, P, t for which the determinant

$$\det \left[\frac{\partial^2 S}{\partial q_i \partial P_j} \right] \neq 0 \tag{7.5}$$

when evaluated at any point (q, P, t) of its domain. Set $p_i = \partial S / \partial q_i$ and invert these equations to solve for P, so that $P = F_2(q, p, t)$ for some function F_2. This inversion is possible by the nonvanishing of the determinant. Now define the coordinates Q by $Q_i = \partial S / \partial P_i$, which may be written as $Q = F_1(q, p, t)$ for some function F_1. Set $\mathcal{H} = H + \partial S / \partial t$, where $\mathcal{H}$ is written as a function of Q, P, and t. Then the discussion above, starting at equations 7.4, when run in reverse guarantees that condition 7.1 holds for $\mathcal{H}$ and therefore that the transformation $(Q, P) = F(q, p, t) = (F_1(q, p, t), F_2(q, p, t))$ is canonical with Hamiltonian $\mathcal{H}$. We will call the procedure just described, which produces canonical coordinates Q and P, the Hamiltonian $\mathcal{H}$, and the canonical transformation F from an appropriate function S, the **Jacobi procedure**.

Choosing sets of $2N$ variables different from q, P as the independent variables yields other generating functions for the canonical transformation; different choices are appropriate in various other settings and applications. Our choice is useful in deriving the Hamilton–Jacobi differential equation, whose solutions provide generating functions for canonical transformations of a particularly useful type, this being the topic of the next section.

7.3 The Hamilton–Jacobi Equation

Notice that if we can find a canonical transformation for which the Hamiltonian $\mathcal{H}(Q, P)$ is identically zero then, as Hamilton's equations, $\dot{Q}_i = \partial\mathcal{H}/\partial P_i$ and $\dot{P}_i = -\partial\mathcal{H}/\partial Q_i$, hold along any stationary trajectory, Q and P will be constant along any such trajectory. Therefore, the system $Q = Q(q(t), p(t), t) = Q_0$ and $P = P(q(t), p(t), t) = P_0$ may be solved for $q(t)$ and $p(t)$ in terms of the constant vectors Q_0 and P_0, the time t, and the stationary trajectory $q = q(t)$; thus the motion of the system is found. From the discussion of the immediately preceding section, we may obtain such a canonical transformation by finding an appropriate generating function S, namely one that satisfies the third of equations 7.4 with $\mathcal{H} = 0$. Combining this requirement with the first of equations 7.4 gives the following partial differential equation in the unknown function S of the $N+1$ variables q, t that the desired generating function must satisfy:

$$\frac{\partial S}{\partial t} + H\left(q_1, \ldots, q_N, \frac{\partial S}{\partial q_1}, \ldots, \frac{\partial S}{\partial q_N}, t\right) = 0.$$

This equation is known as the **Hamilton–Jacobi equation** and may be rendered succinctly as

$$\frac{\partial S}{\partial t} + H\left(q, \frac{\partial S}{\partial q}, t\right) = 0,$$

where

$$\frac{\partial S}{\partial q} = \left(\frac{\partial S}{\partial q_1}, \ldots, \frac{\partial S}{\partial q_N}\right),$$

a notation that we will employ for the remainder of the lecture. The Hamilton–Jacobi equation contains all the information about our dynamical system in the sense that it replaces the N linear-second order Euler–Lagrange equations or the $2N$ linear first-order Hamilton's equations. Part of the price paid for this reduction from N, or $2N$, equations to a single equation is that generally the Hamilton–Jacobi equation is nonlinear, so one must contend with all the pitfalls nonlinearity entails.

A general solution of the Hamilton–Jacobi equation, known as **Hamilton's principal function**, generates the desired canonical coordinates and transformations according to the Jacobi procedure. On the surface it seems that solutions of the Hamilton–Jacobi equation do not quite give the correct form for a generating function, since the momentum coordinates P are nowhere to be found. It is when the full family of solutions is considered that this is remedied. Indeed, as there are $N+1$ independent variables, $N+1$ constants of integration arise when solving for S. Since S does not appear in the Hamilton–Jacobi equation, but only its various partial derivatives, one of these is an additive constant and, since two generating functions that differ by an additive constant generate the same canonical transformation, we may ignore this additive constant. Name the remaining N nonadditive constants $P_1, \ldots, P_N$ and denote the solution to the Hamilton–Jacobi equation with these constants as $S(q, P, t)$, where $P = (P_1, \ldots, P_N)$. Assuming then that the determinant condition 7.5 holds for S, the Jacobi procedure of the immediately preceding section produces a canonical transformation $(Q, P) = F(q, p, t)$ with Hamiltonian $\mathcal{H} = H + \partial S/\partial t$, which is identically zero since S solves the Hamilton–Jacobi equation. We note that the choice of $P_1, \ldots, P_N$ is far from unique, as there are many ways to choose the constants of integration and each leads to a different generating function, which generates a different canonical transformation with $\mathcal{H} = 0$. In fact, the integration constants may be replaced by functions of the integration constants, as long as those functions form an invertible system of equations. A technical complication also arises from the nonlinearity of the Hamilton–Jacobi equation, namely, the lack of uniqueness of solutions even when initial conditions are specified though this feature of nonlinearity causes no practical difficulties for finding some solution.

Consider now the total time derivative of S along a trajectory $q = q(t)$. Since the momenta P are constant in time, we have

$$\frac{dS}{dt} = \sum_{i=1}^{N} \frac{\partial S}{\partial q_i} \dot{q}_i + \frac{\partial S}{\partial t} = p^{\mathsf{Tr}} \dot{q} - H = L.$$

It follows that, along a trajectory, $S(q, P, t) = \int L(q, \dot{q}, t)\, dt$, i.e., S is an indefinite integral of the Lagrangian. This is why we use S for Hamilton's principal function; the same notation is used also for the definite integral of the Lagrangian that defines the action functional.

The Hamilton–Jacobi equation can be difficult to solve but, when the Hamiltonian H has no explicit time dependence, an additive separation of variables offers a possible way to a solution. Indeed, assuming that $\partial H/\partial t = 0$, we may search for a solution of the form $S(q, t) = \sigma(q) + \tau(t)$. The Hamilton–Jacobi equation becomes

$$H\left(q, \frac{\partial \sigma}{\partial q}(q)\right) = -\dot{\tau}(t)$$

and, as the left-hand side of the equation depends only on q while the right-hand side depends only on t, both sides must equal a constant E, whose value is normally the constant, conserved, energy of the system.[3] Thus the Hamilton–Jacobi equation of interest becomes

$$H\left(q, \frac{\partial \sigma}{\partial q}\right) = E, \tag{7.6}$$

and the generating function takes the form

$$S(q, P, t) = \sigma(q, P_1, \ldots, P_{N-1}) - Et,$$

where $P_N = E$ and $P_1, \ldots, P_{N-1}$ are $N - 1$ nonadditive integration constants that arise from solving equation 7.6.

7.4 Some Sample Applications

We close this lecture by solving for the motion of some simple physical systems using the Hamilton–Jacobi formalism.

We begin with the one-dimensional simple harmonic oscillator,[4] with Hamiltonian given by $H(q, p) = p^2/2m + kq^2/2$ where q measures the displacement from equilibrium, m is the mass, $p = \partial L/\partial \dot{q} = m\dot{q}$, and $k = m\omega^2$ is the spring constant. Since the Hamiltonian does not depend explicitly on time, the desired generating function with new momentum

[3] The constant E is the total energy if the equations of transformation from the generalized coordinates to Cartesian coordinates do not depend explicitly on time and the potential depends only on generalized position coordinates. See Goldstein (1980), pp. 348–351, for a discussion of the general case where the Hamiltonian, even if not explicitly dependent on time, may not be the total energy.

[4] Following Goldstein (1980), pp. 442–445.

coordinate $P = E$ is $S(q, E, t) = \sigma(q) - Et$, where σ is a solution to the Hamilton–Jacobi equation

$$\frac{(\sigma'(q))^2}{2m} + \frac{kq^2}{2} = E.$$

In this case, E is the total energy, which is a constant of the motion. The above Hamilton–Jacobi equation is solved quite easily as

$$\sigma(q) = \int \sqrt{2mE - mkq^2}\, dq.$$

This indefinite integral is easy enough to obtain, but it is unneccessary to do so since our interest is in the canonical coordinate $Q = \partial S/\partial E$. We have, for Hamilton's principal function,

$$S(q, E, t) = \sigma(q) - Et = \int \sqrt{2mE - mkq^2}\, dq - Et, \tag{7.7}$$

so that

$$\begin{aligned} Q = \frac{\partial S}{\partial E} &= m \int \frac{1}{\sqrt{2mE - mkq^2}}\, dq - t \\ &= \sqrt{\frac{m}{2E}} \int \frac{1}{\sqrt{1 - kq^2/2E}}\, dq - t. \end{aligned}$$

Integrating and rearranging, we have

$$Q + t = \sqrt{\frac{m}{k}} \arcsin\left(q\sqrt{\frac{k}{2E}}\right) = \frac{1}{\omega} \arcsin\left(q\sqrt{\frac{m\omega^2}{2E}}\right).$$

The easy inversion of this equation gives the position q as a function of time t as

$$q(t) = \sqrt{\frac{2E}{m\omega^2}} \sin(\omega t + \beta),$$

where $\beta = \omega Q$, a constant. From the first of equations 7.4 the conjugate momentum is given by

$$\begin{aligned} p = \frac{\partial S}{\partial q} &= \sqrt{2mE - mkq^2} \\ &= \sqrt{2mE - 2mE\sin^2(\omega t + \beta)} = \sqrt{2mE}\cos(\omega t + \beta) = m\dot{q}(t). \end{aligned}$$

Substituting $q(t)$ in equation 7.7 gives the generating function S along the stationary trajectory as

$$S = \int 2E\cos^2(\omega t + \beta)\, dt - Et = \int E(2\cos^2(\omega t + \beta) - 1)\, dt,$$

and a simple calculation identifies the integrand as the Lagrangian

$$L = \frac{p^2}{2m} - \frac{m\omega^2 q^2}{2} = E(\cos^2(\omega t + \beta) - \sin^2(\omega t + \beta))$$
$$= E(2\cos^2(\omega t + \beta) - 1).$$

This of course is in agreement with the observations of Section 7.3.

As a second example, we solve Kepler's problem[5] for a particle of mass m in an inverse square central force field. The natural coordinates are polar coordinates in the plane of motion, the radial coordinate r and the angular coordinate θ. The Lagrangian is

$$L = L(r, \theta, \dot{r}, \dot{\theta}) = \frac{1}{2}m(\dot{r}^2 + r^2\dot{\theta}^2) + \frac{K}{r},$$

where $V(r) = -K/r$ is the potential. The canonically conjugate momenta are $p_r = \partial L/\partial \dot{r} = m\dot{r}$ and $p_\theta = \partial L/\partial \dot{\theta} = mr^2\dot{\theta}$, and the Hamiltonian is

$$H(r, \theta, p_r, p_\theta) = \frac{1}{2m}\left(p_r^2 + \frac{p_\theta^2}{r^2}\right) - \frac{K}{r}.$$

Since the Hamiltonian does not depend explicitly on time, the desired generating function is $S(r, \theta, P_1, P_2, t) = \sigma(r, \theta, P_1) - Et$, where $P_2 = E$ and σ is a solution of the Hamilton–Jacobi equation

$$\frac{1}{2m}\left\{\left(\frac{\partial \sigma}{\partial r}\right)^2 + \frac{1}{r^2}\left(\frac{\partial \sigma}{\partial \theta}\right)^2\right\} - \frac{K}{r} = E. \tag{7.8}$$

We look for a solution of the form $\sigma(r, \theta) = \sigma_1(r) + \sigma_2(\theta)$. Substituting this into equation 7.8 and rearranging gives

$$\left(\frac{d\sigma_2}{d\theta}\right)^2 = 2mr^2E + 2mrK - r^2\left(\frac{d\sigma_1}{dr}\right)^2.$$

As the left-hand side is a function of θ alone while the right-hand side is a function of r alone, both sides must be constant. Naming this constant P_1^2 gives $\sigma_2(\theta) = P_1\theta$ and

$$\frac{d\sigma_1}{dr} = \sqrt{2mE + \frac{2mK}{r} - \frac{P_1^2}{r^2}}.$$

Hamilton's principal function becomes

$$S(r, \theta, P_1, E, t) = \int \sqrt{2mE + 2mK/r - P_1^2/r^2}\, dr + P_1\theta - Et,$$

[5] Following Spiegel (1967), pp. 326–327.

and the canonical coordinates Q_1 and Q_2 are given by

$$Q_1 = \frac{\partial S}{\partial P_1} = \int \frac{-P_1}{r^2\sqrt{2mE + 2mK/r - P_1^2/r^2}}\,dr + \theta,$$

$$Q_2 = \frac{\partial S}{\partial E} = \int \frac{m}{\sqrt{2mE + 2mK/r - P_1^2/r^2}}\,dr - t.$$

The first equation integrates to give the equation for the orbit as

$$\frac{1}{r} = \frac{mK}{P_1^2}\left(1 + \epsilon \sin(\theta - Q_1)\right),$$

where the eccentricity is

$$\epsilon = \sqrt{1 + \frac{2EP_1^2}{mK^2}}.$$

This orbit is a conic section with one focus at the origin and is elliptic, parabolic, or hyperbolic according to whether the energy E is less than, equal to, or greater than zero, respectively. The coordinate Q_1 represents the orientation of the polar axis with respect to the (major or minor) axis of the conic section. It turns out that the coordinate P_1 represents the constant total angular momentum of the system. The integration of the second equation is quite a bit more difficult[6] and the inversion of the resulting function to yield r as a function of t is quite formidable, and will not be carried out here.

7.5 End Notes

The mathematics of classical mechanics as presented in this and the preceding lecture is that of the calculus of variations, a topic that began its development in the seventeenth century shortly after the advent of the differential and integral calculus of Newton and Leibnitz. The original motivations were geometric – solving for the largest area enclosed by a fixed length or for the curve of fastest descent from a higher to a lower point and later for the surface of minimal area spanning a fixed boundary curve and the shortest paths connecting two points on a surface. By the mid-eighteenth century, under the genius of Euler and Lagrange, the subject was developed further and, in the nineteenth century, Hamilton applied the calculus of variations to the action defined

[6] See Goldstein (1980), pp. 98–102, for a full discussion.

as the path integral of the Lagrangian, demonstrating that the mechanics of Newton follows from his action principle. Jacobi's seminal and deep contributions followed and the development has proceeded up to the present, with significant contributions in the twentieth century by Hilbert, Noether, Morse and, especially, the Russians Kolmogorov and Arnold, and the German-American Moser.

The modern version of Lagrangian, Hamiltonian, and Hamilton–Jacobi mechanics is written in the language of tangent and cotangent bundles on manifolds of configurations or phases. A beautiful modern treatment appears in the book *Foundations of Mechanics* by Abraham and Marsden (1978).

The development of canonical transformations and Hamilton–Jacobi theory presented in Sections 7.2 and 7.3 relies on Chapter 4 of Gelfand and Fomin (1963) and Chapter 8 of van Brunt (2004). Chapter 3 of Pollard (1976) is a gem and presents the best explanation that I have seen of canonical coordinates that do not depend on time. Pollard develops canonical transformations in terms of symplectic matrices, which may serve as a bridge to the modern symplectic manifold accounting of classical mechanics. It is my opinion that Goldstein's telling of the theory of Hamilton and Jacobi, published originally in 1950, is one of the most elegant. Especially noteworthy are Chapter 8 on the Hamilton equations of motion, Chapter 9 on canonical transformations, and Chapter 10 on Hamilton–Jacobi theory in either Goldstein (1980) or Goldstein, Poole, and Safko (2002). The first example of Section 7.4 is taken from Section 10–2 of Goldstein (1980) and the second from pp. 326 and 327 of Spiegel (1967). See Chapter 20 of Greiner (2010) for other examples of applications of Hamilton–Jacobi theory. Dittrich and Reuter (2001) present a fine discussion of canonical coordinates in their Chapter 5, with a good set of examples of generating functions for canonical transformations. In addition, their Chapter 6 presents a nice set of applications of Hamilton–Jacobi theory.

8

Classical Mechanics Regain'd

A fairer Paradise is founded now ...

John Milton
Paradise Regain'd, Book IV, p. 613

In this lecture we present Ehrenfest's theorem in the context of wave mechanics. Essentially, the theorem asserts that classical equations hold for the time rate of change of the average values of quantum observables. Exactly in what sense this holds is uncovered in the present lecture. In the course of the discussion we will see how natural it is that Dirac should have suggested a loose correspondence between the Poisson bracket of two classical observables and the commutator of the self-adjoint operators representing those observables. Indeed, the naturalness of this suggestion follows from the resemblance of certain Poisson bracket relations to the commutation relations for operators and is reinforced by a comparison of the classical evolution equation 6.4 with its quantum mechanical analogue, the derivation of which is our first order of business. This derivation introduces the commutator, which is then studied on a somewhat superficial level, and there follow several applications of Ehrenfest's theorem that, in a loose sense, recover classical equations. We end with a more thorough discussion of what it means for two unbounded operators to commute, which leads naturally into a discussion of noncommuting self-adjoint operators and the Weyl relations.

In the lecture following, a highly nonclassical feature of quantum mechanics is exposed with the aid of commutators; this forms the basis of our discussion of the Heisenberg uncertainty principle.

8.1 The Quantum Evolution Equation

Recall that the time evolution of a state vector is governed by the Schrödinger equation $\mathbf{i}\hbar\, d\psi_t/dt = \mathsf{H}\psi_t$, where ψ_t is the state at time t and H is the relevant Hamiltonian operator. In the wave mechanical instantiation of the quantum axioms, ψ_t is a square-integrable function and is assumed to be a member of $\mathcal{D}(\mathsf{H})$ for all t as well as differentiable in t. The Hamiltonian operator is given as

$$\mathsf{H} = \frac{1}{2m}\mathbf{P}^2 + V(\mathbf{X}) = -\frac{\hbar^2}{2m}\Delta + V(\mathbf{x}), \tag{8.1}$$

where Δ is the Laplacian and V is the classical potential function, which is assumed to be as differentiable as is needed for any calculations. Our objective is to evaluate the time derivative of the expectation value $\langle \mathsf{A}\rangle_{\psi_t}$ of an observable $a = a(t)$ that is represented at time t by the self-adjoint operator $\mathsf{A} = \mathsf{A}(t)$. We assume that the dependence of A on time is differentiable in the sense that, in the usual L^2-norm, the limit

$$\mathsf{A}'(t)\psi = \lim_{h\to 0}\frac{\mathsf{A}(t+h)\psi - \mathsf{A}(t)\psi}{h}$$

exists for each $\psi \in \mathcal{D}(\mathsf{A}(t))$.[1] It is a straightforward exercise to show that the Leibnitz rules,

$$\frac{d}{dt}\langle\alpha(t)|\beta(t)\rangle = \langle\dot\alpha(t)|\beta(t)\rangle + \langle\alpha(t)|\dot\beta(t)\rangle$$

and

$$\frac{d}{dt}\mathsf{A}(t)\alpha(t) = \mathsf{A}'(t)\alpha(t) + \mathsf{A}(t)\dot\alpha(t),$$

hold for differentiable functions α and β of a real variable into $L^2(\mathbb{R}^3)$,[2] provided, for the second of these, that

$$\lim_{h\to 0}\mathsf{A}(t)\frac{\alpha(t+h)-\alpha(t)}{h} = \mathsf{A}(t)\dot\alpha(t). \tag{8.2}$$

Of course formula 8.2 is a consequence of the continuity of $\mathsf{A}(t)$, provided that $\mathsf{A}(t)$ is a bounded operator, but as we assume only the self-adjointness of $\mathsf{A}(t)$, there is something to discuss. The most interesting operators representing observables, such as the position, momentum, and angular momentum operators, are unbounded. Rather than explore general conditions under which equation 8.2 holds for unbounded operators,

[1] We assume that $\psi \in \mathcal{D}(\mathsf{A}(t+h))$ for any sufficiently small time increment h.

[2] We assume that $\alpha(t), \dot\alpha(t) \in \mathcal{D}(\mathsf{A}(t))$ for each t in the domain of α, and that $\alpha(t) \in \mathcal{D}(\mathsf{A}(t+h))$ and $\alpha(t+h) \in \mathcal{D}(\mathsf{A}(t)) \cap \mathcal{D}(\mathsf{A}(t+h))$ for sufficiently small h.

which would take us too far afield from our goal, we will derive the quantum evolution equation under the assumption that equation 8.2 holds, and verify equation 8.2 for any appropriate operators in applications.

Armed with the assumptions of the preceding paragraph, the quantum evolution equation is easy to derive. Indeed, taking the time derivative, indicated again by a prime, of the expectation value for observable a at time t when the system is in the normalized state ψ_t, which, of course, satisfies the Schrödinger equation, yields

$$\begin{aligned}\frac{d}{dt}\langle \mathsf{A}(t)\rangle_{\psi_t} &= \frac{d}{dt}\langle \psi_t|\mathsf{A}(t)\psi_t\rangle = \langle \psi_t'|\mathsf{A}(t)\psi_t\rangle + \langle \psi_t|(\mathsf{A}(t)\psi_t)'\rangle \\ &= \left\langle \frac{1}{\mathbf{i}\hbar}\mathsf{H}\psi_t \,\middle|\, \mathsf{A}(t)\psi_t \right\rangle + \langle \psi_t|\mathsf{A}(t)\psi_t'\rangle + \langle \psi_t|\mathsf{A}'(t)\psi_t\rangle \\ &= -\frac{1}{\mathbf{i}\hbar}\langle \psi_t|\mathsf{HA}(t)\psi_t\rangle + \frac{1}{\mathbf{i}\hbar}\langle \psi_t|\mathsf{A}(t)\mathsf{H}\psi_t\rangle + \langle \psi_t|\mathsf{A}'(t)\psi_t\rangle.\end{aligned}$$

To write this a bit more elegantly, we introduce the **commutator** of the operators A and H as $[\mathsf{A},\mathsf{H}] = \mathsf{AH} - \mathsf{HA}$. Suppressing the explicit time dependence of A and ψ yields the **quantum evolution equation** as

$$\frac{d\langle \mathsf{A}\rangle_\psi}{dt} = \langle \mathsf{A}'\rangle_\psi + \frac{1}{\mathbf{i}\hbar}\langle [\mathsf{A},\mathsf{H}]\rangle_\psi. \tag{8.3}$$

Compare this with the evolution equation 6.4, which gives the time rate of change of $A(q,p)$, a function of the generalized position and momentum coordinates. In applications of the quantum version of the evolution equation, the self-adjoint operator A will be an operator built from the position $\mathbf{X} = (\mathsf{X}_1,\mathsf{X}_2,\mathsf{X}_3)$ and momentum $\mathbf{P} = (\mathsf{P}_1,\mathsf{P}_2,\mathsf{P}_3)$ triads of operators. The obvious resemblance of equation 8.3 to equation 6.4 is one of several resemblances between quantum commutation relations and classical Poisson bracket equations that led Dirac to suggest a loose correspondence between the Poisson bracket and the commutator:

$$\{A(\mathbf{x},\mathbf{p}), B(\mathbf{x},\mathbf{p})\} \longleftrightarrow \frac{1}{\mathbf{i}\hbar}[\mathsf{A},\mathsf{B}],$$

where the operator A is obtained by replacing x_i by X_i and p_i by P_i in the functional expression $A(\mathbf{x},\mathbf{p})$, and similarly for B. To explore this correspondence more fully, we need to discuss in more detail the commutator of operators and derive some explicit commutation relations.

8.2 Commutation Relations and the Ehrenfest Theorems

The commutator $[\mathsf{A}, \mathsf{B}] = \mathsf{AB} - \mathsf{BA}$ is defined only for those states ψ for which AB and BA are defined. Though normally we will not concern ourselves with identifying the domains of these operators and will assume that, in any calculations of commutators, any state vectors that appear are in appropriate domains, we do present in Section 8.3 a discussion of the subtleties that arise from considering the commutators of unbounded self-adjoint operators.

The commutator $[\mathsf{A}, \mathsf{B}]$ is antisymmetric and $\mathbb{C}$-bilinear and satisfies the identity

$$[\mathsf{A}, \mathsf{BC}] = \mathsf{B}[\mathsf{A}, \mathsf{C}] + [\mathsf{A}, \mathsf{B}]\mathsf{C} \tag{8.4}$$

and, like the Poisson bracket, the Jacobi identity

$$[\mathsf{A}, [\mathsf{B}, \mathsf{C}]] + [\mathsf{B}, [\mathsf{C}, \mathsf{A}]] + [\mathsf{C}, [\mathsf{A}, \mathsf{B}]] = 0.$$

The most important commutator computation is that of $[\mathsf{X}, \mathsf{P}]$ for the one-dimensional position and momentum operators. For appropriate states ψ,

$$\begin{aligned}(\mathsf{PX}\psi)(x) = \frac{\hbar}{\mathbf{i}}\frac{d}{dx}(x\psi(x)) &= -\mathbf{i}\hbar\,(\psi(x) + x\psi'(x)) \\ &= -\mathbf{i}\hbar\,\psi(x) + (\mathsf{XP}\psi)(x),\end{aligned}$$

so

$$[\mathsf{X}, \mathsf{P}] = \mathsf{XP} - \mathsf{PX} = \mathbf{i}\hbar\,1, \tag{8.5}$$

where 1 is the identity operator on the appropriate domain. More generally, for the triads $\mathbf{X} = (\mathsf{X}_1, \mathsf{X}_2, \mathsf{X}_3)$ and $\mathbf{P} = (\mathsf{P}_1, \mathsf{P}_2, \mathsf{P}_3)$, we have the relations

$$[\mathsf{X}_j, \mathsf{X}_k] = [\mathsf{P}_j, \mathsf{P}_k] = 0 \quad \text{and} \quad [\mathsf{X}_j, \mathsf{P}_k] = \mathbf{i}\hbar\,\delta_{jk}1.$$

Here we have another striking resemblance between quantum commutation relations and Poisson bracket equations, as the classical dynamical variables $\mathbf{x}$ and $\mathbf{p}$ satisfy

$$\{x_j, x_k\} = \{p_j, p_k\} = 0 \quad \text{and} \quad \{x_j, p_k\} = \delta_{jk}.$$

As we mentioned previously, these sorts of resemblances led Dirac to suggest that each function f on the classical phase space should be replaced in the act of quantizing classical equations by an operator $\mathsf{Q}(f)$, in such a way that

$$[\mathsf{Q}(f), \mathsf{Q}(g)] = \mathbf{i}\hbar\, \mathsf{Q}(\{f, g\}). \tag{8.6}$$

This idea has served as an inspiration for guiding the process of quantizing classical mechanics and works for "deriving" the Schrödinger, Klein–Gordon, and Dirac equations from their classical counterparts that describe the energy of the system being quantized. As successful a guide as it has been, this replacement is at best an heuristic technique that, in fact, cannot be satisfied for all functions simultaneously without violating quantum mechanics. The exercise below bears this out, but first, it should be noted that there are ambiguities that arise from the noncommutation of operators. For example, the quantum analogues of position and momentum are the position operator X and momentum operator P, respectively. What should be the quantum analogue of the classical expression $f(x, p) = px^2$? There are three pairwise unequal candidates, PX^2, XPX, and $\mathsf{X}^2\mathsf{P}$, or some combination thereof.

Exercise. Suppose there is a map Q from functions on x and p to operators that satisfies equation 8.6 with $\mathsf{Q}(x) = \mathsf{X}$ and $\mathsf{Q}(p) = \mathsf{P}$.[3]

(i) Show that $\mathsf{Q}(x^k) - \mathsf{X}^k$ and $\mathsf{Q}(p^k) - \mathsf{P}^k$ commute with both X and P, for all $k \geq 1$.

(ii) Show that $12\{p^3, x^3\} = \{\{p^3, x^2\}, \{x^3, p^2\}\}$.

(iii) Show that $-12\hbar^2[\mathsf{P}^3, \mathsf{X}^3] \neq [[\mathsf{P}^3, \mathsf{X}^2], [\mathsf{X}^3, \mathsf{P}^2]]$.

(iv) Deduce the Groenwald–van Hove theorem, that such a Q cannot exist.

The following exercise might be useful in the preceding one.

Exercise. Show that if f is a polynomial function then the following hold:

(i) $[\mathsf{P}, f(\mathsf{X})] = -\mathbf{i}\hbar f'(\mathsf{X})$,

(ii) $[\mathsf{XP}, f(\mathsf{X})] = -\mathbf{i}\hbar \mathsf{X} f'(\mathsf{X})$,

(iii) $[\mathsf{X}, f(\mathsf{P})] = \mathbf{i}\hbar f'(\mathsf{P})$,

(iv) $[\mathsf{XP}, f(\mathsf{P})] = \mathbf{i}\hbar \mathsf{P} f'(\mathsf{P})$.

We are now in a position to prove the Ehrenfest theorems. The first result recovers the classical relationship between position and momentum, viz., that $p = m\dot{x}$. For simplicity we work with one-dimensional

[3] This exercise has been lifted from Hannabuss (1997), p. 106.

motion. Let X and P be the one-dimensional position and momentum operators, respectively. Let $\psi = \psi_t$ be the normalized state at time t. Note that $\mathsf{X}'(t) = 0$, since X is independent of time, and observe that X, whose effect on an element of $L^2(\mathbb{R})$ is merely to multiply by the coordinate x on $\mathbb{R}$, easily satisfies equation 8.2. Since the Hamiltonian is given by equation 8.1 and $[\mathsf{X}, V(\mathsf{X})] = 0$, we have $[\mathsf{X}, \mathsf{H}] = \frac{1}{2m}[\mathsf{X}, \mathsf{P}^2]$, and an application of formula 8.4 yields

$$\begin{aligned}[\mathsf{X}, \mathsf{P}^2] &= \mathsf{P}[\mathsf{X}, \mathsf{P}] + [\mathsf{X}, \mathsf{P}]\mathsf{P} \\ &= \mathsf{P}(\mathbf{i}\hbar\mathsf{1}) + (\mathbf{i}\hbar\mathsf{1})\mathsf{P} = 2\mathbf{i}\hbar\mathsf{P}.\end{aligned}$$

Thus

$$[\mathsf{X}, \mathsf{H}] = \frac{\mathbf{i}\hbar}{m}\mathsf{P} \tag{8.7}$$

and the quantum evolution equation 8.3 with $\mathsf{A} = \mathsf{X}$ reads

$$\frac{d\langle\mathsf{X}\rangle_\psi}{dt} = \frac{1}{\mathbf{i}\hbar}\langle[\mathsf{X}, \mathsf{H}]\rangle_\psi = \frac{1}{m}\langle\mathsf{P}\rangle_\psi.$$

Setting $x(t) = \langle\mathsf{X}\rangle_\psi$ and $p(t) = \langle\mathsf{P}\rangle_\psi$, the expectation values of the position and momentum operators when the state of the system is $\psi = \psi_t$, we recover the classical equation

$$p = m\dot{x}.$$

The next Ehrenfest result applies equation 8.3 with $\mathsf{A} = \mathsf{P}$ to recover Newton's second law, that $\dot{p}$ is equal to the applied force. Since P is independent of time, $\mathsf{P}'(t) = 0$. If the wave function $\Psi(x, t) = \psi_t(x)$ has continuous second partial derivatives, then equation 8.2 for P is merely the assertion that the mixed second partial derivatives agree: $\partial^2\Psi/\partial x\partial t = \partial^2\Psi/\partial t\partial x$. A quick calculation shows that $[\mathsf{P}, \mathsf{H}] = [\mathsf{P}, V(\mathsf{X})]$. To compute this latter commutator, note that

$$\frac{\partial}{\partial x}V(x)\Psi(x,t) = \frac{\partial V}{\partial x}\Psi + V\frac{\partial \Psi}{\partial x},$$

so that

$$\left(\frac{\partial}{\partial x}V - V\frac{\partial}{\partial x}\right)\Psi = \frac{\partial V}{\partial x}\Psi.$$

It follows that

$$\left[\frac{\partial}{\partial x}, V\right] = \frac{\partial V}{\partial x} \quad \text{or} \quad [\mathsf{P}, V] = \frac{\hbar}{\mathbf{i}}\frac{\partial V}{\partial x}.$$

From the quantum evolution equation 8.3,

$$\frac{d\langle \mathsf{P} \rangle_\psi}{dt} = \frac{1}{\mathrm{i}\hbar}\left\langle \frac{\hbar}{\mathrm{i}}\frac{\partial V}{\partial x} \right\rangle_\psi = -\left\langle \frac{\partial V}{\partial x} \right\rangle_\psi.$$

Again setting $p(t) = \langle \mathsf{P} \rangle_\psi$, we have

$$\dot{p} = \langle F \rangle_\psi, \tag{8.8}$$

where $F(x) = -\partial V/\partial x$ is the classical force exerted on the particle at point x. This is the quantum version of Newton's second law, but note the difference from the classical version; the classical version is decidedly local in both time and space, in that the instantaneous time rate of change of a particle's momentum is equal to the applied force at the location of the particle at an instant in time. A large change in the potential field, and therefore the force field, will have no effect upon the motion of the particle if the change is restricted to regions far removed from the particle's location during the time interval under consideration. The quantum version, on the other hand, expresses the instantaneous time rate of change of the expected value of momentum measurements, when in state ψ_t, being the expected value of the whole of the force field $F(x)$. The expected value of the force is

$$\langle F \rangle_\psi = -\left\langle \frac{\partial V}{\partial x} \right\rangle_\psi = -\left\langle \psi \,\middle|\, \frac{\partial V}{\partial x}\psi \right\rangle = -\int_{\mathbb{R}} \overline{\psi}\frac{\partial V}{\partial x}\psi, \tag{8.9}$$

so the particle motion is affected when the potential V is changed in regions where the modulus of the wave function ψ is not negligible. The position and momentum of the particle are uncertain; perhaps it is even meaningless to speak of the particle as possessing these attributes before a measurement. The wave function provides a probability amplitude for the position measurement, and it may have a significant modulus over a large region of space. In this case, the information contained in the force field spread over a significant amount of space affects the motion and subsequent position and momentum measurements of the particle. The lesson is that, though equation 8.8 has the form of the classical equation, its content is decidedly nonclassical.

A recovery of the classical equation occurs when the particle is highly localized. This means that the wave function is steeply spiked at some point, say at $x = \lambda$, and is zero outside a small neighborhood of λ. This implies, from equation 8.9, that since ψ is normalized, $\langle F \rangle_\psi \approx F(\lambda)$. In this case equation 8.8 reads, approximately, $\dot{p} \approx F(\lambda)$, so the rate of change of the expected value of momentum does approximately equal the applied force at the "location" of the particle.

The reader should notice that the quantum analogue of Hamilton's equations of classical mechanics are just the quantum evolution equations 8.3 for $\mathsf{A} = \mathsf{X}$ and $\mathsf{A} = \mathsf{P}$. In the Poisson bracket formulation, Hamilton's equations along a classical trajectory $x = x(t)$ read $\dot{x} = \{x, H\}$ and $\dot{p} = \{p, H\}$, while the quantum evolution equations along the "quantum trajectory" $\psi = \psi_t$ read

$$\frac{d\langle \mathsf{X}\rangle_\psi}{dt} = \frac{1}{\mathbf{i}\hbar}\langle[\mathsf{X}, \mathsf{H}]\rangle_\psi$$

and

$$\frac{d\langle \mathsf{P}\rangle_\psi}{dt} = \frac{1}{\mathbf{i}\hbar}\langle[\mathsf{P}, \mathsf{H}]\rangle_\psi.$$

8.3 Commuting Self-Adjoint Operators

In this section we will see how very delicate it can be to work with unbounded operators, which are defined only on a dense linear subspace of the Hilbert space $\mathcal{H}$, as compared to bounded operators defined on the whole of $\mathcal{H}$.

Our motivating question is *What does, or should, it mean to say that two self-adjoint operators* A *and* B *commute?* The immediate answer is that A and B commute precisely when $\mathsf{AB} = \mathsf{BA}$. Though this answer is quite adequate when the operators are bounded, it hides an abundance of difficulty when the operators are unbounded. First, there is a problem with forming the composition AB in that $\mathsf{B}\psi$ may fail to be a member of $\mathcal{D}(\mathsf{A})$ even when ψ is a member of $\mathcal{D}(\mathsf{B})$. As, obviously,

$$\mathcal{D}(\mathsf{AB}) = \{\psi \in \mathcal{D}(\mathsf{B}) : \mathsf{B}\psi \in \mathcal{D}(\mathsf{A})\},$$

we find that the equation $\mathsf{AB} = \mathsf{BA}$ makes sense only on the restricted domain

$$\begin{aligned}\mathcal{D} &= \mathcal{D}(\mathsf{AB}) \cap \mathcal{D}(\mathsf{BA}) \\ &= \{\psi \in \mathcal{D}(\mathsf{A}) \cap \mathcal{D}(\mathsf{B}) : \mathsf{B}\psi \in \mathcal{D}(\mathsf{A}) \text{ and } \mathsf{A}\psi \in \mathcal{D}(\mathsf{B})\}.\end{aligned}$$

However, even though A and B are densely defined, the intersection $\mathcal{D}$ may fail to be so and, indeed, may be finite-dimensional and may even reduce to the trivial subspace $\{0\}$.[4] It seems that calling A and B

[4] It may even happen that $\mathcal{D}(\mathsf{AB}) = \{0\}$.

commuting operators when $\mathcal{D} = \{0\}$, even though, necessarily, $\mathsf{AB} = \mathsf{BA}$ on $\mathcal{D}$, is rather uninspiring and vacuous in meaning or implication. Surely the theory begs for a better, more useful, definition of commuting self-adjoint operators than that $\mathsf{AB} = \mathsf{BA}$ on the common domain $\mathcal{D}$.

A possible, even reasonable, path to a useful definition is to say that A and B commute when there exists a dense linear subspace $\mathcal{C}$ of $\mathcal{D}$ that is both A- and B-invariant and on which $\mathsf{AB} = \mathsf{BA}$. Here, to say that $\mathcal{C}$ is A-invariant is to say that $\mathcal{C} \subset \mathcal{D}(\mathsf{A})$ and $\mathsf{A}(\mathcal{C}) \subset \mathcal{C}$. This would preclude any pair of operators from being declared commuting, even though $\mathsf{AB} = \mathsf{BA}$ on $\mathcal{D}$, when the common domain $\mathcal{D}$ is rather meager. It seems that the requirement that there be a *dense* invariant subspace of both domains $\mathcal{D}(\mathsf{A})$ and $\mathcal{D}(\mathsf{B})$ on which the operators commute, in the classical sense, would be enough to capture the appropriate information so that the natural implications of commuting bounded operators would carry over to the unbounded case. This, of course, begs the question as to what might be these natural implications that follow from the commutation of bounded operators. We will focus our discussion on one such implication, arguably the most important.

For a bounded operator (not necessarily self-adjoint) A with, say, $\|\mathsf{A}\| \leq \mu$, and a normalized vector $\psi \in \mathcal{H}$, we have $\|\mathsf{A}^n\psi\| \leq \mu^n$, so that the Weierstrauss M-test implies that the series

$$\sum_{n=0}^{\infty} \frac{\mathsf{A}^n\psi}{n!} \tag{8.10}$$

converges uniformly on the unit ball in $\mathcal{H}$. It follows that the operator e^{A} defined by

$$e^{\mathsf{A}}\psi = \left(\sum_{n=0}^{\infty} \frac{\mathsf{A}^n}{n!}\right)\psi = \sum_{n=0}^{\infty} \frac{\mathsf{A}^n\psi}{n!}$$

is a bounded linear operator whose domain is the whole of $\mathcal{H}$ and for which $\|e^{\mathsf{A}}\| \leq e^{\|\mathsf{A}\|} \leq e^{\mu}$. Moreover, the usual proof that $e^{a+b} = e^a e^b$, using the infinite series expansion for the exponential function, goes through to show that $e^{\mathsf{A}+\mathsf{B}} = e^{\mathsf{A}}e^{\mathsf{B}}$ whenever A and B are bounded operators that commute. It follows then that the operators e^{A} and e^{B} commute since, obviously, $\mathsf{A} + \mathsf{B} = \mathsf{B} + \mathsf{A}$. In fact, a sort of converse of this discussion holds, which goes like this. First observe that if A and B commute then so do the one-parameter families of operators $\{s\mathsf{A}\}_{s\in\mathbb{R}}$ and $\{t\mathsf{B}\}_{t\in\mathbb{R}}$ and, thus, $e^{s\mathsf{A}}$ and $e^{t\mathsf{B}}$ commute for all real s and t. Here is the converse result. The commutation of the exponential operators $e^{s\mathsf{A}}$ and $e^{t\mathsf{B}}$, for all real s and t, implies that of the operators A and B. To see this take the derivative of the equation $e^{s\mathsf{A}}e^{t\mathsf{B}} = e^{t\mathsf{B}}e^{s\mathsf{A}}$, first with respect to s and then with respect to t,[5] and then evaluate it at $s = t = 0$. Thus, when the operators are bounded, the commutativity of the operators may be described equally well by the commutativity of their associated one-parameter exponential families $\{e^{s\mathsf{A}}\}_{s\in\mathbb{R}}$ and $\{e^{t\mathsf{B}}\}_{t\in\mathbb{R}}$. It follows, on replacing A by $c\mathsf{A}$, where c is a complex constant, and similarly for B, that the commutation of A and B is equivalent to the commutation of the associated one-parameter exponential families $\{e^{\mathbf{u}s\mathsf{A}}\}_{s\in\mathbb{R}}$ and $\{e^{\mathbf{v}t\mathsf{B}}\}_{t\in\mathbb{R}}$, where $\mathbf{u}$ and $\mathbf{v}$ are arbitrary unit complex constants.

Now going back to self-adjoint operators, if we are to emulate the discussion of the preceding paragraph in the case when the operators are unbounded, we need to understand something about the exponential of a general self-adjoint operator. The first thing to say is that when A is self-adjoint but unbounded, the series of equation 8.10 generally fails to converge on any substantial subset of the domain $\mathcal{D}(\mathsf{A})$.[6] The way forward is to invoke the functional calculus of the spectral theorem (see p. 54), which allows us to define the exponential of a purely imaginary multiple of A. For any real s, define

$$e^{\mathbf{i}s\mathsf{A}} = \int_{\mathbb{R}} e^{\mathbf{i}s\lambda}\, d\mathsf{E}_\lambda,$$

where $\{\mathsf{E}_\lambda : \lambda \in \mathbb{R}\}$ is the resolution of the identity corresponding to A, on the domain

$$\mathcal{D}(e^{\mathbf{i}s\mathsf{A}}) = \left\{\psi \in \mathcal{H} : \int_{-\infty}^{\infty} |\, e^{\mathbf{i}s\lambda}\,|^2 \; d\langle\psi|\mathsf{E}_\lambda\psi\rangle < \infty\right\}.$$

Now, because we have chosen to multiply A by the purely imaginary parameter "$\mathbf{i}s$" rather than the real parameter "s" the integrand is identically 1 and, as $\int_{-\infty}^{\infty} d\langle\psi|\mathsf{E}_\lambda\psi\rangle = \|\psi\|^2 < \infty$ for all $\psi \in \mathcal{H}$, we find that $\mathcal{D}(e^{\mathbf{i}s\mathsf{A}}) = \mathcal{H}$. Moreover, this exponential operator is bounded. All this is rather interesting in its own right: namely, though A may be defined only on a proper subspace of $\mathcal{H}$, the one-parameter family of exponentials $\{e^{\mathbf{i}s\mathsf{A}}\}_{s\in\mathbb{R}}$ is defined on the whole of $\mathcal{H}$.[7]

[5] The derivative of the one-parameter family of operators $e^{s\mathsf{A}}$ with respect to the parameter s equals what one expects, namely, $\mathsf{A}e^{s\mathsf{A}}$; moreover, the Leibnitz rule for taking derivatives holds when, as here, the operators are bounded.

[6] It converges on the set of so-called analytic vectors, which, though forming a dense subspace of the domain, still may manage to have an infinite-dimensional complement in that domain. Analytic vectors are defined in the next section.

[7] We should point out that on the space of analytic vectors to which the previous footnote refers, this definition of the exponential using the functional calculus does agree with that using the series of equation 8.10.

Returning to the problem at hand, the shortfall with defining general self-adjoint operators A and B to be commuting when there exists a dense linear subspace $\mathcal{C}$ of $\mathcal{D}$ that is both A- and B-invariant and on which $\mathsf{AB} = \mathsf{BA}$ is that there are examples of such operators, necessarily unbounded, where the equation $e^{\mathsf{i}(\mathsf{A}+\mathsf{B})} = e^{\mathsf{iA}}e^{\mathsf{iB}}$ fails and the exponential operators e^{iA} and e^{iB} fail to commute. This proposed definition, then, though seemingly reasonable, is a mere curiosity, for it offers not even the least of the consequences of commutativity found in the bounded operator world, and this the most useful of those consequences that follow from commutativity.

The benefit of our analysis, aside from manifesting the unsatisfactory nature of this seemingly reasonable definition, is that it proposes a definition of commuting operators more than adequate to the task. Indeed, to skirt the whole trouble that domains present, we replace the self-adjoint operator A by its exponential family $e^{\mathsf{i}s\mathsf{A}}$ in discussions of commutativity, as then the problem of domains vanishes since these exponential operators are defined on the whole of $\mathcal{H}$. The definition adequate to express the commutation of self-adjoint operators A and B is then that these operators are said to commute whenever their corresponding one-parameter families of bounded exponential operators, $\{e^{\mathsf{i}s\mathsf{A}}\}_{s\in\mathbb{R}}$ and $\{e^{\mathsf{i}t\mathsf{B}}\}_{t\in\mathbb{R}}$, are found to commute in the usual sense.

This definition can be given in terms of the resolutions of the identity from the spectral theorem. Indeed, if $\{\mathsf{E}_\mu : \mu \in \mathbb{R}\}$ and $\{\mathsf{F}_\nu : \nu \in \mathbb{R}\}$ are the respective resolutions of A and B, the operators commute if and only if each E_μ commutes with each F_ν and, as these projection operators are bounded, their commutation is in the usual sense of the word.

8.4 The Baker–Hausdorff Formula

Having explored the meaning of commuting self-adjoint operators in the preceding section, we now turn our attention to noncommuting operators and a formula that in some sense quantifies the noncommutation. This leads naturally to the Weyl relations on unitary representations. This is a precursor to the C^*-algebra approach to quantum mechanics, a presentation of which may be found in Strocchi (2008), for example.

We begin with a definition. A vector $\psi \in \mathcal{H}$ is **analytic** with respect to the operator A if ψ is in the domain of A^n, for all n, and the series of equation 8.10 converges. The **Baker–Hausdorff formula** expresses the noncommutation of operators A and B as follows. Assume that the domains of A, B, and $\mathsf{A}+\mathsf{B}$ have a common dense subspace $\mathcal{D}$ of vectors that are analytic with respect to all three operators. If A and B both commute with $[\mathsf{A},\mathsf{B}]$ on $\mathcal{D}$ then

$$e^{\mathsf{A}}e^{\mathsf{B}} = e^{\mathsf{A}+\mathsf{B}+\frac{1}{2}[\mathsf{A},\mathsf{B}]} \tag{8.11}$$

on the domain $\mathcal{D}$. As an example, since $[\mathsf{X},\mathsf{P}] = \mathbf{i}\hbar 1$ holds on the domain $\mathcal{D} = C_c^\infty(\mathbb{R})$ in one-dimensional wave mechanics, which is a domain of analytic vectors for X and P, the Baker–Hausdorff formula holds for the position and momentum operators on $\mathcal{D}$.

To see that the Baker–Hausdorff formula holds, define the function $\mathsf{F}(x) = e^{-x\mathsf{B}}\mathsf{A}e^{x\mathsf{B}}$ of the real variable x and take the derivative[8] to get

$$\frac{d\mathsf{F}}{dx}(x) = -e^{-x\mathsf{B}}[\mathsf{A},\mathsf{B}]e^{x\mathsf{B}} = [\mathsf{A},\mathsf{B}]. \tag{8.12}$$

The second equality follows since the commutation of B with the commutator $[\mathsf{A},\mathsf{B}]$ implies the commutation of the exponential operator $e^{x\mathsf{B}}$ with the commutator. Treating equation 8.12 as a differential equation of operators, its solution is $\mathsf{F}(x) = [\mathsf{A},\mathsf{B}]x + \mathsf{F}(0)$ so that $e^{-x\mathsf{B}}\mathsf{A}e^{x\mathsf{B}} = [\mathsf{A},\mathsf{B}]x + \mathsf{F}(0)$, and evaluating at $x = 0$ gives $\mathsf{F}(0) = \mathsf{A}$. This implies that

$$e^{-x\mathsf{B}}\mathsf{A}e^{x\mathsf{B}} - [\mathsf{A},\mathsf{B}]x - \mathsf{A} = 0. \tag{8.13}$$

Consider now the function

$$\mathsf{G}(x) = e^{x\mathsf{A}}e^{x\mathsf{B}}e^{-x(\mathsf{A}+\mathsf{B})}e^{-x^2[\mathsf{A},\mathsf{B}]/2}.$$

Taking the derivative of G using the Leibnitz rule, the commutativity of e^{A} and e^{B} with the commutator $[\mathsf{A},\mathsf{B}]$, and applying equation 8.13, gives

$$\frac{d\mathsf{G}}{dx}(x) = e^{x\mathsf{A}}e^{x\mathsf{B}}\left(e^{-x\mathsf{B}}\mathsf{A}e^{x\mathsf{B}} - [\mathsf{A},\mathsf{B}]x - \mathsf{A}\right)e^{-x(\mathsf{A}+\mathsf{B})}e^{-x^2[\mathsf{A},\mathsf{B}]/2} = 0.$$

This of course implies that G is constant and therefore

$$e^{\mathsf{A}}e^{\mathsf{B}}e^{-(\mathsf{A}+\mathsf{B})}e^{-[\mathsf{A},\mathsf{B}]/2} = \mathsf{G}(1) = \mathsf{G}(0) = 1.$$

The Baker–Hausdorff formula now follows from this and the fact that $e^{\mathsf{A}+\mathsf{B}+[\mathsf{A},\mathsf{B}]/2} = e^{\mathsf{A}+\mathsf{B}}e^{[\mathsf{A},\mathsf{B}]/2}$, which in turn follows from the commutation of $\mathsf{A}+\mathsf{B}$ with the commutator $[\mathsf{A},\mathsf{B}]$.

We now apply this to the one-dimensional wave mechanical operators of position and momentum with commutation relation $[\mathsf{X},\mathsf{P}] = \mathbf{i}\hbar 1$. The unitary exponential operators $\mathsf{U}(s) = e^{\mathbf{i}s\mathsf{X}}$ and $\mathsf{V}(t) = e^{\mathbf{i}t\mathsf{P}}$ for real s

[8] The Leibnitz rule for taking derivatives of products holds in this particular setting because the operators depending on x are exponential operators and therefore bounded.

and t are bounded and defined on all of $L^2(\mathbb{R})$ and satisfy the **Weyl relations**, given by

$$\begin{aligned}\mathsf{U}(s)\mathsf{U}(t) &= \mathsf{U}(t)\mathsf{U}(s) = \mathsf{U}(s+t),\\ \mathsf{V}(s)\mathsf{V}(t) &= \mathsf{V}(t)\mathsf{V}(s) = \mathsf{V}(s+t),\\ \mathsf{U}(s)\mathsf{V}(t) &= e^{-\mathbf{i}\hbar st}\mathsf{V}(t)\mathsf{U}(s).\end{aligned}$$

The first two lines merely say that $\mathsf{U}(s)$ and $\mathsf{V}(t)$ define one-parameter groups of unitary transformations, representations of $\mathbb{R}$ on the group of unitary transformations on $L^2(\mathbb{R})$, while the third describes precisely and quantitatively the failure of commutativity between these two representations. These all follow from the Baker–Hausdorff formula on setting A and B as the various possible combinations of $\mathbf{i}s\mathsf{X}$, $\mathbf{i}t\mathsf{X}$, $\mathbf{i}s\mathsf{P}$, and $\mathbf{i}t\mathsf{P}$. The mathematician Hermann Weyl suggested that much of the physics described by wave mechanics could more easily be encoded mathematically by replacing the self-adjoint operators X and P, with all the concomitant problems that arise from the use of unbounded operators, by the one-parameter groups $\mathsf{U}(s)$ and $\mathsf{V}(t)$ of bounded unitary operators. The closed subalgebra of unitary operators generated by these one-parameter groups is called the **Weyl algebra** and, more generally, any C^*-algebra generated by one-parameter families of elements satisfying the Weyl relations is termed a Weyl algebra.

8.5 End Notes

Much of my development of commutation relations owes a debt to Chapter 7 of Hannabuss (1997). The discussion in Sections 8.3 and 8.4 is derived partially from Reed and Simon (1980), Section VIII.5.

9
Wave Mechanics I: Heisenberg Uncertainty

> Uncertainty and mystery are energies of life. Don't let them scare you unduly, for they keep boredom at bay and spark creativity.
>
> R.I. Fitzhenry

> When one admits that nothing is certain one must, I think, also admit that some things are much more nearly certain than others.
>
> Bertrand Russell

We have, in some sense, recovered some of the equations of classical mechanics with the Ehrenfest results. This has been accomplished by deriving the quantum evolution equation, most naturally formulated in terms of the commutator of operators. We now use commutators to explore one of the foundational insights of quantum mechanics, the justifiably famous **Heisenberg uncertainty principle**, a decidedly nonclassical result that places limits on what the quantum formalism can say about the precision of the measurements of two attributes of a physical system. There has been much confusion about exactly what this principle says and what it fails to say. We will try to let the mathematics speak through to articulate precisely what may be asserted from the formalism and what is open to interpretation. After stating the principle in the first section, we search for a suitable interpretation of its meaning in the next. We then examine in detail so-called **minimal uncertainty states** and end our lecture with a discussion of the role of the Fourier transform in the mathematical expression of Heisenberg's principle.

9.1 Statement of the Principle

Consider self-adjoint operators A and B, and note that $\mathbf{i}[\mathsf{A},\mathsf{B}]$ is self-adjoint.[1] Let ψ be a normalized state vector. A short calculation shows that $\langle \mathbf{i}[\mathsf{A},\mathsf{B}]\rangle_\psi = -2\,\mathrm{Im}\langle \mathsf{A}\psi|\mathsf{B}\psi\rangle$, so that

$$\langle \mathbf{i}[\mathsf{A},\mathsf{B}]\rangle_\psi^2 = 4(\mathrm{Im}\langle \mathsf{A}\psi|\mathsf{B}\psi\rangle)^2 \le 4|\langle \mathsf{A}\psi|\mathsf{B}\psi\rangle|^2 \le 4\|\mathsf{A}\psi\|^2\|\mathsf{B}\psi\|^2;$$

the second inequality holds by an application of the Schwarz inequality. Now, $\|\mathsf{A}\psi\|^2 = \langle \mathsf{A}\psi|\mathsf{A}\psi\rangle = \langle \psi|\mathsf{A}^2\psi\rangle = \langle \mathsf{A}^2\rangle_\psi$, and similarly for B, and we get

$$\frac{1}{4}\langle \mathbf{i}[\mathsf{A},\mathsf{B}]\rangle_\psi^2 \le \langle \mathsf{A}^2\rangle_\psi \langle \mathsf{B}^2\rangle_\psi. \tag{9.1}$$

When is inequality 9.1 an equality? We avoid the trivial case by assuming that $\mathsf{B}\psi$ is nonzero. Then the Schwarz inequality is an equality precisely when the vectors $\mathsf{A}\psi$ and $\mathsf{B}\psi$ are linearly dependent, so that $\mathsf{A}\psi = c\mathsf{B}\psi$ for some complex number c. We must have also that $|\langle \mathsf{A}\psi|\mathsf{B}\psi\rangle| = \pm\mathrm{Im}\langle \mathsf{A}\psi|\mathsf{B}\psi\rangle$. Combining these two conditions gives as a criterion for equality that $|c|\|\mathsf{B}\psi\|^2 = \pm\mathrm{Im}(\bar{c})\|\mathsf{B}\psi\|^2$, or $|c| = \pm\mathrm{Im}\,\bar{c}$, since $\mathsf{B}\psi \neq 0$. This implies that $\mathrm{Re}\,c = 0$, so c is purely imaginary, say $c = s\mathbf{i}$ for some real number s. This may be stated succinctly as: equality holds in inequality 9.1 if, and only if, either $\mathsf{B}\psi = 0$ or $(\mathsf{A} - \mathbf{i}s\mathsf{B})\psi = 0$ for some real number s.

We are now in a position to examine the Heisenberg uncertainty principle in some detail. Recall from Lecture 3 that the variance $\nu_\psi(\mathsf{A})$ of the observable represented by A and the state vector ψ is the expectation value of the operator $(\mathsf{A} - \langle \mathsf{A}\rangle_\psi)^2$, and that the dispersion $\Delta_\psi(\mathsf{A})$, serving in the role of the standard deviation, is the positive square root of the variance. We have

$$\nu_\psi(\mathsf{A}) = \mathrm{Exp}_\psi\left[(\mathsf{A} - \langle \mathsf{A}\rangle_\psi)^2\right] = \langle A^2\rangle_\psi - \langle \mathsf{A}\rangle_\psi^2$$

and

$$\Delta_\psi(\mathsf{A}) = \sqrt{\nu_\psi(\mathsf{A})} = \sqrt{\langle A^2\rangle_\psi - \langle \mathsf{A}\rangle_\psi^2}.$$

Heisenberg uncertainty principle. For the normalized state vector ψ, the dispersions of the position and momentum observables in a one-dimensional quantum system satisfy

$$\Delta_\psi \mathsf{X}\,\Delta_\psi \mathsf{P} \ge \frac{\hbar}{2}. \tag{9.2}$$

[1] It is easy to see that $\mathbf{i}[\mathsf{A},\mathsf{B}]$ is symmetric, and domains must be adjusted to make sure it is self-adjoint. According to standard practice, we happily ignore this.

The inequality is an equality precisely when, and only when,

$$\psi(x) = Ce^{-s(x-\mu)^2/2\hbar},$$

for some positive constant s, complex constant μ, and constant C chosen to normalize ψ.

We will derive this from inequality 9.1 and the discussion of the paragraph following that inequality. Indeed, let ψ be a normalized state vector and let $\mathsf{A} = \mathsf{P} - \langle \mathsf{P} \rangle_\psi 1$ and $\mathsf{B} = \mathsf{X} - \langle \mathsf{X} \rangle_\psi 1$. A straightforward calculation, using the linearity of the expected value function $\langle - \rangle_\psi$ and the easily verified fact that $\langle \mathsf{A} \rangle_\psi = 0$, gives the variance $\nu_\psi(\mathsf{P}) = (\Delta_\psi \mathsf{P})^2 = \langle \mathsf{A}^2 \rangle_\psi$, and, similarly, $\nu_\psi(\mathsf{X}) = (\Delta_\psi \mathsf{X})^2 = \langle \mathsf{B}^2 \rangle_\psi$. This, of course, will give the right-hand side of inequality 9.1, and for the left-hand side we calculate easily that $[\mathsf{A}, \mathsf{B}] = [\mathsf{P}, \mathsf{X}] = -\mathbf{i}\hbar 1$. Putting the pieces together yields

$$\begin{aligned}(\Delta_\psi \mathsf{P})^2 \, (\Delta_\psi \mathsf{X})^2 &= \langle \mathsf{A}^2 \rangle_\psi \langle \mathsf{B}^2 \rangle_\psi \\ &\geq \frac{1}{4} \langle \mathbf{i}[\mathsf{A}, \mathsf{B}] \rangle_\psi^2 = \frac{1}{4} \langle \hbar 1 \rangle_\psi^2 = \frac{\hbar^2}{4} \langle \psi | \psi \rangle^2 = \frac{\hbar^2}{4},\end{aligned}$$

or $\Delta_\psi \mathsf{P} \, \Delta_\psi \mathsf{X} \geq \hbar/2$, the uncertainty relation of Heisenberg.

According to the discussion right after inequality 9.1, equality holds in the Heisenberg relation precisely when either $\mathsf{B}\psi = 0$ or there exists a real number s such that $(\mathsf{A} - \mathbf{i}s\mathsf{B})\psi = 0$. As the function $\mathsf{B}\psi(x) = (x - \langle \mathsf{X} \rangle_\psi)\psi(x)$ cannot be zero almost everywhere if ψ is normalized, we may dispense with the first possibility and assume the second. This second possibility implies that the criterion for equality is that

$$(\mathsf{P} - \mathbf{i}s\mathsf{X})\psi = (\langle \mathsf{P} \rangle_\psi - \mathbf{i}s\langle \mathsf{X} \rangle_\psi)\psi. \tag{9.3}$$

This is an eigenvalue problem of the form $(\mathsf{P} - \mathbf{i}s\mathsf{X})\psi = \lambda\psi$, where $\lambda = \langle \mathsf{P} \rangle_\psi - \mathbf{i}s\langle \mathsf{X} \rangle_\psi$.

This general eigenvalue problem is easy to solve and may be written as

$$-\mathbf{i}\hbar\psi'(x) = (\lambda + \mathbf{i}sx)\psi(x),$$

which integrates immediately to give

$$\log \psi(x) = \frac{\mathbf{i}}{\hbar}\left(\frac{\mathbf{i}sx^2}{2} + \lambda x\right) + \beta$$

for some constant β and branch of the logarithm. By completing the square, we get

$$\log \psi(x) = -\frac{s}{2\hbar}(x - \mu)^2 + \alpha,$$

where $\mu = \mathbf{i}\lambda/s$ and α is an appropriate complex constant. Setting $C = e^{\alpha}$ gives the solution

$$\psi(x) = Ce^{-s(x-\mu)^2/2\hbar}.$$

Note that ψ is in $L^2(\mathbb{R})$ if and only if $s > 0$ and, assuming this, we may assume also that β can be chosen so that $\|\psi\| = 1$. This shows that, when equality holds in the uncertainty relation, ψ must take the form claimed.

Conversely, the form claimed gives equality follows when one calculates the expectation values for position and momentum. Indeed, if we set $\mu = a + ib$ then straightforward calculations show that $\langle \mathsf{X} \rangle_\psi = a$, $\langle \mathsf{P} \rangle_\psi = sb$, and $(\mathsf{P} - \mathbf{i}s\mathsf{X})\psi = -\mathbf{i}s\mu\psi$. It then follows that $-\mathbf{i}s\mu = \langle \mathsf{P} \rangle_\psi - \mathbf{i}s\langle \mathsf{X} \rangle_\psi$, so that equation 9.3 holds, as required.

9.2 Interpretation

Much has been written about the Heisenberg uncertainty principle and the limitations it places on the measurability of the position and momentum of a quantum particle. Unfortunately, misconceptions have arisen that suggest that these limitations are somehow related to the disturbance that a position measurement, for instance, imposes upon the system that is just large enough to cause an uncertainty in a subsequent momentum measurement that yields the Heisenberg relation when the uncertainty in position is taken into account. This "disturbance theory" of uncertainty is not what the quantum formalism says. In the early days of the development of quantum mechanics, the principal architects of the formalism proposed such an explanation for Heisenberg's insight, but as the formalism and orthodox interpretation solidified, this disturbance theory was abandoned as untenable and inconsistent with, for example, the completeness hypothesis of quantum mechanics. Unfortunately, the disturbance theory seems to have a long half-life in the popular mind, long after the recognition that it is unwarranted by the formalism.

What exactly then does the Heisenberg Principle mean? To answer this question, first recall that the dispersion $\Delta_\psi \mathsf{A}$ of the observable a represented by A plays the role of the standard deviation in probability theory. This is understood in the orthodox interpretation to mean that, when a-measurements are performed on numerous systems similarly prepared in the state $\mathfrak{S}$ and represented by the state vector ψ, the statistical distribution of the resulting instrument readings will have a calculated standard deviation that converges to $\Delta_\psi \mathsf{A}$ in the limit of infinitely many measurements. This interpretation of the dispersion as the standard deviation of the statistical distribution of a-measurements does not follow logically from the formalism but is suggested by analogy with classical probability theory, as outlined in Lecture 3. Whatever one may think of the theoretical merits of this interpretation, the fact is that this interpretation is backed up by 80 years of successful experimental verification.

With this understanding of the meaning of the dispersion of an observable, we understand the Heisenberg principle, in practical terms, as follows: it asserts that if numerous systems are similarly prepared in a state represented by ψ and then measured for either position or momentum then the statistical distributions of the measurements will have experimental standard deviations agreeing with the dispersions $\Delta_\psi \mathsf{X}$ and $\Delta_\psi \mathsf{P}$ in the limit of infinitely many measurements, and the product of these dispersions will be found to be of order at least $\hbar/2$. This suggests that a quantum particle described by the state vector ψ possesses, in some sense, an uncertain position and an uncertain momentum, to the degree that the Heisenberg relation asserts, and the more localized the particle or certain its position, the less certain its momentum. Of course, this last statement is subject to philosophical criticism, and one who accepts the completeness of quantum mechanics might have to assert more, viz., that it is not meaningful to refer to the position or momentum of a quantum particle before measurement, as the wave function ψ is an eigenfunction of neither the position nor momentum operator. All one can say is that ψ offers a probability amplitude distribution that allows us to compute probabilities that the position, and through the Fourier transform the momentum, when measured will be found to lie in various intervals. Despite the fact that physicists may speak of position and momentum as if quantum particles possess these attributes even when not measured, when pressed philosophically they will generally assert the orthodox view with its completeness hypothesis.

The Heisenberg principle really is a profound insight into the working of nature. There is nothing like it in classical mechanics where, in principle if not in practice, various attributes can be measured simultaneously with arbitrary accuracy. In practical terms, if numerous systems are prepared in the same classical state, which is characterized by exact position and momentum values, and either a position or momentum measurement is performed on each state, then in principle the experimental statistical dispersions of the measurements may be made arbitrarily small. There is nothing in the theoretical edifice of classical mechanics that forbids this, and the product of dispersions is bounded below only by the ingenuity of the experimentalists. Though we have stated the principle in its original form as an assertion about position and momentum measurements, its derivation from inequality 9.1 immediately generalizes. The **general uncertainty principle** states that, for operators A and B representing respectively observables a and b, the product of the dispersions satisfies

$$\Delta_\psi \mathsf{A}\, \Delta_\psi \mathsf{B} \geq \frac{1}{2}|\langle \mathbf{i}[\mathsf{A}, \mathsf{B}]\rangle_\psi|.$$

When $[\mathsf{A}, \mathsf{B}] \neq 0$ we say that the observables a and b are **incompatible**, and, otherwise, when A and B commute, they are **compatible**. In particular, the uncertainty relation places no limitations on the simultaneous magnitudes of the dispersions of two compatible observables, but limits the smallness of the dispersion of one in relation to that of the other when the observables are incompatible and the state ψ has nonzero expectation with respect to $\mathbf{i}[\mathsf{A}, \mathsf{B}]$.

Another articulation of the uncertainty relation is that it places a fundamental limit on not only the preparation but even the existence of quantum states with specified combinations of attributes. Indeed, the position–momentum relation is really a statement that no quantum system may be prepared that has statistical dispersions for position and momentum that violate the Heisenberg relation. More generally, when the operators A and B represent incompatible observables, so that the commutator $[\mathsf{A}, \mathsf{B}] \neq 0$, no system may be prepared that has statistical dispersions for A and B arbitrarily small simultaneously unless its state vector satisfies $\langle \mathbf{i}[\mathsf{A}, \mathsf{B}]\rangle_\psi = 0$.[2] Any theory from which quantum mechanics emerges, but for which systems actually possess exact values for the observables a and b before measurement, must come to grips with the limitation suggested by the uncertainty relations and must explain how the statistical spreads arise when the system possesses exact values at all times. To warn the reader that this is not an impossible task, Bohmian mechanics, developed for example in Holland (1993), accomplishes precisely this.

[2] It sometimes is not recognized that these uncertainty relations depend not only on the observables, but also on the state vector ψ. When $\mathsf{A} = \mathsf{X}$ and $\mathsf{B} = \mathsf{P}$, then $\langle \mathbf{i}[\mathsf{A}, \mathsf{B}]\rangle_\psi$ is never equal to 0 since $[\mathsf{X}, \mathsf{P}] = \mathbf{i}\hbar\mathbb{1}$. Moreover, this implies that $\langle \mathbf{i}[\mathsf{A}, \mathsf{B}]\rangle_\psi = \hbar$, i.e., $\langle \mathbf{i}[\mathsf{A}, \mathsf{B}]\rangle_\psi$ is constant in ψ. In general though, when A and B are incompatible, the expectation $\langle \mathbf{i}[\mathsf{A}, \mathsf{B}]\rangle_\psi$ may be zero for some states ψ, and nonzero for others and, when nonzero, the expectation $\langle \mathbf{i}[\mathsf{A}, \mathsf{B}]\rangle_\psi$ and therefore the lower bound in the uncertainty may vary with ψ.

9.3 Minimal-Uncertainty States

The states identified as giving equality in the Heisenberg relation are called **minimal-uncertainty states**. Using that

$$\lambda = \langle \mathsf{P} \rangle_\psi - \mathbf{i}s\langle \mathsf{X} \rangle_\psi = -\mathbf{i}s\mu$$

and applying a modest amount of algebraic manipulation allows one to write

$$\psi(x) = C' e^{\mathbf{i}x\langle \mathsf{P} \rangle_\psi/\hbar} e^{-s(x-\langle \mathsf{X} \rangle_\psi)^2/2\hbar}, \tag{9.4}$$

where C' is a normalization constant which may be taken to be real. This expression is a complex Gaussian. The unimodular factor, $u(x) = \exp(\mathbf{i}x\langle \mathsf{P} \rangle_\psi/\hbar)$, is a **phase factor** that depends on x, and the remaining factors, $g(x) = C' \exp(-s(x - \langle \mathsf{X} \rangle_\psi)^2/2\hbar)$, give the position probability distribution as

$$|\psi(x)|^2 = C'^2 \exp\left(-\frac{s(x - \langle \mathsf{X} \rangle_\psi)^2}{\hbar}\right).$$

This is a Gaussian distribution with mean $\langle \mathsf{X} \rangle_\psi$ and standard deviation $\sqrt{\hbar/2s}$. That ψ is normalized implies that $C' = \sqrt[4]{s/\pi\hbar}$.[3] A calculation then reveals that the dispersion $\Delta_\psi \mathsf{X}$ is equal to the classical standard deviation $\sqrt{\hbar/2s}$ of this Gaussian distribution, again justifying the interpretation of the dispersion $\Delta_\psi \mathsf{X}$ as the standard deviation of the statistical spread of position measurements. Another calculation gives the dispersion $\Delta_\psi \mathsf{P}$ as $\sqrt{s\hbar/2}$, confirming that the product of the two dispersions equals $\hbar/2$, as required.

When $\langle \mathsf{P} \rangle_\psi = 0$, the state ψ represents a particle in equilibrium about the point $\mu = \langle \mathsf{X} \rangle_\psi$. Set $s = m\omega$ for positive constants m and ω, and take $\mu = 0$ to get the minimal uncertainty state wave function

$$\psi_0(x) = \left(\frac{m\omega}{\pi\hbar}\right)^{1/4} \exp\left(-\frac{m\omega x^2}{2\hbar}\right), \tag{9.5}$$

representing a particle in equilibrium about the origin, with expected momentum zero and expected position coordinate $x = 0$. The reader may recall from Lecture 1 that this is the $N = 0$ eigenfunction, the **ground state**, of the harmonic oscillator, with energy $E_0 = \hbar\omega/2$ for a particle of mass m in the quadratic potential well $V(x) = m\omega^2 x^2/2$.

[3] The standard form of the Gaussian distribution in probability theory is

$$G(x) = \frac{1}{\sigma\sqrt{2\pi}} e^{-(x-\mu)^2/2\sigma^2},$$

where μ is the mean and σ the standard deviation.

This is no coincidence. Indeed, recall in the derivation of the form of the minimal uncertainty states, from equation 9.3, that ψ_0 is a solution of the eigenvalue equation $(\mathsf{P} - \mathbf{i}s\mathsf{X})\psi = \lambda\psi$, where $\lambda = \langle\mathsf{P}\rangle_\psi - \mathbf{i}s\langle\mathsf{X}\rangle_\psi$. With $\psi = \psi_0$, $\lambda = 0$ so that $(\mathsf{P} - \mathbf{i}s\mathsf{X})\psi_0 = 0$. Applying the adjoint $(\mathsf{P} - \mathbf{i}s\mathsf{X})^* = (\mathsf{P} + \mathbf{i}s\mathsf{X})$ to this gives

$$0 = (\mathsf{P} + \mathbf{i}s\mathsf{X})(\mathsf{P} - \mathbf{i}s\mathsf{X})\psi_0 = (\mathsf{P}^2 + s^2\mathsf{X}^2)\psi_0 - \mathbf{i}s[\mathsf{P},\mathsf{X}]\psi_0.$$

Using $[\mathsf{P},\mathsf{X}] = -\mathbf{i}\hbar\mathbb{1}$, multiplying by $1/2m$, and replacing s with $m\omega$ gives

$$\left(\frac{\mathsf{P}^2}{2m} + \frac{1}{2}m\omega^2\mathsf{X}^2\right)\psi_0 = \frac{1}{2}\hbar\omega\psi_0,$$

or

$$-\frac{\hbar}{2m}\psi_0'' + \frac{1}{2}m\omega^2x^2\psi_0 = \frac{1}{2}\hbar\omega\psi_0,$$

which is recognized as the time-independent Schrödinger equation for the harmonic oscillator ground state.

In Lecture 11 we will study the general solution to the Schrödinger equation for the harmonic oscillator potential, where we will make use of the operators

$$\mathsf{a}_- = \mathsf{P} - \mathbf{i}m\omega\mathsf{X} \quad \text{and} \quad \mathsf{a}_+ = \mathsf{P} + \mathbf{i}m\omega\mathsf{X}$$

to find not only the ground state ψ_0 but all the eigenstates of the harmonic oscillator in an elegant, algebraic, manner. These operators are called **ladder operators**, with a_- the **annihilation operator** and a_+ the **creation operator**.

9.4 The Fourier Transform and Uncertainty

An alternative way in which to understand the position–momentum uncertainty relations is in terms of the Fourier transform. Recall that, while the square modulus $|\psi(x)|^2$ of the state vector ψ gives the position measurement probability distribution for the quantum particle, the square modulus $\left|\widehat{\psi}(p)\right|^2$ of the Fourier transform $\widehat{\psi}$ gives the momentum measurement probability distribution. The Fourier transform on appropriately integrable functions is given by

$$\mathcal{F}(\psi)(p) = \widehat{\psi}(p) = \frac{1}{\sqrt{2\pi\hbar}}\int_{\mathbb{R}} \psi(x)e^{-\mathbf{i}px/\hbar}\,dx.$$

In particular, as the Fourier transform of a Gaussian is a Gaussian,[4] we may read off directly the standard deviation, and therefore the dispersion $\Delta_\psi \mathsf{P}$, from the form that the Gaussian $\mathcal{F}(\psi)$ takes, when ψ is a minimal uncertainty state. A calculation of the Fourier transform of the minimal uncertainty state ψ of equation 9.4 gives

$$\widehat{\psi}(p) = e^{\mathrm{i}\theta} \frac{C'}{\sqrt{s}} e^{-\mathrm{i}p\langle \mathsf{X}\rangle_\psi/\hbar} e^{-(p-\langle \mathsf{P}\rangle_\psi)^2/2s\hbar}, \tag{9.6}$$

where $\theta = \langle \mathsf{X}\rangle_\psi \langle \mathsf{P}\rangle_\psi/\hbar$. The momentum measurement probability distribution $\left|\widehat{\psi}(p)\right|^2$ is then the Gaussian distribution

$$\begin{aligned}\left|\widehat{\psi}(p)\right|^2 &= \frac{C'^2}{s} \exp\left(-\frac{(p-\langle \mathsf{P}\rangle_\psi)^2}{s\hbar}\right) \\ &= \frac{1}{\sqrt{s\hbar\pi}} \exp\left(-\frac{(p-\langle \mathsf{P}\rangle_\psi)^2}{s\hbar}\right).\end{aligned}$$

A comparison of this with the standard form for the Gaussian distribution reported in the footnote in the previous section reveals that the mean is $\langle \mathsf{P}\rangle_\psi$ with standard deviation $\sqrt{s\hbar/2}$, confirming again the results of the preceding section.

In general, when ψ is not necessarily a minimal uncertainty state, it is still true that the position and momentum measurement probability distributions are given by $|\psi(x)|^2$ and $\left|\widehat{\psi}(p)\right|^2$, respectively. The uncertainty relation may then be seen as an instantiation of a famous result in the theory of the Fourier transform that implies that the more concentrated a function, the more spread out its Fourier transform. The exact result is known sometimes as the **Fourier uncertainty principle** and implies that when f is a probability amplitude, meaning that $|f|^2$ is a probability distribution, with Fourier transform given by $F = \mathcal{F}(f)$, which by the Plancherel theorem is also a probability amplitude, then the product of the standard deviations of $|f(x)|^2$ and $|F(p)|^2$ is at least $\frac{1}{2}$. The Heisenberg principle follows immediately when it is observed just from the definitions that the dispersion $\Delta_\psi \mathsf{X}$ is precisely the standard deviation of the probability distribution $|\psi(x)|^2$, and the dispersion $\Delta_\psi \mathsf{P}$ is precisely the standard deviation of the probability distribution $\left|\widehat{\psi}(p)\right|^2$, along with the observation that the factor $\hbar$ arises because our Fourier transform is slightly modified from the usual by insertion of the $\hbar$-terms.

The next lecture probes the Fourier transform and its intimate connection to wave mechanics.

[4] For details see the next lecture.

9.5 End Notes

The approach I have taken in Section 9.1 to develop the Heisenberg uncertainty principle was inspired by the first two sections of Chapter 7 of Hannabuss (1997). The interpretation I present in Section 9.2 is my take on what the mathematics of the uncertainty principle implies. Most explanations I have seen suggest that the Heisenberg principle forbids the simultaneous arbitrarily precise measurement of both the position and the momentum of a quantum particle represented by state vector ψ – the more accurate the measurement of position, the less accurate the measurement of momentum. While it perhaps does suggest this, it seems to me that it implies so much more, viz., that repeated separate single measurements, each measurement highly precise and in each case either a position or a momentum measurement, made serially on a large number of quantum particles that are prepared in exactly the same state represented by ψ, cannot pinpoint the position and momentum of a quantum particle in that state to an accuracy greater than that allowed by inequality 9.2. A distribution of measurement results occurs even though each measurement in theory may be made as precisely as desired, and the experimental distribution of results used to calculate standard deviations always satisfies the Heisenberg principle in the limit of infinitely many measurements. This suggests either that those attributes – position and momentum – of a quantum particle represented by ψ are in some way fuzzy, or that position and momentum are attributes not enjoyed by the quantum object until the very point of measurement of those attributes. For one who accepts the completeness hypothesis for quantum mechanics, this says that, for the position attribute for instance, we should not think of a quantum particle as a point particle of classical mechanics that at all times has a definite position in space, but as an object that in some unknown sense is spread out across space and materializes to a more localized object upon interaction with the position measuring instrument. This discussion again highlights the measurement problem for quantum mechanics. What exactly constitutes a measurement? What about it localizes a quantum object – what is the mechanism of localization? Shouldn't the measuring instrument be covered by quantum mechanics, and if so, how does the quantum object's interaction with another (admittedly rather large) quantum object force a nonunitary evolution that departs from the Schrödinger evolution governed by his equation? How does the act of measurement – so mundane and pedestrian in classical mechanics – become so central to a description of the micro-world?

The attempts to understand better what the quantum formalism says about the ontological nature of a quantum object is one of the major struggles in the history of the development of quantum mechanics. Though we have uncovered the mechanics that underlies the deep reality of the world, the very nature of this deep reality still eludes us.

10

Wave Mechanics II: The Fourier Transform

> The equivalence of Heisenberg's and Schrödinger's theories, however, was soon established.
>
> The whole development gave Hilbert "a great laugh," according to [E.U.] Condon:
>
> > "... when [Born and Heisenberg and the Göttingen theoretical physicists] first discovered matrix mechanics they were having, of course, the same kind of trouble that everybody else had in trying to solve problems and to manipulate and really do things with matrices. So they had gone to Hilbert for help and Hilbert said the only times he had ever had anything to do with matrices was when they came up as a sort of by-product of the eigenvalues of the boundary-value problem of a differential equation. So if you look for the differential equation which has these matrices you can probably do more with that. They thought it was a goofy idea and Hilbert didn't know what he was talking about. So he was having a lot of fun pointing out to them that they could have discovered Schrödinger's wave mechanics six months earlier if they had paid a little more attention to him."
>
> Constance Reid
> *Hilbert*, 1978

Schrödinger's wave mechanics is the model interpretation of the abstract axioms of quantum mechanics presented in Lecture 2. So far in these lectures we have restricted our attention to one-dimensional wave mechanics, which describes the motion of a quantum particle along a line. In this lecture we develop wave mechanics more fully for a quantum particle adrift in three-dimensional space and look carefully at the Fourier transform and its mapping between the "position" and "momentum" representations of the wave function. We have mentioned already in some preceding lectures the importance of the Fourier transform in moving between the position and momentum probability distributions. The first task of the present lecture is to acquaint the reader with a deeper understanding of the Fourier transform, and so we begin with a mathematical section that develops the transform quickly and catalogues its important properties. This is followed by an examination of the eigenvalues and eigenfunctions of the Fourier transform, which leads to a brief glance at an important class of special functions called the Hermite functions. Next come the applications to quantum mechanics, with a more careful study of the two premier observable operators of wave mechanics – the position and momentum operators – and their relationship under the transform. In particular, we examine in greater detail the mathematical symmetry between position and momentum, in the way that wave functions can be represented as probability amplitudes for either position or momentum measurements, and how the Fourier transform maps between these representations by a unitary mapping of the state space. Finally, we examine the construction of wave packets by the superposition of plane waves and see how the Fourier transform arises as a continuum limit of Fourier series.

10.1 The Fourier Transform

We will give a quick outline and follow with details. We first define the Fourier transform on the Schwartz space $\mathcal{S}(\mathbb{R}^n)$ of C^∞ functions of rapid decrease. This is done with the usual integral definition that physicists use, which, however, is inadequate for defining the Fourier transform for an arbitrary square-integrable function. To remedy this, after establishing some salient features of the Fourier transform on $\mathcal{S}(\mathbb{R}^n)$, we extend the definition to the space of tempered distributions $\mathcal{S}'(\mathbb{R}^n)$ (see below) and examine its important properties. Since

$$\mathcal{S}(\mathbb{R}^n) \subset L^2(\mathbb{R}^n) \subset \mathcal{S}'(\mathbb{R}^n), \tag{10.1}$$

this allows us to define the Fourier transform on the Hilbert space $L^2(\mathbb{R}^n)$ and tease out its significant and useful properties. We will see, for example, that the Fourier transform is an isometry of $L^2(\mathbb{R}^n)$ onto itself with the four infinitely degenerate eigenvalues ± 1 and $\pm \mathbf{i}$.

A complex-valued C^∞ function $\psi : \mathbb{R}^n \to \mathbb{C}$ is in the **Schwartz space** $\mathcal{S}(\mathbb{R}^n)$, provided that all its partial derivatives decrease to zero in modulus as $|x| \to \infty$ faster than the reciprocal of any polynomial in the variables $x = (x_1, \ldots, x_n)$.[1] To develop this more precisely, define the partial differential operator D^k by

$$D^k = \frac{\partial^{|k|}}{\partial_1^{k_1} \cdots \partial_n^{k_n}} = \partial_1^{k_1} \cdots \partial_n^{k_n}.$$

Here, $k = (k_1, \ldots, k_n)$ is a **multi-index**, with each k_i a nonnegative integer and $|k| = \sum_{i=1}^n k_i$. We define, for the complex-valued C^∞ function ψ and for each pair of multi-indices j and k,

$$\|\psi\|_{j,k} = \sup_{x \in \mathbb{R}^n} |x^j D^k \psi(x)|.$$

The notation x^k means $x_1^{k_1} \cdots x_n^{k_n}$. The Schwartz space $\mathcal{S}(\mathbb{R}^n)$ is then the set of all such ψ, called **test functions**, for which $\|\psi\|_{j,k} < \infty$ for all multi-indices j and k. Aside from the C^∞ functions of compact support, the Schwartz space also contains the n-dimensional Gaussian functions and, hence, the wave functions describing the minimal uncertainty states, as well as the products of Gaussians with polynomial functions. The natural topology on $\mathcal{S}(\mathbb{R}^n)$ is the weakest topology in which vector addition as well as all the semi-norms $\| - \|_{j,k}$ are continuous, and this makes the Schwartz space into a Fréchet space, a complete metrizable locally convex topological vector space.

We need not concern ourselves too much with the properties of this topology. We mention it only to define the space of **tempered distributions**, $\mathcal{S}'(\mathbb{R}^n)$, the topological dual space of the Schwartz space. This is just the space of *continuous* linear functionals defined on $\mathcal{S}(\mathbb{R}^n)$, and the topology is needed primarily in reference to the continuity of the linear functionals. The upshot is that the linear functional $T : \mathcal{S}(\mathbb{R}^n) \to \mathbb{C}$ is continuous provided there is a constant C and nonnegative integers J and K such that, for all $\psi \in \mathcal{S}(\mathbb{R}^n)$,

$$|T(\psi)| < C \sum_{\substack{|j| \le J \\ |k| \le K}} \|\psi\|_{j,k}.$$

An important case that will be used in the paragraph following is that

[1] In the standard mathematical convention used in this lecture, the unadorned variable x is used for an element of $\mathbb{R}^n$ with subscripts used to denote coordinates. When $n = 3$, we normally use the standard physicists' convention and write an element of $\mathbb{R}^3$ in bold-face type as $\mathbf{x}$. We are hardly consistent in our use, though, as we revert to bold-face type for elements of $\mathbb{R}^n$ in our discussion of the Feynman integral, primarily because we think of elements of $\mathbb{R}^n$ as column vectors and use matrix manipulations.

the continuity of T is guaranteed whenever there is a constant C such that $|T(\psi)| < C\|\psi\|_{0,0} = C\|\psi\|_\infty$, where $\|-\|_\infty$ is the usual sup-norm.

We now describe the natural inclusion of $\mathcal{S}(\mathbb{R}^n)$ into $\mathcal{S}'(\mathbb{R}^n)$. This begins with the observation that each test function ψ is square-integrable, so that $\mathcal{S}(\mathbb{R}^n) \subset L^2(\mathbb{R}^n)$.[2] It follows that, for any test functions φ and ψ, the L^2 inner product $\langle\overline{\varphi}|\psi\rangle = \int_{\mathbb{R}^n} \varphi\psi$ is finite and, hence, with φ fixed, the function T_φ defined by the prescription $\psi \mapsto \langle\overline{\varphi}|\psi\rangle$ defines a linear functional of the Schwartz space into the complex field. Observing also that each test function is integrable and setting $C = \int_{\mathbb{R}^n} |\varphi|$, we see that the functional T_φ is continuous since $|T_\varphi(\psi)| = |\langle\overline{\varphi}|\psi\rangle| \leq C\|\psi\|_\infty$. Using the Dirac bra-ket notation, this functional may be denoted alternatively as $T_\varphi = \langle\overline{\varphi}|$.[3] The prescription $\varphi \mapsto T_\varphi = \langle\overline{\varphi}|$ defines a linear inclusion[4] of the Schwartz space of test functions into the space of tempered distributions, and this allows us to write the containment $\mathcal{S}(\mathbb{R}^n) \subset \mathcal{S}'(\mathbb{R}^n)$.

We now are in a position to define the Fourier transform on the Schwartz space $\mathcal{S}(\mathbb{R}^n)$ and then to extend the definition to its dual $\mathcal{S}'(\mathbb{R}^n)$. For the function $\psi \in \mathcal{S}(\mathbb{R}^n)$, the **Fourier transform** $\mathcal{F}(\psi) = \widehat{\psi}$ and the **inverse Fourier transform** $\mathcal{F}^{-1}(\psi) = \widetilde{\psi}$ are the functions defined as

$$\mathcal{F}(\psi)(p) = \widehat{\psi}(p) = \frac{1}{(2\pi)^{n/2}} \int_{\mathbb{R}^n} e^{-\mathbf{i}p\cdot x}\psi(x)\, dx \tag{10.2}$$

and

$$\mathcal{F}^{-1}(\psi)(x) = \widetilde{\psi}(x) = \frac{1}{(2\pi)^{n/2}} \int_{\mathbb{R}^n} e^{\mathbf{i}x\cdot p}\psi(p)\, dp. \tag{10.3}$$

Here, $x \cdot p$ is just the Euclidean inner product $x_1p_1 + \cdots + x_np_n$. Of course, these integrals exists since every test function is in $L^1(\mathbb{R}^n)$.

Our aim is to review the most important of the properties of the Fourier transform and its inverse without proof, directing the reader to the End Notes of this lecture for good references. First, the **Fourier inversion theorem** asserts that the Fourier transform is a linear bicontinuous bijection of $\mathcal{S}(\mathbb{R}^n)$ onto $\mathcal{S}(\mathbb{R}^n)$ whose inverse mapping is the

[2] In fact, $\mathcal{S}(\mathbb{R}^n)$ is dense in $L^2(\mathbb{R}^n)$ and, moreover, the identity mapping of $\mathcal{S}(\mathbb{R}^n)$ into $L^2(\mathbb{R}^n)$ is continuous, though we will not need this fact here.

[3] Note that this is not the same as $\overline{\langle\varphi|} = |\varphi\rangle$; indeed, $\langle\overline{\varphi}|$ is a linear functional while $\overline{\langle\varphi|}$ is a conjugate-linear functional.

[4] The reader should note the slight difference from the dual correspondence of the preceding lectures, where φ corresponds to $\varphi^* = \langle\varphi|$. This previous correspondence (used by physicists) is conjugate-linear while the correspondence considered here (used by mathematicians) is linear and, in fact, is necessary for the cleanest method of extending the Fourier transform from $\mathcal{S}(\mathbb{R}^n)$ to $L^2(\mathbb{R}^n)$.

inverse Fourier transform and, moreover, preserves the L^2-norm. This latter assertion means that, for any $\psi \in \mathcal{S}(\mathbb{R}^n)$,

$$\|\psi\|_2^2 = \int_{\mathbb{R}^n} |\psi|^2 = \int_{\mathbb{R}^n} |\widehat{\psi}|^2 = \|\widehat{\psi}\|_2^2. \tag{10.4}$$

As a consequence of the inversion theorem, we have

$$\begin{aligned} \psi(x) = \widetilde{\widehat{\psi}}(x) &= \frac{1}{(2\pi)^n} \int_{\mathbb{R}^n} e^{\mathbf{i}x\cdot p} \widehat{\psi}(p)\, dp \\ &= \frac{1}{(2\pi)^n} \int_{\mathbb{R}^n} \int_{\mathbb{R}^n} e^{\mathbf{i}(x-y)\cdot p} \psi(y)\, dy\, dp, \end{aligned} \tag{10.5}$$

Replacing $x - y$ by $y - x$, one obtains an integral formula for $\widehat{\widehat{\psi}} = \psi$. Be careful here. The latter expression in this formula is not an integral over the space $\mathbb{R}^n \times \mathbb{R}^n$; rather, it is an iterated integral, and the order of integration cannot be reversed since it is not absolutely convergent.

The Fourier transform behaves particularly well with derivatives, and this is one of the facts that makes the transform so valuable. Indeed, if j and k are multi-indices then

$$\mathcal{F}(D^j((-\mathbf{i}x)^k \psi(x)))(p) = (\mathbf{i}p)^j (D^k \mathcal{F}(\psi))(p). \tag{10.6}$$

This says, in some loose sense, that the Fourier transform exchanges the taking of derivatives with multiplication by powers of the independent variable. For example, when $j = (0, \ldots, 1, \ldots, 0)$ has a nonzero entry only in the ith slot and $k = (0, \ldots, 0)$ has all zero entries, this becomes

$$\widehat{\frac{\partial \psi}{\partial x_i}} = \mathbf{i} p_i \widehat{\psi}. \tag{10.7}$$

The extension of the Fourier transform to the space of tempered distributions is rather straightforward, though the fact that it does indeed define an *extension* of $\mathcal{F}$ on $\mathcal{S}(\mathbb{R}^n)$ to the larger domain $\mathcal{S}'(\mathbb{R}^n)$ requires some explanation. First the definition: the Fourier transform $\widehat{T}$ of the tempered distribution $T \in \mathcal{S}'(\mathbb{R}^n)$ is defined by

$$\widehat{T}(\psi) = T(\widehat{\psi}) \quad \text{for all} \quad \psi \in \mathcal{S}(\mathbb{R}^n). \tag{10.8}$$

Thus, $\widehat{T} = T \circ \mathcal{F}$. Of course, this makes sense as an extension of the Fourier transform on $\mathcal{S}(\mathbb{R}^n)$ only if $\widehat{T_\varphi} = T_{\widehat{\varphi}}$ for each $\varphi \in \mathcal{S}(\mathbb{R}^n)$.[5] In

[5] This is precisely why we need to use the linear mapping $\varphi \mapsto \langle\overline{\varphi}|$ rather than the conjugate-linear mapping $\varphi \mapsto \langle\varphi|$ for the inclusion of $\mathcal{S}(\mathbb{R}^n)$ into $\mathcal{S}'(\mathbb{R}^n)$. Had we used the latter, we would have obtained $\widehat{T_\varphi} = T_{\widetilde{\varphi}}$, as can be seen by the argument given here for $\widehat{T_\varphi} = T_{\widehat{\varphi}}$.

Dirac notation this condition becomes

$$\langle \overline{\varphi} \mid \widehat{\psi} \rangle = \langle \overline{\widehat{\varphi}} \mid \psi \rangle,$$

and, in terms of integrals,

$$\int_{\mathbb{R}^n} \varphi \widehat{\psi} = \int_{\mathbb{R}^n} \widehat{\varphi} \psi.$$

To confirm this, note that $\widehat{\overline{\varphi}} = \overline{\widetilde{\varphi}}$, by observation, $\widehat{\widetilde{\varphi}} = \varphi$ from the Fourier inversion theorem, and $\langle \widehat{\varphi} \mid \widehat{\psi} \rangle = \langle \varphi | \psi \rangle$ from equation 10.4 and the polarization identities.[6] Putting these together gives

$$\langle \overline{\varphi} \mid \widehat{\psi} \rangle = \left\langle \widehat{\widetilde{\overline{\varphi}}} \,\middle|\, \widehat{\psi} \right\rangle = \langle \overline{\widetilde{\varphi}} \mid \psi \rangle = \langle \overline{\widehat{\varphi}} \mid \psi \rangle.$$

It follows that $\widehat{T_\varphi} = T_{\widehat{\varphi}}$ and so expression 10.8 defines an extension of the Fourier transform, as claimed.

We are now in a position to examine the Fourier transforms of square-integrable functions. Recall first that the topological dual space of the Hilbert space $L^2(\mathbb{R}^n)$ is just $L^2(\mathbb{R}^n)$ itself, under the dual correspondence $\psi \longleftrightarrow \langle \overline{\psi} |$. It follows that the restriction of $T_\psi = \langle \overline{\psi} |$ to $\mathcal{S}(\mathbb{R}^n)$ is a tempered distribution, so that the containments of expression 10.1 hold whenever $L^2(\mathbb{R}^n)$ is identified as its dual. We see then that the Fourier transform of a square-integrable function is defined through the dual correspondence as $\widehat{T_\psi} = T_\psi \circ \mathcal{F}$. The question of importance is whether there exists a square-integrable function φ such that $\widehat{T_\psi} = T_\varphi$. If so then $\widehat{\psi}$ will be defined as this φ. One might be tempted to use equation 10.2 to define such a φ, but of course the problem is that ψ, though square-integrable, may fail to be integrable and in that case the integral of equation 10.2 could diverge. This, though, is but a slight nuisance since the subset of integrable functions is dense in the Hilbert space $L^2(\mathbb{R}^n)$. Indeed, note first that the Fourier transform extends, using equation 10.2, from $\mathcal{S}(\mathbb{R}^n)$ to $L^1(\mathbb{R}^n)$; moreover, it turns out that equation 10.4 holds for functions that are both integrable and square integrable. Let χ_R be the characteristic function of the closed ball of

[6] The polarization identities for Hermitian inner products are given by

$$2\mathrm{Re}(\langle \varphi | \psi \rangle) = \langle \varphi + \psi | \varphi + \psi \rangle - \langle \varphi | \varphi \rangle - \langle \psi | \psi \rangle$$

and

$$-2\mathrm{Im}(\langle \varphi | \psi \rangle) = \langle \varphi + \mathbf{i}\psi | \varphi + \mathbf{i}\psi \rangle - \langle \varphi | \varphi \rangle - \langle \psi | \psi \rangle.$$

Thus, a map that preserves the squared norm $\langle \psi | \psi \rangle$ preserves the Hermitian form $\langle \varphi | \psi \rangle$, and vice-versa.

radius R centered at the origin:

$$\chi_R(x) = \begin{cases} 1 & |x| \leq R; \\ 0 & |x| > R. \end{cases}$$

Then $\chi_R\psi \in L^1(\mathbb{R}^n)$ so that $\widehat{\chi_R\psi}$ is given by the integral of equation 10.2 and, moreover, $\chi_R\psi$ converges to ψ in the L^2-norm as $R \to \infty$. Since equation 10.4 holds for the functions $\chi_R\psi$, this implies that the L^2-limit of $\widehat{\chi_R\psi}$ as $R \to \infty$ exists, and that this limit, called $\widehat{\psi}$, does indeed satisfy $\widehat{T_\psi} = T_{\widehat{\psi}}$. We record, then, using the notation $\lim^2$ to denote the limit in the L^2-norm, the formula for the Fourier transform of the square-integrable function ψ as

$$\widehat{\psi} = \frac{1}{(2\pi)^{n/2}} \lim{}^2_{R\to\infty} \int_{|x|\leq R} e^{-\mathrm{i}p\cdot x}\psi(x)\, dx. \tag{10.9}$$

It turns out that equation 10.4 holds also for square-integrable functions. These results come under the heading of the **Plancherel theorem**, which states further that $\mathcal{F} : L^2(\mathbb{R}^n) \to L^2(\mathbb{R}^n)$ is a unitary map of Hilbert spaces whose adjoint is the inverse Fourier transform $\mathcal{F}^* = \mathcal{F}^{-1} : L^2(\mathbb{R}^n) \to L^2(\mathbb{R}^n)$.

It is important to notice that formula 10.2 defines the Fourier transform $\widehat{\psi}$ pointwise and unambiguously for every p, whenever ψ is in $\mathcal{S}(\mathbb{R}^n)$ (or even in $L^1(\mathbb{R}^n)$). However, when ψ is in $L^2(\mathbb{R}^n)$, the Plancherel theorem defines $\widehat{\psi}$ as an element of $L^2(\mathbb{R}^n)$ and thus is defined only almost everywhere. Whether the integral of expression 10.9 converges pointwise as $R \to \infty$ is a rather difficult and delicate question. If $n = 1$ then the pointwise limit of the integral expression in equation 10.9 does exist almost everywhere and therefore defines $\widehat{\psi}$ as a pointwise function almost everywhere.[7] When $n > 1$, the pointwise almost-everywhere-convergence of the limit of formula 10.9 is still an open question, though if integration over n-balls of increasing radii is replaced by integration over n-cells of increasing side lengths then the almost-everywhere-convergence follows.

We need one more important feature of the Fourier transform, namely, its behavior under the multiplication of functions. The new ingredient is that of the **convolution** of two functions μ and ν. This is denoted as

[7] The fact that the integral formula converges for almost all p when $n = 1$ is exceptionally difficult to prove. It was conjectured by N.N. Luzin in 1915 and finally proved in 1966 by Lennart Carleson in the notoriously difficult paper "On convergence and growth of partial sums of Fourier series," Acta Math., 116: 135–157. There still are no easy proofs of this. See the End Notes of this lecture for more information.

$\mu * \nu$ and defined for suitably integrable functions by

$$\mu * \nu(y) = \int_{\mathbb{R}^n} \mu(y-x)\nu(x)\,dx = \int_{\mathbb{R}^n} \mu(x)\nu(y-x)\,dx = \nu * \mu(y),$$

the equality of the integrals being a consequence of the change of variables $x \to y - x$. This is defined, for example, whenever one of μ and ν is a test function and the growth of the other is polynomially bounded. It is also defined, obviously, when both functions μ and ν are square-integrable. Its importance for the Fourier transform is that the transform turns products into convolutions and vice versa. Precisely,

$$\widehat{\mu\nu} = \frac{1}{(2\pi)^{n/2}} \widehat{\mu} * \widehat{\nu} \quad \text{and} \quad \widehat{\mu * \nu} = (2\pi)^{n/2} \widehat{\mu}\,\widehat{\nu}. \tag{10.10}$$

10.2 Eigenvalues and Eigenfunctions: Hermite Functions

What are the eigenvalues and eigenfunctions of the Fourier transform on $L^2(\mathbb{R}^n)$? Since $\mathcal{F}$ is unitary, the eigenvalues all lie on the unit circle. In fact, we have

$$\begin{aligned}
\mathcal{F}^2(\psi)(x) &= \frac{1}{(2\pi)^{n/2}} \lim^2_{R\to\infty} \int_{|p|\le R} e^{-\mathrm{i}x\cdot p} \mathcal{F}(\psi)(p)\,dp \\
&= \frac{1}{(2\pi)^{n/2}} \lim^2_{R\to\infty} \int_{|p|\le R} e^{\mathrm{i}(-x)\cdot p} \mathcal{F}(\psi)(p)\,dp \\
&= \mathcal{F}^{-1}\mathcal{F}(\psi)(-x) = \psi(-x)
\end{aligned}$$

for almost all x in $\mathbb{R}^n$ so that, in $L^2(\mathbb{R}^n)$, we have $\mathcal{F}^4(\psi) = \psi$. It follows that if ψ is an eigenfunction of the Fourier transform with eigenvalue λ then

$$\psi = \mathcal{F}^4(\psi) = \lambda^4\psi,$$

so that $\lambda^4 = 1$, implying that $\lambda = \pm 1, \pm \mathrm{i}$.

What of the corresponding eigenfunctions? We will show that the eigenfunctions of the one-dimensional Fourier transform are the so-called **Hermite functions**, a class of special functions that find their natural place in quantum theory as the energy eigenfunctions corresponding to the harmonic oscillator potential. In this section we will introduce the Hermite functions and study their relationship to the Fourier transform. The application of the Hermite functions as energy eigenfunctions of the harmonic oscillator Hamiltonian awaits presentation for the next lecture, where we will also confirm their completeness for use as a basis in $L^2(\mathbb{R})$.

The **Hermite function** of **order** n, for $n = 0, 1, \ldots$, is

$$h_n(x) = H_n(x)e^{-x^2/2},$$

where $H_n(x)$ is the nth **Hermite polynomial**, defined by Rodrigues's formula

$$H_n(x) = (-1)^n e^{x^2} \frac{d^n}{dx^n} e^{-x^2};$$

H_n is a polynomial of degree n and, obviously, h_n is a member of the Schwartz space of test functions. Another expression for h_n is as follows:

$$h_n(x) = (-1)^n \left(\frac{d}{dx} - x\right)^n e^{-x^2/2} = (-1)^n \left(\frac{d}{dx} - x\right)^n h_0(x).$$

It is this expression that allows a transparent proof that the Hermite functions are eigenfunctions of the Fourier transform. The pertinent observation is that[8]

$$\mathcal{F} \circ \left(\frac{d}{dx} - x\right) = -\mathbf{i}\left(\frac{d}{dp} - p\right) \circ \mathcal{F},$$

which is verified rather easily using equation 10.6. This gives

$$\mathcal{F}(h_n)(p) = (-1)^n(-\mathbf{i})^n \left(\frac{d}{dp} - p\right)^n \mathcal{F}(h_0)(p). \tag{10.11}$$

Notice that h_0 is a Gaussian function. Its Fourier transform is another Gaussian since, with $G(x) = e^{-ax^2/2}$ where $a > 0$, we have

$$\begin{aligned}\widehat{G}(p) &= \frac{1}{\sqrt{2\pi}} \int_{\mathbb{R}} e^{-\mathbf{i}px} e^{-ax^2/2}\, dx = \frac{1}{\sqrt{a\pi}} \int_{-\infty}^{\infty} e^{-\left(t^2 + \mathbf{i}pt\sqrt{2/a}\right)}\, dt \\ &= \frac{e^{-p^2/2a}}{\sqrt{a\pi}} \int_{-\infty}^{\infty} e^{-(t+\mathbf{i}p/\sqrt{2a})^2}\, dt = \frac{e^{-p^2/2a}}{\sqrt{a\pi}} \int_{-\infty}^{\infty} e^{-s^2}\, ds = \frac{e^{-p^2/2a}}{\sqrt{a}}.\end{aligned} \tag{10.12}$$

With $a = 1$, this shows in fact that $\mathcal{F}(h_0) = h_0$ and, from expression 10.11, that

$$\mathcal{F}(h_n) = (-\mathbf{i})^n h_n.$$

Thus, for $\ell \in \{0, 1, 2, 3\}$, the Hermite functions $h_{4k+\ell}$ are in the $(-\mathbf{i})^\ell$-eigenspace of the Fourier transform $\mathcal{F}$, confirming that each eigenvalue has infinite multiplicity, or infinite degeneracy in a physics context. Since

[8] In this expression, the symbol $\circ$ means the composition of operators.

the collection of Hermite functions forms a complete orthogonal basis of $L^2(\mathbb{R})$, as shown in Section 11.3 of the lecture following, the $(-\mathbf{i})^\ell$-eigenspace in fact is spanned by those Hermite functions whose orders are congruent to ℓ modulo 4.

The reader might think that we have committed one "peccadillo" in the derivation of the Fourier transform of the Gaussian in equation 10.12. In the next-to-last equality, it looks on the surface as if we have performed the change of variables $s = t + \mathbf{i}p/\sqrt{2a}$. This, of course, would not be valid. Indeed, both s and t are real variables and so cannot be related by this simple equation. An ordinary change of variables does not change the domain of integration but merely reparameterizes it. In changing from $t+\mathbf{i}p/\sqrt{2a}$ to s, though, we have changed the integration of the complex function e^{-z^2} from the line in the complex plane parallel to the real axis and passing through $\mathbf{i}p/\sqrt{2a}$ to the real axis itself. The way out of this is through the integration of e^{-z^2} along the contour C_R, a rectangle with vertices $-R$, R, $R+\mathbf{i}p/\sqrt{2a}$, and $-R+\mathbf{i}p/\sqrt{2a}$, oriented counterclockwise. Since e^{-z^2} is entire, Cauchy's theorem implies that $\int_{C_R} e^{-z^2}\,dz = 0$ for every real a and p and $R > 0$. Since e^{-z^2} decreases in modulus rapidly as $|z|$ increases, the integrals of e^{-z^2} over the two vertical sides decrease to zero in the limit $R \to \infty$, which then justifies the next-to-last equality of equation 10.12.

Having derived the Fourier transform of the one-dimensional Gaussian in equation 10.12, it is now easy to compute the transform of the n-dimensional Gaussian. We leave it to the reader to verify that if $G(x) = e^{-a|x|^2/2}$ then

$$\widehat{G}(p) = \frac{e^{-|p|^2/2a}}{a^{n/2}}.$$

10.3 The Position and Momentum Representations

We now examine wave mechanics, where the state space is the Hilbert space $L^2(\mathbb{R}^n)$. When $n = 3k$ this turns out to be useful for describing certain quantum systems of k particles in $\mathbb{R}^3$, with $x_{3j-3+\ell}$ the ℓth position coordinate for the jth particle, where $\ell = 1, 2, 3$. When n is not a multiple of 3, the quantum system might describe several constrained particles.

In Section 3.1.1 the probability interpretation of the wave function ψ of a quantum state $\mathfrak{S}$ was presented for a single quantum particle moving in one dimension. This interpretation, with the wave function

$\psi(x)$ serving as the probability amplitude for position measurements and the Fourier transform $\widehat{\psi}(p)$ serving as the probability amplitude for the corresponding momentum measurements, generalizes to higher dimensions. The first thing we should mention, though, is the minor modification used in quantum mechanics for the Fourier transform. This amounts to a rescaling of measurements in terms of the basic unit $\hbar$. In n-dimensions, the Fourier transform pertinent for quantum mechanics is given for functions in $L^1(\mathbb{R}^n)$ by

$$\mathcal{F}(\psi)(p) = \widehat{\psi}(p) = \frac{1}{(2\pi\hbar)^{n/2}} \int_{\mathbb{R}^n} e^{-\mathbf{i}p\cdot x/\hbar}\psi(x)\,dx, \tag{10.13}$$

then extended to $L^2(\mathbb{R}^n)$ as in the preceding section. This Fourier transform, rescaled by $\hbar$, still defines a unitary mapping on $L^2(\mathbb{R}^n)$.

Equation 10.7 expresses a symmetry of the Fourier transform where differentiation with respect to a variable is interchanged with multiplication by the variable. Using the rescaled transform of expression 10.13, equation 10.7 reads $\hbar\,\widehat{\partial_i\psi}(p) = \mathbf{i}p_i\widehat{\psi}(p)$. In quantum mechanics, this expresses a symmetry between the position and momentum operators and implies that quantum mechanics is symmetric with respect to the representation of a quantum state by either a position amplitude wave function or its symmetric analogue under the Fourier transform, to wit, a momentum amplitude wave function. Letting X_i represent the position operator that multiplies by the ith coordinate and letting $\mathsf{P}_i = -\mathbf{i}\hbar\partial_i$, the corresponding momentum operator, equation 10.7 may be expressed as

$$\mathcal{F}\,\mathsf{P}_i = \mathsf{X}_i\,\mathcal{F}. \tag{10.14}$$

A similar observation gives the relationship

$$\mathcal{F}\mathsf{X}_i = -\mathsf{P}_i\,\mathcal{F}. \tag{10.15}$$

Using the Plancherel theorem, that the Fourier transform is unitary, the relationships above give the expectation values for momentum and position as

$$\langle \mathsf{P}_i\rangle_\psi = \langle\psi|\mathsf{P}_i\psi\rangle = \langle\widehat{\psi}\mid \mathsf{X}_i\widehat{\psi}\,\rangle = \langle \mathsf{X}_i\rangle_{\widehat{\psi}}.$$

and

$$\langle \mathsf{X}_i\rangle_\psi = \langle\psi|\mathsf{X}_i\psi\rangle = \langle\widehat{\psi}\mid -\mathsf{P}_i\widehat{\psi}\,\rangle = -\langle \mathsf{P}_i\rangle_{\widehat{\psi}}.$$

In terms of integrals, and using the variable x in **position space** and p

in **momentum space**, the first of these relations reads

$$-\mathbf{i}\hbar \int_{\mathbb{R}^n} \overline{\psi(x)} \partial_i \psi(x)\, dx = \int_{\mathbb{R}^n} \overline{\widehat{\psi}(p)} p_i \widehat{\psi}(p)\, dp.$$

Thus the expectation of momentum in the ith coordinate direction may be calculated either from the momentum operator applied to the state function or from the position operator applied to its Fourier transform.

Let $\mathsf{P} = (\mathsf{P}_1, \ldots, \mathsf{P}_n)$ and, similarly, $\mathsf{X} = (\mathsf{X}_1, \ldots, \mathsf{X}_n)$ and let $g = g(t_1, \ldots, t_n)$ be a polynomial in n variables. Using property 10.6, it then is easy to see that expressions 10.14 and 10.15 generalize to

$$\mathcal{F} g(\mathsf{P}) = g(\mathsf{X}) \mathcal{F} \tag{10.16}$$

and

$$\mathcal{F} g(\mathsf{X}) = g(-\mathsf{P}) \mathcal{F}. \tag{10.17}$$

This can be generalized to the case where g is an analytic function, but we will not pursue this here.

From our discussion, it is manifest that the Fourier transform is an isomorphism of the quantum theory that uses the state function as a position probability amplitude, the so-called **position representation**, with that one that uses the state function as a momentum probability amplitude, the **momentum representation**. The two representations are exactly symmetric with the Fourier transform carrying each to the other.

According to Axiom 4 of Lecture 2, the time evolution of the wave function ψ in the position representation is governed by the Schrödinger wave equation. What equation describes time evolution in the momentum representation? Our aim is to apply the Fourier transform to the Schrödinger equation to transform it to its momentum representation. On the surface, this is quite transparent, resulting in the evolution equation

$$\frac{|p|^2}{2m} \widehat{\Psi} + \widehat{V\Psi} = \mathbf{i}\hbar \partial_t \widehat{\Psi}, \tag{10.18}$$

where $\widehat{\Psi} = \widehat{\Psi}(t, p)$ denotes the application of the Fourier transform at each instant t. The right-hand side follows from exchanging the order of differentiation and integration, valid for suitably smooth functions. This is an integro-differential equation, involving as it does the term

$$\widehat{V\Psi}(t, p) = \frac{1}{(2\pi)^{n/2}} \int_{\mathbb{R}^n} e^{-\mathbf{i} p \cdot x} V(t, x) \Psi(t, x)\, dx, \tag{10.19}$$

which may admit no further simplification. This sort of equation is generally more difficult to address than a pure differential equation. Our desire then is to simplify this by rewriting $\widehat{V\Psi}$ in terms of $\widehat{\Psi}$, and hopefully to obtain a differential equation of motion. Restricting to a path of suitably smooth wave functions $\Psi(t,x)$ and assuming that the potential function V is a polynomial independent of t, using equation 10.17 we may rewrite the evolution equation for the momentum wave function $\Phi(t,p) = \widehat{\Psi}(t,p)$ as

$$\frac{|p|^2}{2m}\Phi + V(-\mathsf{P})\Phi = \mathbf{i}\hbar\partial_t\Phi. \tag{10.20}$$

This is a pure differential equation since $V(-\mathsf{P})$ is a differential operator. This special case covers many of the most important applications of quantum mechanics.

For example, the one-dimensional harmonic oscillator potential is the square potential $V(x) = m\omega^2 x^2/2$, so that the Hamiltonian operator is of the form $\mathsf{H} = a\mathsf{P}^2 + b\mathsf{X}^2$ for real constants $a = 1/2m$ and $b = m\omega^2/2$. The Fourier-transformed Hamiltonian that applies to momentum space wave functions Φ is then $\widehat{\mathsf{H}} = b\mathsf{P}^2 + a\mathsf{X}^2$, so that no advantage is gained by transforming to the momentum representation. This, of course, is the only case where this phenomenon occurs, and normally the momentum space Schrödinger equation $\widehat{\mathsf{H}}\Phi = \mathbf{i}\hbar\partial_t\Phi$ offers a real alternative that may prove either more or less difficult to handle than the position space Schrödinger equation $\mathsf{H}\Psi = \mathbf{i}\hbar\partial_t\Psi$.

10.4 Wave Packets and Superposition

In this section we consider the Fourier transform as representing a wave packet that arises as an uncountable superposition of stationary plane waves of various frequencies. This is a rather physical picture of the transform, which offers a valuable aid to intuition. The starting point is to understand the physicists' interpretation of the function

$$\psi_p(x) = \frac{1}{(2\pi\hbar)^{n/2}} e^{\mathbf{i}p\cdot x/\hbar}. \tag{10.21}$$

It is not a wave function in the sense of the axioms of Lecture 2 since it is not square-integrable, but it is used to represent a system with a precise value of momentum and fits naturally into the framework of the Dirac calculus and Gelfand triples of Lecture 5. In fact a quick review of pp. 69 and 71 reveals that, in the one-dimensional case, ψ_p represents a generalized eigenvector of the momentum operator with generalized eigenvalue p in the following way. The generalized eigenvector with eigenvalue p is denoted as $|p\rangle$ and, technically, is the conjugate linear functional defined in equation 5.1. This generalized eigenvector $|p\rangle$ is used to represent the idealized quantum state of a particle with precise momentum p and, therefore, from the Heisenberg uncertainty principle, with completely imprecise, or undetermined, position. The relation of ψ_p to $|p\rangle$ is precisely the relation of ψ to $|\psi\rangle$ when $\psi \in L^2(\mathbb{R})$. Indeed, the square-integrable function ψ defines the action of the conjugate linear functional $|\psi\rangle$ by $\varphi \mapsto \langle\varphi|\psi\rangle = \int_{\mathbb{R}} \overline{\varphi}\psi$ for a test function φ. In the same way, the function ψ_p defines the action of the conjugate linear functional $|p\rangle$ on the test function φ by

$$\langle\varphi|p\rangle = \langle\varphi|\psi_p\rangle = \int_{\mathbb{R}} \overline{\varphi}\psi_p = \frac{1}{\sqrt{2\pi\hbar}} \int_{\mathbb{R}} \overline{\varphi(x)} e^{\mathrm{i}px/\hbar}\, dx = \overline{\widehat{\varphi}}(p),$$

or, taking conjugates formally,

$$\langle p|\varphi\rangle = \langle\psi_p|\varphi\rangle = \frac{1}{\sqrt{2\pi\hbar}} \int_{\mathbb{R}} \varphi(x) e^{-\mathrm{i}px/\hbar}\, dx = \widehat{\varphi}(p).$$

Repeating equation 5.8, this may be written as

$$\begin{aligned}\widehat{\varphi}(p) = \langle p|\varphi\rangle = \langle p|1_S|\varphi\rangle &= \langle p| \int_{\mathbb{R}} |x\rangle\langle x|\, dx\, |\varphi\rangle \\ &= \int_{\mathbb{R}} \langle p|x\rangle\langle x|\varphi\rangle\, dx = \frac{1}{\sqrt{2\pi\hbar}} \int_{\mathbb{R}} \varphi(x) e^{-\mathrm{i}px/\hbar}\, dx.\end{aligned}$$

Similarly,

$$\begin{aligned}\varphi(x) = \langle x|\varphi\rangle &= \langle x| \int_{\mathbb{R}} |p\rangle\langle p|\, dp\, |\varphi\rangle \\ &= \int_{\mathbb{R}} \langle x|p\rangle\langle p|\varphi\rangle\, dp = \frac{1}{\sqrt{2\pi\hbar}} \int_{\mathbb{R}} \widehat{\varphi}(p) e^{\mathrm{i}xp/\hbar}\, dp. \qquad (10.22)\end{aligned}$$

Taking the test function φ as a state vector representing a quantum state, these latter two equations may be used to write the wave function φ in the symbolic Dirac calculus variously as

$$|\varphi\rangle = \int_{\mathbb{R}} |x\rangle\langle x|\varphi\rangle\, dx = \int_{\mathbb{R}} \varphi(x)|x\rangle\, dx,$$

in the position basis representation, and

$$|\varphi\rangle = \int_{\mathbb{R}} |p\rangle\langle p|\varphi\rangle\, dp = \int_{\mathbb{R}} \widehat{\varphi}(p)|p\rangle\, dp, \qquad (10.23)$$

in the momentum basis representation.

These formulae are very interesting on several levels. First, note that they encode the Fourier inversion formulae in the language of the Dirac calculus. Indeed, using equation 5.3 we get from the second of these that $\varphi(x) = \langle x|\varphi\rangle = \mathcal{F}^{-1}\widehat{\varphi}(x)$. Second, they symbolically represent the momentum and position state vectors as superpositions of uncountably many "mutually orthogonal" (equation 5.4) position and momentum "eigenstate vectors," respectively. For example, writing $\varphi = \int \widehat{\varphi}(p)|p\rangle\, dp$, the physicist recognizes that the values of the Fourier transform of φ give the coefficients of the idealized quantum states with precise momenta p that are superposed to produce the state vector φ. As such, $\widehat{\varphi}$ is seen to provide the probability amplitude density for momentum measurements. Third, these formulae are seen to arise as a kind of continuum limit of the standard rigorous formulae for generalized Fourier series expansions of functions in $L^2([-\pi, \pi])$. Note the formal similarity between these and the rigorous formula

$$\phi = \sum_{n=-\infty}^{\infty} |n\rangle\langle n|\phi\rangle,$$

which expands an arbitrary vector $\phi \in L^2([-\pi, \pi])$ in terms of a complete orthonormal basis $\{|n\rangle : n \in \mathbb{Z}\}$. In the classical theory of Fourier series, $|n\rangle = e^{\mathbf{i}nx}/\sqrt{2\pi}$ and the nth Fourier coefficient is

$$\langle n|\phi\rangle = \frac{1}{\sqrt{2\pi}} \int_{-\pi}^{\pi} \phi(x) e^{-\mathbf{i}nx}\, dx.$$

By rescaling the interval $[-\pi, \pi]$ to the interval $[-T, T]$ and taking a formal limit as T approaches ∞, the Fourier transform formulae are recovered from the Fourier series formulae. Explicitly, let ϕ be a square-integrable function defined on $\mathbb{R}$ and set $\phi_T = \chi_{[-T,T]}\phi$. The family $|n, T\rangle = e^{\mathbf{i}p_n x}/\sqrt{2T}$, where $p_n = n\pi/T$, is a complete orthonormal basis of $L^2([-T, T])$. It follows that

$$\phi_T = \sum_{n=-\infty}^{\infty} |n, T\rangle\langle n, T|\phi_T\rangle, \tag{10.24}$$

where the Fourier coefficient is

$$\begin{aligned}\langle n, T|\phi_T\rangle &= \frac{1}{\sqrt{2T}} \int_{-T}^{T} \phi_T(x) e^{-\mathbf{i}n\pi x/T}\, dx \\ &= \sqrt{\frac{\pi}{T}} \frac{1}{\sqrt{2\pi}} \int_{-\infty}^{\infty} \phi_T(x) e^{-p_n x}\, dx = \sqrt{\frac{\pi}{T}}\, \widehat{\phi}_T(p_n).\end{aligned}$$

Substituting this into expression 10.24 gives

$$\phi_T = \frac{1}{\sqrt{2\pi}} \sum_{n=-\infty}^{\infty} \widehat{\phi}_T(p_n) e^{\mathrm{i} p_n x} \Delta p_n,$$

where $\Delta p_n = p_n - p_{n-1} = \pi/T$. The sum in this formula is a Riemann-type sum that looks like an approximation to the integral $\int_{\mathbb{R}} \widehat{\phi}_T(p) e^{\mathrm{i}px}\, dp$ and, noting that $\Delta p_n \to 0$ as $T \to \infty$, the formal limit of this expression as $T \to \infty$ gives precisely the Fourier inversion formula $\phi = \mathcal{F}^{-1}(\widehat{\phi})$.

Returning to expression 10.21, ψ_p represents an idealized quantum state with precise momentum p and completely undetermined position. This latter property of undetermined position is reflected by the fact that the square-amplitude $|\psi_p|^2$ is constant, suggesting a constant probability density function for position measurements. Of course, this is quite unwarranted according to the axioms of wave mechanics and their standard interpretation since ψ_p is not a member of $L^2(\mathbb{R}^n)$; nonetheless, it has weight and is acceptable in the Gelfand development of wave mechanics. As the Fourier transform rewrites wave functions in their momentum space representation, the expectation is that the Fourier transform of ψ_p should represent $|p\rangle$ in momentum space and so act somehow as a generalized eigenvector of the transformed momentum operator. What we mean is that, in accordance with equation 10.14, we expect that $\mathsf{X}_i(\mathcal{F}|p\rangle) = \mathcal{F}\mathsf{P}_i|p\rangle = p_i(\mathcal{F}|p\rangle)$, so that in some suitable sense $\mathcal{F}|p\rangle$ is an eigenvector of the position operator X_i applied to momentum-space wave functions. Recalling the definitions of p. 69, we would expect that $\mathcal{F}|p\rangle = \overline{\delta_p}$.

Can the discussion of the preceding paragraph be made rigorous? To answer this, we need to ask what is the Fourier transform of a conjugate-linear functional such as $|p\rangle$. To see how this should be defined, consider the conjugate-linear functional $|\psi\rangle$ corresponding to the square-integrable function ψ. The Fourier transform $\mathcal{F}|\psi\rangle$ should be the conjugate-linear functional defined by $|\mathcal{F}\psi\rangle$. Applying this to the test function φ, we have $\langle\varphi|\widehat{\psi}\,\rangle = \langle\widetilde{\varphi}|\psi\rangle$, which is confirmed easily, so that $\mathcal{F}|\psi\rangle = |\psi\rangle \circ \mathcal{F}^{-1}$. The appropriate definition, then, for $\mathcal{F}|p\rangle$ is that $\mathcal{F}|p\rangle = |p\rangle \circ \mathcal{F}^{-1}$. Applying this now to the test function φ and recalling expressions 5.2, we have

$$\mathcal{F}|p\rangle(\varphi) = |p\rangle(\widetilde{\varphi}) = \langle\widetilde{\varphi}|p\rangle = \overline{\widetilde{\widetilde{\varphi}}}(p) = \overline{\varphi(p)} = \overline{\delta_p}(\varphi),$$

confirming the expectation that $\mathcal{F}|p\rangle = \overline{\delta_p}$. This is just a special case of the definition of the product of an operator with a ket-vector, as the equation $\mathsf{A}|\psi\rangle = |\mathsf{A}\psi\rangle = |\psi\rangle \circ \mathsf{A}^*$ suggests that the correct definition of A acting on any element $\ell \in \overline{\mathcal{D}'}$ is $\mathsf{A}\ell = \ell \circ \mathsf{A}^*$, a functional acting from the left on test functions.

All this is really subsumed under the theory of the Fourier transform applied to distributions that was articulated in the first section of this lecture, and our derivation here is the same as there, modulo that physics is interested in conjugate-linear functionals as well as linear ones. Indeed, a little thinking shows that if the Fourier transform of the tempered distribution T is defined as $\widehat{T} = T \circ \mathcal{F}$, then the Fourier transform of the conjugate-linear distribution $\overline{T}$ should be $\widehat{\overline{T}} = \overline{\widetilde{T}} = \overline{T} \circ \mathcal{F}^{-1}$, which agrees with $\mathcal{F}|p\rangle = |p\rangle \circ \mathcal{F}^{-1}$.

The usual development presented in a physics text goes, at least in spirit, more like the following. How should $\mathcal{F}|p\rangle$ be understood? Obviously, it should be that object represented by the expression $\mathcal{F}\psi_p$ in the way that $|p\rangle$ is represented by ψ_p. As $|p\rangle = |\psi_p\rangle$, we should have $\mathcal{F}|p\rangle = |\mathcal{F}\psi_p\rangle$. The expected result then is that $\overline{\delta_p} = |\mathcal{F}\psi_p\rangle$, or taking conjugates, $\delta_p = \langle\mathcal{F}\psi_p|$. There is a critical problem with this, though, as the Fourier transform of the function ψ_p diverges and so does not represent a function on $\mathbb{R}^n$. Nonetheless, formally applying the divergent term $\langle\mathcal{F}\psi_p|$ to the test function φ gives

$$\begin{aligned}\langle\mathcal{F}\psi_p|\varphi\rangle &= \int_{\mathbb{R}^n} \frac{1}{(2\pi\hbar)^{n/2}} \int_{\mathbb{R}^n} \overline{\psi_p(x)e^{\mathrm{i}q\cdot x/\hbar}}\, dx\, \varphi(q)\, dq \\ &= \frac{1}{(2\pi\hbar)^n} \int_{\mathbb{R}^n} \int_{\mathbb{R}^n} e^{\mathrm{i}(q-p)\cdot x/\hbar} \varphi(q)\, dx\, dq.\end{aligned}$$

Of course, this says nothing for, after all, the integral $\int_{\mathbb{R}^n} e^{\mathrm{i}(q-p)\cdot x/\hbar}\, dx$ diverges so the expression is meaningless. But where the mathematician is at a dead end, the fearless physicist points to the Fourier inversion formula 10.5 and exchanges the order of integration to get

$$\langle\mathcal{F}\psi_p|\varphi\rangle = \frac{1}{(2\pi\hbar)^n} \int_{\mathbb{R}^n} \int_{\mathbb{R}^n} e^{\mathrm{i}(q-p)\cdot x/\hbar} \varphi(q)\, dq\, dx = \widehat{\widetilde{\varphi}}(p) = \varphi(p) = \delta_p(\varphi),$$

so, yes, $\delta_p = \langle\mathcal{F}\psi_p|$ as expected. Though this derivation fails the test of rigor, it is a convenient heuristic for the physicist that has practical use in that it leads to correct calculations. We will see an instance of this sort of derivation of correct information from mathematically suspect reasoning in Lecture 32 when evaluating path integrals that describe propagators at singularities.

The history of physics is rife with example after example of clever reasoning that, though rigorously suspect, nonetheless leads to correct information and to a calculus that works despite its deficiencies in rigor. One task of the mathematician in the endeavor to understand the workings of nature is to examine the heuristics of the physicists and find or construct a rigorous theoretical edifice that supports the physical calculus and, ultimately, to give back to the physicists by providing mathematical tools unknown to them that might be used to further the science. The history of quantum mechanics, with the interplay in the early years between the physicists, on the one hand, and mathematicians such as von Neumann, Noether, Weyl, Hilbert, and Stone on the other, and the current fruitful interplay in an area such as field theory, is a case study in the above thesis. A similar story unfolds in the history of relativity – everyone knows of Einstein's great insights, but his theory would have been impossible to develop in its fullness without the work of the mathematicians Riemann, Poincaré, Minkowski, Ricci, Levi-Civita, Hilbert, and Weyl; furthermore, Roger Penrose's more recent work has contributed to the modern understanding of some intricacies of the theory of singularities in general relativity. Today the most pressing problem in the realm of physics for the mathematician is in understanding the rigorous underpinnings of quantum field theory and how it can incorporate gravity into its edifice.[9]

10.5 End Notes

There are many excellent references in the literature for the Fourier transform. I particularly like the discussions of Reed and Simon (1980) and Strichartz (2003), and the development in this lecture is based on theirs. Chapter IX of Reed and Simon (1980) gives a complete rigorous development and is preceded by a wealth of functional analytic development crucial to a rigorous foundation for quantum mechanics. Strichartz (2003) is a gentle introduction to the theory that is particularly accessible. It concentrates on applications to partial differential equations, which happens to be precisely the material needed for the student studying quantum mechanics. I also recommend Walter Rudin's (1991) presentation of test functions and distributions in his Chapter 6 and of the Fourier transform in his Chapter 7. Those who appreciate austere, economical, explanations will take to Rudin's style of mathe-

[9] For a discussion of this interplay between the disciplines, see the translation of Mikhail Gelfand's 2008 interview with Yuri Manin in the November 2009 issue of *The Notices of the American Mathematical Society*, 56(10):1268–1274.

matical exposition – saying only what needs to be said and doing so with careful attention to detail and accuracy. In his rather terse prose Rudin reminds me of Dirac and his style of writing physics.

There is also a slew of important references for the special functions of mathematics and mathematical physics. I will mention two that have chapters on Hermite polynomials and functions. The first is the classic text of Rainville (1960) and the second is the modern text of Beals and Wong (2010). These texts develop the theory broadly, in the context of complete systems of orthogonal polynomials, and relate them to their corresponding differential equations.

11
Wave Mechanics III: The Quantum Oscillator

The career of a young theoretical physicist consists of treating the harmonic oscillator in ever-increasing levels of abstraction.

Sidney Coleman
in *David Tong: Lectures on Quantum Field Theory*, 2006

We arrive finally at our first full analysis of a system by the quantum mechanical formalism. The results have been reported already in Lecture 1, where we presented both the classical and quantum mechanical treatment of the simple harmonic oscillator, the quintessential physical system. In the present lecture we present an elegant algebraic treatment of the quantum oscillator using the ladder operators introduced in Section 9.3.

After deriving the commutation relations for the creation and annihilation operators – the ladder operators already mentioned – and the Hamiltonian operator, we use these algebraic results to identify the ground state of the oscillator potential and then apply the ladder operators to find the excited states. We then prove that the energy eigenfunctions that correspond to these excited states have nondegenerate eigenvalues and form a complete orthonormal basis of the Hilbert space $L^2(\mathbb{R})$. Finally, we develop the very interesting theory of coherent states of the harmonic oscillator and tease out some beautiful results that these states satisfy, including their relationship to minimal uncertainty states.

In a very real and tangible sense, the results of this lecture are foundational to the whole edifice of quantum mechanics, including quantum field theory, and a careful study and understanding of the content of this lecture is indispensable for the serious student.

11.1 Ladder Operators and the Ground State

By quantizing the classical Hamiltonian for a particle of mass m having angular frequency ω in a square potential well $V(x) = \frac{1}{2}m\omega^2x^2$, we arrive at the Hamiltonian operator for the system, given as

$$\mathsf{H} = \frac{\mathsf{P}^2}{2m} + \frac{1}{2}m\omega^2\mathsf{X}^2 = -\frac{\hbar^2}{2m}\frac{d^2}{dx^2} + \frac{1}{2}m\omega^2x^2.$$

We seek square-integrable solutions to the eigenvalue equation $\mathsf{H}\psi = E\psi$, where E is a real constant. Recall from Section 9.3 that the **ladder operators** $\mathsf{a}_\pm = \mathsf{P} \pm \mathbf{i}m\omega\mathsf{X}$ are adjoints of one another, as $\mathsf{a}_\pm^* = \mathsf{a}_\mp$. We have

$$\mathsf{a}_\pm\mathsf{a}_\mp = \mathsf{P}^2 + m^2\omega^2\mathsf{X}^2 \pm \mathbf{i}m\omega[\mathsf{X},\mathsf{P}] = 2m\mathsf{H} \mp m\omega\hbar\mathbb{1}. \tag{11.1}$$

This shows that $[\mathsf{a}_\pm, \mathsf{a}_\mp] = \mp 2m\hbar\omega\mathbb{1}$. Let ψ be any normalized state vector in $L^2(\mathbb{R})$ and observe that

$$\|\mathsf{a}_\pm\psi\|^2 = \langle\psi|\mathsf{a}_\mp\mathsf{a}_\pm\psi\rangle = 2m\langle\psi|\left(\mathsf{H} \pm \tfrac{1}{2}\hbar\omega\mathbb{1}\right)\psi\rangle$$

so that if ψ is an eigenfunction of H with eigenvalue E then

$$\|\mathsf{a}_\pm\psi\|^2 = 2m\langle\psi|\left(E \pm \tfrac{1}{2}\hbar\omega\right)\psi\rangle = 2m\left(E \pm \tfrac{1}{2}\hbar\omega\right). \tag{11.2}$$

As $\|\mathsf{a}_-\psi\|^2 \geq 0$, we conclude that $E \geq \frac{1}{2}\hbar\omega$ for any eigenvalue E of H. Seeing that $\|\mathsf{a}_-\psi\|^2 = 2m(E - \frac{1}{2}\hbar\omega)$, equality holds, i.e., $E = \frac{1}{2}\hbar\omega$, if and only if $\|\mathsf{a}_-\psi\|^2 = 0$. This is equivalent to $\mathsf{a}_-\psi = (\mathsf{P} - \mathbf{i}m\omega\mathsf{X})\psi = 0$, which is precisely the equation in Section 9.3 satisfied by the minimal uncertainty state ψ_0 given by equation 9.5. This minimal uncertainty state then is the **ground state**, representing the state of minimum energy $\frac{1}{2}\hbar\omega$ for the quantum oscillator. From the discussion of Section 9.3, ψ_0 represents a quantum particle in equilibrium about the origin, with expected momentum zero and expected position coordinate $x = 0$. The position dispersion is $\Delta_{\psi_0}\mathsf{X} = \sqrt{\hbar/2m\omega}$, with momentum dispersion $\Delta_{\psi_0}\mathsf{P} = \sqrt{m\omega\hbar/2}$ and product of dispersions equal to $\hbar/2$ as required for a minimal uncertainty state. Moreover, from the discussion in Sections 9.1 and 9.3, every other state of minimum energy, i.e., every other eigenstate of the Hamiltonian H with eigenvalue $\frac{1}{2}\hbar\omega$, is a constant nonzero multiple of ψ_0. The eigenspace associated with this minimum energy $\frac{1}{2}\hbar\omega$ is then one-dimensional, and we indicate this by saying that the eigenvalue $\frac{1}{2}\hbar\omega$ is **nondegenerate**.

11.2 Higher Energy States

From equation 11.1 the Hamiltonian may be written as

$$\mathsf{H} = \frac{1}{2m}\mathsf{a}_\pm\mathsf{a}_\mp \pm \frac{1}{2}\hbar\omega 1, \tag{11.3}$$

which allows us to calculate the commutator $[\mathsf{H}, \mathsf{a}_\pm]$ as

$$\begin{aligned}[\mathsf{H}, \mathsf{a}_\pm] &= \frac{1}{2m}[\mathsf{a}_\pm\mathsf{a}_\mp, \mathsf{a}_\pm] \pm \frac{1}{2}\hbar\omega[1, \mathsf{a}_\pm] \\ &= \frac{1}{2m}(\mathsf{a}_\pm\mathsf{a}_\mp\mathsf{a}_\pm - \mathsf{a}_\pm\mathsf{a}_\pm\mathsf{a}_\mp) \\ &= \frac{1}{2m}\mathsf{a}_\pm[\mathsf{a}_\mp, \mathsf{a}_\pm] = \pm\hbar\omega\mathsf{a}_\pm,\end{aligned} \tag{11.4}$$

since $[\mathsf{a}_\pm, \mathsf{a}_\mp] = \mp 2m\hbar\omega 1$. Thus,

$$\mathsf{H}\mathsf{a}_\pm = \mathsf{a}_\pm(\mathsf{H} \pm \hbar\omega 1). \tag{11.5}$$

Since $\mathsf{a}_-\psi_0 = 0$, equation 11.3 implies that $\mathsf{H}\psi_0 = \frac{1}{2}\hbar\omega\psi_0$, a fact derived first in Section 9.3. Applying equation 11.5 with the plus sign gives

$$\mathsf{H}\mathsf{a}_+\psi_0 = \frac{3}{2}\hbar\omega\mathsf{a}_+\psi_0,$$

and, since $\|\mathsf{a}_+\psi_0\|^2 = 2m\hbar\omega$ from equation 11.2, $\mathsf{a}_+\psi_0 \neq 0$. Therefore $\mathsf{a}_+\psi_0$ is an eigenfunction of H with eigenvalue $\frac{3}{2}\hbar\omega$ and we write

$$\psi_1 = C_1\mathsf{a}_+\psi_0,$$

where C_1 is a normalization constant making $\|\psi_1\| = 1$. An easy induction argument now shows that

$$\psi_N = C_N\mathsf{a}_+^N\psi_0, \tag{11.6}$$

where C_N is a normalization constant[1] and ψ_N is a normalized eigenfunction of H with eigenvalue $E_N = (N + \frac{1}{2})\hbar\omega$, for $N \in \mathbb{N}$. More generally, if $\mathsf{H}\psi = E\psi$ then, for $N \in \mathbb{N}$,

$$\mathsf{H}\mathsf{a}_\pm^N\psi = (E \pm N\hbar\omega)\mathsf{a}_\pm^N\psi.$$

Moreover, an induction argument using equation 11.2 reveals that if $\psi \neq 0$ then $\mathsf{a}_+^N\psi \neq 0$. Also, again using induction and equation 11.2, along with the fact that the eigenvalue $\frac{1}{2}\hbar\omega$ is nondegenerate, reveals that if $\psi \neq 0$ then $\mathsf{a}_-^N\psi = 0$ if and only if E takes on a value in the list $\frac{1}{2}\hbar\omega, \frac{3}{2}\hbar\omega, \ldots, (N - \frac{1}{2})\hbar\omega$.

[1] Note that equation 11.2 implies that ψ_N is square-integrable and therefore can be normalized.

The results of the preceding paragraph explain why a_+ is called the **creation operator** and a_- the **annihilation operator**: a_+ "creates" a quantum $\hbar\omega$ of energy by mapping an E-eigenstate to an $(E+\hbar\omega)$-eigenstate, and a_- "annihilates" a quantum by mapping an E-eigenstate to an $(E-\hbar\omega)$-eigenstate, provided that $E > \hbar\omega$. The state represented by the Nth energy eigenfunction ψ_N is said to be **excited** above the ground state by the accumulation of N quanta, each of size $\hbar\omega$.

We have seen that, for each $N = 0, 1, 2, \ldots$, the number $(N + \frac{1}{2})\hbar\omega$ is an eigenvalue of the Hamiltonian H and we have identified a corresponding eigenfunction, $\psi_N = C_N \mathsf{a}_+^N \psi_0$, in the $(N+\frac{1}{2})\hbar\omega$-eigenspace. We now verify that this is the complete set of eigenvalues, that each is nondegenerate, so that ψ_N spans the $(N+\frac{1}{2})\hbar\omega$-eigenspace, and that the set $\{\psi_N\} = \{\psi_N : N = 0, 1, 2, \ldots\}$ is a complete orthonormal basis of the Hilbert space $L^2(\mathbb{R})$.

First we verify that the collection $\{E_N = (N+\frac{1}{2})\hbar\omega : N = 0, 1, 2, \ldots\}$ is a complete set of eigenvalues. Assume that ψ is an eigenfunction with eigenvalue E. In particular, $\psi \neq 0$ and $\mathsf{H}\psi = E\psi$. Choose $N \in \mathbb{N}$ such that $N > E/\hbar\omega$, so that $E - N\hbar\omega < 0$. As the minimum eigenvalue is $\frac{1}{2}\hbar\omega$, the expression $E - N\hbar\omega < 0$ cannot be an eigenvalue. From our previous discussion, it follows that $\mathsf{a}_-^N \psi = 0$ and, therefore, $E = (n - \frac{1}{2})\hbar\omega$ for some integer $1 \leq n \leq N$, establishing that E is of the desired form. That the eigenvalue E_N is nondegenerate follows from the nondegeneracy of the minimum eigenvalue $E_0 = \hbar\omega/2$. Indeed, let ψ be an eigenfunction in the E_N-eigenspace. Then $\mathsf{a}_-^N\psi$ is an eigenstate with eigenvalue E_0, and as such is a nonzero multiple of ψ_0. As ψ_N also is an eigenfunction in the E_N-eigenspace, $\mathsf{a}_-^N\psi_N$ too is a nonzero multiple of ψ_0 so that $\mathsf{a}_-^N\psi = \lambda \mathsf{a}_-^N\psi_N$ for some nonzero constant λ. If $\psi - \lambda\psi_N \neq 0$ then it is an eigenfunction in the E_N-eigenspace, for which $\mathsf{a}_-^N(\psi - \lambda\psi_N) = 0$. But then the eigenvalue for the eigenfunction $\psi - \lambda\psi_N$, namely E_N, must come from the list $\frac{1}{2}\hbar\omega, \frac{3}{2}\hbar\omega, \ldots, (N-\frac{1}{2})\hbar\omega$, a contradiction. It follows that $\psi = \lambda\psi_N$, and the eigenvalue E_N is nondegenerate.

In the next section, we develop the idea of a generating function for the oscillator wave functions and use this to prove the orthogonality of the family $\{\psi_N\}$ and, in the section following, the completeness of this family in $L^2(\mathbb{R})$. First, though, we calculate the normalization constants C_N for the eigenstates ψ_N. The normalized ground state from equation 9.5 is

$$\psi_0(x) = \left(\frac{m\omega}{\pi\hbar}\right)^{1/4} \exp\left(-\frac{m\omega x^2}{2\hbar}\right). \tag{11.7}$$

From equation 11.2, for each $N \in \mathbb{N}$, $\|\mathsf{a}_+\psi_{N-1}\|^2 = 2m\hbar\omega N$, so that

$$\psi_N = \frac{1}{\sqrt{N}}\frac{1}{\sqrt{2m\hbar\omega}}\mathsf{a}_+\psi_{N-1}. \tag{11.8}$$

Continuing through $N-1$ more such reductions gives

$$\psi_N = \frac{1}{\sqrt{N!}}\left(\frac{1}{\sqrt{2m\hbar\omega}}\right)^N \mathsf{a}_+^N\psi_0, \tag{11.9}$$

and, comparing with equation 11.6,

$$C_N = \frac{1}{\sqrt{N!}}\left(\frac{1}{\sqrt{2m\hbar\omega}}\right)^N.$$

Thus a normalized E_N-eigenfunction is given as

$$\psi_N(x) = \frac{1}{\sqrt{N!}}\left(\frac{1}{\sqrt{2m\hbar\omega}}\right)^N \left(\frac{m\omega}{\pi\hbar}\right)^{1/4}\left(\frac{\hbar}{\mathbf{i}}\frac{d}{dx} + \mathbf{i}m\omega x\right)^N e^{-m\omega x^2/2\hbar}.$$

We now are going to rewrite ψ_N using a little trick. Application of the creation operator to a function φ may be written as

$$\mathsf{a}_+\varphi = \frac{\hbar}{\mathbf{i}}\left(\frac{d\varphi}{dx} - \frac{m\omega}{\hbar}x\varphi\right) = e^{m\omega x^2/2\hbar}\frac{\hbar}{\mathbf{i}}\frac{d}{dx}\left(e^{-m\omega x^2/2\hbar}\varphi\right).$$

Write $\xi^{\pm} = \exp(\pm m\omega x^2/2\hbar)$ for the operator that multiplies a function by $\exp(\pm m\omega x^2/2\hbar)$. Then the creation operator may be written as $\mathsf{a}_+ = \xi^+\mathsf{P}\xi^-$, and

$$\mathsf{a}_+^N = (\xi^+\mathsf{P}\xi^-)^N = \xi^+\mathsf{P}^N\xi^- = \left(\frac{\hbar}{\mathbf{i}}\right)^N \xi^+\frac{d^N}{dx^N}\xi^-.$$

We have

$$\psi_N = C_N\mathsf{a}_+^N\psi_0 = C_N\left(\frac{\hbar}{\mathbf{i}}\right)^N \xi^+\frac{d^N}{dx^N}\xi^-\psi_0.$$

After absorbing the unimodular term $1/\mathbf{i}^N$ into the normalization, we may write a normalized eigenstate ψ_N with eigenvalue E_N for the harmonic oscillator as

$$\psi_N(x) = \left(\frac{m\omega}{\pi\hbar}\right)^{1/4}\frac{1}{\sqrt{N!}}\left(\frac{\hbar}{2m\omega}\right)^{N/2} e^{m\omega x^2/2\hbar}\frac{d^N}{dx^N}e^{-m\omega x^2/\hbar}. \tag{11.10}$$

Using Rodrigues's formula, that the Nth Hermite polynomial $H_N(u)$ is given by

$$H_N(u) = (-1)^N e^{u^2}\frac{d^N}{du^N}e^{-u^2},$$

substituting $u = \sqrt{m\omega/\hbar}\,x$, and absorbing the unimodular term $(-1)^N$ into the normalization, we may rewrite equation 11.10 as

$$\psi_N(x) = (2^N N!)^{-1/2}\left(\frac{m\omega}{\pi\hbar}\right)^{1/4} \exp\left(-\frac{m\omega x^2}{2\hbar}\right) H_N\left(\sqrt{\frac{m\omega}{\hbar}}x\right),$$

or

$$\psi_N(x) = (2^N N!)^{-1/2} H_N\left(\sqrt{\frac{m\omega}{\hbar}}x\right)\psi_0(x). \tag{11.11}$$

11.3 Generating Function and Completeness

Let D_N be the constant comprising the first three factors on the right-hand side of formula 11.10 and let $\varphi_N(x) = \xi^+ f^{(N)}(x)$, where $f(x) = \exp(-m\omega x^2/\hbar)$. Thus, φ_N comprises the remaining functional factors of formula 11.10, so that $\psi_N = D_N\varphi_N$. Define a **generating function** for the harmonic oscillator energy eigenfunctions by

$$G(s,x) = \sum_{N=0}^{\infty} \frac{s^N}{N!}\varphi_N(x). \tag{11.12}$$

For each real number s, let G_s be the function $G_s(x) = G(s,x)$. Observe that the constant D_N is of the form $D_N = cd^N/\sqrt{N!}$ for positive constants c and d. Therefore the real series

$$\sum_{N=0}^{\infty} \frac{|s|^N}{N!}\|\varphi_N\| = \sum_{N=0}^{\infty} \frac{|s|^N}{N!} D_N^{-1} = \frac{1}{c}\sum_{N=0}^{\infty} \frac{(|s|/d)^N}{\sqrt{N!}}$$

converges, which in turn implies that

$$G_s = \sum_{N=0}^{\infty} \frac{s^N}{N!}\varphi_N, \tag{11.13}$$

where the convergence is in the L^2-norm. To obtain an explicit, closed analytic formula for G, note that

$$G(s,x) = \xi^+ \sum_{N=0}^{\infty} \frac{f^{(N)}(x)}{N!} s^N,$$

and observe that the sum in this formula is the Taylor series expansion for $f(x+s)$, which is pointwise convergent for all real s and x. Therefore,

$$G(s,x) = \xi^+ e^{-m\omega(x+s)^2/\hbar} = \exp\left[-\frac{m\omega}{2\hbar}\left(2s^2 + 4sx + x^2\right)\right],$$

valid for all real s and x.

We now use the generating function to verify that the collection $\{\psi_N\}$ is orthonormal in $L^2(\mathbb{R})$. For fixed s and t we evaluate the product $\langle G_s|G_t\rangle$ in two ways. First, since for fixed s the series in equation 11.13 converges to G_s in the L^2-norm, we have

$$\langle G_s|G_t\rangle = \sum_{M=0}^{\infty}\sum_{N=0}^{\infty} \frac{s^M}{M!}\frac{t^N}{N!}\langle \varphi_M|\varphi_N\rangle.$$

Second,

$$\begin{aligned}\langle G_s|G_t\rangle &= \int_{\mathbb{R}} \overline{G(s,x)}G(t,x)\,dx \\ &= \int_{\mathbb{R}} \exp\left[-\frac{m\omega}{\hbar}\left(s^2+2sx+x^2+2tx+t^2\right)\right]\,dx \\ &= e^{2m\omega st/\hbar}\int_{\mathbb{R}} \exp\left[-\frac{m\omega}{\hbar}(x+s+t)^2\right]\,dx.\end{aligned}$$

On substituting $u = \sqrt{m\omega/\hbar}(x+s+t)$, integrating, and expanding the remaining exponential term in a Taylor series, we get

$$\langle G_s|G_t\rangle = e^{2m\omega st/\hbar}\sqrt{\frac{\hbar}{m\omega}}\int_{\mathbb{R}} e^{-u^2}\,du = \left(\frac{\pi\hbar}{m\omega}\right)^{1/2}\sum_{N=0}^{\infty}\frac{s^N t^N}{N!}\left(\frac{2m\omega}{\hbar}\right)^N.$$

Comparing coefficients in the two expressions for $\langle G_s|G_t\rangle$ gives

$$\langle \varphi_M|\varphi_N\rangle = \delta_{M,N}N!\left(\frac{\pi\hbar}{m\omega}\right)^{1/2}\left(\frac{2m\omega}{\hbar}\right)^N,$$

confirming the orthogonality of the family $\{\varphi_N\}$, and therefore of the family $\{\psi_N = D_N\varphi_N\}$. Note that this also confirms the normalization of ψ_N since this expression shows that

$$D_N^{-1} = \|\varphi_N\| = \sqrt{N!}\left(\frac{\pi\hbar}{m\omega}\right)^{1/4}\left(\frac{2m\omega}{\hbar}\right)^{N/2}.$$

We now turn to the question of completeness of the orthonormal family $\{\psi_N\}$. To verify the completeness, we need to show that, for any $\psi \in L^2(\mathbb{R})$, $\psi = 0$ whenever $\langle\psi_N|\psi\rangle = 0$ for all $N = 0,1,2,\ldots$ This will then demonstrate that $\{\psi_N\}$ is a Schauder basis for $L^2(\mathbb{R})$, so that every square-integrable function ψ may be expanded as a series $\psi = \sum_{N=0}^{\infty}\psi_N\langle\psi_N|\psi\rangle$ convergent in the L^2-norm. The Hamiltonian H is then an example of exactly the type of operator dissected in Sections 4.3 and 5.4. In particular, the spectral decomposition of the Hamiltonian may be written as

$$\mathsf{H} = \sum_{N=0}^{\infty} E_N |\psi_N\rangle\langle\psi_N|,$$

and the resolution of the identity allows us to write

$$1_{L^2(\mathbb{R})} = \sum_{N=0}^{\infty} |\psi_N\rangle\langle\psi_N| \tag{11.14}$$

for the identity mapping on $L^2(\mathbb{R})$.

Let $\psi \in L^2(\mathbb{R})$. Since, for any fixed value of x, equation 11.12 gives $G(s,x)$ as a convergent power series in s, we may replace s by a complex variable z to extend the domain of G so that, for each real x, the function $G(-,x)$ becomes an entire complex function. We have, for all $(z,x) \in \mathbb{C} \times \mathbb{R}$,

$$G(z,x) = \sum_{N=0}^{\infty} \frac{z^N}{N!}\varphi_N(x) = \exp\left[-\frac{m\omega}{2\hbar}\left(2z^2 + 4zx + x^2\right)\right], \tag{11.15}$$

and, moreover, $G_z = \sum_{N=0}^{\infty}(z^N/N!)\varphi_N$, where convergence is in the L^2-norm as before. After setting $z = a + \mathrm{i}b$ for real a and b and after a modest amount of manipulation, we have

$$e^{-m\omega b^2/\hbar} G(a + \mathrm{i}b, x) = e^{-2\mathrm{i}m\omega b(x+a)/\hbar} G(a,x). \tag{11.16}$$

Set $F(x,y) = G(a,x)G(a,y)$ and apply the Fourier inversion theorem for L^1-functions[2] to the function $f(x) = F(t,x)\psi(x)$,[3] where t at first is a fixed number, and then evaluate the resulting equation at $t = x$. The result is

$$F(x,x)\psi(x) = \frac{1}{2\pi\hbar}\int_{-\infty}^{\infty}\int_{-\infty}^{\infty} e^{\mathrm{i}p(x-y)/\hbar} F(x,y)\psi(y)\, dy\, dp.$$

[2] This Fourier inversion theorem states, in part, that if both a function f and its Fourier transform $\widehat{f}$ are in $L^1(\mathbb{R})$ then $f(x) = (\mathcal{F}^{-1}\mathcal{F})(f)(x)$ almost everywhere, where $\mathcal{F}^{-1}$ is the **inverse Fourier transform**:

$$\mathcal{F}^{-1}(f)(x) = \frac{1}{\sqrt{2\pi\hbar}}\int_{\mathbb{R}} e^{\mathrm{i}xp/\hbar} f(p)\, dp.$$

Thus, the inversion theorem gives

$$f(x) = \frac{1}{2\pi\hbar}\int_{-\infty}^{\infty}\int_{-\infty}^{\infty} e^{\mathrm{i}p(x-y)/\hbar} f(y)\, dy\, dp$$

almost everywhere. See Rudin (1991), Theorem 7.7, or Strichartz (2003).

[3] The function $f(x)$ is in $L^1(\mathbb{R})$ by the Cauchy–Schwarz inequality.

Substituting $F(x, y) = G(a, x)G(a, y)$ and $p = 2m\omega b$ in the iterated integral gives

$$G(a,x)^2\psi(x) = \frac{m\omega}{\pi\hbar}\int_{-\infty}^{\infty}\int_{-\infty}^{\infty} e^{2\mathrm{i}m\omega b(x-y)/\hbar}G(a,x)G(a,y)\psi(y)\,dy\,db.$$

From equation 11.16, since $G(a, x)$ is real,

$$G(a,x) = \overline{G(a,x)} = e^{-m\omega b^2/\hbar}e^{-2\mathrm{i}m\omega b(x+a)/\hbar}\overline{G(a+\mathrm{i}b,x)}$$

and

$$G(a,y) = e^{-m\omega b^2/\hbar}e^{2\mathrm{i}m\omega b(y+a)/\hbar}G(a+\mathrm{i}b,y).$$

The integrand in the iterated integral becomes

$$e^{-2m\omega b^2/\hbar}\overline{G(a+\mathrm{i}b,x)}G(a+\mathrm{i}b,y)\psi(y),$$

and integration with respect to y yields

$$G(a,x)^2\psi(x) = \frac{m\omega}{\pi\hbar}\int_{-\infty}^{\infty} e^{-2m\omega b^2/\hbar}\,\overline{G(a+\mathrm{i}b,x)}\langle\overline{G}_{a+\mathrm{i}b}|\psi\rangle\,db. \quad (11.17)$$

Since $G_z = \sum_{N=0}^{\infty}(z^N/N!)\varphi_N$ converges in the L^2-norm, the Hermitian product in the integrand may be expanded as

$$\langle\overline{G}_z|\psi\rangle = \sum_{N=0}^{\infty}\frac{z^N}{N!}\langle\varphi_N|\psi\rangle = \sum_{N=0}^{\infty}\frac{z^N}{N!}D_N^{-1}\langle\psi_N|\psi\rangle.$$

This shows that if $\langle\psi_N|\psi\rangle = 0$ for all $N = 0, 1, 2, \ldots$ then $\langle\overline{G}_z|\psi\rangle = 0$ and, therefore, equation 11.17 guarantees that $G(a, x)^2\psi(x) = 0$, implying $\psi = 0$. It follows that the orthonormal family $\{\psi_N\}$ is complete.

11.4 Coherent States of the Oscillator

In this section, we introduce the very interesting collection of **coherent states** for the harmonic oscillator potential. These may be defined in various ways and have found applications to various physical processes, including processes in the theory of quantum optics and electronics, particularly those articulating a theory of lasers or a theory of the electromagnetic field. Coherent states are, in some sense, the most classical of the possible states for the harmonic oscillator in that they turn out to be minimal uncertainty states that minimize quantum correlations and, for this reason, are sometimes termed **semiclassical states**. Coherent states are defined here as those represented by eigenfunctions of the non-self-adjoint annihilation operator but this definition belies their useful properties, which are teased out in this section.

Before continuing with a discussion of coherent states, we rescale the annihilation and creation operators already introduced by defining

$$\mathbf{a} = \frac{\mathbf{i}}{\sqrt{2\hbar m\omega}}\mathsf{a}_- = \frac{\mathbf{i}}{\sqrt{2\hbar m\omega}}(\mathsf{P} - \mathbf{i}m\omega\mathsf{X})$$

and its adjoint as

$$\mathbf{a}^* = \frac{-\mathbf{i}}{\sqrt{2\hbar m\omega}}\mathsf{a}_+ = \frac{-\mathbf{i}}{\sqrt{2\hbar m\omega}}(\mathsf{P} + \mathbf{i}m\omega\mathsf{X}).$$

These rescalings conform to the more common use in the physics community and offer simpler formulae for the coherent states than those using $\mathsf{a}_\pm$. Equation 11.3 becomes

$$\mathsf{H} = \hbar\omega\left(\mathbf{a}^*\mathbf{a} + \frac{1}{2}\mathbb{1}\right) = \hbar\omega\left(\mathbf{a}\mathbf{a}^* - \frac{1}{2}\mathbb{1}\right) \tag{11.18}$$

and the commutator relation becomes

$$[\mathbf{a}, \mathbf{a}^*] = \mathbb{1}.$$

From equations 11.8 and 11.9, the action of $\mathbf{a}^*$ on the energy eigenfunctions ψ_N becomes

$$\mathbf{a}^*\psi_{N-1} = \sqrt{N}\,\psi_N \tag{11.19}$$

with

$$\psi_N = \frac{1}{\sqrt{N!}}\left(\mathbf{a}^*\right)^N\psi_0, \tag{11.20}$$

provided that ψ_N is renormalized by absorbing a phase factor of $\mathbf{i}^N$ into the normalization. The astute reader will notice that this reproduces the absorptions of $1/\mathbf{i}^N$ to get equation 11.10 and of $(-1)^N$ to get equation 11.11. The result of this is that the formula for ψ_N as defined by equation 11.20 is given, precisely, by equation 11.11. Applying equation 11.18 to ψ_{N-1} and simplifying, again using the renormalized wave functions, yields

$$\mathbf{a}\psi_N = \sqrt{N}\,\psi_{N-1}. \tag{11.21}$$

Putting equations 11.19 and 11.21 together gives the interesting equation

$$\mathbf{a}^*\mathbf{a}\psi_N = N\psi_N,$$

explaining the name **number operator** given to the self-adjoint operator $\mathsf{N} = \mathbf{a}^*\mathbf{a}$: when applied to the eigenfunction representing the Nth excited state of the harmonic oscillator, it returns the eigenfunction with the count of the number of quanta of size $\hbar\omega$ above the ground state energy. This is particularly apt for those applications where the state vector ψ_N is interpreted rather than the Nth excited state of a single particle, the presence of N quanta, each of energy $\hbar\omega$. The expected value $\langle \mathsf{N} \rangle_\psi$ then represents the average number of quanta expected upon a measurement of a state represented by the normalized vector ψ.

Returning our attention to coherent states, we have defined these as states represented by an eigenfunction of the annihilation operator. Thus $0 \neq \chi \in \operatorname{dom} \mathbf{a}$ represents a coherent state precisely when $\mathbf{a}\chi = c\chi$ for some complex eigenvalue c. It is easy to write explicitly a formula representing a coherent state. Indeed, setting

$$\chi_c = e^{c\mathbf{a}^*}\psi_0 = \sum_{N=0}^{\infty} \frac{c^N}{\sqrt{N!}}\psi_N, \tag{11.22}$$

a sum convergent in the L^2-norm, we find by applying equation 11.21 that $\mathbf{a}\chi_c = c\chi_c$, so that χ_c represents a coherent state with eigenvalue $c \in \mathbb{C}$. This is valid even when $c = 0$, for then $\chi_0 = \psi_0$ and, as $\mathbf{a}\psi_0 = 0$, we see that the ground state serves as a coherent state with zero eigenvalue. Of course, since the annihilation operator is not self-adjoint, there is no expectation for orthogonality of these eigenfunctions, and in fact an easy calculation gives, for arbitrary complex numbers c and d,

$$\langle \chi_c | \chi_d \rangle = e^{\bar{c}d}. \tag{11.23}$$

This shows, for instance, that no two of these functions representing coherent states are orthogonal and that all those given by equation 11.22 lie on the hyperplane orthogonal to, and passing through, the ground state eigenfunction. In terms of the projection operator $\mathsf{P} = |\psi_0\rangle\langle\psi_0|$, we have $\mathsf{P}\chi_c = \psi_0$ for all c.

Using equation 11.23,

$$|c\rangle = e^{-|c|^2/2}\chi_c \tag{11.24}$$

is a normalized eigenfunction and the inner product becomes

$$\langle c|d\rangle = e^{\bar{c}d - (|c|^2+|d|^2)/2}$$

with

$$|\langle c|d\rangle|^2 = e^{-|c-d|^2}.$$

This formula may be taken as a measure of how close the normalized vectors $|c\rangle$ and $|d\rangle$ are to being orthogonal. For example, setting $c = 0$

gives the inner product between the ground state vector $|0\rangle = \psi_0$ and $|d\rangle$ as $e^{-|d|^2/2}$, showing that as the complex variable d moves out toward infinity, the vectors $|d\rangle$ come closer to being orthogonal to the ground state. One may think of the family $\mathcal{C} = \{|c\rangle : c \in \mathbb{C}\}$ as a collection of vectors on the unit sphere in $L^2(\mathbb{R})$, indexed by the complex plane, that reside in the positive cone of the ground state. This family is the image of the mapping $\mathbb{C} \to L^2(\mathbb{R})$ defined by $c \mapsto |c\rangle$, and continuity of this mapping is an immediate consequence of the easily derived formula $\||c\rangle - |d\rangle\|^2 = 2(1 - \mathrm{Re}\langle c|d\rangle)$. In what sense $\mathcal{C}$ forms a complete set will be determined subsequently.

We now turn our attention to the functional form of $|c\rangle$. This has already been worked out in Section 9.1, where the criterion for a minimal uncertainty state was derived and expressed in equation 9.3. Indeed, since $\mathbf{a}|c\rangle = c|c\rangle$, we have $\mathsf{a}_-|c\rangle = -\mathbf{i}\sqrt{2\hbar m\omega}\, c\,|c\rangle$, verifying equation 9.3 with s identified as $m\omega$ and the eigenvalue $\lambda = \langle \mathsf{P}\rangle_\psi - \mathbf{i}s\langle \mathsf{X}\rangle_\psi$ identified as $-\mathbf{i}\sqrt{2\hbar m\omega}\, c$. Therefore $\psi = |c\rangle$ is identified as a minimal uncertainty state and, as such, up to a possible unimodular phase factor takes the form of the complex Gaussian given by equation 9.4 as

$$\langle x|c\rangle = \left(\frac{m\omega}{\pi\hbar}\right)^{1/4} \exp\left(\mathbf{i}x\langle \mathsf{P}\rangle_{|c\rangle}/\hbar\right) \exp\left(-m\omega(x - \langle \mathsf{X}\rangle_{|c\rangle})^2/2\hbar\right). \tag{11.25}$$

From the eigenvalue equation

$$\langle \mathsf{P}\rangle_{|c\rangle} - \mathbf{i}m\omega\langle \mathsf{X}\rangle_{|c\rangle} = -\mathbf{i}\sqrt{2\hbar m\omega}\, c, \tag{11.26}$$

the expected position coordinate is $\langle \mathsf{X}\rangle_{|c\rangle} = \sqrt{2\hbar/m\omega}\,\mathrm{Re}\, c$ and the expected momentum is $\langle \mathsf{P}\rangle_{|c\rangle} = \sqrt{2\hbar m\omega}\,\mathrm{Im}\, c$. Notice that this identifies the family $\mathcal{C}$ of coherent states of the harmonic oscillator as exactly the full family of minimal uncertainty states of the oscillator that range over all possible values of position $\langle \mathsf{X}\rangle$ and momentum $\langle \mathsf{P}\rangle$. From Section 9.3, the dispersions of these states are given by $\Delta_{|c\rangle}\mathsf{X} = \sqrt{\hbar/2m\omega}$ and $\Delta_{|c\rangle}\mathsf{P} = \sqrt{m\omega\hbar/2}$ and are independent of c, depending as they do only on the mass m and the natural frequency of the oscillator ω. Thus the position probability distributions for the oscillator determined by the coherent states all have a common Gaussian shape, with common standard deviation and with their means varying over all possibilities. Similarly, the momentum probability distributions determined by the Fourier transforms have a common Gaussian shape with common standard deviations and means varying over all possibilities.

The generating function of the preceding section may be used to give an alternative derivation of the position representation of the coherent state vector $|c\rangle$. A little care must be exercised to take into account the absorption of the factor $(-1)^N$ in going from equation 11.10, on which the generating function $G(z,x)$ is based, to equation 11.11, the exact formula for the ψ_N of equation 11.20 used here to develop coherent states. The result of this care is the replacement of z by $-z$ in equation 11.15, and we may write

$$|c\rangle = \left(\frac{m\omega}{\pi\hbar}\right)^{1/4} e^{-|c|^2/2} G_{-z},$$

where $z = \sqrt{\hbar/2m\omega}\, c$. Using equation 11.15, this becomes

$$\begin{aligned}\langle x|c\rangle &= \left(\frac{m\omega}{\pi\hbar}\right)^{1/4} \exp\left(-|c|^2/2\right) \exp\left(-m\omega(2z^2 - 4zx + x^2)/2\hbar\right) \\ &= \left(\frac{m\omega}{\pi\hbar}\right)^{1/4} \exp\left(-(c^2 + |c|^2)/2\right) \\ &\quad \times \exp\left(-m\omega(x^2 - 2\sqrt{2\hbar/m\omega}\, cx)/2\hbar\right),\end{aligned}$$

which differs from equation 11.25 only by a unimodular phase factor.

The time-evolution of a coherent state in the harmonic oscillator potential is very interesting. We do not address time-evolution in quantum mechanics in general until Lecture 27, but we will take the opportunity now to report on the result in the context of coherent states. The result is that if the initial state vector of a particle in the harmonic oscillator potential is the coherent state vector $\varphi_0 = |c\rangle$, then at time t the state vector has evolved to $e^{-i\omega t/2}|c\, e^{-i\omega t}\rangle$.[4] The term $e^{-i\omega t/2}$ is merely a unimodular phase factor and can be ignored for discussion of the time -evolution of a single state,[5] since both state vectors $e^{-i\omega t/2}|c\, e^{-i\omega t}\rangle$ and $|c\, e^{-i\omega t}\rangle$ represent the same physical state. The first thing to notice is that a coherent state remains a coherent state under time-evolution and thus remains a minimal uncertainty state with the same spread, or dispersion, as the initial state vector. The second thing to notice is that the expression $\varphi_t = |c\, e^{-i\omega t}\rangle$ describes a very classical type of oscillatory motion. The label c of the coherent state evolves by rotating around a circle in the complex plane of radius $|c|$, with the natural frequency ω of the oscillator. We will see in Section 27.3 that for any initial state vector ψ, the classical expression

$$m\omega x(t) = m\omega x_0 \cos\omega t + p_0 \sin\omega t$$

[4] See equation 27.7.

[5] If there were interactions with other particles or potentials, we would be unable to ignore this phase factor.

that describes the motion of the classical oscillator with initial position x_0 and initial momentum $p_0 = m\dot{x}(0)$ holds in the quantum setting if $x(t)$ is interpreted as the expectation value $\langle \mathsf{X} \rangle_{\psi_t}$, $x_0 = \langle \mathsf{X} \rangle_{\psi}$ and $p_0 = \langle \mathsf{P} \rangle_{\psi}$. This is easy to see if the initial state is coherent, for then equation 11.26 applied to $\varphi_t = |ce^{-\mathrm{i}\omega t}\rangle$, after a little algebra, yields precisely

$$m\omega \langle \mathsf{X} \rangle_{\varphi_t} = m\omega \langle \mathsf{X} \rangle_{\varphi_0} \cos \omega t + \langle \mathsf{P} \rangle_{\varphi_0} \sin \omega t.$$

The picture that emerges is that the Gaussian wave packet described by the coherent state vector $|ce^{-\mathrm{i}\omega t}\rangle$ retains its shape throughout its dynamic development, while its mean oscillates about the origin along the precise classical path that a classical particle traverses.

What is the expectation value for the energy of a coherent state, or, essentially the same question, what is the expectation value of the number operator N? The answer is that

$$\langle \mathsf{N} \rangle_{|c\rangle} = \langle c|\mathsf{N}|c\rangle = \langle c|\mathbf{a}^*\mathbf{a}|c\rangle = |c|^2,$$

since $\langle c|\mathbf{a}^* = \langle c|\bar{c}$, and, as $\mathsf{H} = \hbar\omega\left(\mathsf{N} + \frac{1}{2}1\right)$, the expectation value for energy measurements is $\langle \mathsf{E} \rangle_{|c\rangle} = \hbar\omega\left(|c|^2 + \frac{1}{2}\right)$. The probability of finding the oscillator in the energy eigenstate represented by ψ_N and of obtaining the particular value $E_N = \left(N + \frac{1}{2}\right)\hbar\omega$ upon an energy measurement of a coherent state represented by $|c\rangle$ is

$$|\langle \psi_N|c\rangle|^2 = \frac{|c|^{2N}e^{-|c|^2}}{N!}. \tag{11.27}$$

This is precisely a Poisson distribution with both mean and variance equal to $|c|^2$. We can check the expectation value for energy measurements by calculating the expected value, say $\mathbf{E}_{\mathrm{Poisson}}(E)$, of the random variable $E = E(N) = \left(N + \frac{1}{2}\right)\hbar\omega$ governed by the Poisson distribution of equation 11.27. The expected value is

$$\begin{aligned}\mathbf{E}_{\mathrm{Poisson}}(E) &= \sum_{N=0}^{\infty} E(N)\frac{|c|^{2N}e^{-|c|^2}}{N!} = \hbar\omega e^{-|c|^2}\sum_{N=0}^{\infty}\left(N + \frac{1}{2}\right)\frac{(|c|^2)^N}{N!} \\ &= \hbar\omega e^{-|c|^2}\left(|c|^2 + \frac{1}{2}\right)e^{|c|^2} = \hbar\omega\left(|c|^2 + \frac{1}{2}\right) = \langle \mathsf{E} \rangle_{|c\rangle},\end{aligned}$$

in agreement with the previous calculation.

The final task of this section is to explore the completeness of the set $\mathcal{C}$ of coherent state vectors in the Hilbert space $L^2(\mathbb{R})$. Our aim is to show that, in an appropriate sense,

$$1_{L^2(\mathbb{R})} = \frac{1}{\pi} \int_{\mathbb{C}} |c\rangle\langle c| \, dA. \tag{11.28}$$

This is a completeness relation that allows one to write any state vector ψ as $\int_{\mathbb{C}} |c\rangle\langle c|\psi\rangle \, dA/\pi$, the real meaning of this being that the inner product with any other state vector φ may be written as the integral

$$\langle\varphi|\psi\rangle = \frac{1}{\pi} \int_{\mathbb{C}} \langle\varphi|c\rangle\langle c|\psi\rangle \, dA.$$

These integrals are with respect to the area element in the plane and may be performed using the reader's favorite coordinate system on $\mathbb{C}$. These formulae say that the set $\mathcal{C}$ is complete, but it is not a minimal complete set. For example, the countable subset of $\mathcal{C}$ of those $|c\rangle$ for which $c \in \mathbb{G} = \mathbb{Z} + \mathbb{Z}\mathbf{i}$, the set of Gaussian integers, is also complete. The implication is that the representation given in equation 11.28 is not unique.

To derive equation 11.28, we need the integrals

$$\int_0^\infty r^{2n+1} e^{-r^2} \, dr = \frac{n!}{2},$$

for $n = 0, 1, 2, \ldots$ The $n = 0$ case is immediate and the remaining cases follow from mathematical induction, using integration by parts to reduce the nth case to the $(n-1)$th case.

To perform the integration of equation 11.28, we use polar coordinates and parameterize the complex variable c as $c = re^{\mathbf{i}\theta}$, for $r \geq 0$ and $0 \leq \theta \leq 2\pi$, with area element $dA = r \, dr \, d\theta$. Using equations 11.22 and 11.24, the integral in equation 11.28 takes the following form:

$$\int_{\mathbb{C}} |c\rangle\langle c| \, dA = \int_0^{2\pi} \int_0^\infty e^{-r^2} \sum_{m=0}^\infty \sum_{n=0}^\infty \frac{r^{m+n}}{\sqrt{m!n!}} |\psi_m\rangle\langle\psi_n| e^{\mathbf{i}(m-n)\theta} r \, dr \, d\theta.$$

Exchanging the order of integration and summation, this becomes

$$\sum_{m=0}^\infty \sum_{n=0}^\infty \frac{|\psi_m\rangle\langle\psi_n|}{\sqrt{m!n!}} \int_0^\infty r^{m+n+1} e^{-r^2} \, dr \int_0^{2\pi} e^{\mathbf{i}(m-n)\theta} \, d\theta.$$

Now, the θ-integral is just $2\pi\delta_{m,n}$ so, using equation 11.14, the above expression becomes

$$2\pi \sum_{n=0}^\infty \frac{|\psi_n\rangle\langle\psi_n|}{n!} \int_0^\infty r^{2n+1} e^{-r^2} \, dr = \pi \sum_{n=0}^\infty |\psi_n\rangle\langle\psi_n| = \pi 1_{L^2(\mathbb{R})},$$

thus verifying equation 11.28.

11.5 End Notes

The first three sections of this lecture follow closely the development in Sections 7.5–7.7 of Hannabuss (1997) and Section 16.1 of Elbaz (1998). In particular, my development of the generating function in Section 11.3 relies heavily on Section 7.7 of Hannabuss (1997). Merzbacher (1998), Section 10.6, is recommended for further reading. The quantum mechanical analysis of oscillators in higher dimensions and of coupled oscillators is hardly more difficult than the analysis given here. The novel and important feature that emerges is the possibility of degeneracy of the eigenvalues, which occurs when, for example, the natural frequencies of two coupled oscillators agree. See Section 3.3 of Hannabuss (1997) for a discussion of degeneracy. Section 10.7 of Merzbacher (1998) has a good discussion of coherent states.

12

Angular Momentum I: Basics

Turn! Turn! Turn! (To Everything There Is A Season)

Pete Seeger

With our first quantum mechanical calculation under our belt in the treatment of the harmonic oscillator in the previous lecture, we are ready to tackle a more difficult calculation. This calculation concerns the quantum mechanical treatment of the central force problem, one of the most important problems of classical mechanics. Of course, a full classical treatment of a particle in a central inverse-square force field leads to some of the most beautiful and useful results of classical mechanics, for example, the calculation of the orbits of moons about planets and planets about suns. It was perhaps the struggle to give an explanation for the structure of the hydrogen spectrum that, as much as any impetus, spurred the development of quantum mechanics a century ago. On the one hand, the classical treatment of this electrodynamic inverse-square force problem leads to results diametrically opposed to observation, and dramatically so. The treatment by nonrelativistic wave mechanics, on the other hand, derives correctly the gross behavior observed in the hydrogen atom: the stability of the atom, the quantization of the absorption and emission of radiation, a first-order understanding of the observed spectral lines of hydrogen, and an explanation, via shells, of the rough outline of the periodic table. This spectacular success of wave mechanics was then superseded by a fully relativistic quantum mechanical treatment, first by the Dirac equation, which refines the picture and offers an explanation of the fine structure of the hydrogen spectrum and of a mysterious factor of 2 in the observed g-factor for the electron. After that came a quantum field-theoretic treatment that provides radiative corrections to the g-factor calculation and an explanation of the Lamb shift. The remarkable success of quantum mechanics is richly illustrated by its treatment of the central force problem.

Not an insignificant fraction of these lectures is devoted to the quantum mechanical treatment of the central force problem. This lecture, along with the next five, presents a detailed account of the nonrelativistic wave mechanical treatment of the general inverse-square force problem.

In the course of these lectures we will see the application of some beautiful mathematics that arise from a consideration of angular momentum in quantum mechanics. The Hamiltonian for the central force potential naturally separates into a radial part and a transverse, angular, part, and it is this angular part that produces the primary structures that arise from the quantum mechanical treatment. Our attention is focused first on angular momentum operators in quantum mechanics, with four lectures devoted to the topic. The present lecture introduces the basics and the next provides an elegant framework in terms of representation theory for understanding the angular part of the Hamiltonian. The subsequent lecture provides the quantum mechanical treatment of the general central force problem, and the one following specializes to a central inverse-square force field with no electromagnetic interaction. Even with no electrodynamic considerations, the broad outlines of the structure of the hydrogen atom appear. We follow with lectures on the hidden symmetries of solutions to the hydrogenic potential and on the addition of angular momentum, important for later adjustments to the portrait that slowly emerges of an electron in the inverse-square electrostatic field of the proton.

12.1 Angular Momentum Operators

The classical angular momentum of a particle with position vector $\mathbf{x}$ and momentum vector $\mathbf{p}$ is the cross product $\mathbf{x} \times \mathbf{p}$. This motivates the definition of the **orbital angular momentum triad** $\mathsf{L} = (\mathsf{L}_1, \mathsf{L}_2, \mathsf{L}_3)$ as the three-component vector of operators defined formally in terms of the position and momentum triads of wave mechanics as the cross product $\mathsf{L} = \mathsf{X} \times \mathsf{P}$, with components $\mathsf{L}_1 = \mathsf{X}_2\mathsf{P}_3 - \mathsf{X}_3\mathsf{P}_2$, $\mathsf{L}_2 = \mathsf{X}_3\mathsf{P}_1 - \mathsf{X}_1\mathsf{P}_3$, and $\mathsf{L}_3 = \mathsf{X}_1\mathsf{P}_2 - \mathsf{X}_2\mathsf{P}_1$. Using the commutation relations $[\mathsf{X}_i, \mathsf{P}_j] = \mathbf{i}\hbar\delta_{ij}1$, a quick calculation shows that each of L_1, L_2, and L_3 is symmetric and, when the appropriate domains are considered, self-adjoint. In wave mechanics, the self-adjoint operator L_i represents measurements of the component of orbital angular momentum in the direction of the x_i-axis.

We may write the components succinctly using the totally antisymmetric tensor of degree three, ϵ_{ijk}, where $\epsilon_{ijk} = \operatorname{sgn}(i,j,k)$ if $\{i,j,k\} = \{1,2,3\}$ and $\epsilon_{ijk} = 0$ otherwise. Indeed, the three component operators of L may be written as

$$\mathsf{L}_i = \epsilon_{ijk}\mathsf{X}_j\mathsf{P}_k,$$

where we use the convention that repeated indices in an expression are to be summed over as the repeated indices independently take on all values in $\{1,2,3\}$.

Here are some important commutator relations that hold among the orbital angular momentum operators:

$$\begin{aligned} [\mathsf{L}_i, \mathsf{X}_j] &= \mathbf{i}\hbar\epsilon_{ijk}\mathsf{X}_k \\ [\mathsf{L}_i, \mathsf{P}_j] &= \mathbf{i}\hbar\epsilon_{ijk}\mathsf{P}_k \\ [\mathsf{L}_i, \mathsf{L}_j] &= \mathbf{i}\hbar\epsilon_{ijk}\mathsf{L}_k. \end{aligned} \tag{12.1}$$

The verifications of these identities are straightforward, using identity 8.4. For example, we have $[\mathsf{X}_i\mathsf{P}_j, \mathsf{P}_k] = \mathsf{X}_i[\mathsf{P}_j, \mathsf{P}_k] + [\mathsf{X}_i, \mathsf{P}_k]\mathsf{P}_j = \mathbf{i}\hbar\delta_{ik}\mathsf{P}_j$. Then $[\mathsf{L}_i, \mathsf{P}_j] = \epsilon_{isk}[\mathsf{X}_s\mathsf{P}_k, \mathsf{P}_j] = \mathbf{i}\hbar\epsilon_{isk}\delta_{sj}\mathsf{P}_k = \mathbf{i}\hbar\epsilon_{ijk}\mathsf{P}_k$, giving the second identity among the three. As another example, we have

$$\begin{aligned} [\mathsf{L}_1, \mathsf{L}_2] &= [\mathsf{L}_1, \mathsf{X}_3\mathsf{P}_1 - \mathsf{X}_1\mathsf{P}_3] \\ &= \mathsf{X}_3[\mathsf{L}_1, \mathsf{P}_1] + [\mathsf{L}_1, \mathsf{X}_3]\mathsf{P}_1 - \mathsf{X}_1[\mathsf{L}_1, \mathsf{P}_3] - [\mathsf{L}_1, \mathsf{X}_1]\mathsf{P}_3 \\ &= 0 - \mathbf{i}\hbar\mathsf{X}_2\mathsf{P}_1 + \mathbf{i}\hbar\mathsf{X}_1\mathsf{P}_2 - 0 \\ &= \mathbf{i}\hbar\mathsf{L}_3. \end{aligned}$$

Much of the theory of the orbital angular momentum operators is a purely algebraic consequence of the last of these identities, $[\mathsf{L}_i, \mathsf{L}_j] = \mathbf{i}\hbar\epsilon_{ijk}\mathsf{L}_k$; thus it is useful to abstract the study of these operators and derive as much as possible from the algebra before specifically applying the orbital operators to the central force problem. Our foresight will be rewarded when the purely quantum mechanical property of **spin** arises and the spin operators are found to satisfy analogous identities. This motivates the following definition.

Angular momentum operators. Any three self-adjoint operators J_1, J_2, and J_3 on the Hilbert space $\mathcal{H}$ that satisfy the commutation relations

$$[\mathsf{J}_i, \mathsf{J}_j] = \mathbf{i}\hbar\epsilon_{ijk}\mathsf{J}_k \tag{12.2}$$

are called **angular momentum operators**, and the ordered triple $\mathsf{J} = (\mathsf{J}_1, \mathsf{J}_2, \mathsf{J}_3)$ is an **angular momentum triad**.

There are two useful variations of equation 12.2. Given a real 3-tuple $\mathbf{a} = (a_1, a_2, a_3)$ and an angular momentum triad J, define the **angular momentum operator in the direction of the 3-tuple a** as the formal dot product given by

$$\mathbf{a} \cdot \mathsf{J} = a_1\mathsf{J}_1 + a_2\mathsf{J}_2 + a_3\mathsf{J}_3.$$

Then, in place of equation 12.2 we may write

$$[\mathbf{a} \cdot \mathsf{J}, \mathbf{b} \cdot \mathsf{J}] = \mathbf{i}\hbar(\mathbf{a} \times \mathbf{b}) \cdot \mathsf{J}.$$

The second variant is obtained by formally taking the cross product of the triad J with itself. A straightforward calculation shows that

$$\mathsf{J} \times \mathsf{J} = \mathbf{i}\hbar\mathsf{J}.$$

12.2 Eigenvectors and Eigenvalues

The goal in the remainder of this introductory lecture is to identify the eigenvalues of angular momentum operators with the corresponding complete set of eigenvectors. To this end, it is convenient to define auxiliary operators, the **total angular momentum operator**

$$\mathsf{J}^2 = \mathsf{J} \cdot \mathsf{J} = \mathsf{J}_1^2 + \mathsf{J}_2^2 + \mathsf{J}_3^2$$

and the **ladder operators**

$$\mathsf{J}_\pm = \mathsf{J}_1 \pm \mathbf{i}\mathsf{J}_2.$$

The solution to the problem of identifying eigenvalues turns out to be purely algebraic and a consequence of the commutation relations for the operators J_1, J_2, J_3, J^2, and $\mathsf{J}_\pm$. This mirrors the use of the creation and annihilation operators $\mathsf{a}_\pm = \mathsf{P} \pm \mathbf{i}m\hbar\mathsf{X}$ in identifying the eigenvalues and eigenfunctions of the harmonic oscillator Hamiltonian. That derivation was largely algebraic and worked precisely because of the commutation relations for the operators X, P, H, and $\mathsf{a}_\pm$. We first give the final result, and then spend the remainder of the lecture on its derivation. It turns out that the eigenvalues of J^2 are labeled by nonnegative half-integers and are degenerate. Any of the operators $\mathbf{a} \cdot \mathsf{J}$ may be used to further decompose an eigenspace of J^2 into a sum of eigenspaces of $\mathbf{a} \cdot \mathsf{J}$.[1] We will

[1] See Section 12.4.

concentrate on deriving simultaneous eigenstates of $\mathbf{J}^2$ and $(0,0,1)\cdot\mathbf{J} = \mathsf{J}_3$.

Eigenvalues and eigenvectors of $\mathbf{J}^2$ and J_3 *The $\mathbf{J}^2$-eigenspaces of $\mathcal{H}$ are invariant under the action of J_3. Moreover, the following items hold.*

(i) *The eigenvalues of $\mathbf{J}^2$ have the form $j(j+1)\hbar^2$, where j takes on nonnegative half-integer values: $j = n/2$ for $n = 0, 1, 2, \ldots$*

(ii) *For each choice of j, the eigenvalues of J_3 associated with states in the $j(j+1)\hbar^2$-eigenspace of $\mathbf{J}^2$ are $m\hbar$, where m is to be taken from the list $-j, -j+1, \ldots, j-1, j$.*

(iii) *The degeneracies of all the eigenvalues of J_3 from item* (ii) *coincide, meaning that each $m\hbar$-eigenspace has the dimension of the $j\hbar$-eigenspace.*

(iv) *If $j\hbar$ is a nondegenerate J_3-eigenvalue from item* (ii) *then the $j(j+1)\hbar^2$-eigenspace of $\mathbf{J}^2$ is $(2j+1)$-dimensional with an orthonormal basis of the form $\{|m\rangle : m = -j, -j+1, \ldots, j-1, j\}$, where*

$$\mathsf{J}_3|m\rangle = m\hbar\,|m\rangle$$

and

$$\mathsf{J}_\pm|m\rangle = \sqrt{(j \mp m)(j \pm m + 1)}\hbar\,|m \pm 1\rangle. \tag{12.3}$$

Notice that we are using Dirac ket-notation for the eigenstates, the **magnetic quantum number** m labeling an $m\hbar$-eigenstate $|m\rangle$ of J_3. Of course each $|m\rangle$ is also a $j(j+1)\hbar^2$-eigenstate of $\mathbf{J}^2$. To acknowledge this, many authors use both **quantum numbers** j and m to label the simultaneous eigenstates of the operators $\mathbf{J}^2$ and J_3. In summary then, we may write

$$\begin{aligned} \mathbf{J}^2|j,m\rangle &= j(j+1)\hbar^2\,|j,m\rangle & j &= 0, \tfrac{1}{2}, 1, \tfrac{3}{2}, \ldots \\ \mathsf{J}_3|j,m\rangle &= m\hbar\,|j,m\rangle & m &= -j, -j+1, \ldots, j-1, j. \end{aligned} \tag{12.4}$$

12.3 Derivation

We begin with a list of commutation relations, all easily verified.

$$\begin{aligned} [\mathbf{J}^2, \mathsf{J}_i] &= 0 \quad \text{for} \quad i = 1,2,3, \\ [\mathbf{J}^2, \mathsf{J}_\pm] &= 0, \\ [\mathsf{J}_+, \mathsf{J}_-] &= 2\hbar \mathsf{J}_3, \\ [\mathsf{J}_3, \mathsf{J}_\pm] &= \pm\hbar \mathsf{J}_\pm. \end{aligned} \tag{12.5}$$

For example, using the summation convention to write $\mathbf{J}^2 = \mathsf{J}_j\mathsf{J}_j$, we may calculate

$$[\mathsf{J}_i, \mathbf{J}^2] = [\mathsf{J}_i, \mathsf{J}_j\mathsf{J}_j] = \mathsf{J}_j[\mathsf{J}_i, \mathsf{J}_j] + [\mathsf{J}_i, \mathsf{J}_j]\mathsf{J}_j = \mathbf{i}\hbar\epsilon_{ijk}(\mathsf{J}_j\mathsf{J}_k + \mathsf{J}_k\mathsf{J}_j).$$

This last expression is equal to zero since ϵ_{ijk} is antisymmetric in j and k while $\mathsf{J}_j\mathsf{J}_k + \mathsf{J}_k\mathsf{J}_j$ is symmetric in j and k. This gives the first of the identities in equation 12.5, and the remaining identities are even easier to derive. Another quick calculation gives

$$\mathsf{J}_\pm\mathsf{J}_\mp = \mathbf{J}^2 - \mathsf{J}_3^2 \pm \hbar\mathsf{J}_3. \tag{12.6}$$

Since, by the first of the identities in equation 12.5, $\mathbf{J}^2$ and J_3 commute, the $\mathbf{J}^2$-eigenspaces of $\mathcal{H}$ are invariant under the action of J_3. Let ψ be a common eigenvector of $\mathbf{J}^2$ and J_3 that satisfies $\mathbf{J}^2\psi = \lambda\hbar^2\psi$ and $\mathsf{J}_3\psi = m\hbar\psi$, for some constants λ and m. From the second identity of equation 12.5 we have $\mathbf{J}^2\mathsf{J}_\pm\psi = \lambda\hbar^2\mathsf{J}_\pm\psi$, showing that $\mathsf{J}_\pm\psi$ is in the $\lambda\hbar^2$-eigenspace of $\mathbf{J}^2$ whenever ψ is. The last of the identities of equation 12.5 may be applied to get

$$\mathsf{J}_3\mathsf{J}_\pm\psi = \mathsf{J}_\pm\mathsf{J}_3\psi \pm \hbar\mathsf{J}_\pm\psi = (m \pm 1)\hbar\mathsf{J}_\pm\psi,$$

showing that J_+ creates, and J_- annihilates, a quantum of angular momentum, mapping a $m\hbar$-eigenstate of J_3 to a $(m \pm 1)\hbar$-eigenstate, provided that $\mathsf{J}_\pm\psi \neq 0$. Since the creation operator J_+ and the annihilation operator J_- are adjoints of one another, we have from equation 12.6 that

$$\begin{aligned} \|\mathsf{J}_\pm\psi\|^2 &= \langle\psi|\mathsf{J}_\mp\mathsf{J}_\pm\psi\rangle \\ &= \langle\psi|(\mathbf{J}^2 - \mathsf{J}_3^2 \mp \hbar\mathsf{J}_3)\psi\rangle \\ &= (\lambda - m(m \pm 1))\hbar^2\|\psi\|^2. \end{aligned} \tag{12.7}$$

It follows that $\lambda \geq m(m \pm 1)$, with equality holding if and only if $\mathsf{J}_\pm\psi = 0$. This shows that, for a given value of λ, the value of m is bounded both above and below. By iterating for each positive integer N the application of the creation and annihilation operators to ψ, the state $\mathsf{J}_\pm^N\psi$ either vanishes or is another J_3-eigenstate with eigenvalue $(m \pm N)\hbar$. By the preceding observation, the collection of the eigenvalues $(m \pm N)\hbar$ of J_3 must be bounded. It is evident then that there are nonnegative integers p and q for which $\mathsf{J}_+^{p+1}\psi = 0 = \mathsf{J}_-^{q+1}\psi$, and we may assume that these are the least such integers, so that $\mathsf{J}_+^p\psi \neq 0$ and $\mathsf{J}_-^q\psi \neq 0$. Since $\mathsf{J}_+^{p+1}\psi = 0$ while $\mathsf{J}_+^p\psi \neq 0$, the criterion above implies that $(m+p)(m+p+1) = \lambda$ and, similarly, $(m-q)(m-q-1) = \lambda$. A little algebra then shows that $2m(p+q+1) = (q-p)(p+q+1)$ and, as $p+q+1 > 0$, that

$m = (q - p)/2$. Since p and q are nonnegative integers, $j = (p + q)/2$ is a nonnegative half-integer, and the $\mathbf{J}^2$-eigenvalue is given by $\lambda\hbar^2 = (m + p)(m + p + 1)\hbar^2 = j(j + 1)\hbar^2$, all confirming item (i) of the main result.

For item (ii), first note that the quantum number m denoting the J_3-eigenvalue $m\hbar$ associated with the eigenstate ψ satisfies $m - q = -j$ and $m + p = j$. Evidently, from the choices of p and q, the states

$$\mathsf{J}_-^q\psi, \mathsf{J}_-^{q-1}\psi, \ldots, \mathsf{J}_-\psi, \psi, \mathsf{J}_+\psi, \ldots, \mathsf{J}_+^{p-1}\psi, \mathsf{J}_+^p\psi \tag{12.8}$$

are all nonzero, while $\mathsf{J}_-^{q+1}\psi = 0 = \mathsf{J}_+^{p+1}\psi$. There are $q + 1 + p = 2j + 1$ states listed and the leftmost satisfies

$$\mathsf{J}_3\mathsf{J}_-^q\psi = (m - q)\hbar\mathsf{J}_-^q\psi = -j\hbar\mathsf{J}_-^q\psi,$$

while the rightmost satisfies

$$\mathsf{J}_3\mathsf{J}_+^p\psi = (m + p)\hbar\mathsf{J}_+^p\psi = j\hbar\mathsf{J}_+^p\psi.$$

Since the eigenvalues must increase by one quantum, $\hbar$, of angular momentum as we move from left to right, we have, from the existence of the $m\hbar$-eigenstate ψ, demonstrated the existence of J_3-eigenstates with eigenvalues $-j\hbar, (-j + 1)\hbar, \ldots, (j - 1)\hbar, j\hbar$ and, as ψ is in the list, m must be one of $-j, -j + 1, \ldots, j - 1, j$. This verifies item (ii).

The verification of item (iii) is left to the reader, who merely needs to observe that the image of the $m\hbar$-eigenspace under the ladder operators, if not the null space, is an isomorphic copy of the $m\hbar$-eigenspace.

Arriving at item (iv), we normalize each eigenstate in the list in equation 12.8 and use the quantum number m as a label in the Dirac notation, so that the list becomes

$$|-j\rangle, |-j + 1\rangle, \ldots, |j - 1\rangle, |j\rangle, \tag{12.9}$$

where, for each $-j \leq m \leq j$,

$$\mathbf{J}^2|m\rangle = j(j + 1)\hbar^2\,|m\rangle$$

and

$$\mathsf{J}_3|m\rangle = m\hbar\,|m\rangle.$$

If $k \neq m$ then, since the J_3-eigenvalues for $|k\rangle$ and $|m\rangle$ differ, the two states are orthogonal; thus, $\langle k|m\rangle = \delta_{k,m}$, so the eigenstates listed in equation 12.9 are orthonormal. Equation 12.3 follows from equation 12.7.

The $j(j+1)\hbar^2$-eigenspace $\mathcal{H}_j$ of $\mathcal{H}$ determined by $\mathbf{J}^2$ decomposes as the orthogonal direct sum

$$\mathcal{H}_j = \bigoplus_{m=-j}^{j} \mathcal{H}_{j,m},$$

where $\mathcal{H}_{j,m}$ is the $m\hbar$-eigenspace of $\mathcal{H}_j$ determined by J_3. So far, no assumptions as to the degeneracy of eigenvalues has been made. Now making the assumption that $j\hbar$ is a nondegenerate J_3-eigenvalue, item (iii) implies that each $\mathcal{H}_{j,m}$ is one-dimensional, so the direct sum decomposition of $\mathcal{H}_j$ shows it to be $(2j+1)$-dimensional with orthonormal basis given by equation 12.8. Item (iv) stands verified.

12.4 Further Remarks on Angular Momentum

The structure uncovered in this lecture depends on the total angular momentum operator $\mathbf{J}^2$ and one of its component operators, J_3 being conventionally chosen. A close examination of the arguments in the previuos section reveals that all the results are derived from the commutation relations listed in equations 12.2 and 12.5. For any 3-tuple $\mathbf{a}$ in $\mathbb{R}^3$ of unit length, choose unit vectors $\mathbf{v}$ and $\mathbf{w}$ whose cross-product $\mathbf{v} \times \mathbf{w}$ equals $\mathbf{a}$. Set $\mathsf{K}_1 = \mathbf{v} \cdot \mathbf{J}$, $\mathsf{K}_2 = \mathbf{w} \cdot \mathbf{J}$, $\mathsf{K}_3 = \mathbf{a} \cdot \mathbf{J}$, and $\mathsf{K}_\pm = \mathsf{K}_1 \pm i\mathsf{K}_2$. The cross-product version of equation 12.2 shows that $\mathbf{K} = (\mathsf{K}_1, \mathsf{K}_2, \mathsf{K}_3)$ is an angular momentum triad and, hence, $\mathbf{K}$, $\mathsf{K}_\pm$, and K^2 satisfy equations 12.2 and 12.5, and the theorem of Section 12.1 applies to the eigenvalues and eigenvectors of $\mathbf{K}^2$ and K_3. Let M denote the 3×3 matrix whose row vectors are respectively $\mathbf{v}$, $\mathbf{w}$, and $\mathbf{a}$, and observe that $\mathbf{K} = \mathsf{M}\mathbf{J}$, where $\mathbf{K}$ and $\mathbf{J}$ are interpreted as column vectors of operators. As $\mathbf{v}$, $\mathbf{w}$, and $\mathbf{a} = \mathbf{v} \times \mathbf{w}$ form an orthonormal triple, the matrix M is orthogonal with unit determinant, and therefore $\mathsf{M}^{-1} = \mathsf{M}^{\mathsf{Tr}}$. We now have that $\mathbf{K}^2 = \mathbf{K}^{\mathsf{Tr}}\mathbf{K} = \mathbf{J}^{\mathsf{Tr}}\mathsf{M}^{\mathsf{Tr}}\mathsf{M}\mathbf{J} = \mathbf{J}^{\mathsf{Tr}}\mathbf{J} = \mathbf{J}^2$. Since the total angular momentum operators $\mathbf{K}^2$ and $\mathbf{J}^2$ coincide and $\mathsf{K}_3 = \mathbf{a} \cdot \mathbf{J}$, it follows that items (i) through (iv) hold exactly as stated for $\mathbf{J}^2$, but with J_3 replaced by $\mathbf{a} \cdot \mathbf{J}$. Thus the angular momentum operator obtained from $\mathbf{J}$ but in the direction of any nonzero unit vector $\mathbf{a}$ may be substituted for J_3 in the theorem. This is simply a mathematical validation of an obvious physical fact, to wit, that one is free to choose any direction in which to measure a component of angular momentum with the aim of further decomposing the eigenspaces of the total angular momentum operator.

12.5 End Notes

The content of this lecture appears in practically every text on quantum mechanics. My presentation of the content owes a huge debt to Chapter 8 of Hannabuss (1997), which to me is one of the more mathematically clear expositions of angular momentum operators and their eigenvectors and eigenvalues.

13
Angular Momentum II: Representations of $\mathfrak{su}(2)$

> Now, a few words regarding Quaternions. It is known that Sir W. Rowan Hamilton discovered or invented a remarkable system of mathematics, and that since his death the quaternionic mantle has adorned the shoulders of Prof. Tait, who has repeatedly advocated the claims of Quaternions. Prof. Tait in particular emphasizes its great power, simplicity, and perfect naturalness, on the one hand; and on the other tells the physicist that it is exactly what he wants for his physical purposes. It is known that physicists, with great obstinacy, have been careful (generally speaking) to have nothing to do with Quaternions; and, what is equally remarkable, writers who take up the subject of Vectors are (generally speaking) possessed of the idea that Quaternions is not exactly what they want, and so they go tinkering at it, trying to make it a little more intelligible, very much to the disgust of Prof. Tait, who would preserve the quaternionic stream pure and undefiled. Now, is Prof. Tait right, or are the defilers right? Opinions may differ. My own is that the answer all depends upon the point of view.
>
> Oliver Heaviside
> *Electromagnetic Theory,*
> Dover, 1950
> (Reprint of December 9, 1882 issue of *The Electrician*)

It turns out that representation theory is a particularly apt and powerful language in which to express the important features of angular momentum operators. This lecture continues the theme of its predecessor, beginning with the explicit example of an angular momentum triad in the first nontrivial case, viz., for the $j(j+1)\hbar^2$-eigenspace when $j = \frac{1}{2}$. This introduces the **Pauli spin matrices**, which are used to uncover and expose the Lie algebra $\mathfrak{su}(2)$ lurking just beneath the surface of angular momentum in quantum mechanics. Having uncovered this Lie algebra, we then realize angular momentum triads as $\mathfrak{su}(2)$-representations on the Hilbert state space and use this to explicate more extensively the theory. Quaternions make their appearance as a natural setting for examining both the topology of the Lie group SU(2) as well as the relationships between SU(2) and SO(3). All this leads to representations of the Lie group SU(2), whose Lie algebra is $\mathfrak{su}(2)$, and of its quotient group SO(3), which is used in the next lecture in articulating the wave mechanical approach to orbital angular momentum. We find in a later lecture that the orbital angular momentum triad restricts the quantum number j to take on only integer values rather than the half-integer values allowed by the theory for the general abstract triad. This is a reflection of the fact that SO(3) is double-covered by the simply connected group SU(2), which restricts the representation spaces for SO(3). Also, later we will see that the half-integral total angular momentum quantum numbers are certainly not superfluous in that they index the eigenvalues that correspond to the purely quantum mechanical attribute of **spin**, a feature of nature explained ultimately by a relativistic quantum mechanical treatment of particles. The reader who is deficient in the language of Lie groups and Lie algebras, and their representations, may consult the references given in the End Notes to this lecture.

13.1 The Pauli Spin Matrices and the Lie Algebra $\mathfrak{su}(2)$

When $j = \frac{1}{2}$, $2j + 1 = 2$ and, when the eigenvalues are nondegenerate, the $\frac{1}{2}(\frac{1}{2}+1)\hbar^2$-eigenspace of $\mathbf{J}^2$ is spanned by the two eigenvectors $\psi_\pm = |\frac{1}{2}, \pm\frac{1}{2}\rangle$, in the Dirac notation of equation 12.4. We have

$$\mathsf{J}_3\psi_\pm = \pm\tfrac{1}{2}\hbar\psi_\pm \quad \text{and} \quad \mathbf{J}^2\psi_\pm = \tfrac{3}{4}\hbar^2\psi_\pm.$$

In addition, the ladder operators behave as follows:

$$\mathsf{J}_+\psi_- = \hbar\psi_+, \quad \mathsf{J}_-\psi_+ = \hbar\psi_-, \quad \text{and} \quad \mathsf{J}_\pm\psi_\pm = 0.$$

With respect to the basis $\{\psi_+, \psi_-\}$ of the two-dimensional $\frac{3}{4}\hbar^2$-eigenspace of $\mathbf{J}^2$, the operators J_3 and the ladder operators $\mathsf{J}_\pm$ are represented by matrices as

$$\mathsf{J}_3 = \tfrac{1}{2}\hbar\begin{pmatrix}1 & 0\\ 0 & -1\end{pmatrix}, \quad \mathsf{J}_+ = \hbar\begin{pmatrix}0 & 1\\ 0 & 0\end{pmatrix}, \quad \mathsf{J}_- = \hbar\begin{pmatrix}0 & 0\\ 1 & 0\end{pmatrix}.$$

Using the identities $J_1 = \frac{1}{2}(J_+ + J_-)$ and $J_2 = \frac{1}{2i}(J_+ - J_-)$, the other two components of the triad J may be represented as

$$J_1 = \tfrac{1}{2}\hbar\begin{pmatrix}0 & 1\\ 1 & 0\end{pmatrix} \quad \text{and} \quad J_2 = \tfrac{1}{2}\hbar\begin{pmatrix}0 & -\mathbf{i}\\ \mathbf{i} & 0\end{pmatrix}.$$

This matrix representation of **J** is called the **spin representation**, and the matrices

$$\sigma_1 = \begin{pmatrix}0 & 1\\ 1 & 0\end{pmatrix}, \quad \sigma_2 = \begin{pmatrix}0 & -\mathbf{i}\\ \mathbf{i} & 0\end{pmatrix}, \quad \text{and} \quad \sigma_3 = \begin{pmatrix}1 & 0\\ 0 & -1\end{pmatrix}$$

are the **Pauli spin matrices.** We will use the notation **S** to denote the spin representation, so, using $\boldsymbol{\sigma} = (\sigma_1, \sigma_2, \sigma_3)$, the spin representation may be written as

$$\mathbf{S} = \tfrac{1}{2}\hbar\boldsymbol{\sigma} \quad \text{with} \quad \mathbf{a}\cdot\mathbf{S} = \tfrac{1}{2}\hbar\,\mathbf{a}\cdot\boldsymbol{\sigma} = \tfrac{1}{2}\hbar\begin{pmatrix}a_3 & a_1 - \mathbf{i}a_2\\ a_1 + \mathbf{i}a_2 & -a_3\end{pmatrix},$$

where $\mathbf{a} = (a_1, a_2, a_3) \in \mathbb{R}^3$. Notice that $\mathbf{a}\cdot\boldsymbol{\sigma}$ is a Hermitian matrix of trace zero. In fact, every 2×2 trace-free Hermitian matrix is of this form. Indeed, a quick calculation shows that, for any 2×2 matrix M,

$$M = \tfrac{1}{2}\left[\mathrm{tr}(M)1_{2\times 2} + \sum_{i=1}^{3}\mathrm{tr}(M\sigma_i)\sigma_i\right],$$

where $1_{2\times 2}$ is the 2×2 identity matrix. Therefore, when M is trace-free, $M = \mathbf{a}\cdot\boldsymbol{\sigma}$ where $\mathbf{a} = \frac{1}{2}(\mathrm{tr}(M\sigma_1), \mathrm{tr}(M\sigma_2), \mathrm{tr}(M\sigma_3))$, and, when M is Hermitian, each $\mathrm{tr}(M\sigma_i) \in \mathbb{R}$ so that $\mathbf{a} \in \mathbb{R}^3$.

The commutation properties of the Pauli matrices encode the relevant algebraic information which finds application throughout the quantum mechanical analyses of spin and angular momentum. These are derived from the basic observation that

$$\sigma_i\sigma_j = \delta_{ij}1_{2\times 2} + \epsilon_{ijk}\mathbf{i}\sigma_k \tag{13.1}$$

or, equivalently, for $\mathbf{a}, \mathbf{b} \in \mathbb{R}^3$,

$$(\mathbf{a}\cdot\boldsymbol{\sigma})(\mathbf{b}\cdot\boldsymbol{\sigma}) = (\mathbf{a}\cdot\mathbf{b})1_{2\times 2} + \mathbf{i}(\mathbf{a}\times\mathbf{b})\cdot\boldsymbol{\sigma}. \tag{13.1}$$

This implies commutation relations for $\boldsymbol{\sigma}$, given here in two equivalent forms as

$$[\sigma_i, \sigma_j] = 2\mathbf{i}\epsilon_{ijk}\sigma_k \quad \text{or} \quad [\mathbf{a}\cdot\boldsymbol{\sigma}, \mathbf{b}\cdot\boldsymbol{\sigma}] = 2\mathbf{i}(\mathbf{a}\times\mathbf{b})\cdot\boldsymbol{\sigma}, \tag{13.2}$$

and the **anticommutation relations**

$$\sigma_i\sigma_j + \sigma_j\sigma_i = 2\delta_{ij}1_{2\times 2}$$
$$\text{or} \tag{13.3}$$
$$(\mathbf{a}\cdot\boldsymbol{\sigma})(\mathbf{b}\cdot\boldsymbol{\sigma}) + (\mathbf{b}\cdot\boldsymbol{\sigma})(\mathbf{a}\cdot\boldsymbol{\sigma}) = 2(\mathbf{a}\cdot\mathbf{b})1_{2\times 2}.$$

Readers familiar with quaternions or the Lie algebra $\mathfrak{su}(2)$ may see a hint of resemblance between the relations in $\mathfrak{su}(2)$ and among the Pauli matrices, though something is not quite the same, a missing negative sign here, some extra **i**'s there. We will clear this up now. The **special unitary group** SU(2) of complex 2×2 matrices of unit determinant that preserve the standard Hermitian product on $\mathbb{C}^2$ given by

$$\langle (z_1, w_1)|(z_2, w_2)\rangle = \overline{z}_1 z_2 + \overline{w}_1 w_2$$

is given as follows:

$$\mathrm{SU}(2) = \left\{ \mathsf{M} \in \mathrm{M}_2(\mathbb{C}) : \mathsf{M}\overline{\mathsf{M}}^{\mathsf{Tr}} = 1_{2\times 2} \text{ and } \det \mathsf{M} = 1 \right\}.$$

The transpose-conjugate matrix $\overline{\mathsf{M}}^{\mathsf{Tr}}$ in fact serves as the operator adjoint to M, and we will use the adjoint notation $\mathsf{M}^\dagger$ rather than $\overline{\mathsf{M}}^{\mathsf{Tr}}$. The Lie algebra of SU(2) is $\mathfrak{su}(2)$, the set of 2×2 trace-free anti-Hermitian matrices, i.e., those 2×2 matrices of trace 0 with $\mathsf{M} + \mathsf{M}^\dagger = 0$. This is a real Lie algebra whose Lie bracket is the commutator of matrices $[\mathsf{M}, \mathsf{M}'] = \mathsf{M}\mathsf{M}' - \mathsf{M}'\mathsf{M}$. The anti-Hermitian condition differs from the Hermitian condition only in sign: M is Hermitian when $\mathsf{M} - \mathsf{M}^\dagger = 0$. It is immediate, then, that a matrix M is Hermitian if and only if $\mathbf{i}\mathsf{M}$ is anti-Hermitian, and this allows us to express elements of $\mathfrak{su}(2)$ in terms of the Pauli spin matrices. Indeed, define the triad $\boldsymbol{\tau} = (\tau_1, \tau_2, \tau_3)$ by $\boldsymbol{\tau} = -\mathbf{i}\boldsymbol{\sigma}$. Explicitly,

$$\tau_1 = \begin{pmatrix} 0 & -\mathbf{i} \\ -\mathbf{i} & 0 \end{pmatrix}, \quad \tau_2 = \begin{pmatrix} 0 & -1 \\ 1 & 0 \end{pmatrix}, \quad \text{and} \quad \tau_3 = \begin{pmatrix} -\mathbf{i} & 0 \\ 0 & \mathbf{i} \end{pmatrix}.$$

Our claim is that

$$\mathfrak{su}(2) = \{\mathbf{a}\cdot\boldsymbol{\tau} : \mathbf{a} \in \mathbb{R}^3\}.$$

This is a consequence of the following equivalences:

$$\begin{aligned} \mathsf{M} \in \mathfrak{su}(2) &\iff \operatorname{tr}\mathsf{M} = 0 \text{ and } \mathsf{M} \text{ is anti-Hermitian} \\ &\iff \operatorname{tr}(\mathbf{i}\mathsf{M}) = 0 \text{ and } \mathbf{i}\mathsf{M} \text{ is Hermitian} \\ &\iff \mathbf{i}\mathsf{M} = \mathbf{a}\cdot\boldsymbol{\sigma} \text{ for some } \mathbf{a} \in \mathbb{R}^3 \\ &\iff \mathsf{M} = \mathbf{a}\cdot\boldsymbol{\tau} \text{ for some } \mathbf{a} \in \mathbb{R}^3. \end{aligned}$$

From the Pauli commutation relations of equations 13.2, we have

$$[\mathbf{a} \cdot \boldsymbol{\tau}, \mathbf{b} \cdot \boldsymbol{\tau}] = 2(\mathbf{a} \times \mathbf{b})\boldsymbol{\tau}, \tag{13.4}$$

which shows that the mapping $\mathbf{a} \mapsto \mathbf{a} \cdot \boldsymbol{\tau}$ is a real Lie algebra isomorphism of $\mathbb{R}^3$, with Lie bracket given by twice the cross-product $\times$,[1] onto $\mathfrak{su}(2)$.

In the spin representation of $\mathbf{J}$ we have been a bit cavalier in our notation. In terms of $\boldsymbol{\tau}$, the spin representation reads $\mathbf{S} = \frac{1}{2}\mathbf{i}\hbar\boldsymbol{\tau}$. To be precise, we should say that the restriction $\mathbf{J}|\mathcal{H}_{1/2}$ is represented by the matrix triad $\mathbf{S} = \frac{1}{2}\mathbf{i}\hbar\boldsymbol{\tau}$ in the basis $\{\psi_+, \psi_-\}$, where $\mathcal{H}_{1/2}$ is the $\frac{1}{2}(\frac{1}{2}+1)\hbar^2$-eigenspace of the operator $\mathbf{J}^2$ on the Hilbert state space $\mathcal{H}$. This is not mere pedantry, for we will have to take a bit more care with notational accuracy in the remainder of this lecture. We now consider the general case and incorporate the quantum numbers j, as they vary over all nonnegative half-integer values, into the $\mathfrak{su}(2)$ description.

13.2 Angular Momentum and $\mathfrak{su}(2)$-Representations

A **representation** of the Lie algebra $\mathfrak{g}$ on the complex vector space V is a Lie algebra homomorphism ρ of $\mathfrak{g}$ into the Lie algebra $\mathrm{End}(V)$. Here $\mathrm{End}(V)$ is the real vector space of endomorphisms of V, the complex linear maps of V into V. The Lie bracket of $\mathrm{End}(V)$ is just the commutator of operators. When $\mathrm{End}(V)$ is given this Lie algebra structure via the commutator, it is denoted $\mathfrak{gl}(V)$. The representation ρ makes V into a $\mathfrak{g}$-module via $g \cdot v = \rho(g)(v)$, for $g \in \mathfrak{g}$ and $v \in V$, and the mapping $\rho(g)$ is said to represent the element g of $\mathfrak{g}$ as a linear transformation on V. The representation ρ is **irreducible** if there are no proper nonzero $\mathfrak{g}$-invariant subspaces of V. This is all unambiguous and straightforward when V is finite-dimensional, but various amendments may be appropriate when V is infinite-dimensional. For example, questions of continuity arise so that it might be fitting to require that ρ be continuous when $\mathfrak{gl}(V)$ is given some particular topology, or that each transformation $\rho(g)$ be continuous when V is given some particular topology. When V is an infinite-dimensional Hilbert space, the target space of ρ might be adjusted to allow that self-adjoint operators represent elements of $\mathfrak{g}$ rather than linear maps defined on the whole of V. Instead of pursuing the

[1] The usual Lie bracket on $\mathbb{R}^3$ is given by the cross product $\mathbf{a} \times \mathbf{b}$. Many authors avoid the presence of the multiplier 2 by defining the tau-matrices as $\boldsymbol{\tau} = -\frac{1}{2}\mathbf{i}\boldsymbol{\sigma}$, which does make for slightly nicer formulae. However, the natural Lie bracket on the Lie algebra of the 3-sphere group, as presented two sections hence, has the multiplier 2, which is our justification for the development here.

intricacies of infinite-dimensional representations, we will simply place the angular momentum operators in the bare-boned algebraic context of representation theory in the case of infinite dimensions, and then restrict our attention to the well-developed and quite straightforward case of finite-dimensional representations.

We turn the spin representation of the preceding section around and describe an angular momentum triad in terms of $\mathfrak{su}(2)$ representations. Consider, then, an angular momentum triad J on the Hilbert space $\mathcal{H}$. For each $i = 1, 2, 3$, let $\mathsf{T}_i = -(2\mathbf{i}/\hbar)\mathsf{J}_i$ and observe that equation 12.2 implies that

$$[\mathbf{a} \cdot \mathsf{T}, \mathbf{b} \cdot \mathsf{T}] = 2(\mathbf{a} \times \mathbf{b})\mathsf{T}, \tag{13.5}$$

so that the triad $\mathsf{T} = (\mathsf{T}_1, \mathsf{T}_2, \mathsf{T}_3)$ satisfies exactly the relation of equation 13.4. The operators T_i are not quite self-adjoint but, rather, are anti-self-adjoint in that $\langle\psi|\mathsf{T}_i\psi\rangle = -\langle\mathsf{T}_i\psi|\psi\rangle$ for each ψ in the domain of J_i. Hence the mapping $\mathbf{a} \cdot \boldsymbol{\tau} \mapsto \mathbf{a} \cdot \mathsf{T}$, which is linear and preserves commutators, can be thought of as an anti-self-adjoint representation of $\mathfrak{su}(2)$ on the Hilbert space $\mathcal{H}$. The angular momentum triad $\mathsf{J} = \frac{1}{2}\mathbf{i}\hbar\mathsf{T}$ may be thought of as arising from this $\mathfrak{su}(2)$ representation. Conversely, any anti-self-adjoint representation $\mathbf{a} \cdot \boldsymbol{\tau} \mapsto \mathbf{a} \cdot \mathsf{T}$ of $\mathfrak{su}(2)$ on a Hilbert space $\mathcal{H}$ produces an angular momentum triad via $\mathsf{J} = \frac{1}{2}\mathbf{i}\hbar\mathsf{T}$.

We now focus our attention on finite-dimensional representations, assuming the eigenvalues of J_3 to be nondegenerate and concentrating on the individual J^2-eigenspaces. In the case where $j(j+1)\hbar^2$ is an eigenvalue of J^2 for some nonnegative half-integer j, let $\mathcal{H}_j$ denote the $j(j+1)\hbar^2$-eigenspace of J^2 and let $\mathsf{J}^{(j)}$ and $\mathsf{T}^{(j)}$ denote the restrictions of the operator triads to $\mathcal{H}_j$. By the results of the preceding lecture, $\mathcal{H}_j$ is $(2j+1)$-dimensional with an orthonormal basis $\{|j,m\rangle : m = -j, -j+1, \ldots, j-1, j\}$, where $\mathsf{J}_3|j,m\rangle = m\hbar\,|j,m\rangle$. Then $\mathsf{T}_3|j,m\rangle = -2\mathbf{i}m\,|j,m\rangle$, so that this basis consists of eigenstates of T_3. The function $\rho^j : \mathfrak{su}(2) \to \mathfrak{gl}(\mathcal{H}_j)$ defined as $\rho^j(\mathbf{a}\cdot\boldsymbol{\tau}) = \mathbf{a}\cdot\mathsf{T}^{(j)}$ is then a representation of $\mathfrak{su}(2)$ on the vector space $\mathcal{H}_j$, which, by the work of the previous lecture, is seen to be the unique irreducible $(2j+1)$-dimensional representation of $\mathfrak{su}(2)$, up to equivalence.

We turn now to finite-dimensional representations of the Lie groups SU(2) and SO(3), but first we must uncover their topological structure. In the next section, we use quaternions to help realize SU(2) as the simply connected double covering of the rotation group SO(3) as well as to uncover certain useful relationships among other orthogonal groups.

13.3 The Quaternions $\mathbb{H}$, $\mathbf{S}^3$, SU(2), SO(3), and SO(4)

The **quaternions** or the **Hamiltonians** form a four-dimensional real algebra $\mathbb{H}$ generated by $\mathbf{1}, \mathbf{i}, \mathbf{j}$, and $\mathbf{k}$. The multiplication of quaternions is generated by the requirements that $\mathbf{1}$ be the multiplicative identity, $\mathbf{i}^2 = \mathbf{j}^2 = \mathbf{k}^2 = -\mathbf{1}$, $\mathbf{ij} = \mathbf{k} = -\mathbf{ji}$, $\mathbf{jk} = \mathbf{i} = -\mathbf{kj}$, and $\mathbf{ki} = \mathbf{j} = -\mathbf{ik}$. A nice mnemonic for the multiplication of $\mathbf{i}$, $\mathbf{j}$, and $\mathbf{k}$ is

$$\begin{array}{ccc} \mathbf{i} & \longrightarrow & \mathbf{j} \\ & \nwarrow \quad \swarrow & \\ & \mathbf{k} & \end{array} \tag{13.6}$$

where a multiplication of two different ones of these three basis elements gives the third, with a negative sign when one "goes against the arrow." This is a noncommutative algebra whose algebraic operations can be used to encode the rigid motions of the Euclidean spaces $\mathbb{R}^3$ and $\mathbb{R}^4$. This is accomplished by identifying $\mathbb{R}^3$ as $\operatorname{span}\{\mathbf{i}, \mathbf{j}, \mathbf{k}\} = \mathbb{R}\mathbf{i} \oplus \mathbb{R}\mathbf{j} \oplus \mathbb{R}\mathbf{k}$ and observing that conjugation by unit quaternions preserves the usual Euclidean norm in $\mathbb{R}^3$ and by identifying $\mathbb{R}^4$ as $\mathbb{H}$ and observing that left and right unit quaternion multiplications act as rigid motions. To fully develop this, we will expose some of the **geometric algebra** of quaternions.

The starting point is to note that two features of complex numbers – conjugation and norm – extend to quaternions. The **conjugate** of the quaternion $q = q_0\mathbf{1} + q_1\mathbf{i} + q_2\mathbf{j} + q_3\mathbf{k}$ is $\overline{q} = q_0\mathbf{1} - q_1\mathbf{i} - q_2\mathbf{j} - q_3\mathbf{k}$. The **norm** $|q|$ has square $|q|^2 = \overline{q}q = q\overline{q} = q_0^2 + q_1^2 + q_2^2 + q_3^2$. Conjugation is $\mathbb{R}$-linear as well as being an **anti-involution**, meaning that $\overline{\overline{q}} = q$ and $\overline{pq} = \overline{q}\,\overline{p}$ for all $p, q \in \mathbb{H}$. By identifying the real numbers $\mathbb{R}$ with $\operatorname{span}\{\mathbf{1}\} = \mathbb{R}\mathbf{1}$, and thus denoting the identity $\mathbf{1}$ as $1 \in \mathbb{R}$, we may write the quaternion q as $q = q_0 + \mathbf{q}$, where $\mathbf{q} = q_1\mathbf{i} + q_2\mathbf{j} + q_3\mathbf{k} = (q_1, q_2, q_3) \in \mathbb{R}^3$. We define the **real** and **pure parts** of the quaternion q to be, respectively, $\operatorname{Re} q = q_0 \in \mathbb{R}$ and $\operatorname{Pu} q = \mathbf{q} \in \mathbb{R}^3$. The quaternion q is **real** if $q = \operatorname{Re} q$ and is **pure** if $q = \operatorname{Pu} q$. It is easy to see that a quaternion q is in the center of $\mathbb{H}$ if and only if q is real and that q^2 is a nonpositive real number if and only if q is pure. The representation of q as $q_0 + \mathbf{q}$ is called the **real representation** of q and identifies $\mathbb{H}$ as the internal direct sum $\mathbb{H} = \mathbb{R} \oplus \mathbb{R}^3$.

In addition to the real representation of quaternions, there are two other useful representations, the **complex** and the **matrix** representations. For the complex representation, it is easy to see that if $\mathbf{u}$ is any unit pure quaternion, "unit" meaning that $|\mathbf{u}| = 1$, then the real subalgebra generated by $\mathbf{1}$ and $\mathbf{u}$, denoted as $\mathbb{C}_{\mathbf{u}} = \mathbb{R}\mathbf{1} \oplus \mathbb{R}\mathbf{u}$, is isomorphic to the complex numbers $\mathbb{C}$, with $\mathbf{1}$ identified with the complex number 1 and $\mathbf{u}$ identified with the imaginary unit $\mathbf{i}$. The **canonical complex subalgebra** of $\mathbb{H}$ is $\mathbb{C}_{\mathbf{i}}$, the subalgebra generated by the quaternions $\mathbf{1}$ and $\mathbf{i}$, which, when identified with $\mathbb{C}$, provides the canonical $\mathbb{R}$-algebra embedding $\mathbb{C} \subset \mathbb{H}$. By setting $z = q_0 + q_1\mathbf{i} \in \mathbb{C}$ and $w = q_2 + q_3\mathbf{i} \in \mathbb{C}$, the quaternion $q = q_0\mathbf{1} + q_1\mathbf{i} + q_2\mathbf{j} + q_3\mathbf{k}$ may be written as $q = z + w\mathbf{j}$. Quaternion multiplication translates in the complex representation as $pq = (u + v\mathbf{j})(z + w\mathbf{j}) = (uz - v\overline{w}) + (uw + v\overline{z})\mathbf{j}$, since $\mathbf{j}z = \overline{z}\mathbf{j}$ whenever $z \in \mathbb{C}$. Conjugation becomes $\overline{q} = \overline{z} - w\mathbf{j}$ and the norm is $|q| = \sqrt{|z|^2 + |w|^2}$. The complex representation $q = z + w\mathbf{j}$ identifies $\mathbb{H}$ as the internal direct sum $\mathbb{H} = \mathbb{C}_{\mathbf{i}} \oplus \mathbb{C}_{\mathbf{i}}\mathbf{j}$.

For the matrix representation, we define a real algebra isomorphism of $\mathbb{H}$ onto a subalgebra of $\mathrm{M}_2(\mathbb{C})$, the algebra of 2×2 matrices over the complex numbers under the usual matrix addition and multiplication and real scalar multiplication. Define $\varrho : \mathbb{H} \to \mathrm{M}_2(\mathbb{C})$ by

$$\varrho(q) = \varrho(z + w\mathbf{j}) = \begin{pmatrix} z & w \\ -\overline{w} & \overline{z} \end{pmatrix}. \tag{13.7}$$

A quick observation reveals that ϱ is an injective $\mathbb{R}$-linear map that preserves multiplication, meaning that $\varrho(pq) = \varrho(p)\varrho(q)$ for all $p, q \in \mathbb{H}$. If the conjugate of a matrix is defined as the transpose-conjugate or adjoint then ϱ preserves conjugation since then $\varrho(\overline{q}) = \overline{\varrho(q)}^{\mathsf{Tr}} = \varrho(q)^\dagger$. Similarly, if we define the norm of a matrix M as $\sqrt{\det \mathsf{M}}$ then ϱ is a norm-preserving homomorphism of algebras, as we then have $|q|^2 = |z|^2 + |w|^2 = \det \varrho(q)$. Notice that $\varrho(\mathbf{1}) = 1_{2\times 2}$, the 2×2 identity matrix, $\varrho(\mathbf{i}) = -\tau_3$, $\varrho(\mathbf{j}) = -\tau_2$, and $\varrho(\mathbf{k}) = -\tau_1$.[2] It follows that $\mathbb{H}$ may be realized as the real matrix algebra generated by the identity $1_{2\times 2}$ and the triad $\boldsymbol{\tau}$.

We will use these three representations – the real, the complex, and the matrix – of $\mathbb{H}$ interchangeably and without comment as each suits our purpose. We will also use the standard embeddings that realize the

[2] Why not $\varrho(\mathbf{i}) = -\tau_1$, etc.? This is an accident caused by the historical choice of measuring the x_3-component of spin. If the choice had been to call the measured spin-component the x_1-component rather than the x_3-component, the standard spin basis would change and, in that basis, the first and third Pauli matrices would be exchanged and the second multiplied by -1, and we would have $\varrho(\mathbf{i}) = -\tau_1$, $\varrho(\mathbf{j}) = \tau_2$, and $\varrho(\mathbf{k}) = -\tau_3$. This follows from the discussion of Section 12.4 showing how to move from one choice of spin-component measurements to another.

containments $\mathbb{R} \subset \mathbb{C} \subset \mathbb{H}$ given by the identifications

$$\mathbb{R} \ni a \leftrightarrow a + 0\mathbf{i} \in \mathbb{C} \quad \text{and} \quad \mathbb{C} \ni z \leftrightarrow z + 0\mathbf{j} \in \mathbb{H},$$

which are conjugate- and norm-preserving embeddings of real algebras. Each nonzero quaternion q has a multiplicative inverse given by $q^{-1} = \overline{q}/|q|^2$ and, hence, the **group of units** of $\mathbb{H}$ is the multiplicative group $\mathbb{H}^* = \mathbb{H} - \{\mathbf{0}\}$. This implies that the quaternions form a real division algebra.

So far we have concentrated on the algebraic structure of quaternions. The property that makes quaternions so useful for the geometry of Euclidean spaces, and hence in the treatment of angular momentum in quantum mechanics, is the fact that $\mathbb{H}$ forms a **normed algebra**. The requirement for a normed algebra is that the multiplication operation, a purely algebraic feature, be intertwined with the norm, which infuses the underlying vector space with a geometric structure by the equation

$$|pq| = |p||q|, \tag{13.8}$$

for all $p, q \in \mathbb{H}$. That $\mathbb{H}$ is a normed algebra under the norm already defined is a straightforward exercise.[3] The standard **Euclidean inner product** on $\mathbb{H}$ is given by $p \cdot q = p_0q_0 + p_1q_1 + p_2q_2 + p_3q_3$, and p and q are **orthogonal** if $p \cdot q = 0$. Note that even though $p\overline{q} \neq p \cdot q$ in general, nonetheless $q\overline{q} = q \cdot q$, so the norm may be written as $|q| = \sqrt{q \cdot q}$. A real linear transformation T of $\mathbb{H}$ is said to be **orthogonal** if it preserves the inner product, so that $\mathsf{T}p \cdot \mathsf{T}q = p \cdot q$ for all $p, q \in \mathbb{H}$. The **cross product** of pure quaternions $\mathbf{p}$ and $\mathbf{q}$ is the usual cross product $\mathbf{p} \times \mathbf{q}$ in $\mathbb{R}^3$.

In any algebra, the operators L_q and R_q,i.e., left and right multiplication by the element q, respectively, are linear transformations of the underlying vector space. The extra ingredient that equation 13.8 affords is that not only are these transformations invertible but, moreover, L_q and R_q are orthogonal when $|q| = 1$. Of particular importance is the operation of (multiplicative) conjugation $C_q = L_q R_{q^{-1}} = L_q R_{\overline{q}}$ in $\mathbb{H}$ by a unit quaternion q. We will examine the ramifications of this after a quick catalogue of equations, all easily verified, that can be employed when using quaternions to describe Euclidean rotations and reflections. The following formulae hold for all quaternions p and q and all pure

[3] It is an interesting fact that $\mathbb{R}$, $\mathbb{C}$, and $\mathbb{H}$ are the only real unitary normed associative algebras up to isomorphism and that the only additional real unitary normed algebra, nonassociative of course, is the eight-dimensional Cayley algebra $\mathbb{CA}$, also known as the octonian algebra $\mathbb{O}$. See the End Notes to this lecture for references and more detail.

quaternions $\mathbf{p}$, $\mathbf{q}$, and $\mathbf{r}$:

$$p \cdot q = \tfrac{1}{2}(\overline{p}q + \overline{q}p), \tag{13.9}$$

$$\mathbf{p} \cdot \mathbf{q} = -\tfrac{1}{2}(\mathbf{pq} + \mathbf{qp}) = -\mathrm{Re}(\mathbf{pq}), \tag{13.10}$$

$$\mathbf{p} \times \mathbf{q} = \mathrm{Pu}(\mathbf{pq}) = \tfrac{1}{2}(\mathbf{pq} - \mathbf{qp}), \tag{13.11}$$

$$\mathbf{pq} = -\mathbf{p} \cdot \mathbf{q} + \mathbf{p} \times \mathbf{q}, \tag{13.12}$$

$$\mathbf{p}^2 = -\mathbf{p} \cdot \mathbf{p}, \tag{13.13}$$

$$\mathbf{p} \times (\mathbf{q} \times \mathbf{r}) = (\mathbf{p} \cdot \mathbf{r})\mathbf{q} - (\mathbf{p} \cdot \mathbf{q})\mathbf{r}. \tag{13.14}$$

These equations imply that $\mathbf{p}$ and $\mathbf{q}$ are orthogonal if and only if $\mathbf{p}$ and $\mathbf{q}$ anticommute under quaternion multiplication, meaning that $\mathbf{pq} = -\mathbf{qp}$, and in this case $\mathbf{pq} = \mathbf{p} \times \mathbf{q}$.

Equation 13.8 implies that the 3-sphere $\mathbf{S}^3 = \{q \in \mathbb{H} : |q| = 1\}$ is closed under quaternion multiplication and, since $|q| = |\overline{q}|$ and $q^{-1} = \overline{q}$ whenever $|q| = 1$, $q^{-1} \in \mathbf{S}^3$ whenever $q \in \mathbf{S}^3$. Therefore the 3-sphere is a group under quaternion multiplication. In fact, $\mathbf{S}^3$ is a simply connected Lie group and its Lie algebra, the tangent space at the identity 1, denoted as $\mathfrak{s}^3$, is naturally identified with the space of pure quaternions $\mathbb{R}^3$ whose bracket is just the commutator for quaternion multiplication $[\mathbf{p}, \mathbf{q}] = \mathbf{pq} - \mathbf{qp}$. By equation 13.11, $[\mathbf{p}, \mathbf{q}] = 2(\mathbf{p} \times \mathbf{q})$ and, by the discussion surrounding equation 13.4, $\mathfrak{s}^3$ is isomorphic to the Lie algebra $\mathfrak{su}(2)$. In fact, we can say more. Indeed, $\mathfrak{su}(2)$ is just the matrix representation of the space of pure quaternions under the mapping of equation 13.7 restricted to $\mathbb{R}^3$. Since ϱ preserves commutators, which serve as the Lie bracket in both $\mathbb{R}^3$ and in $\mathfrak{su}(2)$, this restricted mapping $\varrho|\mathbb{R}^3$ is a Lie algebra isomorphism of $\mathfrak{s}^3$ onto $\mathfrak{su}(2)$.

What is the relationship between the two Lie groups $\mathbf{S}^3$ and SU(2) that share a common Lie algebra $\mathfrak{su}(2)$? We explore the answer to this query next by showing that $\mathbf{S}^3$ and SU(2) are isomorphic Lie groups that double-cover the group SO(3), the **rotation group** of $\mathbb{R}^3$ that consists of all orthonormal 3×3 matrices of unit determinant. The key to realizing an isomorphism of $\mathbf{S}^3$ with SU(2) is the matrix representation ϱ of equation 13.7. Indeed, consider the restriction $\varrho|\mathbf{S}^3$, a mapping of $\mathbf{S}^3$ into $\mathrm{M}_2(\mathbb{C})$. First note that if $q = z + w\mathbf{j} \in \mathbf{S}^3$ then, using equation 13.7, $\det \varrho(q) = z\overline{z} + w\overline{w} = |q|^2 = 1$ and $\varrho(q)\varrho(q)^\dagger = 1_{2\times 2}$, so that $\varrho(q) \in$ SU(2). Therefore we have a mapping

$$\varrho|\mathbf{S}^3 : \mathbf{S}^3 \to \mathrm{SU}(2), \tag{13.15}$$

which is injective since ϱ is injective and is a group homomorphism since

ϱ preserves multiplication. It remains only to prove that $\varrho|\mathbf{S}^3$ is surjective. For this, straightforward calculations show that the requirements that a 2×2 complex matrix M have unit determinant and that $\mathsf{M}^{-1} = \mathsf{M}^\dagger$ together force M to have the form

$$\begin{pmatrix} z & w \\ -\overline{w} & \overline{z} \end{pmatrix}$$

so that $\mathsf{M} = \varrho(z + w\mathbf{j})$, with $|z + w\mathbf{j}|^2 = \det \mathsf{M} = 1$.

We next describe a surjective homomorphism of Lie groups

$$\vartheta : \mathbf{S}^3 \to \mathrm{SO}(3) \tag{13.16}$$

with kernel $\{\pm 1\}$, which shows the rotation group of $\mathbb{R}^3$ to be, topologically, a three-dimensional real projective space double-covered by the simply connected 3-sphere group $\mathbf{S}^3$. Let $q \in \mathbf{S}^3$ and observe that $C_q = L_q R_{\overline{q}}$ is an orthogonal transformation by equation 13.8, since $|qp\overline{q}| = |q||p||\overline{q}| = |p|$. Thus $C_q \in \mathrm{O}(\mathbb{H})$, the group of orthogonal transformations of $\mathbb{H}$. Since $C_q(1) = q\overline{q} = 1$ and C_q is orthogonal, we have $C_q(\mathbb{R}^3) = \mathbb{R}^3$. Let $c_q = C_q|\mathbb{R}^3$, the restriction of C_q to the space of pure quaternions, and note that $c_q \in \mathrm{O}(\mathbb{R}^3)$. It follows that $c_q(\mathbf{i}) \times c_q(\mathbf{j}) = \pm c_q(\mathbf{k})$, with the plus sign if and only if c_q is orientation-preserving. Calculating with equation 13.11, we have $c_q(\mathbf{i}) \times c_q(\mathbf{j}) = (q\mathbf{i}\overline{q}) \times (q\mathbf{j}\overline{q}) = \mathrm{Pu}(q\mathbf{i}\overline{q}\, q\mathbf{j}\overline{q}) = \mathrm{Pu}(q\mathbf{k}\overline{q}) = \mathrm{Pu}(c_q(\mathbf{k})) = c_q(\mathbf{k})$, so c_q is orientation-preserving and $c_q \in \mathrm{SO}(\mathbb{R}^3)$. The element $\vartheta(q)$ now may be defined as the 3×3 orthogonal matrix, denoted as $[c_q]$, that represents c_q with respect to the basis $\{\mathbf{i}, \mathbf{j}, \mathbf{k}\}$. Thus, $\vartheta(q) = [c_q] \in \mathrm{SO}(3)$ has columns given, respectively, by the components of $q\mathbf{i}\overline{q}$, $q\mathbf{j}\overline{q}$, and $q\mathbf{k}\overline{q}$.

Our claim is that ϑ is a surjective group homomorphism with kernel $\{\pm 1\}$. Indeed, that ϑ is a homomorphism merely results from the observation that $c_p c_q(\mathbf{x}) = p(q\mathbf{x}\overline{q})\overline{p} = pq\mathbf{x}\overline{pq} = c_{pq}(\mathbf{x})$, itself a consequence of the fact that conjugation in $\mathbb{H}$ is an anti-involution. Letting q be an element of the kernel of ϑ implies that $c_q = 1_{\mathbb{R}^3}$. Since C_q fixes $\mathbb{R}$, it follows that $C_q = 1_{\mathbb{H}}$ and therefore $qp\overline{q} = p$ for all $p \in \mathbb{H}$. This implies that $qp = pq$ since $\overline{q}q = 1$ and, therefore, $q \in \mathrm{Center}(\mathbb{H}) \cap \mathbf{S}^3 = \mathbb{R} \cap \mathbf{S}^3 = \{\pm 1\}$. Finally, the fact that ϑ is surjective is a consequence of analyzing the action of c_q on $\mathbb{R}^3$. We will show that when $q = q_0 + \mathbf{q} \in \mathbf{S}^3$, c_q is a rotation of $\mathbb{R}^3$ through an angle of $\theta = 2\cos^{-1} q_0$ about the axis $\ell_{\mathbf{q}} = \mathrm{span}\{\mathbf{q}\}$, the sense of rotation being countercockwise "looking down from $\mathbf{q}$." Assuming this for the moment, let $\mathbf{u}$ be any unit vector in $\mathbb{R}^3$ and define the path $q(t)$ in $\mathbf{S}^3$ by

$$q(t) = \cos t + \mathbf{u} \sin t = q_0(t) + \mathbf{q}(t) \quad \text{for} \quad -\infty < t < \infty.$$

Then $c_{q(t)}$ is a one-parameter family of rotations of $\mathbb{R}^3$ with common axis $\ell_{\mathbf{u}}$ that rotates through the angles $\theta(t) = 2t$, which range from 0 to 2π as t ranges from 0 to π. This shows that ϑ is surjective. Notice that the curve $q(t)$ traces out a great circle in $\mathbf{S}^3$, as its range is the intersection of $\mathbf{S}^3$ with the subalgebra $\mathbb{C}_{\mathbf{u}}$, a two-dimensional plane through the origin. This intersection $\mathbf{S}^1_{\mathbf{u}} = \mathbf{S}^3 \cap \mathbb{C}_{\mathbf{u}}$ is a subgroup of $\mathbf{S}^3$ isomorphic to the circle group $\mathbf{S}^1$ of unit complex numbers. In fact, the mapping $q(t)$ is a one-parameter subgroup of $\mathbf{S}^3$ since $q(s+t) = q(s)q(t)$ for all $s, t \in \mathbb{R}$. This will be used in the following section to understand the exponential mapping $\exp : \mathfrak{s}^3 \to \mathbf{S}^3$.

It remains to verify that c_q is, as claimed, a "counterclockwise" rotation of $\mathbb{R}^3$ through an angle of $\theta = 2\cos^{-1} q_0$ about the axis $\ell_{\mathbf{q}}$. For this, assume that $q_0 \neq \pm 1$ so that $\mathbf{q} \neq \mathbf{0}$. Set $t = \cos^{-1} q_0$ so that $q = q(t) = \cos t + \mathbf{u} \sin t$, where $\mathbf{u} = \mathbf{q}/|\mathbf{q}|$ is the unit vector in the direction of $\mathbf{q}$. Choose unit vectors $\mathbf{v}$ and $\mathbf{w}$ such that $\{\mathbf{u}, \mathbf{v}, \mathbf{w}\}$ is an oriented orthonormal triad with $\mathbf{u} \times \mathbf{v} = \mathbf{w}$. A quick calculation using equation 13.13, that $\mathbf{u}^2 = -1$, shows that $\mathbf{u}$ is fixed by the rotation c_q. It follows that c_q has the line $\ell_{\mathbf{u}} = \ell_{\mathbf{q}}$ as its axis of rotation. The plane $\mathbb{P}_{\mathbf{u}} = \operatorname{span}\{\mathbf{v}, \mathbf{w}\}$ orthogonal to $\ell_{\mathbf{u}}$ is invariant under the action of c_q, which acts as a rotation of $\mathbb{P}_{\mathbf{u}}$ through some angle θ. The image of $\mathbf{v}$ under c_q gives both the sense and the angle of rotation of c_q by its expression as a linear combination of $\mathbf{v}$ and $\mathbf{w}$ as follows:

$$\begin{aligned} c_q(\mathbf{v}) = q\mathbf{v}\overline{q} &= (\cos t + \mathbf{u}\sin t)\mathbf{v}(\cos t - \mathbf{u}\sin t) \\ &= (\cos^2 t)\mathbf{v} + \cos t \sin t(\mathbf{uv} - \mathbf{vu}) - (\sin^2 t)\mathbf{uvu} \\ &= (\cos^2 t - \sin^2 t)\mathbf{v} + (2\sin t \cos t)\mathbf{w} = \mathbf{v}\cos 2t + \mathbf{w}\sin 2t. \end{aligned}$$

Here we have used equation 13.11 to get $\mathbf{uv} - \mathbf{vu} = 2\mathbf{w}$ and equation 13.12 twice to get $\mathbf{uvu} = \mathbf{v}$. This identifies the angle of rotation θ as $2t = 2\cos^{-1} q_0$ and the sense of rotation as counterclockwise "looking down from $\mathbf{u}$," as claimed.

Note that since the map ϱ is an isomorphism and ϑ is a local isomorphism as it is a covering map, the differentials of these maps induce the isomorphisms of Lie algebras

$$\mathfrak{su}(2) \xleftarrow[\cong]{d\varrho} \mathfrak{s}^3 \xrightarrow[\cong]{d\vartheta} \mathfrak{so}(3).$$

We close this section by presenting a surjective homomorphism of Lie groups

$$\Theta : \mathbf{S}^3 \times \mathbf{S}^3 \to \mathrm{SO}(4)$$

with kernel $\{\pm(1,1)\}$, which extends the homomorphism ϑ. Define $\Theta(p,q) = [L_p R_{\overline{q}}]$, the 4×4 orthogonal matrix that represents the othogonal transformation $L_p R_{\overline{q}}$ with respect to the basis $\{\mathbf{1}, \mathbf{i}, \mathbf{j}, \mathbf{k}\}$. Then Θ is, as claimed, a surjective homomorphism with kernel $\{\pm(1,1)\}$, facts we will not verify. Let $\Delta : \mathbf{S}^3 \to \mathbf{S}^3 \times \mathbf{S}^3$ be the diagonal map $\Delta(p) = (p,p)$ and define an embedding $\delta : \mathrm{SO}(3) \to \mathrm{SO}(4)$ by

$$\delta(\mathsf{M}) = \begin{pmatrix} 1 & 0 \\ 0 & \mathsf{M} \end{pmatrix}.$$

Since $C_p = L_p R_{\overline{p}}$, we have

$$\Theta\Delta(p) = \Theta(p,p) = [C_p] = \begin{pmatrix} 1 & 0 \\ 0 & [c_p] \end{pmatrix} = \delta([c_p]) = \delta\vartheta(p),$$

verifying that the following diagram of Lie group homomorphisms, with horizontal embedding maps and vertical double-covering maps, is commutative:

$$\begin{array}{ccc} \mathbf{S}^3 & \xrightarrow{\Delta} & \mathbf{S}^3 \times \mathbf{S}^3 \\ \vartheta \big\downarrow & & \big\downarrow \Theta \\ \mathrm{SO}(3) & \xrightarrow[\delta]{} & \mathrm{SO}(4). \end{array}$$

The map Θ induces a Lie algebra isomorphism

$$\mathfrak{so}(3) \oplus \mathfrak{so}(3) \xleftarrow[\cong]{d\vartheta \oplus d\vartheta} \mathfrak{s}^3 \oplus \mathfrak{s}^3 \xrightarrow[\cong]{d\Theta} \mathfrak{so}(4). \tag{13.17}$$

The isomorphism $\mathfrak{so}(3) \oplus \mathfrak{so}(3) \cong \mathfrak{so}(4)$ will be used in Lecture 16 as we uncover the so-called hidden SO(4) symmetries of the bound electron in a central Coulomb potential.

13.4 Representations of Lie Groups SU(2) and SO(3)

Recall that a **representation** of the Lie group G is just a group homomorphism $R \colon G \to \mathrm{GL}(V)$ for some complex vector space V, and is finite-dimensional if V is finite-dimensional and **irreducible** if there are no proper nonzero G-invariant subspaces of V. The finite-dimensional representation R gives rise to a representation of Lie algebras via the derivative $dR_e : \mathfrak{g} \to \mathfrak{gl}(V)$ evaluated at the identity $e \in G$. In general, though, there may be more representations of the Lie algebra $\mathfrak{g}$ than those arising in this way from the representations of the Lie group G. However, when G is simply connected and connected, every representation of $\mathfrak{g}$ arises from a representation of G and there is in fact a one-to-one correspondence between the representations of G and those of $\mathfrak{g}$, given by $R \leftrightarrow dR_e$. We will see this illustrated in our determination of the irreducible finite-dimensional representations of the simply connected Lie group SU(2), which will be seen to be in one-to-one correspondence with the irreducible finite-dimensional representations of its Lie algebra $\mathfrak{su}(2)$, already found in Section 13.2. We will then use this to determine the finite-dimensional representations of the rotation group SO(3). Rather than invoking and applying general theorems from the theory of representations of Lie groups, we take a parochial view and work with the rich structures of SU(2) and $\mathfrak{su}(2)$ afforded by their realizations as subsets of $\mathbb{H}$. In this section then, we explicitly and directly construct the irreducible representations of SU(2) from the irreducible representations of $\mathfrak{su}(2)$ already identified.

The connection between the representations of a Lie group G and its associated Lie algebra $\mathfrak{g}$ is found in the exponential mapping $\exp : \mathfrak{g} \to G$. In the case of interest for us, observe that there is a commutative diagram that transfers information back and forth between the isomorphic Lie-algebra–Lie-group pairs $\mathfrak{s}^3, \mathbf{S}^3$ and $\mathfrak{su}(2), \mathrm{SU}(2)$, whose horizontal maps are Lie isomorphisms and whose vertical maps are exponential mappings of the Lie algebras:

$$
\begin{array}{ccc}
\mathfrak{s}^3 & \xrightarrow[\cong]{\varrho|\mathbb{R}^3} & \mathfrak{su}(2) \\
{\scriptstyle \exp}\downarrow & & \downarrow{\scriptstyle \exp} \\
\mathbf{S}^3 & \xrightarrow[\varrho|\mathbf{S}^3]{\cong} & \mathrm{SU}(2).
\end{array}
$$

Here, the exponential mappings are given in both cases by the usual infinite series for e^x, with x taking on pure quaternionic values $\mathbf{q}$ in $\mathbb{R}^3$ for the left-hand mapping, and x taking on 2×2 matrix values M in $\mathfrak{su}(2)$ for the right-hand mapping. That the diagram commutes is an exercise left to the reader, in which it will be useful to calculate $\exp(\mathbf{q})$ explicitly. For $\mathbf{q} \in \mathfrak{s}^3 = \mathbb{R}^3$, write $\mathbf{q} = t\mathbf{u}$ where $\mathbf{u} = \mathbf{q}/|\mathbf{q}|$ is the unit vector in the direction of $\mathbf{q}$ and $t = |\mathbf{q}|$. Recalling from equation 13.13

that $\mathbf{u}^2 = -1$, so that $\mathbf{u}^{2n} = (-1)^n$, we have

$$\begin{aligned}\exp(\mathbf{q}) = \sum_{n=0}^{\infty} \frac{\mathbf{q}^n}{n!} &= \sum_{n=0}^{\infty} \mathbf{u}^{2n} \frac{t^{2n}}{(2n)!} + \sum_{n=0}^{\infty} \mathbf{u}^{2n} \frac{t^{2n+1}}{(2n+1)!}\mathbf{u} \\ &= \cos t + \mathbf{u} \sin t = q(t), \end{aligned} \tag{13.18}$$

where $q(t) = \exp(t\mathbf{u})$ is the one-parameter subgroup of the preceding section. We see then that the exponential mapping wraps the line $\mathbb{R}\mathbf{u}$ around the unit circle $\mathbf{S}^1_{\mathbf{u}}$ in the plane $\mathbb{C}_{\mathbf{u}}$ with period 2π.

We now show explicitly how to define an irreducible finite-dimensional representation of $\mathbf{S}^3$ by starting with a representation on $\mathfrak{s}^3$. Let $\rho^j : \mathfrak{s}^3 \to \mathfrak{gl}(V^j)$ be an irreducible representation with $\dim V^j = 2j + 1$ for some nonnegative half-integer j. Our goal is to find a group homomorphism R^j that makes the following diagram commute:

$$\begin{array}{ccc} \mathfrak{s}^3 & \xrightarrow{\rho^j} & \mathfrak{gl}(V^j) \\ \mathbf{exp}\big\downarrow & & \big\downarrow\mathbf{exp} \\ \mathbf{S}^3 & \xrightarrow[R^j]{} & \mathrm{GL}(V^j). \end{array} \tag{13.19}$$

The obvious choice for R^j is the composition $\exp \circ \rho^j \circ \exp^{-1}$, but we must check that this is well defined. Because of the observation in the preceding paragraph, that $\exp : \mathfrak{s}^3 \to \mathbf{S}^3$ is 2π-periodic on lines $\mathbb{R}\mathbf{u}$, this is accomplished if we can show that, for each unit pure quaternion $\mathbf{u}$, the function $\exp(\rho^j(t\mathbf{u}))$ is 2π-periodic in t. From our previous discussion, the linear transformation $\rho^j(\mathbf{u})$ decomposes V^j into $2j+1$ one-dimensional eigenspaces, so that $V^j = \oplus_{m=-j}^{j} V_{\mathbf{u},m}$, where $V_{\mathbf{u},m}$ is the span of the eigenvector $|j,m\rangle_{\mathbf{u}}$ associated with the eigenvalue $-2\mathbf{i}m$. The matrix representation of the transformation $\rho^j(\mathbf{u})$ with respect to the basis

$$\mathcal{B}_{\mathbf{u}} = \{|j,j\rangle_{\mathbf{u}}, |j,j-1\rangle_{\mathbf{u}}, \ldots, |j,-j+1\rangle_{\mathbf{u}}, |j,-j\rangle_{\mathbf{u}}\}$$

is then the diagonal matrix

$$2\mathbf{i}\begin{pmatrix} -j & & & & \\ & -j+1 & & \mathbf{0} & \\ & & \ddots & & \\ & \mathbf{0} & & j-1 & \\ & & & & j \end{pmatrix}.$$

Therefore, the matrix representation of $\exp(\rho^j(t\mathbf{u}))$ with respect to $\mathcal{B}_{\mathbf{u}}$

is the diagonal matrix

$$\begin{pmatrix} e^{-2\mathrm{i}jt} & & & & \\ & e^{2\mathrm{i}(-j+1)t} & & \mathbf{0} & \\ & & \ddots & & \\ & \mathbf{0} & & e^{2\mathrm{i}(j-1)t} & \\ & & & & e^{2\mathrm{i}jt} \end{pmatrix}. \tag{13.20}$$

The diagonal elements are $e^{-2m\mathrm{i}t}$ for $m = j, j-1, \ldots, -j+1, -j$ and, as j is a half-integer, each value of $2m$ is an integer so each diagonal element is of the form $e^{k\mathrm{i}t}$ for an integer k. As each $e^{k\mathrm{i}t}$ is 2π-periodic in t, so too is $\rho^j(t\mathbf{u})$, and $R^j = \exp \circ \rho^j \circ \exp^{-1}$ is a well-defined irreducible representation of SU(2). It is interesting to note the action of the one-parameter subgroup $\mathbf{S}^1_{\mathbf{u}}$ of $\mathbf{S}^3$ on V^j via this representation. Indeed, the matrix in expression 13.20 implies that $R^j(e^{\mathbf{u}t})$ acts on V^j by rotating the complex plane $V_{\mathbf{u},m} = \mathbb{C}|j,m\rangle_{\mathbf{u}}$ through an angle of $-2mt$. Each one-parameter subgroup $\mathbf{S}^1_{\mathbf{u}}$ gives rise to a decomposition of V^j into one-dimensional eigenspaces, with $R^j(e^{\mathbf{u}t})$ acting as described on that decomposition.

We can obtain a useful description of these representations of SU(2) by identifying the $(2j+1)$-dimensional representation space V^j concretely. Let V^j be the complex vector space of homogeneous polynomials of degree $2j$ in the variables z_1 and z_2. The standard basis for V^j is the collection of monomials $z_1^{2j}, z_1^{2j-1}z_2, \ldots, z_1 z_2^{2j-1}, z_2^{2j}$. The unit quaternion $q = z + w\mathbf{j} \in \mathbf{S}^3$ acts on the pair (z_1, z_2) via

$$q \cdot (z_1, z_2) = \varrho(q)\begin{pmatrix} z_1 \\ z_2 \end{pmatrix} = \begin{pmatrix} z & w \\ -\overline{w} & \overline{z} \end{pmatrix}\begin{pmatrix} z_1 \\ z_2 \end{pmatrix} = (zz_1 + wz_2, -\overline{w}z_1 + \overline{z}z_2).$$

Notice that $(pq) \cdot (z_1, z_2) = p \cdot (q \cdot (z_1, z_2))$. If $f \in V^j$ is a homogeneous polynomial of degree $2j$ then so too is $R(q)f = f(\overline{q}\cdot)$. It follows that $R(q) \in \mathrm{GL}(V^j)$ with inverse $R(\overline{q})$. Also, R is a group homomorphism since $R(pq)f = f((\overline{pq})\cdot) = f((\overline{q}\,\overline{p})\cdot) = f(\overline{q} \cdot (\overline{p}\cdot)) = R(q)f(\overline{p}\cdot) = R(p)R(q)f$. Therefore R is an irreducible $(2j+1)$-dimensional representation of $\mathbf{S}^3$ and, hence, $R = R^j$. In fact, we can identify the standard monomial basis for V^j as $\mathcal{B}_\mathbf{i}$ from the discussion in the previous paragraph, with $|j,m\rangle_\mathbf{i} = z_1^{j+m}z_2^{j-m}$. To see this, note that if $q = e^{\mathbf{i}t} \in \mathbf{S}^1_\mathbf{i}$ then $\overline{q} \cdot (z_1, z_2) = (e^{-\mathbf{i}t}z_1, e^{\mathbf{i}t}z_2)$ so that $R(q)z_1^{j+m}z_2^{j-m} = e^{-2m\mathbf{i}t}z_1^{j+m}z_2^{j-m}$. This shows that the matrix that represents $R(q)$ with respect to the monomial basis is given by expression 13.20.

We close this lengthy lecture with a determination of the irreducible finite-dimensional representations of the rotation group SO(3). If r is such a representation of the rotation group then the homomorphism $r \circ \vartheta$ is an irreducible finite-dimensional representation of $\mathbf{S}^3$ and, as such, is equal to R^j for some value of j. It follows that all the irreducible finite-dimensional representations of SO(3) are the maps $r^j = R^j \circ \vartheta^{-1}$ but only for those values of j for which this expression is well defined, i.e., for those j for which $R^j \circ \vartheta^{-1}$ is single-valued. As ϑ is a double-covering map of the rotation group with $\ker \vartheta = \{\pm 1\}$, the expression for the maps is well defined precisely when $R^j(q) = R^j(-q)$ for all unit quaternions q. Setting $q = e^{\mathbf{u}t}$ for a unit pure quaternion $\mathbf{u}$, the matrix representation of $R^j(q)$ with respect to the basis $\mathcal{B}_{\mathbf{u}}$ is given by expression 13.20. Now, $-q = e^{\mathbf{u}(t+\pi)}$ so that the matrix representation of $R^j(-q)$ is given also by expression 13.20, with t replaced by $t + \pi$. It follows then that $R^j(q) = R^j(-q)$ if and only if $e^{2\mathbf{i}m\pi} = 1$ for each m. Since $e^{2\mathbf{i}m\pi} = (-1)^{2m}$, this precisely occurs when each m is an integer, equivalently, when j is an integer. Thus, $R^j \circ \vartheta^{-1}$ is single-valued and $r^j = R^j \circ \vartheta^{-1}$ is a representation if and only if j is integral. In particular, the half-integral values of j do not give rise to representations of the rotation group; only the integral values do this.

As with SU(2), we may describe these representations of the rotation group by identifying useful concrete incarnations of the representation spaces V^j for integral values of j. We will see how to do this in the next lecture when we identify V^j as the space of spherical harmonics of degree j, and use these concrete representation spaces in the study of the central force problem in wave mechanics. The eigenfunctions of the orbital angular momentum operators described in Lecture 12 will be seen to have their most natural expression in terms of these spherical harmonics.

13.5 End Notes

The reader will see the ideas developed in this lecture put to good use in the quantum analysis of the central force problem, in developing the notion of spin in quantum mechanics, in uncovering the structure of hydrogen and the general outline of the periodic table, in exposing the hidden symmetries of the Kepler problem, and in clarifying the nonlocality of quantum mechanics in an analysis of the Bell inequalities. Almost all texts that I have seen develop these ideas using the matrix groups SU(2) and SO(3) and their Lie algebras, and in fact never mention quaternions and the 3-sphere group $\mathbf{S}^3$. I will follow the pack in this respect except when exposing the hidden SO(4) symmetry of the Kepler system and the hydrogenic system in Lectures 16 and 17, where I, like Takhtajan (2008), find that the use of quaternions makes the mathematics more transparent.

See Sternberg (1995) for more on representation theory and its extensive use in physics. A good reference for Lie groups, Lie algebras, and their finite-dimensional representations is the beautifully designed book Fulton and Harris (1991), which targets the beginner in the subject. A more advanced text on the representations of Lie algebras is Humphreys (1972), and a text on the finer points of Lie groups and algebras is that of Knapp (2002). I learned Lie group theory from reading Frank Warner's (1983) excellent text *Foundations of Differentiable Manifolds and Lie Groups*, and I still find his 50-page-long Chapter 3 one of the more concise and well-thought-out presentations of the topic that I have seen.

For the footnote on p. 187, A. Hurwitz in 1898 proved that a real normed algebra can exist only in dimensions 1, 2, 4, and 8, and of course the four examples $\mathbb{R}$, $\mathbb{C}$, $\mathbb{H}$, and $\mathbb{O}$ were known. The actual determination of these as the only real unitary normed algebras was not obtained until the 1940s by A.A. Albert. This is known today as *Hurwitz's theorem*, but in fairness perhaps should be named as the *Hurwitz–Albert theorem*. Good resources for normed algebras are Curtis (1963), pp. 185–187 of Gluck, Warner, and Ziller (1986), and pp. 140–144 of Harvey and Lawson (1982). For a readable historical development of normed algebras, I highly recommend the first two sections of Curtis (1963).

14
Angular Momentum III: The Central Force Problem

> Although Bohr's model opened the way for the later theories, it is important to realize that electrons do not move around the nucleus in circular orbits like the planets orbiting the sun.
>
> Michael Atif
> in a lecture on Atomic Structure

In this lecture we present a method for solving the general central force problem in wave mechanics. Actually, we will show how to use the orbital angular momentum operators to reduce the general problem of a particle in a central potential to a one-dimensional Schrödinger equation whose solutions multiply the angular momentum eigenfunctions to give a complete family of solutions. The orbital angular momentum eigenfunctions are realized as spherical harmonics, special functions defined on the unit sphere $\mathbf{S}^2$ in $\mathbb{R}^3$, and the space of these functions serves as the representation space for the rotation group SO(3) as promised at the end of the previous lecture.

We begin with derivations of the orbital angular momentum operators in the coordinates ideally suited for central force problems, namely, spherical coordinates in $\mathbb{R}^3$ wherein the central symmetry of the potential is manifest. This leads to the spherical eigenvalue problem for the total angular momentum operator $\mathbf{L}^2$, which we solve using some beautiful classical mathematics of special functions. In particular, the Legendre functions, including the polynomials, are seen to supply the raw material needed to construct the eigenfunctions of $\mathbf{L}^2$ as spherical harmonics. We then give the promised description of the irreducible finite-dimensional representation spaces of SO(3) as spaces of these spherical harmonics, and we decompose the Hilbert space $L^2(\mathbf{S}^2)$ into an orthogonal direct sum of irreducible finite-dimensional representations that sits densely in $L^2(\mathbf{S}^2)$. This development provides an example of the important observation that symmetry in the equation of motion is reflected in unitary symmetry in the Hilbert space of solutions expressed in terms of representations. When the Hamiltonian for the central force problem is written in terms of spherical coordinates, the total orbital angular momentum operator condenses out of the calculations. The solutions of the central force problem then are seen to decompose naturally into orbital angular momentum eigenfunctions – spherical harmonics – and a radial term that solves the one-dimensional Schrödinger equation already mentioned.

Up to this point the development applies to all centrally symmetric potentials. This includes not only the hydrogenic potential well with inverse-square force law but also the free particle, the harmonic potential well already treated in Lecture 11, and even the symmetric "particle in a box" problem in elementary quantum mechanics. In the next lecture, we specialize to the hydrogenic potential well, with potential proportional to $1/r$. In particular, the solutions give the first approximations to the orbital structure of an electron in a hydrogen atom and a first approximation to the emission and absorption spectra of atomic hydrogen.

14.1 Orbital Angular Momentum

Recall from Lecture 12 that the orbital angular momentum triad of wave mechanics is defined formally in terms of the position operator triad X and the momentum operator triad P as $\mathsf{L} = \mathsf{X} \times \mathsf{P}$. The component operators are $\mathsf{L}_i = \epsilon_{ijk}\mathsf{X}_j\mathsf{P}_k$, and the commutation relations of formulae 12.1 hold. The **total orbital angular momentum operator** is $\mathsf{L}^2 = \mathsf{L} \cdot \mathsf{L}$, with ladder operators $\mathsf{L}_\pm = \mathsf{L}_1 \pm \mathbf{i}\mathsf{L}_2$. These operators operate on suitable functions in $L^2(\mathbb{R}^3)$ and in Cartesian coordinates are given by the partial differential operators

$$\mathsf{L}_i = \epsilon_{ijk}\mathsf{X}_j\mathsf{P}_k = -\mathbf{i}\hbar\epsilon_{ijk}x_j\frac{\partial}{\partial x_k}$$

for $i = 1, 2, 3$. It is, of course, an implicit assumption when we write these expressions for operators that the functions of $L^2(\mathbb{R}^3)$ are written in terms of Cartesian coordinates. In studying the central force problem, however, we are interested in writing functions in spherical coordinates, the natural coordinates when the potential function is centrally symmetric. It is important, then, to derive the expressions for the orbital angular momentum and ladder operators in terms of spherical coordinates in $\mathbb{R}^3$. For convenience we use the notation ∂_t for the partial derivative operator $\partial/\partial t$, for any variable t except for $t = x_i$, when we use ∂_i for ∂_{x_i}. Standard spherical coordinates are (r, θ, ϕ), where r measures the signed length of the radius vector, θ measures the angle of the radius vector down from the z-axis, and ϕ measures the angle of the projection of the radius vector onto the x_1x_2-plane, measured counterclockwise from the x_1-axis.[1] The coordinate transformations between Cartesian and spherical coordinates are given by

$$\begin{aligned} x_1 &= r\sin\theta\cos\phi, & r^2 &= x_1^2 + x_2^2 + x_3^2, \\ x_2 &= r\sin\theta\sin\phi, & \tan\theta &= \frac{\sqrt{x_1^2 + x_2^2}}{x_3}, \\ x_3 &= r\cos\theta & \tan\phi, &= \frac{x_2}{x_1}. \end{aligned} \tag{14.1}$$

By the chain rule, we have

$$\begin{aligned} \partial_\phi &= (\partial_\phi x_1)\partial_1 + (\partial_\phi x_2)\partial_2 + (\partial_\phi x_3)\partial_3 \\ &= -(r\sin\theta\sin\phi)\partial_1 + (r\sin\theta\cos\phi)\partial_2 = x_1\partial_2 - x_2\partial_1, \end{aligned}$$

and it follows that, when functions in $L^2(\mathbb{R}^3)$ are written with respect to spherical coordinates, the operator L_3 is given by

$$\mathsf{L}_3 = -\mathbf{i}\hbar\,\partial_\phi = -\mathbf{i}\hbar\frac{\partial}{\partial\phi}. \tag{14.2}$$

The chain rule gives

$$\begin{aligned} \mathsf{L}_1 &= -\mathbf{i}\hbar(x_2\partial_3 - x_3\partial_2) \\ &= -\mathbf{i}\hbar\left[(x_2\partial_3 - x_3\partial_2)r\,\partial_r + (x_2\partial_3 - x_3\partial_2)\theta\,\partial_\theta + (x_2\partial_3 - x_3\partial_2)\phi\,\partial_\phi\right]. \end{aligned}$$

Since $\partial_3 r = x_3/r$ and $\partial_2 r = x_2/r$, the coefficient of ∂_r in this expression is zero. Differentiating $\tan\theta = \sqrt{x_1^2 + x_2^2}/x_3$ with respect to both x_2 and x_3 and using the transformation equations yields

$$x_2\partial_3\theta - x_3\partial_2\theta = -\sin\phi$$

for the coefficient of ∂_θ, and differentiating $\tan\phi = x_2/x_1$ with respect to both x_2 and x_3 yields

$$x_2\partial_3\phi - x_3\partial_2\phi = -\cot\theta\cos\phi$$

[1] This is the usual convention among physicists. Mathematicians often reverse the roles of θ and ϕ.

for the coefficient of ∂_ϕ. We arrive at

$$\mathsf{L}_1 = \mathbf{i}\hbar\,(\sin\phi\,\partial_\theta + \cot\theta\cos\phi\,\partial_\phi)\,, \tag{14.3}$$

the expression for L_1 in spherical coordinates. Similarly, the expression for L_2 in spherical coordinates is

$$\mathsf{L}_2 = \mathbf{i}\hbar\,(-\cos\phi\,\partial_\theta + \cot\theta\sin\phi\,\partial_\phi)\,.$$

The ladder operators are easily calculated to be

$$\mathsf{L}_\pm = \hbar e^{\pm\mathbf{i}\phi}(\pm\partial_\theta + \mathbf{i}\cot\theta\,\partial_\phi). \tag{14.4}$$

A straightforward calculation, inserting the spherical expressions we have derived already into either $\mathbf{L}^2 = \mathsf{L}_1^2 + \mathsf{L}_2^2 + \mathsf{L}_3^2$ or $\mathbf{L}^2 = \frac{1}{2}(\mathsf{L}_+\mathsf{L}_- + \mathsf{L}_-\mathsf{L}_+) + \mathsf{L}_3^2$, yields the spherical expression for the total orbital angular momentum operator as

$$\begin{aligned}\mathbf{L}^2 &= -\hbar^2[\partial_\theta^2 + \cot\theta\,\partial_\theta + \csc^2\theta\,\partial_\phi^2] \\ &= -\hbar^2\left[\frac{1}{\sin\theta}\frac{\partial}{\partial\theta}\left(\sin\theta\frac{\partial}{\partial\theta}\right) + \frac{1}{\sin^2\theta}\frac{\partial^2}{\partial\phi^2}\right].\end{aligned} \tag{14.5}$$

14.2 Legendre Functions and Spherical Harmonics

Let $\psi_{\ell,m} = |\ell,m\rangle$, a simultaneous eigenstate of the commuting operators $\mathbf{L}^2$ and L_3, with

$$\mathbf{L}^2\psi_{\ell,m} = \ell(\ell+1)\hbar^2\psi_{\ell,m} \quad \text{and} \quad \mathsf{L}_3\psi_{\ell,m} = m\hbar\psi_{\ell,m}.$$

Recall from Lecture 12 that ℓ must be a nonnegative half-integer and that m, the magnetic quantum number, appears in the list $-\ell, -\ell+1, \ldots, \ell-1, \ell$. From equation 14.2, the eigenvalue equation for L_3 reads

$$-\mathbf{i}\hbar\partial_\phi\psi_{\ell,m} = m\hbar\psi_{\ell,m}$$

whose solution is

$$\psi_{\ell,m}(r,\theta,\phi) = f_{\ell,m}(r,\theta)e^{\mathbf{i}m\phi}$$

for an arbitrary function $f_{\ell,m}(r,\theta)$, subject only to the constraint that $\psi_{\ell,m}$ be square-integrable. Since $\psi_{\ell,m}$ must be single-valued, m must be an integer, which forces ℓ to be an integer. It follows that, for orbital angular momentum operators, the quantum number ℓ that parameterizes the $\mathbf{L}^2$-eigenstates must be a nonnegative integer. It is called, variously, the **azimuthal** or **orbital** quantum number.

We next compute the θ-dependence of $\psi_{\ell,m}$. Recall that the ladder operator L_+ maps $\psi_{\ell,m}$ to a nonzero multiple of $\psi_{\ell,m+1}$, except that $\mathsf{L}_+\psi_{\ell,\ell} = 0$. Applying equation 14.4 to $\psi_{\ell,\ell} = f_{\ell,\ell}(r,\theta)e^{i\ell\phi}$ gives

$$\partial_\theta f_{\ell,\ell} = \ell \cot\theta f_{\ell,\ell}(r,\theta)$$

with solution $f_{\ell,\ell}(r,\theta) = g(r)\sin^\ell\theta$, for an appropriate function g of the radial variable r. Starting with

$$\psi_{\ell,\ell}(r,\theta,\phi) = g(r)e^{i\ell\phi}\sin^\ell\theta \tag{14.6}$$

and applying L_- successively gives the eigenstates $\psi_{\ell,m}$ in the order $m = \ell, \ell-1, \ldots, -\ell+1, -\ell$. Equation 14.4 shows that L_- involves only θ and ϕ derivatives, which implies that the form of the solutions $\psi_{\ell,m}$ is

$$\psi_{\ell,m}(r,\theta,\phi) = C_{\ell,m}g(r)Y_\ell^m(\theta,\phi)$$

for appropriate constants $C_{\ell,m}$. Moreover, a straightforward inductive argument shows that when L_- is applied successively to equation 14.6, the form of Y_ℓ^m is given by the separable expression

$$Y_\ell^m(\theta,\phi) = c_{\ell,m}p_\ell^m(\theta)e^{im\phi}, \tag{14.7}$$

for some functions p_ℓ^m and constants $c_{\ell,m}$. We will show that

$$p_\ell^m(\theta) = P_\ell^m(\cos\theta) = \sin^{|m|}\theta\, P_\ell^{(|m|)}(\cos\theta) \tag{14.8}$$

for certain polynomial functions $P_\ell^{(|m|)}(z)$. Further, we will show that the functions Y_ℓ^m, $-\ell \le m \le \ell$, defined on the 2-sphere $\mathbf{S}^2$ with coordinates $\{(\theta,\phi) : 0 \le \theta \le \pi, 0 \le \phi < 2\pi\}$, span a $(2\ell+1)$-dimensional linear subspace V^ℓ of the Hilbert space $L^2(\mathbf{S}^2) = L^2(\mathbf{S}^2, d\Omega)$ and that the direct sum $\mathcal{V} = \oplus_{\ell=0}^\infty V^\ell$ is an orthogonal direct sum dense in $L^2(\mathbf{S}^2)$. The standard normalization imposed on the basis functions Y_ℓ^m is the orthogonality condition with respect to the spherical volume element $d\Omega = \sin\theta\, d\theta d\phi$, given by

$$\int_{\mathbf{S}^2} \overline{Y_\ell^m} Y_{\ell'}^{m'}\, d\Omega = \int_0^{2\pi}\int_0^\pi \overline{Y_\ell^m} Y_{\ell'}^{m'} \sin\theta\, d\theta d\phi = \delta_{\ell\ell'}\delta_{mm'}. \tag{14.9}$$

The elements of V^ℓ are called **spherical harmonics of degree** ℓ and the set $\{Y_\ell^m : \ell = 0,1,\ldots; m = -\ell, -\ell+1, \ldots, \ell-1, \ell\}$ forms a complete orthonormal basis for $L^2(\mathbf{S}^2)$.

To fill in the details of this discussion, define the operator $\Delta_{\mathbf{S}^2}$ as the restriction of $-\hbar^{-2}\mathsf{L}^2$ to functions defined on $\mathbf{S}^2$. By equation 14.5, the so-called **spherical Laplacian**, in spherical coordinates, is

$$\Delta_{\mathbf{S}^2} = \csc\theta\,\partial_\theta(\sin\theta\,\partial_\theta) + \csc^2\theta\,\partial_\phi^2 \tag{14.10}$$

and operates on functions of the form $h(\theta, \phi)$.[2] Easily, from our discussion thus far, we have

$$\Delta_{\mathbf{S}^2} Y_\ell^m = -\ell(\ell+1) Y_\ell^m.$$

Inserting the expression for Y_ℓ^m given in equation 14.7, substituting $p_\ell^m(\theta) = P_\ell^m(\cos\theta)$, and performing the change of variables $z = \cos\theta$ gives

$$(1-z^2)\partial_z^2 P_\ell^m - 2z\partial_z P_\ell^m - \frac{m^2}{1-z^2} P_\ell^m = -\ell(\ell+1) P_\ell^m.$$

Setting $u(z) = P_\ell^m(z)$, we find that $P_\ell^m(z)$ satisfies **Legendre's associated differential equation**

$$\frac{d}{dz}\left[(1-z^2)u'(z)\right] + \left(\lambda - \frac{m^2}{1-z^2}\right) u(z) = 0, \tag{14.11}$$

with $\lambda = \ell(\ell+1)$. When $m = 0$, we get **Legendre's equation**

$$[(1-z^2)u'(z)]' + \lambda u(z) = 0,$$

which is singular at $z = \pm 1$, i.e., at the north and south poles of $\mathbf{S}^2$. The solutions of Legendre's equation relevant to us are those for which $\lambda = \ell(\ell+1)$ and they may be found by expanding $u(z)$ in a power series about the regular point $z = 0$. Substituting $u(z) = \sum_{n=0}^{\infty} a_n z^n$ in Legendre's equation with $\lambda = \ell(\ell+1)$ gives the recurrence relation

$$(n+1)(n+2)a_{n+2} + [\ell(\ell+1) - n(n+1)]a_n = 0,$$

valid for all $n = 0, 1, \ldots$ When ℓ is even, set $a_0 \neq 0$ and $a_1 = 0$ to obtain the even polynomial solution $P_\ell(z) = a_0 + a_2 z^2 + \cdots + a_\ell z^\ell$, as the recurrence relation shows that $a_n = 0$ for all odd values of n as well as for all $n > \ell$. Similarly, when ℓ is odd, set $a_0 = 0$ and $a_1 \neq 0$ to obtain the odd polynomial solution $P_\ell(z) = a_1 z + a_3 z^3 + \cdots + a_\ell z^\ell$. The functions P_ℓ are known as the **Legendre polynomials** and form, as we shall see, a complete orthogonal set of functions on the interval $[-1, 1]$. The nonzero values of a_0 when ℓ is even and of a_1 when ℓ is odd are chosen so that $P_\ell(1) = 1$, in which case $P_\ell(\pm 1) = (\pm 1)^\ell$ as P_ℓ is an even or odd function according to whether ℓ is even or odd. The Legendre

[2] For aficionados of Riemannian geometry, $\Delta_{\mathbf{S}^2} = \operatorname{div}\operatorname{grad}$ is the Laplace–Beltrami operator on the 2-sphere $\mathbf{S}^2$ expressed in the local coordinates (θ, ϕ).

polynomials are given also by Rodrigues's formula[3] as

$$P_\ell(z) = \frac{1}{2^\ell \ell!}\frac{d^\ell}{dz^\ell}(z^2-1)^\ell.$$

The orthogonality relations for the Legendre polynomials are

$$\int_{-1}^{+1} P_\ell(z)P_{\ell'}(z)\,dz = \frac{2}{2\ell+1}\delta_{\ell\ell'} \tag{14.12}$$

and may be derived by using Rodrigues's formula and performing integration by parts $\min\{\ell,\ell'\}$ times, noting that the mth derivative of $(z^2-1)^\ell$ vanishes at the endpoints ± 1 when $m<\ell$. Since the collection $\{P_\ell : \ell = 0,1,\ldots\}$ of polynomials is orthogonal on $[-1,1]$ with each nonnegative degree represented, the Weierstrauss approximation theorem implies that it is a complete orthogonal set.

Appropriate solutions of Legendre's associated differential equation with $\lambda = \ell(\ell+1)$ may now be obtained from the Legendre polynomials $P_\ell(z)$. For each integer $0 \le m \le \ell$, define the **associated Legendre functions of the first kind** $P_\ell^{\pm m}(z)$ by

$$P_\ell^{\pm m}(z) = (1-z^2)^{m/2}\frac{d^m P_\ell(z)}{dz^m} = \frac{1}{2^\ell \ell!}(1-z^2)^{m/2}\frac{d^{\ell+m}}{dz^{\ell+m}}(z^2-1)^\ell.$$

To see that these functions solve Legendre's associated equation with $\lambda = \ell(\ell+1)$, first substitute $u(z) = (1-z^2)^{m/2}w(z)$ into equation 14.11 and multiply by $(1-z^2)^{-m/2}$ to get, after some simplification, the equation

$$(1-z^2)w'' - 2(m+1)zw' + [\ell(\ell+1) - m(m+1)]w = 0. \tag{14.13}$$

As the Legendre polynomial P_ℓ is a solution to Legendre's equation, we may write

$$(1-z^2)P_\ell'' - 2zP_\ell' + \ell(\ell+1)P_\ell = 0.$$

Differentiating this equation m times reveals that

$$(1-z^2)P_\ell^{(m+2)} - 2(m+1)zP_\ell^{(m+1)} + [\ell(\ell+1) - m(m+1)]P_\ell^{(m)} = 0,$$

[3] See Dunham Jackson (1941), pp. 57–58, for a quick proof, and Chapter 2 of Hochstadt (1971) for an extensive treatment of Rodrigues's formulae for various special functions. Chapter 4 of Williams (2003) presents a modern treatment of the general second order hypergeometric differential equation, which encompasses many classical equations of mathematical physics, including those of Legendre, Hermite, Laguerre, Chebyshev, Bessel, Gauss, Jacobi, and Euler. A general method of solution is presented that writes the solutions in a generalized Rodrigues form. Chapter 7 of Hassani (1999) presents another modern approach due to F.G. Tricomi that produces most classical orthogonal polynomials of interest to physicists in an elegant, compact development; this also produces general formulae for Rodrigues forms, orthogonality and normalization, recurrence relations, and the differential equations that define the polynomials.

meaning that the function $w(z) = P_\ell^{(m)}(z)$ satisfies equation 14.13. Consequently, for nonnegative m,

$$u(z) = (1 - z^2)^{m/2} w(z) = (1 - z^2)^{m/2} P_\ell^{(m)}(z) = P_\ell^m(z)$$

satisfies Legendre's associated differential equation with $\lambda = \ell(\ell + 1)$. Since we define $P_\ell^{-m} = P_\ell^m$ and Legendre's associated equation is even in m, the associated equation is also satisfied by P_ℓ^m when m is negative. With this, $p_\ell^m(\theta)$ may be written in spherical coordinates with $z = \cos\theta$, exactly as in equation 14.8, where, of course, $P_\ell^{(|m|)}$ is the $|m|$th derivative of the Legendre polynomial P_ℓ. We should note in passing that there are other solutions of the associated equation, those of the **second kind**, that are singular at the endpoints but are irrelevant for our purposes as they have logarithmic singularities at the poles of $\mathbf{S}^2$ and hence are not elements of $L^2(\mathbf{S}^2)$.

Putting all this together, the basis for the space V^ℓ of spherical harmonics of degree ℓ takes the form

$$Y_\ell^m(\theta, \phi) = c_{\ell,m} e^{\mathrm{i}m\phi} \sin^{|m|}\theta \, P_\ell^{(|m|)}(\cos\theta). \tag{14.14}$$

As the functions Y_ℓ^m, $-\ell \le m \le \ell$, are linearly independent, the space V^ℓ is $(2\ell + 1)$-dimensional. To derive the orthogonality relations 14.9, we need a partial orthogonality condition for the associated Legendre functions whose verification is left to the reader,[4] namely, that

$$\int_{-1}^{+1} P_\ell^m(z) P_{\ell'}^m(z)\, dz = \frac{2}{2\ell + 1} \frac{(\ell + |m|)!}{(\ell - |m|)!} \delta_{\ell\ell'}. \tag{14.15}$$

This is a *partial* orthogonality relation as it says nothing about Legendre functions with different values of m, which in fact may not be orthogonal. The relations 14.9 follow on inserting the formula for the spherical harmonics of equation 14.14 into equation 14.9 and calculating. This follows first for $m = m'$ and $\ell \neq \ell'$ by a change of variables $z = \cos\theta$ in equation 14.15, and second for $m \neq m'$ just by the ϕ integration in equation 14.9 alone. Finally, when $\ell = \ell'$ and $m = m'$, a calculation using equation 14.15 gives

$$\int_{\mathbf{S}^2} |Y_\ell^m|^2 \, d\Omega = |c_{\ell,m}|^2 \frac{4\pi}{2\ell + 1} \frac{(\ell + |m|)!}{(\ell - |m|)!}.$$

Setting

$$c_{\ell,m} = (-1)^{(m+|m|)/2} \sqrt{\frac{2\ell + 1}{4\pi} \frac{(\ell - |m|)!}{(\ell + |m|)!}}$$

[4] See, for example, pp. 121–126 of Jackson (1941).

then normalizes the spherical harmonics Y_ℓ^m to unit length and finishes the verification of the relations 14.9.

14.3 Radial Symmetry and Representations

We have left unconfirmed one claim made in Section 14.2, namely, that the space $\mathcal{V} = \oplus_{\ell=0}^{\infty} V^\ell$ of spherical harmonics is dense in $L^2(\mathbf{S}^2)$ with $\{Y_\ell^m : \ell = 0, 1, \ldots; m = -\ell, -\ell+1, \ldots, \ell-1, \ell\}$ forming a complete orthonormal basis for $L^2(\mathbf{S}^2)$. To confirm this, we first present an alternative description of the space of spherical harmonics, viz., that $\mathcal{V}$ may be described as precisely the set of complex-valued homogeneous harmonic polynomials in three variables restricted to the 2-sphere. A **harmonic polynomial** is just a polynomial $h(x_1, x_2, x_3)$ with complex coefficients in the kernel of the Laplacian, i.e., for which Δh is identically zero. We will see that V^ℓ consists of those homogeneous harmonic polynomials of degree ℓ restricted to $\mathbf{S}^2$ and that it serves naturally as a $(2\ell+1)$-dimensional irreducible representation space of SO(3).

Before filling in the details of the ideas in the previous paragraph, we pause to paint the broad outline of the general theory of spherical harmonics in an arbitrary number of dimensions.[5] This will reappear in Section 17.1 where a study of the momentum space Schrödinger equation for the hydrogen atom leads naturally to spherical harmonics on the 3-sphere. Let $\mathbf{S}^n$ denote the unit n-sphere in $\mathbb{R}^{n+1}$. For the vector variable $x \in \mathbb{R} \times \mathbb{R}^n = \mathbb{R}^{n+1}$ and a multi-index α, i.e., an $(n+1)$-tuple $\alpha = (\alpha_0, \ldots, \alpha_n)$ of nonnegative integers, define a monomial of degree $|\alpha| = \sum_{i=0}^n \alpha_i$ by

$$x^\alpha = x_0^{\alpha_0} \cdots x_n^{\alpha_n}.$$

A complex-valued polynomial h in $n+1$ variables $x_0, \ldots, x_n$ is a function of the form

$$h(x) = \sum_\alpha c_\alpha x^\alpha,$$

where the sum is over a finite collection of multi-indices α and the coefficients c_α are complex constants. The polynomial h is homogeneous of degree ℓ if $c_\alpha = 0$ whenever $|\alpha| \neq \ell$ or, equivalently, if $h(\lambda x) = \lambda^\ell h(x)$ for any real λ. The function h is harmonic if $\Delta_{n+1} h = 0$, where $\Delta_{n+1} = \sum_{i=0}^n \partial_i^2$ is the $(n+1)$-dimensional Laplace operator. Let $V^{n,\ell}$ denote

[5] A good reference is Section IV.2 of Stein and Weiss (1971) *Introduction to Fourier Analysis on Euclidean Spaces.*

the collection of the restrictions to $\mathbf{S}^n$ of the homogeneous harmonic polynomials on $\mathbb{R}^{n+1}$ of degree ℓ, and let $\mathcal{V}^n = \oplus_{\ell=0}^{\infty} V^{n,\ell}$. The facts are as follows. The direct sum in the definition of $\mathcal{V}^n$ is an orthogonal direct sum in $L^2(\mathbf{S}^n) = L^2(\mathbf{S}^n, d^n\Omega)$ with respect to the inner product $\langle h|k\rangle = \int_{\mathbf{S}^n} \overline{h}k\, d^n\Omega$, where $d^n\Omega$ is the $\mathrm{SO}(n+1)$-invariant spherical measure on the n-sphere. Moreover, $\mathcal{V}^n$ is dense in $L^2(\mathbf{S}^n)$[6] and the subspace $V^{n,\ell}$ is precisely the λ-eigenspace of the Laplace–Beltrami operator $\Delta_{\mathbf{S}^n}$ on the n-sphere with eigenvalue $\lambda = -\ell(\ell+n-1)$. The Laplace–Beltrami operator $\Delta_{\mathbf{S}^n}$ is the angular part of the n-dimensional Laplacian Δ_{n+1}, in that, in spherical coordinates,

$$\Delta_{n+1} = r^{-n}\partial_r(r^n\partial_r) + r^{-2}\Delta_{\mathbf{S}^n}, \tag{14.16}$$

where $r = \|x\|$. Since spherical coordinates on higher-dimensional spaces may be unfamiliar to the reader, we give a brief inductive description. On $\mathbb{R}^2$ these are just the usual polar coordinates (r, ϕ) with ϕ the one-dimensional spherical (circular) coordinate on the 1-sphere $\mathbf{S}^1$. The move to spherical coordinates on $\mathbb{R}^3$ captures the inductive procedure. Add the coordinate θ that measures the angle from the x_3-axis to $x = (x_1, x_2, x_3)$, set $r = \|x\|$, and complete the picture by assigning the one-dimensional spherical coordinate ϕ to the unit vector in the direction of (x_1, x_2), which is a point on $\mathbf{S}^1$. This gives the point x of $\mathbb{R}^3$ with coordinates (r, θ, ϕ), and (θ, ϕ) serve as two-dimensional spherical coordinates on the 2-sphere $\mathbf{S}^2$. In general, starting with $(n-1)$-dimensional spherical coordinates Φ on $\mathbf{S}^{n-1}$, so that (r, Φ) are spherical coordinates on $\mathbb{R}^n$, where $r = \|y\|$ for $y \in \mathbb{R}^n$, assign the spherical coordinates (r, ϑ, Φ) to the point $x \in \mathbb{R}^{n+1}$, where $r = \|x\|$, ϑ is the angle from the x_0-axis to x, and Φ are the $(n-1)$-dimensional spherical coordinates on $\mathbf{S}^{n-1}$ of the unit vector in the direction of $(x_1, \ldots, x_n) \in \mathbb{R}^n$. The coordinate ϑ is the **axial coordinate** determined by the axis $\mathbb{R} \times \{\mathbf{0}\}$ and (ϑ, Φ) serve as n-dimensional spherical coordinates on the n-sphere $\mathbf{S}^n$. The Laplace–Beltrami operator, or spherical Laplacian, on $\mathbf{S}^n$ is defined inductively by

$$\Delta_{\mathbf{S}^n} = \csc^{n-1}\vartheta\, \partial_\vartheta(\sin^{n-1}\vartheta\, \partial_\vartheta) + \csc^2\vartheta\, \Delta_{\mathbf{S}^{n-1}}, \tag{14.17}$$

with $\Delta_{\mathbf{S}^1} = \partial_\phi^2$, and satisfies equation 14.16. Equation 14.10 is this equation for $n = 2$. We confirm the claims of this paragraph only in the case $n = 2$, to which we now turn.

Let $\mathcal{P}^\ell$ be the space of complex-valued polynomials that are homogeneous of degree ℓ in the three variables x_1, x_2, and x_3. The space $\mathcal{P}^\ell$ is

[6] This is an easy application of the Stone-Weierstrauss theorem.

a complex vector space with basis the collection $\{\mathbf{x}^\alpha : |\alpha| = \ell\}$ of monomials of degree ℓ. Here we revert to using bold-face letters to represent three-dimensional vectors. It follows that

$$\dim \mathcal{P}^\ell = \binom{\ell+2}{2} = \frac{(\ell+2)(\ell+1)}{2}.^7$$

The Laplace operator $\Delta = \Delta_3$ maps $\mathcal{P}^\ell$ linearly into $\mathcal{P}^{\ell-2}$ whenever $\ell \geq 2$. The restriction of Δ to $\mathcal{P}^\ell$ is in fact surjective, a fact whose verification is left to the reader. This implies that the space of homogeneous harmonic polynomials of degree ℓ, namely, $\mathcal{H}^\ell = \ker \Delta|\mathcal{P}^\ell$, has dimension

$$\dim \mathcal{H}^\ell = \dim \mathcal{P}^\ell - \dim \mathcal{P}^{\ell-2} = 2\ell + 1. \tag{14.18}$$

Given a function $h \in \mathcal{H}^\ell$, notice that for $\mathbf{0} \neq \mathbf{x} \in \mathbb{R}^3$,

$$h(\mathbf{x}) = \|\mathbf{x}\|^\ell h(\hat{\mathbf{x}}),$$

where $\hat{\mathbf{x}} = \mathbf{x}/\|\mathbf{x}\| \in \mathbf{S}^2$ is the unit vector in the direction of $\mathbf{x}$. It follows that, when written in spherical coordinates, h takes the form

$$h(r, \theta, \phi) = r^\ell Y(\theta, \phi) \tag{14.19}$$

for some function Y defined on the 2-sphere $\mathbf{S}^2$. This means, of course, that the restriction of h to the unit sphere $\mathbf{S}^2$ is precisely the function Y. Let $V^{2,\ell}$ denote the set of these restrictions of the elements of $\mathcal{H}^\ell$ to $\mathbf{S}^2$. Note that $V^{2,\ell}$ is a complex vector space isomorphic to $\mathcal{H}^\ell$ via the correspondence $V^{2,\ell} \ni Y \leftrightarrow r^\ell Y \in \mathcal{H}^\ell$ and, therefore, is a $(2\ell+1)$-dimensional complex vector subspace of the Hilbert space $L^2(\mathbf{S}^2)$. The next task is to show that $V^{2,\ell} = V^\ell$, i.e., that the restrictions to the 2-sphere of the homogenous harmonic polynomials of degree ℓ are precisely the spherical harmonics of degree ℓ defined in the preceding section.

Equations 14.16 and 14.17 may be derived easily by application of the chain rule. Most of the derivation for $n = 2$ has already been worked out in deriving equations 14.5 and 14.10. When the $n = 2$ case of equation 14.16 is applied to expression 14.19, we obtain, since $\Delta h = 0$,

$$0 = \ell(\ell+1)r^{\ell-2}Y + r^{\ell-2}\Delta_{\mathbf{S}^2}Y$$

or

$$\Delta_{\mathbf{S}^2}Y = -\ell(\ell+1)Y, \tag{14.20}$$

[7] Each multi-index α of degree ℓ corresponds to a choice of two "separator" slots in a linear list of $\ell+2$ slots, the two chosen slots separating α_1 slots to the left for the exponent on x_1, α_2 slots in the middle for the exponent on x_2, and α_3 slots to the right for the exponent on x_3.

verifying that $V^{2,\ell}$ is contained in the $-\ell(\ell+1)$-eigenspace of the operator $\Delta_{\mathbf{S}^2}$ on $L^2(\mathbf{S}^2)$. An application of the Stone–Weierstrauss theorem implies that the subspace $\mathcal{V}^2 = \oplus_{\ell=0}^{\infty} V^{2,\ell}$ is dense in $L^2(\mathbf{S}^2)$, which in turn implies that $V^{2,\ell}$ is precisely the $-\ell(\ell+1)$-eigenspace of $\Delta_{\mathbf{S}^2}$. Now the fact that the spaces $V^{2,\ell}$ are eigenspaces of the self-adjoint operator $\Delta_{\mathbf{S}^2}$ with different eigenvalues implies that the sum is orthogonal. It follows also, since V^{ℓ} is a collection of $-\ell(\ell+1)$-eigenfunctions, that $V^{\ell} \subset V^{2,\ell}$ and, since their dimensions agree, that this containment is in fact an equality, confirming the claim of the preceding paragraph.

Recall the promise of the previous lecture that the irreducible finite-dimensional representation spaces of the rotation group SO(3) would be given a concrete incarnation in terms of spherical harmonics. With the identification of the spherical harmonics as the restrictions to $\mathbf{S}^2$ of the homogeneous harmonic polynomials, we are able to realize V^{ℓ} concretely as an irreducible $(2\ell+1)$-dimensional representation space of SO(3). The first task is to demonstrate the symmetry of the Laplacian with respect to rotations. This means that $(\Delta\psi) \circ \mathsf{R} = \Delta(\psi \circ \mathsf{R})$ whenever $\mathsf{R} \in \mathrm{SO}(3)$ is a rotation. Before considering the details, though, we can see that the equation $(\Delta\psi)(\mathsf{R}\mathbf{x}) = \Delta(\psi \circ \mathsf{R})(\mathbf{x})$ should hold for any rigid motion R of $\mathbb{R}^3$, by recalling the descriptive definition of the Laplacian of a function ψ at the point $\mathbf{x}$ as a measure of the infinitesimal difference between the value $\psi(\mathbf{x})$ and the average value of ψ near $\mathbf{x}$. Under a rigid motion, these infinitesimal averages are preserved because there is no compression or expansion of values of ψ by R, so that the distribution of the values of ψ near $\mathsf{R}\mathbf{x}$ is the same as the distribution of the values of $\psi \circ \mathsf{R}$ near $\mathbf{x}$. For those who like to see ideas expressed in equations, here are the details. Several applications of the chain rule yield the general formula for $\Delta(\psi \circ \mathsf{R})(\mathbf{x})$, for any C^2 function $\mathsf{R} = (\mathsf{R}_1, \mathsf{R}_2, \mathsf{R}_3) : \mathbb{R}^3 \to \mathbb{R}^3$, as

$$\Delta(\psi \circ \mathsf{R})(\mathbf{x}) = \sum_{i,j,k=0}^{3} \frac{\partial^2 \psi}{\partial x_i \partial x_j} \frac{\partial \mathsf{R}_i}{\partial x_k} \frac{\partial \mathsf{R}_j}{\partial x_k} + \sum_{i=1}^{3} \frac{\partial \psi}{\partial x_i} \Delta \mathsf{R}_i. \tag{14.21}$$

Here all the terms involving derivatives of ψ are evaluated at $\mathsf{R}\mathbf{x}$ and all the terms involving derivatives of the component functions of R are evaluated at $\mathbf{x}$. Now if the component functions R_i are linear in the variables x_j, the second derivatives of the R_i with respect to the x_j vanish and this implies that the Laplacians $\Delta \mathsf{R}_i$ are zero and the second term of equation 14.21 vanishes. Assume then that

$$\mathsf{R}_i(\mathbf{x}) = \mathsf{R}_{i1} x_1 + \mathsf{R}_{i2} x_2 + \mathsf{R}_{i3} x_3 + a_i \tag{14.22}$$

for $i = 1, 2, 3$. This means that R is obtained by applying the linear mapping with matrix $[\mathsf{R}_{ij}]$ followed by translation by the vector $\mathbf{a} = (a_1, a_2, a_3)$. In this case, then, equation 14.21 becomes

$$\Delta(\psi \circ \mathsf{R})(\mathbf{x}) = \sum_{i,j=0}^{3} \frac{\partial^2 \psi}{\partial x_i \partial x_j}(\mathsf{R}\mathbf{x}) \sum_{k=1}^{3} \mathsf{R}_{ik}\mathsf{R}_{jk}. \tag{14.23}$$

Now, if R is a rigid motion then R is of the form given by equation 14.22 with the matrices $[\mathsf{R}_{ij}]$ rotation matrices and therefore orthonormal. This means that $\sum_{k=1}^{3} \mathsf{R}_{ik}\mathsf{R}_{jk} = \delta_{ij}$, and equation 14.23 now reads

$$\Delta(\psi \circ \mathsf{R})(\mathbf{x}) = \sum_{i,j=0}^{3} \frac{\partial^2 \psi}{\partial x_i \partial x_j}(\mathsf{R}\mathbf{x})\delta_{ij} = \sum_{i=1}^{3} \frac{\partial^2 \psi}{\partial x_i^2}(\mathsf{R}\mathbf{x}) = (\Delta\psi)(\mathsf{R}\mathbf{x}).$$

This of course is the desired result.

Armed with this SO(3)-symmetry of the Laplacian, V^ℓ is realized in short order as a concrete example of an irreducible representation space of the rotation group. Define the function

$$\sigma : \mathrm{SO}(3) \to \mathbf{U}(L^2(\mathbf{S}^2))$$

by $\sigma(\mathsf{R})(\psi) = \psi \circ \mathsf{R}^{-1}$. Here, $\mathbf{U}(L^2(\mathbf{S}^2))$ is the group of unitary operators on $L^2(\mathbf{S}^2)$ and $\sigma(\mathsf{R})$ is seen to be unitary from the SO(3)-invariance of the spherical measure $d^2\Omega$ on the 2-sphere. Obviously, $\sigma(\mathsf{R})^{-1} = \sigma(\mathsf{R}^{-1})$ and $\sigma(\mathsf{RS}) = \sigma(\mathsf{R}) \circ \sigma(\mathsf{S})$ for $\mathsf{R}, \mathsf{S} \in \mathrm{SO}(3)$, so that σ is a representation of SO(3). Moreover, we claim that V^ℓ is an invariant subspace of the representation σ, i.e., that if $Y \in V^\ell$ then $\sigma(\mathsf{R})(Y) = Y \circ \mathsf{R}^{-1} \in V^\ell$ whenever $\mathsf{R} \in \mathrm{SO}(3)$. Indeed, since $Y \in V^\ell$, the function $h = r^\ell Y \in \mathcal{H}^\ell$ so that $\Delta(h \circ \mathsf{R}^{-1}) = (\Delta h) \circ \mathsf{R}^{-1} = 0$ since h is harmonic. Since $h \circ \mathsf{R}^{-1} = r^\ell(Y \circ \mathsf{R}^{-1})$, the argument leading to equation 14.20 shows that $Y \circ \mathsf{R}^{-1}$ is a $-\ell(\ell+1)$-eigenfunction of $\Delta_{\mathbf{S}^2}$ so that $Y \circ \mathsf{R}^{-1} \in V^\ell$, as promised.[8] It follows that there is a homomorphism

$$\sigma^\ell : \mathrm{SO}(3) \to \mathrm{GL}(V^\ell)$$

defined by $\sigma^\ell(\mathsf{R}) = \sigma(\mathsf{R})|V^\ell$ that defines a representation of SO(3) on the space V^ℓ of spherical harmonics of degree ℓ. That V^ℓ is irreducible is a consequence of the main result of Lecture 12, that $\dim V^\ell = 2\ell+1$, and of the construction of the spherical harmonics Y_ℓ^m of Section 14.1; the

[8] Alternatively, this follows from the fact that $h \circ \mathsf{R}^{-1} = r^\ell(Y \circ \mathsf{R}^{-1})$ is homogeneous, which in turn follows from the observation that a linear change of coordinates in a monomial of degree ℓ produces a homogenous polynomial of degree ℓ in the new coordinates.

details are left to the reader. We see then that $L^2(\mathbf{S}^2)$ decomposes as a Hilbert space orthogonal direct sum of the finite-dimensional irreducible representation spaces of SO(3).

14.4 A Single Particle in a General Central Potential

Recall that the Hamiltonian operator of wave mechanics is given by

$$\mathsf{H} = \frac{1}{2m}\mathbf{P}^2 + V = -\frac{\hbar^2}{2m}\Delta + V,$$

where $\mathbf{P}^2 = \mathsf{P}_j\mathsf{P}_j$ (with summation over the repeated index j) is $-\hbar^2$ times the Laplacian operator Δ and V is the potential function. The second of equations 12.1, that $[\mathsf{L}_i, \mathsf{P}_j] = \mathbf{i}\hbar\epsilon_{ijk}\mathsf{P}_k$, may be used to show that the operators L_i and $\mathbf{P}^2$ commute:

$$\begin{aligned}[\mathsf{L}_i, \mathbf{P}^2] = [\mathsf{L}_i, \mathsf{P}_j\mathsf{P}_j] &= \mathsf{P}_j[\mathsf{L}_i, \mathsf{P}_j] + [\mathsf{L}_i, \mathsf{P}_j]\mathsf{P}_j \\ &= \mathbf{i}\hbar\epsilon_{ijk}(\mathsf{P}_j\mathsf{P}_k + \mathsf{P}_k\mathsf{P}_j) \\ &= 2\mathbf{i}\hbar\epsilon_{ijk}\mathsf{P}_j\mathsf{P}_k = 0,\end{aligned}$$

since $\mathsf{P}_j\mathsf{P}_k = \mathsf{P}_k\mathsf{P}_j$ and the final sum over j and k produces only two nonzero terms, equal in value and opposite in sign. It follows that $\mathbf{P}^2$ and $\mathbf{L}^2$ commute since $[\mathbf{P}^2, \mathbf{L}^2] = [\mathbf{P}^2, \mathsf{L}_j\mathsf{L}_j] = \mathsf{L}_j[\mathbf{P}^2, \mathsf{L}_j] + [\mathbf{P}^2, \mathsf{L}_j]\mathsf{L}_j = 0$.

When V is centrally symmetric, the potential function may be written as $V = V(r)$ with $\partial_\theta V = 0 = \partial_\phi V$. Since the component operators L_i involve only the partial derivatives with respect to θ and ϕ and not with respect to r, it follows that $[\mathsf{L}_i, V] = 0 = [\mathbf{L}^2, V]$. This discussion shows that when V is centrally symmetric, each component of the orbital angular momentum triad, as well as the total orbital angular momentum operator, commutes with the Hamiltonian:

$$[\mathsf{L}_i, \mathsf{H}] = 0 = [\mathbf{L}^2, \mathsf{H}], \quad \text{for} \quad i = 1, 2, 3. \tag{14.24}$$

Thus, for centrally symmetric potentials, the operators H, $\mathbf{L}^2$, and L_3 commute pairwise and, when the Hamiltonian has a discrete spectrum, we may expect simultaneous eigenstates $\psi_{n,\ell,m}$ where n parameterizes the eigenvalues of the Hamiltonian. We have already worked out a complete set of simultaneous eigenfunctions on $\mathbf{S}^2$ for $\mathbf{L}^2$ and L_3 as the spherical harmonics, and we may write

$$\psi_{n,\ell,m}(r, \theta, \phi) = g_{n,\ell}(r)Y_\ell^m(\theta, \phi), \tag{14.25}$$

for a radial function $g_{n,\ell}(r)$ dependent on the radial variable r and on the energy eigenvalues, which are indexed by the parameter n, and the azimuthal quantum number ℓ. Often n turns out to be a discrete index, but we are enforcing no such requirement.

To find the radial functions we first write the Hamiltonian in spherical form. Using equations 14.16 and 14.17 with $n = 2$, the spherical form of the three-dimensional Laplacian may be rewritten as

$$\Delta = \frac{1}{r}\frac{\partial^2}{\partial r^2}r + \frac{1}{r^2 \sin\theta}\frac{\partial}{\partial\theta}\left(\sin\theta\frac{\partial}{\partial\theta}\right) + \frac{1}{r^2\sin^2\theta}\frac{\partial^2}{\partial\phi^2}.$$

Comparing this with the spherical form of L^2 in equation 14.5 gives

$$\Delta = \frac{1}{r}\frac{\partial^2}{\partial r^2}r - \frac{1}{r^2\hbar^2}\mathsf{L}^2.$$

The Hamiltonian becomes

$$\mathsf{H} = -\frac{\hbar^2}{2mr}\frac{\partial^2}{\partial r^2}r + \frac{1}{2mr^2}\mathsf{L}^2 + V(r).$$

Letting the energy eigenvalue for $\psi_{n,\ell,m}$ be E_n and recalling that the L^2-eigenvalue for the spherical harmonic Y_ℓ^m is $\ell(\ell+1)\hbar^2$, we have, from $\mathsf{H}\psi_{n,\ell,m} = E_n\psi_{n,\ell,m}$, after some simplification using equation 14.25,

$$-\frac{\hbar^2}{2mr}\frac{d^2}{dr^2}rg_{n,\ell} + \left(\frac{\hbar^2}{2m}\frac{\ell(\ell+1)}{r^2} + V\right)g_{n,\ell} = E_n g_{n,\ell}.$$

Multiplying by r and letting $f_{n,\ell}(r) = rg_{n,\ell}(r)$, we obtain the **radial equation**

$$\left(-\frac{\hbar^2}{2m}\frac{d^2}{dr^2} + V_{\text{eff}}\right)f_{n,\ell} = E_n f_{n,\ell}, \tag{14.26}$$

which is a one-dimensional time-independent Schrödinger equation with **(effective) potential**

$$V_{\text{eff}}(r) = \frac{\hbar^2}{2m}\frac{\ell(\ell+1)}{r^2} + V(r).$$

This reduces the three-dimensional central potential problem to a one-dimensional problem in wave mechanics.

The eigenfunctions for a central potential are found once the radial equation 14.26 is solved for an appropriate potential V. For example, the three-dimensional harmonic well problem is solved with the insertion of $V(r) = \frac{1}{2}m\omega^2r^2$, and the particle in a spherical box of radius R is solved with the insertion of $V(r) = 0$ for $r < R$ and $V(r) = \text{constant} > 0$ for $r > R$. Our interest in the next lecture turns to solving the radial equation for the **hydrogenic potential well** with $V(r) = -Ze^2/r$, where Z is a positive integer (the proton number) and e is a nonzero constant (the unit electrostatic charge). This will give our first approximation to the quantum mechanical description of the hydrogen atom, which is quite successful in deriving the gross features of the spectral characteristics of atomic hydrogen. This represents one of the crowning achievements of nonrelativistic quantum mechanics, an accomplishment detailed in the papers of Schrödinger from 1926 that first presented wave mechanics and introduced the equation that bears his name.

14.5 End Notes

I drew from many deep sources to learn and digest the material for this lecture. Chapters 11 and 12 of Merzbacher (1998) give a thorough treatment of angular momentum and the hydrogenic system that I find appealing. I must hold up Stephanie Singer's 2005 text *Linearity, Symmetry, and Prediction in the Hydrogen Atom*, a book targeting the advanced undergraduate student of mathematics, for particular praise. Singer (2005) is a beautifully developed work whose purpose seems to be to demonstrate how naturally physics and mathematics blend to complement one another. It develops the unitary representation theory of the Lie groups SO(3), SU(2), and SO(4) in the context of understanding the quantum mechanical treatment of the hydrogen atom, and in doing so presents a rigorous treatment valuable to its intended audience of students of physics and chemistry as well as mathematics.

Sternberg (1995) presents a rigorous accounting of representation theory applied to physical problems. I garnered much of what I have relayed in Section 14.3 on harmonic polynomials and their use as representation spaces from Sternberg. See Chapters 7 and 8 of Lebedev (1972) for a detailed accounting of spherical harmonics, and Section III.3 of Stein (1970) for spherical harmonics in higher dimensions. The Carus Mathematical Monograph Jackson (1941) is a nice quick introduction to orthogonal polynomials. For more on Legendre polynomials and functions, see the detailed presentation in Chapter XV of the classic Whittaker and Watson (1958), Chapter 10 of Rainville (1960), or the quick presentation in Appendix E of Fitts (1999), as well as his chapter on the hydrogen atom. More modern treatments of special functions that are encyclopedic in nature may be found in Andrews, Askey, and Roy (1999) *Special Functions* and Beals and Wong (2010) *Special Functions: A Graduate Text.*

We already have seen two families of special functions – the Hermite polynomials and Legendre functions – and we will see several others as these lectures progress, including for example the Laguerre functions, the hypergeometric functions, and Bessel functions. These are primarily topics in the classical analysis of the eighteenth and nineteenth centuries whose development continued through the twentieth. This is a good place to give general references for those special functions of mathematical physics and the classical analysis that has significant applications in the development of quantum mechanics. I do so by recommending three older classics that are rarely studied today but in which I have found good and complete explanations for much of the classical analysis used in the quantum mechanical analysis of real systems. The first of these, Whittaker and Watson (1902, fourth edition 1958) *A Course of Modern Analysis*, is a bible of classical analysis whose range of discourse is related so ably by its subtitle, *An Introduction to the General Theory of Infinite Processes and of Analytic Functions; With an Account of the Principal Transcendental Functions.* It treats in great detail all the classical special functions of mathematical physics. The second of the classics is Jeffreys and Jeffreys (1946) *Methods of Mathematical Physics*, intended "to provide an account of those parts of pure mathematics that are frequently used in modern physics."[9] There is much to appreciate in this book on classical analysis as applied to problems in physics. The emphasis in the text falls decidedly on the rigorous development of the analytic methods that are used in physics, so that the mathematics takes precedence over the physics that motivates it. This attitude is reversed somewhat in the third of the classic texts, Morse and Feshbach (1953a, b) *Methods of Theoretical Physics*, where the physics takes center stage in motivating and directing the presentation of the mathematics. The emphasis here is on developing just enough of the mathematics to analyze various physical systems and theories. The Morse and Feshbach work appears as two volumes, Parts I and II, and in 1978 pages covers almost all the classical mathematics one needs for the whole of physics.

[9] Jeffreys and Jeffreys (1946) p. v.

15

Wave Mechanics IV: The Hydrogenic Potential

> The hydrogen atom – the two-body motion with inverse-square force rendered by quantum mechanics – possesses eigenstates which form a rich collection of three-dimensional structures. Quantum states stand in marked contrast to classical trajectories for which the orbit lies wholly in the plane perpendicular to the angular momentum vector. The eigenstates are intricate geometric shapes formed of exponential, trigonometric, and polynomial functions. They contrast dramatically against the classical motion for which the geometric structures are the circle and ellipse. This collection of eigenstates of the quantum mechanical two-body motion provides the basis for the manifold structure of the elements – structure not contained in the ellipse of classical motions.
>
> David Oliver
> *The Shaggy Steed of Physics*, 1994

The *full* quantum mechanical treatment of the interaction of proton and electron in the hydrogen atom represents, arguably, the pinnacle of success for the explanatory powers of quantum mechanics. Our treatment began with the study of angular momentum operators applied to the central force problem in the last lecture. In this one, we address the radial equation and arrive at our first approximation to the quantum mechanical solution of the hydrogen atom, which, through the incorporation of the hydrogenic potential, models the electrostatic environment of the orbital electron. In the full solution to the Schrödinger equation for the hydrogenic potential, we see the broad outlines of the order that prevails in the periodic table of the elements; this is surely one of the most beautiful and exceptional achievements, not only of the science of quantum mechanics but indeed of the whole of human science. In this lecture and the next, we explore several approaches to solving the radial equation and look again at the importance of symmetry in understanding the workings of nature.

Even with these successes, the quantum mechanical treatment of the hydrogen atom is not complete. When the full electromagnetic environment, including not only the electrostatic but also the magnetic contributions, is included, as well as the relativistic and spin effects described by the Dirac equation, more impressive agreement with the experimental data is obtained. Even then, with the full electromagnetic potential included in the Dirac equation, there are slight discrepancies with experiment. These are traced to the fact that although the particles have been treated in a fully relativistic and quantum mechanical manner, the electromagnetic field itself has been treated classically. And so the full quantum mechanical treatment of hydrogen requires the development of a quantum theory of the electromagnetic field – a **quantum electrodynamics**. The *full* picture that finally emerges from these successively finer theories offers the most precise agreement between theory and experiment found in all the sciences and must be seen as a crowning achievement of humanity's quest for understanding the workings of nature.

The primary work of this lecture is in obtaining closed form expressions for the wave functions of the hydrogenic potential in quantum mechanics. The lecture following continues this one and examines some of its implications. We begin, then, with an algebraic solution of the radial equation, inspired by Pauli, that is reminiscent of our treatment of the harmonic oscillator. Here we obtain a complete listing of the bound state eigenfunctions of the radial equation, using ladder operators, and the corresponding discrete energy spectrum. The section following presents a more classical treatment due to Schrödinger using power series techniques. This rewrites the bound state wave functions in terms of the classical confluent hypergeometric functions and Laguerre polynomials. The two forms of the eigenfunctions obtained, that of Pauli and of Schrödinger, are proved to represent the same quantum states. The results of our examination of the radial equation are then combined with those of the last lecture to give a complete orthonormal basis for the bound state energy subspace of $L^2(\mathbb{R}^3)$ determined by the time-independent Schrödinger equation for the central Coulomb potential. This, of course, entails the introduction of spherical harmonics into the mix. Finally, we examine the positive energy solutions to the radial equation, which govern the unbound states.

In the next lecture, we continue our discussion of the hydrogenic potential in quantum mechanics by examining the implications of this lecture for the spectral lines of atomic hydrogen and for the arrangement of the elements in the periodic table. There the momentum space representation of the Schrödinger equation is examined and group theory is used to rederive the energy levels of atomic hydrogen.

15.1 An Algebraic Approach to the Radial Equation

The goal of this section and the next is to present two different methods for solving the one-dimensional time-independent Schrödinger equation

$$\left(-\frac{\hbar^2}{2m}\frac{d^2}{dr^2} + V_{\text{eff}}\right) f_{n,\ell} = E_n f_{n,\ell} \tag{15.1}$$

on the ray $r \geq 0$, in the presence of the effective Coulomb potential given by

$$V_{\text{eff}}(r) = \frac{\hbar^2}{2m}\frac{\ell(\ell+1)}{r^2} - \frac{Ze^2}{r}. \tag{15.2}$$

This is, of course, the radial equation 14.26 for the hydrogenic potential well $V(r) = -Ze^2/r$, where Z is a positive integer, the **proton number**, and e is the unit electrostatic charge. The positive integer n, known as the **principal quantum number**, is used to index the bound state energy eigenvalues E_n, which turn out to be negative and to form a discrete energy spectrum. Now, it turns out that the spectrum of the Hamiltonian operator of equation 15.1 includes not only the point spectrum consisting of the discrete negative sequence of eigenvalues E_n but also the continuous spectrum consisting of all nonnegative real numbers. When the energy of the hydrogenic system is in the continuous spectrum, the electron is not bound to the nucleus and typically may move freely to or from infinity, subject, of course, to the influence of the infinite-range potential $V(r)$. This section with the two following primarily restrict attention to determining the negative eigenvalues E_n and their bound state eigenfunctions, and the penultimate section then focuses attention on the positive energy states.

The reader will see that the development of this section is reminiscent of our algebraic treatment of the quantum oscillator in Lecture 11 as well as the use of ladder operators in Lectures 12 and 14 in deriving the eigenspaces of angular momentum operators and the spherical harmonics that serve as their eigenfunctions. In those previous developments, the important feature is that the Hamiltonian may be written in terms of appropriate ladder operators that annihilate and create quanta, which do so precisely because of appropriate commutation relations between the ladder operators and the Hamiltonian. Here the matter is a bit more complicated, primarily due to the appearance of the integral parameter ℓ, the azimuthal quantum number, in the Hamiltonian of the radial equation. Define the Hamiltonian operator H_ℓ as

$$\mathsf{H}_\ell = -\frac{\hbar^2}{2m}\frac{d^2}{dr^2} + \frac{\hbar^2}{2m}\frac{\ell(\ell+1)}{r^2} - \frac{Ze^2}{r}.$$

In examining the eigenvalue equation $\mathsf{H}_\ell u = Eu$, we look for a near-factorization of the Hamiltonian of the form of equation 11.3 using a pair of adjoint operators $\mathsf{A}_\ell^\pm$ on the Hilbert space $L^2([0,\infty))$. Following the method of Lecture 11, we might expect to find that we may use the creation operator A_ℓ^+ to generate the higher energy eigenfunctions from a ground state, but this fails as the appropriate ladder operators do not quite satisfy the relations of equations 11.4 and 11.5 of Lecture 11. Rather, it turns out the important feature of $\mathsf{A}_\ell^\pm$ is that these operators intertwine the two successive Hamiltonians H_ℓ and $\mathsf{H}_{\ell+1}$ in a way that allows their use for creating eigenfunctions of H_ℓ from known eigenfunctions of $\mathsf{H}_{\ell+1}$. Ultimately we are able to identify a complete set of eigenfunctions for the bound states of atomic hydrogen by identifying first the ground state eigenfunctions of each H_ℓ and using the ladder operators to generate the remaining eigenfunctions.

What, then, are the appropriate ladder operators for the radial equation with hydrogenic potential? They turn out to be

$$\mathsf{A}_\ell^\pm = \frac{\hbar}{\sqrt{2m}}\left(\mp\frac{d}{dr} + \frac{\ell+1}{r} - \frac{mZe^2}{\hbar^2}\frac{1}{\ell+1}\right). \tag{15.3}$$

A calculation gives

$$\mathsf{A}_\ell^-\mathsf{A}_\ell^+ = \mathsf{H}_\ell + \frac{mZ^2e^4}{2\hbar^2}\frac{1}{(\ell+1)^2} \tag{15.4}$$

and

$$\mathsf{A}_\ell^+\mathsf{A}_\ell^- = \mathsf{H}_{\ell+1} + \frac{mZ^2e^4}{2\hbar^2}\frac{1}{(\ell+1)^2}. \tag{15.5}$$

The important consequence of equations 15.4 and 15.5 is that

$$\mathsf{H}_\ell\mathsf{A}_\ell^- = \mathsf{A}_\ell^-\mathsf{H}_{\ell+1}. \tag{15.6}$$

If u is an eigenfunction of $\mathsf{H}_{\ell+1}$ with eigenvalue E then expression 15.6

implies that $u^- = \mathsf{A}_\ell^- u$ satisfies $\mathsf{H}_\ell u^- = Eu^-$, so that u^- is an eigenfunction of H_ℓ with the same eigenvalue E provided that $u^- \neq 0$. This little observation is the key that will allow us to construct an infinite sequence of eigenfunctions for the operator H_ℓ.

To find our first eigenfunction for H_ℓ, note that if $\mathsf{A}_\ell^+ u = 0$ then equation 15.4 implies that

$$\mathsf{H}_\ell u = \mathsf{A}_\ell^- \mathsf{A}_\ell^+ u - \frac{mZ^2e^4}{2\hbar^2} \frac{1}{(\ell+1)^2} u = -\frac{mZ^2e^4}{2\hbar^2} \frac{1}{(\ell+1)^2} u.$$

Thus if $u \neq 0$ then u is an eigenfunction of H_ℓ with eigenvalue

$$-\frac{Z^2}{(\ell+1)^2} \mathcal{RY}, \tag{15.7}$$

where $\mathcal{RY} = me^4/2\hbar^2$, the natural unit of energy for this problem, which is called the **Rydberg** whenever m is taken as the rest mass of the electron, a requirement we now impose. The first-order equation given by $\mathsf{A}_\ell^+ u = 0$ is easy to solve. Its solutions are square-integrable on $[0, \infty)$ and are given as the real multiples of the solution

$$u_\ell(r) = r^{\ell+1} e^{-Zr/(\ell+1)a_0}, \tag{15.8}$$

where $a_0 = \hbar^2/me^2$ is the **Bohr radius**, the classical radius of the electron orbit in the Bohr model of the hydrogen atom. The implication of equation 15.6, already explicated, implies that, for positive integers $k \leq \ell$, the function

$$u_{\ell;k} = \mathsf{A}_{\ell-k}^- \cdots \mathsf{A}_{\ell-1}^- u_\ell \tag{15.9}$$

is an eigenfunction of $\mathsf{H}_{\ell-k}$ with eigenvalue given by expression 15.7, provided that $u_{\ell;k} \neq 0$. We will demonstrate that $u_{\ell;\ell} \neq 0$, which implies that each of $u_{\ell;0}, \ldots, u_{\ell;\ell}$ is nonzero, where, of course, $u_{\ell;0} = u_\ell$. The key observation is that when the operator A_m^- for $0 \leq m \leq \ell - 1$ is applied to a function of the form $p(r)e^{Ar/(\ell+1)}$, where $p(r)$ is a polynomial of degree n, the result is a function of the form $q(r)e^{Ar/(\ell+1)}$, where q is a polynomial of degree at most n. It follows that $u_{\ell;\ell} = p_\ell(r)e^{Ar/(\ell+1)}$, where $p_\ell(r)$ is a polynomial of degree at most $\ell + 1$ and $A = -Z/a_0$. Our claim is that the degree of the polynomial $p_\ell(r)$ is in fact $\ell + 1$. We will demonstrate this by identifying the coefficient of $r^{\ell+1}$ as follows. An examination of expressions 15.3 and 15.9 shows that the coefficient of $r^{\ell+1}$ is obtained from two sources: first, by applying the derivative d/dr in each A_m^- to the exponential factor of u_ℓ and, second, by applying the constant term of each A_m^-, which may be written as $A/(m + 1)$, successively. The result is that the coefficient of $r^{\ell+1}$ is

$$\left(\frac{\hbar}{\sqrt{2m}}\right)^{\ell} A^{\ell}\left[\frac{1}{(\ell+1)^{\ell}}+\frac{1}{\ell!}\right],$$

which obviously is nonzero.

Taking stock of the development thus far, we have identified $\ell+1$ eigenfunctions, namely, $u_{\ell;0},\ldots,u_{\ell;\ell}$, with common eigenvalue given by equation 15.7. The important observation, though, is that these are eigenfunctions of $\ell+1$ distinct operators, namely, $\mathsf{H}_{\ell-k}$ for $k=0,\ldots,\ell$. Turning this around a bit, we might ask, what eigenfunctions have we identified of H_{ℓ}, for fixed ℓ? The answer is that we have identified an infinite sequence of such eigenfunctions, namely, $u_{\ell+k;k}$ for $k=0,1,\ldots,$ with $u_{\ell+k;k}$ belonging to the eigenvalue

$$Z^2\mathcal{E}_{\ell,k}=-\frac{Z^2}{(\ell+k+1)^2}\mathcal{R}\mathcal{Y}, \tag{15.10}$$

where $\mathcal{E}_{\ell,k}=-\mathcal{R}\mathcal{Y}/(\ell+k+1)^2$. Note that since these are distinct eigenvalues of the fixed self-adjoint operator H_{ℓ}, the family $u_{\ell+k;k}$ as k ranges over the nonnegative integers is pairwise orthogonal.

Our claim is that formula 15.10 identifies all the negative eigenvalues of all the operators H_{ℓ}. To see this, we argue first that $Z^2\mathcal{E}_{\ell,0}$ is the smallest eigenvalue of H_{ℓ}. Indeed, if the function u is an eigenfunction of the operator $\mathsf{A}_{\ell}^{-}\mathsf{A}_{\ell}^{+}$ with eigenvalue λ then in $L^2([0,\infty))$ we have

$$\lambda\|u\|^2=\langle u|\lambda u\rangle=\langle u|\mathsf{A}_{\ell}^{-}\mathsf{A}_{\ell}^{+}u\rangle=\langle\mathsf{A}_{\ell}^{+}u|\mathsf{A}_{\ell}^{+}u\rangle=\|\mathsf{A}_{\ell}^{+}u\|^2\geq 0,$$

which implies that the self-adjoint operator $\mathsf{A}_{\ell}^{-}\mathsf{A}_{\ell}^{+}$ can have no negative eigenvalues. Now equation 15.4 implies that if μ is any eigenvalue of H_{ℓ} then $\lambda=\mu-Z^2\mathcal{E}_{\ell,0}$ is an eigenvalue of $\mathsf{A}_{\ell}^{-}\mathsf{A}_{\ell}^{+}$ and, therefore, $\lambda\geq 0$, or $\mu\geq Z^2\mathcal{E}_{\ell,0}$. As $Z^2\mathcal{E}_{\ell,0}$ is an eigenvalue of H_{ℓ}, it must be the smallest. Assume then that there is an eigenvalue other than those listed in formula 15.10, say μ, an eigenvalue of H_{ℓ} with $Z^2\mathcal{E}_{\ell,k}<\mu<Z^2\mathcal{E}_{\ell,k+1}$. In our argument we will use the fact that

$$\mathsf{A}_{\ell}^{+}\mathsf{H}_{\ell}=\mathsf{H}_{\ell+1}\mathsf{A}_{\ell}^{+}, \tag{15.11}$$

the counterpart to equation 15.6 obtained in like manner from equations 15.4 and 15.5. By a discussion similar to that after equation 15.6, equation 15.11 implies that if u is an eigenfunction of H_{ℓ} with eigenvalue E then $\mathsf{A}_{\ell}^{+}u$ is an eigenfunction of $\mathsf{H}_{\ell+1}$ with the same eigenvalue E, *provided that* $\mathsf{A}_{\ell}^{+}u\neq 0$. Of course, we know the full solution of the first-order differential equation $\mathsf{A}_{\ell}^{+}u=0$, namely the real multiples of the function of equation 15.8, and if u is such a solution then u is an eigenfunction of H_ℓ belonging to the smallest eigenvalue $Z^2\mathcal{E}_{\ell,0}$. All the ingredients are now in place for the argument. Let u now be a μ-eigenfunction for H_ℓ. From our discussion, since μ is not equal to any of the smallest eigenvalues $Z^2\mathcal{E}_{m,0}$ for any nonnegative integer m, the function $\mathsf{A}_\ell^+ u \neq 0$ and therefore is a μ-eigenfunction of $\mathsf{H}_{\ell+1}$. Repeating k more times, the function

$$u^+ = \mathsf{A}_{\ell+k}^+ \cdots \mathsf{A}_\ell^+ u$$

is nonzero and therefore is a μ-eigenfunction of $\mathsf{H}_{\ell+k+1}$. But here we have a problem, for $\mu < Z^2\mathcal{E}_{\ell,k+1} = Z^2\mathcal{E}_{\ell+k+1,0}$ and the latter term in this inequality is the smallest eigenvalue of $\mathsf{H}_{\ell+k+1}$. This contradiction then shows that no such μ exists, and so we have a complete determination of all the negative energy eigenvalues of the bound states of the electron in our hydrogenic system.

15.2 Power Series and Hypergeometric Functions

We now present a derivation of the solutions to the radial equation using hypergeometric functions.[1] This will allow us to give a closed form expression for the eigenfunctions of the preceding section in terms of classical special functions. It also will allow us to examine the positive energy solutions and gain some understanding of their form and qualitative properties.

We begin with a short review of hypergeometric functions. For the verifications of the properties recalled below, see the references in the End Notes to this lecture. The **generalized hypergeometric functions** are defined by the power series expression

$$ {}_pF_q(a_i; b_j; z) = \sum_{n=0}^{\infty} \frac{(a_1)_n \cdots (a_p)_n}{(b_1)_n \cdots (b_q)_n} \frac{z^n}{n!}. \tag{15.12}$$

In this expression, p labels the number of numerator symbols and q the number of denominator symbols. The **Pochhammer symbol** $(c)_n$, also called the **rising factorial**, for the complex number c and nonnegative integer n is a generalization of the factorial and is defined as $(c)_0 = 1$

[1] The famous book Jeffreys and Jeffreys (1946) condenses a complete and very tersely argued examination of this topic to two pages, pp. 618–19; in some ways this discussion is more extensive than our presentation in this section.

and

$$(c)_n = c(c+1)\cdots(c+n-1).$$

For example, $(1)_n = n!$. The **hypergeometric series** on the right-hand side of expression 15.12 is well-defined as long as not any b_j is a nonpositive integer. The series terminates when at least one a_i is a nonpositive integer, but otherwise it is a power series where the radius of convergence becomes relevant, as well as questions of analytic continuation.

Perhaps the most elegant way to define a hypergeometric series is that it is a series $\sum_{n=0}^{\infty} t_n$ defined by the properties that $t_0 = 1$ and the ratio of two consecutive terms is a rational function of the index. This means that there is a rational function $r = P/Q$ for polynomials P and Q for which $t_{n+1}/t_n = r(n)$. The relationship of this description to that of expression 15.12 is that

$$P(x) = z(x+a_1)\cdots(x+a_p)$$

and

$$Q(x) = (x+b_1)\cdots(x+b_q)(x+1).^2$$

Note that the geometric series $\sum_{n=0}^{\infty} z^n$ is the special case where the ratio of consecutive terms is a constant that is independent of the index n, viz., $r(n) = z$ with $P(x) = z(x+1)$ and $Q(x) = x+1$. In terms of expression 15.12, $\sum_{n=0}^{\infty} z^n = {}_1F_0(1; -; z)$.

The classical **hypergeometric functions** are the functions of the form ${}_2F_1(a_1, a_2; b; z)$ while the **confluent hypergeometric functions** are those of the form ${}_1F_1(a; b; z)$. Many of the important functions of mathematics and physics may be realized in a useful form using the hypergeometric functions. This includes the transcendental functions of calculus – the exponential, trigonometric, and binomial functions – as well as the special functions of number theory and mathematical physics, including Bessel functions, elliptic and theta functions, and the classical orthogonal polynomials of Legendre, Chebyshev, Hermite, and Laguerre.

[2] The factor $x+1$ in Q could be absorbed in the notation with $b_{q+1} = 1$, but historical precedence holds sway.

As examples we report that

$$
\begin{aligned}
(1+z)^n &= {}_2F_1(-n,1;1;-z),\\
\log(1+z) &= z\,{}_2F_1(1,1;2;-z),\\
e^z &= {}_0F_0(-;-;z),\\
\sin^{-1} z &= z\,{}_2F_1\left(\frac{1}{2},\frac{1}{2};\frac{3}{2};z^2\right),\\
J_p(z) &= \frac{z^p}{2^p p!}\,{}_0F_1\left(-;p+1;-\frac{z^2}{4}\right),\\
K(z) = \int_0^{\pi/2}(1-z^2\sin^2\theta)^{-1/2}\,d\theta &= \frac{\pi}{2}\,{}_2F_1\left(\frac{1}{2},\frac{1}{2};1;z^2\right),\\
\mathrm{Erf}\, z &= \frac{2z}{\sqrt{\pi}}\,{}_1F_1\left(\frac{1}{2};\frac{3}{2};-z^2\right),
\end{aligned}
$$

where $J_p(z) = \sum_{n=0}^{\infty}(-1)^n z^{2n+p}/(2^{2n+p}n!(n+p)!)$ is the pth Bessel function, $K(z)$ is the complete elliptic integral of the first kind, and $\mathrm{Erf}\, z$ is the error function.

The classical and confluent hypergeometric functions arose originally in Gauss's study of series solutions of second-order ordinary differential equations, and an extensive literature has developed around the subject in the last two centuries. Our interest is that of Gauss's – the solution of a differential equation, the radial equation, using series methods.

Pertinent to the radial equation is the series solution to the second-order equation

$$zw'' + (b-z)w' - aw = 0,$$

known as **Kummer's equation**. A straightforward power series calculation verifies that $w(z) = {}_1F_1(a;b;z)$ is one solution to Kummer's equation. We will connect this development to the radial equation of the preceding section and examine the bound state solutions.

The radial equation 15.1 may be written in the form

$$u'' - \frac{\ell(\ell+1)}{r^2}u + \frac{2mZe^2}{\hbar^2}\frac{1}{r}u = -\frac{2mE}{\hbar^2}u. \tag{15.13}$$

At this point we do not assume a bound state, so that E is not restricted to be negative. When $E < 0$, then already we have seen that E takes on discrete values,

of the form $E_n = -(Z^2/n^2)\mathcal{RY}$. When $E \geq 0$, there is no further restriction on E. Substituting the expression $u(r) = r^{\ell+1}e^{-\kappa r}W(r)$ into equation 15.13, where $\kappa^2 = -2mE/\hbar^2$, and simplifying yields the equation

$$rW'' + 2((\ell+1) - \kappa r)W' - 2\left(\kappa(\ell+1) - \frac{mZe^2}{\hbar^2}\right)W = 0.$$

Making the change of variable $z = 2\kappa r$ and setting $w(z) = W(z/2\kappa) = W(r)$ yields

$$zw'' + (2(\ell+1) - z)w' - \left((\ell+1) - \frac{mZe^2}{\hbar^2\kappa}\right)w = 0.$$

This is Kummer's equation with $b = 2(\ell+1)$ and

$$a = (\ell+1) - \frac{mZe^2}{\hbar^2\kappa}. \tag{15.14}$$

The solution of the radial equation, then, using this method of series and hypergeometric functions, is

$$u(r) = r^{\ell+1}e^{-\kappa r}\,{}_1F_1\left(\ell+1 - \frac{mZe^2}{\hbar^2\kappa}; 2\ell+2; 2\kappa r\right), \tag{15.15}$$

where $\kappa = \sqrt{-2mE}/\hbar$.

We examine this solution in two cases, $E < 0$ in this section and $E \geq 0$ in the penultimate section of the lecture. In the former case, the confluent hypergeometric function that appears in expression 15.15 is of the form ${}_1F_1(a;b;\rho)$ where $\rho = 2\kappa r \geq 0$. If a is not a nonpositive integer, then ${}_1F_1(a;b;\rho)$ grows as e^ρ for large ρ and hence u grows as $e^{\rho/2}$ and is not square-integrable.[3] Square-integrable, and hence appropriate, solutions are obtained only when a is a nonpositive integer, forcing the series to terminate. In equation 15.14, setting a equal to $-k$ for an integer $k \geq 0$ and solving for the energy E gives the quantization condition

$$E = -\frac{Z^2}{(k+\ell+1)^2}\frac{me^4}{2\hbar^2} = -\frac{Z^2}{(k+\ell+1)^2}\mathcal{R}\mathcal{Y},$$

agreeing with expression 15.10. Setting $n = k + \ell + 1$ and recalling that the Bohr radius is $a_0 = \hbar^2/me^2$, the solution may be written in terms of n and ℓ as

$$u(r) = r^{\ell+1}e^{-Zr/na_0}\,{}_1F_1\left(\ell+1-n; 2\ell+2; \frac{2Z}{na_0}r\right). \tag{15.16}$$

Since $k \geq 0$, we must have $n \geq \ell + 1$. The positive integer n is called the

[3] See, for example, Seaborn (1991), p. 189.

principal quantum number of the hydrogenic system and indexes the bound state energy eigenvalues

$$E_n = -\frac{Z^2}{n^2}\mathcal{RY}, \tag{15.17}$$

and the azimuthal quantum number ℓ takes the values $0, 1, \ldots, n-1$. The integer k is called the **radial quantum number**.

For nonnegative integers p and q, the polynomial

$$\mathcal{L}_q^p(z) = \frac{(p+q)!}{p!q!}\, {}_1F_1(-q; p+1; z) \tag{15.18}$$

has degree q and is called the **generalized Laguerre polynomial.**[4] We may write the solution of equation 15.16 up to multiplication by a nonzero constant as

$$f_{n,\ell}(r) = \rho^{\ell+1} e^{-\rho/2}\, \mathcal{L}_{n-\ell-1}^{2\ell+1}(\rho)\,, \text{ where } \rho = \rho(n) = \frac{2Z}{na_0} r. \tag{15.19}$$

Notice that the variable ρ depends on the principal quantum number n.

What is the relationship of $f_{n,\ell}$ to the solutions $u_{\ell;k}$ of the preceding section? First note that $f_{n,\ell}$ is an eigenfunction of the operator H_ℓ with eigenvalue $-(Z^2/n^2)\mathcal{RY}$, while $u_{\ell+k;k}$ is an eigenfunction of H_ℓ with eigenvalue $-(Z^2/(k+\ell+1)^2)\mathcal{RY}$. It is tempting to suggest that the k of the previous section is the k of this section and that $f_{n,\ell}$ and $u_{\ell+k;k}$ are the same, up to a nonzero multiplicative constant, when $n = k + \ell + 1$. Stated differently, we might expect that, in terms of the quantum numbers k and ℓ,

$$f_{k+\ell+1,\ell} = C_{k,\ell}\, u_{\ell+k;k} \tag{15.20}$$

or, in terms of n and ℓ,

$$f_{n,\ell} = D_{n,\ell}\, u_{n-1;n-\ell-1}, \tag{15.20}$$

for appropriate constants $C_{k,\ell}$ and $D_{n,\ell}$. Checking this first whenever ℓ is an arbitrary nonnegative integer and $k = 0$, we ask whether $f_{\ell+1,\ell} = u_{\ell;0}$. Observing that ${}_1F_1(0; 2\ell+2; \rho) = 1$, we have

$$f_{\ell+1,\ell}(r) = \rho^{\ell+1} e^{-\rho/2} = \left(\frac{2Z}{(\ell+1)a_0}\right)^{\ell+1} r^{\ell+1} e^{-Zr/(\ell+1)a_0} = C_{0,\ell} u_{\ell;0}(r),$$

by equation 15.8. We proceed by induction on k. Assume that equation 15.20 holds for all values of ℓ and some value of k. Our goal, then,

[4] Section 4.17 of Lebedev (1972) and Section 6.2 of Andrews, Askey, and Roy (1999) give good treatments of Laguerre polynomials, as well as Chapter 6 of W. Bell (1968).

is to demonstrate that $f_{k+\ell+2,\ell} = C_{k+1,\ell}\, u_{\ell+k+1;k+1}$ for all values of ℓ. For this we will need three relations satisfied by Laguerre polynomials, namely,

$$z\frac{d}{dz}\mathcal{L}_q^p = q\mathcal{L}_q^p - (p+q)\mathcal{L}_{q-1}^p, \tag{15.21}$$

$$\mathcal{L}_q^p = \mathcal{L}_q^{p+1} - \mathcal{L}_{q-1}^{p+1}, \tag{15.22}$$

$$(q+1)\mathcal{L}_{q+1}^p = (p+q+1)\mathcal{L}_q^p - z\mathcal{L}_q^{p+1}, \tag{15.23}$$

where z is the variable. All are proved in a reasonably straightforward way.[5] Continuing the inductive argument, letting ℓ be an arbitrary non-negative integer, and applying the inductive hypothesis yields

$$u_{\ell+k+1;k+1} = \mathsf{A}_\ell^-\, u_{\ell+k+1;k} = \mathsf{A}_\ell^-\, u_{(\ell+1)+k;k} = C_{k,\ell+1}^{-1}\mathsf{A}_\ell^-\, f_{k+\ell+2;\ell+1}.$$

For this calculation, set $n = k+\ell+2$ and $\rho = \rho(n)$ in equation 15.19, and observe that the operator A_ℓ^- in terms of the variable ρ becomes

$$\mathsf{A}_\ell^- = \frac{\hbar}{\sqrt{2m}}\frac{2Z}{na_0}\left(\frac{d}{d\rho} + \frac{\ell+1}{\rho} - \frac{n}{2\ell+2}\right).$$

Setting $C = C_{k,\ell+1}^{-1}2Z\hbar/(na_0\sqrt{2m})$, the calculation continues as follows:

$$u_{\ell+k+1;k+1}(r) = C\left(\frac{d}{d\rho} + \frac{\ell+1}{\rho} - \frac{n}{2\ell+2}\right)\left(\rho^{\ell+2}e^{-\rho/2}\mathcal{L}_k^{2\ell+3}(\rho)\right)$$

$$= C\rho^{\ell+1}e^{-\rho/2}\left[(2\ell+3)\mathcal{L}_k^{2\ell+3}(\rho) - \frac{k+2\ell+3}{2\ell+2}\rho\mathcal{L}_k^{2\ell+3}(\rho) + \rho\frac{d}{d\rho}\mathcal{L}_k^{2\ell+3}(\rho)\right]. \tag{15.24}$$

Applying relation 15.21 to the derivative term in the last line and simplifying yields

$$\frac{k+2\ell+3}{2\ell+2}\left[(2\ell+2)\left(\mathcal{L}_k^{2\ell+3}(\rho) - \mathcal{L}_{k-1}^{2\ell+3}(\rho)\right) - \rho\mathcal{L}_k^{2\ell+3}(\rho)\right]$$

for the term within large square brackets. Applying relation 15.22 to $\mathcal{L}_k^{2\ell+3}(\rho) - \mathcal{L}_{k-1}^{2\ell+3}(\rho)$ and adding and subtracting $(k+1)\mathcal{L}_k^{2\ell+2}$ yields

$$\frac{k+2\ell+3}{2\ell+2}\left[(k+2\ell+3)\mathcal{L}_k^{2\ell+2}(\rho) - \rho\mathcal{L}_k^{2\ell+3}(\rho) - (k+1)\mathcal{L}_k^{2\ell+2}(\rho)\right].$$

Applying relation 15.23 to the first two summands yields

$$\frac{k+2\ell+3}{2\ell+2}(k+1)\left[\mathcal{L}_{k+1}^{2\ell+2}(\rho) - \mathcal{L}_k^{2\ell+2}(\rho)\right].$$

[5] Or see Theorem 6.11, p. 178, in the 2004 Dover republication of Bell (1968).

Finally, applying relation 15.22 again yields

$$\frac{k+2\ell+3}{2\ell+2}(k+1)\mathcal{L}_{k+1}^{2\ell+1}(\rho).$$

Setting

$$C_{k+1,\ell}^{-1} = C\frac{2\ell+k+3}{2\ell+2}(k+1),$$

equation 15.24 becomes

$$u_{\ell+k+1;k+1}(r) = C_{k+1,\ell}^{-1}\,\rho^{\ell+1}e^{-\rho/2}\mathcal{L}_{k+1}^{2\ell+1}(\rho) = C_{k+1,\ell}^{-1}\,f_{k+\ell+2,\ell}(r),$$

finishing off the inductive verification of equation 15.20.

15.3 The Full Solution for Bound State Electrons

We now are in a position to write down a full orthonormal basis for the subspace of $L^2(\mathbb{R}^3)$ that represents the state space for a negative energy electron bound to a central Coulomb potential. The ingredients are the spherical harmonics of Lecture 14 and the negative energy radial eigenfunctions derived above. To normalize the final wave functions, we need to derive the following integral expression involving Laguerre polynomials:

$$\int_0^\infty \rho^{l+1}e^{-\rho}\left[\mathcal{L}_n^l(\rho)\right]^2 d\rho = \frac{(l+n)!}{n!}(2n+l+1). \tag{15.25}$$

For this we need the Rodrigues formula for the Laguerre polynomials, given as[6]

$$\mathcal{L}_n^l(\rho) = \frac{\rho^{-l}e^{\rho}}{n!}\frac{d^n}{d\rho^n}(e^{-\rho}\rho^{l+n}).$$

Making this replacement for one of the $\mathcal{L}_n^l$ factors in the integral, integrating by parts n times, and observing that the boundary values at 0 and ∞ of $e^{-\rho}\rho^{l+m}$ are zero yields

$$\int_0^\infty \rho^{l+1}e^{-\rho}\left[\mathcal{L}_n^l(\rho)\right]^2 d\rho = \frac{(-1)^n}{n!}\int_0^\infty e^{-\rho}\rho^{l+n}\frac{d^n}{d\rho^n}\left[\rho\mathcal{L}_n^l(\rho)\right] d\rho. \tag{15.26}$$

Calculating n derivatives, we have

$$\frac{d^n}{d\rho^n}\left[\rho\mathcal{L}_n^l(\rho)\right] = \rho(\mathcal{L}_n^l)^{(n)}(\rho) + n(\mathcal{L}_n^l)^{(n-1)}(\rho).$$

[6] See the footnote on p. 203, especially the Hussani reference.

Differentiating the power series representation of $\mathcal{L}_n^l$ m times yields the relation

$$\frac{d^m}{d\rho^m}\mathcal{L}_n^l = (-1)^m \mathcal{L}_{n-m}^{l+m}, \tag{15.27}$$

provided that $m \leq n$. In particular, this gives

$$(\mathcal{L}_n^l)^{(n)}(\rho) = (-1)^n \mathcal{L}_0^{l+n}(\rho) = (-1)^n,$$

and

$$(\mathcal{L}_n^l)^{(n-1)}(\rho) = (-1)^{n-1}\mathcal{L}_1^{l+n-1}(\rho) = (-1)^{n-1}(l+n-\rho),$$

by direct evaluation using equations 15.12 and 15.18. The right-hand side of equation 15.26 becomes

$$\begin{aligned}\frac{(-1)^n}{n!}\int_0^\infty & e^{-\rho}\rho^{l+n}(-1)^n\left[(n+1)\rho - n(l+n)\right]d\rho \\ &= \frac{1}{n!}\left[(n+1)\Gamma(l+n+2) - n(l+n)\Gamma(l+n+1)\right] \\ &= \frac{(l+n)!}{n!}(2n+l+1),\end{aligned}$$

where we have used the relation $\Gamma(z+1) = z\Gamma(z)$. This confirms equation 15.25.

Recalling equation 14.25 and the work of the preceding section in deriving the radial equation eigenfunctions, we may write the bound state eigenfunctions for the hydrogenic potential as

$$\psi_{n,\ell,m}(r,\theta,\phi) = g_{n,\ell}(r)Y_\ell^m(\theta,\phi),$$

where Y_ℓ^m are the spherical harmonics of the previous lecture while $g_{n,\ell}(r) = f_{n,\ell}(r)/r$. Recall that the spherical harmonics are orthonormal with respect to the spherical measure $d\Omega = \sin\theta\, d\theta d\phi$ on $\mathbf{S}^2$. To give an orthonormal collection of bound state eigenfunctions for the hydrogenic potential, we need to normalize the radial functions $g_{n,\ell}$ with respect to the radial measure $r^2\,dr$ on the nonnegative real axis. After the change of variable $\rho = (2Z/na_0)r$ in the integral below, expression 15.25 implies that

$$\int_0^\infty [g_{n,\ell}(r)]^2\, r^2\, dr = \int_0^\infty [f_{n,\ell}(r)]^2\, dr = \frac{n^2 a_0}{Z}\frac{(n+\ell)!}{(n-\ell-1)!}.$$

The functions $g_{n,\ell}$ are appropriately normalized if we redefine them as

$$\begin{aligned} g_{n,\ell}(r) &= \frac{1}{n}\sqrt{\frac{Z}{a_0}\frac{(n-\ell-1)!}{(n+\ell)!}}\frac{f_{n,\ell}(r)}{r} \\ &= \frac{2}{n^2}\sqrt{\left(\frac{Z}{a_0}\right)^3\frac{(n-\ell-1)!}{(n+\ell)!}}\,\rho^\ell e^{-\rho/2}\mathcal{L}_{n-\ell-1}^{2\ell+1}(\rho). \end{aligned} \tag{15.28}$$

Concerning the orthogonality of the functions $g_{m,\ell}$ and $g_{n,\ell}$ when $m \neq n$, observe that since $f_{m,\ell}$ and $f_{n,\ell}$ are eigenfunctions of the self-adjoint operator H_ℓ with different eigenvalues, they are orthogonal in $L^2([0,\infty))$ with respect to the standard measure dr. It follows that

$$\int_0^\infty g_{m,\ell}(r)g_{n,\ell}(r)r^2\,dr = C\int_0^\infty f_{m,\ell}(r)f_{n,\ell}(r)\,dr = 0, \tag{15.29}$$

where C is an appropriate constant.

Collecting together the work of this lecture thus far, the bound state eigenfunction $\psi_{n,\ell,m}$ may be written in polar coordinates as[7]

$$\begin{aligned} \psi_{n,\ell,m}(r,\theta,\phi) &= g_{n,\ell}(r)Y_\ell^m(\theta,\phi) \\ &= \frac{2}{n^2}\sqrt{\left(\frac{Z}{a_0}\right)^3\frac{(n-\ell-1)!}{(n+\ell)!}}\,\rho^\ell e^{-\rho/2}\mathcal{L}_{n-\ell-1}^{2\ell+1}(\rho)Y_\ell^m(\theta,\phi), \end{aligned} \tag{15.30}$$

where $\rho = (2Z/na_0)r$. These collectively form an orthonormal basis for the bound state subspace for the hydrogenic potential, since expressions 14.9 and 15.29 with the normalization of equation 15.28 combine

[7] We should mention that some sources use the **associated Laguerre polynomials** $\mathcal{L}_q^{(p)}$ rather than, as we have done, the generalized Laguerre polynomials $\mathcal{L}_q^p$; this changes the indices decorating $\mathcal{L}$. The associated polynomials are defined, appropriately, by $\mathcal{L}_q^{(p)} = d^p\mathcal{L}_q^0/dz^p$ and their relationship to the generalized polynomials may be derived from equation 15.27 as $\mathcal{L}_q^{(p)} = (-1)^p\mathcal{L}_{q-p}^p$ for $p \leq q$. To make matters worse, even their definition and notation are not standard and may vary from source to source, with different normalization constants and indexing. For example, some sources use the notation $\mathcal{L}_q^{(p)}$ for the generalized polynomials, and some have an extra factor of $q!$ in their definition of $\mathcal{L}_q^{(p)}$. This makes for a confusing hodge-podge of forms for equations 15.28 and 15.30 that differ mildly among one another, from an extra factor of $1/(n+\ell)!$ to different indices and notations for the Laguerre polynomial in use. *Caveat emptor!*

to give

$$\langle \psi_{n,\ell,m} | \psi_{n',\ell',m'} \rangle = \int_{\mathbb{R}^3} \overline{\psi}_{n,\ell,m} \psi_{n',\ell',m'} \, d\mathbf{x}$$

$$= \int_0^\infty g_{n,\ell}(r) g_{n',\ell'}(r) r^2 \, dr \int_0^{2\pi} \int_0^\pi \overline{Y_\ell^m} Y_{\ell'}^{m'} \sin\theta \, d\theta d\phi = \delta_{nn'} \delta_{\ell\ell'} \delta_{mm'},$$

where $d\mathbf{x} = r^2 \sin\theta \, dr d\theta d\phi$.

15.4 The Unbound Electron in the Coulomb Potential

Turning now to the case $E \geq 0$, that of an unbound electron under the influence of the Coulomb potential, we can but scratch the surface. The proper venue for this study is that of scattering theory, which is neglected in these lectures, but on which we nevertheless make a few comments. We have found all the eigenvalues of the operators H_ℓ, for $\ell \geq 0$, in expression 15.10, but this requires some explanation. In our considerations we have regarded the differential operators H_ℓ as self-adjoint operators on the Hilbert space $L^2([0, \infty))$. The point spectrum of the operator H_ℓ is the collection of negative numbers given by expression 15.10 for $k \geq 0$. But this is not the whole story, for the continuous spectrum of the operator H_ℓ is the set of nonnegative real numbers, so that there are quantum states of the electron that may have energy measured to be within any interval of the nonnegative reals. These states are the unbound states of the electron in the central Coulomb potential, and our interest in this section is in understanding the building blocks of these unbound states.

The eigenvalue equation 15.13 is relevant to this discussion when $E \geq 0$. The solution given by equation 15.15 when $E > 0$ does not describe an appropriate wave function since, it turns out, it is not square-integrable. The reason why it is not square-integrable, though, is different from the reason why the negative energy solutions are not square-integrable. For the latter, when $E < 0$ is different from the eigenvalues of expression 15.10 the solutions are not square-integrable because ${}_1F_1(a; b; \rho)$ grows for large ρ as e^ρ and hence u grows as $e^{\rho/2}$. But, for the former, when $E > 0$, ${}_1F_1(a; b; \rho)$ remains bounded even though it fails to be square-integrable. This allows the solution to be a generalized eigenfunction of the Hamiltonian operator H_ℓ with generalized eigenvalue $E > 0$. The remainder of this lecture explores these issues in more detail.

It might be helpful for the reader to review Section 10.4, where the generalized eigenvector $|p\rangle$ that represents the idealized quantum state with exact momentum p and completely indeterminate position is represented by the function $\psi_p(x) = \langle x|p\rangle = e^{\mathrm{i}xp/\hbar}/\sqrt{2\pi\hbar}$, a non-square-integrable function that nonetheless may be used under an integral sign with the aim of representing an appropriate wave function. The pertinent expressions are those of equations 10.22 and 10.23, which exhibit a wave function as an uncountable superposition of "mutually orthogonal" momentum "eigenvectors." The Fourier transform $\widehat{\varphi}(p)$ serves as the coefficient of $|p\rangle$ in that superposition. In the same way, the solution of the eigenvalue equation 15.13 when $E > 0$ is a non-square-integrable function that represents a generalized eigenvector $|E\rangle$ of the operator H_ℓ, which, in turn, represents the idealized quantum state of an electron in the Coulomb potential with exact positive energy E.

To explore this further, let

$$\nu = \frac{Z\hbar}{a_0\sqrt{2mE}}.$$

The expression ν is the positive energy version of the principal quantum number n of the bound electron. Of course, n is a discrete quantum number taking positive integer values while ν is a continuous parameter taking all positive real values. The energy E in terms of ν is given by a continuous analogue of equation 15.17, namely,

$$E = \frac{Z^2}{\nu^2}\mathcal{R}\mathcal{Y}. \tag{15.31}$$

Let $h_{\nu,\ell}$ be the solution to the radial equation reported in equation 15.15 when $E > 0$. In this case, the expression for κ is interpreted as

$$\kappa = \frac{\sqrt{2mE}}{\hbar}\,\mathrm{i}.$$

In terms of the positive real variable $\rho = -2\kappa r\mathrm{i}$ the solution reads, up to a nonzero multiplicative constant,

$$h_{\nu,\ell}(r) = \rho^{\ell+1}e^{-\mathrm{i}\rho/2}\,{}_1F_1\left(\ell+1+\nu\mathrm{i}; 2\ell+2; \rho\mathrm{i}\right). \tag{15.32}$$

To gain an idea of the large-scale characteristics of this function, we use the standard asymptotic series for the confluent hypergeometric function derived, for example, as equation 10.45 of Seaborn (1991). Taking the leading terms of this asymptotic expansion gives the approximation, for complex numbers a and b and large values of $\rho > 0$,

$$ {}_1F_1(a;b;\rho\mathbf{i}) \approx \frac{\Gamma(b)}{\Gamma(a)}\frac{1}{\rho^{b-a}}e^{\mathbf{i}[\rho-\pi(b-a)/2]} + \frac{\Gamma(b)}{\Gamma(b-a)}\frac{1}{\rho^{a}}e^{\mathbf{i}\pi a/2}. $$

Applying this with $a = \ell+1+\nu\mathbf{i}$ and $b = 2\ell+2$, so that $b-a = \ell+1-\nu\mathbf{i}$, recalling that $\Gamma(\overline{z}) = \overline{\Gamma(z)}$, and simplifying gives for $h_{\nu,\ell}$ the expression

$$ h_{\nu,\ell}(r) \approx e^{-\nu\pi/2}\frac{2\Gamma(2\ell+2)}{|\Gamma(\ell+1+\nu\mathbf{i})|}\sin\left(\frac{\rho}{2}-\frac{\pi}{2}\ell+\nu\log\rho-\Theta_{\nu,\ell}\right), $$

where $\Gamma(\ell+1+\nu\mathbf{i}) = |\Gamma(\ell+1+\nu\mathbf{i})|e^{\mathbf{i}\Theta_{\nu,\ell}}$. To see better the dependence on ν and r this may be expressed as

$$ h_{\nu,\ell}(r) \approx C_{\nu,\ell}\sin\left(\frac{Z}{\nu a_0}r+\nu\log r-\eta_{\nu,\ell}\right), $$

for some constants $C_{\nu,\ell}$ and $\eta_{\nu,\ell}$ dependent only on ν and ℓ. The function $h_{\nu,\ell}$ behaves for large r like a sinusoidal wave with period attenuated by the logarithm term.

Let $|E\rangle = |\nu,\ell\rangle$ be the conjugate linear functional defined on the space of functions on $[0,\infty)$ that vanish appropriately as $r \to 0+$ and $r \to \infty$ and are given as

$$ \varphi \mapsto \langle\varphi|\nu,\ell\rangle = \int_0^\infty \overline{\varphi}(r)h_{\nu,\ell}(r)\,dr. $$

Since $\mathsf{H}_\ell h_{\nu,\ell} = Eh_{\nu,\ell}$ where E is given by equation 15.31, a couple of integrations by parts show that $|\nu,\ell\rangle$ is a generalized eigenvector of H_ℓ with generalized eigenvalue E. The function $h_{\nu,\ell}$ represents this eigenvector and, after combining $h_{\nu,\ell}/r$ with the appropriate spherical harmonics, may be used in scattering calculations for an unbound electron of positive energy E scattered by a central Coulomb potential.

15.5 End Notes

Any text or reference on special functions will give a treatment of hypergeometric series and functions. Hassani (1999) offers a quick introduction; for more details consult W. Bell (1968) or Beals and Wong (2010). The book Seaborn (1991) is a treasure trove of theory and applications of hypergeometric functions. One of my favorite books is the charmingly titled $A = B$ by Petkovšek, Wilf, and Zeilberger (1996). It uses hypergeometric series and functions to build algorithms for solving very complex binomial coefficient identities, hypergeometric sum and integral identities, and other identities and demonstrates the power of pure mathematics armed with computational strength in tackling hard mathematical identities.

For more details on Laguerre polynomials see Sections 4.17–4.25 of Lebedev (1972), Chapter 12 of Rainville (1960), Chapter 6 of Bell (1968), or Section 6.2 of Andrews, Askey, and Roy (1999). For the scattering of a positive energy electron, see Morse and Feshbach (1953b), p. 1669.

16
Wave Mechanics V: Hidden Symmetry Revealed

> Other experiments showed the finer structure of the hydrogen spectrum ... To describe the results of these experiments, it is useful to introduce the *azimuthal quantum number* ℓ. States corresponding to the "sharp" spectral lines on the photographic plates (often labeled s) have $\ell = 0$; those corresponding to "principal" lines (labeled p) have $\ell = 1$; those corresponding to "diffuse" lines (labeled d) have $\ell = 2$ and those corresponding to "fundamental" lines (labeled f) have $\ell = 3$. The experiments showed that each spectral line of hydrogen with at least one state of azimuthal quantum number ℓ contains $2(2\ell + 1)$ different states with azimuthal quantum number ℓ. Because these spectral lines split in the presence of a magnetic field, the new split lines were labeled by the *magnetic quantum number* m. The magnetic quantum number could take any of the $2\ell + 1$ values $-\ell, 1 - \ell, \ldots, \ell - 1, \ell$. Similarly, the *spin quantum number* s takes either of the values $\pm 1/2$.
>
> Stephanie Singer
> *Linearity, Symmetry, and Prediction in the Hydrogen Atom*, 2005

Having obtained the negative energy levels of the electron in the central Coulomb potential in the previous lecture, this lecture begins with an examination of the degeneracy of the energy eigenvalues of the bound states and then derives the spectral Balmer series for the hydrogen atom from these theoretical considerations. This gives a quantum mechanical explanation of this experimentally determined spectral series and showcases one of the first great theoretical achievements of wave mechanics. Of course, more exacting experimentation gives spectral lines that differ, ever so slightly, from those derived by Schrödinger's 1926 wave mechanical treatment of atomic hydrogen, even in the physically more realistic analysis where the reduced mass of the proton–electron system is used. In fact, it is interesting to note that Sommerfield had derived in 1915 a more accurate expression for the energy levels of atomic hydrogen from the heuristics of the old quantum theory, and it was not until Dirac's 1928 application to atomic hydrogen of his relativistic equation for the electron that Sommerfield's accuracy was superseded. Nevertheless, Schrödinger's derivation was a triumph for the new wave mechanics that validated the equation that bears his name. We report all these details in the first section of this lecture.

Presented in the following section is Pauli's elegant solution, obtained in 1925 before the introduction of the Schrödinger equation in 1926, for the energy levels of atomic hydrogen. This was accomplished by quantizing a classical constant of motion of the **Kepler problem**, which is the problem of determining the motion of a body in an inverse-square force field. Motion around a central Coulomb potential is of this type, and it is then not surprising that the classical analysis of the Kepler problem might help to solve a quantum version of the problem. An extra benefit appears out of the analysis, in that Pauli's solution offers a hint that there is more symmetry to the Coulomb problem in quantum mechanics than first meets the eyes. Of course one's expectation is only SO(3) symmetry, reflecting the symmetry of the central Coulomb potential in $\mathbb{R}^3$. However, Pauli's analysis suggests hidden symmetries of the hydrogenic system that are not manifest without some delicate probing. After looking carefully at the the group theory underlying Pauli's solution in the third section of this lecture, we examine in the fourth section the hidden symmetries of the hydrogenic system that are based on the group SO(4).

16.1 Quantum Numbers, Degeneracy, and Fine Structure

For the central Coulomb potential, the $E_n = -(Z^2/n^2)\mathcal{RY}$-eigenspace $\mathcal{H}_n$ is spanned by the eigenfunctions $\psi_{n,\ell,m}$ for $\ell = 0, \ldots, n-1$ and $m = -\ell, -\ell+1, \ldots, \ell-1, \ell$. It follows that the dimension of this eigenspace is

$$\dim \mathcal{H}_n = \sum_{\ell=0}^{n-1} (2\ell+1) = n^2, \tag{16.1}$$

so that the nth bound state energy level of an electron in the hydrogenic potential is n^2-degenerate. Actually, there is another degree of freedom that a bound electron enjoys, namely, a spin degree of freedom, indexed by a quantum number s that takes the possible values $\pm\frac{1}{2}$. This property of spin is not in any way captured as a natural consequence of the wave mechanics developed thus far, though Pauli did produce a theory within wave mechanics that, in an ad hoc manner, incorporated spin. When the hydrogenic potential is treated with the relativistic Dirac equation rather than the Schrödinger equation, spin is a natural consequence of the analysis, and the energy eigenfunctions then are indexed by the quantum numbers n, ℓ, m, and s. This doubles the degeneracy[1] of the nth energy level to $2n^2$.

Spectroscopists from the late nineteenth to the mid twentieth century performed more and more accurate measurements of the frequencies of the spectral lines of atomic hydrogen, in a variety of circumstances.[2] Not only did they discover emission spectra in the visible, ultraviolet, and infrared ranges, they observed the splitting of these lines under magnetic field and spin influences. The **Balmer series** of spectral lines in the visible range of the spectrum was discovered in 1885 and the wavelengths λ of the emissions were found empirically to obey the **Rydberg formula**,

$$\frac{1}{\lambda} = \left(\frac{1}{n^2_{\text{initial}}} - \frac{1}{n^2}\right) R_\infty \quad \text{where} \quad n > n_{\text{initial}}.$$

Here $n_{\text{initial}} = 2$ and R_∞ is the **Rydberg constant**, defined by

$$R_\infty = \frac{\mathcal{Ry}}{hc}.$$

The **Lyman series** of spectral lines in the ultraviolet range of the spectrum has wavelengths that obey the Rydberg formula for $n_{\text{initial}} = 1$. In the infrared range there are several named series, the **Paschen series** for $n_{\text{initial}} = 3$, the **Brackett series** for $n_{\text{initial}} = 4$, the **Pfund series** for $n_{\text{initial}} = 5$, and the **Humphreys series** for $n_{\text{initial}} = 6$. In fact, all the wavelengths of the emission spectrum of atomic hydrogen are found to obey the Rydberg formula, modulo the very small but significant

[1] This terminology is not quite accurate, as the term *degeneracy* refers to independent states with the same eigenvalue. When including spin, the E_n-energy level splits ever so slightly into levels of different energies depending on the quantum numbers ℓ and s – in fact, on $\ell \pm s$. This is explained below.

[2] For an absorbing account of the history of spectroscopy in the nineteenth century, from the earliest measurements of emission lines in the solar spectrum and the spectra of electric arcs to the articulation of Rydberg's formula, see Appendix F of Sternberg (1995).

shifting of lines due to relativistic effects and the splitting of lines due to spin effects.

Of course, the Rydberg formula is a direct consequence of our work of the previous lecture, where the energy levels of atomic hydrogen were found to satisfy equation 15.17 with $Z = 1$, so that

$$E_n = -\frac{1}{n^2}\mathcal{R}\mathcal{Y}.$$

The Rydberg formula follows from this and the photoelectric formula, $E = h\nu = hc/\lambda$, relating the energy E of a photon of light to its frequency ν and wavelength λ. The **ionization energy** of hydrogen is the energy required to unbind an electron in the innermost orbital and its value is approximately

$$|E_1| = 1\,\mathcal{R}\mathcal{Y} = \frac{m_e e^4}{2\hbar^2} = \frac{e^2}{2a_0} = 13.6 \text{ eV},$$

where m_e is the mass of the electron.

We should point out that our treatment of an electron in a central Coulomb potential has made a serious simplification, in that we have treated the electron as the lone particle in a rigidly fixed electrostatic environment. In particular, we have allowed no influence of the electron on the Coulomb potential, or, more precisely, the source of the Coulomb potential. Essentially, in fixing the location of the Coulomb source at $r = 0$, we have assumed an infinite mass for the source. In reality, the hydrogenic system should be modeled as a two-body problem with a Hamiltonian that incorporates kinetic terms for both the electron and the central nucleus, of proton number Z. When this is done, by moving to center of mass coordinates, the Schrödinger equation for the two-body problem reduces to the one-body Schrödinger equation for a particle in a fixed Coulomb potential field – the problem already solved – but with the mass of the electron replaced by a corresponding "reduced mass." Since the mass m_e of the electron is less than 0.05446% of the mass m_p of the proton, the reduced mass $m_p m_e/(m_p + m_e)$ adjusts the value of the Rydberg $\mathcal{R}\mathcal{Y} = m_e(e^4/2\hbar^2)$ by less than 0.05443%. This amounts to a small, yet measurable, shift in the energy levels E_n of atomic hydrogen.

As reported in the quote that begins this lecture, spectroscopists have developed a nomenclature centered around the azimuthal quantum number ℓ, where states with $\ell = 0$ are known as **s** states (sharp), $\ell = 1$ as **p** states (principal), $\ell = 2$ as **d** states (diffuse), and $\ell = 3$ as **f** states (fundamental). The quantum numbers $n = 2$, $\ell = 1$, and $m = -1$, for example, describe an electron in the $2\mathbf{p}_{-1}$ **orbital.**[3] Electrons with the same principal quantum number n are said to be in the same **shell**, since n fixes the mean distance from the nucleus.

Various structural effects are observed in the hydrogen spectrum in a variety of experimental settings. The **Zeeman effect** is the splitting of the spectral lines of the same azimuthal quantum number ℓ, but different magnetic quantum numbers m, in the presence of a weak constant external magnetic field. A similar splitting as well as a shifting of the spectral lines is observed in the presence of a static electric field; this the **Stark effect**. Even finer structural effects are observed, and can be incorporated and explained by the theory, when relativistic and spin effects are included via the Dirac equation, and again when electromagnetic field effects are included via quantum electrodynamics. This starts with a slight correction of the energy levels due to a more precise formula for the energy eigenvalues that follows from the Dirac equation, which is reviewed in the next paragraph. The extremely small splitting of spectral lines known as the **Lamb shift** occurs in atomic hydrogen and has been traced to the interaction of the electron with the vacuum. This is described successfully when the electromagnetic field itself is quantized.

Though we do not treat here the Dirac equation and its application to the fine structure of the spectrum of atomic hydrogen, we will go ahead and report the results for the energy levels of atomic hydrogen. A relativistic calculation of the energy eigenvalues gives the formula

$$E_{n,j} = m_e c^2 \left(1 + \frac{\alpha^2}{\left\{ n - j - \frac{1}{2} + \left[\left(j + \frac{1}{2}\right)^2 - \alpha^2 \right]^{1/2} \right\}^2} \right)^{-1/2} \tag{16.2}$$

for the energy of an electron in atomic hydrogen. The ingredients of this formula are as follows. The parameter α is a dimensionless constant called the **fine structure constant** and is defined as

$$\alpha = \frac{e^2}{\hbar c} \approx \frac{1}{137.036}.$$

While n is the principal quantum number from the Schrödinger analysis, j is a new quantum number describing the coupling of the orbital angular momentum and the intrinsic spin of the electron; it is related to the

[3] When the relativistic hydrogen atom is analyzed with the Dirac equation, different quantum numbers arise naturally and the spectroscopists' notation is duly modified.

azimuthal quantum number ℓ by $j = \ell \pm \frac{1}{2}$. The values that j may take are the half integer values $\frac{1}{2}, \frac{3}{2}, \frac{5}{2}, \ldots$ A Taylor expansion in the parameter α^2 of the right-hand side of equation 16.2 gives, for the first three terms,

$$E_{n,j} \approx m_e c^2 \left[1 - \frac{\alpha^2}{2n^2} - \frac{\alpha^4}{2n^4}\left(\frac{n}{j+\frac{1}{2}} - \frac{3}{4}\right)\right].$$

The first term is $m_e c^2$, the rest mass of the electron. The second term is

$$-\frac{m_e c^2 \alpha^2}{2n^2} = -\frac{1}{n^2}\frac{m_e e^4}{2\hbar^2} = -\frac{1}{n^2}\mathcal{R}\mathcal{Y} = E_n,$$

the nth energy eigenvalue from the nonrelativistic Schrödinger analysis. To the order α^2, then, the Dirac analysis gives the same energy levels of atomic hydrogen as the Schrödinger analysis. The first correction term is of order α^4 and splits spectral lines with the same value of n but different values of ℓ, while the Schrödinger theory gives them all the same energy and hence a single emission spectral line. The Dirac theory further splits the lines with the same value of ℓ into two lines charaterized by j values that differ by unity. For example, with $n = 2$ and $\ell = 1$, the Schrödinger theory gives an energy level of E_2. In the Dirac theory, we have $j = \ell \pm \frac{1}{2}$ so that $j + \frac{1}{2}$ takes the values 1 and 2, and the E_2 energy level splits into $E_{2,1/2}$ and $E_{2,3/2}$. The difference between the $j + \frac{1}{2} = 1$ and $j + \frac{1}{2} = 2$ energies is given by $\Delta E = E_{2,3/2} - E_{2,1/2} = \alpha^4 m_e c^2/32$ to order α^4, which is approximately 4.53×10^{-5} eV or $3.33 \times 10^{-6}\,\mathcal{R}\mathcal{Y}$, in excellent agreement with spectroscopic observation. The replacement of ℓ by j introduces a two-fold degeneracy[4] to the Schrödinger theory that accounts for the intrinsic spin of the electron and its coupling to the electron's orbital angular momentum. Now the degeneracy of the nth shell is $2n^2$ rather than n^2, but in fact this represents orbitals of slightly differing energies that all have the same principal quantum number n. This gives a wave mechanical explanation for the observations of experimental physicists, who had determined by 1925 the shell structure of electron orbitals, enumerated by the principal quantum number n, as well as much of the detail of the **subshell** structure, enumerated by the azimuthal quantum number ℓ.

[4] The first footnote on p. 235 is pertinent.

16.2 The Laplace–Runge–Lenz Vector

We turn now to Pauli's elegant solution to the energy levels of the hydrogenic system, rederiving expression 15.17. This is accomplished by quantizing the **Laplace–Runge–Lenz (LRL) vector**, a classical constant of motion of the Kepler problem. We begin with a review of the classical treatment that is not so well known, even among physicists.

Consider the classical motion of a particle under the influence of a central inverse-square force $\mathbf{f} = -(K/x^2)\hat{\mathbf{x}}$ centered at the origin in $\mathbb{R}^3$. Here, $\mathbf{x} = (x_1, x_2, x_3)$ denotes the position vector of the particle and $\hat{\mathbf{x}} = \mathbf{x}/x$ is the unit vector in the direction of $\mathbf{x}$, with $x = \|\mathbf{x}\| = \sqrt{x_1^2 + x_2^2 + x_3^2}$. The momentum is $\mathbf{p} = m\dot{\mathbf{x}}$ while the angular momentum, which is a constant of the motion since the force is central, is $\mathbf{L} = \mathbf{x} \times \mathbf{p}$. Our aim is to show that the Laplace-Runge-Lenz vector, defined as

$$\mathbf{A} = \mathbf{L} \times \mathbf{p} + mK\hat{\mathbf{x}}, \tag{16.3}$$

is a constant of the motion. To see this, first note from Newton's second law that $\dot{\mathbf{p}} = \mathbf{f} = -(K/x^2)\hat{\mathbf{x}}$ and, since $\mathbf{L}$ is conserved, $\dot{\mathbf{L}} = \mathbf{0}$. It follows that

$$\frac{d}{dt}(\mathbf{L} \times \mathbf{p}) = \mathbf{L} \times \dot{\mathbf{p}} = -\frac{mK}{x^2}\left[(\mathbf{x} \times \dot{\mathbf{x}}) \times \hat{\mathbf{x}}\right].$$

Applying the triple vector product formula, that $(\mathbf{a} \times \mathbf{b}) \times \mathbf{c} = (\mathbf{a} \cdot \mathbf{c})\mathbf{b} - (\mathbf{b} \cdot \mathbf{c})\mathbf{a}$, this reduces to

$$\frac{d}{dt}(\mathbf{L} \times \mathbf{p}) = -\frac{mK}{x^2}\left[(\mathbf{x} \cdot \hat{\mathbf{x}})\dot{\mathbf{x}} - (\dot{\mathbf{x}} \cdot \hat{\mathbf{x}})\mathbf{x}\right] = -mK\frac{x\dot{\mathbf{x}} - \dot{x}\mathbf{x}}{x^2} = -mK\frac{d}{dt}\hat{\mathbf{x}},$$

implying that $d\mathbf{A}/dt = \mathbf{0}$. Thus $\mathbf{A}$ is conserved. In passing we mention that the vector $\mathbf{A}$ lies along the major axis of the elliptical orbit of the particle and its length is the eccentricity of the orbit.

To derive the energy levels of atomic hydrogen, Pauli quantized equation 16.3. This is accomplished by promoting the angular momentum vector $\mathbf{L}$ to the angular momentum operator triad L, the momentum vector $\mathbf{p}$ to the momentum operator triad P, and the position vector $\mathbf{x}$ to the position operator triad X. One technical point immediately arises. How should the cross product $\mathbf{L} \times \mathbf{p}$ be quantized – as $\mathsf{L} \times \mathsf{P}$ or $-\mathsf{P} \times \mathsf{L}$? After all, even though $\mathbf{L} \times \mathbf{p} = -\mathbf{p} \times \mathbf{L}$, it is not the case that $\mathsf{L} \times \mathsf{P} = -\mathsf{P} \times \mathsf{L}$ since X and P fail to commute. Pauli made the democratic choice $\mathbf{L} \times \mathbf{p} \rightsquigarrow \frac{1}{2}(\mathsf{L} \times \mathsf{P} - \mathsf{P} \times \mathsf{L})$. As for the unit vector $\hat{\mathbf{x}}$, it is quantized by the triad X/x and the Laplace–Runge–Lenz vector $\mathbf{A}$

is correspondingly promoted to the **quantized LRL triad** defined by

$$\mathbf{A} = \frac{1}{2}(\mathbf{L}\times\mathbf{P} - \mathbf{P}\times\mathbf{L}) + \frac{mK}{x}\mathbf{X}. \tag{16.4}$$

Using the second commutator relation of equations 12.1, we find that $-\mathbf{P}\times\mathbf{L} = \mathbf{L}\times\mathbf{P} - 2\mathrm{i}\hbar\mathbf{P}$, and this shows that the LRL operator triad may be written variously as

$$\mathbf{A} = \mathbf{L}\times\mathbf{P} + \frac{mK}{x}\mathbf{X} - \mathrm{i}\hbar\mathbf{P} = -\mathbf{P}\times\mathbf{L} + \frac{mK}{x}\mathbf{X} + \mathrm{i}\hbar\mathbf{P}. \tag{16.5}$$

The components of the LRL triad $\mathbf{A}$ commute with the inverse-square Hamiltonian

$$\mathsf{H} = -\frac{\hbar^2}{2m}\Delta - \frac{K}{x} = \frac{1}{2m}\mathbf{P}^2 - \frac{K}{x} \tag{16.6}$$

and satisfy the dot product relation

$$\mathbf{A}\cdot\mathbf{L} = \mathbf{L}\cdot\mathbf{A} = \mathbf{0} \tag{16.7}$$

and the commutation relations

$$\begin{aligned}[\mathsf{L}_i, \mathsf{A}_j] &= \mathrm{i}\hbar\epsilon_{ijk}\mathsf{A}_k \quad \text{and}\\ [\mathsf{A}_i, \mathsf{A}_j] &= -2\mathrm{i}\hbar m\epsilon_{ijk}\mathsf{H}\mathsf{L}_k,\end{aligned} \tag{16.8}$$

as well as the useful relation

$$\mathbf{A}^2 = \mathbf{A}\cdot\mathbf{A} = 2m\mathsf{H}(\mathbf{L}^2 + \hbar^2 1) + m^2K^2 1. \tag{16.9}$$

The verification of these relations is somewhat involved. First, to verify that $[\mathsf{H}, \mathsf{A}_i] = 0$ for each component operator A_i, observe that

$$\begin{aligned}[\mathsf{H}, \mathsf{A}_i] &= \left[\mathsf{H}, \frac{1}{2}(\mathbf{L}\times\mathbf{P} - \mathbf{P}\times\mathbf{L})_i + \frac{mK}{x}\mathsf{X}_i\right]\\ &= \frac{1}{2}[\mathsf{H}, (\mathbf{L}\times\mathbf{P})_i] - \frac{1}{2}[\mathsf{H}, (\mathbf{P}\times\mathbf{L})_i] + mK\left[\mathsf{H}, \frac{1}{x}\mathsf{X}_i\right]\\ &= \frac{\mathrm{i}\hbar K}{2}\left((\mathbf{L}\times\mathbf{X})_i\frac{1}{x^3} - \frac{1}{x^3}(\mathbf{X}\times\mathbf{L})_i\right) + \frac{K}{2}\left[\mathbf{P}^2, \frac{1}{x}\mathsf{X}_i\right].\end{aligned} \tag{16.10}$$

To see the last line, using the fact that $[\mathsf{H}, \mathsf{L}_k] = 0$ from expression 14.24 and the easily proved fact that

$$\left[\mathsf{P}_s, \frac{1}{x}\right] = \frac{\mathrm{i}\hbar}{x^3}\mathsf{X}_s, \tag{16.11}$$

we have, for example,

$$\begin{aligned}[\mathsf{H}, (\mathbf{P}\times\mathbf{L})_i] &= \epsilon_{ijk}[\mathsf{H}, \mathsf{P}_j\mathsf{L}_k] = \epsilon_{ijk}\left(\mathsf{P}_j[\mathsf{H},\mathsf{L}_k] + [\mathsf{H},\mathsf{P}_j]\mathsf{L}_k\right)\\ &= \epsilon_{ijk}\left(-K\left[\frac{1}{x},\mathsf{P}_j\right]\mathsf{L}_k\right) = \mathbf{i}\hbar K\frac{1}{x^3}\epsilon_{ijk}\mathsf{X}_j\mathsf{L}_k = \mathbf{i}\hbar K\frac{1}{x^3}(\mathbf{X}\times\mathbf{L})_i.\end{aligned}$$

Again using equation 16.11, and with the repeated index s summed from 1 to 3, we have

$$\begin{aligned}\left[\mathbf{P}^2, \frac{1}{x}\mathsf{X}_i\right] &= \left[\mathsf{P}_s\mathsf{P}_s, \frac{1}{x}\mathsf{X}_i\right] = \mathsf{P}_s\left[\mathsf{P}_s, \mathsf{X}_i\frac{1}{x}\right] + \left[\mathsf{P}_s, \frac{1}{x}\mathsf{X}_i\right]\mathsf{P}_s\\ &= \mathsf{P}_s\mathsf{X}_i\left[\mathsf{P}_s, \frac{1}{x}\right] + \mathsf{P}_s\left[\mathsf{P}_s,\mathsf{X}_i\right]\frac{1}{x} + \frac{1}{x}\left[\mathsf{P}_s,\mathsf{X}_i\right]\mathsf{P}_s + \left[\mathsf{P}_s,\frac{1}{x}\right]\mathsf{X}_i\mathsf{P}_s\\ &= \mathbf{i}\hbar\left(\mathsf{P}_s\mathsf{X}_i\mathsf{X}_s\frac{1}{x^3} - \mathsf{P}_i\frac{1}{x} - \frac{1}{x}\mathsf{P}_i + \frac{1}{x^3}\mathsf{X}_s\mathsf{X}_i\mathsf{P}_s\right).\end{aligned}$$

From equation 16.10, the verification that H and A_i commute is finished once it is observed that

$$(\mathbf{L}\times\mathbf{X})_i + \mathsf{P}_s\mathsf{X}_i\mathsf{X}_s - \mathsf{P}_i x^2 = 0 = (\mathbf{X}\times\mathbf{L})_i \;+ \mathsf{X}_s\mathsf{X}_i\mathsf{P}_s - x^2\mathsf{P}_i.$$

The case $i = 1$ for the left-hand expression illustrates why this holds. Indeed, we have

$$\begin{aligned}[(\mathbf{L}\times\mathbf{X})_1 + \mathsf{P}_s\mathsf{X}_1\mathsf{X}_s - \mathsf{P}_1x^2] &= \mathsf{L}_2\mathsf{X}_3 - \mathsf{L}_3\mathsf{X}_2\\ &\quad + \mathsf{P}_1\mathsf{X}_1^2 + \mathsf{X}_1(\mathsf{P}_2\mathsf{X}_2 + \mathsf{P}_3\mathsf{X}_3) - \mathsf{P}_1x^2\\ &= (\mathsf{X}_3\mathsf{P}_1 - \mathsf{X}_1\mathsf{P}_3)\mathsf{X}_3 - (\mathsf{X}_1\mathsf{P}_2 - \mathsf{X}_2\mathsf{P}_1)\mathsf{X}_2\\ &\quad + \mathsf{P}_1\mathsf{X}_1^2 + \mathsf{X}_1(\mathsf{P}_2\mathsf{X}_2 + \mathsf{P}_3\mathsf{X}_3) - \mathsf{P}_1x^2\\ &= \mathsf{P}_1(\mathsf{X}_1^2 + \mathsf{X}_2^2 + \mathsf{X}_3^2) - \mathsf{P}_1x^2 = 0.\end{aligned}$$

The remaining cases are handled similarly and we conclude that each component operator of the LRL triad commutes with the inverse-square Hamiltonian.

As for the dot-product relation 16.7, first note that, since P_i and X_j commute when $i \neq j$,

$$\mathbf{P}\cdot\mathbf{L} = \mathsf{P}_i\mathsf{L}_i = \epsilon_{ijk}\mathsf{P}_i\mathsf{X}_j\mathsf{P}_k = \epsilon_{ijk}\mathsf{X}_j\mathsf{P}_i\mathsf{P}_k = 0.$$

Similarly, $\mathbf{L}\cdot\mathbf{P} = \mathbf{X}\cdot\mathbf{L} = \mathbf{L}\cdot\mathbf{X} = 0$. Using equations 12.1 and 16.5, this implies that

$$\begin{aligned}\mathbf{A}\cdot\mathbf{L} &= (\mathbf{L}\times\mathbf{P})\cdot\mathbf{L} = \epsilon_{ijk}\mathsf{L}_j\mathsf{P}_k\mathsf{L}_i\\ &= \epsilon_{ijk}(\mathsf{P}_k\mathsf{L}_j + \mathbf{i}\hbar\epsilon_{jki}\mathsf{P}_i)\mathsf{L}_i = \epsilon_{ijk}\mathsf{P}_k\mathsf{L}_j\mathsf{L}_i + 2\mathbf{i}\hbar\,\mathbf{P}\cdot\mathbf{L} = 0.\end{aligned}$$

The remaining relations expressed in equations 16.8 and 16.9 are derived in a similar fashion, though perhaps, the calculations turn out to be more strenuous. As nothing really new is needed, we leave their derivations to the energetic reader.

Armed with these relations, Pauli's argument now may be presented. Let $E < 0$ be a negative eigenvalue for the central-force Hamiltonian H and let $\mathcal{H}_E$ be the E-eigenspace of H. Define two new operator triads $\mathbf{J}^{\pm}$ by

$$\mathbf{J}^{\pm} = \frac{1}{2}\left(\mathbf{L} \pm \frac{1}{\sqrt{-2mE}}\mathbf{A}\right). \tag{16.12}$$

Using the fact that the component operators of both $\mathbf{L}$ and $\mathbf{A}$ commute with H, it is easy to verify that the components of the triads $\mathbf{J}^{\pm}$ also commute with H. It follows, then, that the operators $(\mathbf{J}^{\pm})^2$ commute with H. This means that $\mathcal{H}_E$ is invariant under the action of both the components of the triads $\mathbf{J}^{\pm}$ as well as the operators $(\mathbf{J}^{\pm})^2$. Our claim is that the operator triads $\mathbf{J}^{\pm}$ are commuting angular momentum triads on $\mathcal{H}_E$. Indeed, a calculation using relations 16.8 gives, for example,

$$[\mathsf{J}_i^+, \mathsf{J}_j^+] = \frac{\mathbf{i}\hbar}{4}\epsilon_{ijk}\left(\mathsf{L}_k + \frac{2}{\sqrt{-2mE}}\mathsf{A}_k + \frac{1}{E}\mathsf{HL}_k\right). \tag{16.13}$$

Since L_k commutes with H and on $\mathcal{H}_E$ the Hamiltonian reduces to multiplication by E, the operator product $(1/E)\mathsf{HL}_k$ reduces to L_k on $\mathcal{H}_E$. It follows from this and expression 16.13 that, on the subspace $\mathcal{H}_E$, $[\mathsf{J}_i^+, \mathsf{J}_j^+] = \mathbf{i}\hbar\epsilon_{ijk}\mathsf{J}_k^+$. Similar calculations show that on the subspace $\mathcal{H}_E$ $[\mathsf{J}_i^-, \mathsf{J}_j^-] = \mathbf{i}\hbar\epsilon_{ijk}\mathsf{J}_k^-$ and $[\mathsf{J}_i^+, \mathsf{J}_j^-] = 0$, so that, as claimed, the triads $\mathbf{J}^{\pm}$ are commuting angular momentum triads *on the subspace* $\mathcal{H}_E$. Another quick calculation gives

$$(\mathbf{J}^{\pm})^2 = \frac{1}{4}\left(\mathbf{L}^2 - \frac{1}{2mE}\mathbf{A}^2 \pm \frac{2}{\sqrt{-2mE}}\mathbf{A}\cdot\mathbf{L}\right), \tag{16.14}$$

from which, using equation 16.7, we get $(\mathbf{J}^+)^2 = (\mathbf{J}^-)^2$. Using equations 16.9 and 16.14, the fact that $\mathbf{L}^2$ commutes with H and again the fact that $\mathsf{H} = E1_{\mathcal{H}_E}$ when restricted to $\mathcal{H}_E$, we obtain that on the subspace $\mathcal{H}_E$ that

$$2[(\mathbf{J}^+)^2 + (\mathbf{J}^-)^2] + \hbar^2 1_{\mathcal{H}_E} = -\frac{mK^2}{2E}1_{\mathcal{H}_E}. \tag{16.15}$$

As the operators H and $(\mathbf{J}^{\pm})^2$ pairwise commute, we may expect a common eigenvector $\psi \in \mathcal{H}_E$ with $\mathsf{H}\psi = E\psi$ and $(\mathbf{J}^{\pm})^2\psi = j_{\pm}(j_{\pm} +$

$1)\hbar^2\psi$ for some half-integers $j_\pm$. Since $(\mathbf{J}^+)^2 = (\mathbf{J}^-)^2$ we have $j_+ = j_- = j$ and, applying equation 16.15 to ψ, we deduce that

$$[4j(j+1)+1]\hbar^2 = -\frac{mK^2}{2E}$$

or

$$E = -\frac{1}{(2j+1)^2}\frac{mK^2}{2\hbar^2}.$$

Since j may take on half-integer values starting with $j = 0$, $n = 2j+1$ may take positive integer values and, for the hydrogenic system with proton number Z where $K = Ze^2$, we have

$$E = E_n = -\frac{Z}{n^2}\frac{me^4}{2\hbar^2} = -\frac{Z^2}{n^2}\mathcal{R}\mathcal{Y},$$

reproducing equation 15.17 for the bound state energy levels.

At this point, we revert to the notation of the first section and denote the $E = E_n$-eigenspace $\mathcal{H}_E$ as $\mathcal{H}_n$. Note that, since $n = 2j+1$ for precisely one value of the half-integer j, each ψ in $\mathcal{H}_n$ is an eigenfunction of $(\mathbf{J}^\pm)^2$ with eigenvalue $j(j+1)\hbar^2$. Since the component operators $\mathsf{J}_3^\pm$ also commute with H and $(\mathbf{J}^\pm)^2$ and each other, there are linearly independent common eigenvectors in $\mathcal{H}_n$ indexed by the magnetic quantum number pairs (m_+, m_-) with $-j \leq m_\pm \leq j$. Specifically, we may find normalized eigenfunctions $|m_+, m_-\rangle$ in $\mathcal{H}_n$ for which $\mathsf{J}_3^\pm|m_+, m_-\rangle = m_\pm\hbar|m_+, m_-\rangle$ and for which $\langle m_+, m_-|m'_+, m'_-\rangle = \delta_{m_+m'_+}\delta_{m_-m'_-}$. This gives a degeneracy of the nth energy level of at least $(2j+1)^2 = n^2$, which is all we get from the developments of this section. Of course, from the first section of this lecture, based on the work of the previous lecture, we know that the degeneracy is precisely n^2 so that $\{|m_+, m_-\rangle : -j \leq m_\pm \leq j\}$ forms an orthonormal basis of $\mathcal{H}_n$.

16.3 Hidden Symmetry

There is a hint of subtle, hidden symmetry in the hydrogenic system in the advent of the commuting operator triads $\mathbf{J}^\pm$. The remainder of this lecture, as well as the next, will flesh out the details of this symmetry. Of course these operator triads, though defined on a dense subspace of $L^2(\mathbb{R}^3)$, do not act as angular momentum triads on their whole domain; it is only on the E_n-eigenspace $\mathcal{H}_n$, an n^2-dimensional subspace of the infinite-dimensional space $L^2(\mathbb{R}^3)$, that they act in this way. This can be seen easily in equation 16.13 since it is only on $\mathcal{H}_n$ that the operator product $(1/E_n)\mathsf{H}\mathsf{L}_k$ reduces to L_k. Being a bit more careful with the notation, we now use the symbols $\mathbf{J}_n^{\pm}$, for $n \in \mathbb{N}$, to denote the triads of the previous section that act as angular momentum operators when restricted to $\mathcal{H}_n$.

First, consider the bound state energy E_n for a fixed positive integer n. We concentrate on the structure of the n^2-dimensional space $\mathcal{H}_n$. As observed at the beginning of this lecture, the work of the previous lecture gives as an orthonormal basis for $\mathcal{H}_n$ the n^2 functions $\psi_{n,\ell,m}$, where ℓ takes the integer values $0, \ldots, n-1$ and m takes the integer values $-\ell, -\ell+1, \ldots, \ell-1, \ell$. Recalling the discussion of Section 13.2, this gives the decomposition

$$\mathcal{H}_n = \bigoplus_{\ell=0}^{n-1} V^{\ell}, \tag{16.16}$$

where V^{ℓ} is the $(2\ell+1)$-dimensional irreducible representation space of $\mathfrak{su}(2)$ corresponding to the angular momentum triad $\mathbf{L}$ and spanned by $\psi_{n,\ell,m}$ for $-\ell \leq m \leq \ell$. This decomposition arises precisely because the angular momentum triad $\mathbf{L}$ commutes with the Hamiltonian H.

We have in $\mathbf{J}_n^{\pm}$ the presence of two further angular momentum triads on $\mathcal{H}_n$ that commute with H, which should lead to a different decomposition of $\mathcal{H}_n$ in terms of the irreducible representation spaces of $\mathfrak{su}(2)$. Indeed, continuing with the observations of the final paragraph of the previous section, let W^j be a $(2j+1)$-dimensional irreducible representation space for $\mathfrak{su}(2)$ with basis $\{\mathbf{e}_m : -j \leq m \leq j\}$. The bilinear mapping $(\mathbf{e}_{m_+}, \mathbf{e}_{m_-}) \mapsto |m_+, m_-\rangle$ induces an isomorphism $W^j \otimes W^j \cong \mathcal{H}_n$. Noting that $2j+1 = n$, the resulting isomorphism,

$$W^j \otimes W^j \cong \bigoplus_{\ell=0}^{2j} V^{\ell}, \tag{16.17}$$

is a special case of the general result that the reader will find expressed in equation 21.3. Let $\mu : \mathfrak{su}(2) \to \mathrm{End}(W^j)$ be the representation homomorphism. Then $W^j \otimes W^j$ is an irreducible representation space of $\mathfrak{su}(2) \oplus \mathfrak{su}(2)$ via the homomorphisms $\tau_i \oplus 0 \mapsto \mu(\tau_i) \otimes 1_{W^j}$ and $0 \oplus \tau_i \mapsto 1_{W^j} \otimes \mu(\tau_i)$, for $i = 1, 2, 3$, where, the reader may recall, the τ_i are the generators of $\mathfrak{su}(2)$.

The bound state subspace $\mathcal{H}_{<0}$ of $L^2(\mathbb{R}^3)$ now may be realized as the completion of the orthogonal direct sum of irreducible representation spaces of $\mathfrak{su}(2) \oplus \mathfrak{su}(2)$ given by

$$\bigoplus_{n=1}^{\infty} \mathcal{H}_n \cong \bigoplus_{2j+1 \in \mathbb{N}} W^j \otimes W^j.$$

We may define an angular momentum triad that operates on the whole of $\mathcal{H}_{<0}$ by taking the direct sum of the restrictions of the triads $\mathsf{J}_n^{\pm}$ to $\mathcal{H}_n$. Specifically, we define the triads $\mathsf{J}^{\pm}$ operating on $\oplus_{n=1}^{\infty} \mathcal{H}_n$ by $\mathsf{J}_i^{\pm} \sum_n \psi_n = \sum_n (\mathsf{J}_n^{\pm})_i \psi_n$. Under this definition, the triads $\mathsf{J}^{\pm}$ are commuting angular momentum triads defined on a dense subspace of the bound state space $\mathcal{H}_{<0}$. They satisfy $(\mathsf{J}^+)^2 = (\mathsf{J}^-)^2$ and correspond to the standard representation of $\mathfrak{su}(2) \oplus \mathfrak{su}(2)$ as the sum of irreducible representations on $\oplus\{W^j \otimes W^j : 2j+1 \in \mathbb{N}\}$. We point out that $\mathsf{J}^{\pm}$ may be defined directly, precisely as in equation 16.12 if E is replaced by the restriction of the operator H to $\mathcal{H}_{<0}$. Of course, this means that we need to make sense of the expression $1/\sqrt{-2m\mathsf{H}}$ on $\mathcal{H}_{<0}$. This in turn may be accomplished by general theory or, as we will choose, directly by definition.[5] The definition of $1/\sqrt{-\mathsf{H}}$ on the direct sum $\oplus_n \mathcal{H}_n$ is

$$\frac{1}{\sqrt{-\mathsf{H}}} \sum_n \psi_n = \sum_n \frac{1}{\sqrt{-E_n}} \psi_n,$$

where this is, of course, a finite sum.

What does this all mean? After all, it is the SO(3) symmetry of the hydrogenic potential that leads directly to the vanishing of the commutators $[\mathsf{H}, \mathsf{L}_i]$ and the decomposition of the energy eigenspaces into irreducible representation spaces of $\mathfrak{su}(2) \cong \mathfrak{so}(3)$ given by equation 16.16. Moreover, the SO(3) symmetry of the potential translates into both the conservation of angular momentum expectation values and a SO(3) symmetry of the solutions to the Schrödinger equation.[6] In particular, if $\psi_t \in \mathcal{H}_{<0}$ solves the Schrödinger equation then so does $\psi_t \circ R$, where $R \in \mathrm{SO}(3)$ acts as a rotation on $\mathbb{R}^3$. The question posed by the existence of the angular momentum operator triads $\mathsf{J}^{\pm}$ that commute with the Hamiltonian on $\mathcal{H}_{<0}$ and the subsequent $\mathfrak{su}(2) \oplus \mathfrak{su}(2)$ decomposition of $\mathcal{H}_n$ as $W^j \otimes W^j$ is whether this Lie algebra symmetry arises from an action of a corresponding Lie group. According to equation 13.17, the obvious candidates for the Lie group are $\mathrm{SU}(2) \times \mathrm{SU}(2)$ and its quotient $\mathrm{SO}(3) \times \mathrm{SO}(3) \cong \mathrm{SO}(4)$. It turns out that the hidden symmetry

[5] In the general theory, $-\mathsf{H}$ is a positive operator on $\mathcal{H}_{<0}$ and hence has a square root, which then can be inverted on a dense subspace of $\mathcal{H}_{<0}$ that contains $\oplus_n \mathcal{H}_n$.

[6] See the first two sections of Lecture 28 for a fuller discussion of the conservation of angular momentum and a verification of the claims of this sentence.

is an SO(4) symmetry, which may be realized most easily as a higher-dimensional symmetry in momentum space.

16.4 Momentum Representation and SO(4) Symmetry

Naively applying equation 10.10 to the Fourier transform of the time-independent Schrödinger equation gives the Schrödinger equation in momentum space for an inverse-square potential as

$$\left(\frac{1}{2m}\|\mathbf{p}\|^2 - E\right)\widehat{\psi} = K\widehat{\frac{1}{x}\psi} = \frac{K}{(2\pi\hbar)^{3/2}}\widehat{\frac{1}{x}} * \widehat{\psi}, \qquad (16.18)$$

where $*$ denotes the convolution operator, which for functions μ and ν defined on $\mathbb{R}^3$, has the form

$$\mu * \nu(\mathbf{p}) = \int_{\mathbb{R}^3} \mu(\mathbf{p}-\mathbf{q})\nu(\mathbf{q})\,d\mathbf{q}.$$

The problem with this is that the Fourier transform of $v(\mathbf{x}) = 1/x = 1/\sqrt{x_1^2+x_2^2+x_3^2}$ fails to exist. There are two ways to proceed – that of the physicist or that of the mathematician.

The physicist's solution. Since the integral $\int_{\mathbb{R}^3} e^{-i\mathbf{p}\cdot\mathbf{x}/\hbar}v(\mathbf{x})\,d\mathbf{x}$ diverges, the physicist replaces the Coulomb potential $v(\mathbf{x}) = 1/x$ by the **Yukawa potential** $u_\alpha(\mathbf{x}) = e^{-\alpha x}/x$ for $\alpha > 0$, a central potential that looks like the Coulomb potential for $\mathbf{x}$ close to the origin, but which decreases exponentially to zero as $x \to \infty$. Note that as $\alpha \to 0+$, $u_\alpha \to v$ pointwise. The idea is to compute the Fourier transform $\widehat{u}_\alpha$, which does exist, and to recover a meaning for the symbol $\widehat{v}$ by taking the limit as $\alpha \to 0+$. To calculate the Fourier transform $\widehat{u}_\alpha(\mathbf{p})$, we may assume without loss of generality that the vector $\mathbf{p}$ lies along the z-axis. The reason is that an orthogonal transformation can be used to align $\mathbf{p}$ along the z-axis and, since the Yukawa potential is centrally symmetric, this change of variables does not affect the form of the Fourier integral. Set $p = \|\mathbf{p}\|$ and note that the angle between $\mathbf{p}$ and $\mathbf{x}$ is precisely the coordinate θ of the standard spherical coordinates (r, θ, ϕ) in equations 14.1. In passing to spherical coordinates we have $x = r$ and $d\mathbf{x} = r^2 \sin\theta\, dr d\theta d\phi$, and the Fourier integral becomes

$$\begin{aligned}
\widehat{u}_\alpha(\mathbf{p}) &= \frac{1}{(2\pi\hbar)^{3/2}} \int_{\mathbb{R}^3} e^{-\mathbf{i}\mathbf{p}\cdot\mathbf{x}/\hbar} \frac{e^{-\alpha x}}{x}\, d\mathbf{x} \\
&= \frac{1}{(2\pi\hbar)^{3/2}} \int_0^\infty \left(\int_0^\pi e^{-\mathbf{i}pr\cos\theta/\hbar} \sin\theta\, d\theta \right) e^{-\alpha r} r\, dr \int_0^{2\pi} d\phi \\
&= \frac{1}{\sqrt{2\pi\hbar}} \frac{1}{\mathbf{i}p} \int_0^\infty e^{-(\alpha - \mathbf{i}p/\hbar)r} - e^{-(\alpha + \mathbf{i}p/\hbar)r}\, dr = \sqrt{\frac{2\hbar}{\pi}} \frac{1}{p^2 + \hbar^2\alpha^2}.
\end{aligned} \tag{16.19}$$

So, even though $\widehat{v}$ does not exist in the usual sense as a convergent integral, the formula for $\widehat{u}_\alpha$ suggests that we may attach the meaning as follows:

$$\widehat{v}(\mathbf{p}) = \lim_{\alpha \to 0+} \widehat{u}_\alpha(\mathbf{p}) = \sqrt{\frac{2\hbar}{\pi}} \frac{1}{p^2}.$$

Plugging this into the convolution of the right-hand side of equation 16.18 and simplifying yields a momentum space Schrödinger equation:

$$\left(\frac{p^2}{2m} - E \right) \varphi(\mathbf{p}) = \frac{K}{2\pi^2\hbar} \int_{\mathbb{R}^3} \frac{\varphi(\mathbf{q})}{\|\mathbf{p} - \mathbf{q}\|^2}\, d\mathbf{q}, \tag{16.20}$$

where $\varphi = \widehat{\psi}$ must be a function for which the integral exists. Though this derivation fails the test of rigor, it does yield the valid expression 16.20.

The mathematician's solution. Recognizing that the Fourier transform of v is nonexistent, the mathematician also recognizes the introduction of the convolution in equation 16.18 as invalid and so sticks with the expression $\widehat{v\psi}$. A calculation similar to that of expression 16.19, passing again to spherical coordinates in the calculation of the transform, gives the inverse Fourier transform of $w(\mathbf{p}) = 1/p^2$ as

$$\widetilde{w}(\mathbf{x}) = \sqrt{\frac{\pi}{2\hbar}} \frac{1}{x} = \sqrt{\frac{\pi}{2\hbar}} v(\mathbf{x}),$$

a perfectly valid formula. We have from this that $v = \sqrt{2\hbar/\pi}\,\widetilde{w}$ so that

$$\begin{aligned}
\widehat{v\psi}(\mathbf{p}) &= \frac{1}{(2\pi\hbar)^{3/2}} \int_{\mathbb{R}^3} v(\mathbf{s})\psi(\mathbf{s})e^{-\mathbf{i}\mathbf{p}\cdot\mathbf{s}/\hbar}\, d\mathbf{s} \\
&= \frac{1}{(2\pi\hbar)^{3/2}} \sqrt{\frac{2\hbar}{\pi}} \int_{\mathbb{R}^3} \left[\frac{1}{(2\pi\hbar)^{3/2}} \int_{\mathbb{R}^3} w(\mathbf{q})e^{\mathbf{i}\mathbf{s}\cdot\mathbf{q}/\hbar}\, d\mathbf{q} \right] \psi(\mathbf{s})e^{-\mathbf{i}\mathbf{p}\cdot\mathbf{s}/\hbar}\, d\mathbf{s} \\
&= \frac{1}{2\pi^2\hbar} \int_{\mathbb{R}^3} w(\mathbf{q}) \left[\frac{1}{(2\pi\hbar)^{3/2}} \int_{\mathbb{R}^3} \psi(\mathbf{s})e^{-\mathbf{i}(\mathbf{p}-\mathbf{q})\cdot\mathbf{s}/\hbar}\, d\mathbf{s} \right] d\mathbf{q} \\
&= \frac{1}{2\pi^2\hbar} \int_{\mathbb{R}^3} w(\mathbf{q})\widehat{\psi}(\mathbf{p}-\mathbf{q})\, d\mathbf{q} = \frac{1}{2\pi^2\hbar} w * \widehat{\psi}(\mathbf{p}) = \frac{1}{2\pi^2\hbar} w * \varphi(\mathbf{p}).
\end{aligned}$$

Thus, the middle term of equation 16.18 becomes precisely the right-hand-side of equation 16.20. This is rigorous as long as the change in the order of integration in going from line two to line three in the above equation is valid. This is indeed valid provided that the original integrand is suitably well behaved and decreases to zero fast enough toward infinity; it certainly is valid, for example, when ψ is a test function.

Whether one prefers the physicist's or the mathematician's explanation is, perhaps, a test of the reader's temperament. Whatever the case, we now concentrate our study on the momentum space equation 16.20. First, assuming still that $E < 0$, we rewrite that equation by introducing the parameter $p_0 = \sqrt{-2mE}$, the average momentum, and performing the change of variables $\mathbf{u} = \mathbf{p}/p_0$. Using $\mathbf{v} = \mathbf{q}/p_0$ as the new integration variable and observing that $d\mathbf{q} = p_0^3\, d\mathbf{v}$, the momentum space equation becomes

$$\left(1 + u^2\right) \Phi(\mathbf{u}) = \frac{mK}{p_0\pi^2\hbar} \int_{\mathbb{R}^3} \frac{\Phi(\mathbf{v})}{\|\mathbf{u} - \mathbf{v}\|^2}\, d\mathbf{v}, \tag{16.21}$$

where we have written $\Phi(\mathbf{u}) = \varphi(p_0\mathbf{u})$ and $u = \|\mathbf{u}\|$. Our first task is to see how this equation transforms under stereographic projection of $\mathbb{R}^3$ to the 3-sphere $\mathbf{S}^3$ in $\mathbb{R}^4$. The **stereographic projection** $\mathfrak{p}$ of $\mathbb{R}^n$ to the n-sphere $\mathbf{S}^n \subset \mathbb{R}^{n+1} = \mathbb{R} \times \mathbb{R}^n$ is defined by the requirement that $\hat{\mathbf{n}} = (1, \mathbf{0})$, $(0, \mathbf{u})$, and $\mathfrak{p}(\mathbf{u})$ be collinear points in $\mathbb{R}^{n+1}$ with $\|\mathfrak{p}(\mathbf{u})\| = 1$ whenever $\mathbf{u} \in \mathbb{R}^n$. It follows that, for $\mathbf{u} \in \mathbb{R}^n$, there exists $t \in \mathbb{R}$ such that $\mathfrak{p}(\mathbf{u}) = (1 - t, t\mathbf{u})$ and $(1 - t)^2 + t^2u^2 = 1$, where $u = \|\mathbf{u}\|$. Solving for t gives $t = 2/(u^2 + 1)$, so that

$$\mathfrak{p}(\mathbf{u}) = \left(\frac{u^2 - 1}{u^2 + 1}, \frac{2\mathbf{u}}{u^2 + 1} \right),$$

which defines a bijection between $\mathbb{R}^n$ and the sphere $\mathbf{S}^n$ punctured at

$\hat{\mathbf{n}}$. Specializing to the case $n = 3$, we may identify $\mathbb{R}^4$ as the quaternions $\mathbb{H}$, thus identifying $\mathbb{R}^3$ as span$\{\mathbf{i}, \mathbf{j}, \mathbf{k}\} \subset \mathbb{H}$ and $\mathbf{S}^3$ as the group of unit quaternions. Recalling equation 13.13, this admits the elegant quaternionic expressions

$$\mathfrak{p}(\mathbf{u}) = -\frac{1 - \mathbf{u}}{1 + \mathbf{u}} \quad \text{and} \quad \mathfrak{p}^{-1}(\mathsf{s}) = \frac{\mathbf{s}}{1 - s_0}, \tag{16.22}$$

where $\mathsf{s} = s_0 + \mathbf{s} = \operatorname{Re}\mathsf{s} + \operatorname{Pu}\mathsf{s}$ is a quaternionic variable on $\mathbf{S}^3$. The fraction for $\mathfrak{p}(\mathbf{u})$ makes sense since $1/(1 + \mathbf{u})$ commutes with $1 - \mathbf{u}$.

The pull-back to $\mathbb{R}^3$ of the standard spherical volume element $d\Omega$ on $\mathbf{S}^3$ is given by

$$\mathfrak{p}^*(d\Omega) = dV = \frac{8}{(1 + u^2)^3}\, d\mathbf{u}.$$

A quick way to verify this is to note that, since $\mathfrak{p}$ is conformal, the pull-back takes the form $dV = \mu(\mathbf{u})^3 d\mathbf{u}$ where $\mu(\mathbf{u})$ is the local dilatation factor. Since the mapping is conformal, the dilatation at the point $\mathbf{u}$ may be obtained by calculating the dilatation in any direction at $\mathbf{u}$. It is easy to see that since the image under $\mathfrak{p}$ of any circle centered at the origin and of radius u in $\mathbb{R}^3$ is a latitudinal circle in $\mathbf{S}^3$ of radius $2u/(1 + u^2)$, the factor is $\mu(\mathbf{u}) = 2/(1 + u^2)$. This means that a function f on the sphere $\mathbf{S}^3$ may be integrated via the pull-back metric as follows:

$$\int_{\mathbf{S}^3} f(\mathsf{s})\, d\Omega = \int_{\mathbb{R}^3} f(\mathfrak{p}(\mathbf{u}))\, dV = \int_{\mathbb{R}^3} f(\mathfrak{p}(\mathbf{u})) \frac{8}{(1 + u^2)^3}\, d\mathbf{u}. \tag{16.23}$$

As an easy exercise, the reader may want to verify that the volume of the 3-sphere is $2\pi^2$ by evaluating the integral on the right with $f \equiv 1$ using spherical coordinates.

Our aim is to rewrite the integral of equation 16.21 as an integral on $\mathbf{S}^3$. Applying the first of equations 16.22 to $\mathsf{s} = \mathfrak{p}(\mathbf{u})$ and $\mathsf{t} = \mathfrak{p}(\mathbf{v})$ for $\mathbf{u}, \mathbf{v} \in \mathbb{R}^3$ gives

$$\begin{aligned}\frac{1}{2}(1 + \mathbf{u})(\mathsf{s} - \mathsf{t})(1 + \mathbf{v}) &= \frac{1}{2}(1 + \mathbf{u})\left(-\frac{1 - \mathbf{u}}{1 + \mathbf{u}} + \frac{1 - \mathbf{v}}{1 + \mathbf{v}}\right)(1 + \mathbf{v}) \\ &= \frac{1}{2}\left[-(1 - \mathbf{u})(1 + \mathbf{v}) + (1 + \mathbf{u})(1 - \mathbf{v})\right] = \mathbf{u} - \mathbf{v}.\end{aligned}$$

Using the normed algebra structure of quaternions expressed in equation 13.8 and the fact that $|\mathsf{q}|^2 = \mathsf{q}\bar{\mathsf{q}} = \mathsf{q} \cdot \mathsf{q} = \|\mathsf{q}\|^2$ for the quaternion q,

$$\frac{1}{4}(1 + u^2)(1 + v^2)|\mathsf{s} - \mathsf{t}|^2 = \|\mathbf{u} - \mathbf{v}\|^2. \tag{16.24}$$

Setting

$$Y(\mathsf{s}) = (1 + \|\mathfrak{p}^{-1}(\mathsf{s})\|^2)^2 \Phi \circ \mathfrak{p}^{-1}(\mathsf{s})$$

so that $Y(\mathsf{s}) = (1 + u^2)^2 \Phi(\mathbf{u})$ whenever $\mathsf{s} = \mathfrak{p}(\mathbf{u})$, and applying equations 16.23 and 16.24 to equation 16.21, gives

$$Y(\mathsf{s}) = \frac{\lambda}{2\pi^2} \int_{\mathbf{S}^3} \frac{Y(\mathsf{t})}{|\mathsf{s} - \mathsf{t}|^2} \, d\Omega, \tag{16.25}$$

where $\lambda = mK/p_0\hbar$. Equation 16.25 is the momentum space Schrödinger equation on the 3-sphere for the Kepler problem with negative energy. Its solutions, after inverse stereographic projection followed by inverse Fourier transform, give the bound state eigenfunctions for the Kepler problem. In this form, though, we are able to see the emergence of symmetries that are not apparent from the standard position space Schrödinger equation, for which the SO(3) symmetry of the central potential is the only obvious symmetry.

Indeed, we can see already the manifestation of SO(4) symmetry. Recall from Lecture 13 that the orthogonal action of SO(4) on $\mathbf{S}^3$ may be realized as the action of $\mathbf{S}^3 \times \mathbf{S}^3$ by left and right quaternionic multiplication. Specifically, for unit quaternions p and q, the transformation $L_{\mathsf{p}} R_{\overline{\mathsf{q}}}(\mathsf{s}) = \mathsf{p}\mathsf{s}\overline{\mathsf{q}}$ defines a special orthogonal transformation of $\mathbf{S}^3$. All the special orthogonal transformations are obtained this way, with two-to-one ambiguity given by $L_{\mathsf{p}} R_{\overline{\mathsf{q}}} = L_{-\mathsf{p}} R_{-\overline{\mathsf{q}}}$. With $\Upsilon = Y \circ L_{\mathsf{p}} R_{\overline{\mathsf{q}}}$, since the standard volume element $d\Omega$ is $\mathbf{S}^3 \times \mathbf{S}^3$-invariant, the change of variable $\mathsf{t} = \mathsf{p}\mathsf{t}'\overline{\mathsf{q}}$ gives

$$\begin{aligned} \Upsilon(\mathsf{s}) = Y(\mathsf{p}\mathsf{s}\overline{\mathsf{q}}) &= \frac{\lambda}{2\pi^2} \int_{\mathbf{S}^3} \frac{Y(\mathsf{t})}{|\mathsf{p}\mathsf{s}\overline{\mathsf{q}} - \mathsf{t}|^2} \, d\Omega \\ &= \frac{\lambda}{2\pi^2} \int_{\mathbf{S}^3} \frac{Y(\mathsf{p}\mathsf{t}'\overline{\mathsf{q}})}{|\mathsf{s} - \mathsf{t}'|^2} \, d\Omega = \frac{\lambda}{2\pi^2} \int_{\mathbf{S}^3} \frac{\Upsilon(\mathsf{t})}{|\mathsf{s} - \mathsf{t}|^2} \, d\Omega, \end{aligned}$$

since $|\mathsf{p}\mathsf{s}\overline{\mathsf{q}} - \mathsf{t}| = |\mathsf{p}||\mathsf{s} - \mathsf{t}'||\overline{\mathsf{q}}| = |\mathsf{s} - \mathsf{t}'|$ by equation 13.8. This confirms that whenever Y solves equation 16.25, so does the function $\Upsilon = Y \circ L_{\mathsf{p}} R_{\overline{\mathsf{q}}}$. This manifests SO(4) symmetry in the solutions to the stereographically projected momentum space Schrödinger equation 16.25.

We explore this SO(4) symmetry of the Kepler problem in more detail in the next lecture, where the introduction of techniques from representation theory, and in particular that of group characters, will aid our discussion.

16.5 End Notes

We saw at the at the end of Section 16.2 that Pauli's derivation of the energy levels of the hydrogenic system leads to a hint of SO(4) symmetry in the system. Pauli's work of 1925 built upon that of Rünge and Lenz, who had shown, by the end of the nineteenth century, that the classical Kepler problem of planetary orbits exhibits SO(4) symmetry in its action on a portion of the six-dimensional phase space of the planet. More recently, it has been demonstrated that the classical Kepler problem exhibits even more symmetry, with the 15-dimensional group $O(2,4)$ playing a key role. Pauli's analysis followed that of Rünge and Lenz, but with the Poisson brackets of the classical theory replaced by the commutators of the quantum theory. Pauli's theory provides merely a hint of the hidden symmetry, and it was Fock who showed that the symmetry is in fact SO(4) symmetry. A mathematically intense presentation of the higher symmetries exhibited by the classical Kepler problem may be found in Guillemin and Sternberg (1990). Fock's (1935) paper, an English translation of which appears in Stephanie Singer's (2005) beautiful text, thrashes out the details of the SO(4) symmetry. An extensive modern treatment of the Kepler problem and its group-theoretic aspects in both the classical and quantum settings may be found in Cordani (2003).

Hannabuss (1997) gives a quick, short development of hidden symmetry in the hydrogenic system using the matrix group SU(2). I have filled in his development with more details in this and the next lecture. I find that the choice of quaternionic variables instead of matrix variables to describe the momentum space Schrödinger equation offers a more concise language, which is more conducive to performing calculations: these appear in the next lecture. In this I follow Section 5.3 of Takhtajan (2008).

An invaluable resource for a more in-depth development of the hidden symmetries of hydrogen are the two research papers Bander and Itzykson (1966a, b). After a thorough discussion of the internal SO(4) symmetry of the hydrogenic system, the authors expose a more general symmetry based on the de-Sitter group $O(1,4)$. The SO(4) symmetry is used to study bound state problems in paper I and the $O(1,4)$ symmetry is used to study scattering states of the unbound electron in their second paper.

17

Wave Mechanics VI: Hidden Symmetry Solved

> It has long been known that the energy levels of the hydrogen atom are degenerate with respect to the azimuthal quantum number ℓ; one speaks occasionally of an "accidental" degeneracy. But any degeneracy of eigenvalues is linked to the transformation group of the relevant equation: e.g., the degeneracy with respect to the magnetic quantum number m is allied to the usual rotation group. However, until now, the group corresponding to the "accidental" degeneracy of the hydrogen levels was unknown.
>
> In this work we will show that this group is equivalent to the four-dimensional rotation group.
>
> Vladimir Fock
> "On the theory of the hydrogen atom," 1935

This lecture continues our investigation of the hidden SO(4) symmetry of the Kepler problem in wave mechanics. Having uncovered in Section 16.4 the SO(4) symmetry of the solutions to the bound state momentum space wave equation for the Kepler problem on the 3-sphere, we now solve that momentum space equation by presenting two methods of solution – the first using classical analytic arguments and the second, a modern treatment, using group characters. The first method realizes the solutions as spherical harmonics on the three-dimensional sphere, establishing again the SO(4) symmetry of the solutions in a way that mirrors the SO(3) symmetry evident in the standard solutions to the position space Schrödinger equation. The second is more abstract, but the advantage is that it gives a rich source of ready-made solutions to the momentum space equation on the 3-sphere.

In the first section, we present our first solution to equation 16.25.

This is one of Fock's contributions in his 1935 article elucidating the hidden symmetries of atomic hydrogen.[1] We present Fock's original argument, augmented by the work of Bander and Itzykson (1966), which provides an explanation of the fact that equation 16.25 is precisely the integral equation on the 3-sphere for the three-dimensional version of spherical harmonics. Since in our argument we use a higher-dimensional version of Green's second identity, which is not so well known to students of pure mathematics, we include in the first section a short primer on vector calculus and the Laplace operator in n dimensions. After this, a more modern treatment of the momentum space equation is presented that uses the theory of group characters in its solution. This modern treatment is presented after a mathematical interlude on character theory. We round out the discussion by returning to the angular momentum operator triads $\mathbf{J}^{\pm}$ that arise from the LRL-vector and relating them to the SO(4) representations that arise from our analysis of the solutions to the momentum space equation.

17.1 Fock's Treatment of the Momentum Space Equation

As the theory of the Laplace operator is no more difficult in n dimensions, we have chosen not to restrict our development to four dimensions only. For Fock's argument, we need the n-dimensional version of Green's second identity, which reads

$$\int_M (u\Delta v - v\Delta u)\, dx = \int_{\partial M} (u\partial_X v - v\partial_X u)\, dS, \tag{17.1}$$

valid for real-valued functions u and v that are C^2 on a neighborhood of the closure of $M \subset \mathbb{R}^n$. Here, M is an open set with boundary ∂M smooth enough to support almost everywhere an outward-pointing unit normal vector field $X = (X^1, \ldots, X^n)$; $\Delta = \Delta_n$ is the n-dimensional Laplacian, $\nabla u = \operatorname{grad} u = (\partial_1 u, \ldots, \partial_n u)$ is the n-dimensional gradient, and $\partial_X = X \cdot \nabla$ is the derivative in the direction of the outward normal X. We are using $x = (x_1, \ldots, x_n)$ as Cartesian coordinates on $\mathbb{R}^n$, $\partial_i = \partial/\partial x_i$, and dS as the volume element on ∂M inherited from the usual measure dx on $\mathbb{R}^n$.

To derive identity 17.1, we use the n-dimensional divergence theorem,

[1] V. Fock (1935), Zur Theorie des Wasserstoffatoms. *Z. Phys.*, 98: 145–54. I have relied on the English translation on pp. 286–296 in Singer (2005).

which expressed in traditional vector calculus notation reads

$$\int_M \nabla \cdot \mathbf{F}\, dx = \int_{\partial M} \mathbf{F} \cdot X\, dS,$$

for any C^2 vector field $\mathbf{F}$ defined on a neighborhood of the closure of M. First note that since $\nabla \cdot (u\nabla v) = \nabla u \cdot \nabla v + u\Delta v$, and similarly with u and v exchanged, the divergence theorem applied to $\mathbf{F} = u\nabla v - v\nabla u$ gives

$$\begin{aligned}\int_M u\Delta v - v\Delta u\, dx &= \int_M \nabla \cdot (u\nabla v - v\nabla u)\, dx \\ &= \int_{\partial M} (u\nabla v - v\nabla u) \cdot X\, dS.\end{aligned}$$

This is Green's identity 17.1 since $(u\nabla v - v\nabla u) \cdot X = (u\partial_X v - v\partial_X u)$.

Armed with Green's identity, we are ready to examine the Laplace operator and harmonic functions in n dimensions in more detail. Let $\mathcal{H}^\ell$ be the space of complex-valued harmonic polynomials in n variables that are homogeneous of degree ℓ. As reported in Section 14.3, a typical element h of $\mathcal{H}^\ell$ is of the form $h(x) = \|x\|^\ell Y(\hat{x})$, where Y is the restriction of h to the $(n-1)$-sphere $\mathbf{S}^{n-1}$ and $\hat{x} = x/\|x\|$ is the unit vector in the direction of x. Using expression 14.16 it is easy to show that, since $\Delta_n h = 0$ whenever $h \in \mathcal{H}^\ell$, the restriction Y of h to the $(n-1)$-sphere satisfies $\Delta_{\mathbf{S}^{n-1}} Y = -\ell(\ell + n - 2)Y$. The function Y is called a spherical harmonic of degree ℓ. Our aim is to derive an integral equation that the function Y satisfies.

For this, let $R(x) = 1/\|x\|^{n-2}$. A straightforward calculation shows that, for any $s \in \mathbb{R}^n$, $R_s(x) = R(x - s)$ is harmonic on $\mathbb{R}^n \setminus \{s\}$ whenever $n > 2$. We would like to apply Green's identity where the set M is the open n-ball $\mathbf{B}^n$, $u = h \in \mathcal{H}^\ell$, and $v = R_s$, where s is a point fixed on the sphere $\mathbf{S}^{n-1} = \partial\mathbf{B}^n$. The problem with this is that the function R_s is not C^2 on $\mathbf{S}^{n-1}$, since it has a singularity at the point s. To get around this problem, we modify the n-ball in a small neighborhood of the singularity and take the limit as the neighborhood size vanishes. Let $\overline{N}_\varepsilon$ be the closure of the open ε-neighborhood N_ε about s, for small $\varepsilon > 0$, and define M_ε as the set difference $\mathbf{B}^n - \overline{N}_\varepsilon$. Applying Green's identity on M_ε gives

$$0 = \int_{\partial M_\varepsilon} (h\partial_X R_s - R_s\partial_X h)\, dS.$$

By the description of M_ε, the boundary can be written as $\partial M_\varepsilon = \Sigma_\varepsilon \cup \sigma_\varepsilon$, where $\Sigma_\varepsilon = \mathbf{S}^{n-1} - N_\varepsilon$ and $\sigma_\varepsilon = \partial\overline{N}_\varepsilon \cap \overline{\mathbf{B}}^n$. This gives

$$\int_{\sigma_\varepsilon} (h\partial_\rho R_s - R_s\partial_\rho h)\, dS = -\int_{\Sigma_\varepsilon} (Y\partial_r R_s - R_s\partial_r h)\, d\Omega,$$

where ∂_ρ is the derivative in the direction of the *inward* pointing unit normal vector field on $\partial\overline{N}_\varepsilon$, ∂_r is the derivative in the direction of the *outward* pointing unit normal vector field on $\mathbf{S}^{n-1}$, and $d\Omega = d^{n-1}\Omega$ is the rotationally invariant measure on the unit sphere. Eventually we will take the limit of both sides of this expression as ε decreases to zero.

Considering the right-hand side first, since $h(x) = r^\ell Y(\hat{x})$ we have $\partial_r h = \ell r^{\ell-1} Y$, which, evaluated on the unit sphere, gives $\partial_r h|_{r=1} = \ell Y$. A further calculation gives $\partial_r R_s = (2-n)R_s/2$, so that the integrand reduces to $(2-n-2\ell)YR_s/2$. It follows that the right-hand side reduces to

$$-\int_{\Sigma_\varepsilon} (Y\partial_r R_s - R_s\partial_r h)\, d\Omega = \frac{n+2\ell-2}{2}\int_{\Sigma_\varepsilon} \frac{Y(t)}{\|s-t\|^{n-2}}\, d\Omega,$$

where we have used t as a variable on the $(n-1)$-sphere. The limit, then, of the right-hand side as ε decreases to zero is

$$\frac{n+2\ell-2}{2}\int_{\mathbf{S}^{n-1}} \frac{Y(t)}{\|s-t\|^{n-2}}\, d\Omega. \tag{17.2}$$

The argument for the left-hand side is a little more delicate. First, a calculation shows that $\partial_\rho R_s(x) = (n-2)\|x-s\|^{1-n}$, so that $\partial_\rho R_s|_{\sigma_\varepsilon} = (n-2)\varepsilon^{1-n}$. Obviously $R_s|_{\sigma_\varepsilon} = \varepsilon^{2-n}$. Letting h^ε denote the average value of h on σ_ε, and similarly for $\partial_\rho h^\varepsilon$, we may write the left-hand side as

$$\int_{\sigma_\varepsilon} (h\partial_\rho R_s - R_s\partial_\rho h)\, dS = [(n-2)h^\varepsilon - \varepsilon\partial_\rho h^\varepsilon]\, \varepsilon^{1-n}\, \mathrm{Vol}(\sigma_\varepsilon).$$

Noting that the volume of a sphere in $\mathbb{R}^n$ of radius ε is $\varepsilon^{n-1}\,\mathrm{Vol}(\mathbf{S}^{n-1})$ and that σ_ε is, essentially, a hemisphere, we have $\lim_{\varepsilon\to 0+} \varepsilon^{1-n}\,\mathrm{Vol}(\sigma_\varepsilon) = \mathrm{Vol}(\mathbf{S}^{n-1})/2$. The limit, then, of the left-hand side as ε decreases to zero is

$$\frac{n-2}{2}\,\mathrm{Vol}(\mathbf{S}^{n-1})Y(s), \tag{17.3}$$

since the continuity of h implies that $\lim_{\varepsilon\to 0+} h^\varepsilon = h(s) = Y(s)$. The equality of expressions 17.2 and 17.3 now gives the integral equation on the $(n-1)$-sphere $\mathbf{S}^{n-1}$ satisfied by the spherical harmonics of degree ℓ as

$$Y(s) = \frac{n+2\ell-2}{n-2}\,\frac{1}{\mathrm{Vol}(\mathbf{S}^{n-1})}\int_{\mathbf{S}^{n-1}} \frac{Y(t)}{\|s-t\|^{n-2}}\, d\Omega. \tag{17.4}$$

Recalling that in the exercise following equation 16.23 we noted that $\mathrm{Vol}(\mathbf{S}^3) = 2\pi^2$, we see that this integral equation reduces to equation 16.25 when $n = 4$ provided that $\lambda = \ell + 1$. Since $\lambda = mK/p_0\hbar$ and $p_0 = \sqrt{-2mE}$, this gives

$$E = -\frac{mK^2}{2\hbar^2}\frac{1}{(\ell+1)^2}. \tag{17.5}$$

This is precisely equation 15.7 if $K = Ze^2$ represents the hydrogenic potential, and so the above discussion presents yet another derivation of the energy levels of atomic hydrogen. It follows also that the spherical harmonics on $\mathbf{S}^3$ provide a set of solutions to the projected momentum space Schrödinger equation for the Kepler problem of expression 16.25. Moreover, since the space of spherical harmonics is dense in $L^2(\mathbf{S}^3)$, these solutions provide a complete set in which every solution can be expanded.

Recall that equation 16.1 reports the dimension of the space $\mathcal{H}_n$ of E_n-eigenfunctions as n^2. The argument relies on the fact that the elements of $\mathcal{H}_n$ arise from the partial space of two-dimensional spherical harmonics $\oplus_{\ell=0}^{n-1} V^\ell$, where $V^\ell = V^{2,\ell}$ is the $(2\ell+1)$-dimensional space of spherical harmonics of degree ℓ on the 2-sphere $\mathbf{S}^2$. In the setting of the momentum space equation of this section, the E_n-eigenfunctions arise from the n^2-dimensional space $V^{3,n-1}$, $n = \ell + 1$, of spherical harmonics of degree $n - 1$ on the 3-sphere $\mathbf{S}^3$. The dimension count may be seen by an argument exactly like the one which yielded equation 14.18 in Lecture 14. Indeed,

$$\dim V^{3,\ell} = \dim \mathcal{P}^{3,\ell} - \dim \mathcal{P}^{3,\ell-2} = \binom{\ell+3}{3} - \binom{\ell+1}{3} = (\ell+1)^2,$$

where $\mathcal{P}^{3,\ell}$ is the space of complex-valued polynomials in four variables that are homogeneous of degree ℓ. See the derivation of equation 14.18.

We have already recognized the SO(4) symmetry of the solutions to the projected momentum space equation using only the form of the integral equation. Now that we also have identified the solutions as the space of spherical harmonics, $\mathcal{V}^3 = \oplus_{\ell=0}^{\infty} V^{3,\ell}$ in the notation of Section 16.4, we may recognize the symmetry again by observing that $V^{3,\ell}$ is an irreducible representation space of SO(4) just as $V^\ell = V^{2,\ell}$ is an irreducible representation space of SO(3). This will be expanded upon in the penultimate section of this lecture.

17.2 Group Characters and Representations

We pause for a brief mathematical interlude where we calculate the characters of the irreducible representations of the group of unit quaternions[2] $\mathbf{S}^3$. These are then used in the section following to offer an alternative method for solving the Schrödinger equation for the Kepler problem via the integral equation 16.25.

We begin with a definition. Let $R : G \to \mathrm{GL}(V)$ be a finite-dimensional representation of the group G on the (finite-dimensional) Hilbert space V with inner product $\langle - | - \rangle$. The **character** χ_R of the representation R is the complex-valued function defined on G given by

$$\chi_R(g) = \mathsf{Tr}(R(g)) = \sum_i \langle \xi_i | R(g) | \xi_i \rangle,$$

where $\{\xi_i\}$ is an orthonormal basis. The trace functional is, of course, independent of which particular basis is used. Recalling also that the trace functional is invariant under conjugation, we have

$$\chi_R(hgh^{-1}) = \mathsf{Tr}(R(h)R(g)R(h)^{-1}) = \mathsf{Tr}(R(g)) = \chi_R(g),$$

implying that χ_R is constant on conjugacy classes in G.

At this point the reader might want to review Sections 13.3 and 13.4 as the discussion in this section will draw freely from those. The first task is to evaluate the character of the representation R^j, the irreducible $(2j+1)$-dimensional representation of $\mathbf{S}^3$ on $\mathrm{GL}(V^j)$, which is defined by diagram 13.19 for half-integer j. Recall from equation 13.18 that each element of the group $\mathbf{S}^3$ may be written in the form $e^{\mathbf{u}\vartheta} = \cos\vartheta + \mathbf{u}\sin\vartheta$ for a unit pure quaternion $\mathbf{u} \in \mathbf{S}^2 \subset \mathbb{R}^3$ and real number ϑ, which measures the angle between the axis $\mathbb{R}$ of real quaternions and the point $e^{\mathbf{u}\vartheta}$. With this description, it is easy to see that any element $\mathsf{s} \in \mathbf{S}^3$ other than ± 1 may be written as

$$\mathsf{s} = \mathrm{Re}\,\mathsf{s} + \mathrm{Pu}\,\mathsf{s} = s_0 + \mathbf{s} = \cos\vartheta + \hat{\mathbf{s}}\sin\vartheta = e^{\hat{\mathbf{s}}\vartheta},$$

where $\vartheta = \cos^{-1} s_0 = \arg \mathsf{s}$ and $\hat{\mathbf{s}} = \mathbf{s}/\|\mathbf{s}\|$. The conjugacy class of s is

$$C(\mathsf{s}) = \{\mathsf{q}\mathsf{s}\bar{\mathsf{q}} : \mathsf{q} \in \mathbf{S}^3\} = \{e^{\mathbf{u}\vartheta} : \mathbf{u} \in \mathbf{S}^2\} = C_\vartheta, \tag{17.6}$$

where $\vartheta = \arg \mathsf{s}$. This follows from the fact that conjugation by q is

[2] In any references that the reader might consult, the characters are likely to be worked out in the context of the more familiar group SU(2). As expression 13.15 provides an isomorphism of $\mathbf{S}^3$ with SU(2), by the discussion on p. 194 identifying concrete representation spaces for SU(2) that are equivalent to those of $\mathbf{S}^3$ the two approaches are equivalent.

orthogonal, fixes the real axis, and, when restricted to $\mathbf{S}^2$, describes a transitive action of $\mathbf{S}^3$ on the 2-sphere.[3] Since χ_{R^j} is constant on conjugacy classes, the character depends only on the angle ϑ. Denote the value of χ_{R^j} on the conjugacy class C_ϑ by $c^j(\vartheta)$.

To find $c^j(\vartheta)$, we need a matrix representation of an element of C_ϑ, which already appears in expression 13.20 with $t = \vartheta$. Calculating the trace with $m = -j, -j+1, \ldots, j-1, j$ gives

$$\begin{aligned} c^j(\vartheta) = \sum_{m=-j}^{j} e^{2m\mathrm{i}\vartheta} &= e^{-2j\mathrm{i}\vartheta} \sum_{m=0}^{2j} e^{2m\mathrm{i}\vartheta} = e^{-2j\mathrm{i}\vartheta} \frac{e^{(2j+1)2\mathrm{i}\vartheta} - 1}{e^{2\mathrm{i}\vartheta} - 1} \\ &= \frac{e^{(2j+1)\mathrm{i}\vartheta} - e^{-(2j+1)\mathrm{i}\vartheta}}{e^{\mathrm{i}\vartheta} - e^{-\mathrm{i}\vartheta}} = \frac{\sin\left[(2j+1)\vartheta\right]}{\sin\vartheta} \end{aligned} \tag{17.7}$$

for the values of the characters of R^j. Recall from p. 189 that when the unit quaternion s acts by conjugation on $\mathbb{H}$, it restricts to a rotation on $\mathbb{R}^3$ through an angle $\theta = 2\cos^{-1} s_0 = 2\vartheta$ with axis $\ell_{\mathsf{s}} = \operatorname{span}\{\mathbf{s}\}$. The characters calculated here are often reported using the variable θ rather than ϑ, and so read

$$c^j(\theta/2) = \frac{\sin\left[(j+\frac{1}{2})\theta\right]}{\sin\left(\frac{1}{2}\theta\right)}. \tag{17.8}$$

We note in passing that the conjugacy classes in SO(3) are parameterized by θ, since two rotations are conjugate precisely when they rotate through the same angle θ. Expression 17.8 gives the characters of the irreducible representations of the rotation group SO(3) when j takes on integral values.

The characters calculated above satisfy an orthogonality relation that is used later in applications to the Kepler problem. Extending the Kronecker delta symbol to half-integers j and k in an obvious way, we have

$$\frac{2}{\pi} \int_0^\pi c^j(\vartheta) c^k(\vartheta) \sin^2\vartheta \, d\vartheta = \delta_{jk}. \tag{17.9}$$

This is nothing more than the usual orthogonality relation for Fourier sine series, since it reduces to the familiar relation

$$\frac{2}{\pi} \int_0^\pi \sin m\vartheta \sin n\vartheta \, d\vartheta = \delta_{mn} \tag{17.10}$$

for integers $m = 2j+1$ and $n = 2k+1$.

Having calculated the characters of the irreducible finite-dimensional

[3] Recall that this is how we get the double cover of SO(3) by the 3-sphere group described by equation 13.16.

representations of $\mathbf{S}^3$, we may use them to expand any character of $\mathbf{S}^3$ as a Fourier sum. Indeed, let χ be the character of an arbitrary finite-dimensional representation of $\mathbf{S}^3$. Observe first that expression 17.6 implies that the (group) conjugacy classes $C(\mathsf{s})$ are invariant under quaternionic conjugation $\mathsf{s} \mapsto \bar{\mathsf{s}}$, in that $C(\mathsf{s}) = C(\bar{\mathsf{s}})$. This follows from $\bar{\mathsf{s}} = e^{-\hat{\mathsf{s}}\vartheta} \in C_\vartheta$, where $\vartheta = \arg \mathsf{s}$. This implies that $c(\vartheta) = c(-\vartheta)$, where $c(\vartheta)$ is the value of the character χ on the conjugacy class C_ϑ. Thus, $c(\vartheta) \sin \vartheta$ is an odd function of ϑ and may be expanded in a Fourier sine series:

$$c(\vartheta) \sin \vartheta = \sum_{m=1}^{\infty} \mu_m \sin m\vartheta, \quad \mu_m = \frac{2}{\pi} \int_0^\pi c(\vartheta) \sin \vartheta \sin m\vartheta \, d\vartheta,$$

or, equivalently, with $\lambda_j = \mu_{2j+1}$,

$$c(\vartheta) = \sum_{2j+1 \in \mathbb{N}} \lambda_j c^j(\vartheta), \quad \lambda_j = \frac{2}{\pi} \int_0^\pi c(\vartheta) c^j(\vartheta) \sin^2 \vartheta \, d\vartheta.$$

We now turn to the task of rewriting the integral equation 16.25 in terms of the character values of $\mathbf{S}^3$. To get a hint of how to do this, we are going to engage in a bit of mathematical tomfoolery. It starts out rigorously enough. If ϑ is the angle from the origin between the points s and t of $\mathbf{S}^3$, then $|\mathsf{s} - \mathsf{t}| = 2\sin(\vartheta/2) = \sqrt{2(1 - \cos \vartheta)}$, and writing $\cos \vartheta$ in terms of complex exponentials yields

$$\begin{aligned} \frac{1}{|\mathsf{s} - \mathsf{t}|^2} &= \frac{1}{2 - e^{\mathrm{i}\vartheta} - e^{-\mathrm{i}\vartheta}} = \frac{1}{(1 - e^{\mathrm{i}\vartheta})(1 - e^{-\mathrm{i}\vartheta})} \\ &= \frac{1}{e^{\mathrm{i}\vartheta} - e^{-\mathrm{i}\vartheta}} \left(\frac{1}{1 - e^{\mathrm{i}\vartheta}} - \frac{1}{1 - e^{-\mathrm{i}\vartheta}} \right) \\ &\sim \frac{1}{e^{\mathrm{i}\vartheta} - e^{-\mathrm{i}\vartheta}} \sum_{n=0}^{\infty} \left(e^{\mathrm{i}n\vartheta} - e^{-\mathrm{i}n\vartheta} \right) \\ &= \sum_{n=1}^{\infty} \frac{\sin n\vartheta}{\sin \vartheta} = \sum_{2j+1 \in \mathbb{N}} c^j(\vartheta). \end{aligned} \tag{17.11}$$

The tomfoolery comes in at the symbol $\sim$, for it is there that we have used the formula $(1 - z)^{-1} = \sum_{n=0}^{\infty} z^n$ for a convergent geometric series. This is valid only when $|z| < 1$, of course, and we have applied it to the two values $z = e^{\pm \mathrm{i}\vartheta}$ for which the series diverge. The divergence, however, is quite tame in that the partial sums remain bounded. In fact, for any unit complex number $e^{\mathrm{i}\vartheta} = u \neq 1$, the partial sums $s_n = \sum_{k=0}^{n} u^k$ are the vertices of a polygon with side lengths unity and turning angle ϑ, which is a finite polygon when ϑ is a rational multiple of π and an infinite (generalized) polygon when ϑ is an irrational multiple of π. Though $\lim_{n\to\infty} s_n$ fails to exist, an easy calculation shows that the limit of the averages of the partial sums does exist, in that

$$\lim_{n\to\infty} \frac{\sum_{k=0}^{n} s_k}{n+1} = \frac{1}{1-u}. \tag{17.12}$$

The value $1/(1-u)$ is precisely the geometric center, or center of mass, of the polygon whose vertices are given by the sequence of partial sums s_n. The series $\sum u^k$ whose partial sum averages converge in the sense of expression 17.12 is said to be **Cesàro summable** to the value $1/(1-u)$.[4] Cesàro summability is linear in that if $\sum a_n$ and $\sum b_n$ are Cesàro summable to A and B, respectively, then the series $\sum(\lambda a_n + \mu b_n)$ is Cesàro summable to $\lambda A + \mu B$. It follows then that the symbol $\sim$ in expression 17.11 may be replaced by equality if the series sums in the expression, all of which are divergent, are interpreted as Cesàro sums. In particular, we have the rigorous formula

$$\frac{1}{|\mathsf{s}-\mathsf{t}|^2} = \overline{\sum_{2j+1\in\mathbb{N}}} c^j(\vartheta) = \overline{\sum_{2j+1\in\mathbb{N}}} \chi_{R^j}(\bar{\mathsf{s}}\mathsf{t}), \tag{17.13}$$

where $\overline{\sum}$ represents the Cesàro sum of the series. The equality $c^j(\vartheta) = \chi_{R^j}(\bar{\mathsf{s}}\mathsf{t})$ follows, since the angle, ϑ, between s and t is the same as that between 1 and $\bar{\mathsf{s}}\mathsf{t}$, which in turn follows since multiplication by $\bar{\mathsf{s}}$ is an isometry and hence preserves angles.

17.3 The Momentum Space Equation and Characters

Let V be an n-dimensional vector space over the complex numbers and $\mathrm{End}(V)$ the space of all linear mappings of V to V. In this discussion we will need to work with expressions of the form

$$L = \int_{\mathbf{S}^3} F(\mathsf{t})\, d\Omega, \tag{17.14}$$

where $F : \mathbf{S}^3 \to \mathrm{End}(V)$. The integral in this expression may be defined, rather straightforwardly if inelegantly, by using a basis in V to identify

[4] Every convergent series is Cesàro summable to the same value to which the series converges, so Cesàro summability generalizes the convergence of series. An even more general summability method is **Abel summability**. A series $\sum a_n$ is Abel summable to A if $\lim_{z\to 1-} \sum_{n=0}^{\infty} a_n z^n = A$. Any series Cesàro summable to A is Abel summable to A. In the context of geometric series, it is especially easy to see that $\lim_{z\to 1-} \sum_{n=0}^{\infty} u^n z^n = 1/(1-u)$ whenever $u \neq 1$ lies on the unit circle.

End(V) with the space of $n \times n$ complex matrices $\mathrm{M}_n(\mathbb{C})$ in the usual way. Then $\mathrm{M}_n(\mathbb{C})$ in turn may be identified as the Euclidean space $\mathbb{C}^{n^2}$ and the integral is then to be understood component-wise.

We make four observations that will be useful in our solution to the momentum space equation. First, expression 17.14 defines L as a linear mapping of V to V. Indeed, just observe that the componentwise integrals provide the matrix entries of L. Second, if A is any linear mapping of V to V then A commutes with integration in the sense that $A \circ L = \int_{\mathbf{S}^3} A \circ [F(\mathrm{t})]\, d\Omega$. This is an easy exercise in writing out the matrix entries of these linear maps. Third, the trace functional Tr also commutes with integration, an even easier exercise in writing out matrix entries. Fourth, for $B, C \in \mathrm{End}(V)$, if $\mathsf{Tr}(AB) = \mathsf{Tr}(AC)$ for all $A \in \mathrm{End}(V)$ then $B = C$. Indeed, working in the identification of End(V) with $\mathrm{M}_n(\mathbb{C})$, when A is the transformation represented by the matrix with 1 as the ji entry and 0 elsewhere, we get $\mathsf{Tr}(AB) = B_{ij}$. The trace equality now shows that $B_{ij} = C_{ij}$ for all i, j and thus confirms that $B = C$.

We will need one other fact for our derivation of a solution set for the momentum space equation: that for all nonnegative half-integers j,

$$\int_{\mathbf{S}^3} \frac{\chi_{R^j}(\mathrm{t})}{|1-\mathrm{t}|^2}\, d\Omega = 2\pi^2 = \mathrm{Vol}(\mathbf{S}^3). \tag{17.15}$$

To evaluate the integral, note first that the integrand is dependent only on the angle $\vartheta = \arg \mathrm{t}$ between the quaternion unit 1 and t. Indeed, we have seen already that $\chi_{R^j}(\mathrm{t}) = c^j(\arg \mathrm{t}) = c^j(\vartheta)$ and that $|1-\mathrm{t}|^2 = 2(1-\cos\vartheta)$. Recall from Section 14.3 that standard spherical coordinates on $\mathbf{S}^3$ are given as (ϑ, Φ), where $\Phi = (\theta, \phi)$ are the standard spherical coordinates on $\mathbf{S}^2$ that arise from the standard spherical coordinates (r, θ, ϕ) on $\mathbb{R}^3$. The volume element on $\mathbf{S}^3$ is $d\Omega = \sin^2\vartheta \sin\theta\, d\phi\, d\theta\, d\vartheta$. Integrating the ϕ and θ variables and using the fact that the integrand is independent of these variables gives

$$\int_{\mathbf{S}^3} \frac{\chi_{R^j}(\mathrm{t})}{|1-\mathrm{t}|^2}\, d\Omega = 4\pi \int_0^\pi \frac{c^j(\vartheta)}{2(1-\cos\vartheta)} \sin^2\vartheta\, d\vartheta.[5]$$

If we ignore the question of the validity of interchanging integration and

[5] For those who prefer a geometric derivation to calculations and coordinates, the subset of $\mathbf{S}^3$ of all t with constant argument ϑ is precisely the conjugacy class C_ϑ identified in equation 17.6, and it forms a round Euclidean two-dimensional sphere of radius $\sin\vartheta$ with area $4\pi \sin^2\vartheta$. Fixing ϑ and integrating over the conjugacy class C_ϑ then reduces the integral to that appearing on the right-hand side.

Cesàro summation, we may use equation 17.13 and the orthogonality relations 17.9 to get

$$\int_0^\pi \frac{c^j(\vartheta)}{2(1-\cos\vartheta)} \sin^2\vartheta \, d\vartheta = \sum_{2k+1\in\mathbb{N}} \int_0^\pi c^j(\vartheta) c^k(\vartheta) \sin^2\vartheta \, d\vartheta = \frac{\pi}{2},$$

finishing the confirmation of equation 17.15. Now, since the validity of interchanging integration and summation is unconfirmed, we will offer an alternative direct verification that

$$I(2j+1) = \int_0^\pi \frac{c^j(\vartheta)}{(1-\cos\vartheta)} \sin^2\vartheta \, d\vartheta = \pi$$

for any nonnegative half-integer j. Indeed, this follows from the recursion formula

$$I(n+1) = 2I(n) - I(n-1) \quad \text{for integers} \quad n > 1,$$

and the fact $I(1) = I(2) = \pi$, a fact verified by direct calculation. As for the recursion formula, use formula 17.7 to rewrite I as

$$I(n) = \int_0^\pi \frac{\sin n\vartheta}{\sin\vartheta}(1+\cos\vartheta)\, d\vartheta$$

and apply the sine addition formula to write

$$\sin(n+1)\vartheta + \sin(n-1)\vartheta = 2\sin n\vartheta \cos\vartheta.$$

Together these give

$$\begin{aligned} I(n+1) + I(n-1) &= 2\int_0^\pi \frac{\sin n\vartheta}{\sin\vartheta} \cos\vartheta (1+\cos\vartheta)\, d\vartheta \\ &= 2\int_0^\pi \frac{\sin n\vartheta}{\sin\vartheta}(1+\cos\vartheta - \sin^2\vartheta)\, d\vartheta \\ &= 2I(n) - 2\int_0^\pi \sin n\vartheta \sin\vartheta \, d\vartheta. \end{aligned}$$

The recursion formula now follows from equation 17.10, which shows that the functions $\sin n\vartheta$ and $\sin\vartheta$ are orthogonal on the interval $[0, \pi/2]$ whenever $n > 1$.

Armed with the four observations above and equation 17.15, we are now in a position to solve the integral equation 16.25. The result, which we will confirm in the remainder of this section, is that for *any* linear transformation $A \in \mathrm{End}(V^j)$ of the irreducible $(2j+1)$-dimensional representation space V^j of $\mathbf{S}^3$, the function $Y = Y_A^j : \mathbf{S}^3 \to \mathbb{C}$ defined as

$$Y(\mathsf{s}) = \mathsf{Tr}(A \circ [R^j(\mathsf{s})])$$

satisfies the integral equation

$$Y(\mathsf{s}) = \frac{2j+1}{2\pi^2} \int_{\mathbf{S}^3} \frac{Y(\mathsf{t})}{|\mathsf{s}-\mathsf{t}|^2}\, d\Omega. \tag{17.16}$$

Comparing with equation 16.25, we see that Y_A^j is a solution to the momentum space equation provided that $\lambda = 2j+1$. As j is a nonnegative half-integer, $\ell = 2j$ is a nonnegative integer and so we obtain yet another derivation of the energy levels of equations 17.5 and 15.7. Note that the functions Y_A^j give a ready supply of solutions to the momentum space Schrödinger equation as A ranges over the $(2j+1)^2$-dimensional space $\operatorname{End}(V^j)$. With $n = \ell + 1 = 2j+1$, we see again that the E_n-eigenspace is of dimension n^2, the dimension of the space $\operatorname{End}(V^j)$.

It remains only to confirm that the function $Y = Y_A^j$ satisfies equation 17.16. Since conjugation by the unit quaternion s is an isometry of $\mathbf{S}^3$ and the volume element $d\Omega$ is invariant under isometries, and since $|1 - \mathsf{s}\mathsf{t}\bar{\mathsf{s}}| = |\mathsf{s}||1-\mathsf{t}||\bar{\mathsf{s}}| = |1-\mathsf{t}|$, we have

$$\int_{\mathbf{S}^3} \frac{Y(\mathsf{t})}{|1-\mathsf{t}|^2}\, d\Omega = \int_{\mathbf{S}^3} \frac{Y(\mathsf{s}\mathsf{t}\bar{\mathsf{s}})}{|1-\mathsf{t}|^2}\, d\Omega. \tag{17.17}$$

The first of the observations made at the start of this section may be applied to the function $F(\mathsf{t}) = R^j(\mathsf{t})/|1-\mathsf{t}|^2$, to see that the expression

$$B^j = \int_{\mathbf{S}^3} \frac{R^j(\mathsf{t})}{|1-\mathsf{t}|^2}\, d\Omega$$

defines a linear mapping of V^j. Using the second and third of the observations, that integration commutes with both linear mappings and trace, along with the linearity of trace and of the mapping A, equation 17.17 may be written as

$$\operatorname{Tr}\left(A \circ \int_{\mathbf{S}^3} \frac{R^j(\mathsf{t})}{|1-\mathsf{t}|^2}\, d\Omega\right) = \operatorname{Tr}\left(A \circ \int_{\mathbf{S}^3} \frac{R^j(\mathsf{s}\mathsf{t}\bar{\mathsf{s}})}{|1-\mathsf{t}|^2}\, d\Omega\right).$$

Now, since $A \in \operatorname{End}(V^j)$ is quite arbitrary, the fourth observation along with another application of the second, as well as the fact that R^j is a representation, implies that

$$B^j = R^j(\mathsf{s}) \circ B^j \circ R^j(\mathsf{s})^{-1},$$

and this holds for arbitrary $\mathsf{s} \in \mathbf{S}^3$. This merely states the commutativity of the diagrams of linear mappings given as

$$\begin{array}{ccc} V^j & \xrightarrow{R^j(\mathsf{s})} & V^j \\ {\scriptstyle B^j}\big\downarrow & & \big\downarrow{\scriptstyle B^j} \\ V^j & \xrightarrow[R^j(\mathsf{s})]{} & V^j \end{array}, \quad \text{for each} \quad \mathsf{s} \in \mathbf{S}^3.$$

This in turn just states that the linear map B^j is a morphism of irreducible representations, and Schur's lemma then implies that it is a multiple of the identity.[6] It follows that there is a complex constant μ such that $\mu 1_{V^j} = B^j$. Taking the trace of this equation, commuting the trace with integration, and then applying equation 17.15 gives

$$\mu(2j+1) = \mathsf{Tr}(B^j) = \int_{\mathbf{S}^3} \frac{\chi_{R^j}(\mathsf{t})}{|1-\mathsf{t}|^2}\, d\Omega = 2\pi^2.$$

This identifies the constant μ as $\mu = 2\pi^2/(2j+1)$ and gives

$$1_{V^j} = \frac{2j+1}{2\pi^2} \int_{\mathbf{S}^3} \frac{R^j(\mathsf{t})}{|1-\mathsf{t}|^2}\, d\Omega.$$

Composing both sides of this equation with the map $A \circ [R^j(\mathsf{s})]$ on the left, taking the trace, and commuting with integration then gives

$$\begin{aligned} Y(\mathsf{s}) = Y^j_A(\mathsf{s}) = \mathsf{Tr}(A \circ [R^j(\mathsf{s})]) &= \frac{2j+1}{2\pi^2} \mathsf{Tr}\left(A \circ [R^j(\mathsf{s})] \int_{\mathbf{S}^3} \frac{R^j(\mathsf{t})}{|1-\mathsf{t}|^2}\, d\Omega \right) \\ &= \frac{2j+1}{2\pi^2} \int_{\mathbf{S}^3} \frac{\mathsf{Tr}(A \circ [R^j(\mathsf{st})])}{|1-\mathsf{t}|^2}\, d\Omega. \end{aligned}$$

Finally, performing a change of variable by replacing t by $\bar{\mathsf{s}}\mathsf{t}$ and using again that multiplication by $\bar{\mathsf{s}}$ is an isometry, that the volume element $d\Omega$ is invariant under isometries, and that $|1-\bar{\mathsf{s}}\mathsf{t}| = |\mathsf{s}-\mathsf{t}|$, we arrive at equation 17.16.

17.4 The Infinitesimal Generators of SO(4) ⋆

Recall that we uncovered the first hint of SO(4) symmetry in the Kepler problem from Pauli's quantization of the LRL triad in Section 16.2 and

[6] Schur's lemma, that a morphism of a finite-dimensional irreducible representation R to itself is always a multiple of the identity, is left as an exercise. This is done in three steps. First, prove that for any morphism A of an arbitrary representation R to itself, $\ker A$ and $\operatorname{im} A$ are R-invariant. Second, deduce that, when R is irreducible, either $A = 0$ or A is invertible. Finally, let μ be an eigenvalue of A and apply step two to the morphism $A - \mu 1$.

the introduction of the commuting triads $\mathbf{J}_n^{\pm}$ that act as angular momentum operators on the E_n-eigenspace $\mathcal{H}_n$ of the hydrogenic system. At the level of Lie algebras, the triads $\mathbf{J}_n^{\pm}$ generate a representation of $\mathfrak{su}(2) \oplus \mathfrak{su}(2)$, and this suggests a corresponding Lie group symmetry of $\mathrm{SU}(2) \times \mathrm{SU}(2)$ or its factor group SO(4). This was articulated more fully in Section 16.3. In Section 16.4, by transforming to the momentum representation and stereographically projecting, we explicitly exhibited the SO(4) symmetry of the solutions to the momentum space Schrödinger equation 16.25. In the present lecture, we have solved the momentum space equation by two methods, the first of which demonstrates SO(4) symmetry by realizing the solutions as spherical harmonics on the 3-sphere. Note that the $\mathfrak{su}(2) \oplus \mathfrak{su}(2)$ Lie algebra symmetry is a symmetry in position space and the SO(4) symmetry is a symmetry in the stereographically projected momentum space, but of course these are expected to be related. In this section, we make explicit this relationship.

We saw in the first section of the present lecture that $V^{3,\ell}$ serves as the E_n-eigenspace of the stereographically projected momentum space Schrödinger equation 16.25, where $n = \ell + 1$. Recall that $V^{3,\ell}$ is the space of spherical harmonics of degree ℓ on the 3-sphere, the collection of the restrictions to $\mathbf{S}^3$ of the homogeneous harmonic polynomials on $\mathbb{R}^4$ of degree ℓ. Now, the group SO(4) acts linearly and irreducibly on $V^{3,\ell}$ via

$$((\mathsf{p}, \mathsf{q}) \cdot Y)(\mathsf{t}) = Y(L_{\mathsf{p}} R_{\overline{\mathsf{q}}}(\mathsf{t})) = Y(\mathsf{p}\mathsf{t}\overline{\mathsf{q}}), \tag{17.18}$$

for each $Y \in V^{3,\ell}$. Here, each element of the group SO(4) is represented by a pair of elements $\pm(\mathsf{p}, \mathsf{q}) \in \mathbf{S}^3 \times \mathbf{S}^3$ that act on $\mathbf{S}^3$ via left and right multiplication, $L_{\mathsf{p}} R_{\overline{\mathsf{q}}} = L_{-\mathsf{p}} R_{-\overline{\mathsf{q}}}$. We next translate this action to an action on the E_n-eigenspace $\mathcal{H}_n$ of the Schrödinger equation in the position representation. This is accomplished by applying stereographic projection and the inverse Fourier transform in the following manner. Referring to Section 16.4, and especially to equations 16.20–16.23, a spherical harmonic $Y \in V^{3,n-1}$ gives an E_n-eigensolution ψ of the position space Schrödinger equation that may be expressed as an inverse Fourier transform on the 3-sphere via stereographic projection:

$$\psi(\mathbf{x}) = \frac{p_0^3}{4(2\pi\hbar)^{3/2}} \int_{\mathbf{S}^3} e^{\mathrm{i} p_0 \mathbf{x} \cdot \mathbf{s}/(1-s_0)\hbar} \frac{Y(\mathsf{s})}{1 - s_0}\, d\Omega. \tag{17.19}$$

Here, $\mathsf{s} = \mathrm{Re}\,\mathsf{s} + \mathrm{Pu}\,\mathsf{s} = s_0 + \mathbf{s}$ is a quaternionic variable on $\mathbf{S}^3$ and $p_0 = \sqrt{-2mE_n}$. In this derivation, we used the facts that $|\mathsf{s}|^2 = |s_0| + \|\mathbf{s}\|^2 = 1$

and that, for stereographic projection $\mathfrak{p}$, if $\mathsf{s} = \mathfrak{p}(\mathbf{u})$ then

$$1 + \|\mathbf{u}\|^2 = \frac{2}{1 - s_0}.$$

The reader who verifies all the the remaining calculations of this section will need to use these two facts on several occasions.

The representation of equation 17.18 induces the linear action

$$\begin{aligned}((\mathsf{p},\mathsf{q})\cdot\psi)(\mathbf{x}) &= \frac{p_0^3}{4(2\pi\hbar)^{3/2}}\int_{\mathbf{S}^3} e^{\mathbf{i}p_0\mathbf{x}\cdot\mathsf{s}/(1-s_0)\hbar}\frac{Y(\mathsf{p}\mathsf{s}\overline{\mathsf{q}})}{1-s_0}\,d\Omega \\ &= \frac{p_0^3}{4(2\pi\hbar)^{3/2}}\int_{\mathbf{S}^3} e^{\mathbf{i}p_0\mathbf{x}\cdot\mathrm{Pu}(\overline{\mathsf{p}}\mathsf{s}\mathsf{q})/(1-\mathrm{Re}(\overline{\mathsf{p}}\mathsf{s}\mathsf{q}))\hbar}\frac{Y(\mathsf{s})}{1-\mathrm{Re}(\overline{\mathsf{p}}\mathsf{s}\mathsf{q})}\,d\Omega\end{aligned}$$

on the E_n-eigenspace $\mathcal{H}_n$. Our goal is to calculate the infinitesimal generators of this SO(4) action and to verify in the process of doing this that they generate the $\mathfrak{su}(2)\oplus\mathfrak{su}(2)$ representation that gives rise to the pair of angular momentum operators $\mathsf{J}^\pm = \mathsf{J}_n^\pm$ of Sections 16.2 and 16.3. Actually, we will give the detailed calculation for just one of the components of these angular momentum triads by demonstrating that the operator J_1^+ operating on the E_n-eigenspace $\mathcal{H}_n$ generates the part of the SO(4) action arising from the one-parameter subgroup $\alpha : \mathbb{R} \to \mathbf{S}^3 \times \mathbf{S}^3$ given by $\alpha(t) = (e^{-\mathbf{i}t/2}, 1)$. The remaining components generate similarly, J_2^+ generating the action of the subgroup $(e^{-\mathbf{j}t/2}, 1)$, J_3^+ that of the subgroup $(e^{-\mathbf{k}t/2}, 1)$, and the components of J^- generating the actions of the subgroups $(1, e^{-\mathbf{u}t/2})$ for $\mathbf{u} = \mathbf{i}, \mathbf{j}, \mathbf{k}$.

To simplify the calculation, write equation 17.19 as

$$\psi(\mathbf{x}) = C\int_{\mathbf{S}^3} F(\mathbf{x},\mathsf{s})Y(\mathsf{s})\,d\Omega, \tag{17.20}$$

where $C = p_0^3/4(2\pi\hbar)^{3/2}$ and

$$F(\mathbf{x},\mathsf{s}) = \frac{\exp\left[\mathbf{i}p_0\mathbf{x}\cdot\mathrm{Pu}(\mathsf{s})/(1-\mathrm{Re}(\mathsf{s}))\hbar\right]}{1-\mathrm{Re}(\mathsf{s})}.$$

To verify that J_1^+ generates the one-parameter subgroup α, it suffices to show that

$$\mathsf{J}_1^+\psi = \mathbf{i}\hbar\frac{d}{dt}(\alpha(t)\cdot\psi)\,|_{t=0}\,.^7 \tag{17.21}$$

[7] The reader unfamiliar with this should consult Stone's theorem in Section 27.2, the discussion in Lecture 27 on quantum dynamics, and, especially, Lecture 28 on symmetries in quantum mechanics.

First note that

$$
\begin{aligned}
(\alpha(t)\cdot\psi)(\mathbf{x}) &= ((e^{-\mathrm{i}t/2},1)\cdot\psi)(\mathbf{x})\\
&= C\int_{\mathbf{S}^3} F(\mathbf{x},\mathsf{s})Y(e^{-\mathrm{i}t/2}\mathsf{s})\,d\Omega = C\int_{\mathbf{S}^3} F(\mathbf{x},e^{\mathrm{i}t/2}\mathsf{s})Y(\mathsf{s})\,d\Omega.
\end{aligned}
\tag{17.22}
$$

With $\mathsf{s} = s_0 + s_1\mathbf{i} + s_2\mathbf{j} + s_3\mathbf{k}$, we have

$$
\frac{d}{dt}2e^{\mathrm{i}t/2}\mathsf{s}\,|_{t=0} = \mathbf{i}e^{\mathrm{i}t/2}\mathsf{s}\,|_{t=0} = \mathbf{i}\mathsf{s} = -s_1 + s_0\mathbf{i} - s_3\mathbf{j} + s_2\mathbf{k},
$$

from which we get

$$
\frac{d}{dt}\mathrm{Re}(e^{\mathrm{i}t/2}\mathsf{s})\,|_{t=0} = -\frac{s_1}{2} \quad\text{and}\quad \frac{d}{dt}\mathrm{Pu}(e^{\mathrm{i}t/2}\mathsf{s})\,|_{t=0} = \frac{1}{2}\left(s_0\mathbf{i} - s_3\mathbf{j} + s_2\mathbf{k}\right).
$$

Evaluating the right-hand side of equation 17.21 amounts to moving the operator d/dt under the integral sign of equation 17.22 to calculate

$$
\frac{\partial}{\partial t}F(\mathbf{x},\exp\mathrm{i}t/2\mathsf{s}) = \frac{\partial}{\partial t}\frac{\exp\left[\mathbf{i}p_0\mathbf{x}\cdot\mathrm{Pu}(e^{\mathrm{i}t/2}\mathsf{s})/(1-\mathrm{Re}(e^{\mathrm{i}t/2}\mathsf{s}))\hbar\right]}{(1-\mathrm{Re}(e^{\mathrm{i}t/2}\mathsf{s}))}.
$$

Performing this differentiation, evaluating the result at $t = 0$, and then performing some straightforward algebraic manipulations give

$$
\mathbf{i}\hbar\frac{\partial}{\partial t}F(\mathbf{x},e^{\mathrm{i}t/2}\mathsf{s})\,|_{t=0} = \frac{p_0}{1-s_0}F(\mathbf{x},\mathsf{s})G(\mathbf{x},\mathsf{s}),
\tag{17.23}
$$

where G is given by the expression

$$
G(\mathbf{x},\mathsf{s}) = \frac{1}{2}\left(-x_1s_0 + x_2s_3 - x_3s_2 + \frac{\mathbf{x}\cdot\mathbf{s}}{1-s_0}s_1 - \mathbf{i}\frac{\hbar}{p_0}s_1\right).
\tag{17.24}
$$

The right-hand side of equation 17.21 becomes

$$
\mathbf{i}\hbar\frac{d}{dt}(\alpha(t)\cdot\psi)(\mathbf{x})\,|_{t=0} = C\int_{\mathbf{S}^3}\frac{p_0}{1-s_0}F(\mathbf{x},\mathsf{s})G(\mathbf{x},\mathsf{s})Y(\mathsf{s})\,d\Omega.
\tag{17.25}
$$

Focusing attention now on the left-hand side of equation 17.21, we use equation 16.12 defining the triads $\mathsf{J}^{\pm}$ and the alternative form of the quantized LRL triad of equation 16.5 to obtain the formula

$$
\begin{aligned}
\mathsf{J}_1^+ &= \frac{1}{2}\left(\mathsf{L}_1 + \frac{1}{p_0}\mathsf{A}_1\right)\\
&= \frac{1}{2}\left(\mathsf{X}_2\mathsf{P}_3 - \mathsf{X}_3\mathsf{P}_2\right)\\
&\quad + \frac{1}{2p_0}\left[\mathsf{P}_1(\mathsf{X}_2\mathsf{P}_2 + \mathsf{X}_3\mathsf{P}_3) - \mathsf{X}_1(\mathsf{P}_2^2 + \mathsf{P}_3^2) + \frac{mZe^2}{x}\mathsf{X}_1 - \mathbf{i}\hbar\mathsf{P}_1\right].
\end{aligned}
\tag{17.26}
$$

Applying this to ψ on the left-hand side of equation 17.21 amounts to moving the differential operator J_1^+ under the integral sign of equation 17.20 to calculate

$$\mathsf{J}_1^+ F(\mathbf{x}, \mathsf{s}) = \frac{1}{1-s_0}\mathsf{J}_1^+\left(e^{\mathbf{i}p_0\mathbf{x}\cdot\mathsf{s}/(1-s_0)\hbar}\right).$$

After simplification this gives

$$\mathsf{J}_1^+ F(\mathbf{x}, \mathsf{s}) = \frac{p_0}{1-s_0}F(\mathbf{x}, \mathsf{s})\left[G(\mathbf{x}, \mathsf{s}) + (1-s_0)\frac{mZe^2}{2p_0^2}\frac{x_1}{x} - \frac{x_1}{2}\right].$$

From this and expression 17.25, we get

$$\mathsf{J}_1^+\psi(\mathbf{x}) = \mathbf{i}\hbar\frac{d}{dt}(\alpha(t)\cdot\psi)(\mathbf{x})\mid_{t=0} + \frac{mx_1}{2p_0}R(\mathbf{x}),$$

where $R(\mathbf{x})$ is given by the expression

$$\begin{aligned}\frac{Ze^2}{x}C\int_{\mathbf{S}^3}F(\mathbf{x}, \mathsf{s})Y(\mathsf{s})\,d\Omega - \frac{p_0^2}{m}C\int_{\mathbf{S}^3}F(\mathbf{x}, \mathsf{s})Y(\mathsf{s})\frac{1}{1-s_0}\,d\Omega \\ = \frac{Ze^2}{x}\psi(\mathbf{x}) - \frac{p_0^2}{m}C\int_{\mathbf{S}^3}F(\mathbf{x}, \mathsf{s})Y(\mathsf{s})\frac{1}{1-s_0}\,d\Omega. \qquad (17.27)\end{aligned}$$

To accomplish our desired result, i.e., verifying that equation 17.21 holds, we need to demonstrate that the expression $R(\mathbf{x})$ is the zero function in $\mathcal{H} = L^2(\mathbb{R}^3)$. For this, we concentrate on the integral expression in the second line of equation 17.27 and transform it to an integration on $\mathbb{R}^3$ via inverse stereographic projection. Thus, in a way that is very similar to the derivation of equation 17.19 from equations 16.20–16.23, we may write

$$\begin{aligned}\frac{p_0^2}{m}C\int_{\mathbf{S}^3}F(\mathbf{x}, \mathsf{s})Y(\mathsf{s})\frac{1}{1-s_0}\,d\Omega \\ = \frac{1}{2m(2\pi\hbar)^{3/2}}\int_{\mathbb{R}^3}e^{\mathbf{i}\mathbf{x}\cdot\mathbf{p}/\hbar}\widehat{\psi}(\mathbf{p})(p_0^2 + \|\mathbf{p}\|^2)\,d\mathbf{p}.\end{aligned}$$

This last expression is just the sum of the inverse Fourier transforms of $\widehat{\psi}(\mathbf{p})$ and of $\|\mathbf{p}\|^2\widehat{\psi}(\mathbf{p})$ and, using equation 10.6, becomes

$$\frac{p_0^2}{2m}\psi(\mathbf{x}) + \frac{1}{2m}\mathcal{F}^{-1}\left[(p_1^2+p_2^2+p_3^2)\widehat{\psi}(\mathbf{p})\right](\mathbf{x}) = -E_n\psi(\mathbf{x}) - \frac{\hbar^2}{2m}\Delta\psi(\mathbf{x}).^8$$

Referring back to formula 17.27, we have

$$R(\mathbf{x}) = \frac{Ze^2}{x}\psi(\mathbf{x}) + E_n\psi(\mathbf{x}) + \frac{\hbar^2}{2m}\Delta\psi(\mathbf{x}) = -\left(\mathsf{H} - E_n 1_{\mathcal{H}}\right)\psi(\mathbf{x}),$$

[8] Recall that the factor $\hbar^2$ appears because we are using the Fourier transform with the exponential term $\mathbf{x}\cdot\mathbf{p}$ divided by $\hbar$.

where H is the inverse-square Hamiltonian of equation 16.6. Our task is now complete since we have $\psi \in \mathcal{H}_n$, the E_n-eigenspace of H, so that $\mathsf{H}\psi = E_n\psi$, confirming that $R = 0$ and that equation 17.21 holds.

17.5 End Notes

Stephanie Singer (2005) has provided an English translation of the original Fock (1935) article in her wonderful book. The development of the solutions to the momentum space equation in Section 17.1 augments Fock's original argument with significant contributions from Bander and Itzykson (1966a, b). For more on the use of representation theory in the explanation of physical phenomena, the student can do no better than to consult the outstanding book *Group Theory and Physics* by Sternberg (1990) and, for more on the symmetries of the Kepler problem, both the classical and the quantum cases, see the enticing but formidable Guillemin and Sternberg (1990). I first learned of the use of character theory to solve the momentum space equation from Sections 9.9 and 9.10 of Hannabuss (1997), who worked with matrix variables in his development. My Section 17.3 may be thought of as an expansion of Hannabuss's development but reimagined in quaternionic coordinates. I worked out Section 17.4 primarily for my benefit, as I wanted to know the exact relationship between the $\mathfrak{su}(2) \oplus \mathfrak{su}(2)$ symmetry at the Lie algebra level and the SO(4) symmetry at the Lie group level.

18

Angular Momentum IV: Addition Rules and Spin

> Spinors were first used under that name, by physicists, in the field of Quantum Mechanics. In their most general mathematical form, spinors were discovered in 1913 by the author of this work, in his investigations on the linear representations of simple groups; they provide a linear representation of the group of rotations in space with any number of dimensions, each spinor having 2^ν components where $n = 2\nu+1$ or 2ν.
>
> Élie Cartan
> *The Theory of Spinors*, 1966

There are numerous quantum systems in which it is important to consider several independent, i.e., pairwise commuting, angular momentum triads. The first example that we saw were the triads $\mathbf{J}^{\pm}$ defined from the orbital angular momentum triad $\mathbf{L}$ and the LRL-vector $\mathbf{A}$ used in Section 16.2 in Pauli's solution to the hydrogen atom. For another example, two "independent" particles – two electrons for instance – may interact in the common electrostatic environment binding them to a central potential. Each has an orbital angular momentum described by an angular momentum operator that acts on the state space of the system in one of the full-integer representations of $\mathfrak{su}(2)$. Each electron has an intrinsic spin that also is described by an angular momentum operator, this time acting in a half-integer representation. In these cases one may measure in the laboratory individual angular momenta and spins of the constituent electrons as well as the total angular momentum of the system. One task of this lecture is to understand how individual angular momenta or spins interact to produce restrictions on total angular momentum measurements. The interaction is described mathematically by simple linear algebra and gives rise in practice to what physicists call

selection rules. This is covered in the first half of the present lecture. The second half is devoted to a second task, that of delving further into the mathematics of spin-$\frac{1}{2}$ systems. In particular, we recall the standard spin representation of Section 13.1 and derive important formulae that will be used in Lecture 22 to analyze the EPR paradox and Bell's theorem. We continue with an examination of how rotations of a spin-$\frac{1}{2}$ particle in $\mathbb{R}^3$ affect the state vector of the particle. This will be put to use in Lecture 19 in discussing the spinors of Pauli's theory of the electron.

Our four axioms for quantum mechanics presented in Lecture 2 are deficient in that they do not tell us how to take account of the number of particles in a quantum system. Technically, the instantiation of those axioms we presented in that lecture as wave mechanics covers only one-particle systems, and we must look for an extension of the axioms to tell us how to treat multi-particle quantum systems. More generally, the axioms do not tell us anything about the structure of the state space of a quantum system that can be decomposed into a finite number of "independent" subsystems. The proper domain for describing a system of two or more particles or subsystems in quantum mechanics is that of a Hilbert space that decomposes as a tensor product of subspaces, each of which individually serves as a state space for one of the constituent particles or subsystems. Details will be presented in Lecture 21, but the interaction of two independent angular momentum operators can be presented for the most part without explicit use of the tensorial structure of the state space of the system. All it requires is an examination of the linear algebra of two commuting triads. The one gap in our discussion does arise from this tensorial structure and will await Lecture 21 for its resolution, where the crucial ingredient will be the use of tensor products to construct representations of $\mathfrak{su}(2)$.

18.1 Coupled Angular Momenta

Let $\mathbf{K}$ and $\mathbf{L}$ be angular momentum triads on the state space $\mathcal{H}$ of a quantum system for which each component operator K_i commutes with each component operator L_j. The commutator relations governing these six operators are $[\mathsf{K}_i, \mathsf{K}_j] = \mathbf{i}\hbar\epsilon_{ijk}\mathsf{K}_k$ and $[\mathsf{L}_i, \mathsf{L}_j] = \mathbf{i}\hbar\epsilon_{ijk}\mathsf{L}_k$ and the expressions of commutativity given by $[\mathsf{K}_i, \mathsf{L}_j] = 0$ for all values i, j, and k. All the usual machinery described in Lectures 12 and 13 is at work for both triads. Explicitly, the eigenvalues of $\mathbf{K}^2$ have the form

$k(k+1)\hbar^2$ for nonnegative half-integers k and, for each value of k, the eigenvalues of K_3 associated with the eigenstates of $\mathbf{K}^2$ with quantum number k are $p\hbar$, the magnetic quantum number p taking values from the list $-k, -k+1, \ldots, k-1, k$. We use ℓ for the quantum numbers indexing the eigenvalues of $\mathbf{L}^2$ and q for those of L_3. Since the component operators of $\mathbf{K}$ and $\mathbf{L}$ commute, the four operators $\mathbf{K}^2$, K_3, $\mathbf{L}^2$, and L_3 pairwise commute and we expect simultaneous eigenvectors for all four operators. Using the Dirac ket-notation, for nonnegative half-integers k and ℓ and appropriate p and q, $|k,p,\ell,q\rangle$ represents a normalized state with

$$\begin{aligned}
\mathbf{K}^2|k,p,\ell,q\rangle &= k(k+1)\hbar^2|k,p,\ell,q\rangle,\\
\mathbf{L}^2|k,p,\ell,q\rangle &= \ell(\ell+1)\hbar^2|k,p,\ell,q\rangle,\\
\mathsf{K}_3|k,p,\ell,q\rangle &= p\hbar|k,p,\ell,q\rangle,\\
\mathsf{L}_3|k,p,\ell,q\rangle &= q\hbar|k,p,\ell,q\rangle,
\end{aligned}$$

and is said to represent a system with total angular momentum quantum numbers k and ℓ and associated magnetic quantum numbers p and q.

The two triads $\mathbf{K}$ and $\mathbf{L}$ may be combined as a sum to produce a third triad $\mathbf{J} = \mathbf{K} + \mathbf{L}$ with components $\mathsf{J}_i = \mathsf{K}_i + \mathsf{L}_i$. An easy exercise confirms that $\mathbf{J}$ is an angular momentum triad with $[\mathsf{J}_i, \mathsf{J}_j] = \mathbf{i}\hbar\epsilon_{ijk}\mathsf{J}_k$. The quantum numbers indexing the eigenvalues of $\mathbf{J}$ and J_3 will be denoted generically as j and m with $-j \le m \le j$. The component operators of $\mathbf{J}$ represent the components of the total angular momentum of the system, with individual angular momentum contributions represented by the triads $\mathbf{K}$ and $\mathbf{L}$. It is easy to see that the operators $\mathbf{J}^2$, J_3, $\mathbf{K}^2$, and $\mathbf{L}^2$ pairwise commute and again we expect simultaneous normalized eigenstates $|j,m,k,\ell\rangle$ with

$$\begin{aligned}
\mathbf{J}^2|j,m,k,\ell\rangle &= j(j+1)\hbar^2|j,m,k,\ell\rangle,\\
\mathsf{J}_3|j,m,k,\ell\rangle &= m\hbar|j,m,k,\ell\rangle,\\
\mathbf{K}^2|j,m,k,\ell\rangle &= k(k+1)\hbar^2|j,m,k,\ell\rangle,\\
\mathbf{L}^2|j,m,k,\ell\rangle &= \ell(\ell+1)\hbar^2|j,m,k,\ell\rangle.
\end{aligned}$$

Our task is to investigate relationships among the quantum numbers j, k, ℓ, m, p, and q, if any, and the restrictions these relationships impose on their allowed values. Note the ambiguity in notation in that $|-,-,-,-\rangle$ is used to represent simultaneous eigenstates of both the collections $\mathbf{K}^2$, K_3, $\mathbf{L}^2$, L_3 and $\mathbf{J}^2$, J_3, $\mathbf{K}^2$, $\mathbf{L}^2$. This should not cause confusion in that the labeling of the state by the generic symbols used for the quantum numbers of the respective operators will indicate which is which, with $|k, p, \ell, q\rangle$ and $|j, m, k, \ell\rangle$ simultaneous eigenstates of $\mathbf{K}^2, \mathsf{K}_3, \mathbf{L}^2, \mathsf{L}_3$ and $\mathbf{J}^2, \mathsf{J}_3, \mathbf{K}^2, \mathbf{L}^2$, respectively.

A typical question that arises is: suppose a system has quantum numbers k and ℓ. If a $\mathbf{J}^2$-measurement is made on the system, what are the possible outcomes, and with what probability does each occur? More specifically, how are the values of j restricted by the measured values of k and ℓ, and similarly for the magnetic quantum numbers p, q, and m?

18.2 The Selection Rules

Here are two basic **selection rules** that provide a partial answer to the questions posed.

Selection rule 1 (Addition of magnetic momenta): $m = p + q$.

This means that if $p\hbar$ is measured for K_3 and $q\hbar$ for L_3 then a subsequent J_3-measurement yields $(p+q)\hbar$ with certainty. Alternatively, if a J_3-measurement yields $m\hbar$ then subsequent K_3- and L_3-measurements yield $p\hbar$ and $q\hbar$ such that $m = p + q$.

Selection rule 2 (Fundamental addition theorem): $|k - \ell| \leq j \leq k + \ell$.

The first half of the first selection rule is easy to derive. When a system has quantum numbers p and q such that its state vector $|p, q\rangle$ is a simultaneous eigenstate of K_3 and L_3 with respective eigenvalues $p\hbar$ and $q\hbar$ then

$$\mathsf{J}_3|p, q\rangle = (\mathsf{K}_3 + \mathsf{L}_3)|p, q\rangle = p\hbar|p, q\rangle + q\hbar|p, q\rangle = (p + q)\hbar|p, q\rangle,$$

so that $|p, q\rangle$ is a $(p+q)\hbar$-eigenstate of J_3. Therefore, a J_3-measurement yields $(p+q)\hbar$ with certainty.

For the second half, suppose that a J_3-measurement yields $m\hbar$, so that the state $|m\rangle$ immediately after measurement satisfies $\mathsf{J}_3|m\rangle = m\hbar|m\rangle$. Since the operators $\mathbf{K}^2$, K_3, $\mathbf{L}^2$, and L_3 pairwise commute, the state space $\mathcal{H}$ decomposes as an orthogonal direct sum of the simultaneous eigenstates of the four operators. Any state, for example the state $|m\rangle$, may be represented as a (possibly infinite) sum of the form

$$|m\rangle = \sum a_{k,p,\ell,q}|k, p, \ell, q\rangle,$$

for some complex coefficients $a_{k,p,\ell,q}$ where k and ℓ range over some collection of nonnegative half-integers and $-k \leq p \leq k$ and $-\ell \leq q \leq \ell$. Applying $\mathsf{J}_3 = \mathsf{K}_3 + \mathsf{L}_3$ to both sides yields

$$m\hbar|m\rangle = \sum a_{k,p,\ell,q}(p+q)\hbar|k,p,\ell,q\rangle,$$

or

$$\sum(m-p-q)a_{k,p,\ell,q}|k,p,\ell,q\rangle = 0.$$

Since the eigenstates $|k,p,\ell,q\rangle$ are linearly independent, in fact pairwise orthogonal, the above sum is zero precisely when each coefficient $(m - p - q)a_{k,p,\ell,q}$ is zero. It follows that the only nonzero coefficients $a_{k,p,\ell,q}$ of the original expansion of $|m\rangle$ possible are those for which $m - p - q = 0$, and the state $|m\rangle$ may be written as

$$|m\rangle = \sum_{p+q=m} a_{k,p,\ell,q}|k,p,\ell,q\rangle.$$

This confirms that any subsequent K_3- and L_3-measurements on the system will yield $p\hbar$ and $q\hbar$ with $p + q = m$.

For the second selection rule, there are two inequalities to address. We consider the second now, which states that $j \leq k + \ell$. This inequality contains two implications. The first is that if a system is in a state with quantum numbers k and ℓ and a $\mathbf{J}^2$-measurement is performed then $j(j+1)\hbar^2$ is the measured value with $j \leq k + \ell$. The second is that if a system is in a state with quantum number j and subsequent $\mathbf{K}^2$- and $\mathbf{L}^2$-measurements are performed then the results are $k(k+1)\hbar^2$ and $\ell(\ell+1)\hbar^2$, respectively, with $j \leq k + \ell$. Both these implications follow from an examination of the structure that the operators $\mathbf{K}^2$, $\mathbf{L}^2$, and $\mathbf{J}^2$ reveal in the state space $\mathcal{H}$.

Suppose first that a system has quantum numbers k and ℓ, so that the state of the system lies in the subspace $\mathcal{H}(k,\ell)$ that consists of the simultaneous $k(k+1)\hbar^2$-eigenstates of $\mathbf{K}^2$ and $\ell(\ell+1)\hbar^2$-eigenstates of $\mathbf{L}^2$. Since each component operator J_i commutes with both $\mathbf{K}^2$ and $\mathbf{L}^2$, the restriction of each component operator to $\mathcal{H}(k,\ell)$ set-wise fixes $\mathcal{H}(k,\ell)$ and $\mathbf{J}$ acts as an angular momentum triad on $\mathcal{H}(k,\ell)$. Our task is to find the eigenvalues of the restriction of $\mathbf{J}^2$ to $\mathcal{H}(k,\ell)$.

Let j be realized as a quantum number of the restriction of $\mathbf{J}^2$ to $\mathcal{H}(k,\ell)$, so that the $j(j+1)\hbar^2$-eigenspace of $\mathcal{H}(k,\ell)$ decomposes as a direct sum of $2j+1$ subspaces, each an eigenspace of J_3 with magnetic quantum numbers m from the list $-j, -j+1, \ldots, j-1, j$. There exists, then, a state $|j,j\rangle$ in $\mathcal{H}(k,\ell)$ with $\mathbf{J}^2|j,j\rangle = j(j+1)\hbar^2|j,j\rangle$ and $\mathsf{J}_3|j,j\rangle = j\hbar|j,j\rangle$. The first selection rule then implies that $j = p + q$, where p is a magnetic quantum number determined by K_3 and q is a magnetic quantum number determined by L_3. As $|j,j\rangle$ is in $\mathcal{H}(k,\ell)$, this means in particular that $-k \le p \le k$ and $-\ell \le q \le \ell$, so that the maximum values of p and q are, respectively, k and ℓ. It follows that $j = p + q \le k + \ell$, confirming the second inequality of the second selection rule.

The first inequality of the second rule, that $|k - \ell| \le j$, will be derived in Lecture 21 after we have discussed two-particle state spaces and tensor products.

18.3 Spin-$\frac{1}{2}$ Systems

A **spin-$\frac{1}{2}$ system** is an angular momentum system with $j = \frac{1}{2}$ described by an angular momentum triad $\mathbf{S}$ arising from the spin representation of $\mathfrak{su}(2)$. Since the realization came that spin-$\frac{1}{2}$ systems can be used in quantum computation, it has become common to call any quantum system with two degrees of freedom that may be described by the mathematics of a spin-$\frac{1}{2}$ system a **qbit**, short for **quantum bit**. A qbit is, in some sense, the simplest possible quantum system; its pertinent properties for computation are described by the simple linear algebra exposed in this and the following section. Examples of qbits that might be of use in building a working quantum computer are electrons with their spin-$\frac{1}{2}$ states and polarized photons with their two states of helicity.[1]

Recalling the results of Section 13.1, we may use the irreducible representation space $V^{1/2}$ as the state space for a spin-$\frac{1}{2}$ particle; this state space is $(2j+1=2)$-dimensional with basis $\psi_\pm = |j,m\rangle = |\frac{1}{2}, \pm\frac{1}{2}\rangle$, as in Section 13.1. Recall that

$$\mathsf{S}_3\psi_\pm = \pm\tfrac{1}{2}\hbar\psi_\pm \quad \text{and} \quad \mathbf{S}^2\psi_\pm = \tfrac{3}{4}\hbar^2\psi_\pm. \tag{18.1}$$

In addition, the ladder operators behave as follows:

$$\mathsf{S}_+\psi_- = \hbar\psi_+, \quad \mathsf{S}_-\psi_+ = \hbar\psi_-, \quad \text{and} \quad \mathsf{S}_\pm\psi_\pm = 0.$$

We may decompose the state space as the orthogonal direct sum $V^{1/2} = \mathbb{C}\psi_+ \oplus \mathbb{C}\psi_-$, and we will represent all states and operators in matrix notation with respect to this standard basis $\{\psi_\pm\}$. With respect to this basis, then, we have $\mathbf{S} = \frac{1}{2}\hbar\boldsymbol{\sigma}$, where $\boldsymbol{\sigma}$ is the triad of Pauli matrices,

[1] Though a photon is a spin-1 particle, only the $+1$ and -1 spin states are available, and these two helicity states are described by the mathematics of the spin-$\frac{1}{2}$ system.

and

$$\psi_+ = \begin{pmatrix}1\\0\end{pmatrix} \quad \text{and} \quad \psi_- = \begin{pmatrix}0\\1\end{pmatrix}.$$

Each component of the operator triad S represents measurements of the component of **spin** in the appropriate direction, with $\mathbf{n}\cdot\mathsf{S}$ representing measurements of the component of spin in the direction of the unit vector $\mathbf{n}\in\mathbb{R}^3$. Spin itself is a basic attribute possessed by particles in nature and arises naturally in the relativistic treatment of quantum systems. As the mathematics of spin states is governed by an understanding of the simple linear algebra of the triad S, we are able to present an analysis of the basic spin-$\frac{1}{2}$ system without relativistic considerations.

To simplify the presentation, we measure spin in units of $\frac{1}{2}\hbar$ and we use Dirac notation, renaming $\psi_\pm$ as $|\pm\rangle$. This means, in particular, that $\frac{1}{2}\hbar$ is set equal to 1 in the relevant formulae. For example, formulae 18.1 become

$$\mathsf{S}_3|\pm\rangle = \pm 1|\pm\rangle \quad \text{and} \quad \mathsf{S}^2|\pm\rangle = 3|\pm\rangle,$$

and in these units $\mathsf{S} = \boldsymbol{\sigma}$. For a unit vector $\mathbf{n} = (n_1, n_2, n_3) \in \mathbb{R}^3$, spin-component measurements in the direction of $\mathbf{n}$ are represented by

$$\mathbf{n}\cdot\mathsf{S} = \mathbf{n}\cdot\boldsymbol{\sigma} = \begin{pmatrix} n_3 & n_1 - \mathbf{i}n_2 \\ n_1 + \mathbf{i}n_2 & -n_3 \end{pmatrix}; \tag{18.2}$$

in particular, the operator $\mathsf{S}_i = \sigma_i$ represents spin-component measurements along the x_i-axis for $i = 1, 2, 3$. Let S_θ represent spin-component measurements along an axis at an angle θ measured from the positive x_3-axis toward the positive x_1-axis, and in the x_1x_3-plane. The notation is ambiguous since $\mathsf{S}_1 = \mathsf{S}_{\theta=\pi/2} = \mathsf{S}_{\pi/2}$ and $\mathsf{S}_3 = \mathsf{S}_{\theta=0} = \mathsf{S}_0$, but this should cause little confusion. Obviously,

$$\mathsf{S}_\theta = (\sin\theta, 0, \cos\theta)\cdot\boldsymbol{\sigma} = \begin{pmatrix} \cos\theta & \sin\theta \\ \sin\theta & -\cos\theta \end{pmatrix}.$$

We need to understand the action of S_θ on $V^{1/2}$. Either by recalling that spin-component measurements along any direction yield ± 1 spin units or by examining the characteristic equation $\det(\mathsf{S}_\theta - \lambda 1_2) = 0$, it is immediate that the eigenvalues of S_θ are ± 1, each nondegenerate. Let $|\theta\pm\rangle$ denote an orthonormal pair of S_θ-eigenstates with respective eigenvalues ± 1. First we find expressions for $|\theta\pm\rangle$ in terms of the standard basis $|\pm\rangle = |0\pm\rangle$. Note that S_θ is a reflection in the complex vector space $\mathbb{C}|+\rangle \oplus \mathbb{C}|-\rangle$ that fixes $|\theta+\rangle$ and sends $|\theta-\rangle$ to $-|\theta-\rangle$. Note also that S_θ exchanges $|+\rangle = \left(\begin{smallmatrix}1\\0\end{smallmatrix}\right)$ with the vector $\left(\begin{smallmatrix}\cos\theta\\ \sin\theta\end{smallmatrix}\right)$, which lies at an angle θ from the positive $|+\rangle$-axis toward the positive $|-\rangle$-axis in the real $(|+\rangle|-\rangle)$-plane. This suggests that the line at angle $\theta/2$ in the real $(|+\rangle|-\rangle)$-plane is fixed, so that we should be able to take

$$|\theta+\rangle = \begin{pmatrix} \cos(\theta/2) \\ \sin(\theta/2) \end{pmatrix} = \cos(\theta/2)\,|+\rangle + \sin(\theta/2)\,|-\rangle \tag{18.3}$$

and

$$|\theta-\rangle = \begin{pmatrix} -\sin(\theta/2) \\ \cos(\theta/2) \end{pmatrix} = -\sin(\theta/2)\,|+\rangle + \cos(\theta/2)\,|-\rangle. \tag{18.4}$$

To confirm this for the $+1$-eigenvalue, we have

$$\begin{aligned} \mathsf{S}_\theta|\theta+\rangle &= \begin{pmatrix} \cos\theta\cos(\theta/2) + \sin\theta\sin(\theta/2) \\ \sin\theta\cos(\theta/2) - \cos\theta\sin(\theta/2) \end{pmatrix} \\ &= \begin{pmatrix} \cos(\theta - (\theta/2)) \\ \sin(\theta - (\theta/2)) \end{pmatrix} = |\theta+\rangle, \end{aligned} \tag{18.5}$$

which uses the sine and cosine angle addition formulæ. A similar calculation confirms the result for the -1-eigenvalue.

Equations 18.3 and 18.4 provide the statistics for measuring the x_3-component of spin when the system is in one of the S_θ-eigenstates. Indeed, if an x_3-component of spin is measured when the system is in state $|\theta+\rangle$, then the amplitude for measuring $+1$ is the coefficient of $|+\rangle$ in the equation 18.3, or $\langle+|\theta+\rangle = \cos(\theta/2)$, with corresponding probability $\cos^2(\theta/2)$ for measuring $+1$. Similarly the probability for measuring -1 is $\sin^2(\theta/2)$. Applying the resolution of the identity $1 = |\theta+\rangle\langle\theta+| + |\theta-\rangle\langle\theta-|$ to the basis state $|+\rangle$ gives

$$|+\rangle = \cos(\theta/2)\,|\theta+\rangle - \sin(\theta/2)\,|\theta-\rangle, \tag{18.6}$$

showing that when an S_θ-measurement is performed on a system in state $|+\rangle$, $+1$ and -1 are obtained with respective probabilities $\cos^2(\theta/2)$ and $\sin^2(\theta/2)$. This shows that there is symmetry between S_0-measurements on S_θ-eigenstates and S_θ-measurements on S_0-eigenstates. For physical reasons it is expected, and from the fact that the choice of x_1- and x_3-axes is arbitrary it is confirmed, that, for any angle φ, S_φ-measurements on $\mathsf{S}_{\varphi+\theta}$-eigenstates and $\mathsf{S}_{\varphi+\theta}$-measurements on S_φ-eigenstates are symmetric situations, with corresponding probabilities given by $\cos^2(\theta/2)$ and $\sin^2(\theta/2)$.

Notice that the $\mathsf{S}_{\pi/2}$-eigenstates are an equal mixture of the two orthogonal S_0-eigenstates with

$$|\pi/2\pm\rangle = \pm\frac{1}{\sqrt{2}}|+\rangle + \frac{1}{\sqrt{2}}|-\rangle. \tag{18.7}$$

In particular, an x_3-spin-component measurement on a system in state $|\pi/2+\rangle$ yields $+1$ and -1 with equal probabilities, while an x_1-spin-component measurement yields $+1$ with certainty; it never yields -1. In Section 22.2 we will use tensor products to give a quantum mechanical analysis of correlated spins in coupled spin-$\frac{1}{2}$ systems.

We close this section with a generalization of what we have accomplished thus far. The task is to find a general formula for the state vectors in $V^{1/2}$ that represent a spin-$\frac{1}{2}$ particle whose spin is aligned along an arbitrary direction in $\mathbb{R}^3$. Let $\mathbf{n}$ be a unit vector in $\mathbb{R}^3$ and recall that $\mathbf{n}\cdot\boldsymbol{\sigma}$ is the spin operator that corresponds to spin component measurements in the direction $\mathbf{n}$. Using equation 18.2, elementary algebra solves the eigenvalue equation $(\mathbf{n}\cdot\boldsymbol{\sigma})\psi = \pm\psi$ to give the normalized solutions

$$\psi = \psi_\pm(\mathbf{n}) = |\pm\rangle_{\mathbf{n}} = \frac{1}{\sqrt{2(1\pm n_3)}}\begin{pmatrix} n_3 \pm 1 \\ n_1 + \mathrm{i}n_2 \end{pmatrix}, \tag{18.8}$$

as long as $1 \pm n_3 \neq 0$.[2] When $1 \pm n_3 = 0$, the two cases are $\mathbf{n} = (0,0,1)$ with spin down and $\mathbf{n} = (0,0,-1)$ with spin up. In both these cases we may take

$$\psi_+((0,0,-1)) = \begin{pmatrix} 0 \\ 1 \end{pmatrix} = \psi_-((0,0,1)).$$

The reader might want to see how the eigenvectors $|\theta\pm\rangle$ appear when written in the form of equation 18.8 and to use the half-angle formulae to reconcile this with equations 18.3 and 18.4. Recall that if these state vectors are multiplied by an arbitrary phase factor $e^{\mathrm{i}\phi}$ then they still remain normalized and still represent the same quantum state.

Reversing the process, we leave it to the reader to show that if φ is any normalized state vector in $V^{1/2}$ then

$$\mathbf{n}(\varphi) = \begin{pmatrix} \langle\varphi|\sigma_1\varphi\rangle \\ \langle\varphi|\sigma_2\varphi\rangle \\ \langle\varphi|\sigma_3\varphi\rangle \end{pmatrix}$$

is a unit vector in $\mathbb{R}^3$, with φ serving as a $+1$-eigenvector of the operator

[2] For yet another description of these spin eigenvectors see Section 23.3.1, where they are parameterized as a discontinuous section of the Hopf fibration of $\mathbf{S}^3$. This is called the **Bloch sphere** parameterization and is used extensively in the sciences where spin-$\frac{1}{2}$ systems are studied, such as physical chemistry and quantum computing.

$\mathbf{n}(\varphi)\cdot\boldsymbol{\sigma}$. It follows that $\varphi = e^{i\phi}\psi_+(\mathbf{n}(\varphi))$ for some phase factor $e^{i\phi}$, and, conversely, that $\mathbf{n}(\psi_+(\mathbf{u})) = \mathbf{u}$ for any unit vector $\mathbf{u}$ of $\mathbb{R}^3$. Note that the components of $\mathbf{n}(\varphi)$ are the expectation values of spin measurements in the three standard directions of $\mathbb{R}^3$ on a particle represented by the state vector φ.

A quick calculation shows that the projection operator onto the one-dimensional space spanned by the the spin vector $|+\rangle_\mathbf{n}$ representing a particle whose spin is aligned along the vector $\mathbf{n}$ is

$$\mathsf{P}_\mathbf{n} = |+\rangle_{\mathbf{n}\,\mathbf{n}}\langle +| = \tfrac{1}{2}(1_{2\times 2} + \mathbf{n}\cdot\boldsymbol{\sigma})\,. \tag{18.9}$$

18.4 Rotations of Wave Functions and Spin-$\frac{1}{2}$ Particles

In this section we ask: "what is the effect of a spatial rotation of $\mathbb{R}^3$ on the state vectors of the spin-$\frac{1}{2}$ system?" To answer this, it will help to see the effect of a spatial rotation on the state vectors of wave mechanics, which is hinted at in Section 14.3 and will be addressed fully here. We saw in Lecture 13 that angular momentum triads in quantum mechanics may be described in terms of representation theory as arising from anti-self-adjoint representations of $\mathfrak{su}(2)$ on the state space, as described on p. 184. For the orbital angular momentum triad L of wave mechanics, we saw in Lecture 14 that the quantum number ℓ that parameterizes the L^2-eigenstates must be an integer. Our previous observation that it is only the integer representations of $\mathfrak{su}(2)$ that produce representations of the rotation group SO(3) suggests that we should think of the orbital angular momentum triad as arising from a representation not of $\mathfrak{su}(2)$ but rather from its isomorphic cousin $\mathfrak{so}(3)$, the Lie algebra of the rotation group. Exponentiation of this representation then should give a representation of its Lie group, SO(3), on the state space $L^2(\mathbb{R}^3)$. In fact, this is precisely the case, since Stone's theorem implies that the operator triad L gives the infinitesimal generators of the standard unitary representation of SO(3) on $L^2(\mathbb{R}^3)$. The details await Lecture 28,[3] but we will go ahead and give the result. The result is that if $R = R(t,\mathbf{n}) \in \mathrm{SO}(3)$ is the matrix that represents the positive rotation of $\mathbb{R}^3$ through an angle t with axis $\mathbf{n}$, where $\mathbf{n}$ a unit vector, then the operator $\mathsf{U}(R)$ on $L^2(\mathbb{R}^3)$ defined by $\mathsf{U}(R)\psi = \psi\circ R^{-1}$ defines a unitary representation of SO(3) on $L^2(\mathbb{R}^3)$, and this representation is given in terms of the orbital angular

[3] Specifically, p. 439.

momentum triad as

$$\mathsf{U}(R(t,\mathbf{n})) = e^{-it\mathbf{n}\cdot\mathsf{L}/\hbar}. \tag{18.10}$$

This representation is called the **standard representation of the rotation group** on $L^2(\mathbb{R}^3)$ and that it is unitary reflects the fact that the mere spatial rotation of a coordinate system should have no effect on the physics of a quantum particle. Indeed, if ψ is the wave function of a particle then $\mathsf{U}(R)\psi$ is the wave function with respect to a rotated coordinate system and, since $\mathsf{U}(R)$ is unitary, the probabilities of wave mechanics are preserved. Such a rotation of the coordinate system is sometimes said to be a **passive rotation**, since the particle does not rotate but only the system of coordinates used to locate the particle. This is to be contrasted with an **active rotation**, where the coordinate axes are kept fixed while the particle is rotated. In this case, if ψ represents the quantum system before rotation, $\mathsf{U}(R)\psi$ represents the system after rotation.[4]

The standard representation is not irreducible. Indeed, the state space $L^2(\mathbb{R}^3)$ decomposes as the completion of the tensor product space

$$L^2(\mathbf{S}^2, d\Omega) \otimes L^2(\mathbb{R}_{>0}, r^2 dr),$$

where $d\Omega$ is the standard rotation-invariant measure on the 2-sphere. The standard representation acts as the identity on the radial factor $L^2(\mathbb{R}_{>0}, r^2 dr)$, and acts on the spherical factor by $\sigma(R)\psi = \psi \circ R^{-1}$, as on p. 209. As seen there, this decomposes the spherical factor into a Hilbert-space orthogonal direct sum of irreducible representation spaces of SO(3) with $L^2(\mathbf{S}^2, d\Omega)$ equal to the completion of the sum $\mathcal{V} = \oplus_{\ell=0}^{\infty} V^\ell$ where V^ℓ is the space of spherical harmonics of degree ℓ.

With this review of the mathematics of orbital angular momentum in wave mechanics as a background, we now examine the mathematics of spin in the same detail. Again we have an angular momentum triad, S, this time acting on $V^{1/2}$, an irreducible representation space of $\mathfrak{su}(2)$ acting as the state space for the spin degrees of freedom. By analogy with the discussion on orbital angular momentum leading to equation 18.10, we might expect that the effect of a spatial rotation $R = R(t,\mathbf{n})$ on the spin state represented by $\psi_+(\mathbf{u})$, for a unit vector $\mathbf{u}$, would be given by

[4] It should be mentioned that passive and active rotations have opposite orientations – one a positive, the other a negative rotation – but we are unconcerned with the details at this point.

applying the unitary transformation[5]

$$\mathsf{V}(R(t,\mathbf{n})) = e^{-\mathbf{i}t\mathbf{n}\cdot\mathbf{S}/\hbar} = e^{-\mathbf{i}t\mathbf{n}\cdot\boldsymbol{\sigma}/2}.$$

Equivalently, the equation

$$\mathsf{V}(R(t,\mathbf{n}))\psi_+(\mathbf{u}) = e^{-\mathbf{i}t\mathbf{n}\cdot\boldsymbol{\sigma}/2}\psi_+(\mathbf{u}) = e^{\mathbf{i}\phi(t,\mathbf{n})}\psi_+(R(t,\mathbf{n})\mathbf{u}) \qquad (18.11)$$

should hold, where the phase angle ϕ may depend on the rotation angle t and axis $\mathbf{n}$. This we will confirm after a couple of comments. First, while equation 18.10 defines a unitary representation of the rotation group SO(3), equation 18.11 will do so only if the phase angle ϕ is independent of the rotation. It turns out this is not so, and equation 18.11 does not define a unitary representation of SO(3) on $V^{1/2}$; however, equation 18.11 does define a **projective representation** of SO(3) on the projective space $V^{1/2}/\mathbb{C}^\times$, whose members are the rays $\mathbb{C}^\times\psi$ for nonzero $\psi \in V^{1/2}$. Second, a very interesting phenomenon occurs when a spin state is rotated through a full rotation of 2π. We will see shortly that $\phi(2\pi,\mathbf{n}) = \pi$ so that, upon a full rotation, the state vector of the particle is transformed from $\psi_+(\mathbf{u})$ to $-\psi_+(\mathbf{u})$. Of course, a state vector and its negative represent exactly the same quantum state, so this has no effect on the quantum physics of the fully rotated system of this single particle, as expected, but this introduction of a negative sign is a rather curious property of these spin-$\frac{1}{2}$ particles and, indeed, of all particles of half-integer, as opposed to integer, spin. One might think that this introduction of a negative sign upon a full spatial rotation would have no measurable effect, since it merely introduces a phase shift, and indeed this is true if we restrict attention to a single particle. However, when there are two particles, if one is rotated so that a negative sign appears in its state vector while the other remains unrotated, then quantum mechanics predicts an observable effect. Two groups have confirmed conclusively the predictions of quantum mechanics in this case and so confirm this introduction of a negative sign upon a full rotation.[6]

To confirm equation 18.11, set $R = R(t,\mathbf{n})$ and note that

$$\mathsf{V}(R) = e^{-\mathbf{i}t\mathbf{n}\cdot\boldsymbol{\sigma}/2} = \cos(t/2)\,1_{2\times 2} - \mathbf{i}\sin(t/2)\,\mathbf{n}\cdot\boldsymbol{\sigma}. \qquad (18.12)$$

[5] In the units that we are using, $\hbar = 2$ and $\mathbf{S} = \boldsymbol{\sigma}$.

[6] The effect appears when two identical beams of such particles are combined after the particles of one of the beams experience a rotation. Quantum mechanics predicts interference effects that arise from this phase shift when one beam is rotated that differ from the prediction when neither is rotated. See, for example, Sakurai (1985), pp. 162–63, for a discussion of the experiment confirming this and references to the primary reports.

This is an easy consequence of the series expansion of the exponential along with the observation that $(\mathbf{n}\cdot\boldsymbol{\sigma})^2 = 1_{2\times2}$, which follows immediately from the second version of equation 13.1.[7] Observe that equation 18.12 implies that $\mathsf{V}(R(2\pi,\mathbf{n})) = -1_{2\times2}$ so that, as promised, $\mathsf{V}(R(2\pi,\mathbf{n}))\psi_+(\mathbf{u}) = -\psi_+(\mathbf{u})$, verifying that $\phi(2\pi,\mathbf{n}) = \pi$. To complete the confirmation of equation 18.11, it suffices to verify that $\psi = \mathsf{V}(R)\psi_+(\mathbf{u})$ is a $+1$-eigenvector of the spin operator $\mathbf{v}\cdot\boldsymbol{\sigma}$, where $\mathbf{v} = R(t,\mathbf{n})\mathbf{u}$. As for the case where $\mathbf{u} = \pm\mathbf{n}$ is trivial, we assume otherwise and let $\mathbf{a} = \mathbf{u} - (\mathbf{u}\cdot\mathbf{n})\mathbf{n} \neq \mathbf{0}$ be the projection of $\mathbf{u}$ onto $\mathbf{n}^\perp$, the plane perpendicular to $\mathbf{n}$. Let $\mathbf{b}$ be the unique vector perpendicular to $\mathbf{a}$ lying in $\mathbf{n}^\perp$ and satisfying $\mathbf{a}\times\mathbf{b} = \|\mathbf{a}\|^2\mathbf{n}$. The triple $\mathbf{a}$, $\mathbf{b}$, $\mathbf{n}$ is a right-handed coordinate triple of vectors with $\|\mathbf{a}\| = \|\mathbf{b}\|$ and it follows easily that the action of the rotation $R(t,\mathbf{n})$ on $\mathbf{u}$ is given by

$$\mathbf{v} = R(t,\mathbf{n})\mathbf{u} = (\cos t)\mathbf{a} + (\sin t)\mathbf{b} + (\mathbf{u}\cdot\mathbf{n})\mathbf{u}. \tag{18.13}$$

Inserting the right-hand sides of equations 18.13 and 18.12 into the expression $(\mathbf{v}\cdot\boldsymbol{\sigma})\mathsf{V}(R)$, expanding using the second version of equation 13.1 as well as the identities $\mathbf{a}\cdot\mathbf{n} = \mathbf{0} = \mathbf{b}\cdot\mathbf{n}$, $\mathbf{a}\times\mathbf{n} = -\mathbf{b}$, and $\mathbf{b}\times\mathbf{n} = \mathbf{a}$, and applying the sine and cosine addition formulae, gives the expression

$$\begin{aligned}(\mathbf{v}\cdot\boldsymbol{\sigma})\mathsf{V}(R) = \cos(t/2)\ &[\mathbf{a} + (\mathbf{u}\cdot\mathbf{n})\mathbf{n}]\cdot\boldsymbol{\sigma} \\ &- \mathbf{i}\sin(t/2)\ [(\mathbf{u}\cdot\mathbf{n})1_{2\times2} + \mathbf{i}\mathbf{b}\cdot\boldsymbol{\sigma}]\,.\end{aligned}$$

Inserting $\mathbf{a} + (\mathbf{u}\cdot\mathbf{n})\mathbf{n} = \mathbf{u}$ into the first summand and applying the second version of equation 13.1 to the second summand, after observing that

$$\mathbf{b} = \mathbf{n}\times\mathbf{a} = \mathbf{n}\times[\mathbf{a} + (\mathbf{u}\cdot\mathbf{n})\mathbf{n}] = \mathbf{n}\times\mathbf{u},$$

gives

$$(\mathbf{v}\cdot\boldsymbol{\sigma})\mathsf{V}(R) = \cos(t/2)\,\mathbf{u}\cdot\boldsymbol{\sigma} - \mathbf{i}\sin(t/2)\,(\mathbf{n}\cdot\boldsymbol{\sigma})(\mathbf{u}\cdot\boldsymbol{\sigma}).$$

One more application of equation 18.12 gives the very useful formula

$$(\mathbf{v}\cdot\boldsymbol{\sigma})\mathsf{V}(R) = \mathsf{V}(R)(\mathbf{u}\cdot\boldsymbol{\sigma}),$$

or, written in terms of the rotation angle t and axis $\mathbf{n}$,

$$\{[R(t,\mathbf{n})\mathbf{u}]\cdot\boldsymbol{\sigma}\}\,\mathsf{V}(R(t,\mathbf{n})) = \mathsf{V}(R(t,\mathbf{n}))(\mathbf{u}\cdot\boldsymbol{\sigma}).$$

The verification that $\psi = \mathsf{V}(R)\psi_+(\mathbf{u})$ is a $+1$-eigenvector of the spin operator $\mathbf{v}\cdot\boldsymbol{\sigma}$ is now transparent, since

$$(\mathbf{v}\cdot\boldsymbol{\sigma})\psi = (\mathbf{v}\cdot\boldsymbol{\sigma})\mathsf{V}(R)\psi_+(\mathbf{u}) = \mathsf{V}(R)(\mathbf{u}\cdot\boldsymbol{\sigma})\psi_+(\mathbf{u}) = \mathsf{V}(R)\psi_+(\mathbf{u}) = \psi.$$

[7] Or, just square the matrix of equation 18.2 and use $\|\mathbf{n}\| = 1$.

The confirmation of equation 18.11 is now complete.

It is worth summarizing what has been demonstrated here. When a spatial rotation $R(t, \mathbf{n})$ is applied to a quantum system that consists of a spin-$\frac{1}{2}$ particle whose spin is aligned along the direction $\mathbf{u}$ represented by state vector $\psi_+(\mathbf{u})$, the rotated system has its spin aligned along the rotated direction $\mathbf{v} = R(t, \mathbf{n})\mathbf{u}$ and, necessarily, is represented by a state vector of the form $e^{\phi(t,\mathbf{n})}\psi_+(\mathbf{v})$ for some phase factor $e^{\phi(t,\mathbf{n})}$. The important point is that this rotated state vector $e^{\phi(t,\mathbf{n})}\psi_+(\mathbf{v})$ is given by equation 18.11, showing the spin triad S to be the generator of spatial rotations on $V^{1/2}$ just as the orbital angular momentum triad is the generator of the effects of spatial rotations in wave mechanics. Technically, this prescription gives an ill-defined representation of SO(3), since $\mathsf{V}(R(0, \mathbf{n})) = 1_{2\times 2} \neq -1_{2\times 2} = \mathsf{V}(R(2\pi, \mathbf{n}))$ while $R(0, \mathbf{n}) = 1_{3\times 3} = R(2\pi, \mathbf{n})$. The way to repair this is to pass to the quotient, by the action of the multiplicative group $\mathbb{C}^\times$ of nonzero complex numbers, and to consider expression 18.11 as defining a projective representation of SO(3) on the projective space $V^{1/2}/\mathbb{C}^\times$. An alternative repair may be accomplished by considering the effect of a rotation to be encoded not, as usual, by a representation of the rotation group on the state space but rather by a representation of SU(2), the simply connected double cover of the rotation group. This approach will be presented in the next lecture, where **spinors** are introduced. Pauli hit upon this idea of *spinors* in his attempt to incorporate spin into the edifice of nonrelativistic quantum mechanics. Pauli's two-dimensional spinors were introduced in an ad hoc manner to solve the problem of spin, and later Dirac would introduce a fully relativistic quantum theory of particles out of which would naturally fall four-dimensional spinors.

This phenomenon, of the introduction of a negative sign upon a full spatial rotation of a physical system, was the first example in the history of physical science where it was suggested that a particle can be physically moved and, over a period of time, returned to its former position and momentum environment, with its former orientation, and yet the state of the particle retains an imprint of this temporal history.

18.5 End Notes

The pure mathematics behind spin had been uncovered in 1913 by the mathematician Élie Cartan (1913) in his study of the linear representations of simple groups in an arbitrary number of dimensions. Pauli's theory, which is examined more fully in the next lecture, reproduced the lowest-dimensional version of the Cartan theory, and the later Dirac theory reproduced the four-dimensional version. Of course, the physicists Pauli and Dirac derived their insights from physical considerations and, it might be argued, were forced to the discovery of spin representations by nature herself. They certainly were unaware of Cartan's mathematics and, besides that, Cartan's mathematical language was somewhat foreign to the language of the developing quantum mechanical formalism. After the quantum mechanical formalism was established by Pauli and Dirac, Cartan showed how to derive a Dirac equation for any group and extended the equation to general relativity. All this is laid out in the 1937 text *Leçons sur la théorie des spineurs*, which was printed from Cartan's lectures, themselves gathered and arranged by André Mercier. This is translated into English in Cartan (1966).

Thaller (2005) had a great influence on my presentations in Sections 18.3 and 18.4, especially his Chapter 4 on qubits. For an extensive development of the theory of angular momentum in quantum mechanics, see Edmonds (1960).

19

Wave Mechanics VII: Pauli's Spinor Theory

> Pauli himself was trying to incorporate spin into quantum mechanics and introduced the Pauli matrices, but he could not arrive at a complete theory. He also clearly recognized the necessity of making the theory relativistic and yet could not achieve it. He was known to be a perfectionist who severely criticized other people's work, but he had no alternative other than to publish his unsatisfactory ad hoc theory. Now on the other hand, Dirac established his theory with totally unexpected acrobatics and solved all the problems on spin as well as making the theory relativistic ... However, in 1934 Pauli ended up retaliating against Dirac. Pauli showed that Dirac's argument that nature is satisfied only by the Dirac equation and not by the Klein–Gordon equation is incorrect. According to him, the Klein–Gordon equation is not in contradiction with the framework of quantum mechanics, and there is no reason that nature abhors a particle with spin 0.
>
> Sin-Itiro Tomonaga
> *The Story of Spin*, 1997

We have seen how the whole-integer representations of $\mathfrak{su}(2)$ are used in wave mechanics to represent the operators that correspond to orbital angular momentum, and how this leads to an understanding of the irreducible representations of the rotation group SO(3) on the space of spherical harmonics. This was the content of Lecture 14 and was used subsequently in Lecture 15 to investigate the hydrogenic potential in quantum mechanics. Orbital angular momentum is an example of an observable that might be labeled as **extrinsic**, in that it arises from what we might call **external degrees of freedom** for the particle. It is not a property of the quantum particle itself, but rather a property of that particle's placement and motion within its environment and is very much dependent on that environment. For example, it is measured with respect to a center that may be chosen quite arbitrarily. Of course, it becomes important in those problems in which there is a natural center to consider, such as, for example, when the potential is centrally symmetric as in the Kepler problem.

It was found early on in the development of quantum mechanics, and especially in the experimental investigations of the electron at the time, that quantum particles also possess **internal degrees of freedom** that reflect observables **intrinsic** to the particle. In particular, every quantum particle seems to possess an intrinsic angular momentum internal to the particle and having nothing to do with its environment. For example, all electrons at all times and all places possess a single quantum of angular momentum that yields a value of $\pm\hbar/2$ when measured along any direction in space. Just as with orbital angular momentum, this intrinsic angular momentum also is described in quantum mechanics by representations of $\mathfrak{su}(2)$, but this time includes not only the whole-integer but also the half-integer representations. It seems that nature does not let any mathematics of elegance and substance go to waste!

While the incorporation of orbital angular momentum into wave mechanics is quite straightforward, in that the pertinent operators are natural combinations of the two operator triads $\mathbf{X}$ and $\mathbf{P}$, which operate as self-adjoint operators on the state space $L^2(\mathbb{R}^3)$, it seems that there is no natural way to incorporate spin operators that describe these internal degrees of freedom into the edifice of wave mechanics, as incapsulated in our four axioms of Section 2.2. We need an extension of the axioms to incorporate the spin degrees of freedom of a quantum particle. Pauli accomplished this by enhancing the state space to include not only the external degrees of freedom associated with position within $\mathbb{R}^3$ but also these internal spin degrees of freedom, by integrating into the state space a representation of $\mathfrak{su}(2)$. It is the purpose of the present lecture to elucidate Pauli's theory of spinors, which accomplishes this task of weaving spin into nonrelativistic wave mechanics.

It should be said that the original inventors of wave mechanics viewed the enhancements discussed in this lecture as rather ad hoc – additions that address a special problem but that present no general, fundamental, contributions to the theory. This ad hoc inelegance was remedied a year later when the internal spin degrees of freedom were seen to arise naturally from the mathematics of the Dirac equation. Indeed, no ad hoc additions to incorporate spin are necessary when wave mechanics is translated to a Lorentz-invariant theory. Moreover, the Pauli theory is seen to be nothing more than the low-energy limit of the Dirac theory, which offers few advantages even in the study of nonrelativistic particles with spin. Despite these criticisms, the author feels that Pauli's treatment of spin is only seen to be inelegant as an enhancement of wave mechanics, but not as an instantiation of the four abstract minimalist axioms of quantum mechanics since his treatment fits precisely as an interpretation of those axioms. So we find it helpful to present Pauli's theory in the nonrelativistic version of wave mechanics, for a few reasons. First, the modern version of the Pauli theory uses tensor products to include spin in the theory, and this serves as a model for extending and enhancing any interpretation of the minimalist axioms. This will be put to good use in a later lecture in showing the way to include in the quantum formalism states that describe multiple particles. Second, though it presents only an ad hoc inclusion of spin, the theory is effective in describing the physics of spin for nonrelativistic particles, those of low energy. For example, the Pauli theory gives an adequate explanation of certain fine structure adjustments to the spectrum of hydrogen, which represents an early success of the theory. Third, the theory gives the student an opportunity to see the mathematics of spinors in its simplest two-dimensional setting before its appearance in the more complicated Dirac theory. We will try to give an account of the theory that develops and underscores each of these reasons.

In the initial section of the lecture, we show how tensor products may be used in the setting of wave mechanics to build a state space for a particle with internal degrees of freedom. The Pauli theory will be cast in this setting. We then examine how the state vectors of the Pauli theory transform under spatial rotations and describe what Pauli called the "strange two-valuedness" of the electron state under a full rotation. The development presented here reflects the lessons of Section 18.4. We then examine the intrinsic magnetic moment of the electron arising from its intrinsic spin. Finally, we apply the theory to the hydrogenic potential and demonstrate how the spin degrees of freedom of the electron give rise to the double degeneracy of the hydrogen spectrum already mentioned in Section 16.1. We outline a correction to the hydrogen spectrum that arises from the magnetic interaction of the electron's spin with its orbital angular momentum, the so-called **spin–orbit coupling**.

In the mathematically oriented lecture that follows this one, we ask the question "what is a spinor?" and examine the mathematics of spin representations, discovered first in 1913 by the mathematician Élie Cartan, over a decade before Pauli's rediscovery of this in the context of quantum mechanics and Dirac's subsequent use in bridging the gap between special relativity and wave mechanics.

19.1 Tensor Products and Internal Degrees of Freedom

The question we address in this section is how to incorporate into wave mechanics, as economically as possible, measurable attributes intrinsic to quantum particles where the pertinent observables cannot be represented by combinations of the operator triads X and P. The archetypal example is the attribute of spin, which, in isolation from any other observable considerations, is governed by the irreducible representation spaces V^j of $\mathfrak{su}(2)$. The set-up considered is that of a single quantum particle in motion in $\mathbb{R}^3$, whose extrinsic mechanics is governed by the wave mechanics developed thus far, with state space $L^2(\mathbb{R}^3)$ and observables corresponding to various combinations of the components of the triads X and P. We consider that the particle has internal degrees of freedom described in isolation from the extrinsic degrees of freedom by a *finite-dimensional* state space $\mathcal{H}$ with a complete set of operators $\mathsf{B}_1, \ldots, \mathsf{B}_M$ that correspond to observables intrinsic to the particle. It was Pauli's insight that wave mechanics can be extended to accommodate these internal observables by enhancing the state space to square-integrable functions that take values in a finite-dimensional complex vector space rather than in just the complex numbers. In applying this idea to spin, Pauli found that the transformations of the theory under the spatial rotation group SO(3) give rise to nontrivial phase changes of the state vector upon rotations through 2π, implying that the state space has internal structure beyond that described by the Hilbert space axioms. What Pauli had rediscovered was an example of the spin representations that Cartan had uncovered more than a decade earlier.

Rather than using functions with values in a complex vector space, as in Pauli's original theory, we will develop an equivalent theory using tensor products. This use of tensor products serves as an object lesson in extending wave mechanics and will be put to work in a later lecture in developing the mechanics of multi-particle quantum systems. The state space in Pauli's theory may be recast as the Hilbert space $L^2(\mathbb{R}^3) \otimes \mathcal{H}$, whose inner product is a bilinear extension of the rule $\langle \psi \otimes \rho | \varphi \otimes \nu \rangle = \langle \psi | \varphi \rangle \langle \rho | \nu \rangle$. We will take a little time to dissect this and see how we may describe both the external degrees of freedom, via the L^2-factor, and the internal degrees of freedom, via the $\mathcal{H}$-factor.

Let $\Xi = \{\xi_i : i \in \mathbb{N}\}$ be an orthonormal Schauder basis for $L^2(\mathbb{R}^3)$ and $\Sigma = \{\varsigma_1, \ldots, \varsigma_M\}$ an orthonormal basis for $\mathcal{H}$. It is a straightforward exercise to prove that the set $\Xi \otimes \Sigma = \{\xi_i \otimes \varsigma_m : i, m \in \mathbb{N}, m \leq M\}$ forms an orthonormal Schauder basis for $L^2(\mathbb{R}^3) \otimes \mathcal{H}$.[1] This gives an explicit path from the tensor product description of the state space to Pauli's original description as a space of vector-valued square-integrable functions, in the following way. Each state vector $\psi \in L^2(\mathbb{R}^3) \otimes \mathcal{H}$ may be expanded in this basis as $\psi = \sum_{i,m} a_{im}\, \xi_i \otimes \varsigma_m$, for some square-summable sequence of coefficients a_{im}, for $i \in \mathbb{N}$ and $1 \leq m \leq M$. This allows us to write

$$\psi = \sum_{m=1}^{M} \left(\sum_{i=1}^{\infty} a_{im} \xi_i \right) \otimes \varsigma_m = \sum_{m=1}^{M} \psi_m \otimes \varsigma_m,$$

where $\sum_i a_{im} \xi_i = \psi_m \in L^2(\mathbb{R}^3)$, the convergence of these series being guaranteed by the square summability of the coefficients. This describes a vector space isomorphism of $L^2(\mathbb{R}^3) \otimes \mathcal{H}$ to $L^2(\mathbb{R}^3)^M$, the M-fold self-Cartesian product of $L^2(\mathbb{R}^3)$, via

$$L^2(\mathbb{R}^3) \otimes \mathcal{H} \ni \psi = \sum_{m=1}^{M} \psi_m \otimes \varsigma_m \mapsto (\psi_1, \ldots, \psi_M) \in L^2(\mathbb{R}^3)^M. \quad (19.1)$$

This becomes a Hilbert space isometry if the inner product on $L^2(\mathbb{R}^3)^M$ is defined as

$$\langle \varphi | \psi \rangle = \sum_{m=1}^{M} \langle \varphi_m | \psi_m \rangle = \sum_{m=1}^{M} \int_{\mathbb{R}^3} \overline{\varphi}_m \psi_m. \quad (19.2)$$

Of course, each $\psi \in L^2(\mathbb{R}^3)^M$ may be thought of as an element of the Hilbert space $L^2(\mathbb{R}^3, \mathbb{C}^M)$ with the inner product of formula 19.2; i.e., as a the function on $\mathbb{R}^3$ that takes values in $\mathbb{C}^M$ via

$$\psi(\mathbf{x}) = (\psi_1(\mathbf{x}), \ldots, \psi_M(\mathbf{x})).$$

It is convenient to have these various, but equivalent, representations of

[1] It is left as an exercise to prove that the tensor product $L^2(\mathbb{R}^3) \otimes \mathcal{H}$ is complete in the inner product described in the preceding paragraph when $\mathcal{H}$ is finite-dimensional. Were $\mathcal{H}$ infinite-dimensional, this inner product space would be incomplete and we would have to pass to its completion in order to use it as a state space. In this lecture, we restrict to a finite-dimensional state space $\mathcal{H}$ for the internal degrees of freedom, for simplicity and because our primary interest is in spin, which has the finite-dimensional state space V^j. This restriction will be relaxed in later lectures.

a state vector as, alternatively, a tensor product element, a Cartesian product element, and a vector-valued square-integrable function, each having its advantages in different contexts.

How do observables fit into this picture? The answer is quite straightforward. Indeed, any "external" observable a is represented by a self-adjoint operator A on $L^2(\mathbb{R}^3)$ in wave mechanics and is enhanced to the operator $\hat{\mathsf{A}} = \mathsf{A} \otimes 1_{\mathcal{H}}$ operating on the enhanced state space $L^2(\mathbb{R}^3) \otimes \mathcal{H}$, still representing a. Similarly, any "internal" observable b is represented by a self-adjoint operator B on $\mathcal{H}$ and is enhanced to the operator $\hat{\mathsf{B}} = 1_{L^2(\mathbb{R}^3)} \otimes \mathsf{B}$ still representing b. Note that $\hat{\mathsf{A}}$ and $\hat{\mathsf{B}}$ commute and have the same eigenvalues (or, more generally, the same spectrum) as A and B, respectively.

At this point we specialize to the irreducible representations of $\mathfrak{su}(2)$ for describing internal degrees of freedom. We model a quantum particle with $2j+1$ spin degrees of freedom, where j is a nonnegative half-integer, as one whose state is described by a vector in the state space $L^2(\mathbb{R}^3) \otimes V^j$, where V^j serves as an irreducible $(2j+1)$-dimensional representation space of $\mathfrak{su}(2)$. Such a particle is said to have **spin** j. Recall from Lecture 12 that V^j may be decomposed into one-dimensional J_3-eigenspaces giving a basis $\Sigma = \{|m\rangle : -j \leq m \leq j\}$ with $\mathsf{J}_3|m\rangle = m\hbar|m\rangle$. Here we are using $\mathbf{J}$ as the angular momentum triad on V^j induced by the irreducible representation of $\mathfrak{su}(2)$. A state vector ψ corresponding to a state of this particle with spin j may be expanded as $\psi = \sum_m \psi_m \otimes |m\rangle$, with the index m ranging over the $2j+1$ half-integers $-j, -j+1, \ldots, j-1, j$ and where $\psi_m \in L^2(\mathbb{R}^3)$. A state vector of the form $\varphi \otimes |m\rangle$ represents a state of the particle whose position and momentum are governed by the square-integrable function $\varphi \in L^2(\mathbb{R}^3)$ and which has spin component $m\hbar$ when measured in the x_3-direction.

Specializing now to the simplest nontrivial spin representation with $j = \frac{1}{2}$, we arrive at Pauli's state space for an electron, or any particle of spin $\frac{1}{2}$. We will use the more common notation $\mathbf{S}$ for the angular momentum triad in this case, rather than $\mathbf{J}$. Writing the two standard basis vectors of the representation space $V^{1/2}$ as $|\pm\rangle$, we see that an arbitrary state vector for a particle of spin $\frac{1}{2}$ may be written as $\varphi \otimes |+\rangle + \psi \otimes |-\rangle$ for square-integrable functions φ and ψ. Under the equivalence 19.1, this is written in standard presentations of Pauli's theory as the column vector

$$\begin{pmatrix} \varphi \\ \psi \end{pmatrix} \in L^2(\mathbb{R}^3) \times L^2(\mathbb{R}^3). \tag{19.3}$$

The reader should check his or her understanding by noting that the action of $\hat{\mathsf{S}}_3$ on these column vectors is given by

$$\hat{\mathsf{S}}_3 \begin{pmatrix} \varphi \\ \psi \end{pmatrix} = \frac{\hbar}{2} \begin{pmatrix} \varphi \\ -\psi \end{pmatrix}.$$

If we measure spin in units of $\frac{1}{2}\hbar$, as in the previous lecture, the action of $\hat{\mathsf{S}}_3$ on the Pauli state space represented by column vectors of square-integrable functions has a matrix representation given by the Pauli spin matrix

$$\sigma_3 = \begin{pmatrix} 1 & 0 \\ 0 & -1 \end{pmatrix}.$$

The actions of $\hat{\mathsf{S}}_1$ and $\hat{\mathsf{S}}_2$ are given, respectively, by the remaining Pauli spin matrices σ_1 and σ_2.

For any $\psi \in L^2(\mathbb{R}^3)$ let $\psi_\pm = \psi \otimes |\pm\rangle$, so that ψ_+ represents a state of the electron with spin up in the x_3-direction and ψ_- a state with spin down. Note that $\hat{\mathsf{S}}_3\psi_\pm = \pm\frac{1}{2}\hbar\,\psi_\pm$. An arbitrary state vector may now be written as $\varphi_+ + \psi_-$ for $\varphi, \psi \in L^2(\mathbb{R}^3)$. Ehrenfest called the state vectors of Pauli's theory **spinors**. Catchy name – after all, we are dealing with spin – but the reason for the name is deeper than that. Indeed, there is, as mentioned earlier, what Pauli called the "strange two-valuedness" associated with his column state vectors. This is seen by asking what happens to a state vector when a spatial rotation is applied to the theory: this is the topic of the next section.

19.2 Action of the Double Cover of SO(3) on $L^2(\mathbb{R}^3) \otimes V^{1/2}$

It will come as no surprise to the reader who has studied Lecture 18 that already in that lecture we worked out the ingredients needed to describe the effect of a spatial rotation on a Pauli spinor. As discussed in Section 18.4, the rotation group SO(3) acts on the wave mechanical state space $L^2(\mathbb{R}^3)$ via the standard representation $\mathsf{U}(R)\psi = \psi \circ R^{-1}$ for a rotation $R = R(t, \mathbf{n})$, and acts projectively on the spin representation space $V^{1/2}$ via $\mathsf{V}(R)|\pm\rangle = e^{-it\mathbf{n}\cdot\boldsymbol{\sigma}/2}|\pm\rangle$. Just as the orbital angular momentum operator triad $\mathbf{L}$ generates the action of U on $L^2(\mathbb{R}^3)$ according to equation 18.10, so the spin operator triad $\mathbf{S}$ generates the action of V on $V^{1/2}$ by equation 18.11. From this discussion it is clear that the Pauli state space $L^2(\mathbb{R}^3) \otimes V^{1/2}$ comes equipped with the projective representation $\mathsf{U} \otimes \mathsf{V}$, whose action on a Pauli spinor written as a column vector, as in equation 19.3, is

$$[\mathsf{U}\otimes\mathsf{V}(R)]\begin{pmatrix}\varphi\\ \psi\end{pmatrix}=\left[\left(\cos\frac{t}{2}\right)1_{2\times 2}-\mathbf{i}\left(\sin\frac{t}{2}\right)\mathbf{n}\cdot\boldsymbol{\sigma}\right]\begin{pmatrix}\varphi\circ R^{-1}\\ \psi\circ R^{-1}\end{pmatrix}, \tag{19.4}$$

where we have applied equation 18.12. Upon a full 2π rotation, this introduces a phase factor of -1, indicative that $\mathsf{U}\otimes\mathsf{V}$ is ill-defined as a representation of the rotation group on the Pauli state space. While physicists may be comfortable with this and with the introduction of a phase shift of -1 upon a full rotation, mathematicians tend to be fussy about these sorts of loose ends. There are two ways to fix this. The first is mentioned in Section 18.4, where passing to the projective space $V^{1/2}/\mathbb{C}^\times$ gives a well-defined action of SO(3). In the same vein, the multiplicative group $\mathbb{C}^\times$ of nonzero complex numbers acts on the Pauli state space via the multiplication of Pauli spinors, given as

$$z\cdot\begin{pmatrix}\varphi\\ \psi\end{pmatrix}=\begin{pmatrix}z\varphi\\ z\psi\end{pmatrix},$$

and $\mathsf{U}\otimes\mathsf{V}$ induces a well-defined action of SO(3) on the quotient space $L^2(\mathbb{R}^3)\otimes V^{1/2}/\mathbb{C}^\times$. An appropriate and valid criticism of this projective solution to the problem of an ill-defined representation is that, as mentioned in the discussion following equation 18.11, the phase of the state does actually shift by a factor of -1 upon a full spatial rotation, as verified observationally when two particles are analyzed,[2] and the projective representation obscures this mathematically. An alternative resolution of the problem is to lift the action of $\mathsf{U}\otimes\mathsf{V}$ to the simply connected double cover of the rotation group, which seems to the author a solution of greater elegance and which will be pursued next.

At this point there is a choice to make – use the group $\mathbf{S}^3$ of unit quaternions as the double cover of SO(3) or use its isomorphic cousin, the matrix group SU(2). We choose to develop both viewpoints simultaneously. We already have described explicitly the covering map $\vartheta:\mathbf{S}^3\to$ SO(3) in the discussion after expression 13.16. The result is that if we write the unit quaternion in terms of its real part q_0 and its vector part $\mathbf{q}$, so that $\mathsf{q}=q_0+\mathbf{q}$, and then set $\hat{\mathbf{q}}=\mathbf{q}/\|\mathbf{q}\|$, the unit vector in the direction of $\mathbf{q}$, we have

$$\vartheta(\mathsf{q})=R(2\cos^{-1}q_0,\hat{\mathbf{q}}).$$

This defines a surjective homomorphism with kernel $\{\pm 1\}$. Now the use

[2] See the footnote on p. 281.

of the isomorphism of equation 13.15 is not so helpful. This is merely an accident of our choice of ϱ, equation 13.7, which was employed to describe a matrix representation for the quaternions. The result of using ϱ is that $\varrho(\mathbf{i}) = \mathbf{i}\sigma_3$, $\varrho(\mathbf{j}) = \mathbf{i}\sigma_2$, and $\varrho(\mathbf{k}) = \mathbf{i}\sigma_1$. Notice the reverse order of the Pauli spin matrices, which causes, for example, the rotation $\vartheta(\mathbf{i})$ about the first coordinate axis to correspond to the spin operator representing spin component measurements along the third coordinate axis. The correction for this is to choose an equivalent but, for the purposes of spin in quantum mechanics, more appropriate matrix representation of the quaternions.[3] For this, define the injective $\mathbb{R}$-linear mapping ρ of the quaternions $\mathbb{H}$ into the matrix algebra $\mathrm{M}_2(\mathbb{C})$ by extending $\mathbb{R}$-linearly the basis mapping given by

$$\rho(1) = 1_{2\times 2}, \quad \rho(\mathbf{i}) = \tau_1 \quad \rho(\mathbf{j}) = \tau_2, \quad \rho(\mathbf{k}) = \tau_3,$$

where $\boldsymbol{\tau} = -\mathbf{i}\boldsymbol{\sigma}$. Note that this means that the action of ρ on the quaternion $\mathsf{q} = q_0 + q_1\mathbf{i} + q_2\mathbf{j} + q_3\mathbf{k} = q_0 + \mathbf{q}$ is given by

$$\rho(\mathsf{q}) = \begin{pmatrix} q_0 - q_3\mathbf{i} & -q_2 - q_1\mathbf{i} \\ q_2 - q_1\mathbf{i} & q_0 + q_3\mathbf{i}. \end{pmatrix} = q_0 1_{2\times 2} + \mathbf{q}\cdot\boldsymbol{\tau},$$

or, in terms of the Pauli spin matrices, $\rho(\mathsf{q}) = q_0 1_{2\times 2} - \mathbf{i}\mathbf{q}\cdot\boldsymbol{\sigma}$. A straightforward calculation shows that $\rho(\mathsf{pq}) = \rho(\mathsf{p})\rho(\mathsf{q})$, so that ρ is an injective algebra homomorphism.[4]

[3] The footnote on p. 186 is relevant here. There are many different matrix representations of quaternions, ϱ just being in some sense the cleanest mathematically, given the usual way in which quaternions are written. The choice of any right-handed triple of unit vectors $\mathbf{I}$, $\mathbf{J}$, $\mathbf{K}$ with $\mathbf{IJ} = \mathbf{K}$, etc., in place of the standard triple $\mathbf{i}$, $\mathbf{j}$, $\mathbf{k}$ gives an alternative representation of the quaternions by pairs of complex numbers and then an alternative matrix representation using exactly the development for defining ϱ in Section 13.3. In fact, the avenue pursued in this section may be seen as precisely the use of the choice $\mathbf{I} = -\mathbf{k}$, $\mathbf{J} = -\mathbf{j}$, and $\mathbf{K} = -\mathbf{i}$.

[4] The quick way to see this is to write $\mathsf{q} = z + w\mathbf{j}$, where $z = q_0 + q_3\mathbf{k}$ and $w = q_2 - q_1\mathbf{k}$ and both z and w are elements of the complex plane $\mathbb{C}_{\mathbf{k}}$ in $\mathbb{H}$. Quaternion multiplication translates to

$$\mathsf{pq} = (u + v\mathbf{j})(z + w\mathbf{j}) = (uz - v\overline{w}) + (uw + v\overline{z})\mathbf{j},$$

since $\mathbf{j}z = \overline{z}\mathbf{j}$ whenever $z \in \mathbb{C}_{\mathbf{k}}$. The mapping ρ is given by

$$\rho(\mathsf{q}) = \rho(z + w\mathbf{j}) = \begin{pmatrix} \overline{z} & -\overline{w} \\ w & z \end{pmatrix},$$

where the matrix entries z and w are now interpreted as elements of the usual complex plane $\mathbb{C} = \mathbb{C}_{\mathbf{i}}$ by replacing $\mathbf{k}$ by $\mathbf{i}$. One can see how the matrix representation of the quaternions given by ϱ is mathematically cleaner than that given by ρ; hence the common use of the representation surrounding equation 13.7 in mathematics. For the purposes of spin in quantum mechanics, with the choice of x_3-axis measurements to decompose $V^{1/2}$ that gives the Pauli

Recall that the triad $\boldsymbol{\tau}$ generates the Lie algebra $\mathfrak{su}(2) = \{\mathbf{a}\cdot\boldsymbol{\tau} : \mathbf{a} \in \mathbb{R}^3\}$ with $\mathrm{SU}(2) = \{e^{\mathbf{a}\cdot\boldsymbol{\tau}} : \mathbf{a} \in \mathbb{R}^3\}$. Exactly as $\varrho|\mathbf{S}^3$ was shown to be an isomorphism in the discussion following equation 13.15, so $\rho|\mathbf{S}^3$ is seen to be an isomorphism of the unit quaternion group onto SU(2). We are now in a position to describe the natural action of $\mathbf{S}^3$, the simply connected double cover of SO(3), on the Pauli state space; this encodes in a well-defined representation the effects of spatial rotations on Pauli spinors. For this we use the exponential mapping description of the elements of $\mathbf{S}^3$:

$$\mathbf{S}^3 = \{e^{\mathbf{n}t} : t \in \mathbb{R}, \mathbf{n} \in \mathbf{S}^2\}, \tag{19.5}$$

which follows from the description of the one-parameter subgroups of $\mathbf{S}^3$ as in equation 13.18. As a reminder, the relationship to the unit quaternion q is that $\mathsf{q} = e^{\hat{\mathsf{q}}\cos^{-1} q_0}$ and $e^{\mathbf{n}t} = \cos t + \mathbf{n}\sin t$. The homomorphisms ϑ of $\mathbf{S}^3$ to SO(3) and ρ of $\mathbf{S}^3$ to SU(2) take the forms

$$\vartheta(e^{\mathbf{n}t}) = R(2t, \mathbf{n}) \quad \text{and} \quad \rho(e^{\mathbf{n}t}) = (\cos t)1_{2\times 2} - \mathbf{i}(\sin t)\mathbf{n}\cdot\boldsymbol{\sigma} = e^{t\mathbf{n}\cdot\boldsymbol{\tau}}.$$

We now can define a representation of $\mathbf{S}^3$ on the Pauli state space that describes the effect of a spatial rotation on a state vector without the problem that arises when we try to use SO(3) to parameterize the rotations. Indeed, define the action of the element $e^{\mathbf{n}t}$ of $\mathbf{S}^3$ on a Pauli spinor by

$$e^{\mathbf{n}t}\cdot\begin{pmatrix}\varphi\\ \psi\end{pmatrix} = \rho(e^{\mathbf{n}t})\begin{pmatrix}\varphi\circ\vartheta(e^{\mathbf{n}t})^{-1}\\ \psi\circ\vartheta(e^{\mathbf{n}t})^{-1}\end{pmatrix} = e^{t\mathbf{n}\cdot\boldsymbol{\tau}}\begin{pmatrix}\varphi\circ R(-2t,\mathbf{n})\\ \psi\circ R(-2t,\mathbf{n})\end{pmatrix}.$$

The effect of rotation by an angle θ about an axis $\mathbf{n}$ is given by the action of $e^{\mathbf{n}\theta/2}$ under this representation, since $\vartheta(e^{\mathbf{n}\theta/2}) = R(\theta, \mathbf{n})$. This, rather than being ill-defined as for SO(3) in equation 19.4, is a well-defined representation of $\mathbf{S}^3 \cong \mathrm{SU}(2)$; the effect of a full spatial rotation is given by the action of $e^{\pi\mathbf{n}}$, which multiplies a Pauli spinor by -1, while two full rotations, given by the action of $e^{2\pi\mathbf{n}} = 1$, leaves the Pauli spinor unchanged. This is an elegant solution to this vexing and "strange two-valuedness" associated with Pauli spinors.

matrices in the order given by Pauli, ρ is the natural choice as demonstrated by the remainder of this section.

19.3 The Spin and Magnetic Moment of the Electron

The Hamiltonian for the hydrogenic potential for an electron of mass m_e in wave mechanics is

$$\mathsf{H} = -\frac{\hbar^2}{2m_e}\Delta + V(r) = -\frac{\hbar^2}{2m_e r}\frac{\partial^2}{\partial r^2}r + \frac{1}{2m_e r^2}\mathbf{L}^2 - \frac{Ze^2}{r}, \qquad (19.6)$$

and the Schrödinger equation $\mathsf{H}\psi = E\psi$ has bound state solutions $\psi = \psi_{n,\ell,m}$ for $n \in \mathbb{N}$, $\ell = 0, \ldots, n-1$ and $m = -\ell, -\ell+1, \ldots, \ell-1, \ell$. The principal quantum number n labels the eigenvalues of H, with $\psi_{n,\ell,m}$ belonging to the eigenvalue $E_n = -(Z/n^2)\mathcal{RY}$. The azimuthal, or orbital, angular momentum quantum number ℓ labels the $\mathbf{L}^2$-eigenstates of total angular momentum, and the magnetic quantum number m labels the L_3-eigenstates. The operators H, $\mathbf{L}^2$, and L_3 form a complete set of commuting operators in wave mechanics without spin that specify simultaneous eigenstates for the Coulomb potential that are unique up to a scalar multiple. All this was derived in Lectures 14–16. In Section 16.1 it was pointed out that an electron bound to a hydrogenic potential enjoys a spin degree of freedom indexed by the **spin quantum number** s, which takes the two possible values $\pm\frac{1}{2}$. This doubles the number of available states for the bound electron from the number predicted by wave mechanics without spin. It seems that including spin in this analysis should specify for each appropriate quadruple n, ℓ, m, s a simultaneous eigenstate $\psi_{n,\ell,m,s}$ of the complete set of commuting operators H, $\mathbf{L}^2$, L_3, and S_3, unique up to a scalar multiple. In this section we will see how this arises naturally in Pauli's extension of wave mechanics.

Before examining Pauli's analysis of the hydrogenic system with spin, we take a moment to present the facts concerning the magnetic moment of the electron. In classical mechanics, a particle of mass m, charge q, and orbital angular momentum $\mathbf{l}$ generates a magnetic moment that, in appropriate units, may be written as $\boldsymbol{\mu} = q\mathbf{l}/2mc$, where c is the speed of light in vacuum. More generally, a distribution of total charge q on a body of total mass m that spins with angular momentum vector $\mathbf{l}$ generates a magnetic moment

$$\boldsymbol{\mu} = g\frac{q}{2mc}\mathbf{l}, \qquad (19.7)$$

where g is a dimensionless constant, the **Landé g-factor**, that in practice is determined experimentally and integrates into one constant the particularities of the distribution of charge and mass. When placed in a constant magnetic field with field strength $\mathbf{B}$, the magnetic moment experiences a torque given by $\mathbf{t} = \boldsymbol{\mu} \times \mathbf{B}$ that causes the magnetic moment vector to precess about the direction of $\mathbf{B}$, but otherwise has no effect on the center-of-mass motion of the body. However, if the magnetic field is inhomogeneous of strength $\mathbf{B} = \mathbf{B}(\mathbf{x})$ then classical electromagnetic theory predicts that the body will also experience a force $\mathbf{f} = \nabla(\boldsymbol{\mu} \cdot \mathbf{B}(\mathbf{x}))$, which identifies the potential field as $V_{\text{mag}} = -\boldsymbol{\mu} \cdot \mathbf{B}(\mathbf{x})$.

Promoting the magnetic moment observable to its corresponding operator suggests that the magnetic moment operator for an electron orbiting the nucleus in a hydrogenic potential should be given by

$$\boldsymbol{\mu}_{\text{orbital}} = -\frac{e}{2m_e c}\mathbf{L}$$

where $-e$ is the electron charge. This is an extrinsic property of the electron, which has no effect on the energy levels of atomic hydrogen since the classical Hamiltonian includes no term corresponding to $\boldsymbol{\mu}_{\text{orbital}}$. However, the electron also possesses the intrinsic property of spin angular momentum, which has no classical counterpart like that of orbital angular momentum. One might think that this intrinsic angular momentum is generated by the mass distribution of the electron spinning around an axis, but as far as we can tell, the electron is a point mass without internal motions of its mass. The electron also possesses an intrinsic magnetic moment which naively might be thought to arise from a spinning charge distribution, but this position has been found to be quite untenable when analyzed carefully. The lesson from the experimental and theoretical development of the ideas surrounding electron spin is the following. The intrinsic angular momentum – the spin – of an elementary particle, as well as its intrinsic magnetic moment, should be thought of as a basic property the particle possesses that is in no way derivative of any internal structure or motion. It is an irreducible brute property, like charge or rest mass, that allows no reduction to more elementary considerations. Though we cannot think of the electron as a spinning distribution of charge and mass, we might hypothesize that, just as in the classical world where the angular momentum of a charged mass generates a magnetic moment, the intrinsic spin and intrinsic magnetic moment are intimately linked, the former giving rise to the latter. It is reasonable to assume, and it is born out in laboratory observation, that the relationship between the spin angular momentum and the intrinsic magnetic moment of an electron is of the form of equation 19.7, for some value of the g-factor, say $g = g_e$. Upon quantization, this leads to a relationship between the intrinsic magnetic moment operator $\boldsymbol{\mu}_e$ of the electron and its internal angular momentum, or spin, operator $\mathbf{S}$ that is given by

$$\boldsymbol{\mu}_e = -g_e \frac{e}{2m_e c}\mathbf{S} = -\frac{g_e}{2}\frac{e\hbar}{2m_e c}\boldsymbol{\sigma}. \tag{19.8}$$

The Dirac theory of the electron predicts an exact value for g_e, namely, $g_e = 2$. It turns out that g_e can be measured very precisely in the laboratory and the experimental value is

$$g_e = 2.0023193043738 \pm 0.0000000000082. \tag{19.9}$$

This differs from the Dirac value of $g_e = 2$ to first order in the fine structure constant α by $\alpha/\pi \approx 0.00232$. The quantity $(g_e - 2)/2$ is called the **anomalous magnetic moment** of the electron, and calculations in quantum electrodynamics that take into account so-called radiative corrections give a theoretical value for g_e in agreement with the experimental value of equation 19.9. These calculations convinced physicists of the correctness of quantum electrodynamics, which had developed into a reasonably complete quantum theory of the electromagnetic field by the late 1940s.

From equation 19.8, since the eigenvalues of S_3 are $\pm\frac{1}{2}\hbar$, quantum mechanics predicts that the x_3-component of the electron's magnetic moment should have measured values of

$$\pm\frac{g_e}{2}\frac{e\hbar}{2m_e c}.$$

If one ignores the anomalous magnetic moment, the magnitude of the x_3-component of the electron's magnetic moment is

$$\mu_B = \frac{e\hbar}{2m_e c};$$

this is the basic unit for magnetic moment measurements, a quantity referred to as the **Bohr magneton.**

19.4 The Hydrogenic Potential with Spin

We turn our attention now to one of the main aims of this lecture, to understand how the inclusion of spin via Pauli's theory affects the wave mechanical analysis of the hydrogenic potential. The time-independent Schrödinger equation for the system with spin is obtained by promoting the Hamiltonian of equation 19.6 to the operator $\hat{\mathsf{H}} = \mathsf{H} \otimes 1_{V^{1/2}}$. In matrix notation, with spinors as column vectors, the Hamiltonian takes the form

$$\hat{\mathsf{H}} = \mathsf{H}\, 1_{2\times 2} = \begin{pmatrix} \mathsf{H} & 0 \\ 0 & \mathsf{H} \end{pmatrix}.$$

This Hamiltonian commutes with the operators $\hat{\mathbf{L}}^2$, $\hat{\mathsf{L}}_3$, and $\hat{\mathsf{S}}_3$ and it is easy to see that a complete set of eigenfunctions for the bound states is given by

$$\psi_{n,\ell,m,1/2} = \psi_{n,\ell,m,+} = \psi_{n,\ell,m} \otimes |+\rangle = \begin{pmatrix} \psi_{n,\ell,m} \\ 0 \end{pmatrix}$$

and

$$\psi_{n,\ell,m,-1/2} = \psi_{n,\ell,m,-} = \psi_{n,\ell,m} \otimes |-\rangle = \begin{pmatrix} 0 \\ \psi_{n,\ell,m} \end{pmatrix}.$$

The former is the spin up eigenfunction corresponding to $\psi_{n,\ell,m}$ and the latter is the spin down eigenfunction. The inclusion of spin in the Pauli theory has no effect on the energy eigenvalues of the hydrogenic system, but, as evident here, doubles the degeneracy and doubles the available eigenstates in each energy level.

As a first correction to a pure wave mechanical analysis, we see that the inclusion of spin via the Pauli theory explains the doubled degeneracy seen in the spectroscopic data. This, however, is just the beginning of the story, for the experimental data reveals another phenomenon, slight splitting of the energy levels with respect to the quantum numbers ℓ and s, as reported on p. 238. This phenomenon is explained in the Pauli theory as an effect of the electron's magnetic moment. To understand this, recognize first that, in the rest frame of a classical charged particle orbiting a central electrostatic potential, the particle sees the charge center of the potential orbiting around it. Since a current induces a magnetic field according the Biot–Savart law, the particle finds itself experiencing this magnetic field as a corresponding torque on its magnetic moment vector, as well as a force generated by the potential field $V_{\mathrm{mag}} = -\boldsymbol{\mu} \cdot \mathbf{B}(\mathbf{x})$, where $\mathbf{B}(\mathbf{x})$ is the induced magnetic field at the position $\mathbf{x}$ of the particle. The classical Hamiltonian for this particle then has an additional potential term, which translates into an additional potential term in the quantized Hamiltonian. The Biot–Savart law implies that the magnetic field that the particle experiences is proportional to $\mathbf{E} \times \mathbf{p}$, where $\mathbf{p}$ is the momentum of the particle and $\mathbf{E}(\mathbf{x})$ is the electric field vector due to the orbiting charge center. This is valid for a particle whose speed is small with respect to c. As the particle is in a central inverse-square electrostatic field, the field intensity $\mathbf{E}(\mathbf{x})$ is proportional to $\mathbf{x}/\|\mathbf{x}\|^3$, and we find that the magnetic field satisfies

$$\mathbf{B}(\mathbf{x}) \propto \frac{1}{\|\mathbf{x}\|^3}\mathbf{x} \times \mathbf{p} = \frac{1}{\|\mathbf{x}\|^3}\mathbf{L},$$

where $\mathbf{L}$ is the angular momentum vector. It follows that

$$V_{\text{mag}} = -\boldsymbol{\mu} \cdot \mathbf{B}(\mathbf{x}) = \frac{\lambda}{\|\mathbf{x}\|^3}\boldsymbol{\mu} \cdot \mathbf{L}$$

for some proportionality constant λ. This energy contribution to the classical Hamiltonian, from the interaction of the particle's magnetic moment with the magnetic field experienced by the particle in its moving frame, gives rise to an **interaction term**, known as the **spin–orbit term**, in the Hamiltonian operator for an inverse-square Coulomb system, of the form

$$\mathsf{V}_{\text{mag}} = \frac{\lambda}{\|\mathbf{x}\|^3}\boldsymbol{\mu}_e \cdot \mathsf{L} = \frac{\Lambda}{r^3}\boldsymbol{\sigma} \cdot \mathsf{L} = \frac{\Lambda}{r^3}\begin{pmatrix} \mathsf{L}_3 & \mathsf{L}_1 - \mathrm{i}\mathsf{L}_2 \\ \mathsf{L}_1 + \mathrm{i}\mathsf{L}_2 & -\mathsf{L}_3 \end{pmatrix},$$

where we have used equation 19.8 and absorbed all the constants in the constant Λ. Note that the spin–orbit term naturally operates on the Pauli state space $L^2(\mathbb{R}^3) \otimes V^{1/2}$ of spinors, and the Hamiltonian for the hydrogenic system becomes

$$\mathcal{H} = \hat{\mathsf{H}} + \frac{\Lambda}{r^3}\boldsymbol{\sigma} \cdot \mathsf{L} = \mathsf{H}\, 1_{2\times 2} + \frac{\Lambda}{r^3}\boldsymbol{\sigma} \cdot \mathsf{L}.$$

The coupling of the spin and the orbital angular momenta that leads to the splitting of the spectral lines modeled by this Pauli Hamiltonian $\mathcal{H}$ is called the **spin–orbit interaction**.

The spinor wave equation $\mathcal{H}\psi = E\psi$ is very difficult to solve. The Hamiltonian $\mathcal{H}$ fails to commute with both $\hat{\mathsf{L}}_3$ and $\hat{\mathsf{S}}_3$ since these fail to commute with the interaction term. Thus there are no simultaneous eigenstates of the operators $\mathcal{H}$, $\hat{\mathsf{L}}^2$, $\hat{\mathsf{L}}_3$, and $\hat{\mathsf{S}}_3$. Now, the system is nonetheless spherically symmetric and a separation of variables into an angular term and a radial term produces two equations that can be attacked separately, just as in the spin-free analysis. It turns out that a complete set of commuting operators for this equation consists of $\mathcal{H}$, J^2, J_3, and $\boldsymbol{\sigma} \cdot \mathsf{L}$, where $\mathsf{J} = \hat{\mathsf{L}} + \hat{\mathsf{S}}$ is the total angular momentum operator. The angular equation can be solved exactly and gives solutions called **spinor harmonics** that are simultaneous eigenfunctions of J^2, J_3, and $\boldsymbol{\sigma} \cdot \mathsf{L}$. The radial equation, however, offers no exact solution and one must use approximation methods.

If the details of the solution are carried out for the spinor wave equation using first-order perturbation theory, the derived energy level splits are in good agreement with observation. Since the analysis is so difficult and since the more accurate Dirac theory admits exact solutions with energy levels as in formula 16.2, we will be satisfied with the outline of Pauli's theory of the hydrogenic system presented thus far and will not pursue it further here.

19.5 End Notes

The End Notes of the previous lecture are also relevant for this lecture. Any reader interested in a more thorough investigation of the Pauli theory of the electron and a more thorough treatment of spin-$\frac{1}{2}$ particles is referred to Thaller (2005), Chapters 2 and 3. My development of the latter two sections of this lecture follow Thaller's development. In Section 22.2 we will develop the mathematics of the coupled spin-$\frac{1}{2}$ system and use it to articulate and clarify the most philosophically significant implication of quantum mechanics – that nature is nonlocal and multiparticle quantum systems are inseparable. In Lecture 23 we will further dissect the spin-$\frac{1}{2}$ system and in Section 23.3 we will develop more of the matrix algebra associated with the system.

For a thorough treatment of spinors and their appearance in physics, see Hladik (1999). Tomonaga (1997) is a beautifully written (and translated) historical reminiscence of the development of spin in quantum mechanics, written by one of the pioneers of quantum electrodynamics. The chapter on Pauli spinors is apt to our discussion.

20
Clifford Algebras and Spin Representations ⋆

> No one fully understands spinors. Their algebra is formally understood but their general significance is mysterious. In some sense they describe the "square root" of geometry and, just as understanding the square root of -1 took centuries, the same might be true of spinors.
>
> Sir Michael Atiya, quoted in Graham Farmelo's *The Strangest Man: The Hidden Life of Paul Dirac, Mystic of the Atom*, Basic Books, 2009

This lecture presents a general mathematical framework for spinors that was discovered by the mathematician Élie Cartan in 1913 in his study of the representation theory of linear groups. The Pauli theory corresponds in the Cartan theory to the natural representation of the group Spin_3 while the Dirac theory corresponds to that of $\mathrm{Spin}_{3,1}$. The appropriate theoretical setting for a spin group is that of a **Clifford algebra**. We will present a very quick introduction to the abstract theory of Clifford algebras before specializing to the Clifford algebras over the real and complex numbers, sometimes called **geometric algebras** since they encode so much of the classical geometric structure of pseudo-Euclidean spaces. Moving from the abstract to the concrete, we will present a hands-on construction of these algebras and identify the low-dimensional ones in terms of more familiar matrix algebras. We then will define the **pin** and **spin groups** associated with them and examine how these groups act on their generating pseudo-Euclidean spaces. This will bring us to the general theory of spin representations and spinors. We close the lecture with a precise description of Pauli spinors in this context, and an introduction to Dirac spinors. This is a bonus lecture that presents the mathematics that underlies spin in quantum mechanics, but its contents will not appear in subsequent lectures and, hence, it may be skipped without harming the flow of the book.

20.1 Clifford Algebras

A Clifford algebra is defined with respect to a **quadratic space** (V, g), where g is a quadratic form defined on the vector space V. The Clifford algebra $C\ell(V, g)$ is designed to answer affirmatively and with economy whether one may define an associative multiplication operation on V with the property that

$$v^2 = -g(v) \quad \text{for all} \quad v \in V.^1 \tag{20.1}$$

We will examine in detail only those Clifford algebras generated by finite-dimensional vector spaces. When V is a real vector space of dimension n, every nondegenerate quadratic form is equivalent to one of the standard pseudo-Euclidean forms $g_{p,q}$ of **signature** (p, q) defined on $\mathbb{R}^n$ by

$$g_{p,q}(v) = v_1^2 + \cdots + v_p^2 - v_{p+1}^2 \cdots - v_n^2.$$

Here p and q are nonnegative integers with $p + q = n$ and $v = \sum_i v_i \mathbf{e}_i$, where the vector $\mathbf{e}_i = (0, \ldots, 1, \ldots, 0)$ is the standard ith basis vector of $\mathbb{R}^n$ with a "1" in the ith spot and zeroes elsewhere. Notice that $g_{p,q}(\mathbf{e}_i) = +1$ when $i \leq p$ and $g_{p,q}(\mathbf{e}_i) = -1$ otherwise. For this case we denote the Clifford algebra by $C\ell_{p,q}$ and, when $q = 0$, by $C\ell_n$. Notice that in the Clifford algebra $C\ell_n$, equation 20.1 reads $v^2 = -\|v\|^2$ for all $v \in \mathbb{R}^n$. It is not surprising, then, with this intertwining of the algebraic multiplication operation with the Euclidean norm, that much of the Euclidean geometry of $\mathbb{R}^n$ is encoded in the Clifford multiplication of the algebra $C\ell_n$. When V is a complex vector space of dimension n, every nondegenerate quadratic form is equivalent to the standard Euclidean form on the Euclidean space $\mathbb{C}^n$ defined by

$$v_1^2 + \cdots + v_n^2,$$

where again $v = \sum_i v_i \mathbf{e}_i$ and, of course, the coefficients v_i are complex. In this case the Clifford algebra is denoted as $C\ell_n(\mathbb{C})$. Denoting $\mathbb{R}^n$

[1] Some authors drop the minus sign. This is merely a matter of convention: replace g with $-g$ to go from our description to the other.

endowed with the form $g_{p,q}$ by $\mathbb{R}^{p,q}$, we will see that the subsets

$$\mathbb{R}^{p,q} \subset \mathcal{C}\ell_{p,q} \quad \text{and} \quad \mathbb{C}^n \subset \mathcal{C}\ell_n(\mathbb{C})$$

each generate their respective algebra, that all these algebras have dimension 2^n over their ground fields, and that these algebras encode much of the geometry of these pseudo-Euclidean spaces in their Clifford multiplication. In a very precise sense, the Clifford algebra is the smallest and most economical algebra containing its generating vector space for which equation 20.1 holds.

We will shortly give explicit hands-on constructions for the algebras $\mathcal{C}\ell_{p,q}$ and $\mathcal{C}\ell_n(\mathbb{C})$ but, for those schooled in categorical thinking and universal mapping properties, we first take a moment to give the explicit abstract definitions and existence constructions for a Clifford algebra in the full generality of an arbitrary vector space V over the field $\mathbb{K}$ with quadratic form $g : V \to \mathbb{K}$. A **Clifford algebra** $\mathcal{C}\ell(V, g)$ determined by the pair (V, g) is an associative $\mathbb{K}$-algebra with unit $\mathbf{1}$, equipped with a $\mathbb{K}$-linear embedding $\jmath : V \to \mathcal{C}\ell(V, g)$ such that $\jmath(v)^2 = -g(v)\mathbf{1}$ for all $v \in V$, that satisfies the following universal mapping property: if $\jmath : V \to A$ is any $\mathbb{K}$-linear map to any associative $\mathbb{K}$-algebra A with unit $\mathbf{1}_A$ such that $\jmath(v)^2 = -g(v)\mathbf{1}_A$ for all $v \in V$ then there exists a unique $\mathbb{K}$-algebra homomorphism $h : \mathcal{C}\ell(V, g) \to A$ such that $h \circ \jmath = \jmath$, i.e., such that the diagram

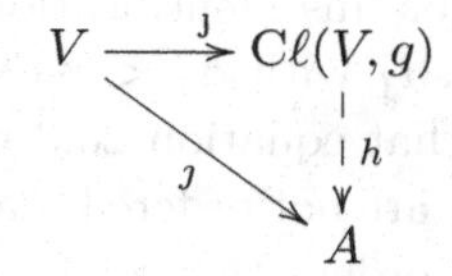

commutes. As the initiate of category theory may guess, $\mathcal{C}\ell$ defines a functor from the category of quadratic $\mathbb{K}$-vector spaces whose morphisms are $\mathbb{K}$-linear maps that preserve the quadratic forms to the category of associative $\mathbb{K}$-algebras with units. As usual for objects defined by universal mapping properties, $\mathcal{C}\ell(V, g)$, if it exists, is unique up to isomorphism and the embedding identifies V as a vector subspace of $\mathcal{C}\ell(V, g)$. Existence follows abstractly by taking the quotient of the tensor algebra $\otimes V$ of V over $\mathbb{K}$ by the two-sided ideal generated by elements of the form $v \otimes v + g(v)\mathbf{1}_{\otimes V}$, as v ranges over V. Actually, we do not need to assume $\jmath$ is injective as this follows from the construction just presented. An immediate but very important consequence of this universal property is that if V is a linear subspace of W then $\mathcal{C}\ell(V, g|V)$ is a subalgebra of $\mathcal{C}\ell(W, g)$, where of course $g|V$ is the restriction of the quadratic form

g to V. For example, this provides the standard inclusion of $C\ell_n$ into $C\ell_{n+1}$ induced from the inclusion $\mathbb{R}^n \subset \mathbb{R}^{n+1}$. One important example of Clifford algebras is the exterior algebra $\wedge V$ over V, which is nothing more than $C\ell(V, 0)$, the Clifford algebra over V with the zero quadratic form.

We now return the discussion to finite-dimensional vector spaces and examine an explicit construction for $C\ell_{p,q}$. Notice from equation 20.1 that in $C\ell_{p,q}$ the relations

$$\mathbf{e}_i^2 = \begin{cases} -1 & \text{for} \quad i = 1, \ldots, p, \\ +1 & \text{for} \quad i = p+1, \ldots, p+q, \end{cases} \tag{20.2}$$

and

$$\mathbf{e}_i \mathbf{e}_j = -\mathbf{e}_j \mathbf{e}_i \quad \text{for} \quad i \neq j \tag{20.3}$$

hold, the latter a consequence of evaluating $(\mathbf{e}_i + \mathbf{e}_j)^2$. The Clifford algebra $C\ell_{p,q}$ is to have dimension 2^n, so we let the underlying real vector space of $C\ell_{p,q}$ be spanned by the set of vectors $\{\mathbf{e}_\alpha\}$ indexed by the subsets α of $\{1, \ldots, p+q = n\}$. To make $C\ell_{p,q}$ into an algebra, we need to define the product of any two elements, and this is accomplished as follows. First, identify $\mathbf{e}_i$ with $\mathbf{e}_{\{i\}}$ and use equation 20.2 to define their squares. Notice that this defines a canonical embedding of $\mathbb{R}^{p,q}$ into $C\ell_{p,q}$. Define $\mathbf{e}_\emptyset$ to be the unit $\mathbf{1}$ of the algebra $C\ell_{p,q}$ and identify its span, $\mathbb{R}\mathbf{e}_\emptyset$, with the reals $\mathbb{R}$. Under this identification, $\mathbf{e}_\emptyset = 1$. For the subset $\alpha = \{\alpha_1, \ldots, \alpha_j\} \subset \{1, \ldots, n\}$ with $\alpha_1 < \cdots < \alpha_j$, define the product $\mathbf{e}_{\alpha_1} \cdots \mathbf{e}_{\alpha_j}$ as $\mathbf{e}_\alpha$. Observe that equation 20.3 tells us how to define the product $\mathbf{e}_{\alpha_1} \cdots \mathbf{e}_{\alpha_j}$ if the α_i are not ordered, namely, as $(\operatorname{sgn} \sigma)\mathbf{e}_\alpha$ where σ is the permutation that orders the α_i. In fact, equations 20.2 and 20.3 tell us how to find all the products of the basis elements. Indeed, from equation 20.3, if $\alpha = \{\alpha_1 < \cdots < \alpha_j\}$ and $\beta = \{\beta_1 < \cdots < \beta_k\}$ are disjoint then $\mathbf{e}_\alpha \mathbf{e}_\beta = (\operatorname{sgn} \sigma)\mathbf{e}_{\alpha \cup \beta}$ where σ is the permutation of $(\alpha_1, \ldots, \alpha_j, \beta_1, \ldots, \beta_k)$ that orders the elements. Generally,

$$\mathbf{e}_\alpha \mathbf{e}_\beta = \pm \mathbf{e}_{\alpha \cup \beta - \alpha \cap \beta},$$

where the sign is uniquely determined by equations 20.2 and 20.3. The final step is to extend this multiplication so defined on the basis vectors of $C\ell_{p,q}$ bilinearly to the whole space. A straightforward exercise now confirms that equation 20.1 holds and that $C\ell_{p,q}$ so defined is the minimal associative algebra with unit containing $\mathbb{R}^{p,q}$ and satisfying equation 20.1 or, more properly, is a solution to the universal mapping property defining Clifford algebras.

The quickest way to obtain the complex Clifford algebras is by complexification of the real Clifford algebras. Recall that a real vector space V can be complexified by tensoring with the complex numbers to yield a real vector space $V_{\mathbb{C}} = V \otimes_{\mathbb{R}} \mathbb{C}$ that contains V as a real subspace via the embedding $v \mapsto v \otimes 1$. This defines $V_{\mathbb{C}}$ as a real vector space, but complex scalar multiplication may now be defined by $\mu(v \otimes \nu) = v \otimes (\mu\nu)$ for $v \in V$ and $\mu, \nu \in \mathbb{C}$. This may be realized more concretely by $V_{\mathbb{C}} \cong V \oplus \mathbf{i}V$, where complex scalar multiplication is given by $(a + b\mathbf{i})(v + w\mathbf{i}) = (av - bw) + (aw + bv)\mathbf{i}$ for $a, b \in \mathbb{R}$ and $v, w \in V$. A real basis $\{\mathbf{e}_i\}$ for V determines the complex basis $\{\mathbf{e}_i \otimes 1\}$ for $V_{\mathbb{C}}$, so that the real dimension of V agrees with the complex dimension for its complexification. When V is an algebra, so too is the complexification via $(v \otimes \mu)(w \otimes \nu) = vw \otimes \mu\nu$, extended bilinearly. The reader may prove that the complex Clifford algebras are given by

$$\mathrm{C}\ell_n(\mathbb{C}) \cong (\mathrm{C}\ell_{p,q})_{\mathbb{C}} = \mathrm{C}\ell_{p,q} \otimes_{\mathbb{R}} \mathbb{C},$$

for any nonnegative pair p and q with $p + q = n$, by verifying the universal mapping property.

In the remainder of this section, let $\mathrm{C}\ell$ be one of the real or complex Clifford algebras. There are three involutions[2] of $\mathrm{C}\ell$ that are a great aid in uncovering the internal structure of Clifford algebras. These take the titles of **main involution**, **transpose**, and **Clifford conjugation**. The main involution is the unique algebra automorphism, which we denote with a tilde, defined on the standard basis vectors $\mathbf{e}_i$ as reflection through the origin: $\mathbf{e}_i \mapsto -\mathbf{e}_i$. Its existence follows from the universal mapping property, but we will give an independent verification at the end of this paragraph. Assuming existence, it follows that the main involution is the linear extension of the rule given on the basis vectors of $\mathrm{C}\ell$ as

$$\tilde{\mathbf{e}}_\alpha = (-1)^{|\alpha|}\mathbf{e}_\alpha, \tag{20.4}$$

where $|\alpha|$ is the cardinality of the set α. Any involution of a real or complex vector space, since it squares to the identity, has only ± 1 as possible eigenvalues. The main involution thus decomposes $\mathrm{C}\ell$ into ± 1-eigenspaces $\mathrm{C}\ell^{\pm}$ and we have $\mathrm{C}\ell = \mathrm{C}\ell^+ \oplus \mathrm{C}\ell^-$. This realizes $\mathrm{C}\ell$ as a $\mathbb{Z}_2$-graded algebra since $\mathrm{C}\ell^s \mathrm{C}\ell^t \subset \mathrm{C}\ell^{st}$, where $\{s, t\} \subset \{+, -\}$.[3] Note that

[2] Recall that an involution of a vector space is a linear self-isomorphism that is its own inverse, i.e., whose square is the identity.

[3] Here we interpret $\mathbb{Z}_2$ as the set that consist of the two symbols $+$ and $-$, with $+$ serving as the identity and $(-)^2 = +$. We point out that the notation for Clifford algebras is not standardized, with some authors writing $\mathrm{C}\ell^0$ or $\mathrm{C}\ell^{\mathrm{even}}$ for our $\mathrm{C}\ell^+$ and $\mathrm{C}\ell^1$ or $\mathrm{C}\ell^{\mathrm{odd}}$ for our $\mathrm{C}\ell^-$, and using our notation for other subspaces.

the **even** subspace $\mathrm{C}\ell^+$ is a subalgebra of $\mathrm{C}\ell$ while the **odd** subspace $\mathrm{C}\ell^-$ is not since it is not closed under Clifford multiplication. The Clifford algebra $\mathrm{C}\ell$ admits another decomposition into a direct sum, as $\mathrm{C}\ell = \oplus_{i=0}^n \mathrm{C}\ell^i$ where $\mathrm{C}\ell^i$ is the vector subspace of dimension $\binom{n}{i}$ spanned by the basis vectors $\mathbf{e}_\alpha$ with $|\alpha| = i$. This is a vector space grading only and not a grading of algebras. The elements of $\mathrm{C}\ell^i$ are said to be **pure** of **degree** i. Notice that the main involution acts as the identity on the even elements and as reflection through the origin on the odd elements. To verify explicitly that the main involution, which obviously is a vector space isomorphism, is in fact an isomorphism of the algebra, we need to show that it preserves Clifford multiplication. Decompose the elements $x, y \in \mathrm{C}\ell$ into their even and odd parts as $x = x^+ + x^-$ and $y = y^+ + y^-$, where $x^\pm, y^\pm \in \mathrm{C}\ell^\pm$, and observe that $\tilde{x} = x^+ - x^-$. Then easily $(xy)^+ = x^+y^+ + x^-y^-$ and $(xy)^- = x^+y^- + x^-y^+$ and we have

$$\widetilde{xy} = (xy)^+ - (xy)^- = x^+y^+ - x^+y^- - x^-y^+ + x^-y^- = \tilde{x}\tilde{y}.$$

This confirms that the main involution is an algebra automorphism.

The two remaining involutions, the transpose and Clifford conjugation, like the main involution in equation 20.4, act on the basis vectors $\mathbf{e}_\alpha$ as either the identity or as reflection through the origin. The transpose is denoted by x^{t} and is defined as the linear extension of the rule $\mathbf{e}_\alpha^{\mathrm{t}} = \mathbf{e}_{\alpha_j} \cdots \mathbf{e}_{\alpha_1}$ when $\alpha = \{\alpha_1 < \cdots < \alpha_j\}$. It follows that $\mathbf{e}_\alpha^{\mathrm{t}} = (\operatorname{sgn}\sigma)\mathbf{e}_\alpha$ where σ is the permutation $(\alpha_j, \ldots, \alpha_1) \mapsto (\alpha_1, \ldots, \alpha_j)$. This may be expressed exactly as

$$\mathbf{e}_\alpha^{\mathrm{t}} = (-1)^{|\alpha|(|\alpha|-1)/2}\mathbf{e}_\alpha, \tag{20.5}$$

and the reader may verify that transposition is an anti-automorphism of algebras, meaning that $(xy)^{\mathrm{t}} = y^{\mathrm{t}}x^{\mathrm{t}}$ for Clifford elements x and y. Finally, Clifford conjugation is just the composition of the main involution with transposition in either order and is denoted with an asterisk. We have $x^* = (\tilde{x})^{\mathrm{t}} = (x^{\mathrm{t}})^{\sim}$, which, easily, is an anti-automorphism. From equations 20.4 and 20.5 we have

$$\mathbf{e}_\alpha^* = (-1)^{|\alpha|(|\alpha|+1)/2}\mathbf{e}_\alpha. \tag{20.6}$$

Equations 20.4–20.6 imply that all three involutions respect the $\mathbb{Z}$-grading in that all three restrict variously as either the identity or as reflection through the origin on the pure elements of a fixed degree, i.e., on $\mathrm{C}\ell^i$. Moreover, all three actions on pure elements depend only on the degree modulo 4, with the action on a pure element x of mod 4 degrees 1, 2, 3, and 4 given by the following table:

degree mod 4	$\tilde{x}$	x^{t}	x^*
1	$-$	$+$	$-$
2	$+$	$-$	$-$
3	$-$	$-$	$+$
4	$+$	$+$	$+$

(20.7)

Recall that the inclusion $\mathbb{R}^n \subset \mathbb{R}^{n+1}$ induces the standard algebra inclusion $C\ell_n \subset C\ell_{n+1}$. More interesting and useful, though, is another embedding of $C\ell_n$ into $C\ell_{n+1}$, namely, as the even subalgebra of $C\ell_{n+1}$. Indeed, our claim is more general, namely, that the even subalgebra $C\ell^+$ of a real or complex Clifford algebra $C\ell$ of dimension 2^{n+1} is itself a Clifford algebra of dimension 2^n and, in the real setting of signature $(p+1, q)$, this becomes $C\ell^+_{p+1,q} \cong C\ell_{p,q}$ while, in the complex setting, $C\ell^+_{n+1}(\mathbb{C}) \cong C\ell_n(\mathbb{C})$. This we verify only for the real case, using the universal mapping property of Clifford algebras. Indeed, let $\{\mathbf{e}_i\}$ be the standard $g_{p,q}$-orthonormal basis for $\mathbb{R}^{p,q}$ with $g_{p,q}(\mathbf{e}_i) = +1$ for $i \le p$ and $g_{p,q}(\mathbf{e}_i) = -1$ for $i > p$. Write $\mathbb{R}^{p+1,q}$ as the internal direct sum $\mathbb{R}\mathbf{e}_0 \oplus \mathbb{R}^{p,q}$ where $g_{p+1,q}(\mathbf{e}_0) = +1$ and $g_{p+1,q}|\mathbb{R}^{p,q} = g_{p,q}$. Define the map $\jmath : \mathbb{R}^{p,q} \to C\ell_{p+1,q}$ on the above basis by

$$\jmath(\mathbf{e}_i) = \mathbf{e}_0\mathbf{e}_i \quad \text{for} \quad 1 \le i \le p+q$$

and extend linearly. Let $v = \sum_{i=1}^{p+q} v_i\mathbf{e}_i$ and use properties 20.2 and 20.3 to calculate that

$$\begin{aligned}\jmath(v)^2 &= \sum_{i,j=1}^{p+q} v_iv_j\mathbf{e}_0\mathbf{e}_i\mathbf{e}_0\mathbf{e}_j = -\sum_{i,j=1}^{p+q} v_iv_j\mathbf{e}_0^2\mathbf{e}_i\mathbf{e}_j \\ &= \sum_{i=1}^{p+q} v_i^2\mathbf{e}_i^2 + \sum_{i<j} v_iv_j\,(\mathbf{e}_i\mathbf{e}_j + \mathbf{e}_j\mathbf{e}_i) = -g_{p,q}(v)\mathbf{1}.\end{aligned}$$

The universal mapping property now implies the existence of an algebra homomorphism

$$h : C\ell_{p,q} \to C\ell_{p+1,q}$$

extending $\jmath$ with $h(\mathbf{e}_i) = \mathbf{e}_0\mathbf{e}_i$, for $1 \le i \le p+q$. To verify that h is an isomorphism of $C\ell_{p,q}$ onto the even subalgebra $C\ell^+_{p+1,q}$, we just observe that the image of the standard basis of $C\ell_{p,q}$ under h is the standard basis of $C\ell^+_{p+1,q}$. Indeed, if $\alpha \subset \{1, \ldots, p+q\}$ then, applying the fact that $\mathbf{e}_0\mathbf{e}_{k-1}\mathbf{e}_0\mathbf{e}_k = \mathbf{e}_{k-1}\mathbf{e}_k$, starting from the right-hand side and moving to the left of the expression $h(\mathbf{e}_\alpha)$ gives

$$h(\mathbf{e}_\alpha) = \begin{cases} \mathbf{e}_\alpha & \text{if } |\alpha| \text{ is even,} \\ \mathbf{e}_0\mathbf{e}_\alpha = \mathbf{e}_{\{0\}\cup\alpha} & \text{if } |\alpha| \text{ is odd.} \end{cases}$$

Now, since the standard basis of the even subalgebra $\mathcal{C}\ell^+_{p+1,q}$ is indexed by the subsets of $\{0,\ldots,p+q\}$ with an even number of elements, which may be described precisely as

$$\{\alpha : |\alpha| \text{ is even}\} \cup \{\{0\} \cup \alpha : |\alpha| \text{ is odd}\}$$

as α ranges over all the subsets of $\{1,\ldots,p+q\}$, we conclude that h is an isomorphism of $\mathcal{C}\ell_{p,q}$ to $\mathcal{C}\ell^+_{p+1,q}$.

We close this section with a description of some symmetries in the collection of the real Clifford algebras. The **main symmetries** for the real Clifford algebras take the forms

$$\mathcal{C}\ell^+_{p,q} \cong \mathcal{C}\ell^+_{q,p} \quad \text{and} \quad \mathcal{C}\ell_{p-1,q} \cong \mathcal{C}\ell_{q-1,p}. \tag{20.8}$$

Moreover, the first isomorphism preserves the degrees of the elements. To obtain these isomorphisms, we use the isomorphism $h : \mathcal{C}\ell_{p,q} \cong \mathcal{C}\ell^+_{p+1,q}$ constructed above as well as the companion isomorphism $h' : \mathcal{C}\ell_{p,q} \cong \mathcal{C}\ell^+_{q,p+1}$ constructed as follows. As before let $\{\mathbf{e}_i\}_{i=1}^n$ be the standard $g_{p,q}$-orthonormal basis for $\mathbb{R}^{p,q}$ with $g_{p,q}(\mathbf{e}_i) = +1$ for $i \le p$ and $g_{p,q}(\mathbf{e}_i) = -1$ for $i > p$. Let $\{\mathbf{f}_i\}_{i=0}^n$ be the $g_{q,p+1}$-orthonormal basis for $\mathbb{R}^{q,p+1}$ listed so that the negative norms appear first, i.e., so that $g_{q,p+1}(\mathbf{f}_i) = -1$ for $i \le p$ and $g_{q,p+1}(\mathbf{f}_i) = +1$ for $i > p$. Define the map $\jmath : \mathbb{R}^{p,q} \to \mathcal{C}\ell_{q,p+1}$ on the basis by

$$\jmath(\mathbf{e}_i) = \mathbf{f}_0\mathbf{f}_i \quad \text{for} \quad 1 \le i \le p+q = n$$

and extend linearly. Then, a calculation shows that

$$\jmath(\mathbf{e}_i)^2 = -\mathbf{f}_i^2 = g_{q,p+1}(\mathbf{f}_i)\mathbf{1} = -g_{p,q}(\mathbf{e}_i)\mathbf{1}$$

for $1 \le i \le n$ and one may apply the universal mapping property and proceed as before. To see the main symmetries, observe that if $p, q \ge 1$ then

$$\mathcal{C}\ell^+_{p,q} \xleftarrow[\cong]{h} \mathcal{C}\ell_{p-1,q} \xrightarrow[\cong]{h'} \mathcal{C}\ell^+_{q,p} \xleftarrow[\cong]{h} \mathcal{C}\ell_{q-1,p},$$

and if $p = n$ and $q = 0$ then

$$\mathcal{C}\ell^+_n = \mathcal{C}\ell^+_{n,0} \xleftarrow[\cong]{h} \mathcal{C}\ell_{n-1} \xrightarrow[\cong]{h'} \mathcal{C}\ell^+_{0,n}.$$

As a final exercise for the student, we invite him or her to verify the symmetry

$$C\ell_{p+4,q} \cong C\ell_{p,q+4}. \tag{20.9}$$

20.2 Low-Dimensional Algebras

The finite-dimensional real and complex Clifford algebras are completely classified, and all are isomorphic either to a matrix algebra over $\mathbb{R}$, $\mathbb{C}$, or $\mathbb{H}$ or to a direct sum of two of these matrix algebras.[4] Since the Pauli and Dirac theories impinge on only a couple of low-dimensional algebras, we are content with identifying only a few select Clifford algebras in terms of these matrix algebras and probing their internal structure. Specifically, we will observe that the following isomorphisms of real algebras hold: $C\ell_0 \cong \mathbb{R}$, $C\ell_1 \cong \mathbb{C}$, $C\ell_2 \cong \mathbb{H}$, $C\ell_3 \cong \mathbb{H} \oplus \mathbb{H}$, $C\ell_4 \cong M_2(\mathbb{H})$, $C\ell_{1,3} \cong M_4(\mathbb{R})$, and $C\ell_{3,1} \cong M_2(\mathbb{H})$. We also will observe that the following complex algebra isomorphisms hold: $C\ell_2(\mathbb{C}) \cong \mathbb{H}_\mathbb{C} \cong M_2(\mathbb{C})$, $C\ell_3(\mathbb{C}) \cong (\mathbb{H} \oplus \mathbb{H})_\mathbb{C} \cong M_2(\mathbb{C}) \oplus M_2(\mathbb{C})$, and $C\ell_4(\mathbb{C}) \cong M_4(\mathbb{C})$. The algebra $C\ell_3$ gives the Clifford algebra framework for the Pauli spinors, while $C\ell_{3,1}$ and $C\ell_4(\mathbb{C})$ give the Clifford algebra framework for Dirac spinors.

Addressing the real algebras first: when $n = 0$, $V = \mathbb{R}^0$ is the trivial vector space with zero quadratic form, and the only subset of the empty indexing set is the empty set. It follows that $C\ell_0$ is spanned by the single basis vector $\mathbf{e}_\emptyset$ and $C\ell_0 = \mathbb{R}\mathbf{e}_\emptyset \cong \mathbb{R}$. The two examples following this curiosity are easy to construct. When $n = 1$, $V = \mathbb{R}$ and the basis for $C\ell_1$ consist of the two elements $\mathbf{e}_\emptyset = \mathbf{1}$ and $\mathbf{e}_1$, the former the multiplicative identity of the algebra and the latter satisfying $\mathbf{e}_1^2 = -\mathbf{1}$. The mapping $a + b\mathbf{i} \mapsto a\mathbf{1} + b\mathbf{e}_1$ defines an $\mathbb{R}$-algebra isomorphism of $C\ell_1$ with the complex field $\mathbb{C}$. When $n = 2$, $V = \mathbb{R}^2$, the basis for $C\ell_2$ consists of the four elements $\mathbf{1}$, $\mathbf{e}_1$, $\mathbf{e}_2$, and $\mathbf{e}_{\{1,2\}} = \mathbf{e}_1\mathbf{e}_2$. We have $\mathbf{e}_1^2 = \mathbf{e}_2^2 = \mathbf{e}_{\{1,2\}}^2 = -\mathbf{1}$. Moreover, the triality of the quaternion units described by diagram 13.6, that

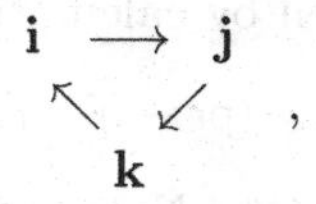

[4] See Theorem 11.3 of Harvey (1990) for a full classification of the algebras $C\ell(p,q)$ as matrix groups and their direct sums.

is valid for these three basis elements in that

$$\begin{array}{ccc} \mathbf{e}_1 & \longrightarrow & \mathbf{e}_2 \\ & \nwarrow \quad \swarrow & \\ & \mathbf{e}_{\{1,2\}} & \end{array} .$$

It follows that the basis mapping $\mathbf{1} \mapsto 1$, $\mathbf{e}_1 \mapsto \mathbf{i}$, $\mathbf{e}_2 \mapsto \mathbf{j}$, and $\mathbf{e}_{\{1,2\}} \mapsto \mathbf{k}$ defines an $\mathbb{R}$-algebra isomorphism of $C\ell_2$ with the real quaternion algebra $\mathbb{H}$.

A little more interesting is the real Clifford algebra when $n = 3$: $C\ell_3$ is an eight-dimensional real algebra spanned by $\mathbf{1}$, $\mathbf{e}_1$, $\mathbf{e}_2$, $\mathbf{e}_3$, $\mathbf{e}_{\{1,2\}}$, $\mathbf{e}_{\{1,3\}}$, $\mathbf{e}_{\{2,3\}}$, and $\mathbf{e}_{\{1,2,3\}}$. We are going to take some time to look at this example through several different lenses. First, perhaps the simplest representation of this Clifford algebra is as the external direct sum algebra $\mathbb{H} \oplus \mathbb{H}$ with component-wise multiplication: $(\mathsf{p}, \mathsf{q})(\mathsf{p}', \mathsf{q}') = (\mathsf{pp}', \mathsf{qq}')$. One isomorphism is given on basis elements by the prescription

$$\begin{aligned} \mathbf{1} &\mapsto (1,1), & \mathbf{e}_1\mathbf{e}_2 &\mapsto (\mathbf{k},\mathbf{k}), \\ \mathbf{e}_1 &\mapsto (\mathbf{i},\mathbf{i}), & \mathbf{e}_1\mathbf{e}_3 &\mapsto (-\mathbf{j},\mathbf{j}), \\ \mathbf{e}_2 &\mapsto (\mathbf{j},\mathbf{j}), & \mathbf{e}_2\mathbf{e}_3 &\mapsto (\mathbf{i},-\mathbf{i}), \\ \mathbf{e}_3 &\mapsto (\mathbf{k},-\mathbf{k}), & \mathbf{e}_1\mathbf{e}_2\mathbf{e}_3 &\mapsto (-1,1). \end{aligned} \tag{20.10}$$

It may easily be checked that this prescription preserves multiplication, and therefore its $\mathbb{R}$-linear vector space extension is an isomorphism of algebras. Notice that, under this isomorphism identifying $C\ell_3$ with $\mathbb{H}\oplus\mathbb{H}$, the standard inclusion $C\ell_2 \subset C\ell_3$ induced from the inclusion $\mathbb{R}^2 \subset \mathbb{R}^3$ identifies $C\ell_2 = \operatorname{span}\{\mathbf{1}, \mathbf{e}_1, \mathbf{e}_2, \mathbf{e}_1\mathbf{e}_2\} \cong \mathbb{H}$ as the diagonal in $\mathbb{H} \oplus \mathbb{H}$, via

$$C\ell_2 \cong \mathbb{H} \ni \mathsf{p} \mapsto (\mathsf{p}, \mathsf{p}) \in \mathbb{H} \oplus \mathbb{H} \cong C\ell_3.$$

The standard inclusion decomposes $C\ell_3$ as a vector space internal direct sum:

$$C\ell_3 = C\ell_2 \oplus C\ell_2\mathbf{e}_3.$$

Notice that, as $C\ell_2 \cong \mathbb{H}$, this offers an alternative description of $C\ell_3$ as a direct sum of quaternions, but this is only a vector space direct sum since the Clifford multiplication does not act component-wise. Indeed, the multiplication can be seen by calculation to be as follows:

$$(\mathsf{p} + \mathsf{q}\mathbf{e}_3)(\mathsf{r} + \mathsf{s}\mathbf{e}_3) = (\mathsf{pr} - \mathsf{q}\tilde{\mathsf{s}}) + (\mathsf{ps} + \mathsf{q}\tilde{\mathsf{r}})\mathbf{e}_3, \tag{20.11}$$

since $\mathbf{e}_3\mathsf{p} = \tilde{\mathsf{p}}\mathbf{e}_3$ for $\mathsf{p} \in \mathbb{H} \cong C\ell_2$. Note that, under this isomorphism of $C\ell_2$, the main automorphism acting on a quaternion p changes the sign of the coefficients of $\mathbf{i}$ and $\mathbf{j}$ and leaves the real term and the coefficient of

k unchanged. This of course describes $C\ell_3$ as isomorphic to an algebra with underlying vector space $\mathbb{H} \oplus \mathbb{H}$, with, instead of component-wise multiplication, the multiplication

$$(\mathsf{p}, \mathsf{q})(\mathsf{r}, \mathsf{s}) = (\mathsf{pr} - \mathsf{q}\tilde{\mathsf{s}}, \mathsf{ps} + \mathsf{q}\tilde{\mathsf{r}})$$

induced from equation 20.11. This may be represented more cleanly as a matrix subalgebra of the real matrix algebra $M_2(\mathbb{H})$, given by the prescription

$$\mathsf{p} + \mathsf{q}\mathbf{e}_3 \mapsto \begin{pmatrix} \mathsf{p} & \mathsf{q} \\ -\tilde{\mathsf{q}} & \tilde{\mathsf{p}} \end{pmatrix}. \tag{20.12}$$

Notice the similarity to equation 13.7 defining the matrix representation of the quaternions; this is hardly an accident.

We see yet another description of $C\ell_3$ by quaternion variables when we examine the even subalgebra $C\ell_3^+$ generated by the two elements $\mathbf{e}_1\mathbf{e}_2$ and $\mathbf{e}_1\mathbf{e}_3$ and which is isomorphic to $C\ell_2 \cong \mathbb{H}$, as is obvious from the triality diagram

$$\begin{array}{ccc} \mathbf{e}_1\mathbf{e}_2 & \longrightarrow & \mathbf{e}_1\mathbf{e}_3 \\ & \nwarrow \quad \swarrow & \\ & \mathbf{e}_2\mathbf{e}_3 & \end{array} . \tag{20.13}$$

This is merely a concrete example of the isomorphism introduced at the end of the previous section.

The Clifford algebra $C\ell_4$ may be seen to be isomorphic to the real matrix algebra $M_2(\mathbb{H})$ by using the universal mapping property to extend the prescription

$$\begin{aligned} \mathbf{e}_1 &\mapsto \begin{pmatrix} \mathbf{i} & 0 \\ 0 & \mathbf{i} \end{pmatrix}, & \mathbf{e}_2 &\mapsto \begin{pmatrix} \mathbf{j} & 0 \\ 0 & \mathbf{j} \end{pmatrix}, \\ \mathbf{e}_3 &\mapsto \begin{pmatrix} \mathbf{k} & 0 \\ 0 & -\mathbf{k} \end{pmatrix}, & \mathbf{e}_4 &\mapsto \begin{pmatrix} 0 & \mathbf{k} \\ \mathbf{k} & 0 \end{pmatrix}, \end{aligned} \tag{20.14}$$

defined on the standard basis of $\mathbb{R}^4$. This offers a common venue from which one can view all the first five Clifford algebras as subalgebras of $M_2(\mathbb{H})$. In fact, the standard inclusions

$$C\ell_0 \subset C\ell_1 \subset C\ell_2 \subset C\ell_3$$

may be viewed as

$$\left\{\begin{pmatrix} r & 0 \\ 0 & r \end{pmatrix}\right\}_{r\in\mathbb{R}} \subset \left\{\begin{pmatrix} z & 0 \\ 0 & z \end{pmatrix}\right\}_{z\in\mathbb{C}} \subset \left\{\begin{pmatrix} \mathsf{p} & 0 \\ 0 & \mathsf{p} \end{pmatrix}\right\}_{\mathsf{p}\in\mathbb{H}} \subset \left\{\begin{pmatrix} \mathsf{p} & 0 \\ 0 & \mathsf{q} \end{pmatrix}\right\}_{\mathsf{p},\mathsf{q}\in\mathbb{H}},$$

all these being subalgebras of $M_2(\mathbb{H}) \cong C\ell_4$. Notice that the matrix representation of $C\ell_3$ at the right-hand end of these containments differs from that of expression 20.12. In fact, the matrix representation of $C\ell_3$ given in expression 20.12 is precisely the representation of the even subalgebra $C\ell_4^+$ under the isomorphism $C\ell_4 \cong M_2(\mathbb{H})$, which can be worked out using representation 20.14 to calculate a matrix basis for this even subalgebra.

We now shift our attention to $C\ell_{1,3}$, which is generated as an $\mathbb{R}$-algebra by $\mathbf{e}_\mu$ with $\mu = 0, 1, 2, 3$ and the relations $\mathbf{e}_0^2 = -1$ and $\mathbf{e}_i^2 = +1$, for $i = 1, 2, 3$, as well as the anticommutation rules $\mathbf{e}_\mu \mathbf{e}_\nu = -\mathbf{e}_\nu \mathbf{e}_\mu$ for $\mu \neq \nu$. Introducing the expression $g^{\mu\nu}$ where $g^{00} = +1$, $g^{ii} = -1$, and $g^{\mu\nu} = 0$ when $\mu \neq \nu$, we may write these relations as

$$\mathbf{e}_\mu \mathbf{e}_\nu + \mathbf{e}_\nu \mathbf{e}_\mu = -2g^{\mu\nu}. \tag{20.15}$$

Our claim is that $C\ell_{1,3}$ is isomorphic to the matrix algebra $M_4(\mathbb{R})$. To prove this we need to find four matrices $\epsilon^\mu \in M_4(\mathbb{R})$ that satisfy the relations

$$\epsilon^\mu \epsilon^\nu + \epsilon^\nu \epsilon^\mu = -2g^{\mu\nu} 1_{4\times4},$$

where $1_{4\times4}$ is an identity matrix. The universal mapping property then guarantees an $\mathbb{R}$-algebra homomorphism $C\ell_{1,3} \to M_4(\mathbb{R})$, which we can then show to be an isomorphism. There are lots of choices for the matrices ϵ^μ. As an example, we will work with the **real Majorana matrices**, given by

$$\epsilon^0 = \begin{pmatrix} 0 & 0 & 0 & -1 \\ 0 & 0 & 1 & 0 \\ 0 & -1 & 0 & 0 \\ 1 & 0 & 0 & 0 \end{pmatrix}, \qquad \epsilon^1 = \begin{pmatrix} 1 & 0 & 0 & 0 \\ 0 & -1 & 0 & 0 \\ 0 & 0 & 1 & 0 \\ 0 & 0 & 0 & -1 \end{pmatrix},$$

$$\epsilon^2 = \begin{pmatrix} 0 & 0 & 0 & 1 \\ 0 & 0 & -1 & 0 \\ 0 & -1 & 0 & 0 \\ 1 & 0 & 0 & 0 \end{pmatrix}, \qquad \epsilon^3 = \begin{pmatrix} 0 & -1 & 0 & 0 \\ -1 & 0 & 0 & 0 \\ 0 & 0 & 0 & -1 \\ 0 & 0 & -1 & 0 \end{pmatrix}.$$

To verify that this assignment gives an isomorphism, it suffices to show that the matrices $\{\epsilon^\alpha\}$, as α ranges over the subsets of $\{1, 2, 3, 4\}$, is linearly independent and therefore a basis for the 16-dimensional algebra $M_4(\mathbb{R})$. Here, of course, when $\alpha = \{\alpha_1 < \cdots < \alpha_j\}$, we have $\epsilon^\alpha = \epsilon^{\alpha_1} \cdots \epsilon^{\alpha_j}$. To show this is a straightforward, albeit unpleasant, exercise. As for $C\ell_{3,1}$, an application of the second of the main symmetries of expression 20.8 with symmetry 20.9 gives the isomorphism $C\ell_{3,1} \cong C\ell_{0,4} \cong C\ell_4 \cong M_2(\mathbb{H})$.

The complex Clifford algebras are the complexifications of the real Clifford algebras and our first example is $C\ell_2(\mathbb{C}) \cong \mathbb{H}_{\mathbb{C}}$. We wish to identify the complexification of the real algebra of quaternions as a complex matrix algebra. Recall that a complex vector space basis for the complexification $V_{\mathbb{C}} = V \otimes_{\mathbb{R}} \mathbb{C}$ of a real vector space V with basis $\{\mathbf{e}_i\}$ is $\{\mathbf{e}_i \otimes 1\}$. It follows that $\{1 \otimes 1, \mathbf{i} \otimes 1, \mathbf{j} \otimes 1, \mathbf{k} \otimes 1\}$ is a complex vector space basis for $\mathbb{H}_{\mathbb{C}}$. Using the matrix representation of the quaternions given in expression 13.7 as a guide, we define a complex vector space homomorphism of $\mathbb{H}_{\mathbb{C}}$ into $M_2(\mathbb{C})$ by $\mathbb{C}$-linearly extending the map of basis elements given as

$$1 \otimes 1 \mapsto \begin{pmatrix} 1 & 0 \\ 0 & 1 \end{pmatrix}, \qquad \mathbf{i} \otimes 1 \mapsto \begin{pmatrix} \mathrm{i} & 0 \\ 0 & -\mathrm{i} \end{pmatrix},$$

$$\mathbf{j} \otimes 1 \mapsto \begin{pmatrix} 0 & 1 \\ -1 & 0 \end{pmatrix}, \qquad \mathbf{k} \otimes 1 \mapsto \begin{pmatrix} 0 & \mathrm{i} \\ \mathrm{i} & 0 \end{pmatrix}.$$

Since these four matrices are linearly independent over $\mathbb{C}$ and $M_2(\mathbb{C})$ is four-dimensional as a complex vector space, they form a basis. It follows that this prescription describes a $\mathbb{C}$-vector space isomorphism. To show that it is an isomorphism of algebras requires only that one observe that multiplication is preserved on the basis elements. This is a quite simple exercise and identifies the complex algebra $C\ell_2(\mathbb{C})$ as the matrix algebra $M_2(\mathbb{C})$. It follows now rather easily that $C\ell_3(\mathbb{C}) \cong (\mathbb{H} \oplus \mathbb{H})_{\mathbb{C}} \cong M_2(\mathbb{C}) \oplus M_2(\mathbb{C})$.

Finally, since it is immediate that $M_n(\mathbb{R})_{\mathbb{C}} \cong M_n(\mathbb{C})$ as complex algebras, we have $C\ell_4(\mathbb{C}) \cong (C\ell_{1,3})_{\mathbb{C}} \cong M_4(\mathbb{R})_{\mathbb{C}} \cong M_4(\mathbb{C})$. The real Majorana matrices serve also as $\mathbb{C}$-algebra generators for $M_4(\mathbb{C})$, but there are other generating sets of importance. The **Dirac gamma matrices** are given in block form by

$$\gamma^0 = \begin{pmatrix} 1_{2\times 2} & \mathbf{0} \\ \mathbf{0} & -1_{2\times 2} \end{pmatrix}, \quad \gamma^i = \begin{pmatrix} \mathbf{0} & \sigma_i \\ -\sigma_i & \mathbf{0} \end{pmatrix}, \tag{20.16}$$

where $\mathbf{0}$ is the 2×2 zero matrix, and they also serve to generate the complex matrix algebra $M_4(\mathbb{C})$. Dirac discovered these matrices in his derivation of a relativistic wave equation for the electron, the setting of which is the Clifford algebra $C\ell_4(\mathbb{C})$. Notice that the Majorana matrices are given in terms of the Pauli matrices by

$$\epsilon^0 = \begin{pmatrix} \mathbf{0} & -\mathrm{i}\sigma_2 \\ -\mathrm{i}\sigma_2 & \mathbf{0} \end{pmatrix}, \qquad \epsilon^1 = \begin{pmatrix} \sigma_3 & \mathbf{0} \\ \mathbf{0} & \sigma_3 \end{pmatrix},$$

$$\epsilon^2 = \begin{pmatrix} \mathbf{0} & \mathrm{i}\sigma_2 \\ -\mathrm{i}\sigma_2 & \mathbf{0} \end{pmatrix}, \qquad \epsilon^3 = \begin{pmatrix} -\sigma_1 & \mathbf{0} \\ \mathbf{0} & -\sigma_1 \end{pmatrix}.$$

We hope that this short survey of some of the low-dimensional Clifford algebras gives the reader some sense of the intricacies of the subject and a healthy respect for the great variety of different ways to view these algebras. Our next task is to show how Clifford multiplication provides a powerful geometric tool for probing the isometries of pseudo-Euclidean spaces.

20.3 The Groups Pin and Spin

The multiplicative group of units of the Clifford algebra $C\ell$ is denoted as $C\ell^\times$ and consists of those elements that have a multiplicative inverse. For example, the groups $C\ell_0^\times$, $C\ell_1^\times$, and $C\ell_2^\times$ are isomorphic to, respectively, the nonzero real numbers, the nonzero complex numbers, and the nonzero quaternions. More interesting is the group $C\ell_3^\times$. Indeed, the Clifford algebra $C\ell_3$ has zero divisors, none of which can have a multiplicative inverse, and so the group $C\ell_3^\times$ is more restrictive than the set of nonzero elements of the algebra. For example, $(\mathbf{1}-\mathbf{e}_1\mathbf{e}_2\mathbf{e}_3)(\mathbf{1}+\mathbf{e}_1\mathbf{e}_2\mathbf{e}_3) = 0$, so that $\mathbf{1} - \mathbf{e}_1\mathbf{e}_2\mathbf{e}_3$ is seen to be a zero divisor and hence not a unit in $C\ell_3$. This is quite transparent when viewed through the lens of the isomorphism $C\ell_3 \cong \mathbb{H} \oplus \mathbb{H}$ of expression 20.10. The element $\mathbf{1} - \mathbf{e}_1\mathbf{e}_2\mathbf{e}_3$ corresponds to $(2, 0)$ while $\mathbf{1} + \mathbf{e}_1\mathbf{e}_2\mathbf{e}_3$ corresponds to $(0, 2)$ under the isomorphism and, obviously, $(2, 0)(0, 2) = (0, 0)$. It is easy to see that, in fact, $(\mathbb{H} \oplus \mathbb{H})^\times = \mathbb{H}^\times \times \mathbb{H}^\times$, so that both elements of the quaternion pair (p, q) must be nonzero to qualify for membership in the group of units. Since $C\ell_3$ is a subalgebra of $C\ell_{p,q}$ when $p \geq 3$, this behavior can be recognized as rather typical. In general, for an arbitrary quadratic space (V, g), we have

$$V \cap C\ell^\times(V, g) = \{v \in V : g(v) \neq 0\},$$

since each v with $g(v) = 0$ satisfies $v^2 = 0$ and is a zero divisor, and each v with $g(v) \neq 0$ has multiplicative inverse $v^{-1} = -v/g(v)$.

There are two subgroups of the group of units of a Clifford algebra that are of primary importance: these are the **pin** and **spin** groups.

Their importance in the real case lies, at least partially, in the way in which the geometry of the pseudo-Euclidean spaces $\mathbb{R}^{p,q}$ is represented, in that these groups capture the isometries of these spaces by their natural pseudo-orthogonal representations. We will give the main definitions in full generality and then specialize to the real Euclidean Clifford algebras $C\ell_n$ before returning to the general pseudo-Euclidean spaces of signature (p, q) and the complex Clifford algebras. Along the way, we will derive the explicit structure of some low-dimensional examples and, in the section following, present the specific representation theory of the low-dimensional examples pertinent to the Pauli and Dirac theories in quantum mechanics.

The group $\mathrm{Pin}(V, g)$ is the subgroup of the group of units $C\ell^\times(V, g)$ generated by the unit sphere $\mathbf{S}(V, g) = \{v \in V : g(v) = \pm 1\}$, and the group $\mathrm{Spin}(V, g)$ is the subgroup of $\mathrm{Pin}(V, g)$ of even elements or, explicitly, $\mathrm{Spin}(V, g) = \mathrm{Pin}(V, g) \cap C\ell^+(V, g)$. It can be seen that $\mathrm{Spin}(V, g)$ has index 2 in $\mathrm{Pin}(V, g)$. The importance of these groups will become evident shortly, but first we will sort out some of their properties. First, note that

$$\mathrm{Pin}(V, g) = \bigcup_{k=1}^{\infty} (\mathbf{S}(V, g))^k \quad \text{and} \quad \mathrm{Spin}(V, g) = \bigcup_{k=1}^{\infty} (\mathbf{S}(V, g))^{2k}, \tag{20.17}$$

where, of course, $(\mathbf{S}(V, g))^k = \{v_1 \cdots v_k : v_i \in \mathbf{S}(V, g)\}$. For each $u = v_1 \cdots v_k \in \mathrm{Pin}(V, g)$, the inverse is just

$$u^{-1} = v_k^3 \cdots v_1^3 = (-1)^k g(v_1) \cdots g(v_k) v_k \cdots v_1.$$

Notice that the net result is that u^{-1} is either $v_k \cdots v_1$ or its negative.

Next, there is the **adjoint action** of $\mathrm{Pin}(V, g)$ on its parent Clifford algebra $C\ell(V, g)$ given by conjugation via

$$\mathrm{Ad}_u(x) = uxu^{-1} \quad \text{for} \quad u \in \mathrm{Pin}(V, g) \quad \text{and} \quad x \in C\ell(V, g),$$

as well as the very important **twisted adjoint action** given by

$$\widetilde{\mathrm{Ad}}_u(x) = \tilde{u}xu^{-1} \quad \text{for} \quad u \in \mathrm{Pin}(V, g) \quad \text{and} \quad x \in C\ell(V, g).$$

We will see that this latter action captures much of the geometry of the pseudo-Euclidean spaces in that it provides a g-orthogonal representation of the pin group in the real vector space setting. Note that $\widetilde{\mathrm{Ad}}_{uv} = \widetilde{\mathrm{Ad}}_u \circ \widetilde{\mathrm{Ad}}_v$ and that $\widetilde{\mathrm{Ad}}_u = \mathrm{Ad}_u$ for any even element $u \in \mathrm{Spin}(V, g)$. In particular, the restrictions to only those $u \in \mathrm{Spin}(V, g)$ of the adjoint and of the twisted adjoint action agree.

At this point we want to examine these adjoint actions restricted to the generating vector space V. We will assume throughout the remainder of this lecture that the quadratic form is nondegenerate and the vector space V is finite-dimensional. The reader might wish to restrict his or her attention further, to only real and complex vector spaces, since these are soon to be our particular interest; however, as the arguments work for a while longer in the generality of an arbitrary vector space over a field $\mathbb{K}$ of characteristic different from 2, this is the only extra assumption that we will make before examining the real and complex cases explicitly. Recall first that the quadratic form g on V induces the bilinear form $g(x, y)$ on V by the polarization identity

$$g(x, y) = \frac{1}{2}\left[g(x + y) - g(x) - g(y)\right] \quad \text{for} \quad x, y \in V,$$

which gives the relationship $g(x) = g(x, x)$. As an example, this takes the form

$$g(x, y) = x_1 y_1 + \cdots + x_p y_p - x_{p+1} y_{p+1} - \cdots - x_{p+q} y_{p+q}$$

in $\mathbb{R}^{p,q}$, and gives the usual Euclidean inner product in $\mathbb{R}^n$. Now let $v \in \mathbf{S}(V, g)$, so that $g(v) = \pm 1$ and observe that, for $x \in V$,

$$\begin{aligned} \mathrm{Ad}_v(x) &= vxv^{-1} = -g(v)vxv \\ &= -g(v)\left(-xvv + xvv + vxv\right) = -x - g(v)(xv + vx)v \\ &= -x - g(v)\left[(x + v)^2 - x^2 - v^2\right]v = -\left[x - 2g(v)g(x, v)v\right]. \end{aligned}$$

Immediately, we obtain the twisted version:

$$\widetilde{\mathrm{Ad}}_v(x) = x - 2g(v)g(x, v)v,$$

which has rather strong geometric content. Notice first that this shows that $\widetilde{\mathrm{Ad}}_v(V) = V$ and in fact that the action of $\widetilde{\mathrm{Ad}}_v$ on V is as a reflection through the subspace that is g-orthogonal to v. To see this, observe that $\widetilde{\mathrm{Ad}}_v$ fixes any x with $g(x, v) = 0$ and sends v to $-v$. Moreover, a quick calculation verifies that $\widetilde{\mathrm{Ad}}_v$ preserves the quadratic form g in that $g(\widetilde{\mathrm{Ad}}_v(x)) = g(x)$ for all $x \in V$, implying also that $g(\widetilde{\mathrm{Ad}}_v(x), \widetilde{\mathrm{Ad}}_v(y)) = g(x, y)$ for all $x, y \in V$. It follows that $\widetilde{\mathrm{Ad}}_v \in \mathrm{O}(V, g)$, the group of g-orthogonal transformations of V, i.e., the linear isomorphisms that preserve the quadratic form. If $u = v_1 \cdots v_k$ is an arbitrary element of $\mathrm{Pin}(V, g)$, then

$$\widetilde{\mathrm{Ad}}_u = \widetilde{\mathrm{Ad}}_{v_1 \cdots v_k} = \widetilde{\mathrm{Ad}}_{v_1} \circ \cdots \circ \widetilde{\mathrm{Ad}}_{v_k}. \tag{20.18}$$

This shows that the action of $\widetilde{\mathrm{Ad}}_u$ on V is as a product of reflections and shows that the mapping

$$\widetilde{\mathrm{Ad}} : \mathrm{Pin}(V, g) \to \mathrm{O}(V, g)$$

given by $u \mapsto \widetilde{\mathrm{Ad}}_u$, which obviously is a homomorphism of groups, defines an orthogonal representation of the pin group on the quadratic space (V, g).

We now specialize to the pin and spin groups of the real and complex Clifford algebras $C\ell_{p,q}$ and $C\ell_n(\mathbb{C})$. In this context, more geometric features are captured than in the general context. The groups are now denoted as $\mathrm{Pin}_{p,q}$ and $\mathrm{Spin}_{p,q}$ in the real case, Pin_n and Spin_n when $p = n$ and $q = 0$, and $\mathrm{Pin}_n(\mathbb{C})$ and $\mathrm{Spin}_n(\mathbb{C})$ in the complex case. We first examine the real case. The first observation is that the image of $\mathrm{Pin}_{p,q}$ under $\widetilde{\mathrm{Ad}}$ is the group generated by reflections and a classical result[5] implies that this is the whole orthogonal group $\mathrm{O}(p, q)$, hence $\widetilde{\mathrm{Ad}}$ is surjective. In fact, we have the following short exact sequences that identify the kernels:

$$1 \longrightarrow \mathbb{Z}_2 \longrightarrow \mathrm{Spin}_{p,q} \xrightarrow{\widetilde{\mathrm{Ad}}} \mathrm{SO}(p, q) \longrightarrow 1$$

and

$$1 \longrightarrow \mathbb{Z}_2 \longrightarrow \mathrm{Pin}_{p,q} \xrightarrow{\widetilde{\mathrm{Ad}}} \mathrm{O}(p, q) \longrightarrow 1 .$$

To verify this, having just observed that the image of $\mathrm{Pin}_{p,q}$ under $\widetilde{\mathrm{Ad}}$ is the orthogonal group $\mathrm{O}(p, q)$, the easy observation that the determinant of each reflection is -1 guarantees that the image of $\mathrm{Spin}_{p,q}$ under $\widetilde{\mathrm{Ad}}$ is the special orthogonal group $\mathrm{SO}(p, q) = \{L \in \mathrm{O}(p, q) : \det L = 1\}$. As for the kernels, note first that equation 20.18 implies that $\det \widetilde{\mathrm{Ad}}_u = +1$ when u is even and $\det \widetilde{\mathrm{Ad}}_u = -1$ when u is odd. It follows that any element u of the kernel must be even and therefore in $\mathrm{Spin}_{p,q}$. The condition for u to be in the kernel is then that u is even and $\widetilde{\mathrm{Ad}}_u(x) = x$ for all $x \in V = \mathbb{R}^{p,q}$, i.e., that $ux = xu$ for all $x \in V$. Let $\mathbf{e}_i$ be any of the standard basis elements of V and write $u = a + b\mathbf{e}_i$ where $a, b \in \mathrm{span}\{\mathbf{e}_\alpha : i \notin \alpha\}$. Since u is even, a is even and b is odd. It follows that $\mathbf{e}_i$ commutes with a and anticommutes with b. This gives $a\mathbf{e}_i + b\mathbf{e}_i^2 = u\mathbf{e}_i = \mathbf{e}_i u = a\mathbf{e}_i - b\mathbf{e}_i^2$, which implies that $b = 0$. It follows that $u \in \mathrm{span}\{\mathbf{e}_\alpha : i \notin \alpha\}$. Applying this with i taking the successive values $1, \ldots, p + q$ gives $u \in \mathrm{span}\{\mathbf{e}_\alpha : \alpha = \emptyset\} = \mathbb{R}\mathbf{1} = \mathbb{R}$. To finish

[5] The Cartan–Dieudonné theorem; see Artin (1957) or Harvey (1990).

the argument, note either from the definition or from table 20.7 that $u^* = u$ and, for each $v \in V$, $vv^* = -v^2 = g(v)$. Writing $u = v_1 \cdots v_k$ for some $v_i \in V$ with $g(v_i) = \pm 1$, the fact that Clifford conjugation is an anti-automorphism implies that

$$u^2 = uu^* = v_1 \cdots v_k v_k^* \cdots v_1^* = g(v_1) \cdots g(v_k) = \pm 1.$$

It follows that $u = \pm 1$ and this verifies the kernel as $\mathbb{Z}_2$.[6]

Specializing now to the Euclidean algebras $C\ell_n$, Spin_n is a connected double cover of the rotation group SO(n) whenever $n > 1$. The connectedness follows from the characterization 20.17 since each $(\mathbf{S}^{n-1})^k$, being the image of the connected product $\mathbf{S}^{n-1} \times \cdots \times \mathbf{S}^{n-1}$ (k factors) under continuous multiplication, is connected and the identity 1 is common to every $(\mathbf{S}^{n-1})^{2k}$. Even more is true. For $n > 2$, the spin group Spin_n is simply connected and so serves as the universal covering space of the rotation group SO(n).[7] Though the higher-dimensional spin groups are distinct from the classical matrix groups, there are so-called accidental isomorphisms of the low-dimensional spin groups to some of the more familiar matrix groups. For example, it is quite straightforward to derive the isomorphisms[8]

$$\begin{aligned}
\mathrm{Spin}_1 &= \mathbf{S}^0 = \mathbb{Z}_2 = \{\pm 1\},\\
\mathrm{Spin}_2 &\cong \mathbf{S}^1 \cong \mathrm{U}(1),\\
\mathrm{Spin}_3 &\cong \mathrm{Sp}(1) = \mathbf{S}^3 \cong \mathrm{SU}(2),\\
\mathrm{Spin}_4 &\cong \mathbf{S}^3 \times \mathbf{S}^3 \cong \mathrm{SU}(2) \times \mathrm{SU}(2),\\
\mathrm{Spin}_5 &\cong \mathrm{Sp}(2),\\
\mathrm{Spin}_6 &\cong \mathrm{SU}(4).
\end{aligned}$$

Let $h_n : C\ell_n \cong C\ell_{n+1}^+ \subset C\ell_{n+1}$ be an isomorphism of $C\ell_n$ to the even subalgebra $C\ell_{n+1}^+$. The isomorphisms of the spin groups are given by the

[6] The argument says a bit more, in that, since u is real, $u^2 = \pm 1$ implies that $u^2 = 1$, which implies that when u is written as a product $v_1 \cdots v_k$ of unit vectors, only an even number of the v_i may satisfy $g(v_i) = -1$.

[7] For a proof, see Helgason (2001).

[8] For those unfamiliar with the **simplectic groups** Sp(n), these are the direct analogues of the unitary groups when the complex vector space $\mathbb{C}^n$ with Hermitian product is replaced by its exact quaternionic extension. Explicitly, define the Hermitian product of two elements $\mathsf{p}, \mathsf{q} \in \mathbb{H}^n$ by $\mathsf{p} \cdot \mathsf{q} = \mathsf{p}\overline{\mathsf{q}}^{\mathrm{Tr}} = \mathsf{p}_1\overline{\mathsf{q}}_1 + \cdots + \mathsf{p}_n\overline{\mathsf{q}}_n$. Then $\mathrm{Sp}(n) = \{L \in \mathrm{M}_n(\mathbb{H}) : L\mathsf{p} \cdot L\mathsf{q} = \mathsf{p} \cdot \mathsf{q} \text{ for all } \mathsf{p}, \mathsf{q} \in \mathbb{H}^n\}$, the set of $n \times n$ quaternionic matrices that preserve the Hermitian product. Notice that $\mathrm{Sp}(1) = \mathbf{S}^3 \subset \mathbb{H}$, the group of unit quaternions.

mappings

$$
\begin{array}{lcl}
\mathbf{S}^0 \subset \mathbb{R} & \xrightarrow[\cong]{} \mathcal{C}\ell_0 \xrightarrow[\cong]{h_0} & \mathcal{C}\ell_1^+ \subset \mathcal{C}\ell_1, \\
\mathbf{S}^1 \subset \mathbb{C} & \xrightarrow[\cong]{} \mathcal{C}\ell_1 \xrightarrow[\cong]{h_1} & \mathcal{C}\ell_2^+ \subset \mathcal{C}\ell_2, \\
\mathbf{S}^3 \subset \mathbb{H} & \xrightarrow[\cong]{} \mathcal{C}\ell_2 \xrightarrow[\cong]{h_2} & \mathcal{C}\ell_3^+ \subset \mathcal{C}\ell_3, \\
\mathbf{S}^3 \times \mathbf{S}^3 \subset \mathbb{H} \oplus \mathbb{H} & \xrightarrow[\cong]{} \mathcal{C}\ell_3 \xrightarrow[\cong]{h_3} & \mathcal{C}\ell_4^+ \subset \mathcal{C}\ell_4, \\
\mathrm{Sp}(2) \subset \mathrm{M}_2(\mathbb{H}) & \xrightarrow[\cong]{} \mathcal{C}\ell_4 \xrightarrow[\cong]{h_4} & \mathcal{C}\ell_5^+ \subset \mathcal{C}\ell_5, \\
\mathrm{SU}(4) \subset \mathrm{M}_4(\mathbb{C}) & \xrightarrow[\cong]{} \mathcal{C}\ell_5 \xrightarrow[\cong]{h_5} & \mathcal{C}\ell_6^+ \subset \mathcal{C}\ell_6.
\end{array}
$$

All the indicated isomorphisms have been discussed previously except $\mathrm{M}_4(\mathbb{C}) \cong \mathcal{C}\ell_5$, which we leave to the interested reader. We will discuss the spin representations in the section following, but note for the moment that the third of these isomorphisms, in combination with the adjoint action $\widetilde{\mathrm{Ad}}$, gives a standard orthogonal representation of $\mathrm{Spin}_3 \cong \mathbf{S}^3$ on $\mathbb{R}^3$ that is equivalent to the one that we have seen so many times in the preceding lectures, given by ϑ of expression 13.16.

We close this section with a listing of low-dimensional examples of indefinite signature as well as low-dimensional examples over the complex field. First, unlike the positive definite case of Spin_n, $\mathrm{Spin}_{p,q}$ is not connected when $p, q \geq 1$ and, in fact, has exactly two connected components. Let $\mathrm{Spin}_{p,q}^0$ denote the connected component containing the identity and note that it is a subgroup of $\mathrm{Spin}_{p,q}$. We list without proof[9] the following isomorphisms:

$$
\begin{aligned}
\mathrm{Spin}_{1,2}^0 &\cong \mathrm{SL}(2,\mathbb{R}) = \{L \in \mathrm{GL}(2,\mathbb{R}) : \det L = +1\}, \\
\mathrm{Spin}_{1,3}^0 &\cong \mathrm{SL}(2,\mathbb{C}) = \{L \in \mathrm{GL}(2,\mathbb{C}) : \det L = +1\}, \\
\mathrm{Spin}_{1,2} &\cong \mathrm{SL}^*(2,\mathbb{R}) = \{L \in \mathrm{GL}(2,\mathbb{R}) : \det L = \pm 1\}, \\
\mathrm{Spin}_{1,3} &\cong \mathrm{SL}^*(2,\mathbb{C}) = \{L \in \mathrm{GL}(2,\mathbb{C}) : \det L = \pm 1\},
\end{aligned}
$$

as well as

$$
\begin{aligned}
\mathrm{Spin}_2(\mathbb{C}) &\cong \mathbb{C}^\times, \\
\mathrm{Spin}_3(\mathbb{C}) &\cong \mathrm{SL}(2,\mathbb{C}), \\
\mathrm{Spin}_4(\mathbb{C}) &\cong \mathrm{SL}(2,\mathbb{C}) \times \mathrm{SL}(2,\mathbb{C}).
\end{aligned}
$$

[9] See Lawson and Michelson (1989) or Harvey (1990).

20.4 Spin Representations and Spinors

The original impetus for Cartan's development of spin representations was to understand the irreducible representations of the orthogonal Lie algebras $\mathfrak{so}(n, \mathbb{K})$, only half of which come from corresponding representations of the orthogonal Lie groups $\mathrm{SO}(n, \mathbb{K})$, for $\mathbb{K} = \mathbb{R}, \mathbb{C}$. The other half come from representations of the the spin groups. This phenomenon is a consequence of the fact that $\mathrm{SO}(n, \mathbb{K})$ is not simply connected; the two-fold covering groups $\mathrm{Spin}_n(\mathbb{K})$, which share Lie algebras with the orthogonal groups, have representations where $-\mathbf{1}$ acts nontrivially and so are not induced by representations of the corresponding orthogonal groups. In this section we present the general definitions and then we specialize to two cases, namely, those of dimensions three and four, where we will work out explicitly the **spin representations** of interest in quantum mechanics.

The general abstract setting for a spinor is as an element of a vector space that has been given a spin structure. So, as an object it is nothing more than a vector, *except* there is an action of a spin group on its underlying vector space that endows the vector with a sort of historical memory of a continuous change of its coordinates or of a continuous traversal through the vector space. What comes to mind here is Pauli's spinors, which represent spin in quantum mechanics and which undergo multiplication by -1 when rotated through a full 2π radians. A **spin structure** on a vector space is an irreducible representation of a spin group whose elements act as linear isomorphisms of the vector space. In the real case, the study is of irreducible representations of the groups $\mathrm{Spin}_{p,q}$ on a complex vector space, and in the complex case the study is of the irreducible representations of the complex spin groups $\mathrm{Spin}_n(\mathbb{C})$. The corresponding representations of the Lie algebras give the representations of $\mathfrak{so}(n, \mathbb{K})$ that do not arise from the orthogonal groups $\mathrm{SO}(n, \mathbb{K})$. As noted already, this was the topic of Cartan's 1913 paper where he "explained" the half-integer representations. Any of the general references on Clifford algebras listed in the End Notes will present a classification of the spin representations and their corresponding spinor spaces. Our aim is not quite so bold. We want only to examine the mathematics behind the Pauli spin representation already presented in the preceding lecture, and the Dirac spin representation that arises in Dirac's theory of the electron. The Pauli spin representation is based on a spin structure defined by an irreducible representation of Spin_3 realized as either $\mathrm{SU}(2)$ or $\mathbf{S}^3$, and the Dirac spin representation is based on a spin structure defined by an irreducible representation of either $\mathrm{Spin}_{3,1}$ or $\mathrm{Spin}_4(\mathbb{C})$.

20.4.1 The Pauli Spin Representation

In the Pauli theory described in the preceding lecture, Pauli spinors arise as vectors in $L^2(\mathbb{R}^3)\otimes V^{1/2}$ equipped with the action of $\mathbf{S}^3 \cong \mathrm{SU}(2)$ given as

$$e^{\mathbf{n}t} \cdot \begin{pmatrix} \varphi \\ \psi \end{pmatrix} = \rho(e^{\mathbf{n}t}) \begin{pmatrix} \varphi \circ \vartheta(e^{\mathbf{n}t})^{-1} \\ \psi \circ \vartheta(e^{\mathbf{n}t})^{-1} \end{pmatrix} = e^{t\mathbf{n}\cdot\boldsymbol{\tau}} \begin{pmatrix} \varphi \circ R(-2t, \mathbf{n}) \\ \psi \circ R(-2t, \mathbf{n}) \end{pmatrix}.$$

Here, we recall that $\mathbf{S}^3 = \{e^{\mathbf{n}t} : t \in \mathbb{R}, \mathbf{n} \in \mathbf{S}^2\}$ and $\mathrm{SU}(2) = \{e^{\mathbf{a}\cdot\boldsymbol{\tau}} : \mathbf{a} \in \mathbb{R}^3\}$, and the isomorphism ρ of $\mathbf{S}^3$ to $\mathrm{SU}(2)$ takes the form

$$\rho(e^{\mathbf{n}t}) = (\cos t)1_{2\times 2} - \mathbf{i}(\sin t)\mathbf{n} \cdot \boldsymbol{\sigma} = e^{t\mathbf{n}\cdot\boldsymbol{\tau}}.$$

We identify the irreducible part of this action on $V^{1/2}$ as a spin representation by identifying the 3-sphere group $\mathbf{S}^3$ with Spin_3 in the following way. The triality diagram 20.13 is only one of many that may be used to identify a real algebra isomorphism of quaternions with $C\ell_3^+$. The isomorphism that we find convenient to use in the present discussion is the real algebra isomorphism $h : \mathbb{H} \to C\ell_3^+$ defined by the triality diagram

$$\begin{array}{ccc} \mathbf{e}_2\mathbf{e}_3 & \longrightarrow & \mathbf{e}_3\mathbf{e}_1 \\ & \nwarrow \quad \swarrow & \\ & \mathbf{e}_1\mathbf{e}_2 & \end{array}.$$

Explicitly, on the standard basis of $\mathbb{H}$ we have $h(1) = \mathbf{e}_\emptyset = 1$, $h(\mathbf{i}) = \mathbf{e}_2\mathbf{e}_3$, $h(\mathbf{j}) = \mathbf{e}_3\mathbf{e}_1$, and $h(\mathbf{k}) = \mathbf{e}_1\mathbf{e}_2$. The image of the 3-sphere group $\mathbf{S}^3$ under h is precisely

$$h(\mathbf{S}^3) = \{a + b\,\mathbf{e}_2\mathbf{e}_3 + c\,\mathbf{e}_3\mathbf{e}_1 + d\,\mathbf{e}_1\mathbf{e}_2 : a^2 + b^2 + c^2 + d^2 = 1\} = \mathbf{S}_3^+.$$

Moreover, since h is an isomorphism of algebras, it restricts to the group $\mathbf{S}^3$ as an isomorphism of groups, implying that the right-hand side, $\mathbf{S}_3^+$, is a group under Clifford multiplication.

Our claim is that $\mathbf{S}_3^+ = \mathrm{Spin}_3$. In the preceding paragraph $\mathbf{S}^2$ was used to denote the set of pure unit quaternions in $\mathbb{H}$. To avoid confusion, in this paragraph only we will use S^2 to denote the unit sphere in $\mathbb{R}^3 = \mathrm{span}\{\mathbf{e}_1, \mathbf{e}_2, \mathbf{e}_3\} \subset C\ell_3$. To see the claim, note that equation 20.17 gives

$$\mathrm{Spin}_3 = \bigcup_{k=1}^{\infty} (\mathsf{S}^2)^{2k},$$

and since the image $h(\mathbf{S}^3) = \mathbf{S}_3^+$ is a group, we obtain $\text{Spin}_3 \subset \mathbf{S}_3^+$ provided we know that $\{uv : u, v \in \mathsf{S}^2\} = (\mathsf{S}^2)^2 \subset \mathbf{S}_3^+$. This in turn follows from a straightforward calculation, expanding the elements u and v in S^2 in the basis $\mathbf{e}_1$, $\mathbf{e}_2$, and $\mathbf{e}_3$ and using straightforward algebra. It remains only to verify the opposite containment, that $\text{Spin}_3 \supset \mathbf{S}_3^+$. For this we use the notation $w^\star$ to mean the Clifford element $w^\star = w_1\mathbf{e}_2\mathbf{e}_3 + w_2\mathbf{e}_3\mathbf{e}_1 + w_3\mathbf{e}_2\mathbf{e}_3 \in C\ell_3^+$ whenever $w = w_1\mathbf{e}_1 + w_2\mathbf{e}_2 + w_3\mathbf{e}_3 \in \mathbb{R}^3$. By writing $u, v \in \mathbb{R}^3$ in terms of the basis $\mathbf{e}_1$, $\mathbf{e}_2$, $\mathbf{e}_3$ and multiplying, one obtains $uv = -u \cdot v + (u \times v)^\star$. Let $s \in \mathbf{S}_3^+$ and observe that s may be written as $s = h(\cos t + \mathbf{w} \sin t)$ for some real t and unit quaternion $\mathbf{w} = w_1\mathbf{i} + w_2\mathbf{j} + w_3\mathbf{k} \in \mathbb{H}$, by expression 19.5. Let $w = w_1\mathbf{e}_1 + w_2\mathbf{e}_2 + w_3\mathbf{e}_3$, so that $s = \cos t + w^\star \sin t$, and choose any $u, v \in \mathbb{R}^3 = \text{span}\{\mathbf{e}_1, \mathbf{e}_2, \mathbf{e}_3\}$ such that u, v, w forms an orthonormal triple. This means in particular that $u^2 = -1$, $u \cdot v = 0$, and $u \times v = w$. Let $x = -u \cos t + v \sin t$ and observe that $x \cdot x = 1$, so that $x \in \mathsf{S}^2 \subset \mathbb{R}^3$. It follows that

$$\begin{aligned} ux = u(-u\cos t + v \sin t) &= -u^2 \cos t + uv \sin t \\ &= \cos t + (u \times v)^\star \sin t = \cos t + w^\star \sin t = s. \end{aligned}$$

Since $u, x \in \mathsf{S}^2$, we have $s = ux \in (\mathsf{S}^2)^2 \subset \text{Spin}_3$, confirming that $\mathbf{S}_3^+ \subset \text{Spin}_3$.

The representation given by $s \cdot \mathsf{v} = h^{-1}(s) \cdot \mathsf{v}$ for $\mathsf{v} \in V^{1/2}$ of Spin_3 on $V^{1/2}$ is the Pauli spin representation written in terms of the spin group and constitutes a spin structure on $V^{1/2}$. The elements of the representation space are spinors. If $R(t)$, $0 \le t \le T$, is a path of rotations in SO(3) starting at the identity then R lifts to a path $\widetilde{R}(t)$ starting at the identity in Spin_3. The effect on a spinor $\mathsf{v} \in V^{1/2}$ of this one-parameter set of rotations is given by the spin structure as $\widetilde{R}(t) \cdot \mathsf{v}$. In particular, if $R(t)$ performs a full rotation of 2π then $\widetilde{R}(T) = -1 \in \text{Spin}_3$, and the effect on v is $(-1) \cdot \mathsf{v} = -\mathsf{v}$, multiplication by a factor of -1. The idea here is that the lift of a path in SO(3) to Spin_3 endows the spinor $\widetilde{R}(T) \cdot \mathsf{v}$ with an imprint of its past history.

20.4.2 The Dirac Spin Representation

Though the Dirac theory of the electron is not covered in this book, we describe the setting of the theory in terms of spinors. Just as a spin-$\frac{1}{2}$ particle is represented by a spinor in the Pauli theory, so too is an electron in the Dirac theory. This time, though, the spin structure of relevance is a $\text{Spin}_{3,1}$ structure as well as its complexification, a $\text{Spin}_4(\mathbb{C})$ structure. The pair $(3, 1)$ arises from the signature of the pseudo-Euclidean space $\mathbb{R}^{1,3}$, the Minkowski space of Einstein's special relativity. In Dirac's attempt to find a relativistic equation governing the electron, he hit upon an equation with coefficients given by "numbers" γ^μ, $\mu = 0, 1, 2, 3$, that satisfy

$$\gamma^\mu \gamma^\nu + \gamma^\nu \gamma^\mu = 2g^{\mu\nu}. \tag{20.19}$$

Notice that this is equation 20.15 *except* for the sign on the right-hand side. Dirac observed that there are no complex numbers that satisfy expression 20.19 but there are matrices that do so, the smallest being 4×4 complex matrices. In fact there are many 4-tuples of matrices that satisfy this expression, and Dirac used the Pauli matrices to form his gamma matrices, which do. Expression 20.19 describes the generating relations for the Clifford algebra $C\ell_{3,1}$, and Dirac's theory is framed rather naturally in the context of this Clifford algebra and its complexification $C\ell_4(\mathbb{C})$.

From Section 20.2 we have $C\ell_{3,1} \cong M_2(\mathbb{H})$ and $C\ell_4(\mathbb{C}) \cong M_4(\mathbb{C})$, but another natural isomorphism is $C\ell_4(\mathbb{C}) \cong M_2(\mathbb{H})_{\mathbb{C}} \cong M_2(\mathbb{H}_{\mathbb{C}})$. Now, as $\mathbb{H}_{\mathbb{C}} \cong M_2(\mathbb{C})$, we may realize $C\ell_4(\mathbb{C})$ as 2×2 matrices whose entries are from $M_2(\mathbb{C})$, i.e., as 4×4 complex matrices written in 2×2 block form, exactly as in the expressions 20.16. These act on $\mathbb{C}^2 \oplus \mathbb{C}^2$ and induce a $\mathrm{Spin}_4(\mathbb{C})$ structure by restriction of the action to the spin group of $C\ell_4(\mathbb{C})$. Identifying $\mathbb{C}^2 \oplus \mathbb{C}^2$ as $V^{1/2} \oplus V^{1/2}$, the Dirac matrices act on the two $V^{1/2}$ entries as Pauli spin matrices. The vectors in $V^{1/2} \oplus V^{1/2}$ are called **Dirac spinors**.

As an alternative to the Dirac matrices, the four Majorana matrices also may serve as $\mathbb{C}$-algebra generators for $M_4(\mathbb{C})$ and the theory may be expressed in this Majorana basis. Of course, the Majorana matrices generate the real Clifford algebra $C\ell_{1,3}$ but not the real Clifford algebra $C\ell_{3,1}$. However, $C\ell_4(\mathbb{C})$, thought of as the complexification of $C\ell_{1,3}$, may use as a basis the Majorana matrices to generate the complex Clifford algebra, as expressions 20.15 and 20.19 are equivalent over the complex numbers.

Dirac hit upon the gamma matrices in trying to find a linear wave equation whose square produces the Klein–Gordon equation for the electron and was led to spin as a natural consequence. In the hands of mathematicians, Dirac's insights into the correct relativistic equation for the electron has led to the theory of Dirac operators on Clifford-valued functions and spin structures on manifolds. This represents a great enrichment of modern pure mathematics that has enjoyed extensive development.

20.5 End Notes

There are several really good resources for both the mathematical theory of Clifford algebras and spinors and their applications to physics. Harvey (1990) is a detailed exposition that presents the classical theory of linear groups and normed algebras in its first half and the theory of Clifford algebras and spinors in its second half. For a high-powered impressive accounting of the modern theory of spin manifolds, spinor fields, and Dirac operators see *Spin Geometry* by Lawson and Michelsohn (1989). I relied heavily on its first chapter, which is one of the more concise and elegantly tailored introductions to Clifford algebras that I have seen. Classifications of the algebras $C\ell_{p,q}$ and $C\ell_n(\mathbb{C})$ may be found in both these volumes. More accessible references are Garling (2011) and Curtis (1984). The first is aimed at graduate students in mathematics and presents a course of study on multilinear algebra, Clifford algebras, the spin groups, and applications. The second is a gentle introduction to matrix groups that includes an excursion into Clifford algebras and is aimed at the undergraduate. Two further sources worth checking out are Benn and Tucker (1987) and Abłamowicz and Sobczyk (2004). All these references are very geometric in flavor. For a more algebraic taste of Clifford algebras, the spin groups, and their representations, see Fulton and Harris (1991).

21
Many-Particle Quantum Systems

> **Some people try to depict and explain the strange in terms of the familiar. Such analogies lead only to paradoxes because quantum phenomena are radically different. Other people reject quantum objects because they are different, but all their argument shows is that there is nothing like classical objects in the quantum realm, not that there is no quantum object.**
>
> Sunny Y. Auyang
> *How is Quantum Field Theory Possible?*, 1995

Multi-particle quantum systems present some of the deepest and most counterintuitive paradoxes of the quantum world. Surely the most profound of these arises from the quantum entanglement of more than one particle, producing systems that exhibit the attributes of nonlocality and inseparability. In the next few lectures we examine in some detail the intricacies of multi-particle quantum systems. In this first lecture on multi-particle states, we present the mathematical framework of tensor products used to describe these states and we extend the axiom system of Lecture 2 to include their treatment. We apply this to two-particle systems and examine in more detail the action of two coupled angular momentum triads on the state space of such a system. This leads to an examination of representations of $\mathfrak{su}(2) \oplus \mathfrak{su}(2)$ on the tensor product of two representation spaces of $\mathfrak{su}(2)$ and fills the gap left from Lecture 18 in confirming the first inequality of the fundamental addition theorem.

The next lecture presents one of the most interesting discussions of quantum mechanics in its articulation of the EPR paradox and the Bell inequalities. Experiments confirm that nature presents us with nonlocal systems that exhibit correlations over vast distances, implying that the Bell inequalities are violated. This is a basic fact of nature. We show how the quantum mechanics of multi-particle states can handle this violation and "explain" these correlations. In four subsequent lectures, we present in turn a discussion of ensembles of particles, bosons and fermions and their state spaces, the Fock space for describing multi-particle states with a non-constant number of particles, and an introduction to the statistical mechanics of aggregates of quantum particles.

In these six lectures, we will introduce entangled and unentangled states, ensembles and density operators, pure states and mixed states, correlated measurements and EPR, realism and local causality, bosons and fermions, particle annihilation and creation operators, Fock space and symmetric and antisymmetric products, and Bose–Einstein and Fermi–Dirac statistics. This will provide a rich tapestry of a quantum world teeming with particles that awaits the insertion of interactions among the particles.

21.1 Multi-Particle States and Tensor Products

How should the quantum mechanical formalism be expanded to represent the state of a quantum system composed of two particles? More specifically, suppose that the two Hilbert spaces $\mathcal{H}_1$ and $\mathcal{H}_2$ are the state spaces of the respective particles p_1 and p_2. Quantum mechanics suggests that we may treat p_1 and p_2 as a single quantum system whose states lie in a single state space $\mathcal{H}$. What relationship do, or should, $\mathcal{H}_1$ and $\mathcal{H}_2$ have with $\mathcal{H}$? We will take some time to convince the reader that the state space $\mathcal{H}$ ought to take the form of the tensor product Hilbert space $\mathcal{H}_1 \overline{\otimes} \mathcal{H}_2$, which is the completion of the algebraic tensor product $\mathcal{H}_1 \otimes \mathcal{H}_2$ with respect to the inner product generated by the rule

$$\langle \varphi_1 \otimes \varphi_2 | \psi_1 \otimes \psi_2 \rangle = \langle \varphi_1 | \psi_1 \rangle \langle \varphi_2 | \psi_2 \rangle.$$

21.1.1 Justification that $\mathcal{H} = \mathcal{H}_1 \overline{\otimes} \mathcal{H}_2$

All state vectors in this discussion are assumed to be normalized. Suppose the state vector $\psi_i \in \mathcal{H}_i$ represents particle p_i, for $i = 1, 2$. For an observable a_i, let A_i be the self-adjoint operator on $\mathcal{H}_i$ associated with a_i. In Section 21.3 we will see an example where both a_1 and a_2 are angular momentum observables. Assume that the two particles are far apart and have had no interaction with one another in the distant past. Think, for instance, of an electron on earth and one in the interior of the sun. For $i = 1, 2$, separate a_i-measurements may be performed on p_i and the expectation values are $\langle \psi_i | \mathsf{A}_i \psi_i \rangle$. As the particles are far removed from one another and have not interacted in any appreciable sense, we would expect that measurements on the one have no effect on measurements of the other.[1] Moreover, if ψ_1 is an eigenstate of A_1 with eigenvalue λ, then a a_1-observation of p_1 yields the result λ with certainty. Further, if $\psi_1 = \sum_j c_j \varphi_j$ is a linear combination of A_1-eigenstates φ_j with respective eigenvalues λ_j then a a_1-measurement of p_1 yields λ_j with probability $|c_j|^2$.

The single state space $\mathcal{H}$ is supposed to represent the quantum system composed of both particles p_1 and p_2. It follows that there should be a state $\psi_{1,2} \in \mathcal{H}$ that describes the two particles when their separate, individual states are ψ_1 and ψ_2, respectively. There should be self-adjoint operators $\widehat{\mathsf{A}}_1$ and $\widehat{\mathsf{A}}_2$ on $\mathcal{H}$ that operate on the ψ_1 and ψ_2 "contributions" to $\psi_{1,2}$ respectively, reproducing the individual expectation values for the observables a_i when considered separately. The operation of $\widehat{\mathsf{A}}_1$ should have no effect on the ψ_2 "contribution" to $\psi_{1,2}$.

Formalizing the expectations of the preceding discussion gives the following five requirements that we might expect of the Hilbert space $\mathcal{H}$ and the operators $\widehat{\mathsf{A}}_i$. The notation $\psi_1\psi_2$ for $\psi_{1,2}$ helps to describe the third and the last of the five requirements succinctly:

1. For $i = 1, 2$, $\langle \psi_{1,2} | \widehat{\mathsf{A}}_i \psi_{1,2} \rangle = \langle \psi_i | \mathsf{A}_i \psi_i \rangle$.
2. If ψ_1 is an eigenstate of A_1 with eigenvalue λ then $\psi_{1,2}$ should be an eigenstate of $\widehat{\mathsf{A}}_1$ with eigenvalue λ.
3. If $\psi_1 = \sum_j c_j \varphi_j$ where $\mathsf{A}_1 \varphi_j = \lambda_j \varphi_j$ then the previous requirement says that each $\varphi_j \psi_2$ is an eigenstate of $\widehat{\mathsf{A}}_1$ with eigenvalue λ_j and, to preserve the probabilities of a_1-measurement outcomes on p_1, we should have the expansion $\psi_{1,2} = \sum_j c_j (\varphi_j \psi_2)$, so that a $\widehat{\mathsf{A}}_1$-measurement observes λ_j with probability $|c_j|^2$.
4. $[\widehat{\mathsf{A}}_1, \widehat{\mathsf{A}}_2] = 0$.
5. $\widehat{\mathsf{A}}_1 \psi_{1,2} = \widehat{\mathsf{A}}_1 (\psi_1 \psi_2) = (\mathsf{A}_1 \psi_1) \psi_2$ and $\widehat{\mathsf{A}}_2 \psi_{1,2} = \widehat{\mathsf{A}}_2 (\psi_1 \psi_2) = \psi_1 (\mathsf{A}_2 \psi_2)$.

Note that the second and fifth requirements in this list imply that $(\lambda \psi_1) \psi_2 = (\mathsf{A}_1 \psi_1) \psi_2 = \widehat{\mathsf{A}}_1 (\psi_1 \psi_2) = \lambda (\psi_1 \psi_2)$ for eigenstates of A_1, and these equalities with the third and fifth requirements imply that $(\sum_j c_j \varphi_j) \psi_2 = \sum_j c_j (\varphi_j \psi_2)$ for eigenstates φ_j of A_1. This observation strongly suggests that the mapping $\mathcal{H}_1 \times \mathcal{H}_2 \to \mathcal{H}$ given by $(\psi_1, \psi_2) \mapsto$

[1] A reasonable assumption, but see the discussion of EPR and nonlocality in Lecture 22.

$\psi_{1,2} = \psi_1\psi_2$ should be bilinear. The universal mapping property for tensor products of vector spaces over $\mathbb{C}$ then implies the existence of a unique linear map $\mathcal{H}_1 \otimes \mathcal{H}_2 \to \mathcal{H}$ with $\psi_1 \otimes \psi_2 \mapsto \psi_1\psi_2$ for all $\psi_i \in \mathcal{H}_i$, for $i = 1, 2$. So it seems that setting $\mathcal{H} = \mathcal{H}_1 \otimes \mathcal{H}_2$ is the simplest, most natural, elegant and economical way to accomplish the goal of constructing a state space for our two-particle system that meets these desirable requirements. Moreover, the fifth requirement suggests that $\widehat{\mathsf{A}}_1 = \mathsf{A}_1 \otimes 1_{\mathcal{H}_2}$ and $\widehat{\mathsf{A}}_2 = 1_{\mathcal{H}_1} \otimes \mathsf{A}_2$, and then the second and fourth requirements both hold automatically. Finally, if we define the inner product of $\mathcal{H}$ to be the bilinear extension of the rule

$$\langle \varphi_1 \otimes \varphi_2 | \psi_1 \otimes \psi_2 \rangle = \langle \varphi_1 | \psi_1 \rangle \langle \varphi_2 | \psi_2 \rangle \tag{21.1}$$

then the first requirement holds automatically and, further, the canonical inclusions $\mathcal{H}_i \hookrightarrow \mathcal{H}_1 \otimes \mathcal{H}_2 = \mathcal{H}$, for $i = 1, 2$, are isometric embeddings: this serves the useful and desirable goal of preserving probabilities.

There is one point, though, that must be made. The tensor product that we have described in the preceding paragraph is the *algebraic* tensor product familiar to students of mathematics. When the Hilbert spaces are not finite-dimensional, the algebraic product will not be complete in the inner product of equation 21.1 and so will not be a Hilbert space that is appropriate as a state space for quantum mechanics. For example, if $\{\xi_i : i \in \mathbb{N}\}$ is an orthonormal basis for the Hilbert space $\mathcal{H}$ then the sequence of partial sums of the infinite series $\sum_i c_i \xi_i \otimes \xi_i$ in $\mathcal{H} \otimes \mathcal{H}$ is Cauchy whenever the real sequence $\{c_i\}$ is square-summable, yet the series fails to converge in $\mathcal{H} \otimes \mathcal{H}$. This, though, is remedied quite easily by defining the product $\mathcal{H}_1 \otimes \mathcal{H}_2$ as the completion of the algebraic tensor product with respect to the inner product of equation 21.1. We still have an interest, though, in the incomplete algebraic tensor product as well as the algebraic symmetric and antisymmetric products, for the relevant algebraic multiplication operations reside in these and do not extend to their completions. It seems, then, that we need notation to indicate whether we are working in the **algebraic tensor product** or in its completion, the **analytic tensor product**. In these lectures we will use the unadorned symbol $\otimes$ to denote the algebraic tensor product, and the adorned symbol $\overline{\otimes}$ to denote the analytic tensor product.[2] When the spaces are finite-dimensional, there is no difference in the two and

[2] This notation is not in common use. Normally only the symbol $\otimes$ is used, with the context making clear which of the several varieties of tensor product one means. We have chosen to distinguish notationally the two varieties – the algebraic and the analytic – used in these lectures.

normally we will revert to the unadorned symbol. Even when the spaces are infinite-dimensional, it is often the case that much of our work is done in the algebraic tensor product, with its completion merely a technical requirement for obtaining an appropriate complete state space.

Of course, these arguments do not require that the tensor product is the inevitable tool for constructing two-particle state spaces, but they do show the naturalness of this choice. Ultimately, this tool is proved correct or not by more practical and mundane matters – does it work to reproduce the results found in the laboratory? The answer is a resounding *yes* and we will crystalize this into a fifth axiom for quantum mechanics that explains how the formalism of Lectures 2 and 4 is extended to handle quantum systems with a fixed, finite number of particles. In Lecture 25 we will extend the setting again to include systems with a variable but finite number of particles in which particles may be created and destroyed.

21.2 The Axiom for Multi-Component Systems

Before we state the axiom, a comment on terminology is in order. We have used the classical term *particle* as if we understand what the term means in the quantum world view, but a careful look at our development of the machinery of quantum mechanics indicates that we have never suggested what we should mean by that term. In the formal development of the minimalist axioms in Lecture 2, no mention is made of *particles*, as the axioms govern *quantum systems*, an undefined term in the formal structure. It was only when giving the Schrödinger interpretation of the axioms as wave mechanics that we used the term to mean something like a classical particle. Rather than worrying about what the meaning of these terms should be, we revert back to our minimalist axiomatic setting and use the general term *system* in place of *particle*.

Axiom 5: Multi-component systems. A quantum system consisting of a finite number of subsystems $S_1, S_2, \ldots, S_N$ has as its state space the N-fold analytic tensor product

$$\mathcal{H} = \mathcal{H}_1 \overline{\otimes} \cdots \overline{\otimes} \mathcal{H}_N,$$

where $\mathcal{H}_i$ is the state space of the single system S_i, for $i = 1, \ldots, N$. This is the completion of the algebraic tensor product $\mathcal{H}_1 \otimes \cdots \otimes \mathcal{H}_N$ with respect to the inner product defined as the multilinear extension of the rule

$$\langle \varphi_1 \otimes \cdots \otimes \varphi_N | \psi_1 \otimes \cdots \otimes \psi_N \rangle = \prod_{i=1}^{N} \langle \varphi_i | \psi_i \rangle \tag{21.2}$$

that is defined on the decomposable elements. If A is a self-adjoint operator on $\mathcal{H}_i$ representing an observable a for system S_i then

$$\mathsf{A}_i = 1_{\mathcal{H}_1} \otimes \cdots \otimes 1_{\mathcal{H}_{i-1}} \otimes \mathsf{A} \otimes 1_{\mathcal{H}_{i+1}} \otimes \cdots \otimes 1_{\mathcal{H}_N}$$

is the self-adjoint operator on $\mathcal{H}$ representing a-measurements on S_i.

The reader will recognize that the multilinear analogues of the five requirements that we have articulated for two-particle systems hold for the multi-component system of N subsystems.

The **decomposable** element $\psi_1 \otimes \cdots \otimes \psi_N$ in the algebraic tensor product $\mathcal{H}_1 \otimes \cdots \otimes \mathcal{H}_N$ is said in the physics community to be **unentangled**. On the other hand, a state ψ in $\mathcal{H}_1 \otimes \cdots \otimes \mathcal{H}_N$ is **indecomposable** or **entangled** if ψ cannot be expressed in the form $\psi_1 \otimes \cdots \otimes \psi_N$, for some $\psi_i \in \mathcal{H}_i$, $i = 1, \ldots, N$. For example, the state $\psi = \psi_1 \otimes \psi_1 + \psi_1 \otimes \psi_2 + \psi_2 \otimes \psi_1 + \psi_2 \otimes \psi_2$ in $\mathcal{H} \otimes \mathcal{H}$ is unentangled since $\psi = (\psi_1 + \psi_2) \otimes (\psi_1 + \psi_2)$, while the state $\psi_1 \otimes \psi_2 + \psi_2 \otimes \psi_1$ is entangled.

While the standard mathematical notation for a decomposable state is $\psi_1 \otimes \cdots \otimes \psi_N$, where $\psi_i \in \mathcal{H}_i$ for $i = 1, \ldots, N$, physicists use several notations:

$$\begin{aligned} \psi_1 \otimes \cdots \otimes \psi_N &= \psi_1 \cdots \psi_N = |\psi_1\rangle \cdots |\psi_N\rangle \\ &= |\psi_1 \cdots \psi_N\rangle = |\psi_1\rangle \otimes \cdots \otimes |\psi_N\rangle. \end{aligned}$$

When an eigenstate is labeled by a quantum number, such as $|j\rangle$ in discussions of angular momentum, this is expressed as

$$|j_1\rangle \otimes \cdots \otimes |j_N\rangle = |j_1\rangle \cdots |j_N\rangle = |j_1, \cdots, j_N\rangle = |j_1 \cdots j_N\rangle.$$

For example, in quantum computing, the state space of the quantum computer is the N-fold tensor product of $\mathbb{C}^2 = \mathbb{C}|0\rangle \oplus \mathbb{C}|1\rangle$, and a typical unentangled basis state referencing each of the N qubits is written variously as

$$|0\rangle|0\rangle|1\rangle \cdots |0\rangle|1\rangle = |001 \cdots 01\rangle.$$

Many of the counter-intuitive aspects of quantum mechanics arise in multi-component systems with entangled states. We will discuss a striking example – the Einstein–Podolski–Rosen paradox – in Lecture 22, as well as the Bell inequalities, which confirm how deeply and subtly strange is the nonclassical quantum world to our sensibilities, trained as they are entirely in the classical world.

21.3 Coupled Angular Momenta, Again

Recall from Lecture 13 that $\mathfrak{su}(2) = \{\mathbf{a} \cdot \boldsymbol{\tau} : \mathbf{a} \in \mathbb{R}^3\}$, where the triad $\boldsymbol{\tau} = (\tau_1, \tau_2, \tau_3)$ is given as $\boldsymbol{\tau} = -\mathbf{i}\boldsymbol{\sigma}$; explicitly,

$$\tau_1 = \begin{pmatrix} 0 & -\mathbf{i} \\ -\mathbf{i} & 0 \end{pmatrix}, \quad \tau_2 = \begin{pmatrix} 0 & -1 \\ 1 & 0 \end{pmatrix}, \quad \text{and} \quad \tau_3 = \begin{pmatrix} -\mathbf{i} & 0 \\ 0 & \mathbf{i} \end{pmatrix}.$$

For a nonnegative half-integer k, let U^k denote the irreducible $(2k+1)$-dimensional representation space of $\mathfrak{su}(2)$. The space U^k may be used as the state space of a system P of total angular momentum $k(k+1)\hbar^2$ when angular momentum is the only degree of freedom of concern. If $\mu : \mathfrak{su}(2) \to \mathrm{End}(U^k)$ is the representation homomorphism then $\mathsf{K} = \mathbf{i}\hbar\mu(\boldsymbol{\tau})$ defines an angular momentum triad on U^k with $\mathsf{K}^2\psi = k(k+1)\hbar^2\psi$ for all $\psi \in U^k$. Further, $\mathsf{K}_3 = \mathbf{i}\hbar\mu(\tau_3)$ decomposes U^k into an orthogonal direct sum

$$U^k = \bigoplus_{p=-k}^{k} U_p^k,$$

where U_p^k is the one-dimensional eigenspace of K_3 with eigenvalue $p\hbar$, for $p = -k, -k+1, \ldots, k-1, k$.

Now consider a second state space V^ℓ, a $(2\ell+1)$-dimensional irreducible representation space of $\mathfrak{su}(2)$ that serves as the angular momentum state space of another system Q with total angular momentum $\ell(\ell+1)\hbar^2$. Let $\mathsf{L} = \mathbf{i}\hbar\nu(\boldsymbol{\tau})$ be the angular momentum triad determined by the representation homomorphism $\nu : \mathfrak{su}(2) \to \mathrm{End}(V^\ell)$. As above, the operator $\mathsf{L}_3 = \mathbf{i}\hbar\nu(\tau_3)$ decomposes V^ℓ into an orthogonal direct sum

$$V^\ell = \bigoplus_{q=-\ell}^{\ell} V_q^\ell,$$

where V_q^ℓ is the one-dimensional eigenspace of L_3 with eigenvalue $q\hbar$, for $q = -\ell, -\ell+1, \ldots, \ell-1, \ell$.

Axiom 5 applies, and it asserts that the tensor product $U^k \otimes V^\ell$ is the appropriate state space of the combined quantum system consisting of the two angular momentum subsystems P and Q. The self-adjoint operators on $U^k \otimes V^\ell$ corresponding to angular momentum measurements in subsystem P make up the angular momentum triad $\mathbf{K} \otimes 1_{V^\ell} = (\mathsf{K}_i \otimes 1_{V^\ell})_{i=1,2,3}$ and in subsystem Q the angular momentum triad $1_{U^k} \otimes \mathbf{L}$. Since each unentangled tensor element $\varphi \otimes \psi \in U^k \otimes V^\ell$ is a simultaneous eigenstate of $(\mathbf{K} \otimes 1_{V^\ell})^2$ with eigenvalue $k(k+1)\hbar^2$ and of $(1_{U^k} \otimes \mathbf{L})^2$ with eigenvalue $\ell(\ell+1)\hbar^2$, the state space $U^k \otimes V^\ell$ consists entirely of such simultaneous eigenstates. The commuting component operators $\mathsf{K}_3 \otimes 1_{V^\ell}$ and $1_{U^k} \otimes \mathsf{L}_3$ decompose $U^k \otimes V^\ell$ into the orthogonal direct sum

$$U^k \otimes V^\ell = \bigoplus_{\substack{p=-k \\ q=-\ell}}^{\substack{q=\ell \\ p=k}} U_p^k \otimes V_q^\ell,$$

where, obviously, $U_p^k \otimes V_q^\ell$ is the subspace of simultaneous eigenstates of $\mathsf{K}_3 \otimes 1_{V^\ell}$ and $1_{U^k} \otimes \mathsf{L}_3$, with respective eigenvalues $p\hbar$ and $q\hbar$. A typical normalized element of $U_p^k \otimes V_q^\ell$ might be written in Dirac notation as $|p, q\rangle$.

This combined system with state space $U^k \otimes V^\ell$ and angular momentum triads $\mathbf{K} \otimes 1_{V^\ell}$ and $1_{U^k} \otimes \mathbf{L}$ is the generic minimal example of a coupled angular momentum system, introduced in Section 18.1. Given the tensorial structure of the state space of two coupled angular momentum triads articulated here we now can fill in the one gap in the discussion of the fundamental addition theorem of Section 18.2 and show that $|k - \ell| \leq j$. Recall that the parameter j serves as the quantum number indexing the eigenvalues of the operator $\mathbf{J}^2$ on $U^k \otimes V^\ell$, where, in this context, $\mathbf{J}$ is the angular momentum triad given as the sum $\mathbf{J} = (\mathbf{K} \otimes 1_{V^\ell}) + (1_{U^k} \otimes \mathbf{L})$. This angular momentum triad $\mathbf{J}$ arises from the **tensor product of representations** μ **and** ν, the homomorphism $\mu \otimes \nu : \mathfrak{su}(2) \to \mathrm{End}(U^k \otimes V^\ell)$ defined on $X \in \mathfrak{su}(2)$ as $(\mu \otimes \nu)(X) = \mu(X) \otimes 1_{V^\ell} + 1_{U^k} \otimes \nu(X)$. The operator $\mathbf{J}^2$ decomposes $U^k \otimes V^\ell$ into an orthogonal direct sum of $\mathbf{J}^2$-eigenspaces indexed by j. We verified in Lecture 18 that the maximum value of j is $j_{\max} = k + \ell$ and our aim now is to confirm that the minimum value of j is given by $j_{\min} = |k - \ell|$. We make the assumption that the values of j are $j_{\min}, j_{\min} + 1, \ldots, j_{\max}$, i.e., that there are no gaps from $j_{\min}$ to $j_{\max}$.

This allows us to write the J^2-decomposition of $U^k \otimes V^\ell$ as

$$U^k \otimes V^\ell = \bigoplus_{j=j_{\min}}^{k+\ell} W^j, \tag{21.3}$$

where W^j is the $j(j+1)\hbar^2$-eigenspace of J^2. With the assumption that each $\mathsf{J}_3|W^j$-eigenvalue is nondegenerate, the work of Section 12.2 implies that the dimension of W^j is $2j+1$. A dimension count on the direct sum decomposition of equation 21.3 gives

$$\begin{aligned}(2k+1)(2\ell+1) &= \sum_{j=j_{\min}}^{k+\ell} (2j+1) \\ &= 2\left(\sum_{j=j_{\min}}^{k+\ell} j\right) + (k+\ell-j_{\min}+1) \\ &= (k+\ell)(k+\ell+1) - (j_{\min}-1)j_{\min} + k + \ell - j_{\min} + 1.\end{aligned}$$

A little algebra gives

$$j_{\min}^2 = k^2 + \ell^2 - 2k\ell = (k-\ell)^2,$$

or

$$j_{\min} = |k-\ell|.$$

This finishes the verification of the fundamental addition theorem modulo the assumption of no gaps in the sequence $j_{\min}, \ldots, j_{\max}$, for which we refer the reader to pp. 33–35 of Edmonds (1960) or Appendix C of Hall (2015); see also Sternberg (1995).

Notice that the tensor product representation $\mu \otimes \nu$ of $\mathfrak{su}(2)$ on $U^k \otimes V^\ell$ is not irreducible. The operator J^2 decomposes $U^k \otimes V^\ell$ into summands W^j of irreducible representation spaces of $\mathfrak{su}(2)$.

21.4 A Mathematical Interlude: Bases for Tensor Products

As it often is useful in quantum mechanics to expand state vectors in terms of specific bases, we will spend some effort in illuminating some standard bases for tensor products. For simplicity we initially make the assumption that each of the tensor product factors $\mathcal{H}_i$ is finite-dimensional, with, say, $\dim \mathcal{H}_i = D_i$. Let $\boldsymbol{\xi}_i$ be an orthonormal basis for

$\mathcal{H}_i$ with D_i elements. The definition of the inner product on the tensor product easily implies that the set

$$\boldsymbol{\xi}_1 \otimes \cdots \otimes \boldsymbol{\xi}_N = \{\xi_1 \otimes \cdots \otimes \xi_N : \xi_i \in \boldsymbol{\xi}_i \text{ for } i = 1, \ldots, N\}$$

is orthonormal, and it is easy to see that it spans $\mathcal{H}_1 \otimes \cdots \otimes \mathcal{H}_N$, hence is a basis. It follows that $\dim \mathcal{H}_1 \otimes \cdots \otimes \mathcal{H}_N = D_1 \cdots D_N$. In particular, when each $\mathcal{H}_i$ is equal to a common Hilbert space $\mathcal{H}$ of dimension D, which is a prevalent special case, the N-fold tensor tensor is denoted variously as $\otimes^N \mathcal{H}$ or $\mathcal{H}^{\otimes N}$ and has dimension D^N. The standard basis then is

$$\otimes^N \boldsymbol{\xi} = \{\xi_1 \otimes \cdots \otimes \xi_N : \xi_i \in \boldsymbol{\xi} \text{ for } i = 1 \ldots, N\}.$$

This discussion applies equally well in infinite dimensions when $\boldsymbol{\xi} = \{\xi_i : i \in \mathbb{N}\}$ is a complete orthonormal basis for the separable Hilbert space $\mathcal{H}$. The only difficulty that must be faced is the question of convergence. To illustrate this, we consider the case where $N = 2$ and $\mathcal{H}_1 = \mathcal{H} = \mathcal{H}_2$. Here, $\boldsymbol{\xi} \otimes \boldsymbol{\xi} = \{\xi_i \otimes \xi_j : i, j \in \mathbb{N}\}$ is clearly an orthonormal subset of the analytic tensor product space $\mathcal{H} \overline{\otimes} \mathcal{H}$. That it is complete and, therefore, serves as a complete orthonormal basis for $\mathcal{H} \overline{\otimes} \mathcal{H}$, follows since, when $\varphi = \sum_{i=1}^{\infty} a_i \xi_i$ and $\chi = \sum_{j=1}^{\infty} b_j \xi_j$, the decomposable element $\varphi \otimes \chi$ may be written as

$$\varphi \otimes \chi = \sum_{i,j \in \mathbb{N}} a_i b_j (\xi_i \otimes \xi_j).$$

The one gap in the discussion is the question of the convergence of the double sum. However, this follows easily, since the square-summability of the sequences $\{a_i\}_{i=1}^{\infty}$ and $\{b_j\}_{j=1}^{\infty}$ implies the square-summability of the double sequence $\{a_i b_j\}_{i,j=1}^{\infty}$, irrespective of the ordering of the sum.[3] Similar remarks apply when $N > 2$ and when the $\mathcal{H}_i$ are pairwise distinct.

It is convenient to collect the algebraic tensor product spaces $\otimes^N \mathcal{H}$ into a single object by taking the direct sum. Defining $\otimes^0 \mathcal{H}$ as $\mathbb{C}$ and $\otimes^1 \mathcal{H}$ as $\mathcal{H}$, we define the **tensor algebra** of $\mathcal{H}$, denoted simply as $\otimes \mathcal{H}$, as an orthogonal direct sum:

$$\bigotimes \mathcal{H} = \bigoplus_{N=0}^{\infty} \left(\otimes^N \mathcal{H}\right).$$

[3] The usual ordering for a double sum of complex numbers c_{ij} is that $\sum_{i,j \in \mathbb{N}} c_{ij}$ is defined to be $\lim_{N \to \infty} \sum_{i,j \leq N} c_{ij}$. When $\sum_{i,j \in \mathbb{N}} |c_{ij}|$ converges, we say the convergence of $\sum_{i,j \in \mathbb{N}} c_{ij}$ is **absolute** and, as for singly indexed series, the sum is independent of the ordering of the summands.

The terminology we are using requires some explanation. The tensor algebra is, first, a direct sum in the category of complex vector spaces and, second, an (incomplete) inner product space where the individual summands are orthogonal. Thus, each vector in $\otimes\mathcal{H}$ is a **finite** sum of the form $\sum_{N=0}^{K} \psi_N$, where $\psi_N \in \otimes^N\mathcal{H}$ and $K \in \mathbb{N}$, and the inner product is given by

$$\left\langle \sum_{N=0}^{K} \psi_N \,\middle|\, \sum_{N=0}^{K} \psi'_N \right\rangle = \sum_{N=0}^{K} \langle \psi_N | \psi'_N \rangle.$$

A multiplication operation is defined on decomposable elements $\psi = \varphi_1 \otimes \cdots \otimes \varphi_M \in \otimes^M\mathcal{H}$ and $\psi' = \chi_1 \otimes \cdots \otimes \chi_N \in \otimes^N\mathcal{H}$ as[4]

$$\psi \otimes \psi' = \varphi_1 \otimes \cdots \otimes \varphi_M \otimes \chi_1 \otimes \cdots \otimes \chi_N \in \otimes^{M+N}\mathcal{H},$$

and extended bilinearly. The tensor algebra $\otimes\mathcal{H}$ is then generated as an algebra by $\mathcal{H} = \otimes^1\mathcal{H}$ and is an associative, noncommutative, graded complex algebra with unit $1 \in \mathbb{C} = \otimes^0\mathcal{H}$.

In Lecture 24 we will explore in more detail algebraic structures defined on appropriate subspaces of $\otimes\mathcal{H}$.

21.5 End Notes

Edmonds (1960) is the standard reference on angular momentum in quantum mechanics and gives a derivation of the fundamental addition theorem and the no-gaps assumption of our Section 21.3 in his Chapter 3. Section 4.2 of Sternberg (1995) presents a concise explanation of how $U^k \otimes V^\ell$ decomposes into irreducible representations, this being the content of Section 21.3.

[4] This is well-defined since the tensor product of elements of $\mathcal{H}$ is associative.

22

The EPR Argument and Bell's Inequalities

> The experimental verification of violations of Bell's inequality for randomly set measurements at space-like separation is the most astonishing result in the history of physics. Theoretical physics has yet to come to terms with what these results mean for our fundamental account of the world. Experimentalists, from Freedman and Clauser and Aspect forward, deserve their share of the credit for producing the necessary experimental conditions and for steadily closing the experimental loopholes available to the persistent skeptic. But the great achievement was Bell's. It was he who understood the profound significance of these phenomena ... Unfortunately, many physicists have not properly appreciated what Bell proved ... What Bell proved, and what theoretical physics has not yet properly absorbed, is that the physical world is non-local.
>
> Tim Maudlin
> *What Bell Did*, 2014

Niels Bohr argued that quantum mechanics is a **complete** theory of nature, i.e., that everything that can be known about a physical system is encoded in the state vector. He, along with many physicists of his day, also asserted a **positivistic** view of the role of science. The job of the physicist is not to discover what nature is really like, not to discover the exact truth of the ontological status of the micro-world, for this is unknowable; rather, the job of the physicist is to discover algorithms that predict correctly and accurately dial readings on laboratory instruments, the outcomes of scientific experiments. Therefore, one should not ask such questions as "which of two slits did that electron pass through in its journey from the cathode ray generator to the phosphorescent screen?," or "what is the x_1-component of the spin of an electron with state vector $|+\rangle$?," or "what is an electron – particle or wave?" These are meaningless questions, for there are no experiments that can answer them, and no such information is encoded in the state vector. Views along these lines, though more developed, have become the dominant view of the physics community. These views evolved into a system referred to as the **Copenhagen interpretation** in a series of sometimes confusing metaphysical writings of Bohr in the decade of the thirties. The Copenhagen hegemony accepts what generations of physicists prior to 1930 would have considered an extremely radical view of nature, namely, that micro-objects cannot be said to possess values for measurable attributes – position, momentum, angular momentum, spin, energy – until those very attributes are measured. To be internally consistent it must accept the belief that nature is, at the basic level of micro-processes, acausal. The statistical aspects of quantum mechanics are not, as is sometimes popularly averred, indications of imperfect knowledge or of measurement influences that limit experimental accuracy, but are instead a basic irreducible "hard fact" of nature.

The most successful critique and criticism of the completeness hypothesis is that of Einstein, Podolsky, and Rosen (1935), the so-called **EPR paradox**, extended by Bohm (1951) and Bell (1964). Bohr's response to the EPR paradox, and the subsequent Bohr–Einstein debates of the 1930s, were the final push towards the "extreme" Copenhagen beliefs about the meaning of quantum mechanics. In this author's opinion, it is unfortunate that Bohr's perplexing and often incoherent views, perhaps formed more by psychology and philosophy than by the exigencies of either the laboratory or logical necessity, prevailed in these debates and stopped in its tracks rational inquiry into these important metaphysical questions. The nails in the coffin of any alternative to Bohr and Copenhagen were hammered in by von Neumann's surprising proof of the impossibility of "hidden variables" theories, actually proved earlier though generally unrecognized previously to EPR.

Some today are less sure about the correctness of the Copenhagen interpretation than were previous generations of quantum physicists, and particularly its claim of completeness. After all, the claim that quantum mechanics is a complete theory clearly is a dogmatic assertion of philosophy, not of physics. Many now hold to a less radical view that is as noncommittal about the status of quantum mechanics as a complete theory as it is practical in its positivistic understanding of its meaning. This **minimal instrumentalist interpretation**, like its Copenhagen cousin, still posits that the meaning of quantum mechanics is found in the calculation of probabilities for various possible instrument readings when experiments are performed to measure system attributes. It tells us that the formalism of the mathematics is related to possible experimental outcomes but does not attempt to construct a correspondence between mathematical objects, such as the wave function, and various physical attributes that in any way models or reflects the real status of those attributes. Instead, the correspondence constructed is between fairly abstract mathematical expressions and very concrete instrument readings. In this view, quantum mechanics is a calculus for determining probabilities that give accurate statistical predictions of physical measurements, not a theory that provides a mental image of the micro-world that to some degree comports with reality. Whereas the positivism of the Copenhagen interpretation is theoretical and pervasive, that of the minimal instrumentalist interpretation is practical and mundane. It is thoroughly positivistic in practice without committing to positivism, nor to any other philosophical position, in theory. It rejects the dogmatism of Copenhagen, being agnostic on the question of completeness. Indeed, it is open to the possibility of a theory that transcends quantum mechanics as long as this agreed with the quantum mechanical predictions in its range of applicability and offered something more – either a prediction not covered by quantum mechanics or a more elegant understanding of how to visualize micro-processes.

There has been renewed interest among some physicists and philosophers in these foundational and metaphysical questions since the work of John Bell in the 1960s and the Aspect experiments of the 1980s. Bell examined EPR in detail and not only dethroned von Neumann's impossibility "proof" but established exactly what types of hidden variables theories are allowed if they are to reproduce the theoretical predictions of quantum mechanics. This should have surprised no one, as there was ample evidence that von Neumann's proof did not apply to all conceivable hidden variables theories; in 1952 David Bohm, building on the work of Louis de Broglie from a generation earlier, had constructed a hidden variables theory that provably reproduced all the predictions of standard nonrelativistic quantum mechanics. This was dismissed by the physics community as at best a mistake and at worst quackery, but no two physicists seemed to agree on exactly where Bohm's mistake lay.

Bell's great contribution to the discussion was a clarification of the roles of **nonlocality** and **inseparability**, not only in quantum mechanics but also in Bohmian mechanics and, indeed, in any theory that purports to reproduce the predictions of quantum mechanics. These insights lay dormant for over 30 years, in Einstein's arguments of the 1930s until Bell carefully reexamined the Bohr–Einstein controversy. He offered an articulation of probably the most anti-classical implication of quantum mechanics in his clarification of what has been called this "spooky action-at-a-distance" that quantum mechanics not only allows but necessitates. Bell and his protégés clarified further the notion of **realism** in physics and the role of experimentation to settle some of these seemingly purely philosophical questions. The Aspect experiments then confirmed nonlocality, with the violation of the Bell inequalities in real systems in nature, and confirmed the predictions of both quantum and Bohmian mechanics in this setting.

In this lecture, we describe a version of the EPR argument as updated by Bohm and Bell that offers a compelling choice between opposing philosophical views. We prove the Bell inequalities, whose violation places strict boundaries and restrictions on any realist hidden variables theory that purports to give exact values at all times to measurable quantities in a quantum system. We also show how quantum mechanics violates the Bell inequalities and reproduces the correct experimentally verified predictions in coupled spin-$\frac{1}{2}$ systems. To make the arguments precise, we need to develop the mathematics of coupled spin-$\frac{1}{2}$ systems, applying the results of Sections 18.3 and 21.1. We end with a striking example of quantum entanglement in a system of three particles.

The author considers these issues to be some of the most profound and interesting in quantum mechanics and hopes that the reader is inspired to examine them in more detail in the references provided in the End Notes.[1]

22.1 The EPR Criticism of Quantum Mechanics

Einstein, working with Boris Podolsky and Nathan Rosen, argued that quantum mechanics is an incomplete description of reality. The clarity of the EPR argument stands in sharp contrast to the obfuscation of Bohr's response. The physics community of the 1930s nonetheless ceded the victory to Bohr,[2] and the EPR argument was ignored by that and

[1] The quote that begins this lecture is the first paragraph of Tim Maudlin's wonderful essay (2014) in *J. Phys. A* **47**, 424010. I highly recommend this special issue of the journal, which was devoted to "50 years of Bell's theorem."

[2] The history of the development, promotion, proselytization, and acceptance of the Copenhagen interpretation presents a fascinating and intriguing case study in the psychology and sociology of science. Perhaps the finest telling of this story

the next generation of the development of physics. We present the EPR argument using spin states rather than the original, which used position and momentum measurements, an improvement due to David Bohm. The argument begins with what appears to be a reasonable definition of physical reality. Quoting from EPR (1935),

> If, without in any way disturbing a system, we can predict with certainty (i.e., with probability equal to unity) the value of a physical quantity, then there exists an element of physical reality corresponding to that quantity.

We now argue that the spin component in any direction for certain spin-$\frac{1}{2}$-particles satisfies this definition of physical reality. Indeed, there are elements that, upon excitation to a high energy state, decay to a lower energy state by emitting two spin-$\frac{1}{2}$-particles A and B in opposite directions, say along the y-axis of a coordinate system whose origin is at the source element. Moreover, the spins of A and B are perfectly correlated in the sense that, for any direction θ perpendicular to the y-axis, spin measurements along that direction always yield $+\frac{1}{2}\hbar$ for one of the particles and $-\frac{1}{2}\hbar$ for the other. In principle, a spin measurement on particle A may be performed such a long time after emission, say when A and B are light years apart, that the measurement of the spin component of A can in no way disturb particle B. If A is measured to have spin component $\pm\frac{1}{2}\hbar$ along direction θ then we are assured that B has spin $\mp\frac{1}{2}\hbar$ without measuring or disturbing particle B in any way. According to the EPR notion of physical reality, there exists an element of physical reality corresponding to the spin component of particle B along every direction θ. The picture that EPR paints is that when particle A and B are emitted, for every possible direction θ one particle has spin $+\frac{1}{2}\hbar$ and the other has spin $-\frac{1}{2}\hbar$ along that direction; we just do not know which has which without measurement. This seems reasonable to minds whose only direct experience is of the world of classical mechanics. The quantum mechanical analysis of this two-particle system is presented in Section 22.2. It knows nothing of A and B possessing spin component values along every direction θ before measurement. The wave functions of A and B can specify at most one spin component with certainty and, concludes EPR, quantum mechanics is incomplete in that there are elements of reality in this simple system that quantum mechanics fails to describe.

is Mara Beller's original and masterful historical reconstruction of the conceptual foundations of nonrelativistic quantum mechanics in Beller (1999), wherein she challenges both the coherence and inevitability of the Copenhagen interpretation.

The Copenhagen interpretation asserts the completeness of quantum mechanics. Its position is that neither A nor B has any values for spin components until at least one is measured. It suggests a certain wholeness in a system of two particles, in fact in an inability to even separate the system into two component subsystems until a measurement is made on the system. At the point of measurement, the wave function of the system collapses into two separate wave functions governing the now-individual particles. The initial system of correlated "particles" is **inseparable** and becomes separable only upon the act of measurement of the spin component of particle A.

The EPR argument points out in clear detail that the orthodox Copenhagen interpretation of quantum mechanics accepts a kind of "action-at-a-distance." The measurement of $+\frac{1}{2}\hbar$ for the θ-component of the spin of A immediately collapses the wave function and forces B to fall into the $-\frac{1}{2}\hbar$ state along the θ-direction, even though they may be space-like separated by light-years. The orthodox view avers that it is incorrect to say that there is an effect at the point of measurement of A that travels faster than light to reach B and change its state; rather, it merely asserts that the act of measurement on the inseparable system has the immediate effect of collapsing the wave function, which does indeed immediately give to the θ-component of the spin of B an exact value. However one parses the cause–effect relationship between measuring a spin component at A and the appearance of an exact value for the spin component at B, all interpreters agree that the collapse of the wave function upon measurement is instantaneous in the state space and has instantaneous repercussions across space-like separated expanses of the universe.

Whereas the Copenhagen interpretation asserts the completeness of quantum mechanics, Bohmian mechanics asserts its incompleteness and offers a more basic theory from which quantum mechanics emerges, in much the same way that statistical mechanics emerges from the classical mechanics of large aggregates of particles. Bohmian mechanics accepts the EPR definition of physical reality and its argument that each of the two particles of the coupled spin-$\frac{1}{2}$ system possesses, for every possible direction θ, a value for the θ-component of its spin from the time of emission of A and B. It asserts that there is a collection of **hidden variables** Ω attached to the wave function ψ of the two-particle system that determines the spin components of the constituent particles along all directions. The possibility of such hidden variables flies in the face of von Neumann's proof asserting that it is impossible to assign hidden variables to a theory in such a way as to reproduce the statistical predictions of quantum mechanics. von Neumann's proof assumes that every reasonable description of reality must have no faster-than-light effects. His argument was made in 1929, six years before EPR and over two decades before David Bohm used the EPR argument and spin to observe that quantum mechanics itself has, if not faster-than-light effects, then instantaneous collapse of the wave function that assures instantaneous, faster-than-light repercussions across space-like separated points. In 1952, Bohm, building on the theoretical work of de Broglie, introduced his hidden variables version of quantum mechanics that: (1) reproduces all predictions of standard quantum mechanics exactly; (2) is a theory that accepts **realism**, i.e., that the micro-world consists of particles, not waves, traveling along definite trajectories with, at all times, definite positions and momenta and definite values of spin, angular momentum, and energy; and (3) has faster-than-light effects. Physicists initially based their criticism of Bohmian mechanics by appealing to special relativity, until it was pointed out that quantum mechanics has instantaneous collapse with exactly the same *measurable* effects as those that arise from the Bohmian faster-than-light effects. The discussion then ceased to be rational and deteriorated into everything from indifference to outrage that someone could still believe in something as old-fashioned as a "classical particle." The history of the response of the physics community to Bohmian mechanics gives an enlightening glance into the psychology of science. Though almost universally criticized and often vilified, Bohm's theory has yet to be refuted successfully.

Our aim in the remainder of this lecture is to present the quantum mechanical treatment of the coupled spin-$\frac{1}{2}$ system, which will lead us into Bell's striking analysis of Bohm's version of the EPR argument where we discover that, rather than being of mere philosophical interest, this "action-at-a-distance" may be tested experimentally.

22.1.1 A Disclaimer

Before we begin the analysis, the author feels it appropriate here to offer a disclaimer. The reader may suspect that, with his fairly harsh criticism of the adoption of the Copenhagen orthodoxy by the physics community, the author actually rejects the completeness hypothesis in favor of Bohmian mechanics. The reader would then be mistaken. What the author rejects is an adoption of the standard Copenhagen interpretation when there is not ample evidence on either physical or logical grounds to make such a clear choice. He rejects what often amounts to dogmatism precisely because it has stopped in its tracks appropriate avenues of research into other possibilities of interpretation. A brief study of Bell's work offers ample historical evidence of this fact. The dogmatic acceptance that Bohr "had slain Einstein" in the debates of the 1930s caused a generation of physicists to miss the most surprising implications of quantum mechanics the in issues surrounding nonlocality and inseparability. The sometimes vicious attacks on Bohm helped to force rational inquiry concerning foundational issues into a backwater, causing someone like Bell to do the work on inequalities which, by all rights, should have been discovered and articulated by the previous generation of physicists, as a "hobby" rather than as part of his work as a professional physicist. As Mara Beller exclaims,[3] "It was the opposition – Einstein and Schrödinger – who in the mid 1930s discovered and mathematically elaborated the basic inseparability of quantum systems ... It was the [Copenhagen] orthodoxy, as I have argued, [which] diffused these arguments by operational stratagems, preventing serious exploration of inseparability until Bell's seminal work. Einstein's characterization of Bohr as a 'Talmudic philosopher' referred precisely to Bohr's circumventing, rather than directly confronting, the most fundamental problem of quantum theory. This deep physical challenge was met with the rhetoric of 'sacrifice' [of classical ideas]. *The 'sacrifice' primarily meant elimination of the opposition's ideas.*"[4]

In truth, this author is agnostic on the issue of interpretation, finding himself in emotional agreement with the completeness hypothesis on Mondays, Wednesdays, and Fridays, and even appreciative of the leeway given in nature by its contention that there is an irreducible, acausal probabilistic aspect to the micro-world. He then finds himself in disagreement on Tuesdays, Thursdays, and Saturdays and open to the possibility of understanding nature at a more basic level than that accorded by quantum mechanics. On the one hand, completeness has its own attraction. On the other hand, the computer-generated diagrams showing the tracks of Bohmian particles in the two-slit experiment[5] are tantalizing in their suggestion that we can indeed understand the microworld using something like our classical sensibilities, albeit corrected by faster-than-light effects. On Sundays he tries not to think about these matters at all!

[3] Beller (1999), p. 285.
[4] Emphasis mine.
[5] See, for example, Holland (1993), pp. 181 and 184; see also the diagrams for interference and tunneling in Holland's Chapter 5.

22.2 The Coupled Spin-$\frac{1}{2}$ System in Quantum Mechanics

The mathematics of the spin-$\frac{1}{2}$ system, developed in detail in Section 18.3, will now be used to give a quantum mechanical treatment of perfectly correlated spin states in a coupled spin-$\frac{1}{2}$ system. We consider a system of two correlated spin-$\frac{1}{2}$ particles emitted in opposite directions from a source that then travel along a line y. Spin-measuring devices S_1 and S_2 are positioned on the line y on opposite sides of the source, far removed from one another, to measure the spin components of the particles along various directions perpendicular to the line of motion. We set up an orthogonal two-dimensional coordinate system perpendicular to the line of motion and centered at the emission source, with axes labeled as x and z. The coordinate system is used to encode with a variable θ the orientation of the spin component detectors. The detector orientation θ is measured from the positive z-axis toward the positive x-axis, and in the xz-plane. If the detectors S_1 and S_2 have different orientations, we will use θ_1 and θ_2 to denote the respective orientations.

We will measure spin components in units of $+\frac{1}{2}\hbar$. To say that the spins are **perfectly correlated** means that a measurement of the spin component in any direction $\theta_1 = \theta_2$ of $+1$ at either detector ensures, with unit probability, a measurement of -1 at the other, and similarly for a measurement of -1 at either detector. In fact, when spin measurements are made by S_1 and S_2, no matter in what order or simultaneously, and no matter where along the line of travel S_1 and S_2 lie as long as the detectors are oriented at the same angle θ, the results are always $+1$ at one detector and -1 at the other. Moreover, the probability of measuring a $+1$ by S_1 is the same as that for measuring a -1, both equal to $\frac{1}{2}$. So, in dual measurements with $\theta_1 = \theta_2$, the outcome is $+1$ at S_1 and -1 at S_2 half the time, and the reverse, -1 at S_1 and $+1$ at S_2, the other half. Perfectly correlated particles do appear in nature and can be generated in the laboratory, but the exact details of such pair production need not concern us. Our first task is to describe the state vector ψ for this perfectly correlated spin pair system.

According to Section 18.3, the irreducible two-dimensional representation space $V^{1/2}$ determined by the spin representation of $\mathfrak{su}(2)$ with angular momentum triad $\mathbf{S}$ and quantum number $j = \frac{1}{2}$ serves as the state space of each of the particles when considered separately. The standard orthonormal basis for $V^{1/2}$ is $\{|\pm\rangle\}$, where $|\pm\rangle$ is the ± 1-eigenstate of the operator S_3 that measures spin components along the positive z

direction, for which $\theta = 0$. When the two particles are allowed to interact, they must be combined into a single system, a coupled spin system whose state space is given, according to Axiom 5 of the previous lecture, as the tensor product $\mathcal{H} = V^{1/2} \otimes V^{1/2}$. Spin component measurements by detector S_1 are described by the self-adjoint operator triad $\mathbf{S}^1 = \mathbf{S} \otimes 1$. In the notation of Section 18.3, where S_θ represents spin component measurements in the direction θ, the operator $\mathsf{S}^1_\theta = \mathsf{S}_\theta \otimes 1$ corresponds to spin component measurements made by S_1 in the direction θ. Similarly, $\mathbf{S}^2 = 1 \otimes \mathbf{S}$ represents spin component measurements by detector S_2, with $\mathsf{S}^2_\theta = 1 \otimes \mathsf{S}_\theta$ representing measurements along the direction θ by S_2.

Since $\mathsf{S}^1_{\theta=0}$ and $\mathsf{S}^2_{\theta=0}$ commute, there is a basis of simultaneous eigenstates corresponding to z-component spin measurements by both the detectors S_1 and S_2. Indeed, an orthonormal basis for the coupled spin state space $\mathcal{H}$ is

$$\{|{+}{+}\rangle, |{+}{-}\rangle, |{-}{+}\rangle, |{-}{-}\rangle\},$$

where $|{\pm}{\pm}\rangle = |\pm\rangle \otimes |\pm\rangle$. The state vector ψ of the perfectly correlated system can therefore be expanded as

$$\psi = a|{+}{+}\rangle + b|{+}{-}\rangle + c|{-}{+}\rangle + d|{-}{-}\rangle,$$

for some complex coefficients a, b, c, d normalized by $|a|^2 + |b|^2 + |c|^2 + |d|^2 = 1$. Our task is to determine the numerical values of these coefficients. Notice then that perfect correlation implies that $a = d = 0$, since a nonzero a would imply a positive probability $|a|^2$ of observing spins of $+1$ at both detectors S_1 and S_2, and similarly for d. It also implies that $|b| = |c|$ since ± 1 are equally likely upon measurement at S_1. It follows that, up to a unimodular phase factor, the form of the normalized state vector ψ is

$$\psi = \frac{1}{\sqrt{2}} \left(|{+}{-}\rangle + e^{\mathrm{i}\phi} |{-}{+}\rangle \right), \tag{22.1}$$

for some phase factor $e^{\mathrm{i}\phi}$.

One may object by suggesting an alternative explanation for the state vector. This argument suggests that when a $+1$ is recorded for a z-component measurement by S_1, we should interpret this to mean that the state before and after measurement is $|{+}{-}\rangle$ and, when -1 is recorded, the state is $|{-}{+}\rangle$. We account for the fact that $+1$ and -1 appear with equal frequency at S_1 by proposing that when the spin pair is emitted, the state is either $|{+}{-}\rangle$ or $|{-}{+}\rangle$ with equal probabilities for each.

This proposition serves to move the explanation of the equal frequency of $+1$ and -1 from the make-up of the state vector to the statistical distribution of different states at the source, and it does indeed offer an alternative explanation for the observed z-component measurements when performed on a large number of perfectly correlated coupled spin systems. The problem with this explanation, though, is that it fails to reproduce the equal frequencies of $+1$ and -1 spin component measurements when the detectors are not aligned along the z-direction. Indeed, in this alternative scenario, what are the frequencies of ± 1 for measurements when both detectors are aligned in the direction θ? The application of the resolution of the identity $1 = |\theta+\rangle\langle\theta+| + |\theta-\rangle\langle\theta-|$ to $|+\rangle$ to get equation 18.6, which states that

$$|+\rangle = \cos(\theta/2)\,|\theta+\rangle - \sin(\theta/2)\,|\theta-\rangle, \tag{22.2}$$

can be applied to $|-\rangle$ to yield

$$|-\rangle = \sin(\theta/2)\,|\theta+\rangle + \cos(\theta/2)\,|\theta-\rangle. \tag{22.3}$$

Taking tensor products gives

$$\begin{aligned} |+-\rangle = |+\rangle \otimes |-\rangle = \cos^2(\theta/2)\,|\theta+-\rangle - \sin^2(\theta/2)\,|\theta-+\rangle \\ + \cos(\theta/2)\sin(\theta/2)\,\left(|\theta++\rangle - |\theta--\rangle\right), \end{aligned} \tag{22.4}$$

where we have used the notation $|\theta\pm\pm\rangle$ for $|\theta\pm\rangle \otimes |\theta\pm\rangle$, and

$$\begin{aligned} |-+\rangle = |-\rangle \otimes |+\rangle = \cos^2(\theta/2)\,|\theta-+\rangle - \sin^2(\theta/2)\,|\theta+-\rangle \\ + \cos(\theta/2)\sin(\theta/2)\,\left(|\theta++\rangle - |\theta--\rangle\right). \end{aligned} \tag{22.5}$$

Suppose, in our alternative scenario, that the particles are emitted in the state $|+-\rangle$ and a S^1_θ-measurement yields $+1$. The projection postulate examined in Lecture 4 then implies that the state after measurement is a multiple of the projection $\cos^2(\theta/2)|\theta+-\rangle + \cos(\theta/2)\sin(\theta/2)|\theta++\rangle$. This presents an immediate problem if θ is not a multiple of π for, to preserve perfect correlation, the coefficient of $|\theta++\rangle$ must be zero in order to guarantee a S^2_θ-measurement of -1. This dooms the alternative scenario as an explanation of perfectly correlated spin pairs.

With equations 22.4 and 22.5, we are now in a position to determine the phase angle ϕ in equation 22.1. Substituting the formulae for $|+-\rangle$ and $|-+\rangle$ in equations 22.4 and 22.5 into equation 22.1 for the state vector ψ gives as the coefficient of the unmixed terms $(|\theta++\rangle - |\theta--\rangle)$ the expression

$$\frac{1}{\sqrt{2}}(1 + e^{\mathrm{i}\phi})\cos(\theta/2)\sin(\theta/2),$$

which must be zero since the spin component measurements are correlated along every orientation angle θ. This happens for every θ precisely when $1+e^{\mathbf{i}\phi} = 0$, i.e., when the phase angle is $\phi = \pi$ and the phase factor is $e^{\mathbf{i}\phi} = -1$. It follows that the state vector for the perfectly correlated coupled spin-$\frac{1}{2}$ system is

$$\psi = \frac{1}{\sqrt{2}}(|+-\rangle - |-+\rangle). \tag{22.6}$$

We now can confirm that ψ does indeed predict perfectly correlated spin component measurements in every direction θ with equal probabilities for recording a $+1$ or a -1 by S_1. Indeed, expanding ψ in the basis $\{|\theta \pm \pm\rangle\}$ by inserting equations 22.4 and 22.5 into equation 22.6 yields

$$\psi = \frac{1}{\sqrt{2}}(|\theta + -\rangle - |\theta - +\rangle). \tag{22.7}$$

This expansion is valid for all direction angles θ and confirms the perfect correlation of the spin components.

The Copenhagen interpretation avers that, when this system is in the state ψ, neither particle of the pair has a value for its z-component of spin, nor for any other component of spin, until the component in question is measured. When a $+1$ in the direction θ is registered at S_1, there is immediate collapse to the state $|\theta + -\rangle$ and an immediate effect is that, since $|\theta + -\rangle$ is a S^2_θ-eigenstate with eigenvalue -1, the second particle has the value -1 for its spin component in the direction θ. The spin detectors S_1 and S_2 may be light years apart, so it is possible to have spacelike-separated regions of the universe where an action in one region has an immediate repercussion in the other.

The fundamental fact that allows this surprising nonclassical behavior is that ψ is an entangled state vector, i.e., it is an indecomposable tensor element. The analysis presented here extends in its essentials to entangled states of many other quantum systems comprising a number of subsystems, and it accounts for some of the most interesting nonclassical aspects of quantum mechanics. Bell took this analysis of the correlated spin system one step further by asking how much correlation there is between spin component measurements by detectors S_1 and S_2 when the detectors are not aligned along the same direction. He analyzed the case where $\theta_1 = 0$ while $0 < \theta_2 < \pi$. His theoretical work, subsequently confirmed by experimental testing, places strict limits on the types of hidden variables theories that are allowed by nature. We turn to this story next.

22.3 Bell Inequalities, Realism, and Nonlocality

A simple, working definition of **realism** suggests that the micro-world of atoms possesses values for physical attributes – position, momentum, angular momentum, spin – at all times, even before measurement. It assumes that the constituents of the micro-world are best thought of in terms of the classical term "localized particles." The nonclassical aspects of this micro-world arise when these particles interact with a nonclassical quantum field. The term "locality" perhaps is better termed **local causality** and asserts that, given complete knowledge of events in their joint causal past, any additional information about the system at spacetime point O_1 should be irrelevant to events at spacetime point O_2. One aspect of this is that local causality suggests that there can be no effects at O_2 dependent on events at O_1 when O_1 and O_2 are spacelike separated. A theory that satisfies local causality is said to be a **local** theory and, otherwise, a **nonlocal** theory. The analysis of entangled, coupled spin states presented previously demonstrates that quantum mechanics violates local causality. Bohr's Copenhagen interpretation, with its completeness hypothesis, asserts that quantum mechanics also violates realism. The orthodox view, then, is that quantum mechanics is a nonlocal anti-realist theory of the micro-world.

Bohmian hidden variables theory, since it gives exactly the same statistical predictions as quantum mechanics, also violates local causality but retains realism. Both quantum and Bohmian mechanics in their violation of local causality seem to offer a conflict with special relativity. These considerations may appear to be of no more than philosophical interest since there seems to be no way to test experimentally such notions as realism and local causality. However, in 1964 Bell presented a careful examination of the EPR argument and gave an inspired and cogent analysis of subtler aspects of the coupled spin system of the previous section that demonstrated that the combination of the hypotheses of realism and local causality is subject to experimental disproof. If local causality turned out not to be experimentally violated in a specific way, then both quantum and Bohmian mechanics would be falsified.

The results of the Aspect (1982) experiments verified that local causality is indeed violated in the specific way required by both quantum and Bohmian mechanics. Some physicists have used the confirmation of the violation of local causality and the quantum mechanical predictions in the Aspect experiments to assert that quantum mechanics is correct and complete and that Bohr's position in his debates with Einstein is vindicated.[6] This inference would be mistaken. The correct inference is that, like both quantum mechanics and Bohmian mechanics, any valid theory of nature must violate local causality in the specific way confirmed by Aspect. No local realist theory can reproduce the correct statistical predictions confirmed by Aspect; however, nonlocal realist theories, such as Bohmian mechanics, cannot be ruled out. At this point in time there seems to be no way to choose experimentally between realism and anti-realism, between quantum and Bohmian mechanics, which leaves the choice to be made by other than physical considerations. For the majority of the physics community, appeals to the supposed simplicity of quantum mechanics, over against the claimed ad hoc nature of the quantum potential of Bohmian mechanics, makes the acceptance of completeness the more compelling option.[7] For the heretical minority who are concerned with these issues, since Bohmian mechanics provably reproduces exactly the statistical predictions of quantum mechanics, the hope is that there will be found a testable prediction from Bohmian mechanics in a realm not treated by quantum mechanics, so that Bohmian mechanics becomes independently falsifiable.[8]

22.3.1 Bell's Argument

The setting for Bell inequalities is the coupled spin-$\frac{1}{2}$ system described above in which repeated spin component measurements are performed by detectors S_1 and S_2 on entangled pairs produced in exactly the same way, so that the state vector for each pair is given by equation 22.6.

[6] A personal anecdote: a physics graduate student at my university expressed great surprise when I asserted that there is a hidden variables theory that reproduces the theoretical predictions of nonrelativistic quantum mechanics. He then suggested that the Bell inequalities and the Aspect experiments proved definitively the impossibility of such theories and verified the completeness of quantum mechanics. And besides, there is the von Neumann impossibility proof ... After further discussion where he admitted to having been taught this in his graduate quantum mechanics courses, he exclaimed, with some incredulity, "why haven't we been told by our professors about the possibility of such hidden variables theories?"

[7] Beller (1999) rejects the view that the physics community arrived at their stance by a careful consideration of simplicity. Her thesis, supported by an historical exegesis, is that this quantum orthodoxy that accepts Bohr's conclusions is an historical accident predicated upon personalities, philosophies, and prejudices, and easily could have had a different outcome.

[8] One such possible realm is that of predicting the average tunneling times of particles through potential barriers that classically forbid transmission of the particles. The average tunneling times for potential barriers of various strengths are subject to experimental measurement, but quantum mechanics does not offer any computation of these averages. See Section 5.5 of Holland (1993) for an enlightening discussion.

Three sets of measurements are performed. The first records repeated S^1_0-measurements by S_1 and S^2_θ-measurements by S_2, and the recorded frequency of outcomes is used to assign an experimental probability $P(0+, \theta-)$ for obtaining a $+1$ by S_1 and a -1 by S_2. The second records repeated S^1_θ-measurements by S_1 and $S^2_{2\theta}$-measurements by S_2 and produces the experimental probability $P(\theta-, 2\theta+)$ for recording a -1 by S_1 and a $+1$ by S_2. Finally, the third records repeated $S^1_{2\theta}$-measurements by S_1 and S^2_0-measurements by S_2 and produces the experimental probability $P(2\theta+, 0+)$ for recording a $+1$ at both detectors. Assuming both realism and local causality, we now argue that the **Bell inequality**

$$P(0+, \theta-) + P(\theta-, 2\theta+) + P(2\theta+, 0+) \leq 1 \qquad (22.8)$$

necessarily holds.[9] Indeed, realism asserts that, for a perfectly correlated spin system, there is a set Ω of variables that encodes all possible spin component values for the pair along every possible direction. Label the particle measured by S_1 as particle A and the other as particle B.[10] Thus, one point ω in the set Ω encodes, for each direction angle θ, a choice of ± 1 for particle A and then, necessarily, ∓ 1 for particle B.[11] How does local causality come into play? The answer is that local causality forbids any change in the "hidden variable" ω attached to the state vector ψ once the particles have been emitted, at least until the measurement of one spin component of one particle is in the causal past of the other. Working in the inertial frame where the particle source and the two detectors S_1 and S_2 are at rest, because the particle spins must remain perfectly correlated it follows that any change in the state that

[9] Many similar inequalities have been derived since Bell's original in 1964. We will examine only this one.

[10] See Lectures 24 and 25 concerning the difficulty in labeling particles. There is no problem in the present discussion since realism assumes particles can be labeled.

[11] The simplest way to realize this mathematically without any hypothesis of correlated spins is to set Ω equal to the set of all functions $\omega : \mathbf{S}^1 \to \{+1, -1\}^2$ from the circle $\mathbf{S}^1$ to the set of pairs

$$\{+1, -1\}^2 = \{(+1, +1), (+1, -1), (-1, +1), (-1, -1)\}$$

of possible spin component values. Then the value $\omega(\theta) = (a, b)$ encodes a θ spin component of a for particle A and a θ spin component of b for B. Of course, the range of a function $\omega \in \Omega$ attached to a perfectly correlated pair is restricted to the set $\{(+1, -1), (-1, +1)\}$. Thus, the variables for the perfectly correlated spin states can be reduced to the set of all functions $\omega : \mathbf{S}^1 \to \{+1, -1\}$, where $\omega(\theta) = a = \pm 1$ means that particle A is in the a state for measurements in the direction θ and B is then necessarily in the $-a$ state for that same direction. The state vector ψ of each production pair would be augmented by a value from Ω, so that each individual entangled pair would have a state vector of the form ψ_ω for some $\omega \in \Omega$.

the hidden variable assigns to A after emission must appear as an instantaneous change, in our inertial frame, in the state of B. Otherwise, perfect correlation would be destroyed during the time intervening from the change of parity at A to the change at B, which presumably would show up in experimental measurements since perfect correlation is independent of the placement of detectors along the line of travel. But such instantaneous changes violate local causality since the change at A is not in the causal past of B at that instant. Even more serious problems arise from special relativity when considering inertial observers moving relative to the rest frame of the source if ω is allowed to fluctuate as the particles separate. Indeed, in the scenario presented where there is an instantaneous change in the hidden variable in the rest frame of the source, other inertial observers moving relative to the rest frame will assert that there are times in their inertial systems when the particles simultaneously share either a $+1$ or a -1 state, which presumably would show up in their simultaneous measurements, violating the perfect correlation of spin component states. This shows that even correlated statistical fluctuations are not allowed.[12] In addition, to preserve local causality the orientation of detector S_1, which can be changed at any time, can have no effect whatsoever on the value of ω from the time of emission to the time of detection, nor until a measurement by S_1 is in the causal past of particle B.

To derive the Bell inequality 22.8, let X be the subset of Ω that consist of all values ω that encode a $+1$ state for S^1_0-measurements for particle A, Y the set that encodes a -1 state for S^1_θ-measurements for particle A, and Z the set that encodes a $+1$ state for $\mathsf{S}^1_{2\theta}$-measurements for particle A. Because the particle spin components are perfectly correlated, Y also may be described as the set of hidden variables that encodes a $+1$ state for S^2_θ-measurements on particle B, and the complement of Y may then be recognized as the set that encodes a -1 state for S^2_θ-measurements. It follows that the set difference $X - Y = \{x \in X : x \notin Y\}$ is precisely the set of hidden variables that encode a $+1$ state for S^1_0-measurements on A and a -1 state for S^2_θ-measurements on B. Similarly,

[12] More serious still is the fact that inertial observers moving relative to the source present problems for quantum mechanics, in the collapse of the wave function for systems extended through spacelike regions, and for Bohmian mechanics, in its nonlocal quantum field with instantaneous nonlocal effects across spacelike separated regions. This points to the fact that neither quantum mechanics, as developed thus far, nor Bohmian mechanics is a relativistic theory. In fact, this problem – one of a Lorentz invariant description of collapse – persists in relativistic quantum theory and still awaits resolution.

the set difference $Y - Z$ encodes a -1 state for S^1_θ-measurements on A and a $+1$ state for $\mathsf{S}^2_{2\theta}$-measurements on B, and, finally, the set difference $Z - X$ encodes a $+1$ state for $\mathsf{S}^1_{2\theta}$-measurements on A and a $+1$ state for S^2_0-measurements on B. Observe that the three sets $X - Y$, $Y - Z$, and $Z - X$ are pairwise disjoint. Now repeated spin component measurements may be used to assign an experimental probability measure to the space Ω of hidden variables with $X - Y$ having measure $P(0+, \theta-)$, $Y - Z$ having measure $P(\theta-, 2\theta+)$, and $Z - X$ having measure $P(2\theta+, 0+)$. Since the total measure is unity and the three sets are pairwise disjoint, the Bell inequality follows.

Observe that our derivation of the Bell inequality is independent of the details of the theory claiming to describe the perfectly correlated, coupled spin system; it assumes only realism – at all times each of A and B has definite values for each spin component – and local causality, implying that these values cannot fluctuate and that the orientation of the detector S_1 can have no effect on measurements by S_2.

22.3.2 The Quantum Mechanical Treatment

The Aspect experiments determined conclusively that the Bell inequality is violated in nature by correlated spin pairs! This means that local realistic theories are invalid, at least in this setting. In particular, no local hidden variables theory can describe perfectly correlated coupled spin systems. How does quantum mechanics fare in its analysis of the correlated spin system? To answer this first note that, for any two direction angles α and β, S^1_α and S^2_β are commuting operators and thus decompose the state space $\mathcal{H}$ into an orthogonal direct sum of eigenspaces. Using the notation $|\pm\pm\rangle_{\alpha,\beta} = |\alpha\pm\rangle \otimes |\beta\pm\rangle$, the basis determined by these two operators is

$$\mathcal{B}_{\alpha,\beta} = \{|++\rangle_{\alpha,\beta}, |+-\rangle_{\alpha,\beta}, |-+\rangle_{\alpha,\beta}, |--\rangle_{\alpha,\beta}\}. \tag{22.9}$$

Since, for example, $|+-\rangle_{\alpha,\beta}$ is a S^1_α-eigenstate with eigenvalue $+1$ and a S^2_β-eigenstate with eigenvalue -1, it represents a state of the system where a spin component measurement by S_1 yields $+1$ in the direction α and a spin component measurement by S_2 yields -1 in the direction β. Remember that the state vector of our perfectly correlated spin-$\frac{1}{2}$ system is the entangled state vector of equation 22.6, given also by equation 22.7. Quantum mechanics predicts that the statistical distribution of dual measurement results, when detector S_1 is aligned in the direction α and S_2 in the direction β, is given by the squares of the moduli of the coefficients in the expansion of the state vector ψ in the basis $\mathcal{B}_{\alpha,\beta}$. In particular, the amplitude for $\pm1,\pm1$ measurement-results is the coefficient of $|\pm\pm\rangle_{\alpha,\beta}$ in the expansion of ψ, which is ${}_{\alpha,\beta}\langle\pm\pm|\psi\rangle$, where we use ${}_{\alpha,\beta}\langle\pm\pm|$ to denote the ket-vector dual to the bra-vector $|\pm\pm\rangle_{\alpha,\beta}$. Thus quantum mechanics predicts the theoretical value for $P(\alpha\pm,\beta\pm)$ as $|{}_{\alpha,\beta}\langle\pm\pm|\psi\rangle|^2$, where $P(\alpha\pm,\beta\pm)$ is the probability for obtaining the respective readings ±1 and ±1 when S_1 is aligned along the direction α and S_2 along the direction β.

Our aim now is to calculate the probabilities $P(\alpha\pm,\beta\pm)$ and see what quantum mechanics predicts for the left-hand side of the Bell inequality 22.8. On physical grounds, since it is the relative difference in the direction angles of the detectors that matters, equations 22.2 and 22.3 can be recast in this setting as

$$|\alpha+\rangle = \cos\left(\frac{\beta-\alpha}{2}\right)|\beta+\rangle - \sin\left(\frac{\beta-\alpha}{2}\right)|\beta-\rangle \tag{22.10}$$

and

$$|\alpha-\rangle = \sin\left(\frac{\beta-\alpha}{2}\right)|\beta+\rangle + \cos\left(\frac{\beta-\alpha}{2}\right)|\beta-\rangle. \tag{22.11}$$

Taking the tensor product of $|\alpha+\rangle$ with both sides of equation 22.11 and of $|\alpha-\rangle$ with both sides of equation 22.10 gives

$$|\alpha+-\rangle = |\alpha+\rangle\otimes|\alpha-\rangle = \sin\left(\frac{\beta-\alpha}{2}\right)|++\rangle_{\alpha,\beta} + \cos\left(\frac{\beta-\alpha}{2}\right)|+-\rangle_{\alpha,\beta},$$

and

$$|\alpha-+\rangle = |\alpha-\rangle\otimes|\alpha+\rangle = \cos\left(\frac{\beta-\alpha}{2}\right)|-+\rangle_{\alpha,\beta} - \sin\left(\frac{\beta-\alpha}{2}\right)|--\rangle_{\alpha,\beta}.$$

Putting this together with the expansion of ψ in equation 22.7 with $\theta=\alpha$ gives the expansion of ψ in the basis $\mathcal{B}_{\alpha,\beta}$ as

$$\begin{aligned}\psi &= \frac{1}{\sqrt{2}}\sin\left(\frac{\beta-\alpha}{2}\right)\left(|++\rangle_{\alpha,\beta} + |--\rangle_{\alpha,\beta}\right)\\ &\quad + \frac{1}{\sqrt{2}}\cos\left(\frac{\beta-\alpha}{2}\right)\left(|+-\rangle_{\alpha,\beta} - |-+\rangle_{\alpha,\beta}\right).\end{aligned}$$

We now may read off the four probabilities $P(\alpha\pm,\beta\pm)$ as

$$P(\alpha+,\beta+) = \frac{1}{2}\sin^2\left(\frac{\beta-\alpha}{2}\right) = P(\alpha-,\beta-)$$

and

$$P(\alpha+, \beta-) = \frac{1}{2}\cos^2\left(\frac{\beta-\alpha}{2}\right) = P(\alpha-, \beta+).$$

In particular, in the experimental set up for the Bell inequality with measurements at angles 0, θ, and 2θ, we have

$$P(0+, \theta-) = \frac{1}{2}\cos^2\left(\frac{\theta}{2}\right) = P(\theta-, 2\theta+)$$

and

$$P(2\theta+, 0+) = \frac{1}{2}\sin^2\left(\frac{-2\theta}{2}\right) = \frac{1}{2}\sin^2\theta.$$

The quantum mechanical prediction for the left-hand side of the Bell inequality, which we now denote as $p(\theta)$, is then

$$p(\theta) = \cos^2\left(\frac{\theta}{2}\right) + \frac{1}{2}\sin^2\theta.$$

Our next observation is that $p(\theta) > 1$ whenever $0 < |\theta| < \pi/2$. To see this, use the cosine half-angle formula and the Pythagorean trigonometric identity to rewrite $p(\theta)$ as

$$p(\theta) = \frac{1+\cos\theta+1-\cos^2\theta}{2} = 1 + \frac{1}{2}(1-\cos\theta)\cos\theta, \tag{22.12}$$

and note that $(1-\cos\theta)\cos\theta > 0$ whenever $0 < |\theta| < \pi/2$. Thus quantum mechanics correctly predicts the violation of the Bell inequality in the perfectly correlated coupled spin system. Moreover, the Aspect experiments confirmed the quantum mechanical prediction of the dependence of p on the angle θ of equation 22.12. This of course does not prove the correctness of the Copenhagen interpretation of quantum mechanics, just its consistency with observation in this particular case. After all, the realist but nonlocal Bohmian hidden variables theory reproduces equation 22.12 and gives precisely the same predictions as does quantum mechanics in this setting.

A final comment: whatever one thinks of the completeness hypothesis and whether there is a more basic theory of which the quantum theory is derivative, it must be admitted that the quantum formalism, with its use of the tensor product state spaces and entangled states for multi-particle systems, offers a calculus that faithfully captures the statistical results of measurements in entangled systems. It is straightforward to apply, quick in its calculations, and beautiful in its simplicity.

22.4 The GHZ Scheme for Spin Triplets

In 1989, D.M. Greenberger, M. Horne, and A. Zeilinger proposed a variation on Bell's inequality that involves an entangled spin triplet rather than a spin doublet. This **GHZ scheme** has the added feature that every possible hidden variables assignment of spin values to the triplet, assuming again realism and locality, would give a spin measurement result that directly contradicts the quantum mechanical prediction in at least one single combination of spin measurements of each of the particles, out of a list of four possible combinations of spin measurement experiments.

To clarify what we are saying, note that, in any hidden variables assignment $\omega \in \Omega$ for a perfectly correlated spin pair, no joint spin measurement can distinguish between standard quantum mechanics and a local realistic hidden variables theory. By this we mean that whether quantum mechanics or a hidden variable ω describes the system, a single joint measurement of the spin of one particle at angle α and the other at angle β, no matter what angles are chosen, cannot decide between the two theories. Indeed, when $\alpha = \beta$, both theories predict perfect correlation and, when $\alpha \neq \beta$, both may yield any of the four possible combinations of spin pair measurements. It is only when many spin pairs are produced and measured at different angles that the two theories differ in their predictions of the statistical distributions of measured spin pair values.

By contrast, the GHZ scheme measures the spin components of three particles in precisely two directions, those of the x- and z- axes, and every choice of hidden variable ω describing the spins of the three particles in these two possible directions offers a prediction that differs from that of the quantum mechanical prediction in at least one combination of x- and z-measurements. In fact, there always is a single combination of spin measurements chosen out of a list of only four possible combinations whose result can distinguish between the quantum mechanical theory and any local realistic hidden variables theory, though we do not have an a priori knowledge of which of the four combinations works.

Before presenting any detailed calculations, we give a complete description of the predictions of the quantum mechanical calculations and compare this with the predictions of any possible local realistic hidden variables theory. The set-up is that of three spin-$\frac{1}{2}$ particles that are allowed to interact and are then separated and carried to three spin measuring devices S_i, $i = 1, 2, 3$. Each device S_i is oriented along either the x- or the z-direction and the spins $s_i \in \{\pm 1\}$ are measured by device S_i and combined in the product $s = s_1 s_2 s_3$. We further restrict our spin measurements to the cases where either (I) all three devices are oriented along the positive z-direction, or (II) two of the devices are oriented along the positive x-direction and one along the positive z-direction. The claim, verified below, is that there is a quantum state of this triplet system with, say, state vector ψ, for which a type-I measurement always yields $s = -1$ and a type-II measurement always yields $s = +1$.

What does a local realistic hidden variables theory predict for the value of s upon a type-I or type-II measurement? To answer this, first note that any realistic theory allows for the separation and labeling of the three particles as particles A, B, and C. Since the hidden variable must assign only two spin values to each particle, a ± 1 along the x and a ± 1 along the z direction, there are only $4^3 = 64$ distinct points ω in the space Ω of hidden variables. In fact, we may identify Ω with the set of functions $\omega : \{A, B, C\} \to \{\pm 1\} \times \{\pm 1\}$, where, for example, $\omega(A) = (1, -1)$ means that A is assigned $+1$ for its x-component spin and -1 for its z-component spin. Recall that the local realism assumption guarantees that once the hidden variable ω is attached to the system, the variable cannot fluctuate and cannot depend on the orientations of the detectors. For the hidden variable ω, let $\omega(A) = (a, a')$, $\omega(B) = (b, b')$, and $\omega(C) = (c, c')$, and let $s' = a'b'c'$. Note that s' is the value of s that a type-I measurement would produce if ω were the hidden variable attached to the system. Thus the requirement that the hidden variables prediction agree with the quantum mechanical prediction for a type-I measurement restricts ω to come from the 32 elements of Ω for which $s' = -1$. Among these hidden variables for which $s' = -1$, what is the value of s upon a type-II measurement? Obviously, the value of s is one of the three products $a'bc$, $ab'c$, or abc'. Multiplying these three possible results gives

$$(a'bc)(ab'c)(abc') = (a'b'c')a^2b^2c^2 = s'a^2b^2c^2 = -1,$$

implying that either exactly one or all three of the products $a'bc$, $ab'c$, or abc' takes the value -1. In particular, when $s' = -1$, at least one of the three possible type-II measurements would yield $s = -1$, contradicting the quantum mechanical prediction of $s = +1$. Then we are left with the fact that, for any $\omega \in \Omega$, the hidden variables prediction of s will differ from the quantum mechanical prediction of s in either a type-I measurement or in at least one of the three type-II measurements. This is striking, in that one experimental run has the potential for deciding for the hidden variables theory over quantum mechanics.

Now, it is to be observed that, as we do not know what hidden variable is attached to the system before measurement, we have no a priori knowledge of which of the four orientation combinations to choose for the devices S_i in order to falsify the quantum mechanical prediction. In practice, then, we would have to perform the experiment several times in order to have a reasonable chance of falsifying the quantum prediction. The thing to notice, though, is that, if the hidden variables theory is correct, one single run of the experiment with the "wrong" result would occur eventually. This may be contrasted with the analysis for the perfectly correlated spin pairs where no single measurement would ever have decided between the two theories, and it is only the statistical distribution of the results of a large number of measurements that can distinguish between them. Of course, the Aspect experiment already confirms a violation of the Bell inequality, so we know that local realistic hidden variables theories are forbidden.

It remains to present a quantum state for which a type-I measurement always yields $s = -1$ and a type-II measurement always yields $s = +1$. The state space of the triplet of spin-$\frac{1}{2}$ particles is the three-fold tensor product $V^{1/2} \otimes V^{1/2} \otimes V^{1/2}$ with orthonormal basis $|\pm\pm\pm\rangle = |\pm\rangle \otimes |\pm\rangle \otimes |\pm\rangle$, where, of course, $|\pm\rangle$ are the normalized ± 1-eigenvectors of S_3, the self-adjoint operator corresponding to z-component spin measurements in the single-particle state space $V^{1/2}$. To flesh this out a bit, let S^i_u be the self-adjoint operator that represents u-component spin measurements by device S_i, for $u = x, z$. Then, for example, $\mathsf{S}^2_z = 1 \otimes \mathsf{S}_{\theta=0} \otimes 1$ and $|+-+\rangle$ is a simultaneous eigenvector of the three operators S^i_z, for $i = 1, 2, 3$, with eigenvalue $+1$ for $i = 1, 3$ and -1 for $i = 2$. Clearly, the state described by the normalized state vector

$$\psi = \frac{1}{2}\left(|---\rangle - |-++\rangle - |+-+\rangle - |++-\rangle\right) \tag{22.13}$$

will always yield $s = -1$ upon a type-I measurement. What does ψ predict for the type-II measurements of s? To answer this we need to expand ψ in a basis of simultaneous eigenvectors of the operators S^i_u, where $u = x$ for two values of i and $u = z$ for the remaining value. Since we have chosen minus signs for all the latter three summands of ψ, the symmetry suggests that we need examine only one of the three type-II measurements, say where S_1 and S_2 are oriented along the x direction and S_3 along the z direction. Using the notation $|\pm\rangle_x$ for the x direction eigenvectors $|\pi/2\pm\rangle$ of the operator $\mathsf{S}_{\theta=\pi/2}$, the basis we need for the expansion of ψ is $|\pm\pm\rangle_x \otimes |\pm\rangle$, where $|\pm\pm\rangle_x = |\pm\rangle_x \otimes |\pm\rangle_x$. To find the expansion in this basis, recall that equations 22.2 and 22.3 give

$$|\pm\rangle = \frac{1}{\sqrt{2}}\left(|+\rangle_x \mp |-\rangle_x\right).$$

A quick calculation now gives

$$\begin{aligned}
|---\rangle &= |-\rangle \otimes |-\rangle \otimes |-\rangle = \frac{1}{2}\left(|+\rangle_x + |-\rangle_x\right) \otimes \left(|+\rangle_x + |-\rangle_x\right) \otimes |-\rangle \\
&= \frac{1}{2}\left(|++\rangle_x + |+-\rangle_x + |-+\rangle_x + |--\rangle_x\right) \otimes |-\rangle, \\
|-++\rangle &= |-\rangle \otimes |+\rangle \otimes |+\rangle = \frac{1}{2}\left(|+\rangle_x + |-\rangle_x\right) \otimes \left(|+\rangle_x - |-\rangle_x\right) \otimes |+\rangle \\
&= \frac{1}{2}\left(|++\rangle_x - |+-\rangle_x + |-+\rangle_x - |--\rangle_x\right) \otimes |+\rangle, \\
|+-+\rangle &= |+\rangle \otimes |-\rangle \otimes |+\rangle = \frac{1}{2}\left(|+\rangle_x - |-\rangle_x\right) \otimes \left(|+\rangle_x + |-\rangle_x\right) \otimes |+\rangle \\
&= \frac{1}{2}\left(|++\rangle_x + |+-\rangle_x - |-+\rangle_x - |--\rangle_x\right) \otimes |+\rangle, \\
|++-\rangle &= |+\rangle \otimes |+\rangle \otimes |-\rangle = \frac{1}{2}\left(|+\rangle_x - |-\rangle_x\right) \otimes \left(|+\rangle_x - |-\rangle_x\right) \otimes |-\rangle \\
&= \frac{1}{2}\left(|++\rangle_x - |+-\rangle_x - |-+\rangle_x + |--\rangle_x\right) \otimes |-\rangle.
\end{aligned}$$

Substituting these into equation 22.13 and using the notation $|\pm_x \pm_x \pm\rangle$ for $|\pm\pm\rangle_x \otimes |\pm\rangle$, we have

$$\psi = \frac{1}{2}\left(-|+_x +_x +\rangle + |+_x -_x -\rangle + |-_x +_x -\rangle + |-_x -_x +\rangle\right).$$

This shows ψ to be a linear combination of only those basis vectors $|\pm_x \pm_x \pm\rangle$ with an even number of minus signs and, as these signs give the parity of the spin components upon the type-II measurement under consideration, the value of s upon measurement is necessarily $+1$. A similar argument, or an observation that symmetry forces the conclusion, proves that the measured value of s will be $+1$ under the two other type-II measurements.

We have completed our objective of finding a state vector ψ with the desired property that type-I measurements yield only $s = -1$ and type-II measurements yield only $s = +1$. As for the real world of experimentation, as of this writing, no measurements have been performed for a GHZ triplet, presumably because of the difficulty of producing triplets with state vector ψ and the fact that the Aspect experiments have already ruled out local realist theories.

22.5 End Notes

The failure of local causality in real systems and the inseparability of two real systems that have interacted in the past are the most surprising and nonclassical aspects of the quantum world. Quantum mechanics handles the description of these systems beautifully in its tensor product formalism and makes the correct quantitative predictions for just how the failure of the Bell inequalities occurs. I would encourage the reader to study the intellectual history of the developments surrounding the identification of the failure of local causality and the inseparability of real systems as necessary implications of the quantum mechanics of the micro-world. The story is fascinating and I think serves to clarify various important issues in the philosophy and psychology of science. Most interesting to me is how this history illuminates issues in the sociology of science – how science is conducted as a human enterprise, rife with large personalities and all too human virtues and vices. This is a history of great ideas (nonlocality and inseparability), explained away so as to be unexplored for a generation, of sometimes vicious attacks on those who would question the received quantum orthodoxy (David Bohm), of dogged brilliance (John Bell) laboring almost in secret to provide a critical analysis of the issues of no interest to the larger community, and finally, after two generations, of the triumph of these ideas and their necessity in understanding just how weird this quantum world really is and their importance in current applications of quantum technology (quantum computers).

The End Notes to Lecture 1 are pertinent to this discussion. There the reader may find the significant references I have found to be effective tellings of these stories with cogent analyses of the history, psychology, and technical mathematics of the issues surrounding nonlocality. Those End Notes give a guide to each reference and so, rather than repeating myself, here I just list the references again with the following additional comments. I recommend rather highly the accounts of the history of the Bohr–Einstein debates of the 1930s and the critical analysis of EPR and nonlocality in Whitaker (2006) and Wick (1995). Beller (1999) is a masterful analytic criticism of quantum orthodoxy that is insightful in its philosophical analysis. Even today many in the physics community get Bell wrong. Tim Maudlin (2014) tries to set the record straight in his tightly argued article "What Bell did," which I cannot recommend highly enough. Maudlin (2011) carefully considers the implications that relativity has for nonlocality and considers in his last chapter a candidate theory for harmonizing the two. Finally, Bell and Gao (2016) is a compilation of scholarly articles centered on the theme of its title, *Quantum Nonlocality and Reality: 50 Years of Bell's Theorem.*

23
Ensembles and Density Operators

> There's no sense in being precise when you don't even know what you're talking about.
>
> John von Neumann

So far we have considered only quantum systems for which we assume complete knowledge of the state vector ψ. What about a quantum system that consists of a mixture of quantum states, each with a perhaps experimentally determined probability of appearance? How can the quantum formalism economically incorporate such **ensembles**, where there is a statistical distribution of various states comprising the system? von Neumann proposed a rather elegant solution to this problem by generalizing the term **state** of a quantum system to include general **density operators**, objects that can encode not only the **pure states** that we have seen thus far but also **mixed states** of statistical distributions of these pure states. We present in this lecture von Neumann's generalization, using once again the quantum mechanical treatment of spin-$\frac{1}{2}$ systems to motivate and elucidate this new setting.

23.1 The Spin-$\frac{1}{2}$ System Revisited

This section should be considered a continuation of Section 18.3, whose notation we largely adopt without review with the one change that we revert to naming the standard orthogonal axes of a fixed rectangular coordinate system with x, y, and z rather than x_1, x_2, and x_3. In particular, we will use $\mathsf{S}_{x,y,z}$ in place of $\mathsf{S}_{1,2,3}$ and will adopt the notation $|\pm\rangle_u$ for a $\pm$-eigenstate of S_u, for $u = x, y, z$. The standard basis for the state space $V^{\frac{1}{2}} = \mathbb{C}|+\rangle_z \oplus \mathbb{C}|-\rangle_z$ consists of two orthogonal eigenstates

$|+\rangle_z$ and $|-\rangle_z$ with respective eigenvalues $+1$ and -1 for the operator S_z, which represents spin-component measurements along the direction of the positive z-axis.

Consider an experimental set-up in which spin-$\frac{1}{2}$ particles are generated by a black box[1] and move along the y-axis to a detector that may be oriented in an arbitrary direction to take spin-component measurements. Suppose that, when the detector is oriented along the positive z-direction, the experimental outcome is that spin $+1$ and spin -1 are equally likely, each having the experimental probability of occurrence of one-half. What can we say about the particles being generated – what state vectors describe these particles? There are two reasonable options that we will consider.

Option 1. Each particle emitted is in an equal superposition of the z-component eigenstates $|+\rangle_z$ and $|-\rangle_z$. This translates into a state vector of the form

$$\psi_\phi = \frac{1}{\sqrt{2}}|+\rangle_z + \frac{e^{i\phi}}{\sqrt{2}}|-\rangle_z,$$

for some phase angle ϕ, which may take different values for different particles. In this scenario, the equal likelihood of the two possible measurement results is an inherent attribute of the state of each single particle.

Option 2. The particles are emitted in various states in a statistical distribution that determines the equal likelihood of the two possible measurement results. Perhaps the easiest arrangement is that each particle is emitted in either state $|+\rangle_z$ or in state $|-\rangle_z$, with an equal probability of each. In this scenario, the equal likelihood of the two possible measurement results is a statistical feature of the distribution of states among many particle emissions, and the state vector of each individual particle carries no imprint of this equal likelihood.

Is there any way to distinguish between these options experimentally? The discussion of Section 22.2, where we examined a similar choice between an explanation in terms of an entangled state versus a statistical ensemble of unentangled states, is suggestive. Consider the

[1] This means that we have no knowledge of the process that creates the specific spin states for the particles and, therefore, we have no a priori knowledge of these spin states.

result of orienting the detector along the positive x-axis and taking x-component spin measurements. Under Option 1, equations 22.2 and 22.3 with $\theta = \pi/2$ give

$$|\pm\rangle_z = \frac{1}{\sqrt{2}}|+\rangle_x \mp \frac{1}{\sqrt{2}}|-\rangle_x,$$

so that

$$\psi_\phi = \frac{1}{2}\left(1 + e^{\mathrm{i}\phi}\right)|+\rangle_x + \frac{1}{2}\left(-1 + e^{\mathrm{i}\phi}\right)|-\rangle_x.$$

Then the theoretical probabilities $P_\phi(x\pm)$ of obtaining ± 1 upon an x-component measurement when the particle is emitted in this state are

$$P_\phi(x\pm) = \frac{1}{4}\left(\pm 1 + e^{\mathrm{i}\phi}\right)\left(\pm 1 + e^{-\mathrm{i}\phi}\right) = \frac{1}{2}\left(1 \pm \cos\phi\right).$$

Assuming that the phase angles are chosen randomly with respect to a probability measure μ on the phase-angle space $\Phi = [0, 2\pi)$, the experimentally determined statistical distribution $P_{\mathrm{exp}}(x\pm)$ of the measurements will be

$$P_{\mathrm{exp}}(x\pm) = \frac{1}{2}\left(1 \pm \mathbf{E}_\mu(\cos\phi)\right), \tag{23.1}$$

where $\mathbf{E}_\mu(\cos\phi)$ is the expected value of the random variable $\cos\phi$, given by $\mathbf{E}_\mu(\cos\phi) = \int_\Phi \cos\phi \, d\mu$. For example, the two extremes occur, on the one hand, when μ is the standard normalized Lebesgue measure in which the μ-measure of an interval $[a, b]$ is $(b - a)/2\pi$, and, on the other, when μ is the atomic measure concentrated at a fixed point $\phi_0 \in \Phi$. In the former case, every phase angle is equally likely to be chosen and the expectation is $\mathbf{E}_\mu(\cos\phi) = \frac{1}{2}\pi^{-1}\int_0^{2\pi} \cos\phi \, d\phi = 0$. The result from equation 23.1 is that $P_{\mathrm{exp}}(x+) = \frac{1}{2} = P_{\mathrm{exp}}(x-)$, so that $+1$ and -1 are equally likely to be recorded upon an x-component measurement. In the latter case, every particle is emitted with phase angle fixed at ϕ_0 and the expectation is $\mathbf{E}_\mu(\cos\phi) = \cos\phi_0$. In this case, equation 23.1 implies that $P_{\mathrm{exp}}(x+) \neq P_{\mathrm{exp}}(x-)$ whenever ϕ_0 is not equal to $\pm\pi/2$.

Under Option 2, half the particles are emitted in state $|+\rangle_z = \frac{1}{\sqrt{2}}|+\rangle_x - \frac{1}{\sqrt{2}}|-\rangle_x$ and half in state $|-\rangle_z = \frac{1}{\sqrt{2}}|+\rangle_x + \frac{1}{\sqrt{2}}|-\rangle_x$, each of these being an equal superposition of the two S_x-eigenstates. The result is that the x-component of spin is measured as ± 1 with equal frequencies, i.e., $P_{\mathrm{exp}}(x+) = \frac{1}{2} = P_{\mathrm{exp}}(x-)$. Comparing with Option 1, we see that when the expected value $\mathbf{E}_\mu(\cos\phi) \neq 0$, Option 1 with its unequal probabilities offers a different experimental outcome than Option 2 and so x-component measurements can decide between the two possibilities.

Even when $\mathbf{E}_\mu(\cos\phi) = 0$, sometimes further measurements can decide between various options. For example, under Option 1, assume we know that the black box emits particles under an extreme case where μ is the atomic measure concentrated at $\phi_0 = \pi/2$. Then $\mathbf{E}_\mu(\cos(\phi)) = \cos(\pi/2) = 0$ and $P_{\text{exp}}(x+) = \frac{1}{2} = P_{\text{exp}}(x-)$. A quick calculation then shows that

$$\psi = \frac{1}{\sqrt{2}}|+\rangle_z + \frac{\mathbf{i}}{\sqrt{2}}|-\rangle_z = |+\rangle_y$$

and, therefore, a measurement of the y-component of spin yields $+1$ with certainty. It follows that $P_{\text{exp}}(y+) = 1$. Under Option 2, though, y-component spin measurements yield $P_{\text{exp}}(y+) = 1/2 = P_{\text{exp}}(y-)$ since half of the particles are emitted in state $|+\rangle_z = \frac{1}{\sqrt{2}}|+\rangle_y + \frac{1}{\sqrt{2}}|-\rangle_y$, and half in state $|-\rangle_z = -\frac{\mathbf{i}}{\sqrt{2}}|+\rangle_y + \frac{\mathbf{i}}{\sqrt{2}}|-\rangle_y$, each of these being an equal superposition of the two S_y-eigenstates. Therefore, y-component spin measurements can decide between these options.

We turn now to von Neumann's economical incorporation of statistical ensembles into the quantum mechanical formalism by the introduction of density operators, which can be used to describe succinctly, and in a way vulnerable to sleek computations, a statistical distribution of states. After articulating the essentials of the theory, we will return to the spin-$\frac{1}{2}$ system and examine the density operators that apply for our various options and, as an illustration, calculate expectation values using the new machinery.

23.2 Density Operators I: Finite-Dimensional Setting

John von Neumann's formalism of density operators is suited particularly well for quantum systems for which the state Hilbert space $\mathcal{H}$ is finite-dimensional. The quintessential example is that of the finite-dimensional irreducible representation spaces of $\mathfrak{su}(2)$, which serve as the state spaces of spin systems. Infinite-dimensional systems introduce many of the usual problems, this time centered on the fact that the trace of an operator is not necessarily defined. Consequently, in this section we first restrict our attention to the finite-dimensional setting and later briefly consider the general setting. Throughout this section $\mathcal{H}$ denotes an N-dimensional state space with orthonormal basis $\boldsymbol{\xi} = \{\xi_1, \ldots, \xi_N\}$, for some $N \in \mathbb{N}$.

Our first task is to recast the notion of a "state" into a more general setting. Let ψ be a normalized state vector in $\mathcal{H}$. The projection operator determined by ψ is, in Dirac notation, $|\psi\rangle\langle\psi|$ and will be denoted also as P_ψ. Since $\mathsf{P}_\psi = \mathsf{P}_{c\psi}$ for any nonzero complex number c, and since every nonzero complex multiple of ψ represents the same quantum state as ψ, von Neumann suggested that the projection operator P_ψ is a more natural choice for an object to represent a quantum state than is the vector ψ. To calculate expectation values of observables directly in terms of projection operators, we need to review the notion of the **trace** of an operator. Recall that the trace Tr is a functional defined for any linear operator A on $\mathcal{H}$ by $\mathsf{Tr}(\mathsf{A}) = \sum_{i=1}^{N} \langle \xi_i|\mathsf{A}|\xi_i\rangle$. An easy exercise verifies that Tr is well-defined in that it does not depend on the choice of orthonormal basis and is complex linear, so that $\mathsf{Tr}(a\mathsf{A}+b\mathsf{B}) = a\mathsf{Tr}(\mathsf{A})+b\mathsf{Tr}(\mathsf{B})$. Using the resolution of the identity associated with the orthonormal basis $\boldsymbol{\xi}$ and given by $1_{\mathcal{H}} = \sum_{i=1}^{N} |\xi_i\rangle\langle\xi_i|$, we calculate the expectation value for the observable a represented by the self-adjoint operator A as

$$
\begin{aligned}
\mathrm{Exp}_\psi(\mathsf{A}) &= \langle\psi|\mathsf{A}\psi\rangle = \langle\psi|\mathsf{A}1_{\mathcal{H}}|\psi\rangle \\
&= \langle\psi|\mathsf{A}\sum_{i=1}^{N} |\xi_i\rangle\langle\xi_i|\psi\rangle = \sum_{i=1}^{N}\langle\xi_i|\psi\rangle\langle\psi|\mathsf{A}|\xi_i\rangle \\
&= \sum_{i=1}^{N}\langle\xi_i|\mathsf{P}_\psi\mathsf{A}|\xi_i\rangle = \mathsf{Tr}(\mathsf{P}_\psi\mathsf{A}). \qquad (23.2)
\end{aligned}
$$

Thus the trace can be used to give a simple expression for the expectation value of an observable when the projection operator notation is used to represent quantum states.

We now apply this to generalize the notion of a quantum state. For the moment we restrict our attention to a quantum system composed of a statistical distribution of finitely many possible states, like that of Option 2. Afterwards, we touch briefly on distributions of states dependent on a continuous parameter, like that of Option 1 where the phase angle ϕ takes values in the parameter space Φ subject to the probability measure μ. Consider, then, an ensemble of particles, any one of which is in the normalized state ψ_k with probability p_k, for $k = 1, \ldots, K$. Here, of course, $\sum_{k=1}^{K} p_k = 1$. The expectation value for the observable a when repeated measurements are performed on this ensemble is the weighted average

$$
\sum_{k=1}^{K} p_k \,\mathrm{Exp}_{\psi_k}(\mathsf{A}) = \sum_{k=1}^{K} p_k\langle\psi_k|\mathsf{A}\psi_k\rangle. \qquad (23.3)
$$

Now define the operator

$$\rho = \sum_{k=1}^{K} p_k \mathsf{P}_{\psi_k} = \sum_{k=1}^{K} p_k |\psi_k\rangle\langle\psi_k|. \tag{23.4}$$

Our claim is that the operator ρ is positive, self-adjoint,[2] and of trace 1. First, since each p_k is nonnegative, the operator ρ is positive since, for any $\chi \in \mathcal{H}$, we have

$$\begin{aligned}\langle\chi|\rho\chi\rangle &= \sum_{k=1}^{K} p_k \langle\chi|\mathsf{P}_{\psi_k}|\chi\rangle \\ &= \sum_{k=1}^{K} p_k \langle\chi|\psi_k\rangle\langle\psi_k|\chi\rangle = \sum_{k=1}^{K} p_k |\langle\psi_k|\chi\rangle|^2 \geq 0.\end{aligned}$$

Second, since each projection operator P_{ψ_k} is self-adjoint and each p_k is nonnegative, ρ is self-adjoint. Finally, since the trace of each projection operator P_{ψ} is 1, which follows from setting $\mathsf{A} = 1_{\mathcal{H}}$ in equation 23.2, we have

$$\mathsf{Tr}(\rho) = \sum_{k=1}^{K} p_k \mathsf{Tr}(\mathsf{P}_{\psi_k}) = \sum_{k=1}^{K} p_k = 1.$$

Using equation 23.2, the expectation value of equation 23.3 is recast in the language of ρ:

$$\mathsf{Tr}(\rho\mathsf{A}) = \sum_{k=1}^{K} p_k \mathsf{Tr}(\mathsf{P}_{\psi_k}\mathsf{A}) = \sum_{k=1}^{K} p_k \mathrm{Exp}_{\psi_k}(\mathsf{A}).$$

It seems reasonable to generalize the notion of the state of a quantum system to include operators such as ρ that encode statistical ensembles. This discussion motivates the following generalizations of Axioms 1 and 3 of the minimal list of Section 2.1, at least in the setting of a finite-dimensional state space $\mathcal{H}$.

Axiom 1. The state of a (finite-dimensional) quantum system is represented by a positive self-adjoint operator of trace 1 on a Hilbert space $\mathcal{H}$. Such an operator is called a **density operator**.

[2] In the context of finite-dimensional Hilbert spaces, a symmetric or Hermitian operator A, defined by the symmetry relation $\langle\varphi|\mathsf{A}\psi\rangle = \langle\mathsf{A}\varphi|\psi\rangle$ for all $\varphi, \psi \in \mathcal{H}$, is exactly the same as a self-adjoint operator. The self-adjoint condition is redundant here since a positive bounded operator on a complex Hilbert space is always symmetric; see Reed and Simon (1980), p. 195.

Axiom 3. For a system represented by a density operator ρ, the expectation value of the observable a represented by the self-adjoint operator A is

$$\mathrm{Exp}_\rho(\mathsf{A}) = \mathsf{Tr}(\rho\mathsf{A}). \tag{23.5}$$

States represented by one-dimensional projections $\mathsf{P}_\psi = |\psi\rangle\langle\psi|$ are called **pure states** while states represented by density operators that are not pure are called **mixed states**. The discussion of this section so far places standard finite-dimensional quantum mechanics in von Neumann's more general setting of density operators and generalizes it to include statistical ensembles composed of finitely many possible pure states. The formalism, though, is more general than this. For instance, suppose that an ensemble of particles is generated for which each particle is in a state ψ_ϕ with probability $d\mu$, where ϕ is some continuous parameter in the space Φ and μ is a probability measure on the parameter space Φ. The density operator ρ is then given by the appropriate integral:

$$\rho = \int_\Phi |\psi_\phi\rangle\langle\psi_\phi|\, d\mu. \tag{23.6}$$

Rather than working out the intricacies and fine details of operators defined by integrals, we will be content with giving some examples in the section following. First, though, we continue von Neumann's development.

How do we recover probabilities for observing various values upon measurement, in the setting of density operators? Let Proj_λ be the orthogonal projection of $\mathcal{H}$ onto the λ-eigenspace of the self-adjoint operator A. Recall from the elaboration of Axiom 3 in Lecture 4 that an A-measurement on a quantum system in state ψ yields the result λ with probability $P(\lambda) = \langle\psi|\mathsf{Proj}_\lambda\psi\rangle$, which, from equation 23.2, may be written as $\mathsf{Tr}(\mathsf{P}_\psi\mathsf{Proj}_\lambda)$. It follows that when the system is in a statistical distribution of states given by the density operator $\rho = \sum_{k=1}^K p_k\mathsf{P}_{\psi_k}$, the probability for obtaining the value λ is

$$\sum_{k=1}^K p_k\langle\psi_k|\mathsf{Proj}_\lambda\psi_k\rangle = \sum_{k=1}^K p_k\,\mathsf{Tr}\left(\mathsf{P}_{\psi_k}\mathsf{Proj}_\lambda\right) = \mathsf{Tr}\left(\rho\mathsf{Proj}_\lambda\right),$$

by the linearity of the trace. This motivates the further embellishment of Axiom 3 in the setting of density operators.

Axiom 3. If a system is represented by the density operator ρ then a measurement of the observable a represented by the self-adjoint

> operator A will record one of the eigenvalues λ of A with probability $\mathsf{Tr}(\rho\mathsf{Proj}_\lambda)$, where Proj_λ is the orthogonal projection onto the λ-eigenspace. The state of the system immediately after measurement is $t\,\mathsf{Proj}_\lambda\rho$, where $t = 1/\mathsf{Tr}(\mathsf{Proj}_\lambda\rho)$.

We can see that this version of the axiom is consistent with the former version by examining the spectral decomposition of A, which is $\mathsf{A} = \sum_{s=1}^{S} \lambda_s \mathsf{Proj}_{\lambda_s}$, where $\lambda_1, \ldots, \lambda_S$ are the distinct eigenvalues of A. Indeed, the expectation value of an A-measurement on a system in state ρ is just the weighted sum $\sum_{s=1}^{S} P(\lambda_s)\lambda_s$, but this is given by

$$\begin{aligned}\mathrm{Exp}_\rho(\mathsf{A}) = \sum_{s=1}^{S} P(\lambda_s)\lambda_s &= \sum_{s=1}^{S} \mathsf{Tr}\left(\rho\mathsf{Proj}_{\lambda_s}\right)\lambda_s \\ &= \mathsf{Tr}\left(\rho\sum_{s=1}^{S}\lambda_s\mathsf{Proj}_{\lambda_s}\right) = \mathsf{Tr}(\rho\mathsf{A}),\end{aligned}$$

again by the linearity of the trace, agreeing with equation 23.5.

We close this section by considering the criteria that identify the pure states among the density operators and explaining further their "pureness." To verify the first of these criteria, we need to know that the product AB of two positive commuting operators A and B is positive. For this, we use the fact that the positive operator B has a square root C, which by definition is a positive and therefore symmetric operator whose square is B and which commutes with every operator that commutes with B.[3] We then have, for $\psi \in \mathcal{H}$, $\langle\psi|\mathsf{AB}\psi\rangle = \langle\psi|\mathsf{AC}^2\psi\rangle = \langle\psi|\mathsf{CAC}\psi\rangle = \langle\mathsf{C}\psi|\mathsf{AC}\psi\rangle \geq 0$, since A is positive.

Density Operators and Pure States. *Let ρ be a density operator on the finite-dimensional Hilbert space $\mathcal{H}$. Then*

(i) $0 \leq \rho^2 \leq \rho \leq 1_{\mathcal{H}}$;

(ii) *ρ represents a pure state if and only if it is a* ***projection,*** *meaning that $\rho = \rho^2$;*

[3] The proof of this **square root lemma** is easy in the case considered in this section, where $\mathcal{H}$ is finite-dimensional, but may be found in the full generality of self-adjoint operators on infinite-dimensional Hilbert spaces in Rudin (1991), p. 349. The proof of the general case makes use of the spectral theorem for self-adjoint operators, which allows for the definition of a function of an operator, as in Section 4.2. For an easier proof in infinite dimensions in the important special case where the operators are bounded, which will be used in the penultimate section of this lecture, see Reed and Simon (1980), p. 196, for example.

(iii) *ρ represents a pure state if and only if it is* **homogeneous**, *meaning that if $\rho = a\sigma + b\tau$ for some $a, b > 0$, $a+b = 1$, and density operators σ and τ then $\sigma = \tau$.*

For item (i), since ρ is self-adjoint, for $\psi \in \mathcal{H}$, we have $\langle\psi|\rho^2\psi\rangle = \langle\rho\psi|\rho\psi\rangle = \|\rho\psi\|^2 \geq 0$, so $0 \leq \rho^2$. Assuming that the state ψ is normalized, the Gram–Schmidt orthogonalization process ensures that we may choose the basis $\boldsymbol{\xi}$ for $\mathcal{H}$ such that $\xi_1 = \psi$. Since ρ is positive, $\langle\xi_i|\rho\xi_i\rangle \geq 0$ for each i, so $\langle\psi|\rho\psi\rangle = \langle\xi_1|\rho\xi_1\rangle \leq \sum_{i=1}^{N}\langle\xi_i|\rho\xi_i\rangle = \mathsf{Tr}(\rho) = 1$, and it follows that $\rho \leq 1_{\mathcal{H}}$. Now both ρ and $1_{\mathcal{H}} - \rho$ are positive commuting operators, and hence the product $\rho(1_{\mathcal{H}} - \rho) = \rho - \rho^2$ is positive, implying that $\rho^2 \leq \rho$. This confirms item (i).

For item (ii), it follows easily that $\rho^2 = \rho$ if $\rho = |\psi\rangle\langle\psi|$. For the converse, suppose that $\rho^2 = \rho$ and let $\{\xi_1, \ldots, \xi_M\}$ be an orthonormal basis for the image $\rho(\mathcal{H})$, where $1 \leq M = \dim\rho(\mathcal{H}) \leq N$, and extend this to an orthonormal basis $\boldsymbol{\xi} = \{\xi_1, \ldots, \xi_N\}$ for $\mathcal{H}$. Note that ρ is the identity on its image, for if $\psi = \rho(\chi) \in \rho(\mathcal{H})$ then $\rho(\psi) = \rho(\rho(\chi)) = \rho^2(\chi) = \rho(\chi) = \psi$. We have, since $\rho(\xi_i) = \xi_i$ whenever $i \leq M$ and $\langle\xi_i|\rho\xi_i\rangle = 0$ whenever $i > M$,

$$1 = \mathsf{Tr}(\rho) = \sum_{i=1}^{N}\langle\xi_i|\rho\xi_i\rangle = \sum_{i=1}^{M}\langle\xi_i|\xi_i\rangle = M.$$

For $i > M = 1$, since ρ is self-adjoint we have $\|\rho\xi_i\|^2 = \langle\rho\xi_i|\rho\xi_i\rangle = \langle\xi_i|\rho^2\xi_i\rangle = \langle\xi_i|\rho\xi_i\rangle = 0$, so $\rho\xi_i = 0$. Since $\rho\xi_1 = \xi_1$ and $\rho\xi_i = 0$ for $i > 1$, it follows that ρ is the pure state $\rho = |\xi_1\rangle\langle\xi_1|$.

Finally, let ρ be a pure state, so that $\rho^2 = \rho$, and assume that $\rho = a\sigma + b\tau$ as in item (iii). Then a calculation using $a + b = 1$ shows that $\rho^2 = a\sigma^2 + b\tau^2 - ab(\sigma - \tau)^2$ and, therefore,

$$0 = \rho - \rho^2 = a(\sigma - \sigma^2) + b(\tau - \tau^2) + ab(\sigma - \tau)^2. \tag{23.7}$$

By item (i), the operators $\sigma - \sigma^2$ and $\tau - \tau^2$ are positive and, since $\sigma - \tau$ is self-adjoint, the operator $(\sigma - \tau)^2$ is positive too. From this observation and the fact that a and b are positive, equation 23.7 implies that $0 = \langle\psi|(\sigma - \tau)^2\psi\rangle$ for every $\psi \in \mathcal{H}$. It follows that, for any $\psi \in \mathcal{H}$, $0 = \langle\psi|(\sigma - \tau)^2\psi\rangle = \langle(\sigma - \tau)\psi|(\sigma - \tau)\psi\rangle = \|(\sigma - \tau)\psi\|^2$, so $(\sigma - \tau)\psi = 0$. We conclude that $\sigma = \tau$, and hence ρ is homogeneous.

For the converse, we show that when ρ is not a pure state, it is not homogeneous. First, expand ρ as a spectral sum $\rho = \sum_{i=1}^{N}\lambda_i|\xi_i\rangle\langle\xi_i|$, where $\boldsymbol{\xi} = \{\xi_1, \ldots, \xi_N\}$ is an orthonormal basis of ρ-eigenvectors with respective, but not necessarily distinct, eigenvalues $\lambda_1, \ldots, \lambda_N$. Since ρ is a density operator, the trace condition implies that $1 = \mathsf{Tr}(\rho) = \sum_{i=1}^{N} \lambda_i$, and item (i) implies that each eigenvalue satisfies $0 \le \lambda_i \le 1$. If ρ is not a pure state then at least two eigenvalues are nonzero, so we may assume without loss of generality that $a = \lambda_1 > 0$ and $b = \sum_{i=2}^{N} \lambda_i > 0$. It follows that $a + b = 1$ and $\rho = a\sigma + b\tau$, where $\sigma = |\xi_1\rangle\langle\xi_1|$ and $\tau = b^{-1}\sum_{i=2}^{N} \lambda_i |\xi_i\rangle\langle\xi_i|$. Obviously σ, a pure state, is a density operator, and an easy examination confirms that τ is a density operator also. This shows that ρ is not homogeneous.

Before we consider density operators on infinite-dimensional state spaces, we pause to see examples of the theory that has already been presented in the context of spin systems.

23.3 Matrix Calculations in the Spin-$\frac{1}{2}$ System

Our matrix calculations are with respect to the basis $\{|+\rangle_z, |-\rangle_z\}$. First observe that the spin state $\psi = a|+\rangle_z + b|-\rangle_z$ is represented by a column matrix and its dual is represented by a row matrix:

$$\psi = |\psi\rangle = \begin{pmatrix} a \\ b \end{pmatrix} \quad \text{and} \quad \langle\psi| = \begin{pmatrix} \overline{a} & \overline{b} \end{pmatrix}.$$

In our first example, under Option 2 where the quantum system is an ensemble where $|+\rangle_z$ and $|-\rangle_z$ occur with equal probability, the **density matrix** is

$$\begin{aligned} \rho_1 &= \frac{1}{2}\mathsf{P}_{|+\rangle_z} + \frac{1}{2}\mathsf{P}_{|-\rangle_z} = \frac{1}{2}|+\rangle_z\,{}_z\langle+| + \frac{1}{2}|-\rangle_z\,{}_z\langle-| \\ &= \frac{1}{2}\begin{pmatrix} 1 \\ 0 \end{pmatrix}\begin{pmatrix} 1 & 0 \end{pmatrix} + \frac{1}{2}\begin{pmatrix} 0 \\ 1 \end{pmatrix}\begin{pmatrix} 0 & 1 \end{pmatrix} = \frac{1}{2}\begin{pmatrix} 1 & 0 \\ 0 & 1 \end{pmatrix} = \frac{1}{2}1_{2\times 2}. \end{aligned}$$

The matrices representing the operators S_z and $\mathsf{Proj}_{+1} = \mathsf{P}_{|+\rangle_z}$ in this basis are

$$\mathsf{S}_z = \begin{pmatrix} 1 & 0 \\ 0 & -1 \end{pmatrix} \quad \text{and} \quad \mathsf{Proj}_{+1} = \begin{pmatrix} 1 & 0 \\ 0 & 0 \end{pmatrix}.$$

Thus, one-half times the identity matrix represents this mixed state and the expectation value for S_z-measurements is $\mathrm{Exp}_{\rho_1}(\mathsf{S}_z) = \mathsf{Tr}(\rho_1 \mathsf{J_z}) = \frac{1}{2}\mathsf{Tr}(\mathsf{J_z}) = 0$, as expected since $+1$ and -1 occur with equal frequencies. The probability of recording a $+1$ upon a z-component spin measurement is $\mathsf{Tr}(\rho_1 \mathsf{Proj}_{+1}) = \frac{1}{2}\mathsf{Tr}(\mathsf{Proj}_{+1}) = \frac{1}{2}$, again as expected.

The standard x- and y-component spin eigenstates are

$$|+\rangle_x = \frac{1}{\sqrt{2}}\begin{pmatrix}1\\1\end{pmatrix} \quad \text{and} \quad |-\rangle_x = \frac{1}{\sqrt{2}}\begin{pmatrix}-1\\1\end{pmatrix},$$

and

$$|+\rangle_y = \frac{1}{\sqrt{2}}\begin{pmatrix}1\\\mathrm{i}\end{pmatrix} \quad \text{and} \quad |-\rangle_y = \frac{1}{\sqrt{2}}\begin{pmatrix}1\\-\mathrm{i}\end{pmatrix}.$$

For our second example, consider an ensemble composed of an equal mixture of $|+\rangle_z$ and $|+\rangle_x$ states. The density matrix is

$$\rho_2 = \frac{1}{2}|+\rangle_{z\,z}\langle+| + \frac{1}{2}|+\rangle_{x\,x}\langle+| = \frac{1}{4}\begin{pmatrix}3 & 1\\1 & 1\end{pmatrix}.$$

For z-component spin measurements, elementary probability theory implies that $P(z+) = \frac{3}{4}$, $P(z-) = \frac{1}{4}$, and $\mathrm{Exp}_\rho(\mathsf{S}_z) = \frac{1}{2}$, agreeing with the trace calculations

$$P(z+) = \mathsf{Tr}(\rho_2 \mathsf{P}_{|+\rangle_z}) = \mathsf{Tr}\frac{1}{4}\begin{pmatrix}3 & 1\\1 & 1\end{pmatrix}\begin{pmatrix}1 & 0\\0 & 0\end{pmatrix} = \frac{1}{4}\mathsf{Tr}\begin{pmatrix}3 & 0\\1 & 0\end{pmatrix} = \frac{3}{4},$$

$$P(z-) = \mathsf{Tr}(\rho_2 \mathsf{P}_{|-\rangle_z}) = \mathsf{Tr}\frac{1}{4}\begin{pmatrix}3 & 1\\1 & 1\end{pmatrix}\begin{pmatrix}0 & 0\\0 & 1\end{pmatrix} = \frac{1}{4}\mathsf{Tr}\begin{pmatrix}0 & 1\\0 & 1\end{pmatrix} = \frac{1}{4},$$

$$\mathrm{Exp}_{\rho_2}(\mathsf{S}_z) = \mathsf{Tr}(\rho_2 \mathsf{S}_z) = \mathsf{Tr}\frac{1}{4}\begin{pmatrix}3 & 1\\1 & 1\end{pmatrix}\begin{pmatrix}1 & 0\\0 & -1\end{pmatrix} = \frac{1}{4}\mathsf{Tr}\begin{pmatrix}3 & -1\\1 & -1\end{pmatrix} = \frac{1}{2}.$$

What is to be expected if y-component spin measurements are made? The matrix representing spin measurements along the y-direction is

$$\mathsf{S}_y = \begin{pmatrix}0 & -\mathrm{i}\\\mathrm{i} & 0\end{pmatrix},$$

and the matrices representing the pertinent projections are

$$\mathsf{P}_{|+\rangle_y} = \frac{1}{2}\begin{pmatrix}1\\\mathrm{i}\end{pmatrix}\begin{pmatrix}1 & -\mathrm{i}\end{pmatrix} = \frac{1}{2}\begin{pmatrix}1 & -\mathrm{i}\\\mathrm{i} & 1\end{pmatrix}$$

and

$$\mathsf{P}_{|-\rangle_y} = \frac{1}{2}\begin{pmatrix}1\\-\mathrm{i}\end{pmatrix}\begin{pmatrix}1 & \mathrm{i}\end{pmatrix} = \frac{1}{2}\begin{pmatrix}1 & \mathrm{i}\\-\mathrm{i} & 1\end{pmatrix}.$$

We get $P(y+) = \mathsf{Tr}(\rho_2 \mathsf{P}_{|+\rangle_y}) = \frac{1}{2} = \mathsf{Tr}(\rho_2 \mathsf{P}_{|-\rangle_y}) = P(y-)$ and $\mathrm{Exp}_{\rho_2}(\mathsf{S}_y) = \mathsf{Tr}(\rho_2 \mathsf{S}_y) = 0$.

23.3.1 The Topology of Spin States

Consider now the general case where spin-component measurements are made in the direction of the unit vector $\mathbf{u} \in \mathbf{S}^2 \subset \mathbb{R}^3$. We may write

$$\mathbf{u} = (\sin\theta\cos\phi, \sin\theta\sin\phi, \cos\theta) \tag{23.8}$$

in rectangular Cartesian coordinates, where $(1, \theta, \phi)$ are the standard spherical coordinates of $\mathbf{u}$ from Section 14.1. The coordinates (θ, ϕ) are the standard spherical coordinates on the 2-sphere $\mathbf{S}^2$ and may be used to locate $\mathbf{u}$. From equation 18.2, the operator representing spin-component measurements in the direction of $\mathbf{u}$ may be written as

$$\mathsf{S}_{\mathbf{u}} = \mathsf{S}_{\theta,\phi} = \begin{pmatrix} \cos\theta & e^{-i\phi}\sin\theta \\ e^{i\phi}\sin\theta & -\cos\theta \end{pmatrix}.$$

Eigenvectors with respective eigenvalues $+1$ and -1 are given by

$$|+\rangle_{\mathbf{u}} = |+\rangle_{\theta,\phi} = \cos\frac{\theta}{2}|+\rangle_z + e^{i\phi}\sin\frac{\theta}{2}|-\rangle_z = \begin{pmatrix} \cos(\theta/2) \\ e^{i\phi}\sin(\theta/2) \end{pmatrix} \tag{23.9}$$

and

$$|-\rangle_{\mathbf{u}} = |-\rangle_{\theta,\phi} = -\sin\frac{\theta}{2}|+\rangle_z + e^{i\phi}\cos\frac{\theta}{2}|-\rangle_z = \begin{pmatrix} -\sin(\theta/2) \\ e^{i\phi}\cos(\theta/2) \end{pmatrix}. \tag{23.10}$$

Notice that $|+\rangle_{-\mathbf{u}} = |+\rangle_{\theta+\pi,\phi} = |-\rangle_{\mathbf{u}}$. Notice also the ambiguity at the south pole $\mathbf{s} = (0, 0, 1)$ with spherical coordinates (π, ϕ) for an arbitrarily chosen ϕ. We make the choice of $\phi = 0$ in this case so that $|+\rangle_{\mathbf{s}} = \binom{0}{1} = |-\rangle_z$. A quick calculation gives the corresponding density matrices as

$$\mathsf{P}_{|+\rangle_{\mathbf{u}}} = \begin{pmatrix} \cos^2(\theta/2) & \frac{1}{2}e^{-i\phi}\sin\theta \\ \frac{1}{2}e^{i\phi}\sin\theta & \sin^2(\theta/2) \end{pmatrix} \tag{23.11}$$

and

$$\mathsf{P}_{|-\rangle_{\mathbf{u}}} = \begin{pmatrix} \sin^2(\theta/2) & -\frac{1}{2}e^{-i\phi}\sin\theta \\ -\frac{1}{2}e^{i\phi}\sin\theta & \cos^2(\theta/2) \end{pmatrix}. \tag{23.12}$$

We may use this development to uncover further topological characteristics of spin states. We will find that the Hopf fibration of the 3-sphere by circles makes a beautiful appearance here. First, the Hilbert space of spin-$\frac{1}{2}$ states is the $\mathfrak{su}(2)$-representation space $V^{\frac{1}{2}} = \mathbb{C}|+\rangle_z \oplus \mathbb{C}|-\rangle_z$. Recall, though, that the vector $c\psi$ represents the same quantum state as the vector $\psi \in V^{\frac{1}{2}} - \{0\}$, for any nonzero complex number c, and thus the state represented by ψ is represented also by every vector in the deleted ray $\mathbb{C}^\times\psi$ in $V^{\frac{1}{2}} = \mathbb{C}^2$. It follows that the complex projective line $\mathbb{P}^1_{\mathbb{C}} = \mathbb{C}^2/\mathbb{C}^\times = V^{\frac{1}{2}}/\mathbb{C}^\times$ may serve more economically as a state space for the spin-$\frac{1}{2}$ system, as there is a one-to-one correspondence between quantum states of the spin system and the points of the projective line.[4] Now the projective line $\mathbb{P}^1_{\mathbb{C}}$ also may be realized as the quotient of the 3-sphere $\mathbf{S}^3$ by an action of the circle group, which fits naturally with the use of $\mathbb{P}^1_{\mathbb{C}}$ as the quantum state space. Indeed, the normalized state vectors in the state space $V^{\frac{1}{2}}$ may be identified as

$$\mathbf{S}^3 = \left\{\psi \in V^{\frac{1}{2}} : \|\psi\| = 1\right\} = \{e^{i\omega}|+\rangle_{\theta,\phi} : \omega, \theta, \phi \in \mathbb{R}\},$$

on which the circle group $\mathbf{S}^1 = \{e^{i\chi} : \chi \in \mathbb{R}\}$ acts by multiplication. As all the vectors in an $\mathbf{S}^1$-orbit represent the same quantum state, there is a one-to-one correspondence between the quantum states of the spin-$\frac{1}{2}$ system and the $\mathbf{S}^1$-orbits, which form the fibers of the Hopf fibration of $\mathbf{S}^3$ with quotient homeomorphic to $\mathbf{S}^2 \cong \mathbb{P}^1_{\mathbb{C}}$. The explicit injection $\mathbf{S}^2 \hookrightarrow \mathbf{S}^3 \subset V^{\frac{1}{2}}$ given by $\mathbf{u} \mapsto |+\rangle_{\mathbf{u}}$ may be thought of as a discontinuous section of the Hopf fibration $\mathbf{S}^1 \hookrightarrow \mathbf{S}^3 \to \mathbf{S}^2$, the discontinuity appearing at the south pole. This identifies the points of the 2-sphere $\mathbf{S}^2$ bijectively with the subset

$$\begin{aligned}\mathbf{B} &= \{|+\rangle_{\mathbf{u}} : \mathbf{u} \in \mathbf{S}^2\} \\ &= \left\{\cos\frac{\theta}{2}|+\rangle_z + e^{i\phi}\sin\frac{\theta}{2}|-\rangle_z : \theta \in [0,\pi], \phi \in [0, 2\pi], \phi = 0 \text{ if } \theta = \pi\right\}.\end{aligned}$$

The set $\mathbf{B}$ is called the **Bloch sphere** and parameterizes the pure states of the spin-$\frac{1}{2}$ system. Topologically the Bloch sphere is an open two-dimensional disk with a single point on the 1-sphere boundary, namely, the point $|+\rangle_{\mathbf{s}}$. This 1-sphere boundary is itself the orbit of $|+\rangle_{\mathbf{s}}$ under the circle action, so every point on the boundary represents the same quantum state. Realizing a pure state by a one-dimensional projection operator using expression 23.11 gives an injection $|+\rangle_{\mathbf{u}} \mapsto \mathsf{P}_{|+\rangle_{\mathbf{u}}}$ of the Bloch sphere into $\mathrm{M}_{2\times 2}(\mathbb{C}) \cong \mathbb{C}^4$, the space of 2×2 complex matrices. A careful examination of equation 23.11 shows that the mapping $\mathbf{u} \mapsto \mathsf{P}_{|+\rangle_{\mathbf{u}}}$ of $\mathbf{S}^2$ to $\mathrm{M}_{2\times 2}(\mathbb{C})$ is continuous and, since it is injective, it is an embedding. It follows that the image of the mapping of the Bloch sphere into $\mathrm{M}_{2\times 2}(\mathbb{C})$ is an embedded 2-sphere. The density operators are the matrices in the convex hull of this image, with pure states represented by matrices on the closure of the image and mixed states represented

[4] Of course this generalizes. If $\mathcal{H}$ is the state space of a quantum system then the projective space $\mathcal{H}/\mathbb{C}^\times$, where $\mathbb{C}^\times$ acts on $\mathcal{H}$ by scalar multiplication, may serve as a state space for the system, with each state of the system represented by precisely one point of $\mathcal{H}/\mathbb{C}^\times$.

by ones that are "interior," the proper convex combinations of the pure states.

One-half times the identity matrix already has been recognized as the density operator ρ_1 of the first example of this section, and it represents a **maximally mixed state**. It is an equal mixture of the states represented by the two basis vectors $|+\rangle_z$ and $|-\rangle_z$. A quick calculation using expressions 23.11 and 23.12 reveals ρ_1 as the equal mixture

$$\rho_1 = \frac{1}{2}1_{2\times 2} = \frac{1}{2}\mathsf{P}_{|+\rangle_{\mathbf{u}}} + \frac{1}{2}\mathsf{P}_{|-\rangle_{\mathbf{u}}} = \frac{1}{2}\mathsf{P}_{|+\rangle_{\mathbf{u}}} + \frac{1}{2}\mathsf{P}_{|+\rangle_{-\mathbf{u}}},$$

for every $\mathbf{u} \in \mathbf{S}^2$. In particular, the same density operator represents many different ensembles. As the probabilities and expectation values are calculated using only the density operator, as recorded in the two versions of Axiom 3 in this lecture, all predictions from quantum mechanics of measurements on ensembles with the same density operator agree. For instance, no experiment on an ensemble with density operator ρ_1 can decide whether the ensemble is composed of an equal mixture of z-component spin eigenstates, an equal mixture of x-component spin eigenstates, or any of a myriad of other possibilities. This property is intrinsic to mixed states. Indeed, measurements can always differentiate two different pure states. For example, when $\mathbf{u} \neq \mathbf{v}$, then repeated spin-component measurements in the directions of $\mathbf{u}$ and $\mathbf{v}$ can decide whether a system is in a state represented by $|+\rangle_{\mathbf{u}}$ or $|+\rangle_{\mathbf{v}}$. Measurements can also differentiate pure states from mixed states, as in the example presented in Section 23.1 with an atomic measure in Option 1 and a maximally mixed state in Option 2. However, the density matrix ρ that represents any mixed state also represents other, different, mixed states, and measurements cannot differentiate the exact composition of the ensemble represented by ρ.

23.3.2 Continuous Ensembles

We end this section by examining Option 1 in more detail. Recall that the ensemble consists of particles emitted in an equal superposition of the z-component spin eigenstates, with each state vector of the form

$$\psi_\phi = \frac{1}{\sqrt{2}}|+\rangle_z + \frac{e^{i\phi}}{\sqrt{2}}|-\rangle_z,$$

where the phase angle ϕ is chosen randomly with respect to a probability measure μ on the phase-angle space $\Phi = [0, 2\pi)$. Note that ψ_ϕ may be identified as precisely the state vector $|+\rangle_{\pi/2,\phi}$, or $|+\rangle_{\mathbf{u}}$ where

$\mathbf{u} = (\cos\phi, \sin\phi, 0)$; this is the $(+1)$-eigenstate of spin-component measurements in the direction of a vector in the xy-plane rotated counterclockwise from the x-axis through angle ϕ. The projection operator corresponding to ψ_ϕ is from expression 23.11

$$\mathsf{P}_{|+\rangle_{\mathbf{u}}} = \frac{1}{2}\begin{pmatrix} 1 & e^{-\mathrm{i}\phi} \\ e^{\mathrm{i}\phi} & 1 \end{pmatrix}. \tag{23.13}$$

According to equation 23.6, the density operator representing this ensemble is

$$\rho = \int_\Phi |\psi_\phi\rangle\langle\psi_\phi|\, d\mu = \frac{1}{2}\int_\Phi \begin{pmatrix} 1 & e^{-\mathrm{i}\phi} \\ e^{\mathrm{i}\phi} & 1 \end{pmatrix} d\mu = \frac{1}{2}\begin{pmatrix} 1 & \mathbf{E}_\mu(e^{-\mathrm{i}\phi}) \\ \mathbf{E}_\mu(e^{\mathrm{i}\phi}) & 1 \end{pmatrix},$$

where $\mathbf{E}_\mu(e^{\pm\mathrm{i}\phi}) = \int_\Phi e^{\pm\mathrm{i}\phi}\, d\mu$ is the expected value of the complex random variable $e^{\pm\mathrm{i}\phi}$.

We discussed two extremes in Section 23.1, one where μ is the standard normalized Lebesgue measure and the other where μ is an atomic measure concentrated at ϕ_0. In the former case, $\mathbf{E}_\mu(e^{\pm\mathrm{i}\phi}) = 0$ and the density matrix representing the ensemble is the maximally mixed state matrix ρ_1. In this case no measurements can distinguish between this ensemble and any other maximally mixed state represented by ρ_1, and Option 1 cannot be distinguished experimentally from Option 2. In the latter case, $\mathbf{E}_\mu(e^{\pm\mathrm{i}\phi}) = e^{\pm\mathrm{i}\phi_0}$ and the ensemble is in the pure state represented by the projection matrix of expression 23.13 with $\phi = \phi_0$.

23.4 Density Operators II: Infinite-Dimensional Setting

We will delve just enough into the intricacies of the theory of the trace of an operator on a separable infinite-dimensional Hilbert space to present a brief outline, so that the reader may appreciate both some of the difficulties and some of the successes of extending the theory of Section 23.2 to infinite dimensions. A difficulty presents itself immediately in the definition. Let A be an operator on the separable infinite-dimensional Hilbert space $\mathcal{H}$ and let $\boldsymbol{\xi} = \{\xi_i : i \in \mathbb{N}\}$ be a complete orthonormal basis. The obvious generalization of the trace concept would be to define the trace of A as

$$\mathsf{Tr}(\mathsf{A}) = \mathsf{Tr}_{\boldsymbol{\xi}}(\mathsf{A}) = \sum_{i=1}^{\infty} \langle \xi_i | \mathsf{A} | \xi_i \rangle, \tag{23.14}$$

but three questions immediately arise. The first asks whether the series is convergent, and the second asks whether the trace so defined is independent of the choice of basis. For general linear operators A, the answers are rather unsatisfactory. Indeed, for many operators the series is divergent and for other operators the series is convergent for some bases and divergent for others, and it converges to different finite values for different bases. The third question that arises is whether the basis vector ξ_i is actually in the domain $\mathcal{D}(\mathsf{A})$ when A is defined only on a dense subset of $\mathcal{H}$. To avoid this latter problem, we will deal only with bounded operators, which, of course, are always defined on the whole of $\mathcal{H}$. In regards to the first two questions, there are operators for which the series that defines trace is always convergent, whatever basis is used, and the value of the sum is independent of the basis. These are the so-called **trace class operators**.

To understand enough of the theory to describe density operators in infinite dimensions, first note that the sum in equation 23.14 is either convergent to a finite value or diverges to ∞ whenever A is a positive, bounded operator, for then the summands $\langle \xi_i | \mathsf{A} | \xi_i \rangle$ are all defined and nonnegative. Assuming then that A is positive and bounded, we show that the trace is independent of basis. First, the square root lemma[5] implies that there is a bounded, positive, self-adjoint operator $\mathsf{B} = \sqrt{\mathsf{A}}$ whose square is A. Letting $\boldsymbol{\varphi}$ be another complete orthonormal basis for $\mathcal{H}$, we will use the two resolutions of the identity determined by the bases $\boldsymbol{\xi}$ and $\boldsymbol{\varphi}$ and given by

$$1_{\mathcal{H}} = \sum_{i=1}^{\infty} |\xi_i\rangle\langle\xi_i| = \sum_{j=1}^{\infty} |\varphi_j\rangle\langle\varphi_j|.$$

We obtain

$$\mathrm{Tr}_{\boldsymbol{\xi}}(\mathsf{A}) = \sum_{i=1}^{\infty} \langle \xi_i | \mathsf{A} | \xi_i \rangle = \sum_{i=1}^{\infty} \langle \mathsf{B}\xi_i | \mathsf{B}\xi_i \rangle = \sum_{i=1}^{\infty} \langle \mathsf{B}\xi_i | 1_{\mathcal{H}} | \mathsf{B}\xi_i \rangle$$

$$= \sum_{i=1}^{\infty}\sum_{j=1}^{\infty} \langle \mathsf{B}\xi_i | \varphi_j \rangle \langle \varphi_j | \mathsf{B}\xi_i \rangle = \sum_{i=1}^{\infty}\sum_{j=1}^{\infty} |\langle \varphi_j | \mathsf{B}\xi_i \rangle|^2.$$

A similar calculation with the resolution of the identity determined by

[5] See the reference to Reed and Simon (1980) in the footnote on p. 368.

$\boldsymbol{\xi}$ gives

$$\mathsf{Tr}_{\boldsymbol{\varphi}}(\mathsf{A}) = \sum_{j=1}^{\infty}\sum_{i=1}^{\infty} |\langle \xi_i | \mathsf{B}\varphi_j \rangle|^2$$

Now since B is self-adjoint, the terms in the two double sums are equal and, since each of these terms is nonnegative, the order of summation can be interchanged, so that

$$\mathsf{Tr}_{\boldsymbol{\xi}}(\mathsf{A}) = \sum_{i=1}^{\infty}\sum_{j=1}^{\infty} |\langle \varphi_j | \mathsf{B}\xi_i \rangle|^2 = \sum_{j=1}^{\infty}\sum_{i=1}^{\infty} |\langle \xi_i | \mathsf{B}\varphi_j \rangle|^2 = \mathsf{Tr}_{\boldsymbol{\varphi}}(\mathsf{A}).$$

Note that this calculation shows that $\mathsf{Tr}(\mathsf{A})$ is a well-defined element of the extended ray $[0, \infty]$, its value being determined by any orthonormal basis. It follows that any positive, bounded, self-adjoint operator with finite trace is a trace class operator.

We will build our density operators out of these, but first, for completeness, we describe without proof the fundamental characterization of trace class operators. If A is now an arbitrary bounded operator then the product $\mathsf{A}^*\mathsf{A}$ is a positive bounded operator and, as such, has a square root $\sqrt{\mathsf{A}^*\mathsf{A}}$, which is often denoted as $|\mathsf{A}|$. The square root lemma implies that $|\mathsf{A}|$ is a positive, bounded, self-adjoint operator, and therefore that $\mathsf{Tr}(|\mathsf{A}|)$ is well defined and can be either finite or ∞. The following conditions[6] on a bounded operator A are equivalent:

1. A is a trace class operator;
2. $\mathsf{Tr}(|\mathsf{A}|) < \infty$;
3. for any orthonormal basis $\boldsymbol{\xi}$, the sum $\sum_{i=1}^{\infty} \langle \xi_i | \mathsf{A} | \xi_i \rangle$ converges absolutely and the limit is independent of the choice of basis.

We now are in a position to extend the density operator formalism of von Neumann to the infinite-dimensional setting. A **density operator**, also called a **statistical operator**, on the separable Hilbert space $\mathcal{H}$ is a trace class operator ρ that is positive, and therefore self-adjoint, and also normalized, meaning that $\mathsf{Tr}(\mathsf{A}) = 1$. In the context of quantum mechanics, the standard example of such an operator is a direct generalization of equation 23.4. Indeed, let $\{\psi_k\}_{k=1}^{\infty}$ be a sequence of normalized state vectors in $\mathcal{H}$ and $\{p_k\}_{k=1}^{\infty}$ a sequence of nonnegative real numbers whose sum $\sum_{k=1}^{\infty} p_k$ is convergent to 1. Then the sum $\rho = \sum_{k=1}^{\infty} p_k \mathsf{P}_{\psi_k}$ is a density operator that represents a quantum ensemble in the state

[6] Detailed proofs may be found in Reed and Simon (1980), Section VI.6, or Blanchard and Brüning (2003), Section 22.4. In these references, our condition 2 is used as the definition of a trace class operator.

represented by ψ_k with probability p_k. This is in fact the archetype of a density operator, since it can be proved[7] that the bounded operator ρ is a density operator if and only if there exists a sequence of nonnegative real numbers $\{p_k\}_{k=1}^{\infty}$ whose sum is convergent to 1 and an orthonormal basis $\boldsymbol{\xi}$ such that $\rho = \sum_{k=1}^{\infty} p_k \mathsf{P}_{\xi_k}$. Notice then that $\rho\xi_k = p_k\xi_k$, so the sequence $\{p_k\}_{k=1}^{\infty}$ is the sequence of eigenvalues of ρ and hence $\sum_{k=1}^{\infty} p_k \mathsf{P}_{\xi_k} = \sum_{k=1}^{\infty} p_k |\xi_k\rangle\langle\xi_k|$ is the spectral decomposition of the density operator ρ.

23.5 End Notes

For more on the trace in the infinite-dimensional setting see Section VI.6 of Reed and Simon (1980), Chapter 30 of Lax (2002), or Section 22.4 of Blanchard and Brüning (2003). von Neumann (1953) contains a wonderful discussion of the trace of an operator. This appears in Section II.11 starting on p. 178. Then in Chapter IV he introduces density matrices to encode statistical ensembles.

[7] See Blanchard and Brüning (2003), p. 308.

24
Bosons and Fermions

> This rule of the 180° phase shift for alternatives involving exchange in identity of electrons is very odd, and its ultimate reason in nature is still only imperfectly understood ... Such particles are called fermions ... Particles for which interchange does not alter phase are called bosons ... All particles are either one or the other, bosons or fermions. These interference properties can have profound and mysterious effects.
>
> R.P. Feynman and A.R. Hibbs
> *Quantum Mechanics and Path Integrals*, 1965

In Lecture 21 we discovered that the state space of a quantum system composed of two subsystems with respective state spaces $\mathcal{H}_1$ and $\mathcal{H}_2$ is the analytic tensor product $\mathcal{H}_1 \overline{\otimes} \mathcal{H}_2$. In this lecture we will refine this picture and try to eliminate redundancies that arise when the two subsystems are indistinguishable. For concreteness, consider a quantum system composed of two particles, the first with state space $\mathcal{H}_1$ and the second with state space $\mathcal{H}_2$. The most general state vector for the system is a sum of the form $\sum_i c_i(\varphi_i \otimes \chi_i)$, where each $\varphi_i \in \mathcal{H}_1$ and $\chi_i \in \mathcal{H}_2$. This description needs no refinement if the two particles in principle are distinguishable. For example, when one particle is a proton and the other an electron, as in atomic hydrogen, the particles are distinguishable at all times and the proton state space differs in kind from the electron state space. There is no chance for confusion, as the state vector $\varphi \in \mathcal{H}_1$ can refer only to the state of, say, the proton, while $\chi \in \mathcal{H}_2$ can refer only to that of the electron. Consider, though, the case where the two particles are indistinguishable, not only in practice but even in principle. For example, both particles may be electrons in the central potential of a helium atom. As far as anyone can establish, there is no way to label the two electrons and keep track of one as "electron no. 1" and the other as "electron no. 2." Once the wave functions of the two electrons have overlapped, any subsequent measurement of any dynamical variable on one of the electrons cannot distinguish one electron from the other, for there will be positive probabilities for either electron to give the measurement. The mathematics and, in particular, the state space for this system should reflect this basic observation.

In developing the mathematics of indistinguishable particles, we will find in this lecture that nature reflects the mathematical dichotomy between symmetry and antisymmetry in the physical dichotomy between the two basic flavors of particles found in nature, viz., **bosons** and **fermions**. The physical implications of this mathematical dichotomy are profound, leading to explanations of the stability of matter articulated most famously in the **Pauli exclusion principle**, and reflected in the statistical mechanics of large aggregates of particles in the dichotomy between **Bose–Einstein** and **Fermi–Dirac** statistics, the subject of Lecture 26. Bosons are particles of integral spin while fermions have half-integral spin. This must be taken as an axiom in quantum mechanics, as developed thus far, but it is a consequence of a deep connection between statistics and spatial rotations expressed in the **spin–statistics theorem** of relativistic quantum mechanics.

We begin with a mathematical discussion of the symmetries that the state vector of a system of indistinguishable particles should satisfy. This leads quickly to the identification of two qualitatively different possible refinements for the state space of the system, one based on the **symmetric** and the other on the **antisymmetric** tensor product. These are found to be satisfactory representation spaces for ensembles of indistinguishable particles whose mathematical properties reflect the physical effects of exchanging two particles among the ensemble. We will find it pedagogically useful to expound the theory in some detail for the special case of two indistinguishable particles before tackling the general case.

As indicated in Section 21.4, the interesting algebraic structure of the state space of a multi-particle system resides in the tensor algebra $\otimes\mathcal{H}$. Here we will construct the **bosonic** and **fermionic state algebras** from elements of the tensor algebra. The building blocks are the finite-rank symmetric and antisymmetric tensor products already mentioned, which in turn embed as dense subspaces of their respective completions, the **N-quanta bosonic** and **fermionic state spaces**. Ultimately these are combined to form various so-called **Fock spaces**, which serve as the state spaces of quantum systems that allow the possibility of a finite, but variable, number of particles. These are constructed toward the end of the present lecture and are explored more fully in the lecture following, where a more physically based recipe is presented.

The term *particle* carries so much classical baggage[1] that we deem it appropriate to change the language to conform better to our (lack of?) understanding of the quantum world.[2] Following Teller (1995), we will use the vague, nondescriptive, and neutral term **quanta** to refer to the "stuff" of the microscopic quantum world. One **quantum** then refers to a system that can be thought of as a single unit in the context of the discussion and described by a single state space, and in practice may refer to what we normally think of as a "particle," such as an electron or proton, or a "composite particle," such as a hydrogen atom, or even non-particle excitations and resonances, such as the quantized vibrational modes of lattices called phonons. In at least one case in this book, when in Lecture 26 we consider the canonical ensemble of statistical mechanics, the term *quantum* will refer to a macroscopic system of a large number of microscopic quanta governed by its own state space and thought of as a unit.

24.1 Bosons and Fermions

First consider the simple case of two indistinguishable quanta where the state space for one quantum of this species is the Hilbert space $\mathcal{H}$. The state space for the two-quanta system is the two-fold analytic tensor product $\overline{\otimes}^2\mathcal{H} = \mathcal{H}\overline{\otimes}\mathcal{H}$. The basic assumption inspired by the indistinguishability of the two quanta is that, if $\varphi\otimes\chi$ is a decomposable (unentangled) summand of the state vector $\psi \in \mathcal{H}\overline{\otimes}\mathcal{H}$ that describes a state of the system, a nonzero multiple of the decomposable vector $\chi\otimes\varphi$ should also appear as a summand of ψ. The natural extension of this idea is that the vector $\mathcal{P}\psi$ should describe exactly the same state as ψ, where $\mathcal{P} : \mathcal{H}\overline{\otimes}\mathcal{H} \to \mathcal{H}\overline{\otimes}\mathcal{H}$ is the **exchange operator** defined on decomposable vectors as $\mathcal{P}(\varphi\otimes\chi) = \chi\otimes\varphi$ and extended linearly. This implies that ψ and $\mathcal{P}\psi$ are nonzero complex multiples of one another or, stated differently, that $\mathcal{P}\psi = \lambda\psi$ for some $\lambda \in \mathbb{C}^*$. Our first observation, then, is that any state vector ψ that describes the state of a system of

[1] The discussion of Merzbacher (1998), pp. 536–7, on the "basic arbitrariness . . . inherent in the definition of the term *particle*" is particularly interesting.

[2] See the first paragraph of Section 21.2.

two indistinguishable quanta should be an eigenstate of the exchange operator.

Observe that $\mathcal{P}^2 = 1_{\mathcal{H}\overline{\otimes}\mathcal{H}}$, which implies that $\lambda = \pm 1$. A basic dichotomy appears here with the two mutually exclusive possibilities that the system is represented by a state vector ψ that satisfies either $\mathcal{P}\psi = \psi$ or $\mathcal{P}\psi = -\psi$. Physicists describe this quite imaginatively by saying, in the former case, that exchanging the two quanta preserves the wave function, while, in the latter, exchanging introduces a phase factor -1. The exchange operator is an isometry, in fact a reflection, that decomposes $\mathcal{H}\overline{\otimes}\mathcal{H}$ into an orthogonal direct sum of closed eigenspaces $\overline{\mathcal{B}}(2) \oplus \overline{\mathcal{F}}(2)$, where $\overline{\mathcal{B}}(2)$ is the $(+1)$-eigenspace fixed by $\mathcal{P}$ and $\overline{\mathcal{F}}(2)$ the (-1)-eigenspace inverted by $\mathcal{P}$.

The $(+1)$-eigenspace $\overline{\mathcal{B}}(2)$ is the space of **bosonic state vectors** and describes the states of pairs of **bosons**, which are, by definition, quanta whose pair-state vectors lie in the kernel of $1_{\mathcal{H}\overline{\otimes}\mathcal{H}} - \mathcal{P}$. We will see that the bosonic state space is precisely the completion of the symmetric product $\mathcal{B}(2) = \odot^2\mathcal{H} = \mathcal{H} \odot \mathcal{H}$.[3]

The (-1)-eigenspace $\overline{\mathcal{F}}(2)$ is the space of **fermionic state vectors** and describes the states of pairs of **fermions**, which are, by definition, quanta whose pair-state vectors lie in the kernel of $1_{\mathcal{H}\overline{\otimes}\mathcal{H}} + \mathcal{P}$. We will see that the fermionic state space is precisely the completion of the antisymmetric, or exterior, product $\mathcal{F}(2) = \wedge^2\mathcal{H} = \mathcal{H}\wedge\mathcal{H}$. As mentioned in the introduction to this lecture, it is a consequence of a deep result of relativistic quantum mechanics that bosons have integral intrinsic spin while fermions have half-integral intrinsic spin.

24.1.1 The State Space for Two Bosons

We will work now to expose some of the mathematical structure that underlies the bosonic state space $\overline{\mathcal{B}}(2)$. For this discussion, we set $1 = 1_{\mathcal{H}\oplus\mathcal{H}}$. Having already described $\overline{\mathcal{B}}(2)$ as the kernel of the operator $1 - \mathcal{P}$, we can also identify it as the image of the operator $1 + \mathcal{P}$. In fact, the operator $\frac{1}{2}(1 + \mathcal{P})$ is precisely the orthogonal projection $\mathsf{Proj}_{\overline{\mathcal{B}}(2)} = \mathsf{Proj}_{+1}$ of $\mathcal{H}\overline{\otimes}\mathcal{H}$ onto $\overline{\mathcal{B}}(2)$. To see this, first note that $\frac{1}{2}(1 + \mathcal{P})$ is a projection operator since it follows easily that this operator is bounded, symmetric,

[3] Unlike the use of the wedge $\wedge$ for the antisymmetric product introduced later, the notation for the symmetric product is not standardized. Our notation $\odot$ for the symmetric product is not in common usage in the mathematical community, and is replaced variously by $\otimes_S$, or $\cdot$, or the juxtaposition of elements in other references.

and idempotent, i.e., $[\frac{1}{2}(1+\mathcal{P})]^2 = \frac{1}{2}(1+\mathcal{P})$, which in turn follows from the fact that $\mathcal{P}^2 = 1$. To identify the image of this projection, note that $\mathcal{P}\frac{1}{2}(1+\mathcal{P}) = \frac{1}{2}(1+\mathcal{P})$, which implies that the image is contained in $\overline{\mathcal{B}}(2)$, and note that $\frac{1}{2}(1+\mathcal{P})$ fixes each element of $\overline{\mathcal{B}}(2)$. Together, these facts imply that the image is precisely $\overline{\mathcal{B}}(2)$, and hence $\frac{1}{2}(1+\mathcal{P}) = \mathsf{Proj}_{\overline{\mathcal{B}}(2)}$. To understand the exact form of the bosonic state vectors in $\overline{\mathcal{B}}(2)$, observe that an arbitrary state vector $\psi \in \mathcal{H}\overline{\otimes}\mathcal{H}$ is given as $\psi = \sum_i c_i(\varphi_i \otimes \chi_i)$, for some $\varphi_i, \chi_i \in \mathcal{H}$, and therefore

$$(1+\mathcal{P})\psi = \sum_i c_i(\varphi_i \otimes \chi_i + \chi_i \otimes \varphi_i) \in \overline{\mathcal{B}}(2). \tag{24.1}$$

Define the **symmetric operator** $\mathcal{S}_2$ by

$$\mathcal{S}_2 = \mathsf{Proj}_{\overline{\mathcal{B}}(2)} = \frac{1}{2}(1+\mathcal{P}),$$

and, for $\varphi, \chi \in \mathcal{H}$, the **symmetric product** $\varphi \odot \chi$ by

$$\varphi \odot \chi = \mathcal{S}_2(\varphi \otimes \chi) = \frac{1}{2}(\varphi \otimes \chi + \chi \otimes \varphi).$$

Notice that the symmetric product is, obviously, symmetric, i.e., $\varphi \odot \chi = \chi \odot \varphi$. It follows now from equation 24.1 that the vector

$$\mathcal{S}_2\psi = \sum_i c_i(\varphi_i \odot \chi_i)$$

is a typical bosonic state vector. The bosonic state space $\overline{\mathcal{B}}(2)$ is precisely the completion of the symmetric tensor product $\mathcal{B}(2) = \mathcal{H} \odot \mathcal{H}$, which is by definition the subspace of $\mathcal{H} \otimes \mathcal{H}$ invariant under the action of the exchange operator.

Our goal now is to describe an orthonormal basis for $\overline{\mathcal{B}}(2)$ that arises naturally from that of $\mathcal{H}\overline{\otimes}\mathcal{H}$. Recall from Section 21.4 that the set $\boldsymbol{\xi} \otimes \boldsymbol{\xi} = \{\xi_i \otimes \xi_j\}_{i,j}$ is an orthonormal basis for $\mathcal{H}\overline{\otimes}\mathcal{H}$ that is complete when $\mathcal{H}$ is separable and infinite-dimensional and when $\boldsymbol{\xi} = \{\xi_i\}_i$ is a complete orthonormal basis for the Hilbert space $\mathcal{H}$. Our goal is to show that the set $\boldsymbol{\xi} \odot \boldsymbol{\xi} = \{\xi_i \odot \xi_j\}_{i \le j}$ is a complete orthogonal basis for $\overline{\mathcal{B}}(2)$. Here we have erased the redundancy that $\xi_i \odot \xi_j = \xi_j \odot \xi_i$ in the set $\boldsymbol{\xi} \odot \boldsymbol{\xi}$ by indexing with the condition that $i \le j$. Now, it is not the case that each $\xi_i \odot \xi_j$ is normalized, so a suitable normalization of $\boldsymbol{\xi} \odot \boldsymbol{\xi}$ will then give a complete orthonormal basis. A discussion of the details finishes this subsection.

For any $\varphi, \chi \in \mathcal{H}$, an easy calculation using equation 21.2 shows that

$\|\varphi \otimes \chi + \chi \otimes \varphi\|^2 = 2\|\varphi\|^2\|\chi\|^2 + 2|\langle\varphi|\chi\rangle|^2$, so that

$$\|\varphi \odot \chi\|^2 = \frac{1}{2}\|\varphi\|^2\|\chi\|^2 + \frac{1}{2}|\langle\varphi|\chi\rangle|^2.$$

In particular, when φ and χ are orthogonal, $\|\varphi \odot \chi\| = \frac{1}{\sqrt{2}}\|\chi\|\|\varphi\|$ and, when orthonormal, $\|\varphi \odot \chi\| = \frac{1}{\sqrt{2}}$. Also, $\|\varphi \odot \varphi\| = \|\varphi \otimes \varphi\| = \|\varphi\|^2$, so that $\|\varphi \odot \varphi\| = 1$ whenever φ is normalized. Another use of equation 21.2 shows that $\varphi_1 \odot \chi_1$ and $\varphi_2 \odot \chi_2$ are orthogonal whenever either pair, φ_1, φ_2 or χ_1, χ_2, is orthogonal. All this goes to verify that the set $\{\xi_i \odot \xi_i\}_i \cup \{\sqrt{2}\,(\xi_i \odot \xi_j)\}_{i<j}$ is orthonormal. That it spans $\overline{\mathcal{B}}(2)$ when the dimension of $\mathcal{H}$ is finite is a straightforward exercise and that it is complete when the dimension is infinite follows similarly to the argument that $\boldsymbol{\xi} \otimes \boldsymbol{\xi}$ is complete for $\mathcal{H} \overline{\otimes} \mathcal{H}$ in Section 21.4. Note that if $\mathcal{H}$ is finite-dimensional, say $D = \dim \mathcal{H}$, then $\dim \mathcal{H} \otimes \mathcal{H} = D^2$ while $\dim \mathcal{H} \odot \mathcal{H} = D(D+1)/2$.

24.1.2 The State Space for Two Fermions

The mathematical structure that underlies the fermionic state space $\overline{\mathcal{F}}(2)$ parallels that of the bosonic state space $\overline{\mathcal{B}}(2)$. The space $\overline{\mathcal{F}}(2)$ is the kernel of the operator $1+\mathcal{P}$ as well as the image of the operator $1-\mathcal{P}$. The operator $\frac{1}{2}(1-\mathcal{P})$ is precisely the projection $\mathsf{Proj}_{\overline{\mathcal{F}}(2)} = \mathsf{Proj}_{-1}$ of $\mathcal{H} \overline{\otimes} \mathcal{H}$ to $\overline{\mathcal{F}}(2)$. Define the **antisymmetric operator** $\mathcal{A}_2$, also known as the **alternating map**, by

$$\mathcal{A}_2 = \frac{1}{2}(1-\mathcal{P}) = \mathsf{Proj}_{\overline{\mathcal{F}}(2)}$$

and, for $\varphi, \chi \in \mathcal{H}$, the **antisymmetric product** $\varphi \wedge \chi$, also called the **exterior** or **wedge** product, by

$$\varphi \wedge \chi = \mathcal{A}_2(\varphi \otimes \chi) = \frac{1}{2}(\varphi \otimes \chi - \chi \otimes \varphi).$$

The antisymmetric product is antisymmetric, i.e., $\varphi \wedge \chi = -\chi \wedge \varphi$, which implies, in particular, that $\varphi \wedge \varphi = 0$. A typical state vector in $\overline{\mathcal{F}}(2)$ is of the form

$$\mathcal{A}_2\psi = \sum_i c_i(\varphi_i \wedge \chi_i).$$

The fermionic state space $\overline{\mathcal{F}}(2)$ is the completion of the antisymmetric tensor product $\mathcal{H} \wedge \mathcal{H}$, by definition the subspace of $\mathcal{H} \otimes \mathcal{H}$ invariant under the action of the negative of the exchange operator and also called the exterior, or wedge product of $\mathcal{H}$ with itself.

A complete orthonormal basis for $\overline{\mathcal{F}}(2)$ is given by the set $\sqrt{2}(\boldsymbol{\xi} \wedge \boldsymbol{\xi}) = \{\sqrt{2}(\xi_i \wedge \xi_j)\}_{i<j}$, where $\boldsymbol{\xi} = \{\xi_i\}_i$ is a complete orthonormal basis for $\mathcal{H}$. To see this, for any $\varphi, \chi \in \mathcal{H}$, an easy calculation again using equation 21.2 shows that $\|\varphi \otimes \chi - \chi \otimes \varphi\|^2 = 2\|\varphi\|^2\|\chi\|^2 - 2|\langle\varphi|\chi\rangle|^2$, so that

$$\|\varphi \wedge \chi\|^2 = \frac{1}{2}\|\varphi\|^2\|\chi\|^2 - \frac{1}{2}|\langle\varphi|\chi\rangle|^2.$$

In particular, when φ and χ are orthogonal, $\|\varphi \wedge \chi\| = \frac{1}{\sqrt{2}}\|\chi\|\|\varphi\|$, and when they are orthonormal, $\|\varphi \wedge \chi\| = \frac{1}{\sqrt{2}}$. This shows that $\sqrt{2}(\boldsymbol{\xi} \wedge \boldsymbol{\xi})$ is an orthonormal set. Again, that it spans when the dimension of $\mathcal{H}$ is finite is a straightforward exercise, and that it is complete when the dimension is infinite follows similarly to the argument that $\boldsymbol{\xi} \otimes \boldsymbol{\xi}$ is complete for $\mathcal{H} \mathbin{\overline{\otimes}} \mathcal{H}$ in Section 21.4. When $\mathcal{H}$ is finite-dimensional with $D = \dim \mathcal{H}$, we have $\dim \mathcal{H} \wedge \mathcal{H} = D(D-1)/2$.

24.1.3 Pauli Exclusion Principle

We already get a glimpse into a deep fact of nature in the antisymmetry of fermionic state vectors. Indeed, the fact that $\varphi \wedge \varphi = 0$ suggests that nature forbids the existence of pairs of indistinguishable fermions each in the same quantum state φ. This is a special case of the **Pauli exclusion principle**, which states that no two quanta in a fermionic gas can occupy the same quantum state. This helps to account for the stability of matter. For example, the ground state of a hydrogenic central potential is doubly degenerate with quantum numbers $n = 0$ (ground state energy), $\ell = 0$ (angular momentum zero), and $s = \pm 1$ (spin, in units of $\frac{1}{2}\hbar$). This ground state then offers just two degrees of freedom for a single electron, and this is captured by the two-dimensional, $j = \frac{1}{2}$, spin representation space $V^{1/2}$, spanned by the two basis vectors $|+\rangle_z$ and $|-\rangle_z$ for example. An electron that occupies this ground state is said to be in the **1s** orbital. However, these two degrees of freedom are not available to each of two electrons in this ground state since electrons are fermions with antisymmetric state vectors. Indeed, the state vector of the system must be antisymmetric and must lie in the one-dimensional exterior product $V^{1/2} \wedge V^{1/2}$ rather than in the four-dimensional tensor product $V^{1/2} \otimes V^{1/2}$ that would be available to two distinguishable fermions. A state vector for the system of two electrons in this potential is any nonzero complex multiple of $\psi = |+\rangle_z \wedge |-\rangle_z$, which guarantees only that one of the electrons is in the $+1$ z-component spin state while the other is in the -1 z-component spin state. Of course there is nothing special about the z-component. If $\mathbf{u}$ is any element of the unit sphere $\mathbf{S}^2$, say with spherical coordinates θ, ϕ, then equations 23.9 and 23.10 imply that $|+\rangle_{\mathbf{u}} \wedge |-\rangle_{\mathbf{u}} = e^{\mathrm{i}\phi}\psi$, representing the same quantum state as ψ. We see here that the degrees of freedom available to two indistinguishable fermions are reduced from four to one in this spin $\frac{1}{2}$-system. In lithium, the **1s** orbital is filled with two electrons and there is a third in the **2s** orbital. Why does this third electron not radiate to the ground state? The explanation is that the antisymmetry of the state vector for a system of three fermions implies that any term of the form $|+\rangle \wedge |-\rangle \wedge |\pm\rangle$ is zero. This is immediate from the generalization of the content of this section to N indistinguishable quanta, which will be presented next.

24.2 N Indistinguishable Quanta

For a state vector ψ representing a system of N indistinguishable quanta, the exchange of any two slots in each of the decomposable tensor product summands should either leave ψ unchanged or introduce a phase factor -1; therefore the state vector ψ should be either symmetric or antisymmetric with respect to such exchanges depending on whether the system is a bosonic or fermionic aggregate. In this section we develop the general versions of the bosonic and fermionic state spaces, but our emphasis is on the incomplete algebraic subsets of these spaces that in their totality make up familiar objects known to mathematicians as the symmetric and exterior algebras. These are graded algebras built up from the finite-rank symmetric and exterior tensor products, respectively, products whose completions form the bosonic and fermionic state spaces for fixed numbers of quanta.

24.2.1 The Bosonic State Algebra

The **N-quanta bosonic state space** $\overline{\mathcal{B}}(N)$ is the completion of the **rank-N symmetric tensor product**

$$\mathcal{B}(N) = \odot^N \mathcal{H} = \mathcal{H} \odot \cdots \odot \mathcal{H} \quad (N \text{ factors}).$$

This is the linear subspace of the N-fold tensor product $\otimes^N \mathcal{H}$ that is invariant under the symmetric action of the symmetric group $\mathfrak{S}_N$. This action is defined on decomposable elements for a permutation $\pi \in \mathfrak{S}_N$

by

$$\pi \cdot (\varphi_1 \otimes \cdots \otimes \varphi_N) = \varphi_{\pi(1)} \otimes \cdots \otimes \varphi_{\pi(N)},$$

and then extended linearly to the whole of $\otimes^N \mathcal{H}$.[4] Define the **symmetric operator** $\mathcal{S}_N$ by

$$\mathcal{S}_N = \frac{1}{N!} \sum_{\pi \in \mathfrak{S}_N} \pi.$$

and, for $\varphi_i \in \mathcal{H}$, the **symmetric product** $\varphi_1 \odot \cdots \odot \varphi_N$ by

$$\begin{aligned} \varphi_1 \odot \cdots \odot \varphi_N &= \mathcal{S}_N(\varphi_1 \otimes \cdots \otimes \varphi_N) \\ &= \frac{1}{N!} \sum_{\pi \in \mathfrak{S}_N} \varphi_{\pi(1)} \otimes \cdots \otimes \varphi_{\pi(N)}. \end{aligned} \tag{24.2}$$

Notice that, for each $\pi \in \mathfrak{S}_N$,

$$\pi \cdot (\varphi_1 \odot \cdots \odot \varphi_N) = \varphi_1 \odot \cdots \odot \varphi_N,$$

which is an expression of symmetry, i.e., of the fact that the order of the factors in the symmetric product is immaterial. From this we see that the symmetric operator maps arbitrary vectors in $\otimes^N \mathcal{H}$ into $\mathcal{B}(N)$ and fixes each element of $\mathcal{B}(N)$. The symmetry operator is a bounded, symmetric, idempotent operator, precisely the projection

$$\mathcal{S}_N = \mathsf{Proj}_{\mathcal{B}(N)}.$$

Applying the symmetry operator to an arbitrary vector ψ in $\otimes^N \mathcal{H}$ gives a bosonic state vector for the N-quanta system as

$$\mathcal{S}_N \psi = \sum_{i=1}^{K} c_i (\varphi_1^i \odot \cdots \odot \varphi_N^i).$$

It is convenient to gather all the finite-rank symmetric products into a single object by taking an orthogonal direct sum[5]

$$\mathcal{B} = \bigoplus_{N=0}^{\infty} \mathcal{B}(N) = \bigoplus_{N=0}^{\infty} (\odot^N \mathcal{H}) = \bigodot \mathcal{H},$$

[4] This is a right action, meaning that, for any $\pi, \varpi \in \mathfrak{S}_N$ and $\psi \in \otimes^N \mathcal{H}$, we have $(\pi\varpi) \cdot \psi = \varpi \cdot (\pi \cdot \psi)$. Right actions are normally written on the right, with $\pi \cdot \psi$ replaced by $\psi \cdot \pi$, which has the notational advantage that the group action equation becomes an instance of associativity, in that $\psi \cdot (\pi\varpi) = (\psi \cdot \pi) \cdot \varpi$, and therefore the expression $\psi\pi\varpi$ is unambiguous. The reader may check this by working out that

$$\varpi \cdot (\pi \cdot (\varphi_1 \otimes \cdots \otimes \varphi_N)) = \varphi_{\pi\varpi(1)} \otimes \cdots \otimes \varphi_{\pi\varpi(N)},$$

a somewhat subtle observation.

[5] See p. 334 for the definition.

which is a linear subspace of the tensor algebra $\otimes\mathcal{H}$ of Section 21.4. Here, $\mathcal{B}(0)$ is defined to be a one-dimensional complex vector space, or just a copy of $\mathbb{C}$, and $\mathcal{B}(1) = \mathcal{H}$. The direct sum $\mathcal{B}$ is not a subalgebra of the tensor algebra since $\psi \otimes \psi'$ is not necessarily symmetric and therefore may not lie in $\mathcal{B}$ whenever ψ and ψ' lie in $\mathcal{B}$. To remedy this, $\mathcal{B}$ is made into an algebra by defining the **symmetric product** of $\psi \in \mathcal{B}(M)$ and $\psi' \in \mathcal{B}(N)$ as

$$\psi \odot \psi' = \mathcal{S}_{M+N}(\psi \otimes \psi') \tag{24.3}$$

and extending bilinearly. This is consistent with the symmetric product already defined in equation 24.2 since, for $\psi = \varphi_1 \odot \cdots \odot \varphi_M$ and $\psi' = \chi_1 \odot \cdots \odot \chi_N$, a straightforward calculation verifies that

$$\psi \odot \psi' = \varphi_1 \odot \cdots \odot \varphi_M \odot \chi_1 \odot \cdots \odot \chi_N.$$

The **bosonic state algebra** $\mathcal{B}$ is then identified as the **symmetric algebra** $\odot\mathcal{H}$ generated by $\mathcal{H} = \mathcal{B}(1)$ and is an associative graded complex algebra with unit $1 \in \mathbb{C} = \mathcal{B}(0)$. Unlike the tensor algebra of $\mathcal{H}$, the symmetric algebra is commutative. In fact, when $\mathcal{H}$ is finite-dimensional with, say, $\dim\mathcal{H} = D$, the symmetric algebra $\odot\mathcal{H}$ is isomorphic to the complex polynomial ring $\mathbb{C}[z_1, \ldots, z_D]$ in the variables z_i via an isomorphism generated by the mapping $\xi_i \mapsto z_i$, where $\{\xi_1, \ldots, \xi_D\}$ is a basis for $\mathcal{H}$.

In terms of the complete orthonormal basis $\boldsymbol{\xi}$ of $\mathcal{H}$, the set

$$\odot^N \boldsymbol{\xi} = \{\xi_{i_1} \odot \cdots \odot \xi_{i_N} : i_1 \leq i_2 \leq \cdots \leq i_N\}$$

is orthogonal and complete in the bosonic state space $\overline{\mathcal{B}}(N)$ and, once normalized, serves as a complete orthonormal basis. From this, when $\dim\mathcal{H} = D$, we see that

$$\dim \odot^N \mathcal{H} = \binom{D-1+N}{N}.$$

A proof of this is given in Section 26.1.2.

24.2.2 The Fermionic State Algebra

The development of the fermionic state algebra and spaces parallels that of the bosonic. The **N-quanta fermionic state space** is the completion of the **rank-N antisymmetric tensor product**

$$\mathcal{F}(N) = \wedge^N \mathcal{H} = \mathcal{H} \wedge \cdots \wedge \mathcal{H} \quad (N \text{ factors}),$$

also known as the N-fold **exterior** or **wedge product**. This is the linear subspace of the N-fold tensor product $\otimes^N \mathcal{H}$ that is invariant under the antisymmetric action of the symmetric group $\mathfrak{S}_N$, which is defined on decomposable elements for a permutation $\pi \in \mathfrak{S}_N$ by

$$\pi * (\varphi_1 \otimes \cdots \otimes \varphi_N) = (\operatorname{sgn} \pi)\, \varphi_{\pi(1)} \otimes \cdots \otimes \varphi_{\pi(N)},$$

then extended linearly to the whole of $\otimes^N \mathcal{H}$. Here sgn, the sign function, is defined by $\operatorname{sgn} \pi = \pm 1$ depending on whether π is an even or an odd permutation. Define the **antisymmetric operator** $\mathcal{A}_N$, also called the **alternating mapping**, by

$$\mathcal{A}_N = \frac{1}{N!} \sum_{\pi \in \mathfrak{S}_N} \pi * = \frac{1}{N!} \sum_{\pi \in \mathfrak{S}_N} (\operatorname{sgn} \pi)\, \pi \cdot,$$

and, for $\varphi_i \in \mathcal{H}$, the **antisymmetric**, or **exterior**, or **wedge product** $\varphi_1 \wedge \cdots \wedge \varphi_N$ by

$$\begin{aligned} \varphi_1 \wedge \cdots \wedge \varphi_N &= \mathcal{A}_N(\varphi_1 \otimes \cdots \otimes \varphi_N) \\ &= \frac{1}{N!} \sum_{\pi \in \mathfrak{S}_N} (\operatorname{sgn} \pi)\, \varphi_{\pi(1)} \otimes \cdots \otimes \varphi_{\pi(N)}. \end{aligned} \tag{24.4}$$

That $\operatorname{sgn} \pi\varpi = \operatorname{sgn} \pi \operatorname{sgn} \varpi$ for two permutations π and ϖ implies that, for each $\pi \in \mathfrak{S}_N$,

$$\pi \cdot (\varphi_1 \wedge \cdots \wedge \varphi_N) = (\operatorname{sgn} \pi)\, \varphi_1 \wedge \cdots \wedge \varphi_N,$$

and, equivalently, that

$$\pi * (\varphi_1 \wedge \cdots \wedge \varphi_N) = \varphi_1 \wedge \cdots \wedge \varphi_N.$$

This is an expression of antisymmetry – of the fact that the transposition of any two factors in the antisymmetric product introduces a phase factor of -1. From this we see that the antisymmetric operator maps arbitrary vectors in $\otimes^N \mathcal{H}$ into the fermionic state space $\mathcal{F}(N)$ and fixes each element of $\mathcal{F}(N)$. The antisymmetry operator is a bounded, symmetric, idempotent operator, precisely the projection

$$\mathcal{A}_N = \mathsf{Proj}_{\mathcal{F}(N)}.$$

Applying the antisymmetry operator to an arbitrary vector ψ in $\otimes^N \mathcal{H}$ gives a fermionic state vector for the N-quanta system as

$$\mathcal{A}_N \psi = \sum_{i=1}^{K} c_i (\varphi_1^i \wedge \cdots \wedge \varphi_N^i).$$

It is convenient to gather all the fermionic state spaces into a single object by taking an orthogonal direct sum

$$\mathcal{F} = \bigoplus_{N=0}^{\infty} \mathcal{F}(N) = \bigoplus_{N=0}^{\infty} (\wedge^N \mathcal{H}) = \bigwedge \mathcal{H},$$

which is a linear subspace of the tensor algebra $\otimes\mathcal{H}$ of Section 21.4. Here, $\mathcal{F}(0)$ is defined to be a one-dimensional complex vector space, or just a copy of $\mathbb{C}$, and $\mathcal{F}(1) = \mathcal{H}$. The direct sum $\mathcal{F}$ is not a subalgebra of the tensor algebra since $\psi \otimes \psi'$ is not necessarily antisymmetric and therefore might not lie in $\mathcal{F}$ whenever ψ and ψ' lie in $\mathcal{F}$. To remedy this, $\mathcal{F}$ is made into an algebra by defining the **antisymmetric product** of $\psi \in \mathcal{F}(M)$ and $\psi' \in \mathcal{F}(N)$ by

$$\psi \wedge \psi' = \mathcal{A}_{M+N}(\psi \otimes \psi') \tag{24.5}$$

and extending bilinearly. This is consistent with the antisymmetric product already defined in equation 24.4, since for $\psi = \varphi_1 \wedge \cdots \wedge \varphi_M$ and $\psi' = \chi_1 \wedge \cdots \wedge \chi_N$, a straightforward calculation verifies that

$$\psi \wedge \psi' = \varphi_1 \wedge \cdots \wedge \varphi_M \wedge \chi_1 \wedge \cdots \wedge \chi_N.$$

The **fermionic state algebra** $\mathcal{F}$ is then identified as the **antisymmetric**, or **exterior algebra** $\wedge\mathcal{H}$ generated by $\mathcal{H} = \mathcal{F}(1)$, and is an associative graded complex algebra with unit $1 \in \mathbb{C} = \mathcal{F}(0)$.

Like the tensor algebra of $\mathcal{H}$, the antisymmetric algebra is noncommutative, but its noncommutativity is somewhat tame with respect to that of the tensor algebra. This is most strikingly apparent in the **skew symmetry** of the product, seen in the expression

$$\psi \wedge \psi' = (-1)^{MN} \psi' \wedge \psi,$$

whenever $\psi \in \wedge^M\mathcal{H}$ and $\psi' \in \wedge^N\mathcal{H}$, which follows by a straightforward calculation.

In terms of the complete orthonormal basis $\boldsymbol{\xi}$ of $\mathcal{H}$, the set

$$\wedge^N \boldsymbol{\xi} = \{\xi_{i_1} \wedge \cdots \wedge \xi_{i_N} : i_1 < i_2 < \cdots < i_N\}$$

is orthogonal and complete in $\overline{\mathcal{F}}(N)$ and, once normalized, serves as a complete orthonormal basis. From this, if $\dim \mathcal{H} = D$, we see that

$$\dim \wedge^N \mathcal{H} = \binom{D}{N}.$$

Notice that $\wedge^N\mathcal{H} = 0$ whenever $N > D$, this being another instance of the Pauli exclusion principle and implying that $D + 1$ indistinguishable fermions cannot occupy a system with only D mutually orthogonal states available. For example, three electrons cannot occupy the ground state of helium since the state space $\mathcal{H} = V^{1/2}$ for a single electron in this ground state is two-dimensional.

Notice that when $\dim \mathcal{H} = D$, the exterior algebra is finite-dimensional of dimension

$$\dim \bigwedge \mathcal{H} = \sum_{N=0}^{D} \binom{D}{N} = 2^D.$$

This contrasts with the symmetric algebra, which is always infinite-dimensional.

24.2.3 Pauli Exclusion Again

We now present an example that illustrates the restraints placed on systems as one goes from distinguishable to indistinguishable quanta and from bosons to fermions. Consider the case where the state space for one quantum of a certain species is the six-dimensional Hilbert space $\mathcal{H}$. For example, the next-to-ground state of a hydrogenic potential well with nonzero orbital angular momentum is triply degenerate in its orbital angular momentum and doubly degenerate in its spin. The quantum numbers are $n = 1$ (energy level), $\ell = 1$ (total orbital angular momentum), $m = 0, \pm 1$ (magnetic quantum number), and $s = \pm 1$ (spin), describing the so-called **2p** orbitals of the second electron shell of a hydrogenic potential. The state space that describes this system is the six-dimensional tensor product space $V^1 \otimes V^{1/2}$, where V^1 is a three-dimensional, $j = 1$, spin representation space for the orbital angular momentum and $V^{1/2}$ is the two-dimensional, $j = \frac{1}{2}$, spin representation space for the intrinsic electron spin.

Were six distinguishable quanta to occupy a system in which each of the quanta had six available degrees of freedom represented by separate six-dimensional state spaces, the number of degrees of freedom available would be $6^6 = 46,656$, the dimension of a six-fold tensor product of six-dimensional spaces. Our interest, though, is in indistinguishable quanta. For six bosons, the state space is the six-fold symmetric product $\odot^6 \mathcal{H}$ of dimension $\binom{6-1+6}{6} = 462$, with 462 mutually orthogonal states available. For fermions, though, like the six electrons that fill the **2p** orbitals of the element neon, the total number of states available is precisely $\dim \wedge^6 \mathcal{H} = \binom{6}{6} = 1$. This reduction in degrees of freedom from 46,656 to 462 to 1 for only six quanta is quite striking and suggests that the statistical mechanics of large aggregates of quanta will have quantitative characteristics with different qualitative implications for their gross behavior depending on the nature of the quanta – whether they are distinguishable or indistinguishable, and whether they are bosons or fermions.

24.3 Algebraic Structure of the Tensor Algebra

In this section we delve a little deeper into the algebraic structure of bosonic and fermionic state spaces and algebras. First we offer an alternate description of the N-fold symmetric and antisymmetric products. In the N-fold tensor product $\otimes^N \mathcal{H}$, where $N \geq 2$, let $\mathfrak{s}_N$ be the vector subspace generated by all elements of the form

$$\varphi_1 \otimes \cdots \otimes \varphi_N - \pi \cdot (\varphi_1 \otimes \cdots \otimes \varphi_N),$$

as π ranges over the symmetric group $\mathfrak{S}_N$, and let $\mathfrak{a}_N$ be the vector subspace generated by all elements of the form

$$\varphi_1 \otimes \cdots \otimes \varphi_N,$$

where $\varphi_i = \varphi_j$ for some pair of indices $i \neq j$. An exercise verifies that $\mathfrak{s}_N$ is the kernel of the symmetry operator $\mathcal{S}_N$ and that $\mathfrak{a}_N$ is the kernel of the antisymmetry operator $\mathcal{A}_N$. It follows that the symmetric product $\odot^N \mathcal{H}$ may be described as the quotient vector space $\otimes^N \mathcal{H}/\mathfrak{s}_N$ and that the antisymmetric product $\wedge^N \mathcal{H}$ may be described as the quotient vector space $\otimes^N \mathcal{H}/\mathfrak{a}_N$.

Let $\mathfrak{s}$ be the ideal in the tensor algebra $\otimes \mathcal{H}$ generated by all elements of the form $\varphi \otimes \chi - \chi \otimes \varphi$ as φ and χ range over all of $\mathcal{H}$. Then it is not hard to see that $\mathfrak{s}$ is precisely the graded ideal

$$\mathfrak{s} = \bigoplus_{N=2}^{\infty} \mathfrak{s}_N.$$

Now, since $\mathfrak{s}$ is not only a vector subspace of the graded algebra $\otimes \mathcal{H}$ but also a graded ideal, the quotient $\otimes \mathcal{H}/\mathfrak{s}$ is a graded algebra with product given by $(\psi + \mathfrak{s}) \cdot (\psi' + \mathfrak{s}) = \psi \otimes \psi' + \mathfrak{s}$. To see that $\otimes \mathcal{H}/\mathfrak{s}$ is isomorphic as a graded algebra to the symmetric algebra $\odot \mathcal{H}$, observe that there is a global symmetric linear mapping $\mathcal{S} : \otimes \mathcal{H} \to \odot \mathcal{H}$ uniquely defined by its restrictions $\mathcal{S}_N$ to $\otimes^N \mathcal{H}$. Obviously, $\mathcal{S}$ is a graded linear map that maps the Nth tensor power $\otimes^N \mathcal{H}$ onto the Nth symmetric power $\odot^N \mathcal{H}$.

From definition 24.3, a calculation shows that, for any $\psi, \psi' \in \otimes\mathcal{H}$,

$$\mathcal{S}(\psi \otimes \psi') = \mathcal{S}(\mathcal{S}\psi \otimes \mathcal{S}\psi') = \mathcal{S}\psi \odot \mathcal{S}\psi'.$$

This shows that the symmetric mapping is not only a linear mapping but also a surjective algebra homomorphism. As the kernel of the graded algebra homomorphism $\mathcal{S}$ is precisely the ideal $\mathfrak{s}$, this demonstrates that $\otimes\mathcal{H}/\mathfrak{s} \cong \odot\mathcal{H}$ via the graded algebra isomorphism $\psi + \mathfrak{s} \mapsto \mathcal{S}\psi$.

A similar development for the antisymmetric algebra, with $\mathfrak{a}$ the ideal generated by the set of all vectors $\psi \otimes \psi$ for $\psi \in \mathcal{H}$, shows that the antisymmetric operator $\mathcal{A} : \otimes\mathcal{H} \to \wedge\mathcal{H}$ is a surjective graded-algebra homomorphism with kernel $\mathfrak{a}$, and hence $\otimes\mathcal{H}/\mathfrak{a} \cong \wedge\mathcal{H}$ via the graded-algebra isomorphism $\psi + \mathfrak{a} \mapsto \mathcal{A}\psi$.

24.4 Analytic Structure of the Tensor Algebra

One of the subtleties that we have left unaddressed in our development of state spaces for indistinguishable quanta occurs when we take a countably infinite direct sum, as in the definitions of the tensor algebra $\otimes\mathcal{H}$, the symmetric bosonic algebra $\mathcal{B}$, and the antisymmetric fermionic algebra $\mathcal{F}$. The subtleties reduce to a mismatch between algebra and analysis. We have taken the direct sum in the algebraic category of complex inner spaces, where the natural construction of a countably infinite orthogonal direct sum does not produce a Hilbert space since it fails to be complete but produces what is normally termed a **pre-Hilbert space**, an incomplete inner product space. Its completion to a Hilbert space remedies this analytically but at the cost of losing the algebraic structure, as the tensor product of two elements of the completion might not be defined. We will sort this out next, and will define the Fock space and its bosonic and fermionic subspaces. In the next lecture we consider in detail the structure of these spaces.

For each nonnegative integer N, let $\mathcal{H}_N$ be a Hilbert space. It will cause no confusion to use a common notation for the inner products and norms in these spaces, the context differentiating in which space the product or norm is applied. We define the **complete orthogonal sum** $\mathcal{H}$ of these spaces as

$$\mathcal{H} = \left\{ (\psi_N) = (\psi_N)_{N=0}^{\infty} : \psi_N \in \mathcal{H}_N \text{ and } \sum_{N=0}^{\infty} \|\psi_N\|^2 < \infty \right\},$$

the set of square-summable infinite sequences (ψ_N) of the elements of the spaces $\mathcal{H}_N$. The inner product on $\mathcal{H}$ is defined by the convergent sum

$$\left\langle \sum_{N=0}^{\infty} \psi_N \,\middle|\, \sum_{N=0}^{\infty} \psi'_N \right\rangle = \sum_{N=0}^{\infty} \langle \psi_N | \, \psi'_N \rangle,$$

which makes the canonical isometrically embedded copies of $\mathcal{H}_M$ and $\mathcal{H}_N$ orthogonal whenever $M \neq N$. The complete orthogonal sum is a Hilbert space, complete in its norm and separable when each $\mathcal{H}_N$ is separable. The vector subspace of sequences that are eventually zero,

$$\mathcal{H}_f = \{(\psi_N) \in \mathcal{H} : \exists K \text{ such that } \psi_N = 0 \text{ for all } N > K\}$$

inherits an inner product from $\mathcal{H}$ and is easily identified algebraically as the direct sum $\mathcal{H}_f = \oplus_{N=0}^{\infty} \mathcal{H}_N$. We call the inner product space $\mathcal{H}_f$ the **orthogonal direct sum**[6] of the $\mathcal{H}_N$ and observe that it is incomplete, and therefore a pre-Hilbert space, when infinitely many of the $\mathcal{H}_N$ are nonzero. Manifestly, the orthogonal direct sum $\mathcal{H}_f$ is a dense subspace of $\mathcal{H}$, which identifies $\mathcal{H}$ as the metric completion of $\mathcal{H}_f$ in its norm. We use the symbol $\overline{\oplus}$ to denote the complete orthogonal sum of Hilbert spaces and reserve the unadorned symbol $\oplus$ for the orthogonal direct sum, and we use an overbar on a pre-Hilbert space to denote its completion. Thus we may write

$$\overline{\bigoplus_{N=0}^{\infty} \mathcal{H}_N} = \overline{\mathcal{H}_f} = \mathcal{H} = \overline{\bigoplus}_{N=0}^{\infty} \mathcal{H}_N.$$

For the categorically minded reader we should warn that though the orthogonal direct sum is a direct sum, or a coproduct, in the category of complex inner product spaces, the complete orthogonal sum is not a coproduct in the category of Hilbert spaces. In fact, the category of Hilbert spaces does not admit coproducts, for infinitely many nontrivial Hilbert spaces.

As a quick example, when, for each N, we have $\mathcal{H}_N = \mathbb{C}$ with inner product $\langle u|v\rangle = \overline{u}v$, the complete orthogonal sum is the Hilbert space $\ell^2(\mathbb{C})$, the space of square summable complex sequences, and the orthogonal direct sum is the pre-Hilbert space $\ell^2_f(\mathbb{C})$ of those sequences that are eventually zero, which may be identified canonically as the increasing union of finite-dimensional complex Euclidean spaces $\cup_{N=1}^{\infty} \mathbb{C}^N$.

[6] This was introduced on p. 334 for the case when all the $\mathcal{H}_N$ coincide. There we used the *internal* direct sum notation of finite sums whereas here we use the *external* direct sum notation of sequences with finitely many nonzero terms.

24.4.1 Fock Spaces for Indistinguishable Quanta

Let $\mathcal{H}$ be a Hilbert space. The **Fock space** over $\mathcal{H}$ is the complete orthogonal sum

$$\mathbf{F}(\mathcal{H}) = \overline{\bigotimes \mathcal{H}} = \bigoplus_{N=0}^{\infty} \left(\overline{\otimes}^N \mathcal{H}\right).$$

Of particular interest, and the subject of the next lecture, are its two Hilbert subspaces, the **symmetric** or **bosonic Fock space** over $\mathcal{H}$,

$$\mathbf{F}_{\odot}(\mathcal{H}) = \overline{\mathcal{B}} = \overline{\bigodot \mathcal{H}} = \bigoplus_{N=0}^{\infty} \left(\overline{\odot}^N \mathcal{H}\right),$$

and the **antisymmetric** or **fermionic Fock space** over $\mathcal{H}$,

$$\mathbf{F}_{\wedge}(\mathcal{H}) = \overline{\mathcal{F}} = \overline{\bigwedge \mathcal{H}} = \bigoplus_{N=0}^{\infty} \left(\overline{\wedge}^N \mathcal{H}\right),$$

where $\overline{\odot}^N \mathcal{H} = \overline{\mathcal{B}}(N)$ and $\overline{\wedge}^N \mathcal{H} = \overline{\mathcal{F}}(N)$ are the N-quanta state spaces. Of course, the bosonic state algebra $\mathcal{B}$ is a dense linear subspace of the bosonic Fock space while the fermionic state algebra $\mathcal{F}$ is a dense linear subspace of the fermionic Fock space. The bosonic and fermionic Fock spaces will serve as the state spaces of aggregates of indistinguishable quanta. The symmetric and wedge products on these algebras do not extend to their completions, but these products are available for the state vectors in the Fock spaces that represent aggregates of finitely many quanta.

24.5 End Notes

The formalism of this heavily algebraic lecture will not find extensive use in the remainder of this book. I have included it in such algebraic detail in order to show how the abstract algebra seen in graduate-level mathematics classes builds the playing fields of the quantum world. I think it is pleasing to see how the mathematical dichotomy of the symmetric and antisymmetric products on the tensor algebra reflects the physical dichotomy between bosons and fermions in the natural world. Nonetheless, more common in the physics literature than the use of the symmetric and antisymmetric products is the use of the Fock space formalism, to be presented in the lecture that follows; this formalism dispenses with the tensor product notation used so heavily here and reworks the material of Section 24.4.1 using physically motivated notation that extends the Dirac calculus to multi-particle states.

Pauli exclusion is of course the primary physical lesson of this lecture. Pauli arrived at his principle in an attempt to understand the results of experimentalists in the decades preceding the articulation of the principle. His explanation relied on the idea that the electron has a two-valuedness that gives the simplest explanation of the Zeeman effect, the splitting of the spectral lines of an element in the presence of a magnetic field. In this he built upon the ideas of Sommerfeld, Stoner, Kossel, and Bohr, who by 1924 had set out a scheme for understanding the periodic table in terms of electron shells. Pauli's principle was an inspired explanation of the purely physical evidence and did not rely on an understanding of the mathematics behind the rule. As the quantum formalism developed and spin became a key feature of the quantum explanation, the mathematics of antisymmetry offered mathematical window dressing for the physical property. A detailed account of this history may be found in Tomonaga (1997).

I recommend Chapter 5, entitled "Identical particles," of Griffiths (1995) for a highly readable account of two-particle systems in quantum mechanics. The level is undergraduate and uses the wave mechanical approach instead of tensor products, but it has a good elementary discussion of atomic helium, the shell structure of atoms, the periodic table, and solids, and how the Pauli exclusion principle is involved in these issues. It ends with a solid discussion of quantum statistical mechanics, the topic of our Lecture 26.

25

The Fock Space for Indistinguishable Quanta

> I have promised to show how Fock space realizes the idea of quanta, understood as entities that can be (merely) aggregated, as opposed to particles, which can be labeled, counted, and thought of as switched.
>
> ...
>
> Our use of particle labels ... to describe multiquanta systems is a historical accident. A Fock space description, free of the surplus structure of labeled tensor product Hilbert space formalism, provides the most parsimonious description of multiquanta states and so the description that arguably gives us the best picture of the nature of quanta.
>
> Paul Teller
> *An Interpretive Introduction to Quantum Field Theory*, 1997

One drawback in the development of the previous lecture is that, while it allows for a finite, and perhaps even an indefinite, number of indistinguishable quanta, it does not incorporate into its formalism in an elegant manner the possibility of a variable number of quanta, in cases where quanta may be created or annihilated in the dynamical development of the system. The proper realm for the creation and annihilation of quanta is that of relativistic quantum field theory, but the machinery describing the state space for indistinguishable quanta, as well as the acts of the creation and annihilation of quanta, fits nicely into the present discussion. We will present a clever reworking of the symmetric and antisymmetric state spaces for bosons and fermions that introduces a mathematically nonstandard, but physically motivated, notation for the bosonic and fermionic states that bears the name of its developer – **Fock space notation**. This uses Dirac ket notation for states, which easily and transparently handles quanta creation and annihilation. Fock space, introduced in skeletal form at the end of the previous lecture and clothed in flesh in this one, can represent in a natural way the evolution of a quantum system from one with N quanta to one with N' quanta, where $N \neq N'$. The formalism of Fock space developed here applies to systems with a countable number of degrees of freedom and is on a sound mathematical footing. It is one of the jumping off points for quantum field theory with its uncountably infinite number of degrees of freedom. Much of the formalism developed here works formally in the field-theoretic setting by mimicking our rigorous development, even though the formalism itself, in the setting of uncountably many degrees of freedom, fails the test of mathematical rigor.

A good case can be made on physical grounds that the Fock space formalism is the most appropriate for describing bosons and fermions, as these quantum objects are, in principle, indistinguishable. Unlike the symmetric and exterior product notation of the preceding lecture, this formalism makes no suggestion that quanta can in any way be labeled and then mentally switched and exchanged multiple times to create formal symmetric or antisymmetric expressions in the appropriate algebra, so that the labels formally disappear. In the previous lecture, we might have the feeling that the indistinguishability of the quanta is a pretense and that, had we more incisive observational powers, we could indeed follow individual quanta as they interact with their kind. But the Fock notation from the outset offers no possibility that quanta can in any way be labeled, and so conforms to the ontology of the quantum world that we see physically in experiment and observation. Yes, the abstract state space of Fock space is described in the previous lecture in terms of tensor products with their distinguishable factors made indistinguishable by (anti)symmetrizing operations, but this, as Teller (1997) says, is more a historical accident than a necessity and we will see that the bosonic and fermionic Fock spaces may be described without any reference to tensor products.

We present this now, not so much because we will make immediate use of these Fock spaces but because it completes our introduction of the various explicit descriptions of Hilbert spaces that will serve as the state spaces in our development of quantum mechanics. In the two lectures following, we continue our discussion of aggregates of quanta by examining briefly the statistical mechanics of ensembles; then in subsequent lectures we increasingly turn our attention to quantum dynamics and the time evolution of state vectors.

25.1 Notation for Fock Space

One should never underestimate the power of well-crafted notation in taming complexity and subtlety in mathematics. The Fock notation for elements of the bosonic and fermionic state algebras is a well-crafted application of Dirac bra-ket notation in the context of aggregates of indistinguishable bosons and fermions. The notation is dependent on the choice of a complete orthonormal basis $\boldsymbol{\xi}$ for the state space $\mathcal{H}$ of a single quantum of the species of interest. This basis is usually determined by a maximal collection of observables represented by pairwise commuting operators whose simultaneous one-dimensional eigenspaces decompose $\mathcal{H}$ into a complete orthogonal sum. For example, if the system of interest consists of electrons in a hydrogenic potential then the four observables, energy, total angular momentum, z-component of angular momentum, and spin, represented respectively by the Hamiltonian H, the angular momentum operators L^2 and L_z, and the spin operator S_z, provide such a maximal collection. The idea is that the basis elements determined by this maximal set of observables provide a complete list of the possible states of a single quantum immediately after a complete set of measurements from the whole list of observables. The state, then, of a system of N indistinguishable quanta immediately after a complete set of measurements can indicate only that there are n_i quanta in the single-quantum state represented by $\xi_i \in \boldsymbol{\xi}$, where $\sum_i n_i = N$.

To describe Fock space notation for indistinguishable quanta, let $\boldsymbol{\xi} = \{\xi_i : i \in \mathbb{N}\}$ be a complete orthonormal basis for the separable Hilbert space $\mathcal{H}$. Given a sequence $n_1, n_2, \dots$ of nonnegative integers with all but finitely many terms equal to zero, let $|n_1, n_2, \dots\rangle$ represent the state of the system in which n_i quanta are in the state represented by ξ_i. Shortly, we will see how to use this **Fock space notation** to manipulate basic state vectors in the bosonic and fermionic algebras, but first we offer some refinements and comments on the notation. The expression $|n_1, n_2, \dots\rangle$ will eventually designate a member of a complete orthonormal basis of either the bosonic or fermionic Fock space, and the expression $N = \sum_i n_i$ will give the total number of quanta represented. Often we will write $|n_1, n_2, \dots, n_K\rangle$ whenever $n_k = 0$ for $k > K$. Notice that the expression $|0, \dots, 0, 1\rangle$, where there are $K - 1$ zeroes, represents the system with one quantum in the state represented by the basis vector ξ_K of $\mathcal{H}$, so that $\boldsymbol{\xi} = \{|1\rangle, |0, 1\rangle \dots, |0, \dots, 0, 1\rangle, \dots\}$. As these expressions will represent basis vectors in a Hilbert space, we will be able to add them, but perhaps at this time a quick caution is in order. The sum $|1\rangle + |0,1\rangle$, for example, represents a superposition of two states for one quantum in which this single quantum has an equal chance of being found in either the state represented by ξ_1 or that represented by ξ_2, upon appropriate measurements. This is different from the expression $|1,1\rangle$, which represents a system of two quanta in which one quantum is in the ξ_1-state while the other is in the ξ_2-state, though which is in which state cannot be determined.

The expression $\mathbf{1} = |\,\rangle = |0\rangle = |0,0,\dots\rangle$ represents the state in which there are no quanta present, called the **vacuum state** and represented by this **vacuum state vector**. The use of the numeral **1** is explained by the fact that the vacuum state vector will be the unit in both the bosonic and fermionic state algebras and will serve as a basis of the rank-zero tensors in $\mathcal{B}(0) = \mathbb{C}\mathbf{1} = \mathcal{F}(0)$.

Instead of going straight to a discussion of the use of this notation for elements of $\mathcal{B}$ and $\mathcal{F}$, we first discuss various operators that may be defined on these expressions and use this to help in refining our discussion of the bosonic and fermionic Fock spaces. For the time being, without identifying further restrictions that may be enforced on this notation, we will assume that we are working in an unnamed Hilbert space that includes vectors labeled with our new notation that represent states of indistinguishable quanta, as noted in the definition of the notation. Moreover, the vectors $\{|n_1, n_2, \dots\rangle\}$, as the n_i range over appropriately chosen values of the nonnegative integers, are assumed to form a complete orthonormal basis for this Hilbert space and, as such, any vector ψ may be expanded uniquely in an appropriate convergent sum of the form

$$\psi = \sum c_{|n_1,n_2,\dots\rangle} |n_1, n_2, \dots\rangle,$$

with $c_{|n_1,n_2,\dots\rangle} = \langle n_1, n_2, \dots |\psi\rangle$, where, of course, $\langle n_1, n_2, \dots |$ is the bra dual to the ket $|n_1, n_2, \dots\rangle$.

It will be convenient to use a compact notation for sequences that names a sequence by a bold-face version of the symbol used for the sequence terms. Thus the sequence $(n_i)_{i=1}^{\infty}$ will be denoted simply as $\mathbf{n}$. This, of course, has the advantage of giving more compact formulæ, though we will still sometimes use the more usual $n_1, n_2, \dots$ for emphasis or necessity.

25.2 Annihilation and Creation Operators

The **creation operator** a_i^+ represents the "creation" of a quantum in a state represented by the ith basis vector ξ_i. Explicitly, $\mathsf{a}_i^+|0\rangle = |0,\ldots,0,1\rangle = \xi_i$ and, generally,

$$\mathsf{a}_i^+|\mathbf{n}\rangle = \mathsf{a}_i^+|n_1,n_2,\ldots\rangle = \lambda_i(|\mathbf{n}\rangle)|\ldots,n_i+1,\ldots\rangle,$$

where the proportionality constant $\lambda_i(|\mathbf{n}\rangle)$ may depend on both i and the state $|\mathbf{n}\rangle = |n_1,n_2,\ldots\rangle$. This constant $\lambda_i(|\mathbf{n}\rangle)$ will be chosen later; here we simply mention that it may take the value zero. The **annihilation operator** a_i^- represents the "destruction" of a quantum in a state represented by ξ_i. Explicitly, $\mathsf{a}_i^-\xi_i = \mathsf{a}_i^-|0,\ldots,0,1\rangle = |0\rangle$ and, generally,

$$\mathsf{a}_i^-|\mathbf{n}\rangle = \mathsf{a}_i^-|n_1,n_2,\ldots\rangle = \mu_i(|\mathbf{n}\rangle)|\ldots,n_i-1,\ldots\rangle. \quad (25.1)$$

Notice there is a negative integer in the ith slot if $n_i = 0$. This is dealt with by defining $|n_1,n_2,\ldots\rangle$ as 0, the zero vector in the Hilbert space and not the vacuum state vector, if any n_i is negative. The same result follows if we define $\mu_i(|\mathbf{n}\rangle) = 0$ whenever $n_i = 0$. Notice, in particular, that $\mathsf{a}_i^-|0\rangle = 0$.

The proportionality constants $\lambda_i(\mathbf{n})$ and $\mu_i(\mathbf{n})$ are usually chosen so that the operators $\mathsf{a}_i^\pm$ are adjoints of one another. To see how this restricts the choice of these constants, first expand $(\mathsf{a}_i^+)^*|\mathbf{n}\rangle$ as

$$(\mathsf{a}_i^+)^*|\mathbf{n}\rangle = \sum c_{|\mathbf{m}\rangle}|\mathbf{m}\rangle,$$

where the coefficients $c_{|\mathbf{m}\rangle}$ are given by

$$\begin{aligned} c_{|\mathbf{m}\rangle} &= \langle\mathbf{m}|(\mathsf{a}_i^+)^*|\mathbf{n}\rangle = \left\langle \mathsf{a}_i^+|\mathbf{m}\rangle \,\middle|\, \mathbf{n}\right\rangle \\ &= \overline{\lambda_i(|\mathbf{m}\rangle)}\,\langle\ldots,m_i+1,\ldots|n_1,n_2\ldots\rangle \\ &= \overline{\lambda_i(|\mathbf{m}\rangle)}\,\delta_{m_1n_1}\delta_{m_2n_2}\cdots\delta_{m_i+1,n_i}\cdots. \end{aligned}$$

It follows that $c_{|\mathbf{m}\rangle} \neq 0$ only when $m_i+1 = n_i$, while $m_k = n_k$ for $k \neq i$, and $\lambda_i(|\mathbf{m}\rangle) \neq 0$. If $n_i = 0$ then $m_i+1 \neq n_i$, so that $(\mathsf{a}_i^+)^*|\mathbf{n}\rangle = 0$. On the other hand, when $n_i \neq 0$, $m_i = n_i - 1 \geq 0$ and

$$(\mathsf{a}_i^+)^*|\mathbf{n}\rangle = \overline{\lambda_i(|\ldots,n_i-1,\ldots\rangle)}\,|\ldots,n_i-1,\ldots\rangle. \quad (25.2)$$

This immediately identifies a condition for $\mathsf{a}_i^\pm$ to be adjoints of one another. Indeed, from equations 25.1 and 25.2, $\mathsf{a}_i^- = (\mathsf{a}_i^+)^*$ if, for each positive integer i and each basis vector $|\mathbf{n}\rangle$, we have

$$\mu_i(|n_1,n_2,\ldots\rangle) = \begin{cases} 0, & \text{when } n_i = 0; \\ \overline{\lambda_i(|\ldots,n_i-1,\ldots\rangle)}, & \text{when } n_i > 0. \end{cases} \quad (25.3)$$

Now, this condition can be satisfied easily by many choices for λ, for instance simply by the choice $\lambda_i(|\mathbf{n}\rangle) = 1$ for all i and all $|\mathbf{n}\rangle$, but we will see that a little patience will lead to a more judicious choice that meshes nicely with other operators of interest to be introduced in the next section. For now, notice that the definitions imply that $\mu_i(\xi_i) = \mu_i(|0,\ldots,0,1\rangle) = 1 = \overline{1} = \overline{\lambda_i(|0\rangle)}$, which satisfies condition 25.3.

25.3 Bosonic Fock Space

Bosons are quanta of integer spin and are characterized by the fact that any number of them may occupy simultaneously the same quantum state. In the setting of the notation introduced in the preceding sections, any state vector of the form $|\mathbf{n}\rangle = |n_1, n_2, \ldots\rangle$, where $n_i \geq 0$ and $\sum_i n_i < \infty$, represents a possible state of the system after a complete set of measurements. The bosonic Fock space $\mathbf{F}_\odot(\mathcal{H})$ may then be described as the complete orthogonal sum

$$\mathbf{F}_\odot(\mathcal{H}) = \overline{\bigoplus_{\mathbf{n}}} \mathbb{C}|\mathbf{n}\rangle,$$

where the sum is over all sequences $\mathbf{n} = (n_i)_{i=1}^\infty$ of nonnegative integers that are eventually zero. An explicit identification of $|n_1, n_2, \ldots\rangle$ with the symmetric product element $\xi_1^{n_1} \odot \xi_2^{n_2} \odot \cdots$ gives an explicit isomorphism between this description of the bosonic Fock space and its description as the complete orthogonal sum $\overline{\bigoplus}_{N=0}^\infty(\overline{\odot}^N \mathcal{H})$ of the preceding lecture. Here, the symbol ξ^n is interpreted as $\xi \odot \cdots \odot \xi$ (n times) and $\xi^0 = \mathbf{1}$ as the identity element of the bosonic state algebra. Note that $\xi_1^{n_1} \odot \xi_2^{n_2} \odot \cdots$ makes sense since the n_i are eventually zero. In the Fock space notation, the symmetric product of two basis vectors in the bosonic state algebra is given by the elegant expression

$$|m_1, m_2, \ldots\rangle \odot |n_1, n_2, \ldots\rangle = |m_1 + n_1, m_2 + n_2, \ldots\rangle,$$

or in the compact notation, $|\mathbf{m}\rangle \odot |\mathbf{n}\rangle = |\mathbf{m} + \mathbf{n}\rangle$. To go back to a previous example, while it is false that $|1\rangle + |0,1\rangle = |1,1\rangle$, it is true that $|1\rangle \odot |0,1\rangle = |1,1\rangle$.

We now introduce the **number operators** and choose the proportionality constants $\lambda_i(|\mathbf{n}\rangle)$ and $\mu_i(|\mathbf{n}\rangle)$ so that the number operators may be factored by the creation and annihilation operators. Whenever a quantum is in the state represented by ξ_i, we will say that it is in state i and call it an i-quantum. The **number operator for the i-quanta**, denoted N_i, is the self-adjoint operator on bosonic Fock space defined on basis state vectors by

$$\mathsf{N}_i|\mathbf{n}\rangle = n_i|\mathbf{n}\rangle.$$

It represents the observable that counts the number of quanta in state i. We will not worry about the precise domain $\mathcal{D}(\mathsf{N}_i)$ of this number operator but note, merely, that it contains the bosonic state algebra $\mathcal{B}$, the dense linear subspace that describes systems with a finite number of quanta.

The **total number operator** N is defined on the basis state vectors by

$$\mathsf{N}|\mathbf{n}\rangle = \left(\sum_{i=1}^{\infty} n_i\right)|\mathbf{n}\rangle,$$

and represents the observable that counts the total number of quanta. We now choose λ and μ by demanding that $\mathsf{N}_i = \mathsf{a}_i^+\mathsf{a}_i^-$. To derive the appropriate requirement, let $\mathbf{n}$ and $\mathbf{n}'$ be sequences that differ only in the ith term, with $n_j = n_j'$ if $i \neq j$ and $n_i' = n_i - 1$. Note that condition 25.3 implies that, when $n_i \geq 1$,

$$n_i|\mathbf{n}\rangle = \mathsf{N}_i|\mathbf{n}\rangle = \mathsf{a}_i^+\mathsf{a}_i^-|\mathbf{n}\rangle = \mu_i(|\mathbf{n}\rangle)\lambda_i(|\mathbf{n}'\rangle)|\mathbf{n}\rangle = |\mu_i(|\mathbf{n}\rangle)|^2|\mathbf{n}\rangle. \quad (25.4)$$

The requirement, then, for $\mathsf{N}_i = \mathsf{a}_i^+\mathsf{a}_i^-$ is that $|\mu_i(|\mathbf{n}\rangle)|^2 = n_i$, which may be accomplished easily by setting $\mu_i(|\mathbf{n}\rangle) = \sqrt{n_i}$. Condition 25.3 then implies that $\lambda_i(|\mathbf{n}\rangle) = \sqrt{n_i + 1}$, giving the final definitions for the creation and annihilation operators on bosonic Fock space as

$$\mathsf{a}_i^-|\mathbf{n}\rangle = \sqrt{n_i}\,|\ldots, n_i - 1, \ldots\rangle$$

and

$$\mathsf{a}_i^+|\mathbf{n}\rangle = \sqrt{n_i + 1}\,|\ldots, n_i + 1, \ldots\rangle.$$

Note that these equations hold even when $n_i = 0$.

We make the following claims for the commutators of the creation and annihilation operators:

$$[\mathsf{a}_i^-, \mathsf{a}_j^+] = \delta_{ij}1_{\mathbf{F}_\odot(\mathcal{H})},$$
$$[\mathsf{a}_i^+, \mathsf{a}_j^+] = 0 = [\mathsf{a}_i^-, \mathsf{a}_j^-].$$

For example, when $i = j$,

$$\langle\mathbf{m}|\mathsf{a}_i^-\mathsf{a}_i^+ - \mathsf{a}_i^+\mathsf{a}_i^-|\mathbf{n}\rangle = (n_i + 1)\langle\mathbf{m}|\mathbf{n}\rangle - n_i\langle\mathbf{m}|\mathbf{n}\rangle = \langle\mathbf{m}|\mathbf{n}\rangle = \delta_{\mathbf{mn}}.$$

This demonstrates that $[\mathsf{a}_i^-, \mathsf{a}_i^+] = 1_{\mathbf{F}_\odot(\mathcal{H})}$, and the remaining calculations are similar.

We may use the creation operators to build any basis state vector from the vacuum state vector. Indeed, since the creation operators a_i^+ all commute with one another, the following expression unambiguously defines the state $|\mathbf{n}\rangle$:

$$|\mathbf{n}\rangle = |n_1, n_2, \ldots\rangle = \left[\prod_{i=1}^{\infty} \frac{\left(\mathsf{a}_i^+\right)^{n_i}}{\sqrt{n_i!}}\right] |0\rangle.$$

We now can get at least a glimpse of how useful Fock space will be for describing processes in nature that lead to the creation and annihilation of quanta. For example, we may start with a system represented by the one-quantum state vector $|1\rangle$. A physical process corresponding to a_2^+ may occur that puts the system in a state represented by $a|1\rangle + b\,\mathsf{a}_2^+|1\rangle = a|1\rangle + b|1,1\rangle$, with $|a|^2 + |b|^2 = 1$, a superposition of a one-quantum and a two-quanta state. A measurement of the total number of quanta, whose representative operator is the number operator N, then yields the answer 2 with probability $|b|^2$. The physical processes that may lead to such creation are modeled in various quantum field theories, where the use of Fock space as the state space is pervasive.

25.4 Fermionic Fock Space

The development of the notation for fermionic Fock space begins in parallel to that for its bosonic analogue but quickly diverges, in no small part owing to the lack of commutativity of the wedge product. As no two fermions may occupy the same quantum state, the only basis vectors in the Fock space notation that can refer to fermions are those $|\mathbf{n}\rangle$ for which each n_i is either 0 or 1, still, of course, with $\sum_{i=1}^{\infty} n_i < \infty$. The fermionic Fock space $\mathbf{F}_\wedge(\mathcal{H})$ then has the alternative description as the complete orthogonal sum

$$\mathbf{F}_\wedge(\mathcal{H}) = \overline{\bigoplus_{\mathbf{n}}} \mathbb{C}|\mathbf{n}\rangle,$$

where the sum is over all **binary** sequences $\mathbf{n} = (n_i)_{i=1}^{\infty}$ that have only finitely many 1s. An explicit identification of $|n_1, n_2, \ldots\rangle$ with the antisymmetric product element $\xi_1^{n_1} \wedge \xi_2^{n_2} \wedge \cdots$ gives an explicit isomorphism between this description of the bosonic Fock space and its description as the complete orthogonal sum $\overline{\bigoplus}_{N=0}^{\infty}(\overline{\wedge}^N \mathcal{H})$ of the preceding lecture. Here again $\xi^0 = \mathbf{1}$; it is now the identity element of the fermionic state algebra. The wedge product of two of these basis vectors is given by the expression

$$|\mathbf{m}\rangle \wedge |\mathbf{n}\rangle = \operatorname{sgn}(i_1, \dots, i_M, j_1, \dots, j_N)|\mathbf{m} + \mathbf{n}\rangle,$$

where $\operatorname{sgn}(i_1, \dots, i_M, j_1, \dots, j_N)$ is the sign of the permutation that orders the concatenated list $i_1, \dots i_M, j_1, \dots, j_N$; here $i_1 < \cdots < i_M$ and $j_1 < \cdots < j_N$ are the indices of the terms of $\mathbf{m}$ and $\mathbf{n}$, respectively, that are nonzero. The above sign is interpreted to be zero if some index i_k equals some index j_ℓ, which is to say that the wedge product of $|\mathbf{m}\rangle$ and $|\mathbf{n}\rangle$ is zero if there is any index i such that $m_i = 1 = n_i$ and otherwise is given by the same formal expression $|\mathbf{m} + \mathbf{n}\rangle$ as the symmetric product, modulo a possible phase shift of -1.

The number operators on the fermionic Fock space are defined exactly as before, with $\mathsf{N}_i|\mathbf{n}\rangle = n_i|\mathbf{n}\rangle$ and $\mathsf{N}|\mathbf{n}\rangle = (\sum_{i=1}^{\infty} n_i)|\mathbf{n}\rangle$. We again choose the proportionality constants $\lambda_i(|\mathbf{n}\rangle)$ and $\mu_i(|\mathbf{n}\rangle)$ so that the number operators factor as $\mathsf{N}_i = \mathsf{a}_i^+\mathsf{a}_i^-$. As an initial observation, since two fermions cannot occupy the same quantum state, we can see immediately that $\lambda_i(|\mathbf{n}\rangle) = 0$ whenever $n_i = 1$, and we know already that $\mu_i(|\mathbf{n}\rangle) = 0$ whenever $n_i = 0$. Let $\mathbf{n}$ and $\mathbf{n}'$ be binary sequences that coincide except at the ith term, where $n_i = 1$ and $n_i' = 0$. Condition 25.3 gives $\mu_i(|\mathbf{n}\rangle) = \overline{\lambda_i(|\mathbf{n}'\rangle)}$, and equation 25.4, which holds here when $n_i = 1$, requires that $|\mu_i(|\mathbf{n}\rangle)|^2 = 1$. At this point it seems that we ought to choose any unimodular constant $e^{i\alpha}$ for $\mu_i(|\mathbf{n}\rangle)$ irrespectively of i as long as $n_i = 1$; this choice would factor the number operators appropriately. However, such a choice would lead to (anti)commutation relations that are a bit more unwieldy than we would like. It turns out that a slightly more complicated choice for $\mu_i(|\mathbf{n}\rangle)$, one dependent on the context, offers cleaner relations that result in a more elegant algebra for the fermionic creation and annihilation operators. We therefore make the following choices for these operators, employing again the notation where $\mathbf{n}$ and $\mathbf{n}'$ are binary sequences for which $n_j = n_j'$ if $j \neq i$ while $n_i = 1$ and $n_i' = 0$. The operators satisfy

$$\mathsf{a}_i^+|\mathbf{n}\rangle = 0 = \mathsf{a}_i^-|\mathbf{n}'\rangle,$$

while

$$\mathsf{a}_i^+|\mathbf{n}'\rangle = \left(\prod_{j=1}^{i-1}(-1)^{n_j'}\right)|\mathbf{n}\rangle \quad \text{and} \quad \mathsf{a}_i^-|\mathbf{n}\rangle = \left(\prod_{j=1}^{i-1}(-1)^{n_j}\right)|\mathbf{n}'\rangle,$$

which is to say that $\mathsf{a}_i^+|\mathbf{n}'\rangle = \pm|\mathbf{n}\rangle$ and $\mathsf{a}_i^-|\mathbf{n}\rangle = \pm|\mathbf{n}'\rangle$, the plus sign being chosen if there is an even number of 1s among $n_1, \ldots, n_{i-1}$, and the minus sign being chosen otherwise. With this choice of phase, the following anticommutation relations hold, where the **anticommutator** $[\,,\,]_+$ is defined on operators by $[\mathsf{A}, \mathsf{B}]_+ = \mathsf{AB} + \mathsf{BA}$:

$$[\mathsf{a}_i^-, \mathsf{a}_j^+]_+ = \delta_{ij} 1_{\mathbf{F}_\wedge(\mathcal{H})},$$
$$[\mathsf{a}_i^+, \mathsf{a}_j^+]_+ = 0 = [\mathsf{a}_i^-, \mathsf{a}_j^-]_+.$$

Again we can use creation operators to build any basis state vector from the vacuum state vector, but the lack of commutativity, since $\mathsf{a}_i^+\mathsf{a}_j^+ = -\mathsf{a}_j^+\mathsf{a}_i^+$ from the second anticommutator, means that the order of creation of quanta matters. If $\mathbf{n}$ is a binary sequence whose only nonzero terms are $n_{i_1}, \ldots, n_{i_N}$ then

$$|\mathbf{n}\rangle = \operatorname{sgn}(i_1, \ldots, i_N)\, \mathsf{a}_{i_1}^+ \cdots \mathsf{a}_{i_N}^+ |0\rangle,$$

where $\operatorname{sgn}(i_1, \ldots, i_N)$ is the sign of the permutation that orders the integers $i_1, \ldots, i_N$. Alternatively, $|\mathbf{n}\rangle = \mathsf{a}_{i_1}^+ \cdots \mathsf{a}_{i_N}^+ |0\rangle$, provided that the indices are ordered as $i_1 < \cdots < i_N$.

25.5 End Notes

The primary reference for this lecture is Teller (1995) *An Interpretive Introduction to Quantum Field Theory.* Written by the philosopher Paul Teller, son of the physicist Edward Teller, it presents only the bare basic ideas of quantum field theory with the precision of language one expects from a philosopher. The book presents a very careful construction of Fock space, the playing field of the theory, and describes in precise language what physicists are doing when they quantize fields. It is short on mathematical detail and neglects the formal tools of calculation in favor of a thorough discussion that helps the reader to think correctly about the theory and to understand the physical reasoning behind field quantization. For the novice who is hardly ready to tackle quantum field theory, its first three chapters in 52 pages laying out the Fock space formalism is worthy of study, even at an early stage of the study of quantum mechanics.

26

An Introduction to Quantum Statistical Mechanics

> There is, essentially, only one problem in statistical thermodynamics: the distribution of a given amount of energy E over N identical systems.
>
> Erwin Schrödinger
> *Statistical Thermodynamics*, 1946

How many distinct, linearly independent, multi-quanta states are available for N quanta when there are N degrees of freedom for each one-quantum state? We saw in Lecture 24 that the answer depends on the species of quanta that populate the system. If the N quanta are pairwise distinguishable from one another then there are N^N independent, mutually orthogonal, states available since each quantum has available any of the N states forming the degrees of freedom. The state space of configurations is the N^N-dimensional N-fold tensor product $\otimes^N \mathcal{H}$, where $\mathcal{H}$ is the N-dimensional state space of a single quantum. On the other hand, if the quanta are indistinguishable bosons, the number of independent states available is the number of basis vectors $|n_1, \ldots, n_N\rangle$ from the bosonic Fock space $\mathbf{F}_{\odot}(\mathcal{H})$ for which $\mathsf{N}|n_1, \ldots, n_N\rangle = n_1 + \cdots + n_N = N$. The answer is the same as that for the combinatorial problem that asks how many ways there are to distribute N indistinguishable balls, the quanta, among N labeled boxes, the possible one-quantum states. This is $\binom{2N-1}{N}$, the dimension of the symmetric product $\odot^N \mathcal{H}$. Finally, if the quanta are fermions then no two quanta may occupy the same state, and so there is exactly one configuration available, the one with state vector $|1, \ldots, 1\rangle$.

This example illustrates in sharp relief that the statistical mechanics of large aggregates must take into account the nature of the quanta and suggests strongly that there will be significant differences between the behavior of aggregates of bosons and fermions. In this lecture, we offer an introduction to the statistical mechanics of quantum aggregates. Our goal is to derive the distribution functions that govern the behavior of the three types of aggregates – ensembles of distinguishable quanta, of bosons, and of fermions – and to determine their most likely configurations.

Statistical mechanics is a vast subject, and a single lecture can accomplish merely an inadequate preamble to this topic. The End Notes to the present lecture will point to various sources for more in-depth study.

26.1 Statistical Mechanics

To avoid technicalities that have no bearing on the final results, we assume in this lecture that the state space $\mathcal{H}$ for a one-quantum system is of an unspecified, possibly very large but nonetheless finite, number of dimensions. This avoids the need to work in the completions of the tensor, bosonic, and fermionic products since the finite products are complete automatically.

Consider, then, a system of N quanta for which the possible one-quantum energies are discrete, say $\varepsilon_1 < \varepsilon_2 < \cdots$. These are the eigenvalues of the Hamiltonian H on the state space $\mathcal{H}$ for the one-quantum system. We assume that the degeneracy of ε_i is d_i, meaning, of course, that the ε_i-eigenspace is d_i-dimensional. Let $\mathbf{n} = (n_1, n_2, \dots)$ denote a multi-quanta **configuration** in which there are n_i quanta in an ε_i energy state.[1] Notice that a single configuration $\mathbf{n}$ may be realized by possibly many different linearly independent state vectors in the state space. We assume that the system is in thermal equilibrium and isolated from the environment with a total constant energy E. For the configuration $\mathbf{n}$, we have $\sum_i n_i = N$ and $\sum_i n_i\varepsilon_i = E$. The fundamental question we ask is, if a measurement is performed that determines the energies of the individual quanta, how likely are we to find the system in configuration $\mathbf{n}$, or, if a measurement is performed on one quantum, how likely

[1] Note that this use of notation is a little different from the discussion of the previous lecture. There, the state space $\mathcal{H}$ was decomposed into one-dimensional subspaces by the basis $\boldsymbol{\xi}$, whereas here we allow for arbitrary finite-dimensional subspaces in our decomposition of $\mathcal{H}$ into energy eigenspaces. There, $\mathbf{n}$ denoted a system in which there were n_i quanta with the same state vector ξ_i, while here it denotes a configuration of our system with n_i quanta distributed among a possibly very large number of linearly independent states, each with energy ε_i.

are we to find it with energy ε_i? The **fundamental assumption of statistical mechanics**[2] is that, in thermal equilibrium, every distinct state of the ensemble with the same total energy is equally probable.[3] The answer, then, to the question depends on how many different ways the configuration $\mathbf{n}$ can be realized as a state of the system, taking $\mathbf{n}$ as more likely than $\mathbf{n}'$ when it has more ways to be realized than $\mathbf{n}'$.

This leads to the question of what precisely the term "different ways" means. After all, the general state is a linear combination of basis states, and a basis is chosen somewhat arbitrarily. To flesh this out a bit, assume that a maximal set of pairwise-commuting observable operators that commute with the Hamiltonian exists whose simultaneous eigenspaces further decompose the ε_i-eigenspaces of the Hamiltonian into one-dimensional subspaces. Then the ε_i-eigenspace is spanned by a basis $\psi_{i,1}, \ldots, \psi_{i,d_i}$ of simultaneous eigenvectors and a further measurement of these observables would place each of the n_i quanta with energy ε_i in exactly one of the basis states from the list $\psi_{i,1}, \ldots, \psi_{i,d_i}$. These basis states then can be used to help count the different ways in which the configuration $\mathbf{n}$ can be achieved, for these form a fine enough sieve to separate out independent ways in which the quantum number n_i can be realized. Thus it is the dimension, d_i, of the ε_i-eigenspace that counts the number of independent ways in which a single quantum of energy ε_i may contribute to the configuration. The implication then is that the number of different ways in which the configuration $\mathbf{n}$ can be realized should be interpreted as the dimension of the **n-configuration space**, defined as the subspace of the N-quanta state space of the system that consists of all state vectors that realize the configuration $\mathbf{n}$. This is the maximal number of linearly independent state vectors for the system for which there are, for each i, n_i quanta in an ε_i-eigenstate; in turn, this depends on the nature of the quanta.

26.1.1 Distinguishable Quanta

For distinguishable quanta, the state space is the N-fold tensor product $\otimes^N \mathcal{H}$, and our task is to find the dimension of the $\mathbf{n}$-configuration space. First, there are $\binom{N}{n_1}$ ways to choose n_1 distinguishable quanta, each with energy ε_1, to occupy the first slot of the configuration. Once the n_1 distinguishable quanta are chosen, there is a d_1-dimensional space, the

[2] Also called the **principle of equal a priori probabilities**.

[3] Phillies (2000), p. 34, has an incisive discussion on this assumption that solidifies its status as an axiom rather than a theorem.

ε_1-eigenspace, for the n_1 distinguishable quanta to occupy independently of one another, with d_1 linearly independent basis states available for each of the n_1 quanta. Thus, the first slot may be filled in

$$\binom{N}{n_1} d_1^{n_1} = \frac{N! d_1^{n_1}}{n_1!(N-n_1)!}$$

linearly independent ways. Continuing in this manner, letting $Q(\mathbf{n})$ denote the dimension of the $\mathbf{n}$-configuration space, i.e., the number of linearly independent ways in which the configuration $\mathbf{n}$ may be realized, we have

$$Q(\mathbf{n}) = \prod_i \binom{N-(n_1+\cdots+n_{i-1})}{n_i} d_i^{n_i} = N! \prod_i \frac{d_i^{n_i}}{n_i!}. \tag{26.1}$$

Note that these are finite products.

26.1.2 Indistinguishable Quanta

The state space for bosons is the N-fold symmetric product $\odot^N \mathcal{H}$. If the quanta are indistinguishable, to realize the configuration $\mathbf{n}$ there is exactly one way to choose n_i quanta to occupy the ith slot – just choose n_i quanta! How many linearly independent ways are there for these n_i quanta to occupy the ε_i-eigenspace? The question is, in how many distinct ways may these n_i indistinguishable quanta occupy the d_i distinguished basis states? In terms of the symmetric product, how many distinct state vectors are there of the form $\psi_{i,1}^{k(1)} \odot \cdots \odot \psi_{i,d_i}^{k(d_i)}$ with $k(1) + \cdots + k(d_i) = n_i$, where each $k(j) \geq 0$? This is the well-known combinatorial "balls in boxes" problem – in how many ways can n_i indistinguishable balls be placed in d_i numbered boxes? For readers who are unfamiliar with this problem, consider as an example a configuration with $n_i = 7$ and $d_i = 5$. One way to realize this data is as $\psi_1^2 \odot \psi_2^1 \odot \psi_3^3 \odot \psi_5^1$, representing two balls in the first box, one in the second, three in the third, none in the fourth, and one in the fifth. This may be represented pictorially as

$$\circ\circ|\circ|\circ\circ\circ||\circ,$$

and it is clear then that the number of different ways to place seven identical balls in five boxes is precisely the number of distinct arrangements of these symbols, seven $\circ$'s and four $|$'s. In general then, the number of linearly independent ways in which the ith slot of $\mathbf{n}$ may be filled is precisely the number of ways in which one can arrange n_i $\circ$'s and $d_i - 1$

|'s in a row. This is given by

$$\binom{n_i + d_i - 1}{n_i}.$$

It follows that the dimension of the $\mathbf{n}$-configuration space for bosons is

$$Q(\mathbf{n}) = \prod_i \binom{n_i + d_i - 1}{n_i} = \prod_i \frac{(n_i + d_i - 1)!}{n_i!(d_i - 1)!}, \tag{26.2}$$

where, again, this is a finite product.

The state space for fermions is the N-fold exterior product $\wedge^N \mathcal{H}$. Again, we need to determine how many linearly independent ways n_i quanta may occupy the ε_i-eigenspace. Now no two quanta may occupy the same quantum state, so, if a basis is chosen for the ε_i-eigenspace, the n_i quanta must be distributed among the d_i basis states, with at most one quantum per state. Thus, there are $\binom{d_i}{n_i}$ distinct ways to fill the ith energy slot of the configuration $\mathbf{n}$, and the dimension of the $\mathbf{n}$-configuration space for fermions is

$$Q(\mathbf{n}) = \prod_i \binom{d_i}{n_i} = \prod_i \frac{d_i!}{n_i!(d_i - n_i)!}. \tag{26.3}$$

26.2 The Most Probable Configuration via the Maximum-Term Method

We wish now to find, among all possible configurations of N quanta with total energy E, the one which is most likely. This amounts to maximizing the function $Q(\mathbf{n})$ subject to the constraints that $\sum_i n_i = N$ and $\sum_i n_i \varepsilon_i = E$. The importance of this stems from the experimental observation that, when N is extremely large, the most probable configuration becomes overwhelmingly more likely than its competitors. More precisely, the probability distribution on the space of configurations derived from Q is sharply spiked at the most probable configuration, and approaches an atomic, or delta, distribution as $N \to \infty$. In terms of state spaces, if $\mathcal{H}_E^N$ is the total energy eigenspace of the appropriate N-quanta state space – $\otimes^N \mathcal{H}$, $\odot^N \mathcal{H}$, or $\wedge^N \mathcal{H}$ – with eigenvalue E, each configuration $\mathbf{n}$ chooses the $\mathbf{n}$-configuration space, a subspace $\mathcal{H}_{\mathbf{n}}$ of $\mathcal{H}_E^N$ of dimension $Q(\mathbf{n})$. The space $\mathcal{H}_E^N$ splits as the orthogonal direct sum

$$\mathcal{H}_E^N = \bigoplus_{\mathbf{n}} \mathcal{H}_{\mathbf{n}},$$

where the sum is over all configurations $\mathbf{n}$ subject to the constraints. Notice that this is a finite sum since there are only finitely many configurations with total energy E. For extremely large values of N the dimension of $\mathcal{H}_E^N$ is very close to that of $\mathcal{H}_{\mathbf{n}}$, where $\mathbf{n}$ is the configuration that maximizes Q, so that the state of the system upon the measurement of total energy is overwhelmingly likely to be represented by a vector in the $\mathbf{n}$-configuration space $\mathcal{H}_{\mathbf{n}}$.

This preceding plausibility argument is an example of the **maximum-term method** that is used extensively in statistical mechanics. The method replaces a sum by the maximum term in the sum and argues that this replacement is statistically accurate under appropriate conditions. The appropriate conditions occur when the maximum term of the sum is at least on the order of an exponential in the number of terms. Specifically, consider the sum

$$S = \sum_{i=1}^{M} q_i,$$

where each term q_i is nonnegative and $Q = \max\{q_1, \ldots, q_M\} \geq O(e^M)$. Then $Q \leq S \leq MQ$ and

$$\log Q \leq \log S \leq \log Q + \log M.^4$$

Since $Q \geq O(e^M)$ we have $\log Q \geq O(M)$, and so the error term $\log M$ is negligible compared with $\log S$ when M is large. We conclude that $\log S = O(\log Q)$ and the estimate $S = Q$ is exponentially good. As an example, a typical macroscopic system that has $M = 6.02214 \times 10^{23}$ particles, a mole's worth, has $\log M \approx 55$, ten billion trillion times smaller than M. In the example of the preceding paragraph, the sum we are estimating is

$$\dim \mathcal{H}_E^N = \sum_{\mathbf{n}} Q(\mathbf{n}),$$

where the sum is over all configurations $\mathbf{n}$ subject to the constraints. The maximum of $Q(\mathbf{n})$ is exponentially larger than the number of configurations $\mathbf{n}$ subject to the constraints. Approximation by the maximum term is even more compelling when, as in this case, the terms are sharply spiked at the maximum.

Turning to the mathematical problem of maximizing Q subject to the

[4] As is common in the mathematics community, we use the symbol "log" for the natural logarithm to base e rather than "ln," which is used more commonly in the science and engineering communities.

constraints, we will maximize the function $\log Q$. Let

$$G = \log Q + \alpha \left(N - \sum_i n_i \right) + \beta \left(E - \sum_i n_i \varepsilon_i \right).$$

Here, α and β are Lagrange multipliers, and maximizing $\log Q$ subject to the constraints amounts to finding the critical points of $G(\mathbf{n}; \alpha, \beta)$. The equation $\partial G/\partial \alpha = 0 = \partial G/\partial \beta$ merely reproduces the constraints and we are left to identify where the partial derivatives $\partial G/\partial n_i$ are zero. This will be explored in the separate contexts of distinguishable and indistinguishable quanta, and of bosons and fermions.

Our analysis, though, in each case will apply only when we may expect that the overwhelming number of occupation numbers n_i is either zero or extremely large. This is necessary for the use of Stirling's approximation,

$$\log n! \approx n \log n - n, \tag{26.4}$$

which is valid for $n = 0$ if the expression $0 \log 0$ is interpreted as the limit $\lim_{n \to 0^+} n \log n = 0$, and is a very close approximation when $n \gg 1$.[5] If one is studying a system where this is not true, our analysis is irrelevant.

26.3 The Mechanics of Maximizing $Q(\mathbf{n})$

As the dimension function Q differs between distinguishable and indistinguishable quanta, and also between bosons and fermions, we will need to apply the machinery of Lagrange multipliers to the three cases separately.

26.3.1 Configurations of Distinguishable Quanta

Using equation 26.1 and approximation 26.4, we may write the approximate equation

$$G = \sum_i (n_i \log d_i - n_i \log n_i + n_i - \alpha n_i - \beta n_i \varepsilon_i) + \log N! + \alpha N + \beta E.$$

[5] The expression $\log n!$ has an asymptotic series expansion, the **Stirling series**,

$$\log n! \sim n \log n - n + \log \sqrt{2\pi n} + \frac{1}{12n} - \frac{1}{360n^3} + \frac{1}{1260n^5} - \cdots,$$

so that $\log n! = n \log n - n + \frac{1}{2} \log 2\pi n + O(1/n)$. Thus the error in approximation 26.4 is on the order of $\log n$, which, relative to n, is negligible whenever n is large.

Recall that this is a finite sum. Taking the partial derivative with respect to n_i gives

$$\frac{\partial G}{\partial n_i} = \log d_i - \log n_i - \alpha - \beta \varepsilon_i$$

and, setting this equal to zero, gives $\log(n_i/d_i) = -\alpha - \beta\varepsilon_i$ as the condition for maximizing $\log Q$ subject to the constraints. In exponential form, this gives

$$n_i = d_i e^{-(\alpha+\beta\varepsilon_i)}$$

as the most probable occupation number in energy slot i of the configuration. Of course, we have used a continuous approximation to a discrete problem in finding this expression for the maximum of Q and, technically, n_i should be taken as the floor of the expression $d_i e^{-(\alpha+\beta\varepsilon_i)}$, which will be zero for large enough values of i.

The value of the constant β for systems of macroscopic size is determined by thermodynamic considerations that link the microscopic statistical mechanics of ensembles of quanta to the gross, macroscopic, behavior of gases described by classical thermodynamics. The derivation would take us too far afield,[6] but the constant is determined to be $1/kT$, where k is Boltzmann's constant and T is the absolute temperature. Setting $C = e^{-\alpha}$, which is a constant independent of i, the most probable occupation number in slot i becomes

$$n_i = C d_i e^{-\varepsilon_i/kT}. \tag{26.5}$$

Assume that the system is in the most probable configuration $\mathbf{n}$, with n_i given by expression 26.5. The probability of finding a quantum with energy ε_i upon measurement is n_i/N, and the probability of finding this quantum in a particular measurement basis state is n_i/d_iN. The term n_i/d_i represents the average number of quanta per basis state with energy ε_i. Let

$$Z = \sum_{\varepsilon_i \leq E} d_i e^{-\varepsilon_i/kT}; \tag{26.6}$$

the quantity Z is known as the **partition function**. Then $N = CZ$ and the probability of finding a quantum with energy ε_i upon measurement becomes

$$\frac{n_i}{N} = \frac{d_i e^{-\varepsilon_i/kT}}{Z};$$

[6] The derivation may be found in McQuarrie (2000), Sections 2 and 3.

the probability of finding a quantum in a particular measurement basis state with energy ε_i is

$$\frac{n_i}{d_i N} = \frac{e^{-\varepsilon_i/kT}}{Z}. \tag{26.7}$$

This is known as the **Boltzmann distribution** and describes a gas of nonrelativistic particles with no quantum effects, meaning that we have derived this distribution under the assumption of the pairwise distinguishability of particles, which is unrealistic for a gas of quanta of a single species. For example, a gas of helium atoms or of electrons would always have quantum effects as the constituent quanta are bosons or fermions, which in principle are indistinguishable. This discussion is useful nonetheless since the distribution functions that govern ensembles of bosons and fermions, derived in the next section, both limit to the Boltzmann distribution under conditions of high temperature and low density, when the wave functions of individual quanta have no overlap and the gas acts like an ensemble of distinguishable quanta. It will also be used subsequently to derive the fundamental law of statistical mechanics.

Thus, the Boltzmann distribution is the classical distribution that governs a gas of distinguishable classical particles and is derived exactly as above. The only difference is that the degrees of freedom, d_i, denote not the dimension of a continuous state space but the number of distinct, discrete, classical states available at energy ε_i. The general statistics derived from the Boltzmann distribution are known as **Maxwell–Boltzmann statistics.**

26.3.2 Configurations of Indistinguishable Quanta

Using equation 26.2 and the Stirling approximation 26.4, we may write the approximate equation

$$G = \sum_i \Big[(n_i + d_i - 1)\log(n_i + d_i - 1) - n_i \log n_i - \log(d_i - 1)! \\ - d_i + 1 - \alpha n_i - \beta n_i \varepsilon_i \Big] + \alpha N + \beta E;$$

hence,

$$\frac{\partial G}{\partial n_i} = \log(n_i + d_i - 1) - \log n_i - \alpha - \beta \varepsilon_i = 0$$

is the condition for maximizing $\log Q$ subject to the constraints. Solving for the occupation number n_i gives

$$n_i = \frac{d_i - 1}{e^{(\alpha + \beta\varepsilon_i)} - 1}.$$

The constant β is equal to $1/kT$ but this time we cannot eliminate the constant α. Again, we will not take a side trip to derive its value but are content to report it as $\alpha = -\mu/kT$, where μ is the **chemical potential** of the system, defined as the marginal change in energy of the system upon the addition of one more quantum while keeping the volume and entropy constant. With these values for the Lagrange multipliers, the occupation number for a bosonic gas becomes

$$n_i = \frac{d_i - 1}{e^{(\varepsilon_i - \mu)/kT} - 1}.$$

For most real systems the degeneracy d_i is very large, especially for high values of i, and the term -1 can be dropped from the numerator. In some sense, μ just relocates the zero-point energy by subtracting from the energy of a quantum the amount of energy needed to place that quantum in the system. The statistics derived from this expression for the occupation number are known as **Bose–Einstein statistics.** Letting n_i^{MB} denote the occupation number for Maxwell–Boltzmann statistics and n_i^{BE} for Bose–Einstein statistics, it is easy to see that $n_i^{MB} = O(n_i^{BE})$ as $\varepsilon_i \to \infty$ with T held constant. In particular, when $\varepsilon_i - \mu \gg kT$, the Bose–Einstein expression for the distribution n_i/N is asymptotically close to the Maxwell–Boltzmann expression.

Turning to fermions, using equation 26.3 and again the Stirling approximation 26.4 in a derivation similar to that for boson occupation numbers, we find that the occupation number for fermions is

$$n_i = \frac{d_i}{e^{(\varepsilon_i - \mu)/kT} + 1}.$$

One caveat is in order, viz., to derive this we have used Stirling's approximation for the expression $\log(d_i - n_i)!$, and so it is valid only under the extra assumption that $d_i \gg n_i$. The statistics derived from this expression for the occupation number are known as **Fermi–Dirac statistics** and revert to Maxwell–Boltzmann statistics when $\varepsilon_i - \mu \gg kT$.

26.4 The Fundamental Law and the Canonical Ensemble

The preceding sections have concentrated on finding the distribution of the occupation numbers for a system of N quanta in thermal equilibrium with total energy E. This is a "frog's level view" of an idealized system and we now take a "bird's eye view." The frog's level view[7] asks, as its fundamental question, if a measurement is performed that determines the energy of an individual quantum in an isolated system with fixed energy E, how likely are we to find that quantum in an ε_i-energy eigenstate? The bird's eye view asks as its fundamental question, if a non-isolated system in thermal equilibrium can be in total-energy eigenstates $|1\rangle, \ldots, |\kappa\rangle, \ldots$ with respective total energies $E(1), \ldots, E(\kappa), \ldots$, and a measurement is performed of the total energy of the system, how likely are we to find the system in the state represented by $|\kappa\rangle$ with energy $E(\kappa)$? More precisely, the system that we will consider now, called the **canonical ensemble**, is a macroscopic system in thermal equilibrium with a heat bath in which the number of quanta N, the volume V, and the temperature T are held constant. There is a spectrum of total-energy eigenstates available, the eigenvalues being determined by the Hamiltonian H for the system, and the question is the likelihood of finding the system in a particular eigenstate upon measurement. In this development, we assume that the state vectors $|1\rangle, \ldots, |\kappa\rangle, \ldots$ form an orthonormal basis of energy eigenstates for the state space of the ensemble. In particular, if a certain energy eigenvalue E has degeneracy d, then E is repeated d times in the list $E(1), E(2), \ldots$

The **fundamental law of statistical mechanics** posits that the probability p_κ that the system will be found in the state represented by $|\kappa\rangle$ with energy $E(\kappa)$ is

$$p_\kappa = \frac{e^{-E(\kappa)/kT}}{\sum_\lambda e^{-E(\lambda)/kT}}. \tag{26.8}$$

The denominator

$$Z = Z(T) = \sum_\lambda e^{-E(\lambda)/kT}$$

is called, as before, the **partition function** and merely normalizes the relative frequencies given by the numerator. Once the partition function is known for a particular system, all the thermodynamic properties –

[7] See the wonderful essay "Birds and frogs" by Freeman Dyson in *Notices AMS* 56(2), 2009, pp. 212–223.

pressure, energy, entropy, free energy, specific heat – of the system may be found. Notice that p_κ attains a maximum for the smallest value of $E(\kappa)$, so the lowest energy states are the most probable. We may think of the system as being in the statistical state represented by the density operator $\rho = \sum_\kappa p_\kappa |\kappa\rangle\langle\kappa|$. By equations 23.3 and 23.5, the expectation value for the observable a represented by A is

$$\mathrm{Exp}_\rho(\mathsf{A}) = \mathsf{Tr}(\rho\mathsf{A}) = \frac{1}{Z}\sum_\kappa \langle\kappa|\mathsf{A}|\kappa\rangle e^{-E(\kappa)/kT}.$$

For example, the expected value of the energy is

$$\mathrm{Exp}_\rho(\mathsf{H}) = \mathsf{Tr}(\rho\mathsf{H}) = \frac{1}{Z}\sum_\kappa \langle\kappa|\mathsf{H}|\kappa\rangle e^{-E(\kappa)/kT} = \frac{1}{Z}\sum_\kappa E(\kappa) e^{-E(\kappa)/kT}.$$

Quoting Feynman,

> This fundamental law is the summit of statistical mechanics, and the entire subject is either the slide-down from this summit, as the principle is applied to various cases, or the climb-up to where the fundamental law is derived and the meaning of the concepts of thermal equilibrium and temperature T clarified.[8]

The fundamental law is just that – fundamental! It has the status of a law of nature and its validity is attested by strong experimental evidence. Despite Feynman's use of the verb *derive*, the law cannot in fact be deduced from more basic postulates and does not follow as a consequence of either the classical Hamiltonian laws of motion or the quantum laws, admitting only plausibility arguments[9] to buttress its theoretical validity or a derivation from an equivalent formulation of the law. We present one such derivation of the fundamental law from the fundamental assumption of the first section of this lecture. But the fundamental assumption is just an equivalent formulation of the fundamental law, each derivable from the other – either may be taken as an axiom, and the other can then be derived as a theorem.[10] Indeed, one might notice that if $E(\kappa) = E(\kappa')$ then the fundamental law equation 26.8 implies that it is equally probable for the system to be in states κ and κ', as $p_\kappa = p_{\kappa'}$. This, of course, merely confirms the fundamental assumption.

That the fundamental law is a consequence of the fundamental assumption follows almost as easily. Indeed, consider a large number, say Ω, of identical systems, each of volume V, each with N quanta and at

[8] Feynman (1998), p. 1.

[9] Two renditions of essentially the same plausibility argument appear in Feynman (1998), pp. 1–3, and Pathria (1996), pp. 44–45.

[10] See the discussion in Sections 3.3 and 3.4 of Phillies (2000).

temperature T. These are inserted into a heat bath at constant temperature T and the whole ensemble, which is usually called the **canonical ensemble**, of Ω systems plus heat bath is allowed to stabilize to thermal equilibrium. The energy eigenstates $|1\rangle, \ldots, |\kappa\rangle, \ldots$ are available for each system, and the total energy is given by $\mathcal{E} = \sum_{\omega=1}^{\Omega} E_\omega$. Here E_ω is the energy of the ωth system of the ensemble and hence appears as one of the entries in the list $E(1), \ldots, E(\kappa), \ldots$ The heat bath merely serves to fix the temperature and to allow for the exchange of energy between systems, but, on average, since the whole ensemble of Ω systems and heat bath is in thermal equilibrium the total energy $\mathcal{E}$ may be treated as a constant. What is the probability that any particular system chosen at random in this ensemble will be found in state κ with energy $E(\kappa)$?

This question is precisely the question answered in Section 26.3.1. If we interpret a "quantum" of the canonical ensemble as one of the Ω systems of volume V composed of N "microscopic quanta" then the question we ask concerns the number of ways to distribute the total energy $\mathcal{E}$ among the Ω distinguishable quanta or systems. The available energy eigenstates are $|1\rangle, \ldots, |\kappa\rangle, \ldots$ with respective energies $E(1), \ldots, E(\kappa), \ldots$ Listed in terms of available energies, we would have energy eigenvalues ε_i with degeneracy d_i, so that ε_i would appear precisely d_i times in the list $E(1), \ldots, E(\kappa), \ldots$ Stated differently, there are exactly d_i energy eigenstates from among the list $|1\rangle, \ldots, |\kappa\rangle, \ldots$ with energy ε_i. Equation 26.7 gives the probability p_κ of finding one system of the ensemble in eigenstate $|\kappa\rangle$ as

$$p_\kappa = \frac{e^{-E(\kappa)/kT}}{Z},$$

where the partition function Z is given by equation 26.6:

$$Z = \sum_i d_i e^{-\varepsilon_i/kT} = \sum_\lambda e^{-E(\lambda)/kT}.$$

Thus we see that the fundamental law 26.8 is a logical consequence of the fundamental assumption.

In closing this lecture we will point out that similar reasoning allows one to analyze other configurations besides the canonical ensemble. There are several that are studied in statistical mechanics. For example, the grand canonical ensemble generalizes the canonical ensemble by allowing the number of micro-quanta, N, in a system to vary. The set-up again assumes a large number, say Ω, of identical systems, each of volume V and at temperature T, but with the number N of quanta in each system variable. These are inserted into a heat bath at constant temperature T, but the walls of the systems are permeable to allow for the exchange of quanta. This **grand canonical ensemble** of Ω systems plus heat bath is allowed to stabilize to thermal equilibrium. Again, the energy eigenstates $|1\rangle, \ldots, |\kappa\rangle, \ldots$ are available for each system, and the total energy is given by $\mathcal{E} = \sum_{\omega=1}^{\Omega} E_\omega$. This time, though, the eigenstates of the total number operator N determine the allowed values of N for any one system. If N_ω denotes the number of quanta in the ωth system, then the total number of quanta in all the systems is $\mathcal{N} = \sum_{\omega=1}^{\Omega} N_\omega$. The heat bath merely serves to fix the temperature and to allow for the exchange of energy and quanta between systems but, on average, since the whole ensemble of Ω systems and heat bath is in thermal equilibrium, the total energy $\mathcal{E}$ and the total number of quanta $\mathcal{N}$ may be treated as constants. The question is as to the probability that any particular system chosen at random in this ensemble will be found with N quanta and total energy E. We emphasize that the state $|\kappa\rangle$ satisfies $\mathsf{H}|\kappa\rangle = E(\kappa)|\kappa\rangle$ and $\mathsf{N}|\kappa\rangle = N(\kappa)|\kappa\rangle$, for some energy value $E(\kappa)$ and nonnegative integer $N(\kappa)$. The analysis of the grand canonical ensemble is left in the hands of the interested reader, with the recommended references described in the End Notes as appropriate helpmates.

26.5 End Notes

The discussion of the first two sections of this lecture is standard but the flow follows closely that of Griffiths (1995), Section 5.4. Statistical mechanics is a vast subject that requires at least a full year course to do it justice. This lecture provides but the briefest of hints as to the importance and techniques of the science. There are several nice treatments of statistical mechanics, including Schwabl (2002), Pathria (1996), Chandler (1987), McQuarrie (2000), and Phillies (2000). The last of these is an elementary introduction at graduate level to what every student of physics or mathematics should know about statistical mechanics. While not comprehensive, it treats the most important topics in the discipline carefully and thoroughly. At the other end of the spectrum is Feynman (1998), an intensive work on quantum statistical mechanics that makes high demands of the student. Even today it holds up as a classic exposition of quantum statistical mechanics.

27
Quantum Dynamics

> If anything moves, it moves either in a place in which it is or in a place in which it is not,
> But neither in a place in which it is (for there it is at rest) nor in a place in which it is not (for how could it do anything in a place in which it simply is not?).
> Therefore, it is not the case that anything moves.
>
> Diodorus Cronus
> Fourth century BC

> Having accumulated rubbish of this sort [the dialectician] frowns and takes out his dialectic and solemnly tries to establish for us by his deductive proofs that ... some things move ... although if we set in opposition to these arguments what appears evidently, that is no doubt enough to shatter their positive affirmation with the equipollent disconfirmation given by what is apparent.
>
> Sextus Empericus
> *Outlines of Skepticism*, Second century AD

For the most part, our development of quantum mechanics thus far has concentrated on understanding the static structure of the theory. Though we have introduced the dynamic aspects with Axiom 4 and the Schrödinger equation, we have yet to investigate the implications of time evolution in quantum mechanics. Aside from two brief excursions into dynamics – in Lecture 8 in deriving the quantum equation that governs the time evolution of expectation values, and again in Lecture 10 where the Fourier transform was used to obtain free particle solutions to the Schrödinger equation – we have spent most of our efforts in understanding the state spaces of quantum systems and the eigenstates of the time-independent Schrödinger equation, which give the stationary states for central potentials.

In this lecture we will study the time evolution of quantum systems. We begin with a description of the Schrödinger picture, where quantum state vectors evolve according to the time-dependent Schrödinger equation of Axiom 4. We present a derivation of the Schrödinger equation from a short list of natural assumptions that culminates in an articulation of Stone's theorem on unitary operator representations. This is explored in some detail in a subsequent lecture where symmetries and conservation laws are introduced into the quantum formalism. After Stone's theorem, time evolution is repackaged in a picture named for Heisenberg, where states remain constant while the operators associated with observables evolve. We then derive the Heisenberg equation of motion, which complements that of Schrödinger. Finally, the pictures of quantum evolution championed by Schrödinger and Heisenberg are seen to be two poles of a single interaction picture due to Dirac.

This interaction picture is derived by generalizing that of Schrödinger by allowing Schrödinger observables to depend explicitly on time and by introducing a reference Hamiltonian. Appropriate choices of the reference Hamiltonian recover the Schrödinger and Heisenberg pictures. The resulting evolution equations in the general interaction picture offer a wide latitude in applications and lead directly to a perturbation approximation that can sometimes be useful for understanding time evolution in cases where the Schrödinger equation has no closed form solution. This may be explored by the interested reader by consulting almost any graduate-level quantum mechanics text written for students of physics; see the Endnotes for more discussion.

27.1 The Schrödinger Picture

According to Axiom 4 of Lecture 2, the time evolution of a quantum state vector is governed by the Schrödinger equation. If $\psi = \psi_t$ is the state vector at time t then

$$\mathbf{i}\hbar\frac{d\psi}{dt} = \mathsf{H}\psi, \tag{27.1}$$

where H is the Hamiltonian operator for the system. This is a basic postulate of quantum mechanics. The operator H is a self-adjoint operator densely defined on the state space $\mathcal{H}$ of the system, but the exact nature of the operator H is left undetermined in the axioms.

In the wave mechanics of a one-particle system described in Section 2.2 the state space $\mathcal{H}$ is the Hilbert space $L^2(\mathbb{R}^3)$, and the Hamiltonian H arises by quantization of the classical Hamiltonian, which describes the total energy of the system in terms of the canonically conjugate position and momentum variables. To review the simplest case, with $\mathbf{x}$ the classical position and $\mathbf{p}$ the classical momentum of the particle, the energy E is given by the classical Hamiltonian as

$$E = H(\mathbf{x}, \mathbf{p}) = \frac{\mathbf{p} \cdot \mathbf{p}}{2m} + V(\mathbf{x}).$$

Quantization proceeds by replacing x_j by X_j and p_j by P_j, from which comes the prescription

$$V(\mathbf{x}) \rightsquigarrow V(\mathsf{\mathbf{X}}),$$
$$\mathbf{p} \cdot \mathbf{p} \rightsquigarrow \mathsf{\mathbf{P}} \cdot \mathsf{\mathbf{P}} = -\hbar^2 \Delta.$$

The operator $V(\mathsf{\mathbf{X}})$ multiplies a function of the variable $\mathbf{x}$ by the potential function $V(\mathbf{x})$. Setting $\Psi(\mathbf{x}, t) = \psi_t(\mathbf{x})$ gives the Schrödinger equation as

$$\mathrm{i}\hbar \frac{\partial \Psi}{\partial t} = -\frac{\hbar^2}{2m} \Delta \Psi + V(\mathbf{x}) \Psi.$$

Thus, for the wave mechanics of relatively simple quantum systems, the Hamiltonian operator is given by the expression $\mathsf{H} = -(\hbar^2/2m)\Delta + V(\mathsf{\mathbf{X}})$, which operates on a suitable dense subspace of $L^2(\mathbb{R}^3)$. Though we have not stated it explicitly, the potential function $V(\mathbf{x})$ may depend on time, in which case it would be more appropriate to write $V(\mathbf{x}, t)$ for the function and the Hamiltonian for the system is then written as $\mathsf{H}(t) = -(\hbar^2/2m)\Delta + V(\mathsf{\mathbf{X}}, t)$. This offers a difficulty that can be quite tricky to handle. Further complications arise with more complicated Hamiltonians as, for example, in the case of a charged particle in an electromagnetic field, which is not covered in these lectures.

The Schrödinger equation is notoriously difficult to solve except in special cases, many of which are presented in the present work. One simplifying assumption that is quite important occurs when the Hamiltonian H is constant in time. In this case, the formal solution to equation 27.1 given by $\psi_t = \mathsf{U}_t \psi_0$, where

$$\mathsf{U}_t = e^{-\mathrm{i}t\mathsf{H}/\hbar},$$

turns out to be correct. Here the exponential expression for U_t is given as follows:

$$e^{-\mathrm{i}t\mathsf{H}/\hbar} = \sum_{k=0}^{\infty}(-1)^k \frac{\mathbf{i}^k t^k}{\hbar^k k!}\mathsf{H}^k, \tag{27.2}$$

an expression that converges once it is applied to appropriate test functions on $\mathbb{R}^3$. It turns out, though, that this exponential expression extends, via the functional calculus of Section 4.2, to a linear operator defined on the whole of $L^2(\mathbb{R}^3)$. Indeed, the spectral theorem for self-adjoint operators of that section guarantees a resolution of the identity $\{\mathsf{E}_\lambda : \lambda \in \mathbb{R}\}$ for H, and the exponential is given as

$$e^{-\mathrm{i}t\mathsf{H}/\hbar} = \int_{\mathbb{R}} e^{-\mathrm{i}t\lambda/\hbar}\, d\mathsf{E}_\lambda.$$

The operator U_t is called the **time evolution operator**. If H is not constant, so that $\mathsf{H} = \mathsf{H}(t)$, then a closed form solution may be impossible to obtain. The situation improves somewhat if the Hamiltonians at various times commute, so that $[\mathsf{H}(s), \mathsf{H}(t)] = 0$ for all s and t. In this case, the time evolution operator takes the form

$$\mathsf{U}_t = \exp\left[-\frac{\mathbf{i}}{\hbar}\int_0^t \mathsf{H}(s)\, ds\right]. \tag{27.3}$$

When the Hamiltonians at different times fail to commute, an array of clever perturbation techniques have been invented that may possibly be used to approximate U_t, but it is just as likely that the problem is intractable.

To avoid difficulties we will assume in the remainder of this section and in the next two sections that H is constant. The fuller picture of the time evolution of a quantum system that emerges upon close study of this case is that the time evolution operator U_t satisfies some quite appealing properties, three of which we expose next. First, time evolution is **unitary**. Indeed, since H is self-adjoint, with $\mathsf{H} = \mathsf{H}^*$, we have that

$$\mathsf{U}_t^* = \left(e^{-\mathrm{i}t\mathsf{H}/\hbar}\right)^* = e^{\mathrm{i}t\mathsf{H}^*/\hbar} = \mathsf{U}_{-t}. \tag{27.4}$$

As $\mathsf{U}_t^{-1} = \mathsf{U}_{-t}$, we get the unitary condition that $\mathsf{U}_t^{-1} = \mathsf{U}_t^*$. It follows that the time evolution of a quantum state vector ψ preserves norms, so that $\|\psi\|_2 = \|\mathsf{U}_t\psi\|_2$, which is to be expected since this condition is a statement of probability conservation: once ψ is normalized, its time-evolved state $\mathsf{U}_t\psi$ remains normalized. Second, the time evolution operator satisfies the **group condition** that $\mathsf{U}_{s+t} = \mathsf{U}_s\mathsf{U}_t$. This is a basic property of the exponentials of operators provided that the operators commute, i.e., $e^{\mathsf{A}+\mathsf{B}} = e^{\mathsf{A}}e^{\mathsf{B}}$ for commuting operators A and B. Finally, U_t is **continuous** in t, as expected for the physical evolution of a system.

These observations together say only that the time evolution of a quantum system with constant Hamiltonian is given by a continuous unitary representation of $\mathbb{R}$ on $\mathcal{H}$. Specifically, the mapping

$$\mathsf{U} : \mathbb{R} \to \mathbf{U}(\mathcal{H})$$

given by $t \mapsto \mathsf{U}_t$ is a continuous group homomorphism of $\mathbb{R}$ into the group $\mathbf{U}(\mathcal{H})$ of unitary operators on $\mathcal{H}$.

An important special case occurs when the energy eigenvalues $E_1 \leq E_2 \leq \cdots$ are discrete and the corresponding normalized eigenvectors $|1\rangle, |2\rangle, \ldots$ form a complete orthonormal set. Notice that we allow for degeneracy of the energy levels by allowing equalities in the sequence of energy eigenvalues. In this special case, since $\mathsf{H}|n\rangle = E_n|n\rangle$, the time evolution of eigenvectors proceeds according to equation 27.2, as a time-periodic phase change of period $2\pi\hbar/E_n$ via $\mathsf{U}_t|n\rangle = e^{-\mathrm{i}E_n t/\hbar}|n\rangle$. Since $\{|n\rangle : n \in \mathbb{N}\}$ is complete and orthonormal, the time evolution representation may be written as

$$\mathsf{U}_t = \mathsf{U}_t \sum_{n=1}^{\infty} |n\rangle\langle n| = \sum_{n=1}^{\infty} e^{-\mathrm{i}E_n t/\hbar}|n\rangle\langle n|,$$

and the time evolution of the arbitrary state vector $\psi = \sum_{n=1}^{\infty} c_n|n\rangle$ proceeds as follows:

$$\psi_t = \mathsf{U}_t\psi = \sum_{n=1}^{\infty} c_n e^{-\mathrm{i}E_n t/\hbar}|n\rangle, \tag{27.5}$$

where $c_n = \langle n|\psi\rangle$. For example, if $|n\rangle = \psi_n$, where ψ_n is the nth excited state of the harmonic oscillator potential, given by equation 11.11, then $E_n = \left(n + \frac{1}{2}\right)\hbar\omega$ is a nondegenerate eigenvalue and equation 27.5 becomes

$$\psi_t = \mathsf{U}_t\psi = e^{-\mathrm{i}\omega t/2} \sum_{n=1}^{\infty} c_n e^{-\mathrm{i}n\omega t}|n\rangle. \tag{27.6}$$

Note that, ignoring the phase factor $e^{-\mathrm{i}\omega t/2}$, this is periodic of period $2\pi/\omega$. Thus the main part of the evolving wave function oscillates with angular frequency ω, the natural frequency of the classical oscillator. The oscillation of the phase at angular frequency $\omega/2$ is a purely quantum phenomenon that arises precisely because the ground state energy is nonzero. This has no measurable effects when attention is restricted to a single oscillator, but does have consequences when the oscillator interacts with other quanta.

To round out our discussion of the harmonic oscillator, we now determine the time evolution of the coherent states of the oscillator. The reader should review the contents of Section 11.4, whose notation we freely use. The coherent state $|c\rangle$ evolves according to equation 27.6 as

$$\begin{aligned} \mathsf{U}_t|c\rangle &= e^{-|c|^2/2}\mathsf{U}_t \sum_{n=0}^{\infty} \frac{c^n}{\sqrt{n!}}\psi_n \\ &= e^{-\mathrm{i}\omega t/2} e^{-|c|^2/2} \sum_{n=0}^{\infty} \frac{\left(c\,e^{-\mathrm{i}\omega t}\right)^n}{\sqrt{n!}}\psi_n = e^{-\mathrm{i}\omega t/2}|c\,e^{-\mathrm{i}\omega t}\rangle. \end{aligned} \tag{27.7}$$

This says that a coherent state remains a coherent state under time evolution and oscillates with angular frequency ω, the natural frequency of the oscillator, though the phase oscillates at half the natural frequency.

27.2 Deriving the Schrödinger Equation: Stone's Theorem

We now turn this discussion around and see how to derive the Schrödinger equation for a constant Hamiltonian from some natural assumptions about time evolution. We use the Dirac notation $|t\rangle$ for the state vector of a system at time t. The starting assumption is that the state vector at time t_2 is obtained from that at time t_1 by a linear map $\mathsf{U}(t_2, t_1) : \mathcal{H} \to \mathcal{H}$, so that $|t_2\rangle = \mathsf{U}(t_2, t_1)|t_1\rangle$. Here is a list of a further four assumptions about $\mathsf{U}(t_2, t_1)$. The first three are reasonable assumptions based on physical grounds that one would expect to hold for almost any fundamental theory, and the fourth is a useful mathematical assumption that is neither unwarranted nor unexpected.

(i) For $t_1 < t_2 < t_3$, $\mathsf{U}(t_3, t_1) = \mathsf{U}(t_3, t_2)\mathsf{U}(t_2, t_1)$. This states that time evolution from t_1 to t_3 should be time evolution from t_1 to t_2 followed by time evolution from t_2 to t_3.

(ii) $\mathsf{U}(t_2, t_1) = \mathsf{U}(t_2 - t_1, 0)$. Time evolution depends only on the difference in time between events and not on an absolute choice of zero time. This assumes a constant Hamiltonian.

(iii) $\langle t_2|t_2\rangle = \langle t_1|t_1\rangle$. This says that time evolution preserves probabilities.

(iv) $\mathsf{U}(t, t) = 1_{\mathcal{H}}$, and U is smooth, meaning that $\mathsf{U}(t, 0)$ may be expanded in a power series in t.

Setting $\mathsf{U}_t = \mathsf{U}(t, 0)$, assumptions (i) and (ii) imply that $\mathsf{U}_{s+t} = \mathsf{U}_s \mathsf{U}_t$ for all s and t. Assumption (iii) implies that U_t is unitary, since $\langle 0|0\rangle = \langle t|t\rangle = \langle \mathsf{U}_t|0\rangle|\mathsf{U}_t|0\rangle\rangle = \langle 0|\mathsf{U}_t^* \mathsf{U}_t|0\rangle$ for all $|0\rangle \in \mathcal{H}$, so $\mathsf{U}_t^* \mathsf{U}_t = 1_{\mathcal{H}}$. Thus, U_t is a unitary representation of $\mathbb{R}$ on $\mathcal{H}$. Assumption (iv) allows us to determine the form of U_t. Expanding in a power series about $t = 0$ gives

$$\mathsf{U}_t = 1_{\mathcal{H}} + \Omega t + \cdots .$$

The operator Ω is defined on an appropriate domain in $\mathcal{H}$, but normally not on the whole of $\mathcal{H}$. As U_t is unitary, we have

$$1_{\mathcal{H}} = \mathsf{U}_t^* \mathsf{U}_t = 1_{\mathcal{H}} + (\Omega^* + \Omega)t + \cdots .$$

It follows that $\Omega^* + \Omega = 0$, so that $\Omega^* = -\Omega$. Setting $\mathsf{H} = \mathbf{i}\hbar\Omega$, we have $\mathsf{H}^* = -\mathbf{i}\hbar\Omega^* = \mathsf{H}$, so that H is self-adjoint, modulo identification of the appropriate domain for Ω, which will be addressed subsequently. Up to first order in t, then, U_t takes the form

$$\mathsf{U}_t = 1_{\mathcal{H}} - \frac{\mathbf{i}}{\hbar}\mathsf{H}t.$$

We have

$$|t + \Delta t\rangle = \mathsf{U}_{\Delta t}|t\rangle = \left(1_{\mathcal{H}} - \frac{\mathbf{i}}{\hbar}\mathsf{H}\Delta t\right)|t\rangle,$$

or

$$\mathbf{i}\hbar \frac{|t + \Delta t\rangle - |t\rangle}{\Delta t} = \mathsf{H}|t\rangle,$$

which is valid to first order in Δt. Taking the limit as Δt approaches zero gives

$$\mathbf{i}\hbar \frac{d|t\rangle}{dt} = \mathsf{H}|t\rangle,$$

which is the Schrödinger equation with constant Hamiltonian H.

For a fixed time t, apply $\mathsf{U}_{t/N}$ N times successively, using the group property, to get

$$\mathsf{U}_t = \left(\mathsf{U}_{t/N}\right)^N \approx \left(1_{\mathcal{H}} - \frac{\mathbf{i}\mathsf{H}t}{\hbar N}\right)^N .$$

Taking the limit as N approaches infinity gives $\mathsf{U}_t = e^{-\mathbf{i}\mathsf{H}t/\hbar}$. The discussion of this section can be made rigorous and precise and is really an outline for the proof of Stone's theorem regarding unitary representations on Hilbert spaces.

Stone's theorem *Let* $\mathsf{U} : \mathbb{R} \to \mathbf{U}(\mathcal{H})$ *be a unitary representation of* $\mathbb{R}$ *on the Hilbert space* $\mathcal{H}$. *Assume that* $\langle\psi|\mathsf{U}_t\psi\rangle$ *is a continuous function of* t *for each* $\psi \in \mathcal{H}$. *Then there exists a unique self-adjoint operator* H, *called the* **infinitesimal generator** *of* U, *such that*

$$\mathsf{U}_t = e^{-\mathbf{i}t\mathsf{H}/\hbar} \tag{27.8}$$

for all $t \in \mathbb{R}$. *The operator* H *is defined on the dense subspace*

$$\operatorname{dom}\mathsf{H} = \{\psi \in \mathcal{H} : (\mathbf{i}\hbar/t)[\mathsf{U}_t - 1_{\mathcal{H}}]\psi \text{ converges as } t \to 0\}.$$

For each $\psi \in \operatorname{dom}\mathsf{H}$, *we have*

$$\mathsf{H}\psi = \mathbf{i}\hbar \lim_{t\to 0} \frac{(\mathsf{U}_t - 1_{\mathcal{H}})\psi}{t}; \tag{27.9}$$

furthermore, from this, the function $\psi(t) = \mathsf{U}_t\psi$ *satisfies the Schrödinger equation* $\mathbf{i}\hbar\psi'(t) = \mathsf{H}\psi(t)$. *Finally, the operator* U_t *commutes with every self-adjoint operator that commutes with the infinitesimal generator* H.

We should note once again that expression 27.8 gives U_t as the exponential of an operator defined by the functional calculus of Section 4.2 using the spectral theorem. It is defined on the whole of $\mathcal{H}$ but is given as a convergent infinite series, as in equation 27.2, only on a proper subdomain of $\mathcal{H}$.[1]

A final note on U_t ends this section. Expression 27.9 allows us to calculate the time derivative of U_t as an operator on $\operatorname{dom}\mathsf{H}$:

$$\frac{d\mathsf{U}_t}{dt} = \lim_{h\to 0} \frac{\mathsf{U}_{t+h} - \mathsf{U}_t}{h} = \lim_{h\to 0} \frac{\mathsf{U}_h - 1_{\mathcal{H}}}{h}\mathsf{U}_t = -\frac{\mathbf{i}}{\hbar}\mathsf{H}\mathsf{U}_t, \tag{27.10}$$

which is exactly as expected for an operator that takes the form $e^{-\mathbf{i}\mathsf{H}t/\hbar}$.

27.3 The Heisenberg Picture and the Heisenberg Equation

In the Schrödinger picture of time evolution, state vectors evolve in time while the operators representing observables remain constant. For the

[1] For those uncomfortable with this behavior, just think of the familiar case of the geometric series $\sum_{n=0}^{\infty} z^n$ for $z \in \mathbb{C}$. This converges to the expression

$$\sum_{n=0}^{\infty} z^n = \frac{1}{1-z} \quad \text{if} \quad |z| < 1,$$

and diverges otherwise. Nonetheless, the function $g(z) = 1/(1-z)$ is defined and analytic on the punctured plane $\mathbb{C} \setminus \{1\}$, though the series defining $g(z)$ in the unit disk diverges when $|z| \geq 1$.

state vector ψ and operator A corresponding to the observable a, the expectation value for a-measurements is

$$\text{Exp}_{\psi}(\mathsf{A}) = \langle \psi | \mathsf{A}\psi \rangle.$$

Evolving ψ to $\psi_t = \mathsf{U}_t\psi$, the expectation becomes

$$\text{Exp}_{\psi_t}(\mathsf{A}) = \langle \psi_t | \mathsf{A}\psi_t \rangle = \langle \mathsf{U}_t\psi | \mathsf{A}\mathsf{U}_t\psi \rangle = \langle \psi | \mathsf{U}_t^* \mathsf{A}\mathsf{U}_t\psi \rangle.$$

This offers an alternative way to think about the time evolution of quantum systems. The only physically measurable quantities in quantum mechanics are expectation values. This equation shows that we get exactly the same expectation values, and therefore an equivalent formulation of the theory, if we think of the operators representing observables as evolving in time, rather than the state vectors as evolving. This **Heisenberg picture** is really just a mathematical change of view, where we consider that a quantum system is represented by an unchanging state vector ψ but the operator A representing the observable a evolves in time via unitary conjugation according to

$$\mathsf{A} \mapsto \mathsf{U}_t^* \mathsf{A}\mathsf{U}_t = \mathsf{A}_t.$$

It is important to recall that we are continuing to assume that the Hamiltonian H has no explicit time dependence. This allows the use of the fact that time evolution is given by the unitary representation $t \mapsto \mathsf{U}_t$ where U_t is given by equation 27.8, and of the fact that $d\mathsf{U}_t/dt$ is given by equation 27.10.

The Schrödinger equation is the differential equation that the time-evolving state vector ψ_t must satisfy. What differential equation must a time-evolving observable operator satisfy? We use the Leibnitz rule for products of operator-valued functions, easily confirmed by the reader, and the fact that $[\mathsf{H}, \mathsf{U}_t] = 0$ in the next calculation:

$$\begin{aligned}\frac{d\mathsf{A}_t}{dt} &= \frac{d}{dt}(\mathsf{U}_t^* \mathsf{A}\mathsf{U}_t) = \frac{d\mathsf{U}_t^*}{dt}\mathsf{A}\mathsf{U}_t + \mathsf{U}_t^*\mathsf{A}\frac{d\mathsf{U}_t}{dt} \\ &= \frac{\mathbf{i}}{\hbar}\mathsf{H}\mathsf{U}_t^*\mathsf{A}\mathsf{U}_t + \mathsf{U}_t^*\mathsf{A}\frac{-\mathbf{i}}{\hbar}\mathsf{H}\mathsf{U}_t = -\frac{\mathbf{i}}{\hbar}(-\mathsf{H}\mathsf{A}_t + \mathsf{A}_t\mathsf{H}) = -\frac{\mathbf{i}}{\hbar}[\mathsf{A}_t, \mathsf{H}].\end{aligned}$$

Thus, the time evolution of the operator representing an observable is governed by the **Heisenberg equation**:

$$\mathbf{i}\hbar\frac{d\mathsf{A}_t}{dt} = [\mathsf{A}_t, \mathsf{H}].$$

Despite its name, this was derived by Dirac in the context of a generalization both to observables that depend explicitly on time and to a so-called *reference evolution*, presented as the interaction picture in the next section.

For now, we examine the one-dimensional harmonic oscillator with Hamiltonian $\mathsf{H} = (1/2m)\mathsf{P}^2 + (m\omega^2/2)\mathsf{X}^2$ in the Heisenberg picture. First,

$$\mathbf{i}\hbar\dot{\mathsf{X}}_t = \mathbf{i}\hbar\frac{d\mathsf{X}_t}{dt} = [\mathsf{X}_t, \mathsf{H}] = \mathsf{U}_t^*\mathsf{X}\mathsf{U}_t\mathsf{H} - \mathsf{H}\mathsf{U}_t^*\mathsf{X}\mathsf{U}_t.$$

Since $[\mathsf{H}, \mathsf{H}] = 0$, Stone's theorem implies that U_t commutes with H and the preceding equation becomes

$$\mathbf{i}\hbar\dot{\mathsf{X}}_t = \mathsf{U}_t^*\,[\mathsf{X}, \mathsf{H}]\,\mathsf{U}_t.$$

Recalling from equation 8.7 that $[\mathsf{X}, \mathsf{H}] = (\mathbf{i}\hbar/m)\mathsf{P}$, we get

$$\mathsf{P}_t = m\dot{\mathsf{X}}_t, \tag{27.11}$$

looking suspiciously like the classical relationship $p = m\dot{x}$ describing the momentum as the product of mass and velocity. In words, the operator representing momentum is the product of the mass and the rate of change of the operator representing position, which is reminiscent of the Ehrenfest results of Lecture 8.

For the momentum operator we have, since $[\mathsf{P}, \mathsf{P}] = 0$,

$$\mathbf{i}\hbar\dot{\mathsf{P}}_t = [\mathsf{P}_t, \mathsf{H}] = \mathsf{U}_t^*\,[\mathsf{P}, \mathsf{H}]\,\mathsf{U}_t = \frac{m\omega^2}{2}\mathsf{U}_t^*\left[\mathsf{P}, \mathsf{X}^2\right]\mathsf{U}_t.$$

Equation 8.4 implies that $[\mathsf{P}, \mathsf{X}^2] = \mathsf{X}[\mathsf{P}, \mathsf{X}] + [\mathsf{P}, \mathsf{X}]\mathsf{X}$, which, with the standard position–momentum commutation relation 8.5, gives

$$\dot{\mathsf{P}}_t = -m\omega^2\mathsf{U}_t^*\mathsf{X}\mathsf{U}_t = -m\omega^2\mathsf{X}_t. \tag{27.12}$$

This, of course, reflects the classical equation of motion, $\dot{p} = m\ddot{x} = -kx = -m\omega^2 x$, for the classical oscillator.

We are now in a position to solve the Heisenberg equations 27.11 and 27.12 exactly as we would solve the classical equations, since their forms agree. Combining the two equations gives

$$\frac{d}{dt}(\mathsf{P}_t \pm \mathbf{i}m\omega\mathsf{X}_t) = -m\omega^2\mathsf{X}_t \pm \mathbf{i}\omega\mathsf{P}_t = \pm\mathbf{i}\omega(\mathsf{P}_t \pm \mathbf{i}m\omega\mathsf{X}_t),$$

whose solution is $(\mathsf{P}_t \pm \mathbf{i}m\omega\mathsf{X}_t) = (\mathsf{P} \pm \mathbf{i}m\omega\mathsf{X})e^{\pm\mathbf{i}\omega t}$. Multiplying and equating real and imaginary parts gives

$$\mathsf{P}_t = \cos\omega t\,\mathsf{P} - m\omega\sin\omega t\,\mathsf{X} \tag{27.13}$$

and

$$m\omega \mathsf{X}_t = m\omega \cos \omega t\, \mathsf{X} + \sin \omega t\, \mathsf{P}. \tag{27.14}$$

Notice that equations 27.13 and 27.14 are obtained from one another by differentiation with respect to time. After noting that $\mathsf{X}_0 = \mathsf{X}$ and $\mathsf{P}_0 = \mathsf{P}$, compare equation 27.14 with the classical expression

$$m\omega x(t) = m\omega x_0 \cos \omega t + p_0 \sin \omega t \tag{27.15}$$

for the displacement of a harmonic oscillator with initial position x_0 and initial momentum $p_0 = m\dot{x}(0)$. The form is the same because the equations of motion in the classical and quantum settings have the same form, but it is striking nonetheless to see this formal agreement between the classical and quantum theories. Equation 27.14 translates to precisely equation 27.15 whenever $x(t)$ is interpreted as the expectation value $x(t) = \langle \mathsf{X}_t \rangle_\psi = \langle \psi | \mathsf{X}_t \psi \rangle$, and similarly for $x_0 = \langle \psi | \mathsf{X} \psi \rangle$ and $p_0 = \langle \psi | \mathsf{P} \psi \rangle$; this is an instantiation of Ehrenfest's theorem.

27.4 Synthesis: The Dirac, or Interaction, Picture

In this section we present a generalization of both the Schrödinger and Heisenberg pictures of time evolution in quantum mechanics. Schrödinger observables are allowed to depend explicitly on time, and the time evolution of the quantum system is measured as the deviation from a reference evolution determined by a reference Hamiltonian. There is wide latitude in the choice of reference Hamiltonian, the Schrödinger and Heisenberg pictures being obtained as polar opposite choices of reference Hamiltonian. Mathematically, the construction of this **interaction** or **Dirac picture** is a straightforward manipulation of the equations that govern evolution in the standard Schrödinger picture, but the resulting change of viewpoint allows for the construction of sophisticated perturbation techniques that can be used to approximate the time evolution of systems in which the Hamiltonian forbids closed form solutions.

Throughout this section, we allow the Hamiltonian $\mathsf{H} = \mathsf{H}(t)$ to depend explicitly on time, but we do require that $[\mathsf{H}(s), \mathsf{H}(t)] = 0$ for all s and t so that the time evolution operator U_t takes the form of equation 27.3. The time evolution of the quantum system will be measured with respect to a constant **reference Hamiltonian** K whose unitary time evolution proceeds via the **reference evolution operator** $\mathsf{V}_t = e^{-it\mathsf{K}/\hbar}$. We allow either for a fixed observable a to be represented by different operators at different times or for measurements of different observables to be made at different times. This situation is represented by a function $\mathsf{A}(t)$, where $\mathsf{A}(t)$ is the self-adjoint operator representing the observable $a(t)$ at time t. The basic requirement for the operator-valued function $\mathsf{A}(t)$ is that it be differentiable in the sense of Section 8.1, and we use $\mathsf{A}'(t)$ to represent the derivative.

In the interaction picture, both states and observables evolve in a way that preserves the expectation values of the observables. This is accomplished by evolving the initial state $\psi = \psi_0$ by

$$\psi_t = \mathsf{V}_t^* \mathsf{U}_t \psi_0$$

and the operator $\mathsf{A}(t)$ according to

$$\mathsf{A}_t = \mathsf{V}_t^* \mathsf{A}(t) \mathsf{V}_t. \tag{27.16}$$

When $\mathsf{K} = 0$, so that $\mathsf{V}_t = 1_{\mathcal{H}}$, the Schrödinger picture is recovered and when $\mathsf{K} = \mathsf{H}$, so that $\mathsf{V}_t = \mathsf{U}_t$, the Heisenberg picture is recovered. To see that this interaction evolution gives an equivalent theory to those of Schrödinger and Heisenberg, we will calculate expectation values. Since

$$\langle \psi_t | \mathsf{A}_t \psi_t \rangle = \langle \mathsf{V}_t^* \mathsf{U}_t \psi_0 | \mathsf{V}_t^* \mathsf{A}(t) \mathsf{V}_t \mathsf{V}_t^* \mathsf{U}_t \psi_0 \rangle = \langle \mathsf{U}_t \psi_0 | \mathsf{A}(t) \mathsf{U}_t \psi_0 \rangle,$$

the interaction picture gives precisely the same expectation values as the Schrödinger. Continuing the calculation,

$$\langle \psi_t | \mathsf{A}_t \psi_t \rangle = \langle \mathsf{U}_t \psi_0 | \mathsf{A}(t) \mathsf{U}_t \psi_0 \rangle = \langle \psi_0 | \mathsf{U}_t^* \mathsf{A}(t) \mathsf{U}_t \psi_0 \rangle,$$

confirming equivalence with the Heisenberg picture.

To derive the equations of motion, first we need

$$\frac{d\psi_t}{dt} = \frac{d}{dt} \mathsf{V}_t^* \mathsf{U}_t \psi_0 = \left(\frac{\mathbf{i}}{\hbar} \mathsf{K}^* \mathsf{V}_t^* \mathsf{U}_t + \mathsf{V}_t^* \left(-\frac{\mathbf{i}}{\hbar} \mathsf{H} \right) \mathsf{U}_t \right) \psi_0,$$

which, upon rearrangement, yields

$$\mathbf{i}\hbar \frac{d\psi_t}{dt} = \left(\mathsf{V}_t^* \mathsf{H} \mathsf{U}_t - \mathsf{K} \mathsf{V}_t^* \mathsf{U}_t \right) \psi_0 = \left(\mathsf{V}_t^* \mathsf{H} \mathsf{V}_t - \mathsf{K} \right) \mathsf{V}_t^* \mathsf{U}_t \psi_0.$$

Setting $\widetilde{\mathsf{H}} = \mathsf{V}_t^* \mathsf{H} \mathsf{V}_t - \mathsf{K}$ gives the equation of motion for the state vector in the interaction picture as the Schrödinger equation

$$\mathbf{i}\hbar \frac{d\psi_t}{dt} = \widetilde{\mathsf{H}} \psi_t, \tag{27.17}$$

with Hamiltonian $\widetilde{\mathsf{H}}$.

To derive now the equation of motion for the operators representing observables, note that

$$\frac{dA_t}{dt} = \frac{d}{dt} V_t^* A(t) V_t = V_t^* A'(t) V_t + \frac{i}{\hbar}(K V_t^* A(t) V_t - V_t^* A(t) K V_t),$$

giving the equation of motion as the Heisenberg equation with Hamiltonian K:

$$\frac{dA_t}{dt} = \frac{\partial A_t}{\partial t} - \frac{i}{\hbar}[A_t, K], \tag{27.18}$$

where $\partial A_t / \partial t = V_t^* A'(t) V_t$. Compare this result with the quantum evolution equation 8.3, which gives the time rate of change of expectation values.

These equations show that, in the interaction picture with Hamiltonian H and reference Hamiltonian K, states evolve via Schrödinger evolution with Hamiltonian $\widetilde{H}$, which is the difference between the K-evolved Hamiltonian H and the reference Hamiltonian, and operators evolve via Heisenberg evolution with respect to the reference Hamiltonian K. The interaction picture is used, for example, when the Hamiltonian H takes the form $H = K + \Delta K$, where ΔK is a small perturbation away from a Hamiltonian K of a system that is well understood and solvable, this being the rationale for perturbation expansions in quantum mechanics. We will not cover the various perturbation theories for approximating solutions to the Schrödinger equation in these lectures but instead refer the reader to the references in the End Notes.

A nice way to think of this interaction picture is that it is Schrödinger evolution taking place in a rotating frame of reference in the state space. Rather than fixing our frame of reference, we use the frame rotated by the representation V^*. This is accomplished at time t by multiplying all states by V_t^* and applying unitary conjugation by V_t to move back and forth between the rotated and nonrotated frames. Specifically, at time t, the initial state vector ψ_0 has rotated to $V_t^* \psi_0$. The time-evolved state vector $U_t \psi_0$ has rotated to $\psi_t = V_t^* U_t \psi_0$, which, when written as $V_t^* U_t V_t \cdot V_t^* \psi_0$, can be seen as the rotated state vector $V_t^* \psi_0$ rotated back by V_t, then time-evolved by U_t, and finally rotated again by V_t^*. The observable a is represented in the rotated frame by the operator A_t, as given by equation 27.16, which, when applied to ψ_t, just rotates ψ_t back via V_t, applies $A(t)$, and then rotates the result by V_t^*.

27.5 End Notes

This lecture is an expansion of Hannabuss's (1997) discussion of the Schrödinger and Heisenberg pictures of quantum mechanics, found in his short Sections 11.1 and 11.2. The interaction picture of Dirac is the basis of time-dependent perturbation theory. In these lectures I neglect the very important topic of perturbation theory in its various forms, which is of great use in approximating solutions to quantum mechanical problems that have no closed form analytic solutions; these constitute the majority of the problems that arise for the practicing physicist. The interested reader may find perturbation treatments of quantum mechanical problems in most graduate texts on quantum mechanics. Hannabuss covers stationary perturbation theory in his Chapter 12 and iterative perturbation theory in his Chapter 13.

Stone's theorem may be found in most books on functional analysis, but here I mention the classic text of Riesz and Sz.-Nagy (1955) and the modern work of Blanchard and Brüning (2003).

28

Unitary Representations and Conservation Laws

> When the ancients declared the circle the most perfect shape, they meant that it was the most symmetric. Each point on the orbit is the same as any other. The principles that are hardest to give up are those that appeal to our need for symmetry and elevate an observed symmetry to a necessity. Modern physics is based on a collection of symmetries, which are believed to enshrine the most basic principles. No less than the ancients, many modern theorists believe instinctively that the fundamental theory must be the most symmetric possible law. Should we trust this instinct, or should we listen to the lesson of history, which tells us that (as in the example of the planetary orbits) nature becomes less rather than more symmetric the closer we look?
>
> Lee Smolin
> *The Trouble with Physics*, 2006

Stone's theorem, discussed in Section 27.2, is a powerful statement about continuous unitary representations of $\mathbb{R}$ on the state Hilbert space $\mathcal{H}$. In Lecture 27 it was used to derive the Schrödinger equation for the evolution of a quantum state vector when the Hamiltonian is constant, using only the reasonable assumption that time evolution should be continuous, unitary, and additive. This is just one application of a theorem that has far reaching implications in mathematics and physics. In the present lecture we will see illustrations of the use of Stone's theorem in deriving quantum conservation laws from symmetries of the Hamiltonian.

In the first section, we use Stone's theorem to determine the infinitesimal generators of unitary representations other than those derived from the time evolution of a quantum system. In particular, we examine unitary representations that arise from symmetries under Euclidean motions – translations and rotations. In the next section we examine how conservation laws arise from symmetries and link this to our discussion in Section 6.4 of Noether's Theorem in classical mechanics. It will be seen that the emergence of conservation laws from symmetries is more direct and transparent in quantum mechanics than in classical mechanics. We then examine symmetry under global and local phase transformations, which are purely quantum mechanical symmetries not present in classical mechanics, and we will find lurking there the first hint of the inevitability and necessity of the electromagnetic interaction in the development of quantum mechanics. Developing this idea further, we will see how the enforcement of local phase symmetry necessarily gives rise to a mediating field that has the precise properties that define the electromagnetic field. The ideas introduced in this section are developed more fully in courses on relativistic quantum mechanics in which this "gauge principle" is explored and the electromagnetic interaction is seen as a necessary consequence of local phase symmetry. This is surely one of the most profound implications of quantum mechanics and provides a model for introducing other interactions into the theory by requiring local symmetries based on groups other than the phase group.

28.1 Euclidean Symmetries

Throughout this lecture, we work in the context of wave mechanics where the state space of a single quantum particle is $\mathcal{H} = L^2(\mathbb{R}^3)$, the space of square-integrable complex functions on $\mathbb{R}^3$. Define for each $\mathbf{a} \in \mathbb{R}^3$ a unitary transformation $\mathsf{U}(\mathbf{a}) : \mathcal{H} \to \mathcal{H}$ by

$$(\mathsf{U}(\mathbf{a})\psi)(\mathbf{x}) = \psi(\mathbf{x} - \mathbf{a}). \tag{28.1}$$

Obviously $\mathsf{U}(\mathbf{a} + \mathbf{b}) = \mathsf{U}(\mathbf{a})\mathsf{U}(\mathbf{b})$, confirming that $\mathsf{U} : \mathbb{R}^3 \to \mathbf{U}(\mathcal{H})$ is a unitary representation of $\mathbb{R}^3$ on $\mathcal{H}$. Calculating the expectation value for the position operator triad $\mathbf{X} = (\mathsf{X}_1, \mathsf{X}_2, \mathsf{X}_3)$, we have, using a normalized wave function ψ,

$$\begin{aligned}
\langle \mathsf{X}_i \rangle_{\mathsf{U}(\mathbf{a})\psi} &= \int_{\mathbb{R}^3} \overline{\psi}(\mathbf{x} - \mathbf{a}) x_i \psi(\mathbf{x} - \mathbf{a})\, d\mathbf{x} \\
&= \int_{\mathbb{R}^3} \overline{\psi}(\mathbf{x})(x_i + a_i)\psi(\mathbf{x})\, d\mathbf{x} = \langle \mathsf{X}_i \rangle_\psi + a_i.
\end{aligned}$$

The expected position is then

$$\langle \mathbf{X} \rangle_{\mathsf{U}(\mathbf{a})\psi} = \langle \mathbf{X} \rangle_\psi + \mathbf{a},$$

which is obvious from the fact that ψ and $\mathsf{U}(\mathbf{a})\psi$ are position probability amplitudes, the second being a translation of the first by the vector $\mathbf{a}$. It is easy to see also that momentum expectation values remain unchanged, with $\langle \mathsf{P} \rangle_{\mathsf{U}(\mathbf{a})\psi} = \langle \mathsf{P} \rangle_\psi$.

The expression $\langle \psi | \mathsf{U}(t\mathbf{a})\psi \rangle$ is continuous in the real parameter t whenever ψ is continuous and, as the continuous square-integrable functions are dense in $L^2(\mathbb{R}^3)$, it is continuous for all ψ. It follows that Stone's theorem applies to give the infinitesimal generator of the one-parameter **translation representation** $\mathsf{U}_\mathbf{a}(t) = \mathsf{U}(t\mathbf{a})$ in the direction of $\mathbf{a}$. We denote this infinitesimal generator by $\mathsf{P}_\mathbf{a}$ and evaluate it from the Schrödinger equation $\mathbf{i}\hbar\psi'(t) = \mathsf{P}_\mathbf{a}\psi(t)$, where $\psi(t) = \mathsf{U}_\mathbf{a}(t)\psi$:

$$\begin{aligned}(\mathsf{P}_\mathbf{a}\psi)(\mathbf{x}) &= \mathbf{i}\hbar\frac{d}{dt}\mathsf{U}(t\mathbf{a})\psi\Big|_{t=0}(\mathbf{x}) = \mathbf{i}\hbar\frac{d}{dt}\psi(\mathbf{x}-t\mathbf{a})\Big|_{t=0} \\ &= \mathbf{i}\hbar\left(\nabla\psi(\mathbf{x}-t\mathbf{a})\right)\cdot(-\mathbf{a})\big|_{t=0} = -\mathbf{i}\hbar\,\mathbf{a}\cdot\nabla\psi(\mathbf{x}) = (\mathbf{a}\cdot\mathsf{P})\psi(\mathbf{x}).\end{aligned} \tag{28.2}$$

Here, $\mathsf{P} = -\mathbf{i}\hbar(\partial_1, \partial_2, \partial_3)$ is the momentum operator triad and we find that $\mathsf{P}_\mathbf{a} = \mathbf{a}\cdot\mathsf{P}$, the angular momentum operator in the direction of $\mathbf{a}$, scaled by $\|\mathbf{a}\|$. Scaling to unity by requiring that $\|\mathbf{a}\| = 1$, we find that the infinitesimal generator of the translation representation in the direction of $\mathbf{a}$ is the momentum operator in the direction of $\mathbf{a}$.

The argument just given was a little sloppy concerning the domains of operators. Stone's theorem gives precise information about the domain of the infinitesimal generator. Indeed in this case, we can identify the domain of $\mathsf{P}_\mathbf{a}$ as

$$\begin{aligned}\operatorname{dom}\mathsf{P}_\mathbf{a} &= \left\{\psi\in\mathcal{H} : \lim_{t\to 0}\frac{(\mathsf{U}(t\mathbf{a}) - 1_\mathcal{H})\psi}{t}\text{ exists}\right\} \\ &= \left\{\psi\in\mathcal{H} : \lim_{t\to 0}\frac{\psi(\mathbf{x}-t\mathbf{a}) - \psi(\mathbf{x})}{t}\text{ exists}\right\}.\end{aligned}$$

At first sight it might seem that this identifies the domain as those ψ for which the directional derivative in the direction of $\mathbf{a}$ at all points $\mathbf{x}$ of $\mathbb{R}^3$ exists, since the limit is that of the difference quotient of calculus. However, remember that the limit is not the pointwise limit of calculus but is taken in the L^2-norm of $\mathcal{H} = L^2(\mathbb{R}^3)$. Though this domain does include those ψ that are pointwise differentiable, it also includes those ψ that have a generalized derivative, in the sense that there exists $\varphi\in\mathcal{H}$ for which

$$\lim_{t\to 0}\int_{\mathbb{R}^3}\left|\frac{\psi(\mathbf{x}-t\mathbf{a}) - \psi(\mathbf{x})}{t} - \varphi(\mathbf{x})\right|^2 d\mathbf{x} = 0.$$

This includes in the domain of $\mathsf{P}_{\mathbf{a}}$ the functions in $\mathcal{H}$ that are (pointwise) differentiable almost everywhere as well as more pathological examples. The derivation of expression 28.2 is valid for those ψ for which the directional derivative along $\mathbf{a}$ does in fact exist.

Having determined the infinitesimal generator of the translation representation $\mathsf{U}(t\mathbf{a})$, we may write

$$\mathsf{U}(t\mathbf{a}) = e^{-\mathbf{i}t\mathsf{P}_{\mathbf{a}}/\hbar}, \tag{28.3}$$

where of course the exponential is defined by the functional calculus and has as its domain the whole of $\mathcal{H}$. It is interesting to explore the effect of this expression when $t = -1$ on an infinitely differentiable function $\psi \in \mathcal{H}$. In this case, when the exponential is expanded in a power series and applied to ψ, it yields

$$\psi(\mathbf{x}+\mathbf{a}) = \mathsf{U}(-\mathbf{a})\psi(\mathbf{x}) = e^{\mathbf{i}\mathsf{P}_{\mathbf{a}}/\hbar}\psi(\mathbf{x}) = \sum_{n=0}^{\infty}\frac{1}{n!}(\mathbf{a}\cdot\nabla)^n\psi(\mathbf{x}). \tag{28.4}$$

This is nothing more than the Taylor series expansion of the C^∞ function ψ about the point $\mathbf{x}$. When ψ is smooth, so that the Taylor series converges to $\psi(\mathbf{x}+\mathbf{a})$, expression 28.4 is valid and the equality of equation 28.3 can be interpreted precisely as a statement of Taylor's theorem. Of course the series expression of equation 28.4 is valid only for smooth functions in $\mathcal{H}$, while the expression of formula 28.3 applies to all functions in $\mathcal{H}$.

Consider now the unitary representation $\widehat{\mathsf{U}} : \mathbb{R}^3 \to \mathbf{U}(\mathcal{H})$ defined by

$$\left(\widehat{\mathsf{U}}(\mathbf{a})\psi\right)(\mathbf{x}) = e^{-\mathbf{i}\mathbf{a}\cdot\mathbf{x}/\hbar}\psi(\mathbf{x}). \tag{28.5}$$

Evidently, the form of equation 28.5 implies that the infinitesimal generator of the real representation $\widehat{\mathsf{U}}_{\mathbf{a}}(t) = \widehat{\mathsf{U}}(t\mathbf{a})$ is the generalized position operator $\mathsf{X}_{\mathbf{a}}$ defined by the expression $(\mathsf{X}_{\mathbf{a}}\psi)(\mathbf{x}) = (\mathbf{a}\cdot\mathbf{x})\psi(\mathbf{x})$. We may write this as $\mathsf{X}_{\mathbf{a}} = \mathbf{a}\cdot\mathbf{X}$, where $\mathbf{X}$ is the position operator triad. This is known as a one-parameter **momentum representation** because it corresponds to a one-parameter translation representation by the Fourier transform. Indeed, $\widehat{\mathsf{U}}_{\mathbf{a}}$ is a translation representation when viewed in momentum space, in the sense that the Fourier transform conjugates $\widehat{\mathsf{U}}_{\mathbf{a}}$ to the one-parameter translation representation $\mathsf{U}_{\mathbf{a}}$. This may be confirmed by the observation that

$$\mathsf{U}_{\mathbf{a}} = \mathcal{F}^{-1}\widehat{\mathsf{U}}_{\mathbf{a}}\mathcal{F}, \tag{28.6}$$

which follows either by direct calculation or from the fact that $\mathsf{P}_{\mathbf{a}} =$

$\mathcal{F}^{-1}\mathsf{X}_{\mathbf{a}}\mathcal{F}$. We see here the symmetry between the position and momentum representations, guaranteed by the fact that the Fourier transform $\mathcal{F}$ is an isometry that carries the position operator to the momentum operator, and vice versa; This is a standard and powerful characteristic of the Fourier transform as already articulated in Lecture 10. This relationship between the position and momentum operators and their induced unitary representations is a special case of the following fact. If U is unitary and A is self-adjoint with domain $\mathcal{D}(\mathsf{A})$ then the operator $\mathsf{B} = \mathsf{U}^*\mathsf{A}\mathsf{U}$ is self-adjoint on $\mathcal{D}(\mathsf{B}) = \mathsf{U}^*\mathcal{D}(\mathsf{A})$ and $e^{-it\mathsf{B}/\hbar} = \mathsf{U}^* e^{-it\mathsf{A}/\hbar}\mathsf{U}$.

We now turn our attention to Euclidean rotations. Let $R = R(t, \mathbf{n}) \in \mathrm{SO}(3)$ be the 3×3 orthonormal matrix that represents a positive rotation of $\mathbb{R}^3$ through angle t about the axis $\mathbf{n}$, a unit vector in $\mathbb{R}^3$. Define the operator $\mathsf{U}(R)$ on $\mathcal{H}$ by $(\mathsf{U}(R)\psi)(\mathbf{x}) = \psi(R^{-1}\mathbf{x})$ and observe that, since rotations preserve volume, $\mathsf{U}(R)$ is unitary. We have $\mathsf{U}(R)\mathsf{U}(S) = \mathsf{U}(RS)$, so that $\mathsf{U} : \mathrm{SO}(3) \to \mathbf{U}(\mathcal{H})$ is seen to be a unitary representation of $\mathrm{SO}(3)$ on $\mathcal{H}$. With $\mathbf{n}$ a fixed unit vector, let $R_{\mathbf{n}}(t) = R(t, \mathbf{n})$, so that the mapping $R_{\mathbf{n}} : \mathbb{R} \to \mathrm{SO}(3)$ is a homomorphism and $\mathsf{U}_{\mathbf{n}} = \mathsf{U} \circ R_{\mathbf{n}}$ is a continuous unitary representation of $\mathbb{R}$ on $\mathcal{H}$, termed a one-parameter **rotation representation**. Let $\mathsf{J}_{\mathbf{n}}$ be its infinitesimal generator. Applying Stone's theorem and the vector operator identity $(\mathbf{n} \times \mathbf{x}) \cdot \nabla = \mathbf{n} \cdot (\mathbf{x} \times \nabla)$ we have, for appropriately differentiable ψ,

$$\begin{aligned} \mathsf{J}_{\mathbf{n}}\psi(\mathbf{x}) &= \mathbf{i}\hbar \frac{d}{dt}\psi(R_{\mathbf{n}}(-t)\mathbf{x}) \bigg|_{t=0} \\ &= \mathbf{i}\hbar \left(\nabla\psi(R_{\mathbf{n}}(-t)\mathbf{x}) \cdot (-\mathbf{n} \times \mathbf{x})\right) \bigg|_{t=0} \\ &= -\mathbf{i}\hbar\, \mathbf{n} \cdot (\mathbf{x} \times \nabla)\psi(\mathbf{x}) = (\mathbf{n} \cdot \mathsf{L})\psi(\mathbf{x}), \end{aligned}$$

where $\mathsf{L} = \mathsf{X} \times \mathsf{P}$ is the orbital angular momentum triad of operators defined in Section 12.1. This identifies the infinitesimal generator $\mathsf{J}_{\mathbf{n}}$ as the orbital angular momentum operator $\mathbf{n} \cdot \mathsf{L}$ along the axis $\mathbf{n}$.

This development reflects the classical fact that the infinitesimal generator for the orthogonal representation $R_{\mathbf{n}} : \mathbb{R} \to \mathrm{SO}(3)$ of $\mathbb{R}$ on $\mathbb{R}^3$ is $J_{\mathbf{n}} = \mathbf{n} \cdot \boldsymbol{J}$, where $\boldsymbol{J} = (J_{\mathbf{i}}, J_{\mathbf{j}}, J_{\mathbf{k}})$ is the triad of generators of the Lie algebra $\mathfrak{so}(3)$ given by

$$J_{\mathbf{k}} = \frac{d}{dt}R_{\mathbf{k}}(t)\bigg|_{t=0} = \frac{d}{dt}\begin{pmatrix} \cos t & \sin t & 0 \\ -\sin t & \cos t & 0 \\ 0 & 0 & 1 \end{pmatrix}\Bigg|_{t=0} = \begin{pmatrix} 0 & -1 & 0 \\ 1 & 0 & 0 \\ 0 & 0 & 0 \end{pmatrix}$$

and, similarly,

$$J_{\mathbf{i}} = \begin{pmatrix} 0 & 0 & 0 \\ 0 & 0 & -1 \\ 0 & 1 & 0 \end{pmatrix} \quad \text{and} \quad J_{\mathbf{j}} = \begin{pmatrix} 0 & 0 & 1 \\ 0 & 0 & 0 \\ -1 & 0 & 0 \end{pmatrix}.$$

The commutation relation $[J_{\mathbf{m}}, J_{\mathbf{n}}] = J_{\mathbf{m}\times\mathbf{n}}$ in $\mathfrak{so}(3)$ is mirrored by the commutation relation $[\mathsf{J}_{\mathbf{m}}, \mathsf{J}_{\mathbf{n}}] = \mathbf{i}\hbar \mathsf{J}_{\mathbf{m}\times\mathbf{n}}$ between angular momentum operators. The equivalence of these two relations follows from a straightforward computation using the position–momentum commutation relations and the fact that $\mathsf{J}_{\mathbf{n}} = -\mathbf{X}^{\mathrm{Tr}} J_{\mathbf{n}} \mathbf{P}$, where the triads are interpreted as column vectors.

Now that we have determined the infinitesimal generators of the representations induced by Euclidean rigid motions, in the next section we examine their use in articulating conservation laws.

28.2 Conservation Laws

Recall from Section 6.4 that symmetries of the Lagrangian under a local differentiable one-parameter family of diffeomorphisms give rise to conserved quantities in classical mechanics. We will see now that symmetries of the Hamiltonian H give rise to conserved quantities in quantum mechanics. In the quantum mechanical case the conservation laws follow more easily than in the classical case, because the classical theory is not linear and the conserved quantities must be teased out by linearizing the nonlinear equations. Quantum theory as a whole is linear, and this makes for an easy, straightforward, quantum version of Noether's theorem.

Let $\mathsf{U}_t = e^{-\mathbf{i}t\mathsf{H}/\hbar}$ be the time-evolution operator for the constant Hamiltonian H and let $\psi \in \mathcal{D}(\mathsf{H})$. Assume that V is a unitary operator on $\mathcal{H}$ that commutes with the Hamiltonian in the sense that $\mathsf{V}\mathcal{D}(\mathsf{H}) = \mathcal{D}(\mathsf{H})$ and $\mathsf{V}\mathsf{H}\mathsf{V}^* = \mathsf{H}$ on $\mathcal{D}(\mathsf{H})$. Note that time evolution under H satisfies $\mathsf{V}\mathsf{U}_t\mathsf{V}^* = \mathsf{U}_t$ and, as the vector function $\psi(t) = \mathsf{U}_t\psi$ is a solution to the Schrödinger equation with Hamiltonian H and initial condition ψ, an easy calculation shows that the vector function $\varphi(t) = \mathsf{V}\psi(t) = \mathsf{U}_t\mathsf{V}\psi$ is a solution with initial condition $\varphi = \mathsf{V}\psi$. We say that the Hamiltonian is invariant under conjugation by V and call V a **symmetry** of H.

Now assume that each unitary transformation $\mathsf{V}(s)$, for each $s \in \mathbb{R}$, commutes with the Hamiltonian in the sense of the preceding paragraph, where $\mathsf{V} : \mathbb{R} \to \mathbf{U}(\mathcal{H})$ is a continuous unitary representation. Thus, V

serves as a **one-parameter group of symmetries** of the Hamiltonian. Our aim is to show that the expectation value of the observable represented by the infinitesimal generator A of V is constant in time, giving a quantum conservation law reminiscent of the classical Noether's theorem. The first thing to notice is that the self-adjoint operator A commutes with the time-evolution operator U_t. Modulo a proper consideration of the appropriate domains, which we will ignore, this is confirmed as follows. Using equation 27.10 and the fact that $[\mathsf{V}(s), \mathsf{U}_t] = 0$ for all s and t, we have

$$\begin{aligned} 0 = \frac{d}{ds}[\mathsf{V}(s), \mathsf{U}_t] &= -\frac{\mathbf{i}}{\hbar}\left(\mathsf{A}\mathsf{V}(s)\mathsf{U}_t - \mathsf{U}_t\mathsf{A}\mathsf{V}(s)\right) \\ &= -\frac{\mathbf{i}}{\hbar}\mathsf{V}(s)\left(\mathsf{A}\mathsf{U}_t - \mathsf{U}_t\mathsf{A}\right), \end{aligned}$$

confirming that $[\mathsf{A}, \mathsf{U}_t] = 0$ on the appropriate domain. To see that the expectation values of A are constant, just note that

$$\langle\psi(t)|\mathsf{A}\psi(t)\rangle = \langle\mathsf{U}_t\psi|\mathsf{A}\mathsf{U}_t\psi\rangle = \langle\mathsf{U}_t\psi|\mathsf{U}_t\mathsf{A}\psi\rangle = \langle\psi|\mathsf{A}\psi\rangle,$$

since U_t is unitary. As an example, let V be the translation representation $\mathsf{U}_\mathbf{a}$ in the direction of $\mathbf{a} \in \mathbb{R}^3$, so that $\mathsf{V} = \mathsf{U}_\mathbf{a}$ is a one-parameter group of symmetries of the Hamiltonian H. In this case, we say that H is invariant under translations in the direction $\mathbf{a}$. A quick calculation shows that this occurs precisely when the potential function $V(\mathbf{x})$ is invariant under translations in the direction of $\mathbf{a}$, in the sense that $V(\mathbf{x} - s\mathbf{a}) = V(\mathbf{x})$ for all $s \in \mathbb{R}$. Since the infinitesimal generator of $\mathsf{U}_\mathbf{a}$ is the momentum operator $\mathsf{P}_\mathbf{a} = \mathbf{a} \cdot \mathsf{P}$, we find that the expectation value of momentum in the direction of $\mathbf{a}$ is conserved. In particular, when V is identically zero, so that H describes a free quantum particle, momentum is conserved in all directions.

As a second example, let V be the rotation representation $\mathsf{U}_\mathbf{n}$ with axis $\mathbf{n} \in \mathbb{R}^3$, so that $\mathsf{V} = \mathsf{U}_\mathbf{n}$ is a one-parameter group of symmetries of the Hamiltonian H. In this case we say that H is invariant under positive rotations with axis $\mathbf{n}$. This occurs precisely when the potential function $V(\mathbf{x})$ is invariant under rotations about the axis $\mathbf{n}$, in the sense that $V(R_\mathbf{n}(s)\mathbf{x}) = V(\mathbf{x})$ for all $s \in \mathbb{R}$. This in turn follows from the fact that the Laplacian operator is symmetric with respect to rotations in that $(\Delta\psi)(R\mathbf{x}) = \Delta(\psi \circ R)(\mathbf{x})$ whenever R is a rotation.[1] Since the infinitesimal generator of $\mathsf{U}_\mathbf{n}$ is the angular momentum operator $\mathsf{J}_\mathbf{n} = \mathbf{n} \cdot \mathsf{L}$, the expectation value of angular momentum along the axis $\mathbf{n}$ is

[1] See p. 208.

conserved. In particular, when V is a centrally symmetric potential, angular momentum is conserved.

28.3 Phase Transformations

In this section we consider a purely quantum mechanical symmetry, which has no classical analogue. Let $\mathrm{U}(1)$ be the group of unit complex numbers under multiplication, the **circle group**, and let $\exp : \mathbb{R} \to \mathrm{U}(1)$ be the exponential mapping $\exp(\chi) = e^{\mathrm{i}\chi}$. Define the representation $\mathsf{E} : \mathrm{U}(1) \to \mathbf{U}(\mathcal{H})$ by $\mathsf{E}(\mathbf{u})\psi = \mathbf{u}\psi$, so that the composition $\mathsf{E} \circ \exp$ is a continuous unitary representation of $\mathbb{R}$ on $\mathcal{H}$. Explicitly, $\mathsf{E} \circ \exp(\chi)\psi = e^{\mathrm{i}\chi}\psi$. The representation E is called the **global** $\mathrm{U}(1)$ **representation** on $\mathcal{H}$ and the state vector $e^{\mathrm{i}\chi}\psi$ is said to be obtained from ψ by a **global phase transformation**. The physics described by the Hamiltonian H and the state vector ψ is invariant under such phase transformations. Indeed, all expectation values are preserved since $\langle e^{\mathrm{i}\chi}\psi|\mathsf{A}e^{\mathrm{i}\chi}\psi\rangle = \langle\psi|\mathsf{A}\psi\rangle$ for any linear operator A and, as $\mathsf{E}\circ\exp(\chi)$ commutes with H, $e^{\mathrm{i}\chi}\psi(t)$ is a solution to the Schrödinger equation whenever $\psi(t)$ is. Of course, this is built into the framework of quantum mechanics in Axiom 1, which states, in part, that two vectors represent the same state whenever they are nonzero complex multiples of one another. We say that quantum mechanics is symmetric under global phase transformations.

The discussion gets interesting when we relax the definition of a phase transformation to allow for local influences in both time and space. Rather than requiring that χ be a real constant, what effect does allowing χ to depend explicitly on the time t and the spatial location $\mathbf{x}$ have on the wave function? In exploring this question, first consider the case where χ depends only on spatial location, so that $\chi = \chi(\mathbf{x})$. The question before us is how the wave function $\psi(\mathbf{x})$ differs in physical content from $\varphi(\mathbf{x}) = e^{\mathrm{i}\chi(\mathbf{x})}\psi(\mathbf{x})$. The latter wave function is said to be obtained from the former by a **local phase transformation**. The wave functions differ merely by a unimodular factor that depends on the spatial location $\mathbf{x}$, and it is obvious that the position probability distribution functions agree in that $|\psi(\mathbf{x})|^2 = |\varphi(\mathbf{x})|^2$. In particular, the expectation values of the position observables agree, as $\langle\mathsf{X}_{\mathbf{n}}\rangle_\psi = \langle\mathsf{X}_{\mathbf{n}}\rangle_\varphi$ for all directions $\mathbf{n}$. This occurs because the local phase transformation commutes with the position operator $\mathsf{X}_{\mathbf{n}}$. This generalizes, since $\langle\mathsf{A}\rangle_\psi = \langle\mathsf{A}\rangle_\varphi$ for any linear operator A that commutes with the local phase transformation. However, there are observables whose corresponding self-adjoint operators fail to commute with the local phase transformation. Indeed, any operator whose application involves coordinate differentiation would not be expected to commute with the local phase transformation. For example, the momentum operator $\mathsf{P}_{\mathbf{a}}$ in the direction $\mathbf{a}$ does not commute with the local phase transformation since a straightforward calculation gives

$$(\mathsf{P}_{\mathbf{a}} - \hbar\mathbf{a}\cdot\nabla\chi)\, e^{\mathrm{i}\chi}\psi = e^{\mathrm{i}\chi}\mathsf{P}_{\mathbf{a}}\psi. \tag{28.7}$$

It follows easily that

$$\langle\mathsf{P}_{\mathbf{a}}\rangle_{\varphi} = \langle\mathsf{P}_{\mathbf{a}}\rangle_{\psi} + \hbar\mathbf{a}\cdot\nabla\chi. \tag{28.8}$$

When $\mathbf{a}\cdot\nabla\chi = 0$, the expectation values of the momenta in the direction $\mathbf{a}$ with respect to ψ and φ coincide. This occurs for instance when χ is constant, meaning that the local phase transformation is actually global, and also when the vector $\mathbf{a}$ is orthogonal to the gradient $\nabla\chi$ at each point of $\mathbb{R}^3$.

We have already seen an example of a local phase transformation in the unitary representation $\widehat{\mathsf{U}}$ of $\mathbb{R}^3$ on $\mathcal{H}$, where $\widehat{\mathsf{U}}(\mathbf{b})\psi = e^{-\mathrm{i}\mathbf{b}\cdot\mathbf{x}/\hbar}\psi$. Setting $\chi(\mathbf{x}) = -\mathbf{b}\cdot\mathbf{x}/\hbar$ gives $\mathbf{a}\cdot\nabla\chi = -\mathbf{a}\cdot\mathbf{b}/\hbar$ and equation 28.8 reads $\langle\mathsf{P}_{\mathbf{a}}\rangle_{\widehat{\mathsf{U}}(\mathbf{b})\psi} = \langle\mathsf{P}_{\mathbf{a}}\rangle_{\psi} - \mathbf{a}\cdot\mathbf{b}$. Setting $\mathbf{a}$ as the unit vector $\mathbf{b}/\|\mathbf{b}\|$ in the direction of $\mathbf{b}$, we see that the effect of $\widehat{\mathsf{U}}(\mathbf{b})$ on ψ is to produce a state vector $\varphi = \widehat{\mathsf{U}}(\mathbf{b})\psi$ that has the same position distribution as ψ but whose momentum distribution is changed in such a way that the expectation value of the momentum in the direction of $\mathbf{b}$ is decreased by the amount $\|\mathbf{b}\|$. A bit more is true. Recall that the Fourier transform of a state vector gives the amplitude distribution for momentum measurements. An easy calculation shows that $\mathcal{F}\varphi = \mathcal{F}\widehat{\mathsf{U}}_{\mathbf{b}}\psi = \mathsf{U}_{-\mathbf{b}}\mathcal{F}\psi$, similar to equation 28.6; thus the probability distribution for momentum measurements on a state represented by φ is precisely that for a state represented by ψ, except translated by the vector $-\mathbf{b}$. In agreement with the preceding discussion, this shifts the mean of the distribution by $-\mathbf{b}$.

More generally, the local phase transformation may vary in time as well as in position. Explicitly, suppose that $\psi(t)$ is a "path" of wave functions in $\mathcal{H}$ and set $\Psi(t,\mathbf{x}) = \psi(t)(\mathbf{x})$. We assume now that $\chi = \chi(t,\mathbf{x})$ and consider the time-dependent state vector $\Phi = e^{\mathrm{i}\chi}\Psi$. Note that at time t, the wave function $\varphi(t)$ defined by $\varphi(t)(\mathbf{x}) = \Phi(t,\mathbf{x})$ is obtained from $\psi(t)$ by applying the local phase transformation $e^{\mathrm{i}\chi(t,-)}$, where $\chi(t,-)$ is the function of the space variable $\mathbf{x}$ defined by $\mathbf{x}\mapsto\chi(t,\mathbf{x})$. The time derivative interacts with the local phase transformation, in a manner similar to the space derivatives of equation 28.7, as

$$(\partial_t - \mathbf{i}\partial_t \chi)\, e^{\mathbf{i}\chi}\Psi = e^{\mathbf{i}\chi}\partial_t \Psi. \tag{28.9}$$

Equation 28.7 may be written in this context without the **a**-dependence, as

$$(\nabla - \mathbf{i}\nabla \chi)\, e^{\mathbf{i}\chi}\Psi = e^{\mathbf{i}\chi}\nabla \Psi. \tag{28.10}$$

We have noted already that local phase transformations do not preserve the physics in that the transformed state vector typically represents a state different from that represented by the original state vector, though with the same position probability distribution. The quantum theory as developed thus far is not symmetric under local phase transformations. We now ask what may seem to be a rather strange question. How would our development have to change if we were to require that the physics remain invariant under, not just global, but local phase transformations? Explicitly, how must the quantum theory as developed thus far be adjusted if the transformation of the wave function from ψ to $\varphi = e^{\mathbf{i}\chi}\psi$ is to have no observable physical effects? The answer to this question is explored in the next section, where we argue that the requirement that the theory be invariant under local U(1) symmetries may be met by the introduction of a field on $\mathbb{R}^3$ with particular properties that mediates the invariance.

28.4 Local Phase Symmetry Requires a Mediating Field

This section is a prelude to a thorough discussion of electromagnetic interactions in quantum mechanics, which the student may pursue in a subsequent study of relativistic quantum mechanics. Our aim here is to demonstrate how local symmetries of the wave functions can be made into symmetries of the theory provided that fields are introduced to mediate the symmetry.

The facts expressed in equations 28.9 and 28.10 suggest that the wave function $\varphi = e^{\mathbf{i}\chi}\psi$ cannot imply the same physics as that of ψ, since these equations may be taken to mean that, typically, the energies and the momentum probability distribution functions of the two wave functions differ. It would seem that the only way to remedy this is to allow local phase transformations to change not only the wave function from ψ to φ but also the operators ∂_t and ∇ to those acting on φ on the left-hand sides of equations 28.9 and 28.10. This would in turn cause familiar operators that correspond to observables to change in form. For example, the momentum operator in the direction of **a** would become $\mathsf{P}_{\mathbf{a}} = -\mathbf{i}\hbar\,\mathbf{a}\cdot(\nabla - \mathbf{i}\nabla\chi)$ in the presence of the local phase transformation corresponding to χ. This immediately raises several problems. First, redefining the operators corresponding to observables is falsified by experimental observations of quantum particles subject only to a conservative force arising from the scalar potential V. Second, if Ψ satisfies the Schrödinger equation with initial value ψ then usually $e^{\mathbf{i}\chi}\Psi$ fails to satisfy the Schrödinger equation with initial value φ. Thus, even though ψ and φ may describe the same quantum state, their time-evolved wave functions may not. It would seem then that the theory requires drastic adjustments to achieve local phase invariance, including adjustments to the Schrödinger equation that governs time evolution.

The hint that rescues our desire for local phase invariance from this damning assessment comes, maybe unexpectedly, from a study of the electromagnetic interaction. The remedy does change quite naturally the relevant operators of the theory in a nontrivial way, so that ultimately they depend on parameters that give rise to local phase transformations. The development is very interesting and arises from the addition of nonconservative potentials to the mix.

Thus far we have not dealt with nonconservative forces and have looked only at those interactions arising from a classical scalar potential V. The electromagnetic interaction too involves a scalar potential ϕ but in addition has a vector potential $\mathbf{A}$ inducing a velocity dependent, nonconservative, force into the classical theory. The result of this in the Hamiltonian formulation of mechanics is that the momentum $\mathbf{p}$ that is canonically conjugate to the position variable $\mathbf{x}$ is not the kinematic momentum $m\dot{\mathbf{x}}$ itself, but rather the kinematic momentum corrected by the vector potential. In particular, for a particle of electric charge q in a field determined by ϕ and $\mathbf{A}$, the canonical momentum conjugate to $\mathbf{x}$ is

$$\mathbf{p} = m\dot{\mathbf{x}} + q\mathbf{A}$$

and the Hamiltonian for the particle is[2]

$$H(\mathbf{x}, \mathbf{p}) = \frac{1}{2m}\left(\mathbf{p} - q\mathbf{A}\right)^2 + q\phi, \tag{28.11}$$

which is the expression for the total energy of the particle in terms

[2] Remember that the square means the dot product.

of the canonically conjugate position and momentum variables. Axiom 4 of wave mechanics then quantizes this by replacing the canonically conjugate momentum $\mathbf{p}$, and not the kinematic momentum, by $-\mathbf{i}\hbar\nabla$ to get the Schrödinger equation for a particle of mass m and charge q in a field determined by the potentials ϕ and $\mathbf{A}$. This may be written as

$$-\frac{\hbar^2}{2m}\left(\nabla - \frac{q}{\hbar}\mathbf{i}\mathbf{A}\right)^2 \Psi = \mathbf{i}\hbar\left(\partial_t + \frac{q}{\hbar}\mathbf{i}\phi\right)\Psi.$$

A useful way to think about this is that the presence of the electromagnetic field induced by the potentials ϕ and $\mathbf{A}$ induces, in turn, a change in the free particle Schrödinger equation

$$-\frac{\hbar^2}{2m}\nabla^2\Psi = \mathbf{i}\hbar\partial_t\Psi$$

by recalibrating the differential operators ∇ and ∂_t, the effect being to replace ∇ by $D_{\mathbf{A}} = \nabla - \mathsf{q}\mathbf{i}\mathbf{A}$ and ∂_t by $D_\phi = \partial_t + \mathsf{q}\mathbf{i}\phi$. Here $\mathsf{q} = q/\hbar$ is a **coupling constant** that measures the magnitude of this recalibration in units of $\hbar$. The momentum triad of operators is also recalibrated, becoming dependent on the vector potential as $\mathbf{P}_{\mathbf{A}} = -\mathbf{i}\hbar D_{\mathbf{A}} = -\mathbf{i}\hbar\nabla - q\mathbf{A}$. The Schrödinger equation becomes

$$-\frac{\hbar^2}{2m}D_{\mathbf{A}}^2\Psi = \mathbf{i}\hbar D_\phi\Psi, \tag{28.12}$$

which is reminiscent of the free particle Schrödinger equation. This amounts to a change in perspective. Rather than considering the introduction of a potential field as introducing an extra potential term, $V\Psi$, into the Schrödinger equation, we think of the introduction of a potential field as changing the geometry of the environment of the quantum particle and, in so doing, the particle is thought to move "freely" in the potential field with its natural **covariate derivative operators** $D_{\mathbf{A}}$ and D_ϕ guiding the particle in the same way that the operators ∇ and ∂_t guide the free particle.

What has this discussion to do with local phase changes? The answer comes from the nonuniqueness of the vector and scalar potentials in determining the electric and magnetic field vectors. In classical electromagnetic theory, the electric field $\mathbf{E} = -\nabla\phi - \partial_t\mathbf{A}$ and the magnetic field $\mathbf{B} = \nabla \times \mathbf{A}$ are physical fields in that they have measurable consequences when acting on a charged particle. However, the potential fields $\mathbf{A}$ and ϕ are not physically measurable quantities and, as such, are considered more as abstract mathematical devices that aid calculations. Though $\mathbf{A}$ and ϕ uniquely determine $\mathbf{E}$ and $\mathbf{B}$, the favor is not returned. The nonuniqueness of the potentials is quantified in the notion of a **gauge transformation**, defined as follows. We say that potentials $\mathbf{A}'$ and ϕ' are obtained by applying the gauge transformation determined by the function $\chi : \mathbb{R} \times \mathbb{R}^3 \to \mathbb{R}$, called a **gauge**, if $\phi' = \phi - \partial_t \chi$ and $\mathbf{A}' = \mathbf{A} + \nabla\chi$.[3] The point is that the potentials $\mathbf{A}'$ and ϕ' determine exactly the same physical fields, $\mathbf{E}$ and $\mathbf{B}$, that $\mathbf{A}$ and ϕ determine, and hence the Maxwell equations are invariant under a gauge transformation and no physical observation can decide whether the primed or the unprimed potential fields are in use. This suggests that this **gauge invariance** or **gauge symmetry** of classical electromagnetic theory should be reflected in the quantized equations. In particular, this means that the expectation values of the quantum theory determined by the use of $\mathbf{A}$ and ϕ must be identical to the expectation values of the theory determined by the gauge-transformed fields $\mathbf{A}'$ and ϕ'.

Since the quantization procedure uses as input the classical Hamiltonian, and the classical Hamiltonian of the electromagnetic interaction given in equation 28.11 explicitly contains the vector and scalar potentials $\mathbf{A}$ and ϕ, these potentials appear in the equations that govern the mechanics and observations of the quantum theory. In particular, if the vector $\psi \in \mathcal{H}$ describes a particular initial quantum state $\mathfrak{S}$, equation 28.12 is applied to find the wave function $\psi(t)$ representing the time-evolved state, and self-adjoint operators $\mathsf{Q}_\mathbf{A}$ are used to calculate expectation values then the theory must accommodate the change in the equations due to a gauge transformation. Specifically, if the potential fields $\mathbf{A}'$ and ϕ' are used in place of $\mathbf{A}$ and ϕ then the Schrödinger equation changes to

$$-\frac{\hbar^2}{2m} D^2_{\mathbf{A}'}\Psi = \mathbf{i}\hbar D_{\phi'}\Psi, \tag{28.13}$$

and the operator $\mathsf{Q}_\mathbf{A}$ changes to $\mathsf{Q}_{\mathbf{A}'}$. Immediately there appears a problem. In general if Ψ solves equation 28.12, then it does not solve equation 28.13 and, moreover, it is generally the case that $\langle \mathsf{Q}_\mathbf{A}\rangle_\psi \neq \langle \mathsf{Q}_{\mathbf{A}'}\rangle_\psi$. The resolution of this problem is found in examining Axiom 1 of Lecture 2. This axiom states in part that a state $\mathfrak{S}$ of a quantum system is represented by a ray in $\mathcal{H}$, since ψ and $c\psi$ represent the same state for any nonzero complex number c. But the axiom does not restrict us to this condition as the only one for which two different vectors may

[3] This can be presented more elegantly in the context of relativistic quantum mechanics, when one has the advantage of the language of contravariant and covariant 4-vectors and 4-vector differential operators.

represent the same state. To fit the electromagnetic interaction into the wave mechanical framework, we need to allow vectors other than nonzero multiples of ψ to represent the state $\mathfrak{S}$ and, in fact, in a way that depends on the choice of gauge. The question now becomes whether there is a vector $\psi' \in \mathcal{H}$ such that the solution Ψ' to equation 28.13 with initial value ψ' gives identical expectation values to Ψ, in the sense that $\langle \mathsf{Q}_{\mathbf{A}} \rangle_{\psi(t)} = \langle \mathsf{Q}_{\mathbf{A}'} \rangle_{\psi'(t)}$, where $\psi(t)$ and $\psi'(t)$ are defined, respectively, by $\psi(t)(\mathbf{x}) = \Psi(t, \mathbf{x})$ and $\psi'(t)(\mathbf{x}) = \Psi'(t, \mathbf{x})$.

The reader may have anticipated the answer to the last question. Note that the covariant derivative operators for the primed and unprimed potentials are related by

$$D_{\mathbf{A}'} = D_{\mathbf{A}} - \mathfrak{q}\mathbf{i}\nabla\chi \quad \text{and} \quad D_{\phi'} = D_{\phi} - \mathfrak{q}\mathbf{i}\partial_t\chi,$$

which are more than reminiscent of equations 28.9 and 28.10. Indeed, except for the innocuous (but important!) appearance of the coupling constant $\mathfrak{q}$, these are the operators that appear in equations 28.9 and 28.10 when $\mathbf{A} = \mathbf{0}$ and $\phi = 0$. A moment's reflection on these equations will confirm that $D_{\mathbf{A}'} e^{\mathbf{i}\mathfrak{q}\chi}\Psi = e^{\mathbf{i}\mathfrak{q}\chi} D_{\mathbf{A}}\Psi$ and $D_{\phi'} e^{\mathbf{i}\mathfrak{q}\chi}\Psi = e^{\mathbf{i}\mathfrak{q}\chi} D_{\phi}\Psi$, implying by a short calculation that

$$\Psi'(t, \mathbf{x}) = e^{\mathbf{i}\mathfrak{q}\chi(t,\mathbf{x})}\Psi(t, \mathbf{x})$$

is a solution to the Schrödinger equation 28.13 whenever Ψ is a solution to the Schrödinger equation 28.12. Moreover, the expectation values of those observables whose operators are written in terms of X and P coincide. For example, $\langle \mathbf{a} \cdot \mathsf{P}_{\mathbf{A}} \rangle_{\psi(t)} = \langle \mathbf{a} \cdot \mathsf{P}_{\mathbf{A}'} \rangle_{\psi'(t)}$ for any direction $\mathbf{a}$.

It will be useful to collect our thoughts and review what we have discovered about the quantum mechanical treatment of the electromagnetic interaction. First, the quantum mechanical axioms for wave mechanics produce a Schrödinger equation that governs time evolution by quantizing a classical Hamiltonian. In electromagnetic theory, this Hamiltonian explicitly contains, rather than the physically real electric and magnetic fields, vector and scalar potential fields that are not observable. This means that these nonobservable potential fields appear in the quantized equations and, since the potential fields are not unique, this amounts to some ambiguity in the quantum formalism. This ambiguity is resolved by observing that the gauge invariance of classical electromagnetic theory descends to a gauge invariance of the quantized version of the theory in the following way. The gauge transformation

$$\begin{aligned}\mathbf{A} \rightsquigarrow \mathbf{A}' &= \mathbf{A} + \nabla\chi \\ \phi \rightsquigarrow \phi' &= \phi - \partial_t\chi\end{aligned} \tag{28.14}$$

in the classical theory induces a gauge transformation of the quantum theory by recalibrating the operators and state vectors via

$$\begin{aligned}D_{\mathbf{A}} \rightsquigarrow D_{\mathbf{A}'} &= D_{\mathbf{A}} - \mathfrak{q}\mathrm{i}\nabla\chi, \\ D_{\phi} \rightsquigarrow D_{\phi'} &= D_{\phi} - \mathfrak{q}\mathrm{i}\partial_t\chi, \\ \Psi \rightsquigarrow \Psi' &= e^{\mathrm{i}\mathfrak{q}\chi}\Psi, \\ \mathsf{P}_{\mathbf{A}} \rightsquigarrow \mathsf{P}_{\mathbf{A}'} &= \mathsf{P}_{\mathbf{A}} - q\nabla\chi.\end{aligned} \tag{28.15}$$

In this setting, the Schrödinger equation using the potential fields $\mathbf{A}$ and ϕ is obtained from the free particle Schrödinger equation on replacing the differential operators ∇ and ∂_t by the covariant derivatives induced by the potentials, $D_{\mathbf{A}}$ for ∇ and D_ϕ for ∂_t. In addition, and of utmost importance, the expectation values of observables are preserved under these gauge transformations.

Notice that this quantum theory is invariant under local phase transformations! This gives a way to realize the goal at the end of the previous section, i.e., to understand how quantum theory needs to be adjusted to satisfy local U(1) invariance. The lesson of this section is that local U(1) invariance may be achieved by introducing a field to mediate this invariance – a field with exactly the properties of the classical electromagnetic field. Indeed, the local phase transformation induced by $e^{\mathrm{i}\chi}$ is a symmetry of the theory provided that $\chi/\mathfrak{q}$ is used as a gauge to transform the potential fields in use. This is surely one of the most beautiful and profound observations of modern physics and serves as a model for introducing other symmetries into the theory. In fact, both the weak and the strong nuclear interactions may be understood as arising from fields that mediate nonabelian gauge symmetries of the quantum theory – nonabelian because the relevant symmetry groups, unlike U(1), are nonabelian. The group SU(2) × U(1) serves as the gauge symmetry group of the electro-weak interaction, which combines the electromagnetic interaction with its U(1) gauge symmetry with the weak nuclear interaction, while the group SU(3) serves as the gauge symmetry group for the strong nuclear interaction. The **gauge principle** asserts that all elementary interactions arise as fields that mediate local symmetries.[4]

[4] The idea of a gauge was introduced originally by the mathematician Hermann Weyl in 1918 in a failed attempt to build a unified geometric theory of

28.5 End Notes

Noether's observation that conservation laws arise from the infinitesimal symmetries of the Lagrangian in classical mechanics is one of the most beautiful principles in the study of the natural world. Her observation, extended rather easily to linear quantum theory, replicates the conservation laws of classical mechanics and provides for conservation laws of a purely quantum mechanical nature. The idea of the symmetry of a theory is central in the development of advanced quantum mechanics, which describes quantum fields and the weak and strong nuclear interactions. This lecture has presented just a hint of how the notion of symmetry guides the development of fundamental physics.

The discussion and development of how a local symmetry is mediated by a field in the previous section arose from my study of Aitchison and Hey (1989, 2003: a, b). Their 1989 Chapter 2, or their 2003 Chapter 3, presents the best introductory discussion of the gauge principle in quantum mechanics that I have seen. Worthy of a careful read is their 1989 Chapter 1 and their 2003 Chapters 1 and 2 for a not too technical overview of the the particles of nature and the standard model that describes the electro-weak and strong nuclear interactions.

electromagnetism and gravitation. It was modified to a local theory, in more or less the form presented here, in 1927 by Weyl and the physicists Vladimir Fock and Fritz London, producing the first example of a so-called **gauge theory** to explain the effect of the electromagnetic field on a charged quantum particle. In 1954, Chen Ning Yang and Robert Mills introduced nonabelian gauge theories to explain nuclear interactions.

29
The Feynman Formulation of Quantum Mechanics

> Thirty-one years ago [1949], Dick Feynman told me about his "sum over histories" version of quantum mechanics. "The electron does anything it likes," he said. "It just goes in any direction at any speed, forward or backward in time, however it likes, and then you add up the amplitudes and it gives you the wave-function." I said to him, "You're crazy." But he wasn't.
>
> Freeman Dyson, 1980

The development of quantum mechanics presented in the preceding lectures is that formulated from 1925 through the early 1930s by the physicists Heisenberg, Schrödinger, Dirac, Born, Pauli, Jordan, Fermi, de Broglie, Fock, and Einstein and the mathematicians von Neumann, Stone, and Weyl, among others. This nonrelativistic version was in essentially complete form with the publication of Dirac's *The Principles of Quantum Mechanics* in 1930 and von Neumann's *The Mathematical Foundations of Quantum Mechanics* in 1932. The former presented a version favored by physicists that introduced the Dirac calculus and synthesized the two approaches of Heisenberg and Schrödinger into a single formalism based on Hilbert spaces. What it lacked in rigor, it made up for in elegance and clarity. The latter presented a version favored by mathematicians, a precise, rigorous, axiomatic quantum mechanical formalism, formulated by generalizing Hilbert's spectral theory of integral operators to the very general setting of self-adjoint operators.

For two decades these two monographs represented the canonical version of quantum mechanics to physicists and mathematicians and it seemed that there was nothing more to say about the foundations of the nonrelativistic version of the theory. The Dirac calculus provided the practical calculational tools sufficient to the task and the von Neumann rigor placed the theory on a firm mathematical foundation. Then came Richard Feynman.

Richard Feynman was one of the most original physicists of the last century. Idiosyncratic and unconventional, he rarely accepted the conventional wisdom but looked to rework it into a version he could understand. This held especially for quantum mechanics with its wave–particle duality, which seemed to be uncommitted to the ontological existence of particles at the atomic level, a position Feynman rejected. He in fact supported the idea of the electron as particle – not a wave, not a wave and a particle, but a particle – and explained the emergence of wave-like phenomena as the interference of probability amplitudes attached to possible events. In this reworking of quantum mechanics, Feynman identified complex probability amplitudes as the crucial ingredients of quantum mechanics that make it nonclassical and give rise to the familiar quantum mechanical phenomena.

In the first section of this lecture, we articulate a new version of quantum mechanics that takes the probability amplitude as the fundamental concept, and we generalize the rules of classical probability to apply, *mutatis mutandis*, to quantum probability amplitudes. This is summarized as three new axioms for quantum mechanics. In the section following, we look at how Feynman interpreted Axiom 2 of this new rendition to arrive at original formulæ for the Feynman propagator, which gives the amplitude for a particle to be at position $\mathbf{x}_1$ at time t_1 given that it is at position $\mathbf{x}_0$ at time t_0. The propagator is the Green's function for the Schrödinger operator and, in Feynman's hands, ultimately takes the form of the famous Feynman path integral, which is expressed in terms of the classical Lagrangian for the particle.

We continue in the next section with an examination of the form that the path integral takes whenever the Lagrangian is quadratic in the position and velocity variables, this being one of the most useful results in practical calculations. At this point we are almost ready to perform our first path integral calculation, but for that we need to know the propagator for the free particle as this is needed to determine the normalizing factors that appear in the calculations. The free particle is examined in the section following and the results are used in later lectures, where the path integral is presented as a limit of integrations over dense subspaces of the full path space.

Unlike conventional nonrelativistic quantum mechanics, which von Neumann and Stone placed on a firm mathematical footing, there are technical mathematical difficulties with making the path integral rigorous in the full generality that Feynman envisioned. Physicists get around this by looking for natural ways in which to interpret Feynman's idea in the context of the problem at hand; nonetheless, the mathematician is a bit nonplussed by the fact that there is no possible measure-theoretic way to construct an integral in the full generality and with all the properties demanded by Feynman. In some sense, the best answer to the question of exactly what the path integral is may be this: it is an heuristic tool encoded in a formal expression that produces formulae appropriate to the application at hand to derive a Green's function, but these formulae may change from application to application.

We do not want to give the impression that the Feynman approach is in all respects satisfactory, even for the physicist. Aside from the significant mathematical difficulties, there are physical defects of the approach, the most glaring of which is that spin operators are rather difficult to incorporate into the theory. As Feynman and Hibbs assert,[1] "it is a serious limitation that the half-integral spin of the electron does not find a simple and ready representation." On the other hand, one of the great advantages to Feynman's formulation is the aid it brings to our physical intuition in understanding the workings of the quantum world. It is often the case that the path integral method suggests approaches to a problem that canonical quantum mechanics misses; moreover, the path integral heuristic finds critical applications to the quantization of fields, offering a powerful organizing tool for understanding this difficult topic.

29.1 A New Look at Quantum Mechanics

This new version of quantum mechanics arises out of a change in the point of view. It posits that the mechanics of the micro-world are governed by a quantum probability theory that arises by attaching not real but complex numbers to quantum events. Indeed, the classical probability associated with an event E is a real number in the interval $[0, 1]$ describing the proportion of outcomes for which the event E appears as the outcome. More specifically, classical probability attaches to each possible alternative e_i resulting in some possible event a real number $p(e_i)$ in $[0, 1]$ that represents the probability that that alternative is followed. Then the probability of the particular event E is $p(E)$, equal to

[1] Feynman and Hibbs (1965), p. 355

the sum of only those numbers $p(e_i)$ for which E is the outcome of following alternative e_i. In the event of equal probabilities, where $p(e_i)$ is the reciprocal of the total number of alternative paths to all possible outcomes, the calculation of the probability that event E occurs reduces to counting the number of alternatives leading to event E. The essential, crucial, difference between classical probability and quantum probability is that, while classical probability arises in this way from counting the number of ways in which E may occur, quantum probability arises from the assignment of a complex number to E, and the fundamental concept in quantum mechanics becomes that of a **complex probability amplitude**.

This assignment of a complex probability amplitude to an event E descends to the assignment of a complex probability amplitude to each possible route to E. Associated to each quantum event E is a probability amplitude $\phi(E)$, a complex number, and the probability for the event is just the modulus squared: $p(E) = |\phi(E)|^2$. The determination of $\phi(E)$ mimics the determination of $p(E)$ in classical probability theory when E occurs via, say, independent alternatives e_1 or e_2. Indeed in this case, the total amplitude for E is the sum $\phi(E) = \phi(e_1 \vee e_2) = \phi(e_1) + \phi(e_2)$, where $\phi(e_i)$ is the amplitude attached to alternative e_i. This is a very simple axiom but has profound consequences leading to interference effects in quantum mechanics. The essential difference between this and classical probability assignments is captured in the general inequality

$$|\phi(e_1)|^2 + |\phi(e_2)|^2 \neq |\phi(e_1) + \phi(e_2)|^2,$$

the left-hand side representing the classical probability law for combining probabilities and the right-hand side representing the new quantum probability law. This, of course, is the simplest situation and must be generalized to include the case where the number of independent alternatives is not only infinite, but uncountably so. Of course, in dealing with continuous spectra the probability amplitude is interpreted as a **probability amplitude density** in exactly the same way that one moves from discrete probability to continuous probability by replacing discrete probabilities of events with probability densities.

The effect of moving from probabilities to complex amplitudes may be illustrated most simply by the well-known two-slit experiment, where an event is the detection of a particle, emitted from a source, at point x after passage through two slits in an opaque screen. The amplitude density for detection at x is $\phi(x) = \phi_1(x) + \phi_2(x)$, where $\phi_i(x)$ is the amplitude density for detection at x if the particle passes through slit

i, for $i = 1, 2$. Each ϕ_i should be given by a function that produces, at least approximately, a gaussian probability distribution upon calculation of $|\phi_i|^2$, with mean shifted parallel to the screen and towards the direction of slit i. The most naive amplitudes might be given as $\phi_1(x) = e^{i\varphi(x)}e^{-a^2(x+h)^2}$ and $\phi_2(x) = e^{-i\varphi(x)}e^{-a^2(x-h)^2}$. If the slits are very close then h is very small and, for the purposes of illustrating interference effects, will be ignored. The total amplitude density for detection at x then becomes $\phi(x) = (e^{i\varphi(x)} + e^{-i\varphi(x)})e^{-a^2x^2} = 2\cos(\varphi(x))\, e^{-a^2x^2}$, which describes an interference pattern. This pattern arises precisely from the phase difference $e^{\pm i\varphi(x)}$ in the amplitude densities ϕ_1 and ϕ_2.

In classical probability, when an event E may occur via two independent, alternative routes with respective probabilities $p(e_1)$ and $p(e_2)$, the total probability for E is the sum $p(E) = p(e_1 \vee e_2) = p(e_1) + p(e_2)$. Quantum mechanics replaces the classical probability with complex amplitudes and combines independent amplitudes exactly as classical probability combines independent probabilities. Recall another rule from classical probability. When an event E may be realized as a two-step process – alternative e_1 followed by alternative e_2 – the probability of $E = e_1 \wedge e_2$ is the product $p(E) = p(e_1 \wedge e_2) = p(e_1)p(e_2|e_1)$, where $p(e_2|e_1)$ is the probability of e_2 given that e_1 has occurred. Quantum mechanics likewise replaces this classical law with its exact analogue for amplitudes: $\phi(E) = \phi(e_1 \wedge e_2) = \phi(e_1)\phi(e_2|e_1)$, where $\phi(e_2|e_1)$ is the amplitude for e_2 given that e_1 has occurred. Of course, none of this is obvious – why should nature behave according to quantum amplitude laws rather than classical probability laws, and how do the complex amplitudes get attached to events? The answer is that no one knows, but the world does so behave. These laws of amplitudes are fundamental laws of nature, irreducible and nonderivable from deeper or more basic principles, so it seems. They are the brute facts of the physical world.

Following Felsager (1998), we write more formally the laws of quantum mechanics à la Feynman. It should be stated that the term *quantum event* is normally associated with the event of measuring the exact value of an observable of a quantum system, and the term *alternative* is associated with a particular way in which event E may be realized, whether or not there is any way, even in principle, to determine that a particular alternative e has occurred. Feynman argues the following metaprinciple: when particular alternatives e_i are subject to experimental verification, then the outcome of the quantum calculation of $p(E)$ using the amplitudes $\phi(e_i)$ yields the same outcome as the classical expectations, and when the alternatives e_i cannot even in principle be observed to have occurred, the quantum amplitudes $\phi(e_i)$ interfere and produce nonclassical results for $p(E)$, such as the fringe pattern of the double slit experiment.

Axiom 1. With each quantum event E is associated a complex number $\phi(E)$, the **probability amplitude** for that event. In the case of continuous spectra, this is to be interpreted as a probability amplitude density. The probability (or probability density) for the event is $p(E) = |\phi(E)|^2$. Moreover, any particular alternative e with outcome E is assigned a complex probability amplitude $\phi(e)$, and conditional alternatives, $e|d$, are assigned a complex probability amplitude $\phi(e|d)$.

Axiom 2. If E may be realized via one of several independent alternatives e_i, the total amplitude for E may be written as

$$\phi(E) = \phi(\vee_i e_i) = \sum_i \phi(e_i).$$

Axiom 3. If E may be decomposed into several individual steps e_j, the total amplitude may be written as

$$\phi(E) = \phi(\wedge_j e_j) = \phi(e_0) \prod_j \phi(e_j | e_{j-1}).$$

These axioms are simple enough in theory, but how do they work in practice? There are several problems that must be addressed before they can be of use. How does one assign the complex number $\phi(e)$ to the route e? How does the canonical version of quantum mechanics follow from these axioms? How are Axioms 2 and 3 to be interpreted when the index parameters i and j are uncountably infinite rather than discrete? In particular, how is the sum in Axiom 2 to be interpreted? The immediate answer is that it is to be interpreted as an integral, which works rigorously in some cases but leads to great difficulties in others. Many view these axioms of quantum mechanics as intuitive and heuristic, supplying some tentative physical insights, but not to be taken too seriously. It is a tribute to the genius of Feynman that he did take them seriously[2] and produced a calculus for calculating with them even in cases where rigorous argument fails. He reproduced much of canonical quantum mechanics from this approach and even used his powerful methods to go

[2] This is not a statement of history – Feynman did not articulate a set of axioms and reason from the axioms – but he did reason from the assignment of amplitudes to events to the sum of Axiom 2. The arrangement of Feynman's ideas presented here follows Chapter 2 of Bjørn Felsager's 1998 book *Geometry, Particles, and Fields*, and we highly recommend Feynman's and Hibb's presentation of these ideas in their 1965 book *Quantum Mechanics and Path Integrals*.

beyond the canonical. Our particular interest is in Feynman's use of Axiom 2 in calculating propagators for particles moving through spacetime.

29.2 Feynman's Propagator Calculus

We consider a quantum particle in spacetime. For spacetime points $(t_0, \mathbf{x}_0)$ and $(t_1, \mathbf{x}_1)$, the Feynman **propagator** $\mathcal{K}(t_1, \mathbf{x}_1; t_0, \mathbf{x}_0)$ is defined as the amplitude for a particle to be at position $\mathbf{x}_1$ at time t_1 given that it is at position $\mathbf{x}_0$ at time t_0. Let $\Psi(t, \mathbf{x})$ be the amplitude for the particle to be at spacetime point $(t, \mathbf{x})$. Obviously, $|\Psi(t, \mathbf{x})|^2$ is the position probability density at time t, so that $\Psi(t, \mathbf{x})$ is the Schrödinger wave function of wave mechanics. By Axiom 3, the product $\mathcal{K}(t, \mathbf{x}; t_0, \mathbf{x}_0)\Psi(t_0, \mathbf{x}_0)$ is the amplitude density for a particle to be at position $\mathbf{x}$ at time t and position $\mathbf{x}_0$ at time t_0. Applying Axiom 2, using the natural interpretation of the sum as an integral, we get

$$\Psi(t, \mathbf{x}) = \int_{\mathbb{R}^3} \mathcal{K}(t, \mathbf{x}; t_0, \mathbf{x}_0)\Psi(t_0, \mathbf{x}_0)\, d\mathbf{x}_0. \tag{29.1}$$

This is a case where the axioms lead to rigorous results. In the parlance of mathematicians, $\mathcal{K}(t, \mathbf{x}; t_0, \mathbf{x}_0)$ is known as the Green's function for the Schrödinger operator.

One of the more important formulae for propagators that follows, at least formally, from the axioms is the **group property**:

$$\mathcal{K}(t_1, \mathbf{x}_1; t_0, \mathbf{x}_0) = \int_{\mathbb{R}^3} \mathcal{K}(t_1, \mathbf{x}_1; t', \mathbf{x}')\mathcal{K}(t', \mathbf{x}'; t_0, \mathbf{x}_0)\, d\mathbf{x}'. \tag{29.2}$$

Here a particle travels from position $\mathbf{x}_0$ at time t_0 to position $\mathbf{x}_1$ at time t_1. At a time t' intermediate between t_0 and t_1, the particle passes through point $\mathbf{x}'$. The amplitude for this event, by Axiom 3, is the integrand in equation 29.2. The equation itself then follows by applying Axiom 2. There are rigorous mathematical methods for finding propagators, this being one of the central topics in modern applied mathematics. Feynman gives an ingenious method that makes this attempt.

29.2.1 The Path Integral

Taking the content of Axiom 2 seriously, the intrepid Feynman reasoned as follows. The crux of time evolution in quantum mechanics is embodied in equation 29.1: if the propagator $\mathcal{K}(t_1, \mathbf{x}_1; t_0, \mathbf{x}_0)$ is known then the wave function at time t_1 can be calculated from that at time t_0. Let Γ

be the collection of continuous paths $\mathbf{x}\colon [t_0, t_1] \to \mathbb{R}^3$ with $\mathbf{x}(t_0) = \mathbf{x}_0$ and $\mathbf{x}(t_1) = \mathbf{x}_1$. To get from $\mathbf{x}_0$ at time t_0 to $\mathbf{x}_1$ at time t_1, the particle must first be localized at $\mathbf{x}_0$ at time t_0 and then traverse some path $\mathbf{x}$ in Γ. Let $\phi(\mathbf{x})$ be the amplitude for the traversal of the path $\mathbf{x} \in \Gamma$ given that the particle is localized at $\mathbf{x}_0$ at time t_0. According to Axiom 2,

$$\mathcal{K}(t_1, \mathbf{x}_1; t_0, \mathbf{x}_0) = \sum_{\mathbf{x} \in \Gamma} \phi(\mathbf{x}), \tag{29.3}$$

which is usually written as a formal integral, the **path integral**,

$$\mathcal{K}(t_1, \mathbf{x}_1; t_0, \mathbf{x}_0) = \int_\Gamma \phi(\mathbf{x})\, \mathcal{D}(\mathbf{x}(t)). \tag{29.4}$$

The central challenge of the Feynman formulation of quantum mechanics is to determine what this sum, or integral, means!

At this point there are two problems to overcome in this **sum over histories** approach to quantum mechanics. The first is to determine what form the amplitude $\phi(\mathbf{x})$ should take; the second is to determine how to perform the "integration" given that it is known[3] that there is no measure on Γ that has all the properties expected of this formal integral. For the former, Feynman, inspired by Dirac, defines the amplitude as

$$\phi(\mathbf{x}) = \exp\left(\frac{\mathbf{i}}{\hbar} S[\mathbf{x}]\right) = \exp\left(\frac{\mathbf{i}}{\hbar} \int_{t_0}^{t_1} L(\mathbf{x}, \dot{\mathbf{x}}, t)\, dt\right), \tag{29.5}$$

where $S[\mathbf{x}]$ is the classical action associated with the path and $L(\mathbf{x}, \dot{\mathbf{x}}, t)$ is the Lagrangian for the particle. Of course, this choice for ϕ immediately restricts the path family to piecewise differentiable paths – or at least to those with a generalized derivative – rather than general continuous paths. In fact, we will go ahead and restrict our consideration to paths that are at least piecewise continuously differentiable. One interesting fact about the choice given in equation 29.5 is that it is democratic with respect to the possible paths the particle may take – each is equally likely, as the probability density is the constant unit density on Γ. But this is precisely where the move from probabilities to complex amplitudes creates a huge qualitative change in the expected behavior. This may be seen in Feynman's physical plausibility explanation for the choice of formula 29.5 as the amplitude. The idea is that the complex amplitude density should reflect the fact that the classical path traversed by the particle is the one overwhelmingly favored by the probabilities when the particle mass, the energies, and the dimensions are macroscopic – large in relation to the size of $\hbar$.

[3] See, for example the discussion in Cameron (1960).

To flesh out Feynman's argument, let $\mathbf{x} \in \Gamma$ be a possible path that the particle may traverse. Suppose that the masses and dimensions of the problem at hand are so large that the variation in the action, $\delta S_{\mathbf{x}}$, is very large in relation to $\hbar$ when applied to suitable variations $\mathbf{h}$ of the path that are small on a classical scale. Then the change in S, though caused by a small classical change in the path $\mathbf{x}$, is likely to be huge when measured with respect to $\hbar$. Thus, as paths in a small "classical" neighborhood of $\mathbf{x}$ are sampled and the amplitude for this set of paths is calculated by the sum in equation 29.3, these small changes in path will produce huge changes in phase and the complex exponential expression 29.5 will oscillate rapidly about the unit circle. The net effect will be destructive interference, giving an approximately zero amplitude for the particle to traverse a path in this neighborhood of $\mathbf{x}$, since the contribution by one path in this neighborhood will be canceled out by one that is very close but out of phase by π radians.

More precisely, the idea is that $S[\mathbf{x}+\mathbf{h}] = S[\mathbf{x}] + \delta S_{\mathbf{x}}\mathbf{h}$, so that the phase change is $e^{\mathbf{i}\delta S_{\mathbf{x}}\mathbf{h}/\hbar}$. As $\mathbf{h}$ varies over a small classical neighborhood, this phase change oscillates randomly about the unit circle, forcing the sum in equation 29.3 to be close to zero. Thus the amplitude for the particle to take a path near $\mathbf{x}$ is virtually zero when the variational derivative is large in relation to $\hbar$. However, this argument fails for a classical path. Indeed, let $\mathbf{x}_{\text{cl}}$ be the classical path traversed by the classical particle. Recall from Chapter 6 that $\mathbf{x}_{\text{cl}}$ gives a stationary value for the action, so that the variational derivative $\delta S_{\mathbf{x}_{\text{cl}}}$ is zero and, up to first order, $S[\mathbf{x}_{\text{cl}}+\mathbf{h}] = S[\mathbf{x}_{\text{cl}}]$ for small variations $\mathbf{h}$ of the path. Now, when the contributions to the sum in equation 29.3 are added, constructive interference is the overriding effect since a small change in $\mathbf{x}_{\text{cl}}$ produces no change in S. It follows that the amplitude for traversing a path near the classical path is large in comparison with that for a typical path $\mathbf{x}$, and the probability becomes overwhelming that the particle will traverse the classical path in the limit as $\hbar$ approaches zero.

Whatever one may think of Feynman's plausibility argument, it must be admitted that even though it fails the test of mathematical rigor, it is an inspired example of physical intuition. Of course, any physical theory is judged ultimately neither by its cleverness, beauty or inspiration, nor by its standard of mathematical rigor, but by its success in calculation and prediction, and it turns out that Feynman's inspired guess for the form of $\phi(x)$ works in many settings to produce correct formulae for the Green's functions of the Schrödinger operator. More importantly, in the setting of quantum field theories and quantum statistical mechanics, the ideas that arise from the path integral formalism are fundamental ingredients that guide the development of the theories.

There is still the problem of the integration over the possible paths, and this latter problem is tougher to solve. Feynman indeed knows well the difficulties in providing rigor to his definition of the path integral.

> There may be other cases where ... the present definition of a sum over all paths is just too awkward to use. Such a situation arises in ordinary integration in which the Riemann definition ... is not adequate and recourse must be had to some other definition, such as that of Lebesgue.
>
> The necessity to redefine the method of integration does not destroy the concept of integration. So we feel that the possible awkwardness of the special definition of the sum over all paths ... may eventually require new definitions to be formulated. Nevertheless, the concept of the sum over all paths, like the concept of an ordinary integral, is independent of the special definition and valid in spite of the failure of such definitions.[4]

Feynman gives a general prescription, the "present definition" of the quote, for constructing the sum in equation 29.3 by partitioning the time interval and approximating paths by piecewise linear ones. A variation of Feynman's prescription is presented in Section 31.1. His scheme involves the introduction of unknown normalizing constants that are determined by the requirement that the path integral approach must lead to the Schrödinger equation. Feynman's argument for this is very clever but glosses over difficult points. He replaces the propagator between space-time points using the action for a single path, with the justification that the time variables are close, but then integrates over space variables. Moreover, he approximates the integrand by a quadratic Taylor approximation in the space variable, which he points out is good as long as the space variable varies over a small interval, but then immediately integrates over all space. In defense of this argument Feynman suggests that, as one integrates over the space variable, it will be only a small variance of the space variable around the classical path that contributes to the integral, similar to his plausibility argument for expression 29.5. Though we are sympathetic to this last argument and are extremely impressed with Feynman's physical intuitions, the development at best is a formal plausibility argument that, as a whole, lacks rigor.

[4] Feynman and Hibbs (1965), p. 34.

29.3 Evaluating Path Integrals: Quadratic Lagrangians

In this section, we examine the one-dimensional case where the particle moves along a line, so that its spacetime location is (t, x) for $x \in \mathbb{R}$. Though the argument can be made in more generality, we will make the simplifying assumption that the possible paths $x(t)$ are continuously differentiable. One of the most beautiful and useful results for evaluating path integrals in this setting is Feynman's observation that when the Lagrangian is quadratic in x and its derivative $\dot{x}$, the path integral reduces to a calculation for which $x_0 = x_1 = 0$. Indeed, if the Lagrangian is of the form

$$L(x, \dot{x}, t) = a(t)x^2 + b(t)x\dot{x} + c(t)\dot{x}^2 + d(t)x + e(t)\dot{x} + f(t), \tag{29.6}$$

then we have for the path integral

$$\begin{aligned}\mathcal{K}(t_1, x_1; t_0, x_0) \\ = e^{\mathrm{i}S_{\mathrm{cl}}/\hbar} \int_{\Gamma_0} \exp\left\{ \frac{\mathrm{i}}{\hbar} \int_{t_0}^{t_1} [a(t)x^2 + b(t)x\dot{x} + c(t)\dot{x}^2]\, dt \right\} \mathcal{D}(x(t)),\end{aligned} \tag{29.7}$$

where S_{cl} is the action associated with the classical path that the particle traverses between spacetime points (t_0, x_0) and (t_1, x_1) and Γ_0 is the space of loops based at the origin.[5] Thus, for quadratic Lagrangians, the phase $e^{\mathrm{i}S_{\mathrm{cl}}/\hbar}$ is expressible exactly in terms of the classical path. This is probably the single most useful result that aids in the evaluation of path integrals. The following comments of Lawrence Schulman are particularly appropriate.

> For all path integrals evaluated above, the result was expressible entirely in terms of the classical path. It turns out that every propagator that anyone has ever been able to evaluate exactly and in closed form is a sum over "classical paths" only. There has even appeared in the literature the claim that *all* propagators are given exactly as a sum over classical paths.[6]

Before we derive this result on quadratic Lagrangians, we make an interesting observation about calculating the action S_{cl} associated with the classical path x_{cl}. Formula 29.7 makes clear that, at least for quadratic Lagrangians – and according to Schulman, for all known propagators –

[5] A loop h based at the origin is a piecewise continuously differentiable closed path $h : [t_0, t_1] \to \mathbb{R}$ with $h(t_0) = 0 = h(t_1)$.

[6] Schulman (1981), p. 39.

it will be crucial to evaluate not only the classical path $x_{\text{cl}}(t)$ but also the value of the action,

$$S_{\text{cl}} = S[x_{\text{cl}}] = \int_{t_0}^{t_1} L(x_{\text{cl}}(t), \dot{x}_{\text{cl}}(t), t)\, dt,$$

for that path. This is to be contrasted with the use of the action in classical mechanics, where its importance is to provide a functional whose stationary arguments, the paths $x_{\text{cl}}(t)$ that solve the Euler–Lagrange equation of motion, are the desired products. These stationary paths give the possible tracks of the classical particle, from which all the physical implications emerge. In contrast, the actual stationary value S_{cl} is of no importance. The progression is from the action functional S to the Euler–Lagrange equation of motion 6.3 to its solution x_{cl}. In quantum mechanics, it now becomes important to add a fourth step to this progression – the actual calculation of S_{cl}.

We invite the reader to derive the stationary values for the action functional for a free particle, for a particle in a constant force field, and for a harmonic oscillator. We will calculate the propagators in these three settings in the course of this and the following lectures. For the record, the Lagrangian for a free particle is $L(x, \dot{x}) = m\dot{x}^2/2$ and the action for the classical path turns out to be

$$S_{\text{cl}} = \frac{m(x_1 - x_0)^2}{2(t_1 - t_0)} = \frac{m(x_1 - x_0)^2}{2T} = \frac{1}{2} m v^2 T, \tag{29.8}$$

where $T = t_1 - t_0$, the time of traversal, and $v = (x_1 - x_0)/T$, the average velocity of the particle as it traverses the classical path. The Lagrangian for a particle in a constant external field K generated by a linear potential is $L(x, \dot{x}) = m\dot{x}^2/2 + Kx$ and the action for the classical path is

$$S_{\text{cl}} = \frac{1}{2} m v^2 T + \frac{1}{2} K T (x_0 + x_1) - \frac{K^2 T^3}{24m}. \tag{29.9}$$

Finally, the Lagrangian for a harmonic oscillator is $L(x, \dot{x}) = m\dot{x}^2/2 - m\omega^2 x^2/2$ and the action for the classical path is

$$S_{\text{cl}} = \frac{m\omega}{2 \sin \omega T} \left[(x_0^2 + x_1^2) \cos \omega T - 2 x_0 x_1 \right]. \tag{29.10}$$

It is time to give the argument for expression 29.7. We write the generic path $x \in \Gamma$ as a variation from the classical path x_{cl}, so that $x = x_{\text{cl}} + h$ where h is a loop based at the origin. Specifically, h is defined by $h(t) = x(t) - x_{\text{cl}}(t)$ and $h(t_0) = h(t_1) = 0$. The space of such loops, denoted now as Γ_0, is the normed linear space of increments $\mathfrak{D}_0$

of Section 6.2 with norm $\| - \|_1$. By equation 6.2, the action functional may be written as

$$S[x] = S[x_{\text{cl}} + h] = S[x_{\text{cl}}] + \delta S_{x_{\text{cl}}} h + \epsilon(h), \tag{29.11}$$

where $\delta S_{x_{\text{cl}}}$ is the variational derivative of S at x_{cl} and $\epsilon(h)/\|h\|_1 \to 0$ as $\|h\|_1 \to 0$. By inserting $x = x_{\text{cl}} + h$ in equation 29.6, expanding, and integrating, the action functional may be written also as

$$S[x] = S[x_{\text{cl}} + h] = S[x_{\text{cl}}] + S_1 h + S_2(h), \tag{29.12}$$

where S_1 is the functional on Γ_0 defined by

$$S_1 h = \int_{t_0}^{t_1} \left[2a(t) x_{\text{cl}} h + b(t)(x_{\text{cl}} h)^{\dot{}} + 2c(t) \dot{x}_{\text{cl}} \dot{h} + d(t) h + e(t) \dot{h} \right] dt$$

and S_2 is the functional on Γ_0 defined by

$$S_2(h) = \int_{t_0}^{t_1} \left[a(t) h^2 + b(t) h \dot{h} + c(t) \dot{h}^2 \right] dt.$$

Notice that S_1 is linear in h and $S_2(h)/\|h\|_1 \to 0$ as $\|h\|_1 \to 0$. Uniqueness of the variational derivative then identifies the summands of equation 29.12: S_1 as $\delta S_{x_{\text{cl}}}$ and S_2 as ϵ. But, since x_{cl} is the classical path and, therefore, a stationary path for the action functional, the variational derivative $\delta S_{x_{\text{cl}}}$ is the zero functional, and we get

$$S[x] = S_{\text{cl}} + \int_{t_0}^{t_1} \left[a(t) h^2 + b(t) h \dot{h} + c(t) \dot{h}^2 \right] dt. \tag{29.13}$$

Notice that this expresses the action $S[x]$ as a constant stationary value S_{cl} plus a term that involves only the increment h. As x varies over the path space Γ, the increment $h = x - x_{\text{cl}}$ varies over the loop space Γ_0. It follows that

$$\begin{aligned} \mathcal{K}(t_1, x_1; t_0, x_0) &= \int_{\Gamma} e^{\mathrm{i}S[x]/\hbar} \mathcal{D}(x(t)) \\ &= e^{\mathrm{i}S_{\text{cl}}/\hbar} \int_{\Gamma_0} \exp\left\{ \frac{\mathrm{i}}{\hbar} \int_{t_0}^{t_1} \left[a(t) x^2 + b(t) x \dot{x} + c(t) \dot{x}^2 \right] dt \right\} \mathcal{D}(x(t)), \end{aligned} \tag{29.14}$$

verifying equation 29.7. The path integral over the space of loops depends only on the times t_0 and t_1, so we may write

$$\mathcal{K}(t_1, x_1; t_0, x_0) = e^{\mathrm{i}S_{\text{cl}}/\hbar} F(t_0, t_1) \tag{29.15}$$

where $F(t_0, t_1)$ denotes the second factor of equation 29.14. This expression leaves only the task of finding the dependence of $\mathcal{K}$ on the temporal endpoints, as the dependence on the spatial endpoints x_0 and x_1 is incorporated in the factor $e^{\mathrm{i}S_{\mathrm{cl}}/\hbar}$. For many problems, including that of the free particle, the particle in a constant force field, and the harmonic oscillator, the dependence of $\mathcal{K}$ on time is in terms of only the time difference $T = t_1 - t_0$, and F takes the form $F = F(T)$.

Quoting Feynman and Hibbs, these results seem "to be characteristic of various methods of doing path integrals; a great deal can be worked out by general methods, but often a multiplying factor is not fully determined. It must be determined by some other known property of the solution"[7] or, we might add, by a brute force calculation. We will see the results of this section illustrated as we evaluate several examples of path integrals by different methods.

29.4 The Free Particle Propagator

There is a well-developed theory of Green's functions for differential operators in mathematics, which can be used to give a rigorous derivation of the propagator for the free particle in quantum mechanics. The free particle Schrödinger equation is

$$-\frac{\hbar}{2m}\Delta\Psi = \mathrm{i}\frac{\partial\Psi}{\partial t}$$

and the Green's function, or propagator, for this equation turns out to be

$$\mathcal{K}_{\mathrm{free}}(t_1, x_1; t_0, x_0) = \left[\frac{m}{2\pi\mathrm{i}\hbar(t_1 - t_0)}\right]^{1/2} \exp\left[\frac{\mathrm{i}m(x_1 - x_0)^2}{2\hbar(t_1 - t_0)}\right]. \quad (29.16)$$

From equation 29.8, this is precisely of the form $\mathcal{K} = F(T)e^{\mathrm{i}S_{\mathrm{cl}}/\hbar}$, as promised in equation 29.15

Rather than reproduce a rigorous derivation from the theory of Green's functions,[8] we use the Dirac calculus and the rigged Hilbert space formalism of Lecture 5 to give a formal derivation that a physicist might like, but which would make a mathematician cringe. Recall, though, that

[7] Feynman and Hibbs (1965), pp. 60–61.
[8] See, for example, pp. 442–447, ending with Equation (7.128), of Byron and Fuller (1969).

the Dirac calculus provides a notation for linear and anti-linear functionals on the Schwarz space $\mathcal{S}$ and, ultimately, a shorthand notation that encodes the substance of the spectral theorem in a compact form.

Since the Hamiltonian $\mathsf{H} = \mathsf{P}^2/2m$ does not depend explicitly on time, the time-evolution operator depends only on the time difference. In particular, $\mathsf{U}(t_1, t_0) = \mathsf{U}_T = e^{-\mathbf{i}T\mathsf{H}/\hbar}$ where $T = t_1 - t_0$. Assuming that it makes sense to time-evolve generalized eigenkets $|x\rangle$ of the position operator, $\psi = \mathsf{U}_T|x_0\rangle$ is the state vector at time t_1 of a particle that is localized at position x_0 at time t_0. The amplitude for the particle to be at x_1 at time t_1, given that it is at x_0 at time t_0, is then

$$\mathcal{K}_{\mathsf{free}}(t_1, x_1; t_0, x_0) = \psi(x_1) = \langle x_1|\psi\rangle = \langle x_1|\mathsf{U}_T|x_0\rangle.$$

We present an evaluation of the expression $\langle x_1|\mathsf{U}_T|x_0\rangle$ in its entirety and then offer an explanation for each line of the computation:

$$\begin{aligned}\langle x_1|\mathsf{U}_T|x_0\rangle = \langle x_1|\mathsf{U}_T 1_{\mathcal{S}}|x_0\rangle &= \langle x_1|\mathsf{U}_T \int_{-\infty}^{\infty} |p\rangle\langle p|\, dp\, |x_0\rangle \\ &= \int_{-\infty}^{\infty} \langle x_1|\mathsf{U}_T|p\rangle\langle p|x_0\rangle\, dp \\ &= \int_{-\infty}^{\infty} e^{-\mathbf{i}Tp^2/2m\hbar}\langle x_1|p\rangle\langle p|x_0\rangle\, dp \\ &= \int_{-\infty}^{\infty} e^{-\mathbf{i}Tp^2/2m\hbar}\frac{1}{2\pi\hbar}e^{\mathbf{i}p(x_1-x_0)/\hbar}\, dp \\ &= \frac{1-\mathbf{i}}{\sqrt{2}}\left(\frac{m}{2\pi\hbar T}\right)^{1/2} e^{\mathbf{i}m(x_1-x_0)^2/2\hbar T} \\ &= \left(\frac{m}{2\pi\mathbf{i}\hbar T}\right)^{1/2} e^{\mathbf{i}m(x_1-x_0)^2/2\hbar T}, \qquad (29.17)\end{aligned}$$

which is precisely expression 29.16. In the first line, the identity operator in the momentum representation, $1_{\mathcal{S}} = \int_{-\infty}^{\infty} |p\rangle\langle p|\, dp$ from expression 5.9, is applied, and the second line then follows from the linearity of time evolution. The third line is obtained from evaluating $\mathsf{U}_T|p\rangle$ by expanding $\mathsf{U}_T = e^{-\mathbf{i}T\mathsf{H}/\hbar}$ in the usual exponential power series and using $\mathsf{P}^m|p\rangle = p^m|p\rangle$, which follows from the eigenvalue equation 5.5. The fourth line follows from two applications of equation 5.3, and the fifth is the result of evaluating the integral of line four. Physicists generally use the symbol $\sqrt{\mathbf{i}}$ for the principal square root of $\mathbf{i}$. Thus $\sqrt{\mathbf{i}} = e^{\mathbf{i}\pi/4} = (1+\mathbf{i})/\sqrt{2}$. The last line of the calculation is just notational, with $\sqrt{1/\mathbf{i}} = \sqrt{2}/(1+\mathbf{i}) = (1-\mathbf{i})/\sqrt{2} = e^{-\mathbf{i}\pi/4}$.

The integral in line four is representative of a class of integrals widely used in physics and applied mathematics. These are known as **Fresnel integrals** in the optics literature and, more generally, as **Gaussian integrals**. They arise not only in optics and path integration but also in the evaluation of Feynman diagrams in quantum field theory. In fact, complex Gaussian integrals will surface in the calculations in the remaining lectures in this book. As this topic is unfamiliar to many mathematicians, in the next lecture we will present derivations of all the Gaussian integral formulae used in this book, including equation 30.3, the one needed to evaluate the integral in line four.

29.5 End Notes

The overall tenor of this lecture owes much to Chapter 2 of Bjørn Felsager's wonderful book (1998) and, of course, Feynman's classic book with Hibbs (1965). Feynman and Hibbs (1965) is the canonical text on the Feynman approach to quantum mechanics through the use of path integrals and remains as one of the better developments of the theory even today. It is strong especially in explanations that give good heuristic reasoning behind the theory, and what it lacks in rigor it makes up for in deep physical insight.

Another treatment I recommend, with several worked-out examples, is that of Dittrich and Reuter (2001) in their Chapters 16–25. This represents a very nice short course in the subject. Finally, Chapter 5 of Felsager (1998) and Chapters 8 and 21 of Shankar (1994) are good concise introductions to the topic.

30
A Mathematical Interlude: Gaussian Integrals

> So, what's an interesting integral ...? I suppose the honest answer ... is sort of along the lines of Supreme Court Associate Justice Potter Stewart's famous 1964 comment on the question of "what is pornography?" He admitted it was hard to define "but I know it when I see it." It's exactly the same with an interesting integral!
>
> Paul J. Nahin
> *Inside Interesting Integrals*, 2015.

Gaussian integrals arise naturally in many areas of physics and applied mathematics. We have seen already a complex Gaussian integral in the derivation of the free particle propagator in the preceding lecture and, in fact, saw another type of Gaussian integral in Section 9.4, when taking the Fourier transform of a Gaussian wave packet. In the next lecture, even more sophisticated Gaussian integrals are needed when we are evaluating propagators using path integrals. As important as these integrals are, they are unfamiliar to many mathematicians and even many physicists who use them regularly have never seen a rigorous derivation of a closed form expression for them. The derivations of these formulae are interesting enough to warrant a lecture dedicated to the evaluation of, at least, all the Gaussian integrals that are used in this book. In fact, anyone who works with path integrals in any significant way would be well served by learning as much as possible about the various types of Gaussian integrals, their rigorous derivations, and the evaluation of specific examples.

In the first section, the classical Fresnel definite integrals of optics are derived, as is the formula already applied in Section 29.4 in the derivation of the free particle propagator. This is generalized in Section 30.2 to an

N-variable complex Gaussian integral in which the squared variable in the exponential in the one-dimensional integrand is replaced by a N-dimensional quadratic form of full rank. The N-variable integral formula is applied in that section and the two sections following to derive specific Gaussian integrals that will be used in the next lecture to evaluate path integrals. The reader who is content with merely using the Gaussian integral formulae and is unconcerned with their derivations may safely skip the details of this mathematical lecture and simply refer to the results.

30.1 Fresnel and Gaussian Integrals

The reader may have seen the derivation of the Gaussian integral

$$\int_{-\infty}^{\infty} e^{-ax^2}\, dx = \sqrt{\frac{\pi}{a}}, \tag{30.1}$$

valid for any $a > 0$ and proved by writing

$$\left(\int_{-\infty}^{\infty} e^{-ax^2}\, dx\right)^2 = \int_{-\infty}^{\infty}\int_{-\infty}^{\infty} e^{-a(x^2+y^2)}\, dxdy$$

and then passing to polar coordinates. If we replace a by $\mathbf{i}a$ in formula 30.1, we obtain

$$\int_{-\infty}^{\infty} e^{-\mathbf{i}ax^2}\, dx = \sqrt{\frac{\pi}{\mathbf{i}a}}. \tag{30.2}$$

Of course there is no particular reason to expect that this naïve move has any validity, but the fact is that this formula, properly interpreted, is correct. Our aim in this section is to give a rigorous proof of this and similar generalizations of formula 30.1. First, we verify that the formula

$$\int_{-\infty}^{\infty} e^{\pm\mathbf{i}(ax^2+bx+c)}\, dx = \frac{1\pm\mathbf{i}}{\sqrt{2}}\sqrt{\frac{\pi}{a}}\, e^{\pm\mathbf{i}(4ac-b^2)/4a} \tag{30.3}$$

is valid for any $a > 0$, $b \in \mathbb{R}$, and $c \in \mathbb{C}$. If we use the physicist's convention that $\sqrt{\mathbf{i}}$ is to be interpreted as the principal square root of $\mathbf{i}$, then $\sqrt{\mathbf{i}} = e^{\mathbf{i}\pi/4} = (1+\mathbf{i})/\sqrt{2}$ and $\sqrt{-\mathbf{i}} = \sqrt{1/\mathbf{i}} = \sqrt{2}/(1+\mathbf{i}) = (1-\mathbf{i})/\sqrt{2} = e^{-\mathbf{i}\pi/4}$, and we may write this formula as

$$\int_{-\infty}^{\infty} e^{\mathbf{i}(ax^2+bx+c)}\, dx = \sqrt{\frac{\mathbf{i}\pi}{a}}\, e^{\mathbf{i}(4ac-b^2)/4a}, \tag{30.4}$$

valid for any $a \neq 0$, $b \in \mathbb{R}$, and $c \in \mathbb{C}$.

The reader should recall that the integrals of these expressions require separate verification that they converge at both $-\infty$ and $+\infty$; i.e., the integrals are interpreted as

$$\int_{-\infty}^{\infty} f(x)\,dx = \lim_{R\to\infty}\int_{-R}^{0} f(x)\,dx + \lim_{R\to\infty}\int_{0}^{R} f(x)\,dx.$$

We begin with the evaluation of the integral

$$\int_0^{\infty} e^{\mathrm{i}x^n}\,dx = \int_0^{\infty} \cos x^n\,dx + \mathrm{i}\int_0^{\infty} \sin x^n\,dx,$$

where $n > 1$ is an integer. Let $C = C_1 + C_2 + C_3$ be an oriented contour in $\mathbb{C}$, where C_1 is the oriented line segment from the origin 0 to 1 along the real axis, C_2 is the oriented circular arc on the unit circle moving counterclockwise from 1 to $e^{\mathrm{i}\pi/2n}$, and C_3 is the oriented segment from $e^{\mathrm{i}\pi/2n}$ back to the origin. Then $RC = RC_1 + RC_2 + RC_3$ is the contour C dilated by the positive factor R. Since the function $e^{\mathrm{i}z^n}$ is entire, Cauchy's theorem implies that

$$0 = \int_{RC} e^{\mathrm{i}z^n}\,dz = \int_{RC_1} e^{\mathrm{i}z^n}\,dz + \int_{RC_2} e^{\mathrm{i}z^n}\,dz + \int_{RC_3} e^{\mathrm{i}z^n}\,dz. \tag{30.5}$$

Parameterizing RC_2 via $z = Re^{\mathrm{i}\theta/n}$ for $0 \le \theta \le \pi/2$ gives

$$\int_{RC_2} e^{\mathrm{i}z^n}\,dz = \frac{\mathrm{i}R}{n}\int_0^{\pi/2} e^{\mathrm{i}(\theta/n + R^n\cos\theta)} e^{-R^n\sin\theta}\,d\theta,$$

and using the inequality $2\theta/\pi \le \sin\theta$ on the interval $[0, \pi/2]$ gives

$$\begin{aligned}\left|\int_{RC_2} e^{\mathrm{i}z^n}\,dz\right| &\le \frac{R}{n}\int_0^{\pi/2} e^{-R^n\sin\theta}\,d\theta \\ &\le \frac{R}{n}\int_0^{\pi/2} e^{-2R^n\theta/\pi}\,d\theta = \frac{\pi}{2nR^{n-1}}\left(1 - e^{-R^n}\right).\end{aligned}$$

Using the notation $\int_{\infty C}$ for $\lim_{R\to\infty}\int_{RC}$, since $n > 1$, this latter estimate implies that

$$\int_{\infty C_2} e^{\mathrm{i}z^n}\,dz = 0.$$

Equation 30.5 now implies that

$$\int_{\infty C_1} e^{\mathrm{i}z^n}\,dz = -\int_{\infty C_3} e^{\mathrm{i}z^n}\,dz, \tag{30.6}$$

provided either, and therefore both, of these limits exist. Parameterizing

$-RC_3$ via $z = te^{\mathrm{i}\pi/2n}$ for $0 \le t \le R$, the second integral is

$$\int_{\infty C_3} e^{\mathrm{i}z^n}\, dz = \lim_{R\to\infty} e^{\mathrm{i}\pi/2n} \int_R^0 e^{\mathrm{i}t^n e^{\mathrm{i}\pi/2}}\, dt = -e^{\mathrm{i}\pi/2n} \int_0^\infty e^{-t^n}\, dt.$$

The substitution $u = t^n$ in this latter integral identifies it as $\Gamma(1/n)/n$. Finally, parameterizing RC_1 via $z = x$ for $0 \le x \le R$, this and equation 30.6 give

$$\int_0^\infty e^{\mathrm{i}x^n}\, dx = e^{\mathrm{i}\pi/2n} \frac{1}{n}\Gamma\left(\frac{1}{n}\right). \tag{30.7}$$

Identifying real and imaginary parts gives the two formulae

$$\int_0^\infty \cos x^n\, dx = \frac{1}{n}\Gamma\left(\frac{1}{n}\right)\cos\frac{\pi}{2n}$$

and

$$\int_0^\infty \sin x^n\, dx = \frac{1}{n}\Gamma\left(\frac{1}{n}\right)\sin\frac{\pi}{2n}.$$

When $n = 2$, these are the classical **Fresnel definite integrals**

$$\int_0^\infty \cos x^2\, dx = \frac{1}{2}\sqrt{\frac{\pi}{2}} = \int_0^\infty \sin x^2\, dx.$$

As $\cos x^2$ and $\sin x^2$ are even functions, we have

$$\int_{-\infty}^\infty \cos x^2\, dx = \sqrt{\frac{\pi}{2}} = \int_{-\infty}^\infty \sin x^2\, dx. \tag{30.8}$$

Performing the change of variables $u = \sqrt{a}x$ for $a > 0$ in the integrals below now gives

$$\begin{aligned}\int_{-\infty}^\infty e^{\pm\mathrm{i}ax^2}\, dx &= \int_{-\infty}^\infty \cos ax^2\, dx \pm \mathrm{i}\int_{-\infty}^\infty \sin ax^2\, dx\\ &= \sqrt{\frac{\pi}{2a}} \pm \mathrm{i}\sqrt{\frac{\pi}{2a}} = \frac{1\pm\mathrm{i}}{\sqrt{2}}\sqrt{\frac{\pi}{a}} = e^{\pm\mathrm{i}\pi/4}\sqrt{\frac{\pi}{a}},\end{aligned}$$

our first generalization of formula 30.1. Again using the physicist's convention that $\sqrt{\pm\mathrm{i}} = e^{\pm\mathrm{i}\pi/4}$, we may write this as

$$\int_{-\infty}^\infty e^{\mathrm{i}ax^2}\, dx = \sqrt{\frac{\mathrm{i}\pi}{a}} = e^{(\operatorname{sgn} a)\mathrm{i}\pi/4}\sqrt{\frac{\pi}{|a|}}, \tag{30.9}$$

valid for any real $a \ne 0$, verifying equation 30.4 in the case where $b = c = 0$ and verifying equation 30.2 when $a < 0$.

The equivalent formulae 30.3 and 30.4 follow by translation of the line of integration by an amount $-b/2a$. Indeed, after substituting $u =$

$x + (b/2a)$ in the second integral below and applying formula 30.9, we have

$$\int_{-\infty}^{\infty} e^{\mathbf{i}(ax^2+bx+c)}\, dx = \int_{-\infty}^{\infty} e^{\mathbf{i}a(x+(b/2a))^2+\mathbf{i}(4ac-b^2)/4a}\, dx$$
$$= e^{\mathbf{i}(4ac-b^2)/4a} \int_{-\infty}^{\infty} e^{\mathbf{i}au^2}\, du = \sqrt{\frac{\mathbf{i}\pi}{a}}\, e^{\mathbf{i}(4ac-b^2)/4a},$$

valid for $a \neq 0$, $b \in \mathbb{R}$, and $c \in \mathbb{C}$.

30.2 The Complex Gaussian Integral in N Dimensions

In this section, we generalize the complex Gaussian integral 30.4 to N dimensions by replacing the quadratic in the exponential by a general quadratic form of full rank. First, we simplify by assuming that the quadratic form is positive definite and confirm in that case the formula

$$\int_{\mathbb{R}^N} e^{\mathbf{i}\left(a\mathbf{x}^{\mathsf{Tr}}\mathbf{A}\mathbf{x}+J\mathbf{b}^{\mathsf{Tr}}\mathbf{x}\right)}\, d\mathbf{x} = \left(\frac{\mathbf{i}\pi}{a}\right)^{N/2} (\det \mathbf{A})^{-1/2}\, e^{-\mathbf{i}J^2\mathbf{b}^{\mathsf{Tr}}\mathbf{A}^{-1}\mathbf{b}/4a}. \tag{30.10}$$

In this formula, $\mathbf{x} \in \mathbb{R}^N$ is represented as a column vector and its transpose is the row vector $\mathbf{x}^{\mathsf{Tr}} = [x_1\ x_2\ \cdots\ x_N]$. The real constant a satisfies $a \neq 0$, $J \in \mathbb{R}$, $\mathbf{b} \in \mathbb{R}^N$, and $\mathbf{A}$ is a positive definite symmetric $N \times N$ matrix. The first term on the right-hand side of the formula is interpreted as follows:

$$\left(\frac{\mathbf{i}\pi}{a}\right)^{N/2} = e^{(\operatorname{sgn} a)\mathbf{i}N\pi/4} \left(\sqrt{\frac{\pi}{|a|}}\right)^N. \tag{30.11}$$

To confirm this formula, let $\mathbf{O}$ be an $N \times N$ orthogonal matrix that diagonalizes $\mathbf{A}$, so that

$$\mathbf{\Lambda} = \mathbf{O}^{\mathsf{Tr}}\mathbf{A}\mathbf{O} = \mathbf{diag}[\lambda_1\ \lambda_2\ \cdots\ \lambda_N], \tag{30.12}$$

where the λ_i are the eigenvalues of $\mathbf{A}$, which are all positive since $\mathbf{A}$ is positive definite and symmetric. Performing the change of variables $\mathbf{x} = \mathbf{O}\mathbf{y}$ gives the integral as

$$\int_{\mathbb{R}^N} e^{\mathbf{i}\left(a\mathbf{y}^{\mathsf{Tr}}\mathbf{\Lambda}\mathbf{y}+J\mathbf{b}^{\mathsf{Tr}}\mathbf{O}\mathbf{y}\right)} |\mathbf{J}(\mathbf{O})|\, d\mathbf{y} = \prod_{n=1}^{N} \int_{-\infty}^{\infty} e^{\mathbf{i}\left(a\lambda_n y_n^2 + J\left[\mathbf{b}^{\mathsf{Tr}}\mathbf{O}\right]_n y_n\right)}\, dy_n, \tag{30.13}$$

where we have used the fact that the Jacobian determinant $|\mathbf{J}(\mathbf{O})|$ is equal to 1, since $\mathbf{O}$ is orthogonal, and have used the notation $[\mathbf{b}^{\mathsf{Tr}}\mathbf{O}]_n$ for the nth entry of the row vector $\mathbf{b}^{\mathsf{Tr}}\mathbf{O}$. Applying formula 30.4 N times in the right-hand side of equation 30.13 gives its value as

$$\left(\sqrt{\frac{\mathbf{i}\pi}{a}}\right)^N \sqrt{\frac{1}{\lambda_1\cdots\lambda_N}}\,\exp\left(-\mathbf{i}\frac{J^2}{4a}\sum_{n=1}^{N}\frac{[\mathbf{b}^{\mathsf{Tr}}\mathbf{O}]_n^2}{\lambda_n}\right).$$

Formula 30.10 now follows from the observations that $\det\mathbf{A} = \det\mathbf{\Lambda} = \lambda_1\cdots\lambda_N$ and

$$\sum_{n=1}^{N}\frac{1}{\lambda_n}[\mathbf{b}^{\mathsf{Tr}}\mathbf{O}]_n^2 = \mathbf{b}^{\mathsf{Tr}}\mathbf{O}\mathbf{\Lambda}^{-1}\mathbf{O}^{\mathsf{Tr}}\mathbf{b} = \mathbf{b}^{\mathsf{Tr}}\mathbf{A}^{-1}\mathbf{b}.$$

The general case follows just as easily. Let $\mathbf{A}$ now be a symmetric matrix of full rank N and of index p, where $0 \le p \le N$. This means that $\mathbf{A}$ may be diagonalized exactly as in equation 30.12, the only difference being that λ_n is positive when $1 \le n \le p$ and negative when $p < n \le N$. Each integration in the product in equation 30.13 contributes a term

$$\sqrt{\frac{\mathbf{i}\pi}{a\lambda_n}} = e^{\operatorname{sgn}(a\lambda_n)\mathbf{i}\pi/4}\sqrt{\frac{\pi}{|a\lambda_n|}}.$$

This gives the following modification of formula 30.10:

$$\int_{\mathbb{R}^N} e^{\mathbf{i}\left(a\mathbf{x}^{\mathsf{Tr}}\mathbf{A}\mathbf{x}+J\mathbf{b}^{\mathsf{Tr}}\mathbf{x}\right)}\,d\mathbf{x} = e^{(\operatorname{sgn} a)(2p-N)\mathbf{i}\pi/4} \times\left(\frac{\pi}{|a|}\right)^{N/2}|\det\mathbf{A}|^{-1/2}\,e^{-\mathbf{i}J^2\mathbf{b}^{\mathsf{Tr}}\mathbf{A}^{-1}\mathbf{b}/4a}. \tag{30.14}$$

Compare the leading exponential factor with equation 30.11, and notice that formula 30.14 reduces to formula 30.10 when $p = N$.

Our first application of the N-dimensional formula 30.10 is in the nonstandard evaluation of the complex Gaussian integral

$$F_N(a,c,d) = \int_{\mathbb{R}^N} e^{\mathbf{i}a\left[(x_1-c)^2+(x_2-x_1)^2+\cdots+(x_n-x_{n-1})^2+(d-x_n)^2\right]}\,d\mathbf{x}, \tag{30.15}$$

where $a \neq 0$ and $c, d \in \mathbb{R}$. The reader is invited to perform a standard evaluation of this integral accomplished by an inductive argument after the one-dimensional case has been derived. Instead of this, we will begin by writing the integral expression in matrix form as

$$F_N(a,c,d) = e^{\mathbf{i}a(c^2+d^2)}\int_{\mathbb{R}^N} e^{\mathbf{i}\left[a\mathbf{x}^{\mathsf{Tr}}(2\mathbf{I}-\mathbf{J})\mathbf{x}-2a\mathbf{b}^{\mathsf{Tr}}\mathbf{x}\right]}\,d\mathbf{x}, \tag{30.16}$$

where $\mathbf{I} = \mathbf{I}_N$ is the $N \times N$ identity matrix, $\mathbf{J} = \mathbf{J}_N$ is the $N \times N$ matrix whose only nonzero entries are along both the off diagonals, where all entries are 1, and $\mathbf{b}^{\mathsf{Tr}}$ is the row vector $[c\ 0\ 0\ \cdots\ 0\ d]$. The positive definite matrix[1] $\mathbf{A} = \mathbf{A}_N = 2\mathbf{I}_N - \mathbf{J}_N$ is

$$\mathbf{A}_N = \begin{pmatrix} 2 & -1 & & & & \\ -1 & 2 & -1 & & \mathbf{0} & \\ & -1 & 2 & -1 & & \\ & & \ddots & \ddots & \ddots & \\ & \mathbf{0} & & -1 & 2 & -1 \\ & & & & -1 & 2 \end{pmatrix}$$

and its determinant may be evaluated by expanding the determinant of $\mathbf{A}_{N+1}$ along the first row, to get the recursive relation

$$\det \mathbf{A}_{N+1} = 2 \det \mathbf{A}_N - \det \mathbf{A}_{N-1}.$$

Direct evaluation gives $\det \mathbf{A}_1 = 2$, $\det \mathbf{A}_2 = 3$, and $\det \mathbf{A}_3 = 4$ and induction now gives $\det \mathbf{A}_N = N+1$. Applying formula 30.10, we obtain

$$F_N(a,c,d) = e^{\mathrm{i}a(c^2+d^2)} \frac{1}{\sqrt{N+1}} \left(\frac{\mathrm{i}\pi}{a}\right)^{N/2} e^{-\mathrm{i}a\mathbf{b}^{\mathsf{Tr}}\mathbf{A}^{-1}\mathbf{b}}. \tag{30.17}$$

To compute the term $\mathbf{b}^{\mathsf{Tr}}\mathbf{A}^{-1}\mathbf{b}$, we let $\mathbf{z} = \mathbf{A}^{-1}\mathbf{b}$ and solve $\mathbf{b} = \mathbf{A}\mathbf{z}$ for the unknown vector $\mathbf{z}$. Define the elements $\boldsymbol{\xi}^N$ and $\boldsymbol{\xi}_N$ in $\mathbb{R}^N$ by

$$\boldsymbol{\xi}^N = \begin{pmatrix} N \\ N-1 \\ \vdots \\ 2 \\ 1 \end{pmatrix} \quad \text{and} \quad \boldsymbol{\xi}_N = \begin{pmatrix} 1 \\ 2 \\ \vdots \\ N-1 \\ N \end{pmatrix}$$

and observe that

$$\mathbf{A}\boldsymbol{\xi}^N = \begin{pmatrix} N+1 \\ 0 \\ \vdots \\ 0 \\ 0 \end{pmatrix} \quad \text{and} \quad \mathbf{A}\boldsymbol{\xi}_N = \begin{pmatrix} 0 \\ 0 \\ \vdots \\ 0 \\ N+1 \end{pmatrix}.$$

[1] The matrix $\mathbf{A}$ is positive definite since

$$\mathbf{x}^{\mathsf{Tr}}\mathbf{A}\mathbf{x} = x_1^2 + \sum_{n=2}^{N} (x_n - x_{n-1})^2 + x_N^2.$$

It follows easily that

$$\mathbf{z} = \frac{c}{N+1}\boldsymbol{\xi}^N + \frac{d}{N+1}\boldsymbol{\xi}_N$$

and, therefore,

$$\mathbf{b}^{\mathsf{Tr}}\mathbf{A}^{-1}\mathbf{b} = \mathbf{b}^{\mathsf{Tr}}\mathbf{z} = \frac{Nc^2 + 2cd + Nd^2}{N+1}.$$

This gives $(c^2 + d^2) - \mathbf{b}^{\mathsf{Tr}}\mathbf{A}^{-1}\mathbf{b} = (c-d)^2/(N+1)$ and, substituting this into equation 30.17,

$$\begin{aligned} F_N(a,c,d) &= \int_{\mathbb{R}^N} e^{\mathbf{i}a\left[(x_1-c)^2+(x_2-x_1)^2+\cdots+(x_n-x_{n-1})^2+(d-x_n)^2\right]}\,d\mathbf{x} \\ &= \frac{1}{\sqrt{N+1}}\left(\frac{\mathbf{i}\pi}{a}\right)^{N/2} e^{\mathbf{i}a(c-d)^2/(N+1)} \\ &= e^{(\operatorname{sgn} a)\mathbf{i}N\pi/4}\sqrt{\frac{\pi^N}{(N+1)|a|^N}}\,\exp\left(\mathbf{i}a\frac{(c-d)^2}{N+1}\right). \end{aligned} \tag{30.18}$$

Formula 30.18 will be employed in the next lecture to normalize the measure used in the evaluation of propagators for the case when the paths in the path integral formalism are piecewise linear. To apply that measure when calculating the propagator of a particle in a constant force field we will need a generalization of this integral, which is presented next.

30.3 Generalizing $F_N(a, c, d)$

We now add a linear term to the exponent in integral 30.15. Our interest is in the integral

$$\begin{aligned} &F_N(a,c,d,k) \\ &\quad = \int_{\mathbb{R}^N} \exp\left\{\mathbf{i}\left[a\left(\sum_{n=1}^{N+1}(x_n - x_{n-1})^2\right) + k\left(\sum_{n=1}^{N} x_n\right)\right]\right\} d\mathbf{x}, \end{aligned} \tag{30.19}$$

where $a \neq 0$, $c = x_0$, $d = x_{N+1}$, and $k \in \mathbb{R}$. We will verify that

$$\begin{aligned} &F_N(a,c,d,k) \\ &\quad = F_N(a,c,d)\exp\left[\mathbf{i}\left(\frac{N}{2}k(c+d) - \frac{N(N+1)(N+2)}{48a}k^2\right)\right], \end{aligned} \tag{30.20}$$

where $F_N(a,c,d) = F_N(a,c,d,0)$ is given by equation 30.18.

One approach is to write expression 30.19 in matrix form as

$$F_N(a,c,d,k) = e^{\mathbf{i}a(c^2+d^2)} \times \int_{\mathbb{R}^N} \exp\left\{\mathbf{i}\left[a\mathbf{x}^{\mathsf{Tr}}\left(2\mathbf{I}-\mathbf{J}\right)\mathbf{x} + (\mathbf{k}-2a\mathbf{b})^{\mathsf{Tr}}\mathbf{x}\right]\right\}\, d\mathbf{x},$$

where the symbols have the same meaning as in formula 30.16 and $\mathbf{k} \in \mathbb{R}^N$ is the column vector all of whose entries are equal to k. Setting $\mathbf{A} = 2\mathbf{I} - \mathbf{J}$, the N-dimensional Gaussian integral formula 30.10 implies that

$$F_N(a,c,d,k) = e^{\mathbf{i}a(c^2+d^2)} \frac{1}{\sqrt{N+1}}\left(\frac{\mathbf{i}\pi}{a}\right)^{N/2} \times \exp\left[-\frac{\mathbf{i}}{4a}(\mathbf{k}-2a\mathbf{b})^{\mathsf{Tr}}\mathbf{A}^{-1}(\mathbf{k}-2a\mathbf{b})\right].$$

We now have the unenviable task of computing the term

$$(\mathbf{k}-2a\mathbf{b})^{\mathsf{Tr}}\mathbf{A}^{-1}(\mathbf{k}-2a\mathbf{b}),$$

which is a tedious calculation. Rather than pursuing this line of attack, we make use of our previous work and verify expression 30.20 by mathematical induction.

The basis of induction, when $N = 1$, is an easy application of formula 30.4. Assume now that formula 30.20 holds for some $N \in \mathbb{N}$. Then a straightforward calculation with $c = x_0$ and $d = x_{N+2}$ gives

$$\begin{aligned} F_{N+1}&(a,c,d,k) \\ &= \int_{-\infty}^{\infty} F_N(a,c,x_{N+1},k) \exp\left\{\mathbf{i}\left[a(d-x_{N+1})^2 + kx_{N+1}\right]\right\} dx_{N+1} \\ &= \frac{1}{\sqrt{N+1}}\left(\frac{\mathbf{i}\pi}{a}\right)^{N/2} \exp\left[\mathbf{i}\left(\frac{1}{2}Nck - \frac{N(N+1)(N+2)}{48a}k^2\right)\right] \\ &\quad \times \int_{-\infty}^{\infty} \exp\left\{\mathbf{i}\left[a\left(\frac{(x-c)^2}{N+1} + (d-x)^2\right) + \frac{N+2}{2}kx\right]\right\} dx. \end{aligned} \tag{30.21}$$

The last integral may be written as

$$\int_{-\infty}^{\infty} e^{\mathbf{i}(Ax^2+Bx+C)}\, dx$$

where

$$A = a\left(\frac{N+2}{N+1}\right), \qquad C = a\left(\frac{c^2}{N+1} + d^2\right),$$

and

$$B = k\left(\frac{N+2}{2}\right) - \frac{2ac}{N+1} - 2ad.$$

We now apply formula 30.4 to integrate this, obtaining

$$\begin{aligned}
&F_{N+1}(a,c,d,k) \\
&\quad = \frac{1}{\sqrt{N+1}}\left(\frac{\mathrm{i}\pi}{a}\right)^{N/2} \exp\left[\mathrm{i}\left(\frac{1}{2}Nck - \frac{N(N+1)(N+2)}{48a}k^2\right)\right] \\
&\quad \times \sqrt{\frac{\mathrm{i}\pi}{A}}\, e^{\mathrm{i}(4AC-B^2)/4A}.
\end{aligned}$$

A lengthy but straightforward algebraic computation simplifies this to give formula 30.20 with N replaced by $N+1$, confirming the inductive step. Formula 30.20 is thus verified for all $N \in \mathbb{N}$.

30.4 Hilbert Matrices and Cauchy Determinants

Complex N-dimensional Gaussian integrals like that of formula 30.10 arise naturally in the evaluation of path integrals. We have just derived the Gaussian formulae 30.18 and 30.20 for use in the evaluation of path integrals using piecewise linear path families. In this section, we evaluate $\det \mathbf{B}$ for a matrix $\mathbf{B}$ that arises in the computation of propagators using polynomial path families and derive a corresponding Gaussian formula. This example is related rather closely to a class of matrices known as **Hilbert matrices**, whose determinants are special cases of a class known as **Cauchy determinants**.

The matrix $\mathbf{B}$ is the $N \times N$ positive definite symmetric matrix given as

$$\mathbf{B} = \mathbf{B}_N = \left[\frac{ij}{i+j+1}T^{i+j+1}\right]_{i,j=1}^{N},$$

where $T > 0$, and our interest is in evaluating the Gaussian integral

$$\int_{\mathbb{R}^N} e^{\mathrm{i}a\mathbf{x}^{\mathrm{Tr}}\mathbf{B}\mathbf{x}}\, d\mathbf{x} = \left(\frac{\mathrm{i}\pi}{a}\right)^{N/2} (\det \mathbf{B})^{-1/2}. \tag{30.22}$$

In computing $\det \mathbf{B}$, the **Hilbert matrices**

$$\mathbf{H}_N = \left[\frac{1}{i+j-1}\right]_{i,j=1}^{N} \quad \text{and} \quad \mathbf{C}_N = \left[\frac{1}{i+j+1}\right]_{i,j=1}^{N} \tag{30.23}$$

play an important role. First note that $\mathbf{B}_N = T\mathbf{T}_N\mathbf{C}_N\mathbf{T}_N$ where $\mathbf{T}_N = \mathbf{diag}\left[T\ 2T^2\ \cdots\ NT^N\right]$, so that

$$\begin{aligned}\det \mathbf{B}_N &= T^N\,(\det \mathbf{T}_N)^2 \det \mathbf{C}_N \\ &= T^N (N!)^2 T^{N(N+1)} \det \mathbf{C}_N = (N!)^2 T^{N(N+2)} \det \mathbf{C}_N. \end{aligned} \tag{30.24}$$

The determinant of the Hilbert matrix $\mathbf{C}_N$ may be evaluated as a special case of a **Cauchy determinant**[2]

$$\det\left[\frac{1}{s_i + t_j}\right] = \frac{\prod_{i>j}[(s_i - s_j)(t_i - t_j)]}{\prod_{i,j}(s_i + t_j)}. \tag{30.25}$$

Taking $s_i = i - 1$ and $t_j = j$ for $\mathbf{H}_N$ and $s_i = i + 1$ and $t_j = j$ for $\mathbf{C}_N$ gives, after some manipulation,

$$\det \mathbf{H}_N = \frac{\left(\prod_{n=1}^{N-1} n!\right)^4}{\prod_{n=1}^{2N-1} n!}, \quad \det \mathbf{C}_N = \frac{\left(\prod_{n=1}^{N-1} n!\right)^4 [N!(N+1)!]^2}{\prod_{n=1}^{2N+1} n!}. \tag{30.26}$$

It then easily follows that

$$\det \mathbf{C}_N = (N+1)^2 \det \mathbf{H}_{N+1}. \tag{30.27}$$

We can get an idea of the size of these determinants by applying Stirling's formula to derive the approximate identity

$$\det \mathbf{H}_N \approx \frac{\pi^{N-1}}{2^{2N^2 - N + 1}}.$$

Putting together equations 30.24, 30.26, and 30.27, we get

$$\begin{aligned}\det \mathbf{B}_N &= ((N+1)!)^2 T^{N(N+2)} \det \mathbf{H}_{N+1} \\ &= ((N+1)!)^2 T^{N(N+2)} \frac{\left(\prod_{n=1}^{N} n!\right)^4}{\prod_{n=1}^{2N+1} n!}. \end{aligned} \tag{30.28}$$

From this we get the Gaussian integral formula

$$\int_{\mathbb{R}^N} e^{\mathbf{i}a\mathbf{x}^{\mathrm{Tr}}\mathbf{B}\mathbf{x}}\, d\mathbf{x} = \left(\frac{\mathbf{i}\pi}{aT^{N+2}}\right)^{N/2} \frac{(N+1)!\sqrt{\prod_{n=1}^{2N+1} n!}}{\left(\prod_{n=1}^{N+1} n!\right)^2}.$$

[2] See Melzak (1976), pp. 152–153 for a proof of the Cauchy determinant formula.

30.5 End Notes

A good resource for the Fresnel definite integrals and some of the Gaussian integrals in this lecture is Chapter 11 of Williams (2003). I worked out all the integrals in this lecture from scratch, with the aid of Hilbert matrices and Cauchy determinants for those in Section 30.4. The reference for Hilbert matrices and Cauchy determinants that I used is Melzak (1976), and this gives me the opportunity to recommend highly this book and its predecessor, Melzak (1973). These companion books are a delightful compilation of problems, techniques, formal manipulations, intuitive discussions, recreational mathematics, beautiful classical geometry, applications of linear algebra, computing, manipulative combinatorics, and entertaining mathematical amusements that span an eclectic array of geometry, analysis, combinatorics, computations and applications. As the title of the first and the subtitle of the second suggest, these are meant to be companions to concrete mathematics, supplementing the standard education of the student in the traditional curriculum. They are a joy to browse!

31
Evaluating Path Integrals I

> There are many misunderstandings between mathematicians and physicists on the place of mathematical rigor in physics ... The physicist cannot understand the mathematician's care in solving an idealized problem. The physicist knows that the real problem is much more complicated. It has already been simplified by intuition, which discards the unimportant and often approximates the remainder.
>
> R.P. Feynman and A.R. Hibbs
> *Quantum Mechanics and Path Integrals*, 1965

> In situations like this, mathematicians are not satisfied just to get an approximation ... mathematicians want to know it *exactly*, if they can. Not just because they are weird obsessives, but because they know from experience that getting that *exact* value often opens unexpected doors and throws light on the underlying math.
>
> John Derbyshire
> *Prime Obsession: Bernhard Riemann and the Greatest Unsolved Problem in Mathematics*, 2003

We now are in position to calculate propagators using path integrals, despite the fact that we do not have a precise definition of exactly what the expression

$$\mathcal{K}(t_1, \mathbf{x}_1; t_0, \mathbf{x}_0) = \int_\Gamma \phi(\mathbf{x})\,\mathcal{D}(\mathbf{x}(t)) = \int_\Gamma e^{iS[\mathbf{x}]/\hbar}\mathcal{D}(\mathbf{x}(t))$$

means. We begin with Feynman's original prescription for such calculations by partitioning the time interval and approximating the actual paths by piecewise linear paths. For some applications this works well, but for others this leads to calculations that are more difficult than necessary, as different choices for approximating a given path, for example by sinusoidal paths, work to give more tractable calculations. The key to Feynman's scheme for interpreting the path integral is to replace the vague idea of some sort of integration over the space Γ of all paths with integration over a dense subspace of paths in Γ. This is accomplished by parameterizing finite-dimensional families of paths and integrating over these families, and then taking the limit as the dimension increases without bound. This method always requires the insertion of appropriate normalizing factors in the finite-dimensional integrations to force convergence of the limit. Feynman himself states that the determination of these factors seems "to be a very difficult problem and we do not know how to do it in general terms."[1] In this lecture we apply Feynman's scheme using his suggested piecewise linear families in the first section, and polynomial families of paths in the third. It is here that the free particle propagator derived in Section 29.4 is used to determine the normalizing factors needed to force convergence. We derive the propagator for a particle in a constant force field in two different calculations, one for each of these two dense families of paths. This suggests that something quite general is afoot, which will be explored in the lecture following.

Restricting our attention to one dimension, we begin with a **filtration** of the space Γ of all paths between (t_0, x_0) and (t_1, x_1). This is a union $\cup_{N=1}^{\infty}\Gamma_N \subset \Gamma$, where Γ_N is finite-dimensional with a natural parameterization by a Euclidean space and the union $\cup\Gamma_N$ is dense in Γ. Using the known free particle propagator, the Euclidean measure is adjusted to a measure $d\mu_N$ such that the free particle propagator may be written as

$$\lim_{N\to\infty} \int_{\Gamma_N} \phi(x)\, d\mu_N.$$

It is in the requirement that this expression yield the free particle propagator that judicious choices must be made for the normalizing factors that force convergence. We then use the **filtration measures** $d\mu_N$ and the limit above to evaluate propagators of particles moving in nonzero potential fields. In this lecture we restrict our attention to the calculation of the propagator for a single example – a particle in a constant force field. In the following and final lecture, we formalize the scheme presented here and apply it to the harmonic oscillator and the forced harmonic oscillator potentials.

[1] Feynman and Hibbs (1965), p. 33.

Other methods have been advanced to evaluate path integrals in various specific cases.[2] In this lecture we do not claim to provide a rigorous definition of the path integral but rather to give an interpretation of the integral that allows for rigorous calculations within that interpretation. It is satisfying to see the correct propagators crystalize from these rigorous computations even though we still do not have a rigorous definition for the general path integral.

31.1 The Piecewise Linear Family

In this section we develop Feynman's original scheme, though packaged differently from the original, for evaluating path integrals using a piecewise linear path family. Our concern is with deriving the propagator – the amplitude for the spacetime coordinates (t_1, X_1) to occur given that initially it has coordinates (t_0, X_0) – for a particle moving along the real line with coordinate x. From Lecture 29, the propagator is given by the Feynman path integral, which heuristically sums over all possible paths between the spacetime points (t_0, X_0) and (t_1, X_1); this is symbolically given as the expression

$$\mathcal{K}(t_1, X_1; t_0, X_0) = \int_\Gamma \exp\left(\frac{\mathbf{i}}{\hbar} S[x]\right) \mathcal{D}(x(t)), \tag{31.1}$$

where Γ is the space of all paths between spacetime points (t_0, X_0) and (t_1, X_1). As recorded in Section 29.1, the challenge of this approach is to make sense of this sum over all paths. Feynman's original scheme was to replace the space of all paths Γ with a space Γ_N of piecewise linear paths, defined below, and use the resulting integration over this finite-dimensional path family as an Nth approximation to the full path integral. We then interpret the full path integral 31.1 as the limit

$$\mathcal{K}(t_1, X_1; t_0, X_0) = \lim_{N\to\infty} \int_{\Gamma_N} \exp\left(\frac{\mathbf{i}}{\hbar} S[x]\right) d\mu_N, \tag{31.2}$$

where the integration over Γ_N is accomplished using the filtration measure $d\mu_N$, yet to be determined. Notice the slight change of notation from Lecture 29, where the initial location of the particle at time t_0 is labeled as X_0 instead of x_0, and similarly for the location at time t_1.

It will not escape the reader that the development presented here is

[2] See, for example, Albeverio and Brzeźniak (1993), Elworthy and Truman (1984), Gustafson and Sigal (2003), and Vassili (1999).

very close in spirit to the development of the classical Riemann integral using Riemann sums. Let $X = X_1 - X_0$ and $T = t_1 - t_0$, and replace the paths $x(t)$ that the particle may traverse by $x(t+t_0)$ to assume, without loss of generality, that $t_0 = 0$ and $t_1 = T$. For a positive integer N, define the time increment $\epsilon_N = T/(N+1)$, and for integers $0 \le n \le N+1$, let $\tau_n = n\epsilon_N$ so that $0 = \tau_0 < \tau_1 < \cdots < \tau_{N+1} = T$ partitions the interval $[0, T]$ into $N+1$ subintervals of equal length; let $x(\tau_n) = x_n$. Let Γ_N be the collection of continuous paths $x(t)$ defined on the closed time interval $[0, T]$ with $x(0) = x_0 = X_0$ and $x(T) = x_{N+1} = X_1$ and such that x is linear on each subinterval $[\tau_{n-1}, \tau_n]$. Observe that a path $x \in \Gamma_N$ is completely determined by its values $x(\tau_n) = x_n$ for $1 \le n \le N$, since its endpoints are fixed. This gives a natural identification of Γ_N with the Euclidean space $\mathbb{R}^N$, via the parameterization

$$\Gamma_N \ni x \leftrightarrow (x_1, \ldots, x_N) \in \mathbb{R}^N. \tag{31.3}$$

With this identification, define the filtration measure $d\mu = d\mu_N$ to be $\Delta(N)d\mathbf{x}$ where $d\mathbf{x}$ is the usual N-dimensional measure on $\mathbb{R}^N$ and $\Delta(N)$ is a presently unknown normalization factor, to be determined subsequently.

We will use the known free particle propagator of equation 29.16 to determine the factor $\Delta(N)$. Writing the free particle action for a path $x \in \Gamma_N$ in terms of the identification 31.3 gives

$$\begin{aligned} S_{\text{free}}[x] &= \int_0^T \frac{1}{2} m\dot{x}^2\, dt = \sum_{n=1}^{N+1} \frac{m}{2} \int_{\tau_{n-1}}^{\tau_n} \frac{(x_n - x_{n-1})^2}{\epsilon_N^2}\, dt \\ &= \frac{m}{2\epsilon_N} \sum_{n=1}^{N+1} (x_n - x_{n-1})^2. \end{aligned} \tag{31.4}$$

We now use expression 31.2 to calculate the free particle propagator. The integration over Γ_N is accomplished by using the filtration measure $d\mu_N$ and inserting the free particle action of equation 31.4 for $x \in \Gamma_N$, to obtain

$$\begin{aligned} \int_{\Gamma_N} \exp\left(\frac{\mathbf{i}}{\hbar} S_{\text{free}}[x]\right) d\mu_N &= \int_{\mathbb{R}^N} \exp\left[\mathbf{i} \frac{m}{2\hbar\epsilon_N} \sum_{n=1}^{N+1} (x_n - x_{n-1})^2\right] d\mu_N \\ &= \Delta(N) F_N\left(\frac{m}{2\hbar\epsilon_N}, X_0, X_1\right), \end{aligned} \tag{31.5}$$

where F_N is the Gaussian integral of equation 30.15. Applying formula 30.18 with $a = m/2\hbar\epsilon_N$, $c = X_0$, and $d = X_1$ gives

$$F_N\left(\frac{m}{2\hbar\epsilon_N}, X_0, X_1\right) = \frac{1}{\sqrt{N+1}}\left(\frac{2\pi\mathbf{i}\hbar\epsilon_N}{m}\right)^{N/2} \exp\left(\frac{\mathbf{i}mX^2}{2\hbar T}\right).$$

The factor $\Delta(N)$ should be chosen so that the limit of expression 31.2, with the approximate propagator given by equation 31.5, gives the value of the propagator derived in Section 29.4 and stated in equation 29.16. There are several ways to do this, but the easiest is to choose $\Delta(N)$ so that the approximate propagator given in expression 31.5 is exact for the free particle. This is accomplished by equating the expression on the right-hand side of equation 29.16 with the second line of equation 31.5, giving

$$\Delta(N) = \left(\frac{m}{2\pi\mathbf{i}\hbar\epsilon_N}\right)^{(N+1)/2}.$$

Now that we have determined the filtration measure $d\mu = d\mu_N = \Delta(N)d\mathbf{x}$ for the piecewise linear family Γ_N of paths, we will use it in integrations over Γ_N in the case when the particle moves in a nonzero potential field $V(x)$. The Lagrangian is $L(x, \dot{x}) = m\dot{x}^2/2 - V(x)$ and the action along the path $x = x(t)$ is $S[x] = S_{\text{free}}[x] - \int_0^T V(x)\,dt$, where $S_{\text{free}}[x]$ is the action associated with the free particle. The prescription for the propagator is given by equation 31.2, where the approximate propagator is

$$\begin{aligned}\int_{\Gamma_N} \exp\left(\frac{\mathbf{i}}{\hbar}S[x]\right) d\mu_N &= \int_{\mathbb{R}^N} \exp\left[\frac{\mathbf{i}}{\hbar}\left(S_{\text{free}}[x] - \int_0^T V(x)\,dt\right)\right] d\mu_N \\ &= \left(\frac{m}{2\pi\mathbf{i}\hbar\epsilon_N}\right)^{(N+1)/2} \\ &\quad \times \int_{\mathbb{R}^N} \exp\left[\frac{\mathbf{i}}{\hbar}\left(\frac{m}{2\epsilon_N}\sum_{n=1}^{N+1}(x_n - x_{n-1})^2 - \int_0^T V(x)\,dt\right)\right] d\mathbf{x}.\end{aligned} \tag{31.6}$$

Generally, expression 31.6 is difficult to evaluate and usually requires perturbation techniques. One further approximation replaces the integral of the potential by a Riemann sum, for example by the right-hand sum $\sum_{n=1}^{N+1} V(x_n)\epsilon_N$, so that the approximate propagator is written as

$$\left(\frac{m}{2\pi \mathrm{i}\hbar\epsilon_N}\right)^{(N+1)/2}$$
$$\times \int_{\mathbb{R}^N} \exp\left\{\frac{\mathrm{i}}{\hbar}\sum_{n=1}^{N+1}\left[\frac{m}{2\epsilon_N}(x_n - x_{n-1})^2 - \epsilon_N V(x_n)\right]\right\} d\mathbf{x}. \quad (31.7)$$

For quadratic Lagrangians the integration may be carried out exactly, though even then the calculations can be quite involved. To illustrate this, in the next section we derive the propagator for a particle moving in a constant force field.

31.2 The Constant Force Propagator

The standard calculation for the propagator for the particle in a constant external field uses the results recorded in Section 29.3 and, once the free particle propagator is known, the calculation reduces to nothing more than evaluating the classical action along the classical path. Indeed, the potential is $V(x) = -Kx$ for the positive constant K, and an application of equation 29.7 with $a(t) = b(t) = 0$ and $c(t) = m/2$ gives the propagator as

$$\mathcal{K}(t_1, X_1; t_0, X_0) = \mathcal{K}_{\text{free}}(T, 0; 0, 0)e^{\mathrm{i}S_{\text{cl}}/\hbar}$$
$$= \left(\frac{m}{2\pi\mathrm{i}\hbar T}\right)^{1/2} \exp\left[\frac{\mathrm{i}}{\hbar}\left(\frac{1}{2}mv^2T + \frac{1}{2}KT(X_0 + X_1) - \frac{K^2T^3}{24m}\right)\right], \quad (31.8)$$

where we have used equations 29.9 and 29.16.

Recall the notation we are using, that $v = X/T$ so that $v^2T = X^2/T$, and observe from equation 29.16 that

$$\mathcal{K}_{\text{free}}(t_1, X_1; t_0, X_0) = \left(\frac{m}{2\pi\mathrm{i}\hbar T}\right)^{1/2} \exp\left[\frac{\mathrm{i}}{\hbar}\left(\frac{1}{2}mv^2T\right)\right].$$

Thus, the propagator for a particle in a constant field may be written as

$$\mathcal{K}(t_1, X_1; t_0, X_0)$$
$$= \mathcal{K}_{\text{free}}(t_1, X_1; t_0, X_0) \times \exp\left[\frac{\mathrm{i}}{\hbar}\left(\frac{1}{2}KT(X_0 + X_1) - \frac{K^2T^3}{24m}\right)\right].$$

We now reevaluate the propagator for the particle in a constant external field from scratch, without using the results of Section 29.3 on quadratic Lagrangians, and demonstrate the way in which the propagator expressed in equation 31.6 crystalizes from Feynman's prescription using piecewise linear paths. For this we first evaluate the integral of the potential $V(x) = -Kx$ for a path $x \in \Gamma_N$:

$$-\int_0^T V(x)\,dt = K\sum_{n=1}^{N+1}\int_{\tau_{n-1}}^{\tau_n} x(t)\,dt = \frac{K\epsilon_N}{2}(X_0 + X_1) + K\epsilon_N\sum_{n=1}^{N} x_n.$$

An application of expression 31.6 gives the approximate propagator that arises from the integration over Γ_N as

$$\left(\frac{m}{2\pi\mathbf{i}\hbar\epsilon_N}\right)^{(N+1)/2} e^{\mathbf{i}K\epsilon_N(X_0+X_1)/2\hbar}$$
$$\times \int_{\mathbb{R}^N} \exp\left\{\mathbf{i}\left[a\left(\sum_{n=1}^{N+1}(x_n - x_{n-1})^2\right) + k\left(\sum_{n=1}^{N} x_n\right)\right]\right\} d\mathbf{x},$$

where $a = m/2\hbar\epsilon_N$ and $k = K\epsilon_N/\hbar$. This may be expressed as

$$\int_{\Gamma_N} \exp\left(\frac{\mathbf{i}}{\hbar}S[x]\right) d\mu_N$$
$$= \left(\frac{m}{2\pi\mathbf{i}\hbar\epsilon_N}\right)^{(N+1)/2} e^{\mathbf{i}KT(X_0+X_1)/2(N+1)\hbar} F_N\left(\frac{m}{2\hbar\epsilon_N}, X_0, X_1, \frac{\epsilon_N K}{\hbar}\right)$$

where F_N is given by expression 30.20. Applying expression 30.20 and simplifying gives the approximate propagator as

$$\left(\frac{m}{2\pi\mathbf{i}\hbar T}\right)^{1/2} \exp\left[\frac{\mathbf{i}}{\hbar}\left(\frac{1}{2}mv^2T\right)\right]$$
$$\times \exp\left\{\frac{\mathbf{i}}{\hbar}\left[\frac{1}{2}KT(X_0 + X_1) - \left(\frac{N(N+2)}{(N+1)^2}\right)\frac{K^2T^3}{24m}\right]\right\}.$$

To obtain the propagator of the particle in a constant force field, we now take the limit as N increases without bound to obtain, from equation 31.2, precisely the second line of equation 31.8.

31.3 The Filtration Measures for the Polynomial Family

Feynman's use of piecewise linear approximations to paths is just one possible tactic for evaluating path integrals. Presumably his strategy for fleshing out the meaning of expression 31.1 may be accomplished via other tactics, ones using dense families of paths other than the piecewise linear family. There are, of course, many different path families $\{\Gamma_N\}$, where Γ_N is finite-dimensional and the union $\cup_{N=1}^{\infty}\Gamma_N$ is dense in the full path family Γ, that have been found useful in various applications of mathematics. One need only browse any book on special functions to find references to the classical polynomial families – Laguerre, Legendre, Hermite, Jacobi – as well as the classical transcendental families – trigonometric, Bessel, various types of hypergeometric families – that have found currency in applied mathematics and physics.[3] In the next lecture, we will present a formal setting in which to explore the generality of Feynman's scheme and develop it to some degree by concentrating on two tactics, one using the trigonometric path family and the other using the unrestricted polynomial path family. As a preparation for this, in this section we derive the filtration measures for the general, unrestricted, polynomial path family, and apply them in the next section to evaluate once again the propagator for a particle in a constant force field using this path family. In the next lecture, we will derive the propagators for the harmonic and forced harmonic oscillators using the trigonometric and polynomial path families.

For a positive integer N, let Γ_N denote the collection of polynomial paths of degree $N+1$ that connect spacetime points (t_0, X_0) and (t_1, X_1). Again let $X = X_1 - X_0$ and $T = t_1 - t_0$, and replace $x(t)$ by $x(t+t_0)$ to assume, without loss of generality, that $t_0 = 0$ and $t_1 = T$. For the path $x(t)$ we have $X_0 = x(0)$, so we may write $x(t) = X_0 + \sum_{n=1}^{N+1} x_{n-1}t^n$. The reason for this choice of notation will become clear as the development progresses. Note that $X_1 = x(T) = X_0 + \sum_{n=1}^{N+1} x_{n-1}T^n$. The classical average velocity is

$$v = \frac{X}{T} = \sum_{n=1}^{N+1} x_{n-1}T^{n-1} = \sum_{n=0}^{N} x_n T^n,$$

so that

$$v^2 T = \sum_{i,j=0}^{N} x_i x_j T^{i+j+1} = x_0^2 T + 2x_0 \sum_{n=1}^{N} x_n T^{n+1} + \sum_{i,j=1}^{N} x_i x_j T^{i+j+1}. \tag{31.9}$$

[3] Rainville (1960) is a classic reference and Andrews, Askey, and Roy (1999) is a modern one. Whittaker and Watson (1902), still available as a 1996 4th edition, remains as an icon in the field of special functions.

The free particle action along the path x can be written as

$$S_{\text{free}}[x] = \int_0^T \frac{1}{2} m\dot{x}^2\, dt = \int_0^T \frac{1}{2} m \left[\sum_{n=0}^{N} (n+1) x_n t^n \right]^2 dt$$

Integrating and substituting from equation 31.9 gives

$$\begin{aligned}
S_{\text{free}}[x] &= \frac{1}{2} m \sum_{i,j=0}^{N} \frac{(i+1)(j+1)}{i+j+1} x_i x_j T^{i+j+1} \\
&= \frac{1}{2} m x_0^2 T + m x_0 \sum_{n=1}^{N} x_n T^{n+1} + \frac{1}{2} m \sum_{i,j=1}^{N} \frac{(i+1)(j+1)}{i+j+1} x_i x_j T^{i+j+1} \\
&= \frac{1}{2} m v^2 T + \frac{1}{2} m \sum_{i,j=1}^{N} \left(\frac{(i+1)(j+1)}{i+j+1} - 1 \right) x_i x_j T^{i+j+1} \\
&= \frac{1}{2} m v^2 T + \frac{1}{2} m \sum_{i,j=1}^{N} \frac{ij}{i+j+1} x_i x_j T^{i+j+1} \\
&= \frac{1}{2} m v^2 T + \frac{1}{2} m \mathbf{x}^{\text{Tr}} \mathbf{B}_N \mathbf{x},
\end{aligned} \tag{31.10}$$

where $\mathbf{x} = [x_1 \; x_2 \; \cdots \; x_N]^{\text{Tr}} \in \mathbb{R}^N$ and $\mathbf{B}_N$ is the positive definite symmetric matrix

$$\mathbf{B}_N = \left[\frac{ij}{i+j+1} T^{i+j+1} \right]_{i,j=1}^{N}. \tag{31.11}$$

As x_0 has been eliminated using the boundary conditions the path $x(t)$ is completely determined by the coefficients $x_1, \ldots, x_N$, and again we have an identification of Γ_N with $\mathbb{R}^N$. As before, the known free particle propagator is used to determine the normalizing factor $\Delta(N)$ for the filtration measure $d\mu_N = \Delta(N) d\mathbf{x}$. The Nth approximation to the free particle propagator is

$$\begin{aligned}
\int_{\Gamma_N} \exp\left(\frac{\mathbf{i}}{\hbar} S_{\text{free}}[x] \right) d\mu_N &= \Delta(N) \int_{\mathbb{R}^N} \exp\left[\frac{\mathbf{i}m}{2\hbar} \left(v^2 T + \mathbf{x}^{\text{Tr}} \mathbf{B}_N \mathbf{x} \right) \right] d\mathbf{x} \\
&= \Delta(N) \exp\left[\frac{\mathbf{i}}{\hbar} \left(\frac{1}{2} m v^2 T \right) \right] \int_{\mathbb{R}^N} \exp\left[\frac{\mathbf{i}m}{2\hbar} \left(\mathbf{x}^{\text{Tr}} \mathbf{B}_N \mathbf{x} \right) \right] d\mathbf{x}.
\end{aligned}$$

Setting this equal to the free particle propagator given by equation 29.16 and applying formula 30.22 gives the normalization factor as

$$\Delta(N) = \left(\frac{m}{2\pi \mathbf{i}\hbar} \right)^{(N+1)/2} \sqrt{\frac{\det \mathbf{B}_N}{T}}. \tag{31.12}$$

The determinant formula 30.28 could be used to write $\Delta(N)$ in terms of N, but equation 31.12 will suffice for our purposes.

31.3.1 The Constant Force Propagator via Polynomial Sums

Armed with the filtration measures for the polynomial path family, we now perform again the calculation of the propagator for a particle in a constant force field. The reader may note that the spatial dependence separates out early in the calculation, with the full time dependence requiring all the heavy lifting.

As the Lagrangian for a particle in a constant external field K generated by a linear potential is

$$L(x, \dot{x}) = \frac{1}{2}m\dot{x}^2 + Kx,$$

the action for the polynomial path $x(t) = X_0 + \sum_{n=1}^{N+1} x_{n-1}t^n$ becomes

$$S[x] = S_{\text{free}}[x] + KX_0T + K\sum_{n=1}^{N+1}\frac{x_{n-1}}{n+1}T^{n+1},$$

where $S_{\text{free}}[x]$ is the action for the free particle from equation 31.10. By using $x(T) = X_1$ we may write, after some manipulation,

$$S[x] = S_{\text{free}}[x] + \frac{1}{2}KT(X_0 + X_1) - \frac{1}{2}K\sum_{n=1}^{N}\frac{n}{n+2}x_nT^{n+2}. \tag{31.13}$$

Setting $\mathbf{b} = [b_1\ b_2\ \cdots\ b_N]^{\mathsf{Tr}}$, where $b_n = nT^{n+2}/(n+2)$, we obtain from equations 31.10 and 31.13 that

$$S[x] = \frac{1}{2}mv^2T + \frac{1}{2}KT(X_0 + X_1) + \frac{1}{2}m\mathbf{x}^{\mathsf{Tr}}\mathbf{B}_N\mathbf{x} - \frac{1}{2}K\mathbf{b}^{\mathsf{Tr}}\mathbf{x}. \tag{31.14}$$

The approximate propagator that arises from the integration over Γ_N with respect to the filtration measure $d\mu_N = \Delta(N)d\mathbf{x}$ becomes

$$\begin{aligned}\int_{\Gamma_N}\exp\left(\frac{\mathbf{i}}{\hbar}S[x]\right)d\mu_N &= \left(\frac{m}{2\pi\mathbf{i}\hbar}\right)^{(N+1)/2}\sqrt{\frac{\det\mathbf{B}_N}{T}}\\ &\times\exp\left[\frac{\mathbf{i}}{\hbar}\left(\frac{1}{2}mv^2T + \frac{1}{2}KT(X_0+X_1)\right)\right]\int_{\mathbb{R}^N}e^{\mathbf{i}\left(a\mathbf{x}^{\mathsf{Tr}}\mathbf{B}_N\mathbf{x} - J\mathbf{b}^{\mathsf{Tr}}\mathbf{a}\right)}d\mathbf{x},\end{aligned} \tag{31.15}$$

where $a = m/2\hbar$ and $J = -K/2\hbar$. Applying the N-dimensional Gaussian integral formula 30.10 gives

$$\left(\frac{m}{2\pi \mathbf{i}\hbar T}\right)^{1/2} \exp\left[\frac{\mathbf{i}}{\hbar}\left(\frac{1}{2}mv^2T + \frac{1}{2}KT(X_0 + X_1)\right)\right] e^{-\mathbf{i}K^2\mathbf{b}^{\mathsf{Tr}}\mathbf{B}_N^{-1}\mathbf{b}/8m\hbar} \tag{31.16}$$

for the approximate propagator. Notice that the only dependence on N for this expression is through the form $\mathbf{b}^{\mathsf{Tr}}\mathbf{B}_N^{-1}\mathbf{b}$. To evaluate this form, recall from Section 30.4 that $\mathbf{B}_N$ decomposes as the product $T\,\mathbf{T}_N\mathbf{C}_N\mathbf{T}_N$, where $\mathbf{C}_N$ is the Hilbert matrix given in equations 30.23 and $\mathbf{T}_N = \mathbf{diag}[T\ 2T^2\ \cdots\ NT^N]$ is the diagonal matrix whose nth entry is nT^n. Setting $\mathbf{c}^{\mathsf{Tr}} = [c_1\ c_2\ \cdots\ c_N]$ where $c_n = 1/(n+2)$, we get

$$\begin{aligned}\mathbf{b}^{\mathsf{Tr}}\mathbf{B}_N^{-1}\mathbf{b} &= \frac{1}{T}\left(\mathbf{b}^{\mathsf{Tr}}\mathbf{T}_N^{-1}\right)\mathbf{C}_N^{-1}\left(\mathbf{T}_N^{-1}\mathbf{b}\right)\\ &= \frac{1}{T}\left(T^2\mathbf{c}^{\mathsf{Tr}}\right)\mathbf{C}_N^{-1}\left(T^2\mathbf{c}\right) = T^3\mathbf{c}^{\mathsf{Tr}}\mathbf{C}_N^{-1}\mathbf{c}.\end{aligned} \tag{31.17}$$

Instead of calculating the inverse of the Hilbert matrix $\mathbf{C}_N$, we finesse the problem by observing that $\mathbf{c}$ is exactly the first column of the matrix $\mathbf{C}_N$, which implies that $\mathbf{c} = \mathbf{C}_N\mathbf{d}$ where $\mathbf{d}^{\mathsf{Tr}} = [1\ 0\ 0\ \cdots\ 0]$ is the standard unit basis vector whose only nonzero entry is a 1 in the first position. Thus $\mathbf{C}_N^{-1}\mathbf{c} = \mathbf{d}$ and

$$\mathbf{b}^{\mathsf{Tr}}\mathbf{B}_N^{-1}\mathbf{b} = T^3\mathbf{c}^{\mathsf{Tr}}\mathbf{C}_N^{-1}\mathbf{c} = T^3\mathbf{c}^{\mathsf{Tr}}\mathbf{d} = T^3c_1 = \frac{T^3}{3}. \tag{31.18}$$

This is rather interesting, in that our form $\mathbf{b}^{\mathsf{Tr}}\mathbf{B}_N^{-1}\mathbf{b}$ reduces to the constant $T^3/3$ independently of N, and hence the dependence of the approximate propagator of expressions 31.15 and 31.16 on N is seen to be illusionary. Applying equation 31.18 to expression 31.16 gives the approximate propagator as equal to precisely the constant force propagator reported in formula 31.8, independently of N.

31.4 From the Dirac Calculus to the Path Integral

In this section the Dirac calculus will be used to derive path integrals and along the way the equivalence of the Feynman and Schrödinger versions of quantum mechanics will become apparent. Our main purpose in choosing the Dirac calculus and rigged Hilbert spaces in the derivations that follow is to demonstrate the power of this well-crafted notation. The derivations are simple, direct, and elegant – one might even claim that they are transparent once digested – and constitute a shorthand for more intricate arguments, which, with care, can be made rigorous. It may be useful for the reader to review the basics of the Dirac calculus in Lecture 5 before proceeding.

31.4.1 The Configuration and Phase Space Path Integrals

Having had success with the Dirac calculus in finding the free particle propagator, we will continue its use in developing the understanding that Feynman's formulation of quantum mechanics is equivalent to the wave mechanical formulation of Schrödinger. Our reasoning begins by evaluating in the Dirac calculus the expression

$$\langle x_j | e^{-\mathrm{i}\epsilon \mathsf{P}^2/2m\hbar} e^{-\mathrm{i}\epsilon V(\mathsf{X})/\hbar} | x_i \rangle$$

by inserting the resolution of the identity in the position representation between the exponentials. Here, of course, P and X are the momentum and position operators of wave mechanics and ϵ is a real parameter. This gives

$$\begin{aligned}\int_{-\infty}^{\infty} \langle x_j | e^{-\mathrm{i}\epsilon \mathsf{P}^2/2m\hbar} | x \rangle \langle x | e^{-\mathrm{i}\epsilon V(\mathsf{X})/\hbar} | x_i \rangle \, dx \\ = e^{-\mathrm{i}\epsilon V(x_i)/\hbar} \int_{-\infty}^{\infty} \langle x_j | e^{-\mathrm{i}\epsilon \mathsf{P}^2/2m\hbar} | x \rangle \langle x | x_i \rangle \, dx \\ = e^{-\mathrm{i}\epsilon V(x_i)/\hbar} \langle x_j | e^{-\mathrm{i}\epsilon \mathsf{P}^2/2m\hbar} | x_i \rangle,\end{aligned}$$

where we have used relation 5.4, proved by equation 5.7. Now, the expression $\langle x_j | e^{-\mathrm{i}\epsilon \mathsf{P}^2/2m\hbar} | x_i \rangle$ has already been identified in equation 29.17 as the free particle propagator $\mathcal{K}_{\mathsf{free}}(\epsilon, x_j; 0, x_i)$ given by equation 29.16. This gives the desired evaluation

$$\langle x_j | e^{-\mathrm{i}\epsilon \mathsf{P}^2/2m\hbar} e^{-\mathrm{i}\epsilon V(\mathsf{X})/\hbar} | x_i \rangle = \left(\frac{m}{2\pi \mathrm{i}\hbar\epsilon} \right)^{1/2} e^{\mathrm{i}m(x_j - x_i)^2/2\hbar\epsilon} e^{-\mathrm{i}\epsilon V(x_i)/\hbar}, \tag{31.19}$$

which will be used to give another derivation of the approximate propagator in equation 31.7.

The propagator for a particle moving in a potential field $V(x)$ to be at spacetime points $(0, X_1)$ and (T, X_1) is

$$\mathcal{K}(T, X_1; 0, X_0) = \langle X_1 | \mathsf{U}_T | X_0 \rangle,$$

where U_T is the time evolution operator

$$\mathsf{U}_T = e^{-\mathbf{i}T\mathsf{H}/\hbar} = \exp\left[-\frac{\mathbf{i}}{\hbar}T\left(\frac{\mathsf{P}^2}{2m} + V(\mathsf{X})\right)\right].$$

In this development, the assumption is that the Hamiltonian H is time-independent. This time-evolution operator U_T is of the form $e^{\lambda(\mathsf{A}+\mathsf{B})}$ for a complex number λ with nonpositive real part and self-adjoint operators A and B. We would like to replace $e^{\lambda(\mathsf{A}+\mathsf{B})}$ by the product $e^{\lambda\mathsf{A}}e^{\lambda\mathsf{B}}$ so that we may use the result of equation 31.19, but the equality of $e^{\lambda(\mathsf{A}+\mathsf{B})}$ and $e^{\lambda\mathsf{A}}e^{\lambda\mathsf{B}}$ fails to hold whenever the operators A and B fail to commute. Fortunately, though, the **Trotter product formula**[4] will suffice for our needs. It states that though $e^{\lambda(\mathsf{A}+\mathsf{B})}$ may fail to equal $e^{\lambda\mathsf{A}}e^{\lambda\mathsf{B}}$, there is a sense in which equality does hold. Precisely, it is the case that if in addition the operator $\mathsf{A}+\mathsf{B}$ is self-adjoint (as is the Hamiltonian H) then

$$e^{\lambda(\mathsf{A}+\mathsf{B})} = \lim_{n\to\infty}\left(e^{\lambda\mathsf{A}/n}e^{\lambda\mathsf{B}/n}\right)^n, \tag{31.20}$$

the limit taken in the strong sense.[5] Assuming that there is no problem with exchanging the order of limit and the application of the inner product,[6] the propagator takes the form

$$\mathcal{K}(t, X_1; 0, X_0) = \lim_{N\to\infty}\langle X_1|\left(e^{\lambda\mathsf{A}/(N+1)}e^{\lambda\mathsf{B}/(N+1)}\right)^{N+1}|X_0\rangle,$$

where $\lambda = -\mathbf{i}T/\hbar$, $\mathsf{A} = \mathsf{P}^2/2m$, and $\mathsf{B} = V(\mathsf{X})$, so that $\mathsf{A}+\mathsf{B} = \mathsf{H}$. The reason for indexing at $N+1$ rather than n will become clear. There are $N+1$ factors in the middle slot of the approximate propagator $\langle X_1|\left(e^{\lambda\mathsf{A}/(N+1)}e^{\lambda\mathsf{B}/(N+1)}\right)^{N+1}|X_0\rangle$. Placing the resolution of the identity $\int_{-\infty}^{\infty}|x_n\rangle\langle x_n|\,dx_n$ between the nth and $(n+1)$th factors, counted from the right, for $n = 1, \ldots, N$ and setting $\epsilon_N = T/(N+1)$, $x_0 = X_0$, and $x_{N+1} = X_1$, gives the formula

$$\begin{aligned}\langle X_1| & \left(e^{\lambda\mathsf{A}/(N+1)}e^{\lambda\mathsf{B}/(N+1)}\right)^{N+1}|X_0\rangle \\ &= \int_{\mathbb{R}^N}\prod_{n=1}^{N+1}\langle x_n|e^{-\mathbf{i}\epsilon_N\mathsf{P}^2/2m\hbar}e^{-\mathbf{i}\epsilon_N V(\mathsf{X})/\hbar}|x_{n-1}\rangle\,d\mathbf{x}\end{aligned} \tag{31.21}$$

[4] For proofs and generalizations, see Reed and Simon (1980), pp. 295–297 and 377–382.

[5] Precisely, for operators A_n and A with $\mathcal{D}(\mathsf{A}_n) = \mathcal{D}(\mathsf{A})$, $\lim_{n\to\infty}\mathsf{A}_n = \mathsf{A}$ means that $\|\mathsf{A}_n\psi - \mathsf{A}\psi\| \to 0$ for all $\psi \in \mathcal{D}(\mathsf{A})$.

[6] This exchange is a statement about distributions couched in the language of the Dirac calculus and can be verified rigorously.

for the approximate propagator. Here, $d\mathbf{x} = dx_1 \cdots dx_N$. Applying formula 31.19 $N+1$ times gives precisely equation 31.7 for the approximate propagator, except for the minor difference that the integral of the potential is replaced by the left-hand sum $\sum_{n=1}^{N+1} V(x_{n-1})\epsilon_N$ rather than the right-hand sum, as in equation 31.7.

We have demonstrated that wave mechanics and the Dirac calculus reproduce formula 31.7. It follows that the Feynman path integral approach to quantum mechanics that leads to expressions 31.6 and 31.7 is equivalent to Schrödinger's wave mechanical approach.

Inserting the approximate propagator given by expression 31.7 into equation 31.2 defines the **configuration space path integral**, distinguished from the **phase space path integral** to be developed next. The phase space integral is obtained from equation 31.21 by inserting a resolution of the identity in the momentum representation between each of the two exponential terms in the product. This gives for the approximate propagator the formula

$$\langle X_1| \left(e^{\lambda \mathsf{A}/(N+1)} e^{\lambda \mathsf{B}/(N+1)}\right)^{N+1} |X_0\rangle$$

$$= \int_{\mathbb{R}^{N+1}\times\mathbb{R}^N} \prod_{n=1}^{N+1} \langle x_n| e^{-\mathrm{i}\epsilon_N \mathsf{P}^2/2m\hbar} |p_n\rangle\langle p_n| e^{-\mathrm{i}\epsilon_N V(\mathsf{X})/\hbar} |x_{n-1}\rangle \, d\mathbf{p}\, d\mathbf{x},$$

where $d\mathbf{p} = dp_1 \cdots dp_{N+1}$. Using equation 5.3, this becomes

$$\int_{\mathbb{R}^{N+1}\times\mathbb{R}^N} \prod_{n=1}^{N+1} \langle x_n|p_n\rangle\langle p_n|x_{n-1}\rangle e^{-\mathrm{i}\epsilon_N {p_n}^2/2m\hbar} e^{-\mathrm{i}\epsilon_N V(x_{n-1})/\hbar} \, d\mathbf{p}\, d\mathbf{x}$$

$$= \left(\frac{1}{2\pi\hbar}\right)^{N+1}$$

$$\times \int_{\mathbb{R}^{N+1}\times\mathbb{R}^N} \prod_{n=1}^{N+1} e^{\mathrm{i}p_n(x_n - x_{n-1})/\hbar} e^{-\mathrm{i}\epsilon_N {p_n}^2/2m\hbar} e^{-\mathrm{i}\epsilon_N V(x_{n-1})/\hbar} \, d\mathbf{p}\, d\mathbf{x}.$$

Combining the exponential factors into a single exponential and writing the domain of integration as $\mathbb{D} = \mathbb{R}^{N+1} \times \mathbb{R}^N$ gives

$$\left(\frac{1}{2\pi\hbar}\right)^{N+1}$$

$$\times \int_{\mathbb{D}} \exp\left\{\frac{\mathrm{i}}{\hbar} \sum_{n=1}^{N+1} \left[p_n \frac{(x_n - x_{n-1})}{\epsilon_N} - \frac{{p_n}^2}{2m} - V(x_{n-1})\right] \epsilon_N \right\} d\mathbf{p}\, d\mathbf{x}$$

for the approximate propagator. When this is used in equation 31.2 for the propagator it is known as the phase space path integral. Notice that the sum in the exponential of this expression is a Riemann-type sum for the integral

$$\int_0^T (p\dot{x} - H(x,p))\, dt,$$

where $H(x,p)$ is the classical Hamiltonian for the system. This expresses the Lagrangian $L(x,\dot{x},t)$ in terms of the canonically conjugate variables x and $p = \partial L/\partial \dot{x}$. This path integral is written symbolically as

$$\begin{aligned}\mathcal{K}(T,X_1;0,X_0) &= \langle X_1|\mathsf{U}_T|X_0\rangle \\ &= \int \exp\left[\frac{\mathbf{i}}{\hbar}\int_0^T (p\dot{x} - H(x,p))\, dt\right] \mathcal{D}(p(t))\mathcal{D}(x(t)),\end{aligned}$$

the integral being taken over the appropriate space of pairs of paths $(p(t), x(t))$. This is the **Hamiltonian form** of the path integral, as opposed to the usual **Lagrangian form** first introduced by Feynman and with which we have been occupied up to this point.

In some senses we have not gained much by the discussion of this section other than a verification of the equivalence of Feynman's and Schrödinger's approaches to quantum mechanics. We have accomplished a rederivation of equation 31.7, Feynman's original interpretation of the Lagrangian form of the path integral, and also a derivation of the Hamiltonian form of the path integral as the phase space path integral, by using the Dirac calculus and the insertion of resolutions of the identity. The primary intention of this section, though, was to illustrate how powerful this method is in deriving new forms for the path integral. To be clear, the method that has been introduced in this section is to write the propagator $\langle X_1|\mathsf{U}_T|X_0\rangle$ as a limit using the Trotter product formula given in equation 31.20 and then to expand each of the factors in the Trotter formula using appropriate resolutions of the identity. Though we will not consider other formulations of path integrals that are derived in this manner, a perusal of the literature will convince the reader of the power of this method; see for example the final chapter of Shankar (1994), which is dedicated to using this method to derive path integrals that incorporate spin and describe fermions.

31.5 End Notes

The filtration measure using the piecewise linear family is Feynman's original prescription for giving meaning to his path integral heuristic.

The computation of Section 31.2 is standard, though the reported propagator in Feynman and Hibbs (1965) contains a typo in its last term.[7] The filtration measure for the general polynomial family and its use in the calculation of the constant force propagator appeared first in Bouch and Bowers (2007). Many worked-out examples, including several involving the derivation of versions of path integrals as in Section 31.4, appear on pp. 181–264 of Dittrich and Reuter (2001).

[7] Feynman and Hibbs (1965) reports $-fT^3/24$ rather than the correct $-f^2T^3/24m$.

32
Evaluating Path Integrals II

> An effort to extend the path integral approach beyond its present limits continues to be a worthwhile pursuit; for the greatest value of this technique remains in spite of its limitations, i.e., the assistance which it gives one's intuition in bringing together physical insight and mathematics analysis.
>
> R.P. Feynman and A.R. Hibbs
> *Quantum Mechanics and Path Integrals*, 1965

In this lecture we formalize the scheme for evaluating path integrals that was introduced in the previous lecture, proposing a general setting in which to perform path integral calculations. The proposal is presented in the first section and then tested by applying it in three settings, to the computation of propagators for harmonic and forced harmonic oscillators using different dense subspaces in the path family. There is no confirmed guarantee that the proposed scheme is independent of the choice of dense subspace, but the examples presented illustrate the method as being successful at least for quadratic Lagrangians.

The harmonic oscillator is one of the most important physical examples to study in quantum mechanics. To see why, consider an arbitrary smooth conservative force $F = F(x) = -\nabla V$ applied to a particle at position x. If x is small, the force may be approximated by its linearization $F \approx F_0 + F_1 x$ and, if no constant force is applied, to first order $F = F_1 x$. This is the restorative force of the harmonic oscillator when $F_1 < 0$, which occurs routinely, for example, in the modeling of the vibrations of atoms in a lattice. We have already derived the eigenfunctions for the harmonic oscillator potential, in Lecture 11. In this lecture we derive the propagator and carefully examine it at its countable set of singularities in the time variable. An interesting phenomenon occurs at a singularity, namely, as the time variable crosses the singularity, a phase factor of $-\mathbf{i}$ appears in the propagator. Ultimately, the propagator at a singularity is itself interpreted as a Dirac delta distribution, generalizing the notion of a propagator. We conclude with an examination of the forced harmonic oscillator, a generalization of the harmonic oscillator potential that finds important applications in quantum field theory, for example in quantum electrodynamics where the electromagnetic field may be represented as a collection of forced harmonic oscillators for which the forcing terms couple the oscillators to one another.

Perhaps a warning is in order. Readers who are interested in the results that path integrals can give but who are not too concerned with the generality of the techniques might do well to skip the third section of this lecture. It gives a detailed computation of the harmonic oscillator propagator using the polynomial path family. The derivation is extremely challenging, and the details provide no physical insights in addition to those given by the much simpler derivation of the same propagator using the Fourier path family presented in the second section of this lecture. The difficult derivation in the third section is presented as a test case of the general proposal and will be of interest to those concerned with the mathematical generality of that proposal, but it has no practical physical value.

32.1 The Proposal

In this first section we generalize the development of Section 31.3. The development here differs in that we expand an arbitrary path in Γ about the classical straight-line path connecting the spacetime points $(0, X_0)$ and (T, X_1). Recall that $v = X/T$ is the average velocity between these points, where $X = X_1 - X_0$. Restricting our attention to paths that are at least of class C^1, Γ_0 is the linear space of loops based at the origin, the set of all continuously differentiable paths x with $x(0) = 0 = x(T)$.

Now, the family of C^1 paths connecting spacetime points $(0, X_0)$ and (T, X_1) may be expressed as

$$\Gamma = \{X_0 + vt + x : x \in \Gamma_0\},$$

which realizes Γ as the set of variations around the classical straight-line path $\ell(t) = X_0 + vt$ connecting $(0, X_0)$ and (T, X_1). The scheme for interpreting the path integral depends on the choice of a countable subset $\mathbf{f} = \{f_n\}$ of loops in Γ_0, which will be used to parameterize subspaces of Γ as Euclidean spaces to perform the path integration. We assume that the linear span of $\mathbf{f}$ is dense in Γ_0. Of course, the question arises to which topology the denseness refers. One possibility is the topology that arises from the norm $\|-\|_1$ of equation 6.2, making $\Gamma_0 = \mathcal{D}_0$, the space of increments. This, though, is a little too restrictive, and we will require only that span $\mathbf{f}$ be dense with respect to the coarser topology generated by the inner product

$$\langle f, g\rangle = \int_0^T f(t)g(t)\,dt,$$

which makes the loop space Γ_0 into a pre-Hilbert space that is not complete. It would be possible to require that the family $\mathbf{f}$ be orthonormal with respect to this inner product. When met, this requirement fosters easier and cleaner computations than otherwise, but the machinery works without it and we prefer to develop the program in greater generality without requiring orthonormality. We do require, though, that the members of the family $\mathbf{f}$ be linearly independent.

For an integer $N \geq 1$, let Γ_N denote that subset of Γ containing all paths of the form $X_0 + vt + x$, where $x \in \operatorname{span}\{f_1, \ldots, f_N\}$. There is an obvious identification of the Euclidean space $\mathbb{R}^N$ with Γ_N via

$$\mathbf{x} = (x_1, \ldots, x_N) \longleftrightarrow X_0 + vt + \sum_{n=1}^{N} x_n f_n = y.$$

We interpret the full path integral over Γ as the limit

$$\mathcal{K}(T, X_1; 0, X_0) = \lim_{N\to\infty} \int_{\Gamma_N} \exp\left(\frac{\mathrm{i}}{\hbar} S[y]\right) d\mu_N \tag{32.1}$$

$$= \lim_{N\to\infty} \Delta(N) \int_{\mathbb{R}^N} \exp\left(\frac{\mathrm{i}}{\hbar} S[y]\right) d\mathbf{x}, \tag{32.2}$$

where, as usual, $d\mu_N = \Delta(N) d\mathbf{x}$ is the filtration measure to be determined from the requirement that this limit must converge to the free particle propagator when the free action is inserted.

The free action for the path $y = X_0 + vt + x \in \Gamma_N$ becomes

$$S_{\text{free}}[y] = \frac{1}{2} m \int_0^T (v + \dot{x})^2\,dt = \frac{1}{2} m v^2 T + \frac{1}{2} m \int_0^T \dot{x}^2\,dt$$

$$= \frac{1}{2} m v^2 T + \frac{1}{2} m \mathbf{x}^{\mathrm{Tr}} \mathbf{B}_N \mathbf{x}, \tag{32.3}$$

where $\mathbf{B}_N$ is the symmetric matrix whose i, j entry is

$$B_{ij} = \int_0^T \dot{f}_i \dot{f}_j \, dt. \tag{32.4}$$

Here we make one more requirement on the family $\mathbf{f}$, viz., that the matrices $\mathbf{B}_N$ be of full rank. Inserting equation 32.3 into the Nth approximate propagator gives

$$\int_{\Gamma_N} \exp\left(\frac{\mathbf{i}}{\hbar} S_{\text{free}}[y]\right) d\mu_N$$
$$= \Delta(N) \exp\left[\frac{\mathbf{i}}{\hbar}\left(\frac{1}{2}mv^2T\right)\right] \int_{\mathbb{R}^N} e^{\mathbf{i}m(\mathbf{x}^{\mathsf{Tr}}\mathbf{B}_N\mathbf{x})/2\hbar} \, d\mathbf{x}.$$

Setting this equal to the free particle propagator given by equation 29.16 and applying formula 30.10 gives the normalization factor as

$$\Delta(N) = \left(\frac{m}{2\pi\mathbf{i}\hbar}\right)^{(N+1)/2} \sqrt{\frac{\det \mathbf{B}_N}{T}} \tag{32.5}$$

whenever the matrices $\mathbf{B}_N$ are positive definite, this being the typical scenario. In the remainder of this discussion, unless otherwise explicitly stated we assume that these matrices are positive definite.[1]

Putting equation 32.5 with equation 32.1 gives our final prescription for evaluating propagators using the path family $\mathbf{f}$ as

$$\mathcal{K}(T, X_1; 0, X_0)$$
$$= \lim_{N\to\infty} \left(\frac{m}{2\pi\mathbf{i}\hbar}\right)^{(N+1)/2} \sqrt{\frac{\det \mathbf{B}_N}{T}} \int_{\mathbb{R}^N} \exp\left(\frac{\mathbf{i}}{\hbar} S[X_0 + vt + \mathbf{x}^{\mathsf{Tr}}\mathbf{f}_N]\right) d\mathbf{x}, \tag{32.6}$$

where $\mathbf{f}_N$ is the column vector $[f_1 \cdots f_N]^{\mathsf{Tr}}$.

Before applying expression 32.6 to evaluate the propagator for the harmonic oscillator in the next section, we pause to see how the development of Section 31.3, which motivated the development of this section, fits into this scheme. Rewriting the polynomial expression $x(t) = X_0 + \sum_{n=1}^{N+1} x_{n-1}t^n$ of Section 31.3 in the form

$$x(t) = X_0 + vt + \sum_{n=1}^{N} x_n t\left(t^n - T^n\right)$$

[1] When positive definite, formula 30.10, rather than the more complicated formula 30.14, is used in determining the normalization factor $\Delta(N)$. More generally, though, when the matrices are not positive definite but still of full rank, formula 30.14 must be used and this introduces a unimodular phase factor. We will see this shortly in our analysis of the singularities of the harmonic oscillator propagator.

identifies the family $\mathbf{f}$ as being given by

$$f_n(t) = t\,(t^n - T^n)\,.$$

The linear span of the linearly independent family $\mathbf{f}$ is dense in Γ_0, and the matrix entry B_{ij} of $\mathbf{B}_N$ is

$$B_{ij} = \int_0^T ((i+1)t^i - T^i)((j+1)t^j - T^j)\,dt = \frac{ij}{i+j+1}T^{i+j+1}, \quad (32.7)$$

in agreement with equation 31.11.

32.2 The Harmonic Oscillator and Fourier Sums

In this section we apply formula 32.6 to evaluate the propagator for the harmonic oscillator using the Fourier family $\mathbf{f}$, where

$$f_n(t) = \sin\left(\frac{\pi n t}{T}\right).$$

Besides the fact that the linear span of $\mathbf{f}$ is dense in the space of loops Γ_0, the family itself is pairwise orthogonal as is the family of the derivatives $\dot{\mathbf{f}} = \{\dot{f}_n\}$. These facts make some of the computations very easy and serve to illustrate the power of orthogonality.

Since the harmonic oscillator Lagrangian is the quadratic form

$$L(x,\dot{x}) = \frac{1}{2}m\dot{x}^2 - \frac{1}{2}m\omega^2 x^2,$$

the results of Section 29.3 apply, and an examination of formula 29.7 shows that the general propagator for the harmonic oscillator may be written as

$$\mathcal{K}(T, X_1; 0, X_0) = \mathcal{K}(T, 0; 0, 0)e^{\mathrm{i}S_{\mathrm{cl}}/\hbar}, \quad (32.8)$$

where the action S_{cl} over the classical path is given by equation 29.10 as

$$S_{\mathrm{cl}} = \frac{m\omega}{2\sin\omega T}\left[(X_0^2 + X_1^2)\cos\omega T - 2X_0X_1\right].$$

This leaves us with the evaluation of $\mathcal{K}(T, 0; 0, 0)$.

The first task is to evaluate the normalization factor given by equation 32.5. As the members of the family $\dot{\mathbf{f}}$ are orthogonal, the matrix $\mathbf{B}_N$ is diagonal, with the diagonal entries given by equation 32.4 as

$B_{nn} = \pi^2 n^2/2T$. It follows that $\det \mathbf{B}_N = (N!)^2 \pi^{2N}/(2T)^N$ and, from equation 32.5,

$$\Delta(N) = N! \left(\frac{m\pi}{4\mathbf{i}\hbar T}\right)^{N/2} \left(\frac{m}{2\pi\mathbf{i}\hbar T}\right)^{1/2}. \tag{32.9}$$

In evaluating $\mathcal{K}(T, 0; 0, 0)$, we note that $X_0 = 0$ and $v = 0$ so that the action used in the integral expression in equation 32.6 is

$$S\left[\mathbf{x}^{\mathsf{Tr}}\mathbf{f}_N\right] = S_{\text{free}}[x] - \frac{1}{2}m\omega^2 \int_0^T x^2\, dt, \tag{32.10}$$

where $x(t) = \sum_{n=1}^N x_n \sin(n\pi t/T)$. Using the orthogonality of the family $\mathbf{f}$, the integral is evaluated easily and the action may be written as

$$S[x] = \frac{m\pi^2}{4T}\sum_{n=1}^N (n^2 - z^2)x_n^2 = \frac{m\pi^2}{4T}\mathbf{x}^{\mathsf{Tr}}\mathbf{A}_N\mathbf{x}$$

where $z = \omega T/\pi$ and $\mathbf{A}_N$ is the diagonal matrix with $A_{nn} = n^2 - z^2$. With the simplifying assumption that $0 < \omega T < \pi$, so that $z^2 < n^2$ for all n, $\mathbf{A}_N$ is positive definite and the approximate propagator becomes, with $a = m\pi^2/4\hbar T$,

$$\begin{aligned}\int_{\Gamma_N} \exp\left(\frac{\mathbf{i}}{\hbar}S[x]\right) d\mu_N &= \Delta(N)\int_{\mathbb{R}^N} e^{\mathbf{i}a\left(\mathbf{x}^{\mathsf{Tr}}\mathbf{A}_N\mathbf{x}\right)}\, d\mathbf{x} \\ &= \Delta(N)\left(\frac{\mathbf{i}\pi}{a}\right)^{N/2} \left(\det \mathbf{A}_N\right)^{-1/2}.\end{aligned}$$

Inserting $\Delta(N)$ from equation 32.9, simplifying, and using the easy observation that $\det \mathbf{A}_N = \prod_{n=1}^N (n^2 - z^2)$, gives

$$N!\left(\frac{m}{2\pi\mathbf{i}\hbar T}\right)^{1/2} \prod_{n=1}^N \left(\frac{1}{n^2 - z^2}\right)^{1/2} = \left(\frac{m}{2\pi\mathbf{i}\hbar T}\right)^{1/2} \prod_{n=1}^N \left(\frac{n^2}{n^2 - z^2}\right)^{1/2}$$

for the approximate propagator. Taking the limit gives the propagator for the harmonic oscillator, valid when $0 < \omega T < \pi$, as

$$\begin{aligned}\mathcal{K}(T, 0; 0, 0) &= \left(\frac{m}{2\pi\mathbf{i}\hbar T}\right)^{1/2} \left(\prod_{n=1}^\infty \frac{n^2}{n^2 - z^2}\right)^{1/2} \\ &= \left(\frac{m}{2\pi\mathbf{i}\hbar T}\right)^{1/2} \left(\frac{\pi z}{\sin \pi z}\right)^{1/2} = \left(\frac{m\omega}{2\pi\mathbf{i}\hbar \sin \omega T}\right)^{1/2},\end{aligned} \tag{32.11}$$

where we have used the sine product formula

$$\sin \pi z = \pi z \prod_{n=1}^\infty \left(1 - \frac{z^2}{n^2}\right). \tag{32.12}$$

It is interesting to see that, in this calculation using Fourier families, $\sin \omega T$ arises from equation 32.12, the classical infinite product formula for the sine function. In the next section we will see that in the calculation of the propagator for the harmonic oscillator using polynomial families, $\sin \omega T$ arises as a Taylor series.

What happens if the simplifying assumption $0 < \omega T < \pi$ does not hold? First note that the propagator given in equation 32.11 has a singularity at each value of $T = n\pi/\omega$ and so is undefined when $\omega T = n\pi$, for any integer n. In this case the symmetric matrix $\mathbf{A}_N$ fails to have full rank – it is then singular and the Gaussian integral diverges. What about away from the singularities? Suppose that $n\pi < \omega T < (n+1)\pi$ for an integer n, $0 \leq n \leq N$. Then $q^2 - z^2$ is negative when $1 \leq q \leq n$ and positive otherwise. This means that the matrix $\mathbf{A}_N$ has index $p = N - n$ and the more general Gaussian integral formula 30.14 must be used in place of formula 30.10 in the derivation of the propagator. The result is the expression given in equation 32.11 except that the sine term is in absolute values and a phase factor of $(-\mathbf{i})^n$ appears. Indeed, the factor $\Delta(N)$ from equation 32.9 contributes the term $(1/\mathbf{i})^{N/2} = e^{-N\mathbf{i}\pi/4}$ and the use of formula 30.14 contributes the term $e^{(2p-N)\mathbf{i}\pi/4}$, to give the unimodular phase factor

$$e^{-N\mathbf{i}\pi/4}e^{(2p-N)\mathbf{i}\pi/4} = e^{(p-N)\mathbf{i}\pi/2} = e^{-n\mathbf{i}\pi/2} = (-\mathbf{i})^n,$$

since $p - N = -n$. The final result for the propagator $\mathcal{K}(T, 0; 0, 0)$, observing that $n = \lfloor \omega T/\pi \rfloor$, is

$$\mathcal{K}(T, 0; 0, 0) = (-\mathbf{i})^{\lfloor \omega T/\pi \rfloor} \left(\frac{m\omega}{2\pi \mathbf{i}\hbar\, |\sin \omega T|} \right)^{1/2}. \tag{32.13}$$

This formula says that when T is increased by π/ω, since then n increases by one the effect is to multiply the expression by a phase factor of $-\mathbf{i}$. Indeed, since $|\sin \omega T|$ is periodic of period π/ω, we have

$$\mathcal{K}\left(T + \frac{\pi}{\omega}, 0; 0, 0\right) = -\mathbf{i}\,\mathcal{K}(T, 0; 0, 0). \tag{32.14}$$

The general propagator, from formula 29.10 and equation 32.8, is

$$\begin{aligned} \mathcal{K}(T, X_1; 0, X_0) = (-\mathbf{i})^{\lfloor \omega T/\pi \rfloor} &\left(\frac{m\omega}{2\pi \mathbf{i}\hbar\, |\sin \omega T|} \right)^{1/2} \\ &\times \exp\left\{ \frac{\mathbf{i}}{\hbar} \frac{m\omega}{2 \sin \omega T} \left[(X_0^2 + X_1^2) \cos \omega T - 2X_0 X_1 \right] \right\}. \end{aligned} \tag{32.15}$$

This is known as the **Feynman–Soriau formula** and was discovered by Soriau in 1974.

Another way to understand the appearance of the phase factor $-\mathbf{i}$ in equation 32.14 when the time is increased by π/ω is as a consequence of analytic continuation. Under the simplifying assumption used to derive equation 32.11, that $0 < t = \omega T < \pi$, the function $s(t) = \sqrt{\sin t}$ that appears in equation 32.11 is defined on the open segment $(0, \pi)$. It is natural to expect that the analytic continuation of $s(t)$ to a function $s(z)$ of the complex variable z is the correct function to appear in the propagator in the general, unrestricted, setting. The function $s(t)$ is real analytic in the open interval $(0, \pi)$ and so does analytically continue, to a function $s(z)$ defined on a Riemann surface lying over the complement of the zero set of $\sin t$, the countable set $\mathbb{Z}\pi$ in $\mathbb{C}$, which is the singular set of $s(z)$. The value of $s(z)$ on the open interval $(n\pi, (n+1)\pi)$, for $n \in \mathbb{Z}$, is

$$s(z) = \mathbf{i}^n s(z - n\pi) = \mathbf{i}^n \sqrt{(-1)^n \sin z} = \mathbf{i}^n \sqrt{|\sin z|}, \tag{32.16}$$

which is precisely what appears in formula 32.13. To see that equation 32.16 holds, we need only observe that the relation $s(z+\pi) = \mathbf{i}s(z)$ holds for the function $s(z)$. This may be seen by a careful study of $s(z)$ in the neighborhood of a singularity $n\pi$, a study we leave to the interested reader.

We return now to the singularities of the propagator that occur at times $T = n\pi/\omega$, for $n \in \mathbb{Z}$. Recall that the value of the propagator $\mathcal{K}(T, X_1; 0, X_0)$ is the complex amplitude for the particle to have spacetime locations $(0, X_0)$ and (T, X_1) and its primary use is to time-evolve the wave function according to equation 29.1. In the context of the present discussion, if the wave function of the harmonic oscillator is given at time $t = 0$ by ψ then the wave function at time t is given as

$$\psi_t(x) = \mathsf{U}_t\psi(x) = \int_{-\infty}^{\infty} \mathcal{K}(t, x; 0, x_0)\psi(x_0)\, dx_0.$$

In particular, the primary use of the propagator is as the kernel of an integral operator. This, of course, is a special case of a linear functional on a linear function space, and it should not be surprising that we may generalize the propagator to include more general functionals such as distributions. We can see this at first hand in examining the wave function at any of the singularities $n\pi/\omega$. Indeed, the initial state vector ψ may be expanded as $\sum_{n=0}^{\infty} c_n|n\rangle$, where $|n\rangle$ is the normalized state represented by the eigenfunction of the harmonic oscillator Hamiltonian given by equation 11.11 with energy eigenvalue $E_n = \left(n + \frac{1}{2}\right)\hbar\omega$. According to equation 27.5, ψ_t is given as

$$\psi_t = \mathsf{U}_t\psi = e^{-\mathbf{i}\omega t/2}\sum_{n=0}^{\infty} c_n e^{-\mathbf{i}n\omega t}|n\rangle. \tag{32.17}$$

Notice that ψ_t is periodic of period $4\pi/\omega$ so that, to understand the propagator at the singularities $t = n\pi/\omega$, all we need is to consider ψ_t for $t = \pi/\omega, 2\pi/\omega, 3\pi/\omega$, with $\psi_{4\pi/\omega} = \psi$. Inserting $t = \pi/\omega$ into equation 32.17, we find that

$$\psi_{\pi/\omega} = -\mathbf{i}\sum_{n=0}^{\infty}(-1)^n c_n|n\rangle. \tag{32.18}$$

By equation 11.11, the eigenfunction $|n\rangle$ is the product of a constant, the ground state eigenfuntion $|0\rangle$, which is an even function of x, and an nth degree Hermite polynomial. Since the nth degree Hermite polynomial is even or odd according to whether n is even or odd, we may rewrite equation 32.18 as

$$\psi_{\pi/\omega}(x) = -\mathbf{i}\sum_{n=0}^{\infty} c_n\langle -x|n\rangle = -\mathbf{i}\psi(-x) = -\mathbf{i}\delta_{-x}(\psi). \tag{32.19}$$

Here, of course, δ_{-x} is the Dirac distribution concentrated at $-x$. Physicists usually write the Dirac distribution using Dirac's fictitious delta function as the kernel in an integral operator equation and would write equation 32.19 as

$$\psi_{\pi/\omega}(x) = -\mathbf{i}\int_{-\infty}^{\infty} \delta(x_0 + x)\psi(x_0)\,dx_0.$$

This discussion identifies the propagator at the singularity π/ω as

$$\mathcal{K}\left(\frac{\pi}{\omega}, X_1; 0, X_0\right) = -\mathbf{i}\delta(X_0 + X_1). \tag{32.20}$$

Notice that the phase factor $-\mathbf{i}$ arises precisely from the ground state energy $\hbar\omega/2$. Similar arguments show that $\psi_{2\pi/\omega}(x) = -\psi(x) = (-\mathbf{i})^2\delta_x(\psi)$ and $\psi_{3\pi/\omega}(x) = \mathbf{i}\psi(-x) = (-\mathbf{i})^3\delta_{-x}(\psi)$. As ψ_t is $(4\pi/\omega)$-periodic, this together with $\psi_{4\pi/\omega}(x) = \psi(x) = (-\mathbf{i})^4\delta_x(\psi)$ gives the formula

$$\psi_{n\pi/\omega}(x) = (-\mathbf{i})^n\delta_{(-1)^n x}(\psi) = (-\mathbf{i})^n\int_{-\infty}^{\infty} \delta\left(x_0 - (-1)^n x\right)\psi(x_0)\,dx_0,$$

and the propagator on the singular set as

$$\mathcal{K}\left(\frac{n\pi}{\omega}, X_1; 0, X_0\right) = (-\mathbf{i})^n\delta(X_0 - (-1)^n X_1). \tag{32.21}$$

We will now give another argument, one that a physicist might present, that the propagator on the singular set should be interpreted as a delta function. On the surface it lacks rigor, but rigor may be supplied with a careful development of distribution theory. Using the group property of equation 29.2, we may write

$$\mathcal{K}\left(\frac{\pi}{\omega}, X_1; 0, X_0\right) = \int_{-\infty}^{\infty} \mathcal{K}\left(\frac{\pi}{\omega}, X_1; \frac{\pi}{2\omega}, x\right) \mathcal{K}\left(\frac{\pi}{2\omega}, x; 0, X_0\right) dx.$$

By time-translation invariance, $\mathcal{K}(\pi/\omega, X_1; \pi/2\omega, x) = \mathcal{K}(\pi/2\omega, X_1; 0, x)$ and, substituting from equation 32.15 for the two factors in the integrand in the first line of equation 32.22, the formula for the propagator at the first singularity π/ω simplifies to

$$\begin{aligned}\mathcal{K}\left(\frac{\pi}{\omega}, X_1; 0, X_0\right) &= \int_{-\infty}^{\infty} \mathcal{K}\left(\frac{\pi}{2\omega}, X_1; 0, x\right) \mathcal{K}\left(\frac{\pi}{2\omega}, x; 0, X_0\right) dx \\ &= -\mathbf{i}\frac{1}{2\pi\hbar}\int_{-\infty}^{\infty} e^{-\mathbf{i}(X_0+X_1)x/\hbar}\, dx, \end{aligned} \tag{32.22}$$

where the factor $m\omega$ that arises from the two substitutions of equation 32.15 has been absorbed into the integration variable. Now, on the face of it this last integral is divergent, but it may be understood, at least formally, as a representation of a delta function in the following way. Formally, the inverse Fourier transform of the Dirac delta function $\delta(p)$ may be written as

$$\mathcal{F}^{-1}(\delta)(x) = \frac{1}{\sqrt{2\pi\hbar}}\int_{-\infty}^{\infty} e^{\mathbf{i}xp/\hbar}\delta(p)\, dp = \frac{1}{\sqrt{2\pi\hbar}}.$$

Taking the Fourier transform and using the Fourier inversion theorem, which technically does not apply as the integral diverges, we have, formally,

$$\begin{aligned}\delta(z) &= \mathcal{F}(\mathcal{F}^{-1}(\delta))(z) \\ &= \frac{1}{\sqrt{2\pi\hbar}}\int_{-\infty}^{\infty} e^{-\mathbf{i}zx/\hbar}\frac{1}{\sqrt{2\pi\hbar}}\, dx = \frac{1}{2\pi\hbar}\int_{-\infty}^{\infty} e^{-\mathbf{i}zx/\hbar}\, dx.\end{aligned}$$

Applying this formal result to equation 32.22 gives precisely equation 32.20. Similar calculations at the remaining singular points give equation 32.21 for the propagator on the singular set.

Mathematicians, like the author, are always amazed at how physicists get away with these sorts of arguments. Physicists often have a certain chutzpah that goes against the grain of the pure mathematician. The arguments are far from rigorous, often nonsensical in their details, but in the end give the correct formulae. The history of physics is rife with example after example of this phenomenon. The appropriate response of the mathematician is to be inspired to understand the physicists' calculations in a rigorous setting, as the success of the nonrigorous methods of physicists provides a strong indication of deep mathematical truths yet to be uncovered.

32.3 The Harmonic Oscillator and Polynomial Sums ⋆

Were this simply a physics text, we would have dispensed with the use of polynomial path families in evaluating propagators altogether, since Feynman's original piecewise linear path family and the Fourier family are sufficient for evaluating all path integrals with quadratic Lagrangians. However, our interest covers also the mathematical generality of Feynman's idea for path integral evaluation, that of using path families to evaluate a sequence of approximate propagators. This section thus presents a case study of this question of generality by examining the evaluation of the harmonic oscillator propagator using the polynomial path family rather than the Fourier path family. The calculations in this setting turn out to be quite formidable and are not for the faint of heart. Nonetheless, the polynomial path family is found to give the same results as the Fourier family.

The calculation presented here has some interesting twists and turns, but ultimately gets complicated enough that the interested reader is referred to Bouch and Bowers (2007) for the final steps of the calculation. We present enough of the calculation here to see the emergence of the term $\sin \omega T$ in the evaluation of $\mathcal{K}(T, 0; 0, 0)$. In the calculations for the free particle and for the constant force field, the classical path that the particle traverses is in fact a polynomial path. This is why the free particle action for the classical path and two of the three terms in the constant force action for the classical path appear quickly in the evaluation of the propagator using the polynomial path family. This will not happen for the harmonic oscillator when the polynomial path family is used, since the classical path in this case is a sinusoid that can only be approximated by polynomial paths, and the action associated with the classical path contains sinusoidal terms.

Recall that the filtration measure for the polynomial path family $\mathbf{f}$, where $f_n(t) = t(t^n - T^n)$, is given by equation 32.5; the matrix $\mathbf{B}_N$

is given by equations 31.11 or 32.7. As in the preceding section, using equation 32.8, we need compute only $\mathcal{K}(T,0;0,0)$. The action is given by expression 32.10 where $x(t) = \sum_{n=1}^{N} x_n f_n(t)$. A straightforward[2] calculation after simplification gives the action as

$$S[x] = \frac{1}{2} m \mathbf{x}^{\mathsf{Tr}} \mathbf{B}_N \mathbf{x} - \frac{1}{2} m\omega^2 T^2 \mathbf{x}^{\mathsf{Tr}} \mathbf{J}_N \mathbf{x},$$

where the matrix $\mathbf{J}_N$ is given by

$$\mathbf{J}_N = \left[\frac{1}{3} \frac{i}{i+3} \frac{j}{j+3} \frac{i+j+6}{i+j+3} T^{i+j+1} \right]_{i,j=1}^{N}.$$

Using the general Gaussian integral formula 30.14 with $a = m/2\hbar$, the approximate propagator becomes, up to a unimodular constant,

$$\int_{\Gamma_N} e^{\mathrm{i}a\mathbf{x}^{\mathsf{Tr}}(\mathbf{B}_N - \omega^2 T^2 \mathbf{J}_N)\mathbf{x}} \, d\mu_N = \left(\frac{m}{2\pi \mathrm{i}\hbar T} \right)^{1/2} \sqrt{\frac{\det \mathbf{B}_N}{|\det(\mathbf{B}_N - \omega^2 T^2 \mathbf{J}_N)|}}.$$

To verify that the polynomial path family gives the correct propagator, we need to prove that the limit of the square root term is $(\omega T/|\sin \omega T|)^{1/2}$. This is equivalent to the statement that

$$\lim_{N\to\infty} \det(\mathbf{I}_N - \omega^2 T^2 \mathbf{B}_N^{-1} \mathbf{J}_N) = \frac{\sin \omega T}{\omega T}. \tag{32.23}$$

To verify this, first rewrite the matrices as $\mathbf{B}_N = T\mathbf{T}_N \mathbf{C}_N \mathbf{T}_N$ and $\mathbf{J}_N = T\mathbf{T}_N \mathbf{K}_N \mathbf{T}_N$, where $\mathbf{T}_N = \mathbf{diag}\left[T \; 2T^2 \; \cdots \; NT^N \right]$, $\mathbf{C}_N$ is the Hilbert matrix of equation 30.23 and the matrix $\mathbf{K}_N$ is given by

$$\mathbf{K}_N = \left[\frac{1}{3} \frac{1}{i+3} \frac{1}{j+3} \frac{i+j+6}{i+j+3} \right]_{i.j=1}^{N}.$$

Obviously, $\det(\mathbf{I}_N - \omega^2 T^2 \mathbf{B}_N^{-1} \mathbf{J}_N) = \det(\mathbf{I}_N - \omega^2 T^2 \mathbf{C}_N^{-1} \mathbf{K}_N)$. The evaluation of the latter determinant is quite involved and takes up the remainder of this section.

Recall that the Hilbert matrix $\mathbf{C}_N$ is an example of a Cauchy matrix, one whose i,j entry is of the form $1/(s_i + t_j)$. We already have applied the general formula for the determinant of a Cauchy matrix, formula 30.25, to find the determinant of $\mathbf{C}_N$. There is also a general formula for the inverse of a Cauchy matrix.[3] Applying this with $s_i = i$ and $t_j = j + 1$

[2] Straightforward means here neither simple nor short. The calculation of the matrix $\mathbf{J}_N$ is rather lengthy and requires several pages of care, but no cleverness.

[3] See, for example, Knuth (1997), Problem 41, p. 38.

gives

$$\mathbf{C}_N^{-1} = \left[(-1)^{i+j}\frac{\prod_{k=1}^{N}(i+k+1)(j+k+1)}{(i+j+1)(i-1)!(j-1)!(N-i)!(N-j)!}\right]_{i,j=1}^{N}$$

$$= \left[(-1)^{i+j}\frac{ij}{i+j+1}\binom{N+i+1}{N}\binom{N}{i}\binom{N+j+1}{N}\binom{N}{j}\right]_{i,j=1}^{N}.$$

It follows that the i,k entry of the matrix product $\mathbf{C}_N^{-1}\mathbf{K}_N$ is

$$\lambda_{ik} = (-1)^i\frac{1}{3}\frac{i}{k+3}\binom{N+i+1}{N}\binom{N}{i}$$
$$\times\sum_{j=1}^{N}(-1)^j\binom{N+j+1}{N}\binom{N}{j}\frac{j}{i+j+1}\frac{j+k+6}{(j+3)(j+k+3)}. \tag{32.24}$$

The key to evaluating these sums is to write explicitly the i,k entry of the matrix product $\mathbf{I}_N = \mathbf{C}_N^{-1}\mathbf{C}_N$. With δ_{ik} as the Kronecker delta, we get

$$\delta_{ik} = (-1)^i i\binom{N+i+1}{N}\binom{N}{i}$$
$$\times\sum_{j=1}^{N}(-1)^j\binom{N+j+1}{N}\binom{N}{j}\frac{j}{i+j+1}\frac{1}{j+k+1}. \tag{32.25}$$

Expand the term $(j+k+6)/(j+3)(j+k+3)$ in formula 32.24 as

$$\frac{j+k+6}{(j+3)(j+k+3)} = \frac{k+3}{k}\frac{1}{j+2+1} - \frac{3}{k}\frac{1}{j+(k+2)+1}. \tag{32.26}$$

Substituting this into equation 32.24 and comparing with equation 32.25 gives, for $1 \le i \le N$ and $1 \le k \le N-2$,

$$\lambda_{i,k} = \frac{1}{3(k+3)}\left(\frac{k+3}{k}\delta_{i,2} - \frac{3}{k}\delta_{i,k+2}\right). \tag{32.27}$$

The reader will notice that the only nonzero elements in the first $N-2$ columns of $\mathbf{C}_N^{-1}\mathbf{K}_N$ are

$$\left.\begin{aligned}\lambda_{2,k} &= \frac{1}{3k}\\ \lambda_{k+2,k} &= \frac{-1}{k(k+3)}\end{aligned}\right\}\ k = 1,\ldots,N-2. \tag{32.28}$$

The form of the matrix $\mathbf{I}_N - \omega^2 T^2 \mathbf{C}_N^{-1} \mathbf{K}_N$ for $N = 8$ is illustrated by the example

$$\left(\begin{array}{cccccc|cc} 1 & 0 & 0 & 0 & 0 & 0 & \mu_{1,7} & \mu_{1,8} \\ * & * & * & * & * & * & \mu_{2,7} & \mu_{2,8} \\ * & 0 & 1 & 0 & 0 & 0 & \mu_{3,7} & \mu_{3,8} \\ 0 & * & 0 & 1 & 0 & 0 & \mu_{4,7} & \mu_{4,8} \\ 0 & 0 & * & 0 & 1 & 0 & \mu_{5,7} & \mu_{5,8} \\ 0 & 0 & 0 & * & 0 & 1 & \mu_{6,7} & \mu_{6,8} \\ 0 & 0 & 0 & 0 & * & 0 & \mu_{7,7} & \mu_{7,8} \\ 0 & 0 & 0 & 0 & 0 & * & \mu_{8,7} & \mu_{8,8} \end{array}\right),$$

where each asterisk represents a nonzero entry calculated using equations 32.28, and where we have written the i, k entry of $\mathbf{I}_N - \omega^2 T^2 \mathbf{C}_N^{-1} \mathbf{K}_N$ as $\mu_{ik} = \delta_{ik} - \omega^2 T^2 \lambda_{ik}$. The determinant will be evaluated after a sequence of column operations yielding a matrix $\mathbf{M}_N$, followed by a sequence of row operations yielding a matrix $\mathbf{A}_N$.

We begin with a sequence of column operations using the entry 1 in column k on the diagonal to eliminate the $k, k-2$ entry in column $k-2$ successively, starting with column $k = N-2$ and moving leftwards one column at a time until column $k = 3$ is used to eliminate the third term from the first column. The result is a matrix $\mathbf{M}_N$, illustrated for $N = 8$ by

$$\left(\begin{array}{cccccc|cc} 1 & 0 & 0 & 0 & 0 & 0 & \mu_{1,7} & \mu_{1,8} \\ * & p & * & * & * & * & \mu_{2,7} & \mu_{2,8} \\ 0 & 0 & 1 & 0 & 0 & 0 & \mu_{3,7} & \mu_{3,8} \\ 0 & 0 & 0 & 1 & 0 & 0 & \mu_{4,7} & \mu_{4,8} \\ 0 & 0 & 0 & 0 & 1 & 0 & \mu_{5,7} & \mu_{5,8} \\ 0 & 0 & 0 & 0 & 0 & 1 & \mu_{6,7} & \mu_{6,8} \\ * & 0 & * & 0 & * & 0 & \mu_{7,7} & \mu_{7,8} \\ 0 & q & 0 & * & 0 & * & \mu_{8,7} & \mu_{8,8} \end{array}\right), \tag{32.29}$$

where we have written p for the $2, 2$ entry and q for the $8, 2$ entry. More generally, we will write p_N for the $2, 2$ entry and q_N for the $S, 2$ entry of $\mathbf{M}_N$, where $S = N$ if N is even and $S = N - 1$ if N is odd. Now, the $2, 2$ entry in the matrix 32.29 is very interesting – we will show that it is the mth Taylor polynomial expanded about the origin of the function $\sin \omega T / \omega T$, for $m = \lfloor \frac{N-2}{2} \rfloor$. Indeed, the reader is advised to check that

$$p_N(\omega T) = \sum_{j=0}^{\lfloor \frac{N-2}{2} \rfloor} \frac{(-1)^j}{(2j+1)!} (\omega T)^{2j}. \tag{32.30}$$

We now continue to reduce the matrix $\mathbf{M}_N$ using row operations, but we shall do nothing that changes the 2, 2 entry p_N or the $S, 2$ entry q_N. We apply row reduction using the diagonal 1 to eliminate the two nonzero entries in column k, for $k = 1, 3, \ldots, N-2$. The result is a matrix $\mathbf{A}_N$, illustrated for $N = 8$ by

$$\left(\begin{array}{cccccc|cc} 1 & 0 & 0 & 0 & 0 & 0 & \mu_{1,7} & \mu_{1,8} \\ 0 & p & 0 & 0 & 0 & 0 & e & f \\ 0 & 0 & 1 & 0 & 0 & 0 & \mu_{3,7} & \mu_{3,8} \\ 0 & 0 & 0 & 1 & 0 & 0 & \mu_{4,7} & \mu_{4,8} \\ 0 & 0 & 0 & 0 & 1 & 0 & \mu_{5,7} & \mu_{5,8} \\ 0 & 0 & 0 & 0 & 0 & 1 & \mu_{6,7} & \mu_{6,8} \\ 0 & 0 & 0 & 0 & 0 & 0 & a & b \\ 0 & q & 0 & 0 & 0 & 0 & c & d \end{array}\right), \tag{32.31}$$

where a, b, c, d, e, and f are appropriate linear combinations of the $\mu_{i,7}$ and the $\mu_{i,8}$. In the general case, the only entries in the last two columns needed for the determinant are the second-row terms e_N and f_N and the lower right 2×2 block $\left(\begin{smallmatrix} a_N & b_N \\ c_N & d_N \end{smallmatrix}\right)$. We arrive at

$$\det(\mathbf{I}_N - \omega^2 T^2 \mathbf{C}_N^{-1}\mathbf{K}_N) = \begin{cases} p(ad-bc) + q(eb-fa) & \text{if } N \text{ is even,} \\ p(ad-bc) - q(ed-fc) & \text{if } N \text{ is odd,} \end{cases} \tag{32.32}$$

where we have suppressed the subscripts N. The limit in equation 32.23 is now within our grasp. The details are quite involved, but the result is that the limit of the term $ad - bc$ is 1 while the limits of the remaining summands, $q(eb - fa)$ and $q(ed - fc)$, are both 0, giving the limit of equation 32.23 as

$$\begin{aligned} \lim_{N\to\infty} \det(\mathbf{I}_N - \omega^2 T^2 \mathbf{B}_N^{-1}\mathbf{J}_N) &= \lim_{N\to\infty} p_N(\omega T) \\ &= \sum_{j=0}^{\infty} \frac{(-1)^k}{(2j+1)!}(\omega T)^{2j} = \frac{\sin \omega T}{\omega T}. \end{aligned}$$

Modulo very complicated computations, then, this confirms that the polynomial path family gives the same result as that for the Fourier path family in evaluating the propagator for the harmonic oscillator.

As for the complicated computations, we need explicit formulae for the terms a, b, c, d, e, f, p, and q and for this we first need to identify explicit formulae for the $\mu_{ik} = \delta_{ik} - \omega^2 T^2 \lambda_{ik}$, i.e., for λ_{ik}, in the last two columns. We then need to confirm the limit claims that $(ad - bc) \to 1$,

$q(eb-fa) \to 0$, and $q(ed-fc) \to 0$ as $N \to \infty$. As these computations are quite involved, they are omitted from this presentation and the interested reader is referred to Bouch and Bowers (2007) for the details.

32.4 The Forced Harmonic Oscillator

It is rare to find physical systems completely isolated from their surroundings. More usually one finds systems that interact, at least weakly, with the environment, and physics must have a way to model these interactions even when the exact nature of the external potentials is unknown. Our first and most important example of this is a harmonic oscillator that interacts with its surroundings via a time-varying external force. In actual physical systems, examples arise as interactions between neighboring atoms in a lattice, or as atomic interactions in molecules in an external magnetic field, or as the interactions of the atoms in a crystal with a passing electron. In this section we want to calculate the propagator for a one-dimensional harmonic oscillator that is subject to a time-varying external force $K(t)$. The Lagrangian is

$$L(x, \dot{x}, t) = \frac{1}{2}m\dot{x}^2 - \frac{1}{2}m\omega^2 x^2 + K(t)x.$$

The term $K(t)x$ is called the **interaction term** and serves to integrate all the external forces on the oscillator, regardless of their sources, that arise from its interaction with its surroundings. This is a quadratic potential and the results of Section 29.3 apply, to give the propagator as

$$\mathcal{K}(T, X_1; 0, X_0) = \mathcal{K}_{\mathrm{HO}}(T, 0; 0, 0)e^{\mathrm{i}S_{\mathrm{cl}}/\hbar}, \tag{32.33}$$

where $\mathcal{K}_{\mathrm{HO}}$ is the propagator for the unforced oscillator of equation 32.13 and S_{cl} is the action over the classical path for the forced oscillator. We will show that this action is

$$\begin{aligned} S_{\mathrm{cl}} = S_{\mathrm{HO}} &+ \frac{1}{\sin\omega T}\int_0^T K(t)\left[X_1 \sin\omega t + X_0 \sin\omega(T-t)\right] dt \\ &- \frac{1}{m\omega\sin\omega T}\int_0^T K(t)\sin\omega(T-t)\left[\int_0^t K(s)\sin\omega s\, ds\right] dt. \end{aligned} \tag{32.34}$$

Here, S_{HO} is the classical action of equation 29.10 for the unforced oscillator.

To verify this last equation, first recall the classical path of the forced oscillator as the solution to the boundary-value problem

$$m\ddot{x} + m\omega^2 x = K(t), \quad x(0) = X_0, \quad x(T) = X_1.$$

A standard variation of parameters handles this equation of motion and supplies the solution. The usual route is to vary the parameters u and v in $u\cos\omega t + v\sin\omega t$, a linear combination of the fundamental set $\{\cos\omega t, \sin\omega t\}$ of solutions for the homogenous equation. While this is appropriate for an initial value problem, more appropriate for a boundary value problem is to use the solutions given by the two functions $\sin\omega t$ and $\sin\omega(T-t)$, which form a fundamental set provided that $\omega T \neq n\pi$ for any $n \in \mathbb{N}$.[4] Assuming this last condition, a standard variation of parameters now gives the solution to the initial-value problem as

$$x_{\text{cl}}(t) = \left(X_1 - \frac{1}{m\omega}\int_0^T K(s)\sin\omega(T-s)\,ds \right)\frac{\sin\omega t}{\sin\omega T} + X_0\frac{\sin\omega(T-t)}{\sin\omega T}$$
$$+ \frac{\sin\omega t}{m\omega\sin\omega T}\int_0^t K(s)\sin\omega(T-s)\,ds - \frac{\sin\omega(T-t)}{m\omega\sin\omega T}\int_0^t K(s)\sin\omega s\,ds.$$

This may be written as

$$x_{\text{cl}}(t) = \left(X_0 - \frac{1}{m\omega}\int_0^t K(s)\sin\omega s\,ds \right)\frac{\sin\omega(T-t)}{\sin\omega T}$$
$$+ \left(X_1 - \frac{1}{m\omega}\int_t^T K(s)\sin\omega(T-s)\,ds \right)\frac{\sin\omega t}{\sin\omega T}, \tag{32.35}$$

and its derivative simplifies to

$$\dot{x}_{\text{cl}}(t) = -\,\omega\left(X_0 - \frac{1}{m\omega}\int_0^t K(s)\sin\omega s\,ds \right)\frac{\cos\omega(T-t)}{\sin\omega T}$$
$$+\,\omega\left(X_1 - \frac{1}{m\omega}\int_t^T K(s)\sin\omega(T-s)\,ds \right)\frac{\cos\omega t}{\sin\omega T}. \tag{32.36}$$

This looks a bit complicated, but it is easy to see that the boundary conditions hold in equation 32.35 and a straightforward calculation of

[4] Obviously, when $\omega T = n\pi$, $\sin\omega(T-t) = (-1)^{n+1}\sin\omega t$ and the two solutions are not linearly independent. In general, the Wronskian for these two functions is constant, as $W(t) = W(\sin\omega t, \sin\omega(T-t); t) = -\omega\sin\omega T$. Since the two solutions $\sin\omega t$ and $\sin\omega(T-t)$ form a fundamental set for the homogeneous equation precisely when the Wronskian is nonzero on the interval $[0, T]$, we see that this occurs exactly when $\omega T \notin \mathbb{Z}\pi$.

the derivative of equation 32.36 used with equation 32.35 verifies that x_{cl} satisfies the equation of motion.

Continuing with the verification of equation 32.34, apply integration by parts to the $\dot{x}^2$-term in the integral expression of the action and insert $m\ddot{x}_{\text{cl}} = -m\omega^2 x_{\text{cl}} + K$ from the equation of motion to get, after simplification,

$$\begin{aligned} S_{\text{cl}} &= \int_0^T \frac{1}{2} m\dot{x}_{\text{cl}}^2 - \frac{1}{2} m\omega^2 x_{\text{cl}}^2 + K(t) x_{\text{cl}}\, dt \\ &= \frac{1}{2} m \left(X_1 \dot{x}_{\text{cl}}(T) - X_0 \dot{x}_{\text{cl}}(0) \right) + \frac{1}{2} \int_0^T K(t) x_{\text{cl}}\, dt. \end{aligned} \tag{32.37}$$

After inserting the expressions in equations 32.35 and 32.36 into the second line of equation 32.37, a straightforward simplification yields the classical action of equation 32.34.

Now that the classical action and propagator have been worked out, they may be used to evaluate the **transition amplitudes** for an oscillator initially in energy eigenstate $|i\rangle$ to end up in energy eigenstate $|j\rangle$ after experiencing a disturbance described by the forcing term $K(t)$ that acts for a time period T. Here $|i\rangle$ is the normalized state vector represented by the eigenfunction of the harmonic oscillator Hamiltonian, given by equation 11.11, with energy eigenvalue $E_i = \left(i + \frac{1}{2}\right) \hbar\omega$. We think of the oscillator as initially unforced and in state $|i\rangle$, but then experiencing a time-varying force $K(t)$ over the time interval from $t = 0$ to $t = T$, at which point the forcing is cut off and the energy of the oscillator is measured. Letting U_t be the time-evolution operator determined by the forced oscillator Hamiltonian, the state vector of the oscillator at time T is given by

$$\langle x|\mathsf{U}_T|i\rangle = \int_{-\infty}^{\infty} \mathcal{K}(T, x; 0, x') \langle x'|i\rangle\, dx',$$

where $\mathcal{K}$ is given by equation 32.33. The transition amplitude is then

$$\begin{aligned} \tau_{ij} = \langle j|\mathsf{U}_T|i\rangle &= \langle j| \left(\int_{-\infty}^{\infty} |x\rangle\langle x|\, dx \right) |\mathsf{U}_T|i\rangle \\ &= \int_{-\infty}^{\infty} \int_{-\infty}^{\infty} \langle j|x\rangle \mathcal{K}(T, x; 0, x') \langle x'|i\rangle\, dx' dx. \end{aligned} \tag{32.38}$$

This is how one quantizes a field of operators, for example, when the electromagnetic field is quantized. A proper explanation requires more background, in special relativity first, then covariant electrodynamics, and finally the Dirac equation. These topics form the basic prerequisites for the study of quantum electrodynamics and would be thoroughly explored in a course on relativistic quantum mechanics, which would be the logical next step for the student.

32.5 End Notes

This lecture probably contains the most difficult calculations in this book. The analysis of the harmonic oscillator propagator of Section 32.2 follows Feynman's original analysis via Fourier polynomials for the range $0 < \omega T < \pi$. The analysis at the singularities, where ωT takes on the values given by integral multiples of π, is my own with help from Felsagar (1998), Section 5.2. I recommend Felsager's treatment for further insight. In it he uses the Feynman propagator to derive the energy spectrum of the harmonic oscillator, as well as the eigenfunctions in terms of Gaussian exponentials and Hermite polynomials. The calculation of the propagator for the harmonic oscillator using the polynomial family in Section 32.3, though extremely challenging, reveals something of the depth of Feynman's propagator calculus. The analysis of the forced oscillator in Section 32.4 follows that of Feynman and Hibbs (1965). Feynman and Hibbs actually go much further and obtain precise formulae for the transition probabilities by performing the integration of equation 32.38. I recommend highly Chapter 8 of Feynman and Hibbs for a complete explanation of the path integral approach to analyzing harmonic oscillators.

Epilogue

The writing of this book has been an exciting journey for me. I hope that the student using it will come away from its study with a keen appreciation for the tight bonds that exist between theoretical physics and pure mathematics and will be inspired to explore further the realms of quantum mechanics. The next step for the student is to include the relativistic theory of particles in the mix. This begins with a reimagining of electrodynamics in covariant relativistic form, and then a study of the Dirac equation and the gauge principle follows. After that come a study of the quantization of the electromagnetic field – quantum electrodynamics – and of the quantization of other fields, culminating in a study of the standard model of particle physics.

For those who wish to continue their pursuit of quantum mechanics, I have included as a final section of this book a selected guide to the literature that I have found useful in gaining some understanding.

Resources for Individual Exploration

> I think I can safely say that nobody understands quantum mechanics.
>
> R.P. Feynman
> *The Character of Physical Law*, 1965

I am of the strong opinion that the student should be referencing a variety of sources as he or she learns quantum mechanics, and this guide consists of my recommendations for enrichment and expansions of what the student will find in any one source. There are hundreds of sources that I have used over the years in support of my quantum mechanics obsession, including textbooks and original research articles, lectures by physicists, discussions with colleagues, historical accounts, critical philosophical writings, television shows and documentaries, written popular accounts, and web-based resources. In selecting samples of the writings that I have found most useful, many will be left out that really deserve inclusion. To their authors I apologize. Each reference discussed below is referred to by author (date of publication) and title, as in Dirac (1958) *The Principles of Quantum Mechanics*, or by author (date of publication), especially after its first occurrence, the requisite publisher information and full reference being given in the bibliography at the end of the book. This guide is arranged topically, with brief comments adorning each reference.

Standard Approaches to Quantum Mechanics

The texts recommended in this section were written by physicists and present standard approaches to the subject. The student will find a wide range of styles of presentation and differences in notation and emphases among them.

Dirac (1958) *The Principles of Quantum Mechanics* is one of the classics of the field. In it Dirac synthesized all the approaches to the then fledgling quantum mechanics into a single interaction picture and presented his bra-ket notational calculus. Its 1957 revision includes a chapter on quantum electrodynamics and can be studied profitably even 85 years after its first edition, and 46 years after the last revision of its fourth edition. Fock (1978) *Fundamentals of Quantum Mechanics* is the English translation of a Russian work of V.A. Fock of 1932. It presents a mathematically detailed development of quantum mechanics, including Dirac's theory of the electron, using classical mathematics. It is out of print but, if one can get hold of a copy, it is a great resource for complete mathematical derivations of some important physical applications of the subject.

Merzbacher (1998) *Quantum Mechanics* is the third edition of a book first published in 1961, whose purpose was "to provide a carefully structured and coherent exposition of quantum mechanics; to illuminate the essential features of the theory without cutting corners, and yet without letting technical details obscure the main storyline; and to exhibit wherever possible the common threads by which the theory links many different phenomena and subfields."[5] Merzbacher has a genius for presenting the physics front and center but with enough mathematical honesty to satisfy even this most unrepentant mathematician. I personally have learned a great deal from this engaging text.

Peebles (1992) *Quantum Mechanics* is a beautifully written exposition of the subject, which I would describe as a "student's book." A pedagogically sound presentation, it presents the basic theory in four chapters before turning to applications in a final four chapters. After developing perturbation theory in some detail, the book explains many physical consequences of the theory, including the Zeeman effect, the Stark effect, the hyperfine structure of atomic hydrogen, transitions between energy levels, details of atomic and molecular structure, and spin–orbit coupling.

Elbaz (1998) *Quantum: the Quantum Theory of Particles, Fields, and Cosmology* is a highly original presentation of the subject, remarkable in its breadth and rather democratic when it comes to interpretative issues. It is a modern rendition of the subject using diagrammatic symbolism

[5] Merzbacher (1998), p. vii.

to indicate certain mathematical operations. This book lends itself to browsing, and I have found it gratifying on many occasions when studying a particular topic to read its take on the subject.

As one might infer from the subtitle, Schwinger and Englert (2001) *Quantum Mechanics: Symbolism of Atomic Measurement* is a highly positivistic presentation of the quantum mechanical calculus that includes a wonderful philosophical essay at the beginning on causality, determinism, and measurement. It develops the formal machinery that nature demands without regard for mathematical underpinnings, and does so quite successfully.

Weinberg (2012) *Lectures on Quantum Mechanics* came across my desk as I was writing this essay, and I include it as an example of a telling of the subject by one of its master practitioners. The book covers many topics not normally presented in standard texts. The directness and conciseness of the prose is appealing, as is Weinberg's willingness to engage with discussions on interpretations of the theory. The first chapter, a historical introduction to the subject, is beautifully written and reminiscent of the fine historical introduction that begins his earlier three-volume work on quantum field theory.

Atypical Approaches to Quantum Mechanics

Feynman and Hibbs (1965) *Quantum Mechanics and Path Integrals* is, in the opinion of this author, still the best place to learn of Feynman's highly original approach to quantum mechanics. It develops many standard results of quantum mechanics using the path integral approach, which necessarily entails some significant mathematical sleights of hand. What the presentation lacks in rigor it more than makes up for in painting a physical description of quantum processes that has a strong intuitive appeal.

The question whether quantum mechanics is a complete theory or whether it is emergent from a more basic theory is addressed in different ways in the next three books. Holland (1993) *The Quantum Theory of Motion* is described accurately by its subtitle, *An Account of the de Broglie–Bohm Causal Interpretation of Quantum Mechanics.* It is a beautifully written and illustrated book, which gives a thorough account of quantum mechanics as an emergent theory that describes statistical ensembles of quantum systems having accurate individualized descriptions in terms of hidden variables. Despite von Neumann's "impossibility proof," it is a nonlocal hidden variables theory that reproduces exactly within its domain all the predictions of standard quantum mechanics. The additional feature of the theory is the quantum potential, a nonclassical nonlocal field that permeates space and guides atomic particles along precise trajectories.

Another theory that describes standard quantum mechanics as an emergent theory is the consistent histories approach detailed in Griffiths (2001) *Consistent Quantum Theory.* Griffiths is the originator of this interpretation, which constructs a sample space of possible "histories" for a quantum system and then describes an appropriate calculus for assigning a probability measure to the space. This allows the author to reproduce the predictions of standard quantum mechanics without including measurement as a fundamental ingredient of the theory and at the same time to resolve many of the famous paradoxes of the standard approach.

Adler (2004) *Quantum Theory as an Emergent Phenomenon* details a third nonstandard approach, asserting that quantum mechanics is not a complete theory. Its premise is that quantum mechanics emerges as the statistical mechanics of matrix models with a global unitary invariance. To say a bit more, the matrix models are analogous to classical Lagrangian and Hamiltonian dynamics but here the phase space variables are noncommutative, and quantum mechanics emerges as the statistical mechanics of these models.

Our final reference in this section is Aharonov and Rohrlich (2005) *Quantum Paradoxes: Quantum Theory for the Perplexed*, which takes the novel approach of beginning each chapter with a paradox that motivates the rest of the chapter. Quantum mechanics is developed as a remedy to paradoxes, whose resolutions help the reader understand the rather unintuitive quantum calculus.

Mathematically Oriented Approaches to Quantum Mechanics

All the references in this section offer sound mathematical developments of quantum theory and are highly recommended by this author for attaining mathematical expertise in differing aspects of the subject.

John von Neumann (1953, 1996) *Mathematical Foundations of Quantum Mechanics* is an English translation of the original 1931 German text that first placed quantum mechanics on a sound mathematical footing. Revolutionary in its time, this work employs the spectral theory of self-adjoint operators, primarily worked out by von Neumann himself as well as the American mathematician Stone, to bring rigor to the theory. Its study is profitable even today for any student interested in the foundations of the subject.

Hall (2013) *Quantum Theory for Mathematicians* is a recent addition to the genre that carefully develops the mathematics of Hilbert spaces and the spectral theorem, as well as that of group and algebra representations, as needed for the precise development of nonrelativistic quantum mechanics.

Takhtajan (2008) *Quantum Mechanics for Mathematicians* assumes a basic knowledge of both the spectral theory of unbounded operators and representation theory and offers a comprehensive and precisely presented mathematical exposition of the subject.

Isham (1995) *Lectures on Quantum Theory: Mathematical and Structural Foundations* is a beautifully written short treatise that presupposes a lower level of mathematical sophistication than that expected of readers of the three books of the preceding paragraphs. Isham avoids the mathematical subtleties of infinite-dimensional spaces by presenting a thorough treatment of the finite-dimensional theory with the goal of illuminating the structural foundations of modern quantum theory.

Gustafson and Sigal (2011) *Mathematical Concepts of Quantum Mechanics* gives a streamlined introduction to quantum mechanics and makes its way swiftly to topics of current interest in which mathematics plays an important role.

Williams (2003) *Topics in Quantum Mechanics* offers a certain mathematical clarity, to particular topics, lacking in many expositions. A noteworthy, unique, and very interesting feature is Williams's use of relatively recent developments in special function theory (the Nikiforov–Uvarov theory of generalized hypergeometric differential equations) to solve the Schrödinger equation with various potentials and obtain energy eigenvalues from a single unified viewpoint.

There are approaches to rigorous quantum mechanics that provide alternatives to the von Neumann version with its use of the spectral theorem. We mention two of these: the first is based on the quantum logic of projection operators and is detailed in Cohen (1989) *An Introduction to Hilbert Space and Quantum Logic*, and the second is based on Weyl's C^*-algebra of operators and is detailed in Strocchi (2008) *An Introduction to the Mathematical Structure of Quantum Mechanics: A Short Course for Mathematicians.* The Cohen book is dual purpose in that it presents a short course on Hilbert spaces as it develops the quantum logic of projection operators. Strocchi presents a sophisticated development of the C^*-algebra approach and develops a dictionary to translate from that to the standard approach.

Singer (2005) *Linearity, Symmetry, and Prediction in the Hydrogen Atom* is a beautifully developed text at the senior undergraduate level whose purpose seems to be to demonstrate how naturally physics and mathematics blend to complement one another. It develops the unitary representation theory of the Lie groups SO(3), SU(2), and SO(4) in the context of understanding the quantum mechanical treatment of the hydrogen atom, and in doing so presents a rigorous treatment valuable to its intended audience of students of physics and chemistry as well as of mathematics. In the book, Singer provides a valuable service to the student by including an English translation of Fock's 1935 article in which he exposes the hidden symmetry of the hydrogen atom.

Hannabuss (1997) *An Introduction to Quantum Theory* is an engaging text that presents a rigorous treatment of quantum mechanics at the upper division undergraduate level. Like Isham (1995) it also avoids the mathematical subtleties of infinite-dimensional Hilbert spaces; nevertheless, its choice of topics includes what would be considered rather advanced material in a physics curriculum, and the mathematical requirements become more significant as the book progresses, touching upon graduate level algebra. This would be my recommendation to any student for where to start their study of rigorous quantum mechanics.

Finally, Sternberg (1995) *Group Theory and Physics* is a highly original and exceptionally well written and arranged book that develops quantum mechanics through the lens of group theory, especially representation theory. The ease with which Sternberg mixes the facts of the physical world with sophisticated mathematics and uses the mathematics to provide rigorous explanations and perceptive insight is impressive. He flows seamlessly from a physical description of a problem to a development of the rigorous mathematics needed to attack that problem and then back to the problem with a beautiful application. Included as an appendix to the book is a rather wonderful essay on the history of nineteenth century spectroscopy. It is amazingly interesting that the seemingly pedestrian act of measuring the characteristic spectral lines of nature's elements leads invariably to an explanation in which group representation theory takes a central role. Were I marooned on a desert island and given access to a single book on quantum mechanics of my choice, this would be it.

Quantum Field Theory

Books on quantum field theory are notoriously opaque to the mathematician. It is difficult for the mathematician even to figure out the definitions of the mathematical terms employed by the physicists, and the theory abounds with unrigorous arguments, ad hoc algorithms for quantum calculations, seemingly arbitrary constructs that are rather suspect upon close examination, and reasoning by analogy; however, the resulting field theory calculus works to produce results that agree with experiment to a degree unattained by any other science and so demands our attention. Despite the difficulty for the mathematician, there are four physically oriented texts written by physicists that I have found valuable in trying to get a handle on the subject.

The two-volume set Aitchison and Hey (2003a, b) *Gauge Theories in Particle Physics, Volumes I and II* begins with a very good introduction to relativistic quantum mechanics and develops field theory with a strong gauge-theoretic emphasis. Its introductory survey on the standard model, gauge theories, and the Dirac equation alone is worth the price of the first volume.

Itzykson and Zuber (2005) *Quantum Field Theory* is one of the classic references that has a complete development of the Dirac equation and its consequences, followed by a detailed description of field quantization with many applications.

Zee (2003) *Quantum Field Theory in a Nutshell* offers a user-friendly introduction to the topic that might serve as a warm-up for more advanced treatments. Whereas the two preceding references are written from the perspective of particle theorists, this one is from that of a condensed-matter physicist. It is a collection of very short chapters, each detailing one aspect of the subject.

Finally, no list of references to quantum field theory would be complete without the mention of the three-volume set Weinberg (1995, 1996, 2000) *The Quantum Theory of Fields, Volumes I, II, III.* In his typically distinct style Weinberg offers a unique perspective, building the theory with representations of the Lorentz group as the starting point. The introductory chapter of the first volume presents an engaging historical account of the birth and early development of quantum field theory.

As much as I like the four references described above, the detailed discussion of field quantization therein remains rather mysterious to this mathematician. The three references below may to some extent ease the anxiety of anyone of a similar mathematical disposition.

Paul Teller (1995) *An Interpretive Introduction to Quantum Field Theory* presents only the bare basic ideas of quantum field theory but with the precision of language one expects from a philosopher. Written by the son of the physicist Edward Teller, the book presents a very careful construction of Fock space, the playing field of the theory, and describes in precise language what the physicists are doing when they quantize fields. It is short on mathematical detail and neglects the formal tools of calculation in favor of a thorough discussion that helps the reader to think correctly about the theory and to understand the physical reasoning behind field quantization.

After Teller, some rigorous mathematics is in order, ably supplied by the next two references. A thorough and rigorous mathematical explanation appears in Folland (2008) *Quantum Field Theory: A Tourist Guide for Mathematicians.* I cannot praise this book highly enough. Folland has done mathematics a huge service in unravelling what physicists have done in quantizing fields; he presents a mathematically sound exposition of the theory as far as is possible, and then pushes further into unrigorous parts of the theory with beautifully written explanations and plausibility arguments.

Equally impressive in its mathematical explanation of quantum field theory is de Faria and de Melo (2010) *Mathematical Aspects of Quantum Field Theory*, which presents a rigorous construction of quantized free fields after a thoroughly sound and complete description of quantum mechanics proper. The theory of interacting fields, where no rigorous treatment is known, is handled by the perturbative approach of Tomonaga, Schwinger, Feynman, and Dyson but is presented in a way more suited for mathematicians than physicists. One strength of the book is in its use of modern geometry in describing the theory. The mathematical community owes a debt of gratitude to the authors of these latter two texts for the great pedagogical aides they have provided for the mathematician struggling to understand this theory.

History and Interpretation of Quantum Mechanics

Jammer (1974) *The Philosophy of Quantum Mechanics: The Interpretations of Quantum Mechanics in Historical Perspective* is a standard work on the interpretations of quantum mechanics and their philosophical implications and interrelations that were current up to the year of its publication. The point of view of this study is historical and its goal is a critical analysis of all the major interpretive proposals. The dominant interpretation of quantum mechanics is the Copenhagen interpretation of Bohr, Born, Heisenberg, and others, and it is used routinely in physics textbooks to provide meaning to the quantum mathematical machinery. However, it is criticized today more than ever, and physicists generally are less committed to it than were those of previous generations. John Bell's analyses of the EPR paradox, the von Neumann impossibility proof of hidden variables, and the mechanics of Bohm and de Broglie have done more to overthrow the dominance of the Copenhagen interpretation than anything else. Beller (1999) *Quantum Dialogue: The Making of a Revolution* is a critical-historical study of the evolution and acceptance of the Copenhagen interpretation that challenges the coherence of the interpretation. This scholarly work demonstrates how contingent was the acceptance of the dominant interpretation on historical accident, rhetorical skill, forceful personalities, scientific ambition, and philosophical beliefs.

Wick (1995) *The Infamous Boundary: Seven Decades of Controversy in Quantum Physics* recounts the major controversial interpretive features of quantum mechanics in an engaging historical retelling of the development of the theory. Wick's use of biographical vignettes of the principle players is especially effective in capturing the interest of the reader. Whitaker (2006) *Einstein, Bohr, and the Quantum Dilemma* thrashes out the arguments for and against the status of quantum mechanics as a complete theory through the historical lens of the famous Bohr–Einstein debates of the 1930s. This book reviews the physical arguments on both sides, updating them with the insights of Bohm, Bell, and others, and in doing so penetrates deeply into the conceptual problems of quantum theory.

Laloë (2012) *Do We Really Understand Quantum Mechanics?* is a recent entry in this genre that surveys current understanding of the meaning of quantum mechanics and compares and contrasts the various interpretations. It covers not only those covered by Jammer (1974) but includes in its discussion the more recent entries into the field, for example, various modal interpretations, continuous spontaneous localizations (GRW model), the transactional interpretation, and the consistent histories approach as well as the impact of quantum computing on the conceptual basis of the theory.

It no an exaggeration to say that since the 1960s John Bell's work has had the greatest impact on the critical analysis of the conceptual foundations of quantum mechanics. Though Einstein and Schrödinger had pointed out the nonlocality of quantum mechanics and Bohr had responded with the idea of the nonseparability of quantum systems, it was not until Bell's work that physicists really began to appreciate the radical nature of these ideas and how nonlocality and nonseparability are quantum mechanics's most profound break with classical mechanics. Bell's assertion that locality is subject to experimental falsification came as a surprise to the physics community, and the Aspect experiments of the 1980s that ruled out locality and affirmed that nature is nonlocal led to a reexamination of the conceptual basis of the theory by many researchers. J.S. Bell (1987) *Speakable and Unspeakable in Quantum Mechanics* is a compilation of all Bell's published and unpublished papers on the conceptual basis of the subject. Cushing and McMullin (1989) *Philosophical Consequences of Quantum Theory: Reflections on Bell's Theorem* is a collection of papers by philosophers and physicists who wrestled with the meaning and implications of Bell's insights and the experimental verification of nonlocality. Baggott (1992) *The Meaning of Quantum Theory* presents a historically based, detailed, and lucid account at the advanced undergraduate level of the development of quantum theory, the Copenhagen interpretation, the Bohr–Einstein debates, the EPR paradox, Bell's theorem, locality and the Aspect experiments, and various interpretations.

The two references following are sophisticated philosophical treatises that concentrate attention on incompleteness, nonlocality, and measurement and are written by philosophers with a clear and deep understanding of quantum mechanics, including its mathematical machinery. Both are philosophical studies that are precise in their explications and claims. After carefully developing the formalism of quantum mechanics, Redhead (1987) *Incompleteness, Nonlocality, and Realism: A Prolegomenon to the Philosophy of Quantum Mechanics* explores the question of its completeness in the context of the EPR paradox, Bell's theorem, and the Kochen–Specker theorem. Dickson (1998) *Quantum Chance and Non-Locality: Probability and Non-Locality in the Interpretations of Quantum Mechanics* gives a careful examination of determinism and nonlocality, concentrating on the measurement problem. The nonlocality of the physical world poses problems for physicists, since how to harmonize nonlocality with relativity is in no way obvious.

The work of yet another philosopher merits mention in the context of Bell's theorem, namely that of Tim Maudlin, who examines the incompatibility of nonlocality with relativity in Maudlin (2011) *Quantum Non-Locality and Relativity: Metaphysical Intimations of Modern*

Physics, recognized as "the premier philosophical study of Bell's theorem and its implication for the relativistic account of space and time."[6] To close this section I mention a nonphilosophical work that presents a purely historical account of the development of the idea of spin in quantum mechanics. Detailed in its mathematical explanations, Tomonaga (1997) *The Story of Spin*[7] presents an enlightening historical study of the main protagonists and their attempts to understand this nonclassical, purely quantum mechanical, attribute of elementary particles.

History and Interpretation of Quantum Field Theory

The philosophical foundations and implications of both quantum mechanics and special relativity have been studied rather thoroughly since the advent of these respective theories, and have stimulated much discussion and reflection. In contrast, a philosophical evaluation of the marriage of these two foundational theories in the form of quantum field theory has been largely ignored by the scientific and philosophical communities. Auyang (1995) *How is Quantum Field Theory Possible?* aims to remedy this with a philosophical analysis of quantum field theory.

Turning to the history of the field, Schweber (1994) *QED and the Men Who Made It: Dyson, Feynman, Schwinger, and Tomonaga* is a masterful recounting of the history and development of quantum electrodynamics by a first rate historian of science who is also a fine theoretical physicist. It is a detailed examination of the history of the subject, through the lens of its chief architects, that does not shy away from mathematics just because its chief concern is historical. Kaiser (2005) *Drawing Theories Apart: The Dispersion of Feynman Diagrams in Postwar Physics* is a really interesting study of how Feynman's personal diagrams for the calculation of amplitudes evolved in the hands of the physics community to become the indispensable tools that they are today. It is a fascinating study in the sociology of science that showcases how differently physicists understood and used Feynman's diagrams during this evolution. Pickering (1984) *Constructing Quarks: A Sociological History of Particle Physics* presents yet another fascinating study in the sociology of science by examining the conceptual development of particle physics in the context of recounting the history of the standard model of particle interactions.

[6] From the back cover of the 2011 third edition.
[7] A translation by Tekeshi Oka of the 1974 Japanese original.

Mathematics

There are almost too many good books on the mathematics that finds great use in quantum mechanics to narrow the focus to just a few for a reading guide like this one. Nonetheless, I will mention here some that have been found especially useful in my studies. I begin with three older classics that are rarely studied today, but in which I have found good and complete explanations for much of the classical analysis that is used in the quantum mechanical analysis of real systems.

The first of these, Whittaker and Watson (fourth edition 1958) *A Course of Modern Analysis*, is a bible of classical analysis whose range of discourse is related ably by its subtitle, *An Introduction to the General Theory of Infinite Processes and of Analytic Functions; With an Account of the Principal Transcendental Functions.* It treats in great detail all the classical special functions of mathematical physics. More modern treatments that are encyclopedic in nature may be found in Andrews, Askey, and Roy (1999) *Special Functions* and in Beals and Wong (2010) *Special Functions: A Graduate Text.*

The second classic is Jeffreys and Jeffreys (1946) *Methods of Mathematical Physics*, intended "to provide an account of those parts of pure mathematics that are frequently used in modern physics."[8] There is much to appreciate in this book on classical analysis as applied to problems in physics. The emphasis in the text falls decidedly on the rigorous development of the analytic methods that are used in physics, so that the mathematics takes precedence over the physics that motivates it.

This attitude is reversed somewhat in the third of the classic texts, Morse and Feshbach (1953a, b) *Methods of Theoretical Physics*, where the physics takes center stage in motivating and directing the presentation of the mathematics. The emphasis is on developing just enough of the mathematics to analyze various physical systems and theories. The Morse and Feshbach work actually appears as two volumes, I and II, and, in 1978 pages, pretty much covers every bit of classical mathematics one needs for the whole of physics.

The (hypothetical) student who masters the classics Whittaker and Watson, Jeffreys and Jeffreys, and Morse and Feshbach will know more classical analysis and how it applies to physics than just about any practicing physicist or mathematician. Nonetheless, he or she still will be deficient in much of the modern analytic machinery that has found such exciting and effective currency in the sound articulation of quantum

[8] Jeffreys and Jeffreys (1946) p. v.

mechanics. This is remedied first by a study of functional analysis, a vast area of modern analysis. The parts of functional analysis needed for quantum mechanics center around the various spectral theorems for operators and unitary representations. These are beautifully covered in another classic text, Riesz and Sz.-Nagy (1990) *Functional Analysis.*

I have a weakness for the classics and consider that the student will not find better treatments than mentioned in these, but it is good to have a variety that includes expositions at a more modest level as well as treatments from a modern vantage and from a variety of viewpoints.

Fano (1971) *Mathematical Methods of Quantum Mechanics* presents a gentler introduction to the functional analysis useful in quantum mechanics, prefaced by a nicely designed course in linear spaces, measure theory, and Hilbert spaces. It also includes as a capstone a final brief chapter on axiomatic quantum mechanics.

Lax (2002) *Functional Analysis* is a modern treatment from one of the foremost applied mathematicians of our age, and Rudin (1991) *Functional Analysis* is a modern treatment from one of the finest expositors of pure mathematics of the last half century.

Zeidler (1995) *Applied Functional Analysis: Applications to Mathematical Physics* gives a rather unique presentation of the functional analysis of quantum mechanics. Applications-oriented in its telling of rigorous mathematics, Zeidler offers a valuable bilingual dictionary that translates between the language of mathematicians and that of physicists. This mathematician has found this dictionary very helpful in getting a handle on the mathematical liberties taken by the physicists.

Strichartz (2003) *A Guide to Distribution Theory and Fourier Transforms* is a highly accessible account of the modern theory of distributions and the Fourier transform. The book is remarkable for its clarity and sufficient to the needs of the student of quantum mechanics. Blanchard and Brüning (2003) *Mathematical Methods in Physics: Distributions, Hilbert Space Operators, and Variational Methods* gives a detailed account of the subjects of its subtitle and is one of the more complete accounts of the modern analytic methods needed for a sound treatment of quantum mechanics. Finally, Teschl (2009) *Mathematical Methods in Quantum Mechanics: With Applications to Schrödinger Operators* presents a careful development of the spectral theory of unbounded operators as its central objective in Part 1 and applies this to understand the technical mathematics of Schrödinger operators in Part 2.

Special mention belongs to the four-volume set entitled *Methods of Modern Mathematical Physics* by the mathematical physicists Michael

Reed and Barry Simon. The first volume, Reed and Simon (1980) *Functional Analysis*, is one of the best available expositions of functional analysis as it is used in modern physics. The subject matter of the remaining volumes is indicated by their titles: (1975) *Volume II: Fourier Analysis, Self-Adjointness*; (1979) *Volume III: Scattering Theory*; and (1978) *Volume IV: Analysis of Operators.* Taken together, these represent a definitive exposition and thoroughgoing analysis of the operators of quantum mechanics, the Schrödinger equation, and its spectral analysis. They are not for the faint-hearted as the level of detail and mathematical sophistication is quite high. A final entry in this guide in the subject of modern analysis is Schechter (1981) *Operator Methods in Quantum Mechanics.* This nicely written work presents operator theory in the context of a single quantum particle moving along a straight line. Its style is to ask physical questions of this one-particle system and then to develop the powerful mathematical methods that help answer the questions. The student can learn a lot of modern hard core analysis by a careful study of this text.

The articulation of a mathematically honest version of quantum mechanics is abetted by the use of modern algebra and geometry. The algebra that finds great present currency centers around those topics that have strong geometric content – Lie algebras and Lie groups as well as Clifford algebras and spin groups. The more sophisticated applications necessarily involve a good understanding of topology, manifold theory, and modern differential geometry. Curtis (1984) *Matrix Groups* is a beautifully presented introduction to the theory of Lie groups via a concrete study of the classical matrix groups. After absorbing this elementary material, the student who wants more about Lie groups can do no better than to study Warner (1983) *Foundations of Differentiable Manifolds and Lie Groups.* It has the added bonus of presenting a course in differentiable manifolds, including differential forms and integration, as well as de Rham theory and Hodge theory.

Another option for the study of differentiable manifolds and Lie groups with the inclusion of a very good course on Riemannian geometry is Boothby (1986) *An Introduction to Differentiable Manifolds and Riemannian Geometry.* A comprehensive treatment of Lie groups and their representations is given in Knapp (2002) *Lie Groups Beyond an Introduction.* The physicist Mikio Nakahara has written a text, Nakahara (2003) *Geometry, Topology, and Physics*, that introduces modern geometry and topology to physicists, and I do think that mathematics students will find much to appreciate here.

Curtis (1984) includes an elementary introduction to Clifford algebras and spin groups, which can be reinforced by a reading of the more sophisticated treatment of these topics found in Chapter I of Lawson and Michelsohn (1989) *Spin Geometry.* Two good follow-up books on Clifford algebras and spin groups are Benn and Tucker (1987) *An Introduction to Spinors and Geometry with Applications to Physics* and Ablamowicz and Sobczyk (2004) *Lectures on Clifford (Geometric) Algebras and Applications.*

For a readable introduction to modern mathematical physics, Geroch (1985) *Mathematical Physics* covers a broad range of subjects, including vector spaces, algebras and their representations, algebraic topology, measure theory and distributions, and the spectral theory of operators. A very good elementary text that covers basic topology, simplicial complexes and manifolds, Hodge and de Rahm theory, and elementary Riemannian geometry is Singer and Thorpe (1967) *Lecture Notes on Elementary Topology and Geometry.* Fifty-three years after its publication, I still consider Singer and Thorpe (1967) to be unsurpassed in its telling of the topics found between its covers. One of my favorite books is the rather unique Frankel (2011) *The Geometry of Physics: An Introduction.* The range of physical topics, especially those of quantum mechanics, is impressive, and equally so is Frankel's applications of modern differential geometry and topology to produce a deeper understanding of physics than is possible by classical analytic approaches.

To finish off this section, the really well written Gelfand and Fomin (1963) *Calculus of Variations* presents a thorough, rigorous study of the topic of its title. A more recent addition to the literature that the student will find useful is van Brunt (2004) *The Calculus of Variations*, which includes chapters on Lagrangian, Hamiltonian, and Jacobian mechanics as well as Noether's Theorem. Finally, Byron and Fuller (1969, 1970) *Mathematics of Classical and Quantum Physics* deserves inclusion on this list of references. Though it is a book intended for graduate students in physics, its level of rigor is comparable with that of mathematics texts; it covers a broad range of topics in modern applied mathematics and its applications to physics.

Physics

Any serious student of quantum mechanics will need some background in general physics and, eventually, may want to obtain expertise in the basic graduate physics curriculum. Besides courses in quantum mechanics, that basic curriculum consists of courses in classical mechanics, electromagnetic theory, statistical mechanics, and the theory of relativity. From these courses the student specializes in a variety of disciplines – condensed matter physics, optics, particle theory, nuclear physics, gravitational physics, materials science, etc. To close this list of references for further study, I will list texts that I have found most beneficial for my edification on these basic topics. Goldstein (1980) *Classical Mechanics* sets the standard for expositions of classical mechanics. It was updated in Goldstein, Poole, and Safko (2002) *Classical Mechanics*, which included new material while maintaining Goldstein's conciseness of language and high standard of prose. I find the presentation of Lagrangian, Hamiltonian, and Jacobian mechanics particularly well written. A very quick and skillfully clear introduction to generalized coordinates and the mechanics of Lagrange and Hamilton appears in the very short Byerly (1944) *An Introduction to the Use of Generalized Coordinates in Mechanics and Physics.* A study of the variational principles of mechanics, written in a lovely old-fashioned style of prose and which does not neglect the philosophical issues surrounding the principles is found in Lanczos (1970) *The Variational Principles of Mechanics.* This book deserves more study than it has engendered, though it might be somewhat difficult to approach by today's students, bereft as they are of a general philosophical education.

Two further recommendations are Greiner (2010) *Classical Mechanics: Systems of Particles and Hamiltonian Dynamics*, which offers clear explanations and a wealth of examples and exercises, many of which are worked out in the text, and Scheck (2010) *Mechanics: From Newton's Laws to Deterministic Chaos*, which emphasizes geometry and includes a very nice introduction to mechanics on symplectic manifolds.

For the classical theory of electromagnetism, Jackson (1999) *Classical Electrodynamics* is the unsurpassed standard for the English-speaking world. Two dual-purpose texts deserve mention here. It is fortunate that we have the English translation of the Russian classic text Landau and Lifshitz (1975) *The Classical Theory of Fields*, the first half of which offers a concise but detailed course in electrodynamics that is elegantly presented; this is followed by a course on the gravitational field. Low (1997) *Classical Field Theory: Electromagnetism and Gravitation* offers another dual-purpose text in the two topics of the subtitle that, though perhaps a bit easier for the student than Landau and Lifshitz, is nonetheless quite complete and self-contained.

There are several nice treatments of statistical mechanics, including Pathria (1996) *Statistical Mechanics*, Chandler (1987) *Introduction to Modern Statistical Mechanics*, McQuarrie (2000) *Statistical Mechanics*, and Phillies (2000) *Elementary Lectures in Statistical Physics.* The last of these is an elementary introduction at the graduate level of what every physics student should know about statistical mechanics. While not comprehensive, it treats the most important topics in the discipline carefully and thoroughly. At the other end of the spectrum is Feynman (1998) *Statistical Mechanics: A Set of Lectures*, an intensive work on quantum statistical mechanics that makes high demands of the student.

Finally, for the theory of relativity, one would be hard pressed to name a better resource than Rindler (2006) *Relativity: Special, General, and Cosmological.* While there are many texts that are more mathematically oriented, there are few that compare favorably with this one in terms of its clarity of explanation, deriving the mathematical form of relativity from physical reasoning, physical observation, physical data, and physical principles. For a swift introduction to the special theory only, and from the physical point of view, Rindler (1991) *Introduction to Special Relativity* is hard to beat. For a modern, global, mathematically intense presentation of the general theory, one might look to Grøn and Hervik (2007) *Einstein's General Theory of Relativity: With Modern Applications in Cosmology.* Perhaps the best advice to a mathematically and philosophically astute student for learning the general theory is to read Rindler (2006) concurrently with the excellent O'Neill (1983) *Semi-Riemannian Geometry With Applications to Relativity.* The former gives deep physical insights as it develops the coordinate point of view while the latter builds the mathematical theory of semi-Riemannian geometry and applies it to give a coordinate-free, global description of Einstein's great accomplishment.

Advice to the Student

Realizing that even the most avid of students must start somewhere, I think it wise to close this essay with a pared down listing of references for the novice who has had little exposure to quantum mechanics. Besides perusing the book in hand, where is a talented mathematics student to begin? My advice centers around seven of the the 99 or so references covered in this guide. I can think of no better resource to serve as an anchor for the student than Hannabuss (1997), supplemented by Peeples

(1992) and Isham (1995). Hannabuss is a mathematician whose expertise is in mathematical physics, and he has given in his text a beautiful introduction to standard quantum mechanics that hits all the right notes, balancing nicely physical explanations with clear and concise rigorous mathematical explanations and derivations. The level is perfect for the student who has been trained in pure mathematics in an undergraduate program at a good US university and has followed this with a taste of mathematics in some first-year courses at graduate level. Peebles is a theoretical physicist and cosmologist whose text presents a streamlined view of the theory from the point of view of a practitioner of the science. What it lacks in mathematical rigor it makes up for in its clear physical explanations and appropriate choice of topics.

Isham is a theoretical physicist who has a strong grasp of pure mathematics. His book is a student-oriented text that attempts to explain to the physics student the subtleties of some of the mathematics of quantum mechanics. These latter two texts serve as a more physical counterweight to the more mathematical Hannabuss. To prepare the student for the representation theory that arises in quantum mechanics, Singer (2005) would be a good book to study concurrently with that of Hannabuss. The mathematical level is undergraduate, but the mathematical topics are graduate level. I strongly urge the student to supplement his or her mathematical studies with a text that stresses the big ideas behind the theory and delves into the controversies surrounding the various interpretations of the quantum calculus. To fulfill this urge and bring a clearer understanding of the meaning behind quantum mechanics, the lucidly written Baggott (1992) is hard to beat.

All five of these references avoid von Neumann's treatment of quantum mechanics, which uses the spectral theory of operators to make the theory mathematically sound. At some point the student will need to obtain a good grasp of functional analysis and will need to become acquainted with the basics of distribution theory and a rigorous treatment of the Fourier transform. These goals can be accomplished with a study of Zeidler (1995) for functional analysis and of Strichartz (1994) for distributions and the Fourier transform.

A mastery of the material in the seven references of this section would provide a strong foundation for further studies in quantum mechanics. To continue feeding his or her quantum habit, the student who has mastered these books may wish to branch out to the more sophisticated treatments of standard quantum mechanics that utilize spectral theory, representation theory, and C^*-algebras, and to bring special relativity into the mix with an exploration of the relativistic theory of particles and fields, including electromagnetic theory using gauge fields. This would require a study of deeper, higher-level, mathematics. The remaining references should be an aid in guiding the student's study in this endeavor.

I wish the reader the same excitement and pleasure I have experienced over the years in learning the workings of nature at the micro-level. May he or she be inspired to learn more deeply the mathematics needed to make quantum theory rigorous and to appreciate more profoundly the interconnectedness of pure mathematics and theoretical physics as studies progress.

And finally, may the student never be satisfied! – always curious and eager to understand more about the quantum world.

Bibliography

Abłamowicz, Rafal, and Sobczyk, Garret. 2004. *Lectures on Clifford (Geometric) Algebras and Application.* Birkhäuser.

Abraham, Ralph, and Marsden, Jerrold E. 1978. *Foundations of Mechanics.* 2nd edn. Benjamin/Cummings.

Adler, Stephen L. 2004. *Quantum Theory as an Emergent Phenomenon: The Statistical Mechanics of Matrix Models as the Precursor of Quantum Field Theory.* Cambridge University Press.

Aharonov, Yakir, and Rohrlich, Daniel. 2005. *Quantum Paradoxes: Quantum Theory for the Perplexed.* Wiley-VCH.

Aitchison, I.J.R., and Hey, A.J.G. 1989. *Gauge Theories in Particle Physics.* 2nd edn. Graduate Student Series in Physics. Institute of Physics Publishing.

Aitchison, I.J.R., and Hey, A.J.G. 2003a. *Gauge Theories in Particle Physics, Volume I: From Relativistic Quantum Mechanics to QED.* 3rd edn. Graduate Student Series in Physics. Institute of Physics Publishing.

Aitchison, I.J.R., and Hey, A.J.G. 2003b. *Gauge Theories in Particle Physics, Volume II: QCD and the Electroweak Theory.* 3rd edn. Graduate Student Series in Physics. Institute of Physics Publishing.

Akhiezer, N.I., and Glazman, I.M. 1993. *Theory of Linear Operators in Hilbert Space.* Reprint of 1961 Book 1 and 1963 Book 2, Frederick Unger Editions. Dover.

Albeverio, S., and Brzeźniak, Z. 1993. Finite dimensional approximation approach to oscillatory integrals and stationary phase in infinite dimensions. *J. Functional Anal.*, **113**, 177–244.

Andrews, George E., Askey, Richard, and Roy, Ranjan. 1999. *Special Functions.* Encyclopedia of Mathematics and its Applications, vol. 71. Cambridge University Press.

Artin, Emil. 1957. *Geometric Algebra.* Interscience.

Aspect, Alain, Grangier, Philippe, and Roger, Gérard. 1982. Experimental realization of Einstein–Podolsky–Rosen–Bohm Gedankenexperiment: a new violation of Bell's inequalities. *Phys. Rev. Lett.*, **49**(2), 91–94.

Auyang, Sunny Y. 1995. *How is Quantum Field Theory Possible?* Oxford University Press.

Baggott, Jim. 1992. *The Meaning of Quantum Theory.* Oxford Science Publications. Oxford University Press.

Bander, M., and Itzykson, C. 1966a. Group theory and the hydrogen atom (I). *Rev. Mod. Phys.*, **38**(2), 330–345.

Bander, M., and Itzykson, C. 1966b. Group theory and the hydrogen atom (II). *Rev. Mod. Phy.*, **38**(2), 346–358.

Beals, Richard, and Wong, Roderick. 2010. *Special Functions: A Graduate Text.* Cambridge Studies in Advanced Mathematics, vol. 126. Cambridge University Press.

Bell, J.S. 1964. On the Einstein–Podolsky–Rosen paradox. *Physics*, **1**, 195–200.

Bell, J.S. 1987. *Speakable and Unspeakable in Quantum Mechanics.* Cambridge University Press.

Bell, Mary, and Gao, Shan (eds). 2016. *Quantum Nonlocality and Reality: 50 Years of Bell's Theorem.* Cambridge University Press.

Bell, W.W. 1968. *Special Functions for Scientists and Engineers.* Dover.

Beller, Mara. 1999. *Quantum Dialogue: The Making of a Revolution.* Science and Its Conceptual Foundations Series. University of Chicago Press.

Benn, I.M., and Tucker, R.W. 1987. *An Introduction to Spinors and Geometry with Applications to Physics.* Adam Hilger, IOP Publishing.

Blanchard, Philippe, and Bruening, Erwin. 2003. *Mathematical Methods in Physics: Distributions, Hilbert Space Operators, and Variational Methods.* Progress in Mathematical Physics, vol. 26. Birkhäuser.

Bohm, David. 1951. *Quantum Theory.* Prentice Hall.

Boothby, William M. 1986. *An Introduction to Differentiable Manifolds and Riemannian Geometry.* 2nd edn. Pure and Applied Mathematics series, vol. 120. Academic Press.

Bouch, Gabriel, and Bowers, Philip L. 2007. Evaluating Feynman path integrals via the polynomial path family. *Math. Phys. Electronic J.*, **13**, 1–29.

Brunt, Bruce van. 2004. *The Calculus of Variations.* Universitext. Springer.

Bub, Jeffrey. 1997. *Interpreting the Quantum World.* Cambridge University Press.

Byerly, William Elwood. 1944. *An Introduction to the Use of Generalized Coordinates in Mechanics and Physics.* Ginn and Co.

Byron, Frederick W., and Fuller, Robert W. Vol. 1 1969, Vol. 2 1970 (Dover edition 2012). *Mathematics of Classical and Quantum Physics.* Addison-Wesley Series in Advanced Physics. Addison-Wesley.

Cameron, R.H. 1960. A family of integrals serving to connect the Wiener and Feynman integrals. *J. Math. Phys.*, **39**, 126–140.

Cartan, Élie. 1913. Les groupes projectifs qui ne laissent invariante aucune multiplicité plane. *Bull. Soc. Math. France*, **41**, 53–96.

Cartan, Élie. 1966. *The Theory of Spinors.* Dover.

Chandler, David. 1987. *Introduction to Modern Statistical Mechanics.* Oxford University Press.

Cohen, David W. 1989. *An Introduction to Hilbert Space and Quantum Logic.* Problem Books in Mathematics. Springer.

Cordani, Bruno. 2003. *The Kepler Problem: Group Theoretic Aspects, Regularization and Quantization, with Application to the Study of Permutations.* Progress in Mathematical Physics, vol. 29. Birkhäuser Verlag.

Curtis, Charles W. 1963. The four and eight square problem and division algebras. Pages 100–125 of: Albert, A.A. (ed), *Studies in Modern Algebra.* MAA Studies in Mathematics, vol. 2. Mathematical Association of America.

Curtis, M.L. 1984. *Matrix Groups.* 2nd edn. Universitext. Springer.

Cushing, James T., and McMullin, Ernan. 1989. *Philosophical Consequences of Quantum Theory: Reflections on Bell's Theorem.* University of Notre Dame Press.

Derbyshire, John. 2003. *Prime Obsession: Bernhard Riemann and the Greatest Unsolved Problem in Mathematics.* Joseph Henry Press.

Dickson, W. Michael. 1998. *Quantum Chance and Non-Locality: Probability and Non-Locality in the Interpretations of Quantum Mechanics.* Cambridge University Press.

Dirac, P.A.M. 1958 (first edition 1930). *The Principles of Quantum Mechanics.* 4th edn. International Series of Monographs on Physics, vol. 27. Oxford University Press.

Dittrich, W., and Reuter, M. 2001. *Classical and Quantum Dynamics: From Classical Paths to Path Integrals.* 3rd edn. Springer.

Dyson, Freeman. 2009. Birds and frogs. *Notices AMS*, **56**(2), 212–223.

Edmonds, A.R. 1960. *Angular Momentum in Quantum Mechanics.* 2nd edn. Princeton Landmarks in Physics. Princeton University Press.

Einstein, A., Podolsky, B., and Rosen, N. 1935. Can quantum-mechanical description of physical reality be considered complete? *Phys. Rev.*, **47**, 777–780.

Elbaz, Edgard. 1998. *Quantum: the Quantum Theory of Particles, Fields, and Cosmology.* Theoretical and Mathematical Physics. Springer.

Elworthy, D., and Truman, A. 1984. Feynman maps, Cameron–Martin formulae, and anharmonic oscillators. *Ann. Inst. H. Poincaré, Section A*, **41**(2), 115–142.

Fano, Guido. 1971. *Mathematical Methods of Quantum Mechanics.* McGraw-Hill.

Faria, Edson de, and de Melo, Welington. 2010. *Mathematical Aspects of Quantum Field Theory.* Cambridge Studies in Advanced Mathematics, vol. 127. Cambridge University Press.

Farmelo, Graham. 2009. *The Strangest Man: the Hidden Life of Paul Dirac, Mystic of the Atom.* Basic Books.

Felsager, Bjørn. 1998. *Geometry, Particles, and Fields.* Graduate Texts in Contemporary Physics. Springer.

Feynman, Richard. 1998. *Statistical Mechanics: A Set of Lectures.* 2nd edn. Advanced Books Classics. Westview Press.

Feynman, Richard P., and Hibbs, A.R. 1965. *Quantum Mechanics and Path Integrals.* International Series in Pure and Applied Physics. McGraw-Hill.

Fitts, Donald D. 1999. *Principles of Quantum Mechanics as Applied to Chemistry and Chemical Physics.* Cambridge University Press.

Fock, Vladimir A. 1935. Zur Theorie des Wasserstoffatoms. *Z. Phys.*, **98**, 145–154.

Fock, Vladimir A. 1978. *Fundamentals of Quantum Mechanics.* Mir Publishers.

Folland, Gerald. 2008. *Quantum Field Theory: A Tourist Guide for Mathematicians.* Mathematical Surveys and Monographs, vol. 149. American Mathematical Society.

Folland, Gerald. 2010. Speaking with the natives: reflections on mathematical communication. *Not. Amer. Math. Soc.*, **57**, 1121–1124.

Frankel, Theodore. 2011. *The Geometry of Physics: An Introduction.* 3rd edn. Cambridge University Press.

Fulton, William, and Harris, Joe. 1991. *Representation Theory: A First Course.* Graduate Texts in Mathematics, vol. 129. Springer.

Garling, D.J.H. 2011. *Clifford Algebras: An Introduction.* London Mathematical Society Student Texts, vol. 78. Cambridge University Press.

Gelfand, I.M., and Fomin, S.V. 1963 (Dover edition 2000). *Calculus of Variations.* Prentice-Hall.

Geroch, Robert. 1985. *Mathematical Physics.* Chicago Lectures in Physics. University of Chicago Press.

Gluck, H., Warner, F., and Ziller, W. 1986. The geometry of Hopf fibrations. *L' Enseignement Math.*, **32**, 173–198.

Goldstein, Herbert. 1980. *Classical Mechanics.* 2nd edn. Addison-Wesley.

Goldstein, Herbert, Jr., Charles P. Poole, and Safko, John L. 2002. *Classical Mechanics.* 3rd edn. Addison-Wesley.

Greiner, Walter. 2010. *Classical Mechanics: Systems of Particles and Hamiltonian Dynamics.* 2nd edn. Springer.

Griffiths, David J. 1995. *Introduction to Quantum Mechanics.* Prentice Hall.

Griffiths, Robert B. 2001. *Consistent Quantum Theory.* Cambridge University Press.

Grøn, Øyvind, and Hervik, Sigbjorn. 2007. *Einstein's General Theory of Relativity: With Modern Applications in Cosmology.* Springer.

Gustafson, Stephen J., and Sigal, Israel Michael. 2011. *Mathematical Concepts of Quantum Mechanics.* 2nd edn. Universitext. Springer.

Hall, Brian C. 2013. *Quantum Theory for Mathematicians.* Graduate Texts in Mathematics, vol. 267. Springer.

Hall, Brian C. 2015. *Lie Groups, Lie Algebras, and Representations.* Graduate Texts in Mathematics, vol. 222. Springer.

Hannabuss, Keith. 1997. *An Introduction to Quantum Theory.* Oxford Graduate Texts in Mathematics, vol. 1. Oxford University Press.

Harvey, F. Reese. 1990. *Spinors and Calibrations.* Perspectives in Mathematics, vol. 9. Academic Press.

Harvey, R., and H. Lawson, Jr. 1982. Calibrated geometries. *Acta Math.*, **148**, 47–157.

Hassani, Sadri. 1999. *Mathematical Physics: A Modern Introduction to Its Foundations.* Springer.

Helgason, Sigurdur. 2001. *Differential Geometry, Lie Groups, and Symmetric Spaces.* Graduate Studies in Mathematics, vol. 34. American Mathematical Society.

Hladik, Jean. 1999. *Spinors in Physics.* Graduate Texts in Contemporary Physics. Springer.

Hochstadt, Harry. 1971. *The Functions of Mathematical Physics.* Dover.

Holland, Peter R. 1993. *The Quantum Theory of Motion: An Account of the de Broglie–Bohm Causal Interpretation of Quantum Mechanics.* Cambridge University Press.

Humphreys, James E. 1972. *Introduction to Lie Algebras and Representation Theory.* Graduate Texts in Mathematics, vol. 9. Springer.

Isham, C.J. 1995. *Lectures on Quantum Theory: Mathematical and Structural Foundations.* World Scientific.

Itzykson, Claude, and Zuber, Jean-Bernard. 2005 (republication of 1980 McGraw-Hill edition). *Quantum Field Theory.* Dover.

Jackson, Dunham. 1941. *Fourier Series and Orthogonal Polynomials.* The Carus Mathematical Monographs. Mathematical Association of America.

Jackson, John David. 1999. *Classical Electrodynamics.* 3rd edn. Wiley.

Jammer, Max. 1974. *The Philosophy of Quantum Mechanics: The Interpretations of Quantum Mechanics in Historical Perspective.* Wiley.

Jeffreys, Harold, and Jeffreys, Bertha. 1946. *Methods of Mathematical Physics.* Cambridge University Press.

Kaiser, David. 2005. *Drawing Theories Apart: The Dispersion of Feynman Diagrams in Postwar Physics.* University Of Chicago Press.

Knapp, Anthony W. 2002. *Lie Groups Beyond an Introduction.* 2nd edn. Birkhäuser.

Laloë, Franck. 2012. *Do We Really Understand Quantum Mechanics?* Cambridge University Press.

Lanczos, Cornelius. 1970 (Dover edition 1986). *The Variational Principles of Mechanics.* 4th edn. Mathematical Expositions, vol. 4. University of Toronto Press.

Landau, L.D., and Lifshitz, E.M. 1972. *Mechanics and Electrodynamics.* A Shorter Course in Theoretical Physics, vol. 1. Pergamon Press.

Landau, L.D., and Lifshitz, E.M. 1975. *The Classical Theory of Fields.* 4th edn. Theoretical Physics series, vol. 2. Butterworth–Heinemann.

Lawson, H. Blaine, and Michelsohn, Marie-Louise. 1989. *Spin Geometry.* Princeton Mathematical Series, vol. 38. Princeton University Press.

Lax, Peter D. 2002. *Functional Analysis.* Pure and Applied Mathematics, vol. 55. Wiley-Interscience.

Lebedev, N.N. 1972. *Special Functions and Their Applications.* Dover.

Low, Francis E. 1997. *Classical Field Theory: Electromagnetism and Gravitation.* Wiley-Interscience.

Madrid, Rafael de la. 2005. The role of the rigged Hilbert space in Quantum Mechanics. *European J. Phys.*, **26**(2), 287–312.

Maudlin, Tim. 2011. *Quantum Non-Locality and Relativity: Metaphysical Intimations of Modern Physics.* 3rd edn. Wiley–Blackwell.

Maudlin, Tim. 2014. What Bell did. *J. Phys. A*, **47**(42), 424010.

McQuarrie, Donald Allan. 2000. *Statistical Mechanics.* University Science Books.

Melzak, Z.A. 1973. *Companion to Concrete Mathematics: Mathematical Techniques and Various Applications.* Wiley-Interscience.

Melzak, Z.A. 1976. *Mathematical Ideas, Modeling and Applications: Volume II of Companion to Concrete Mathematics.* Wiley-Interscience.

Merzbacher, Eugen. 1998. *Quantum Mechanics.* 3rd edn. Wiley.

Morse, Philip McCord, and Feshbach, Herman. 1953a. *Methods of Theoretical Physics, Part I.* International Series in Pure and Applied Physics. McGraw-Hill.

Morse, Philip McCord, and Feshbach, Herman. 1953b. *Methods of Theoretical Physics, Part II.* International Series in Pure and Applied Physics. McGraw-Hill.

Nahin, Paul J. 2015. *Inside Interesting Integrals.* Springer.

Nakahara, Mikio. 2003. *Geometry, Topology, and Physics.* 2nd edn. Graduate Student Series in Physics. Taylor and Francis.

Neumann, John von. 1953, paperback 1996. *Mathematical Foundations of Quantum Mechanics.* Princeton Landmarks in Mathematics and Physics. Princeton University Press.

Oliver, David. 1994. *The Shaggy Steed of Physics: Mathematical Beauty in the Physical World.* Springer-Verlag.

O'Neill, Barrett. 1983. *Semi-Riemannian Geometry With Applications to Relativity.* Pure and Applied Mathematics, vol. 103. Academic Press.

Parshall, Karen Hunger. 2009. Marshall Stone and the internationalization of the American mathematical research community. *Bull. Amer. Math. Soc.*, **46**, 459–482.

Pathria, R.K. 1996. *Statistical Mechanics.* 2nd edn. Elsevier.

Peebles, Phillip James Edwin. 1992. *Quantum Mechanics.* Princeton University Press.

Petkovšek, Marko, Wilf, Hilbert S., and Zeilberger, Doron. 1996. $A = B$. A.K. Peters.

Phillies, George D.J. 2000. *Elementary Lectures in Statistical Physics.* Graduate Texts in Contemporary Physics. Springer.

Pickering, Andrew. 1984. *Constructing Quarks: A Sociological History of Particle Physics.* University of Chicago Press.

Pollard, Harry. 1976. *Celestial Mechanics.* The Carus Mathematical Monographs, vol. 18. Mathematical Association of America.

Rainville, Earl D. 1960. *Special Functions.* McMillan.

Redhead, Michael. 1987. *Incompleteness, Nonlocality, and Realism: A Prolegomenon to the Philosophy of Quantum Mechanics.* Clarendon Paperbacks. Oxford University Press.

Reed, Michael, and Simon, Barry. 1975. *Methods of Modern Mathematical Physics, Volume II: Fourier Analysis, Self-Adjointness.* Academic Press (imprint of Elsevier).

Reed, Michael, and Simon, Barry. 1978. *Methods of Modern Mathematical Physics, Volume IV: Analysis of Operators.* Academic Press (imprint of Elsevier).

Reed, Michael, and Simon, Barry. 1979. *Methods of Modern Mathematical Physics, Volume III: Scattering Theory.* Academic Press (imprint of Elsevier).

Reed, Michael, and Simon, Barry. 1980. *Methods of Modern Mathematical Physics, Volume I: Functional Analysis.* Academic Press (imprint of Elsevier).

Reid, Constance. 1996. *Hilbert.* Copernicus Press.

Riesz, Frigyes, and Sz.-Nagy, Bela. 1955 (Dover edition 1990). *Functional Analysis.* Frederick Ungar Publishing.

Rindler, Wolfgang. 1991. *Introduction to Special Relativity.* 2nd edn. Oxford University Press.

Rindler, Wolfgang. 2006. *Relativity: Special, General, and Cosmological.* 2nd edn. Oxford University Press.

Rovelli, Carlo. 2001. Quantum spacetime: What do we know? In: Callender, Craig, and Huggett, Nick (eds), *Physics Meets Philosophy at the Planck Scale: Contemporary Theories in Quantum Gravity.* Cambridge University Press.

Rudin, Walter. 1991. *Functional Analysis.* 2nd edn. International Series in Pure and Applied Mathematics. McGraw-Hill.

Sakurai, J.J. 1985. *Modern Quantum Mechanics.* Addison Wesley.

Schechter, Martin. 1981 (Dover edition 2003). *Operator Methods in Quantum Mechanics.* Elsevier.

Scheck, Florian. 2010. *Mechanics: From Newton's Laws to Deterministic Chaos.* 5th edn. Graduate Texts in Physics. Springer.

Schrödinger, Erwin. 1989. *Statistical Thermodynamics.* Dover (Republication of 1952 2nd edition of Cambridge University Press).

Schulman, L.S. 1981. *Techniques and Applications of Path Integration.* John Wiley and Sons.

Schwabl, Franz. 2002. *Statistical Mechanics.* Advanced Texts in Physics. Springer.

Schweber, Silvan S. 1994. *QED and the Men Who Made It: Dyson, Feynman, Schwinger, and Tomonaga.* Princeton Series in Physics. Princeton University Press.

Schwinger, Julian, and Englert, Berthold-Georg. 2001. *Quantum Mechanics: Symbolism of Atomic Measurement.* Springer.

Seaborn, James B. 1991. *Hypergeometric Functions and Their Applications.* Texts in Applied Mathematics, vol. 8. Springer.

Shankar, R. 1994. *Principles of Quantum Mechanics.* 2nd edn. Plenum Press.

Singer, I.M., and Thorpe, J.A. 1967. *Lecture Notes on Elementary Topology and Geometry.* Undergraduate Texts in Mathematics. Springer.

Singer, Stephanie Frank. 2005. *Linearity, Symmetry, and Prediction in the Hydrogen Atom.* Undergraduate Texts in Mathematics. Springer.

Smolin, Lee. 2006. *The Trouble with Physics: The Rise of String Theory, the Fall of Science, and What Comes Next.* Houghton Mifflin Co.

Spiegel, Murray R. 1967. *Theory and Problems of Theoretical Mechanics.* Schaum's Outline Series. Schaum.

Spivak, Michael. 2010. *Physics for Mathematicians: Mechanics I.* Publish or Perish, Inc.

Stein, Elias M. 1970. *Singular Integrals and Differentiability Properties of Functions.* Princeton University Press.

Stein, Elias M., and Weiss, Guido. 1971. *Introduction to Fourier Analysis on Euclidean Spaces.* Princeton University Press.

Sternberg, S. 1995. *Group Theory and Physics.* Cambridge University Press.

Stone, Marshell Harvey. 1932. *Linear Transformations in Hilbert Space and Their Applications to Analysis.* AMS Colloquium Publications, vol. XV. The American Mathematical Society.

Strichartz, Robert S. 2003. *Guide to Distribution Theory and Fourier Transforms.* World Scientific.

Strocchi, F. 2008. *An Introduction to the Mathematical Structure of Quantum Mechanics: A Short Course for Mathematicians.* 2nd edn. Advanced Series in Mathematical Physics. World Scientific.

Takhtajan, Leon A. 2008. *Quantum Mechanics for Mathematicians.* Graduate Studies in Mathematics, vol. 95. American Mathematical Society.

Teller, Paul. 1995. *An Interpretive Introduction to Quantum Field Theory.* Princeton University Press.

Teschl, Gerald. 2009. *Mathematical Methods in Quantum Mechanics: with Applications to Schrödinger Operators.* Graduate Studies in Mathematics, vol. 99. American Mathematical Society.

Thaller, Bernd. 2000. *Visual Quantum Mechanics.* Springer.

Thaller, Bernd. 2005. *Advanced Visual Quantum Mechanics.* Springer.

Tomonaga, Sin-itiro. 1997. *The Story of Spin.* University Of Chicago Press.

Vassili, K. 1999. Complex measures on path space: an introduction to the Feynman path integral applied to the Schrödinger equation. *Methodol. Comput. Appl. Probab.*, **1**(3), 349–365.

Warner, Frank W. 1983. *Foundations of Differentiable Manifolds and Lie Groups.* Graduate Texts in Mathematics, vol. 94. Springer.

Weinberg, Stephen. 1995. *The Quantum Theory of Fields, Volume I.* Cambridge University Press.

Weinberg, Stephen. 1996. *The Quantum Theory of Fields, Volume II.* Cambridge University Press.

Weinberg, Stephen. 2000. *The Quantum Theory of Fields, Volume III.* Cambridge University Press.

Weinberg, Stephen. 2012. *Lectures on Quantum Mechanics.* Cambridge University Press.

Whitaker, Andrew. 2006. *Einstein, Bohr, and the Quantum Dilemma.* 2nd edn. Cambridge University Press.

Whittaker, E.T., and Watson, G.N. 1958. *A Course of Modern Analysis.* 4th edn. Cambridge University Press.

Wick, David. 1995. *The Infamous Boundary: Seven Decades of Controversy in Quantum Physics.* Birkhäuser.

Wigner, Eugene P. 1960. The unreasonable effectiveness of mathematics in the natural sciences. *Commun. Pure Appl. Math.*, **XIII**, 1–14.

Williams, Floyd. 2003. *Topics in Quantum Mechanics.* Progress in Mathematical Physics, vol. 27. Birkhäuser.

Zee, A. 2003. *Quantum Field Theory in a Nutshell.* Princeton University Press.

Zeidler, Eberhard. 1995. *Applied Functional Analysis: Applications to Mathematical Physics.* Applied Mathematical Sciences, vol. 108. Springer.

Index

action
  Hamilton's principle, 82
  principle of stationary action, 82
adjoint equation, 29
  for densely defined operator, 31
  for finite-dimensional space, 30
adjoint operator, 29
alternating map, 384, 389
amplitude, 40
analytic vector, 120, 121
angular momentum
  commutator relations, 172
  coupled, 271, 331, 344
  ladder operators, 173
  operators, 172
  orbital triad, 171
  selection rules, 273
  total, 173
  total orbital, 198
  triad, 172
anticommutation relations for Pauli matrices, 182
anticommutator, 406
antisymmetric
  algebra, 390
  operator, 384, 389
  product, 384, 390
  tensor product, 388
Aspect experiments, 18, 348

Baker–Hausdorff formula, 121, 122
Bell, J.S., 17
  analysis of EPR, 17
  inequality, 350
Bloch sphere, 278, 373
Bohm, 17, 95
Bohmian mechanics, 95, 341
Bohr, 12, 13, 17
Bohr magneton, 297
Bohr radius, 218
Boltzmann distribution, 415
Born, 12, 40
boson, 380
  Fock space for, 395
bosonic
  Fock space, 402
  state algebra, 388
  state space, 386
  state vector, 382
bra-ket notation, 68
bra-space, 68
bra-vector, 68, 72

canonical coordinates, 96, 100
canonical ensemble, 417
canonical transformation, 96, 100
Cauchy determinant, 477
causality, 18
characteristic frequency, 3
Clifford algebra, 301
  categorical definition, 303
  complex, 305
  even and odd subspaces, 306
  explicit construction of $C\ell_{p,q}$, 304
  involutions of, 305
  main symmetries for real algebras, 308
  pin group, 301, 314
  spin group, 301, 314
coherent state, 162
collapse, 11
  of the wave function, 44
commutation relations for Pauli matrices, 181

commutator, 113
complete
  Schauder basis, 43
  set of eigenstates, 43
configuration, 408
  most probable, 411
conservation laws
  for classical mechanics, 91
  for quantum mechanics, 440
convolution, 141
coupling constant, 446
covariant derivative, 446

degree of freedom
  external, 285
  internal, 286
density operator, 361, 366
  finite-dimensional setting, 364
  infinite-dimensional setting, 375, 377
Dirac, 65
  spinor, 323
  theory of the electron, 322
Dirac calculus, 65, 67, 73
Dirac gamma matrices, 313
dispersion, 42
distribution, 68
  tempered, 68, 137
dual correspondence, 29
dual space, 67
  anti-dual space, 67
dynamics
  Dirac picture, 431
  Heisenberg picture, 428
  Schrödinger picture, 422

effective potential, 211
Ehrenfest's theorems, 80, 115, 431
eigenvalue
  degenerate for atomic hydrogen, 234
  generalized, 69
  nondegenerate, 155
eigenvector, generalized, 69
Einstein, 13, 17
Einstein–Podolsky–Rosen (EPR) paradox, 17, 336
ensemble, 361
  continuous, 374
entangled state, 330
Euclidean symmetries, 436
Euler–Lagrange equation, 81, 87
evolution equation
  classical, 90
  quantum, 112, 113
evolution operator, 424
  reference, 431
exchange operator, 381
excited state for harmonic oscillator, 157
expectation value, 26
exterior algebra, 390
exterior product, 384, 389

fermion, 380
  Fock space for, 395
fermionic
  Fock space, 404
  state algebra, 390
  state space, 388
  state vector, 382
Feynman
  axioms for quantum mechanics, 456
  formulation of quantum mechanics, 451
  path integral, 457
  propagator, 457
    constant force, 484, 488
    free particle, 464
    group property, 457
Feynman–Soriau formula, 502
filtration measure, 480
  for general dense family, 497
  for piecewise linear family, 483
  for polynomial family, 485
filtration of path space, 480
fine structure, 234, 237
fine structure constant, 237
Fock space, 380, 395
  antisymmetric or fermionic, 395
  notation, 397, 399
  symmetric or bosonic, 395
Fourier
  inverse transform, 138
  inversion theorem, 138
  transform, 131, 136, 138
  uncertainty principle, 132
Fresnel integral, 466, 468, 470
functional calculus, 55

g-factor
  for the electron, 296
    measured value, 297
  general definition, 295
gauge, 447
gauge principle, 449
gauge transformation, 447
Gaussian integral, 466

complex N-dimensional, 471
complex one-dimensional, 468
Gelfand triples, 67
generalized
coordinate, 83
momentum, 88
velocity, 83
generating function, 159
geometric algebra, 185, 301
GHZ scheme, 355
Green's second identity, 253
Groenwald–van Hove theorem, 115
ground state of harmonic oscillator, 130, 155
group character, 257
group condition, 424
Hamilton, 82
equations of motion, 88, 89
principal function, 105
principle of stationary action, 82
Hamilton–Jacobi
equation, 104
mechanics, 95
Hamiltonian, 4
classical, 89
in Hamilton–Jacobi theory, 100
mechanics, 88
operator, 26
reference, 431
Hamiltonians or quaternions, 185
harmonic oscillator, 2
classical treatment, 4
coherent state, 162
time evolution of, 426
Feynman–Soriau formula, 502
generating function, 159
in Heisenberg picture, 430
quantum treatment, 7, 154
time evolution, 425
harmonic polynomial, 205
Heisenberg, 12
equation, 429
uncertainty principle, 125
interpretation, 127
Hellinger–Toeplitz theorem, 32
Hermite
function, 142
polynomial, 143
hidden symmetry
Fock's treatment, 253
in the hydrogenic system, 243
SO(4) symmetry, 246
$\mathfrak{su}(2) \oplus \mathfrak{su}(2)$ symmetry, 245
Hilbert matrix, 476
Hilbert space, 6, 25
complete orthogonal sum, 393
dual correspondence, 29
orthogonal direct sum, 394
Hopf fibration, 372
hydrogenic potential well, 211, 214
bound state eigenfunctions, 228
ladder operators, 217
momentum space equation, 247
in quaternionic coordinates, 250
unbound electrons, 229
hypergeometric function, 220
classical, 221
confluent, 221
hypergeometric series, 221
infinitesimal generator, 428
inseparable, 18, 338, 341
interaction term, 299, 510
interpretation
completeness hypothesis, 17, 336
Copenhagen, 12, 337
hidden variables, 17, 341
local causality, 348
minimal instrumentalist, 11, 337
positivist, 336
realism, 339, 342, 348
ionization energy, 236
Jacobi identity, 91
for commutators, 114
procedure, 104
Kepler problem, 234
Hamilton–Jacobi approach, 108
ket
complete collection, 72
eigenket, 69
ket-space, 68
ket-vector, 68, 72
Kirkegaard, 14
Kummer's equation, 222
ladder operator
for angular momentum, 173
for hydrogenic potential, 217
Lagrangian, 4, 81
in Hamilton–Jacobi theory, 100
mechanics, 84
Laguerre polynomial
associated, 228

generalized, 224
Lamb shift, 237
Landé g-factor, 295
Laplace–Beltrami operator, 202
Laplace–Runge–Lenz
classical vector, 239
quantized triad, 240
Laplacian, spherical, 201
n-dimensional, 206
Legendre
associated differential equation, 202
equation, 202
functions, 200
polynomial, 202
Legendre transformation, 89
Lie algebra
$\mathfrak{s}^3$, 188
$\mathfrak{su}(2)$, 182
Lie group
$\mathbf{S}^3$, 188
$\mathbf{S}^3 \times \mathbf{S}^3$, 191
SO(3), 188
SO(4), 191
SU(2), 182
local causality, 348
local theory, 348

Mach, 13
magnetic moment
anomalous, 297
intrinsic, 296
operator, 297
magnetic quantum number, 174
Majorana matrices, 312
in terms of Pauli spin matrices, 314
measurement, 14
measurement problem, 16, 46, 133
minimal uncertainty state, 130, 165
minimalist axioms, 25
mixed state, 361, 367
maximally mixed, 374
momentum operator, 27, 32
momentum representation, 70, 146, 438
momentum space equation for the Kepler problem, 247
in quaternionic coordinates, 250

Noether's theorem, 91
quantum version, 440
nonlocality, 18, 338, 348
normed algebra, 187
number operator, 163, 402, 405

observable, 4, 25
compatible, 129
extrinsic, 285
incompatible, 129
intrinsic, 286
operator, 31
annihilation, 7, 131, 157
for angular momentum, 175
in Fock space, 401
commuting operators, 118
creation, 7, 131, 157
for angular momentum, 175
in Fock space, 401
densely defined, 31
density operator, 361, 366
Hamiltonian, 26
intrinsic magnetic moment, 297
ladder, 131, 155
for angular momentum, 173
for the hydrogenic potential, 217
momentum triad, 27
multiplication by a function, 60
number, 163, 402, 405
position triad, 27
self-adjoint, 29
resolvent set of, 58
spectrum, 58
spin, 276
statistical, 377
symmetric, 32
time evolution, 424
total orbital angular momentum, 198
in spherical coordinates, 200
trace, 365
trace class, 376
orbital, 237

partition function, 414, 417
path integral, 457
configuration space version, 492
phase space version, 492
Pauli
derivation of hydrogen energy levels, 239
exclusion principle, 380, 385, 390, 391
Hamiltonian, 299
spin matrices, 181
spinor, 283, 321
perfect correlation, 344
phase factor, 6, 130
phase transformation
global, 442
local, 442

pin group, 301, 314
adjoint action on Clifford algebra, 315
as double cover of orthogonal group, 317
general definition, 315
twisted adjoint action, 315
Plancherel theorem, 141
Planck constant, 4
Pochhammer symbol, 220
Podolsky, 17
Poisson bracket, 90
relation to commutator, 113
position operator, 27, 32
position representation, 70, 146
pre-Hilbert space, 393
probability
amplitude, 9, 40
complex, 454
amplitude density, 454
amplitude for an event, 456
density, 40
projection, 44
projection operator, 53
projection postulate, 24, 45
physicists' version, 50
von Neumann's version, 52
von Neumann–Lüders version, 45
propagator
constant force, 484, 488
forced harmonic oscillator, 510
free particle, 464
harmonic oscillator, 499, 501
proton number, 216
pure state, 361, 367

qbit, 275
quantization, 4, 28
quantum, 381
quantum number
azimuthal, 200
magnetic, 174, 200
orbital, 200
principal, 216, 224
spin, 295
quaternion, 185
conjugate, 185
norm, 185
real and pure, 185
representations of, 185

radial equation for central potential, 211
realism, 339, 342, 348
Redhead, 11
reduction of dimension for central force problem, 210
reference Hamiltonian, 431
representation
global U(1), 442
irreducible, 183, 191
of a Lie algebra, 183
of a Lie group, 191
of $\mathbf{S}^3$, 193
of SO(3), 195
of SU(2), 193
of $\mathfrak{su}(2)$, 184
projective, 281
spin, 320
tensor product of, 332
unitary, 425
resolution of the identity, 52
in the Dirac calculus, 75
resolvent set, 58
rigged Hilbert space, 65, 67
rising factorial, 220
Rodrigues's formula
for Hermite polynomials, 143
for Laguerre polynomials, 226
for Legendre polynomials, 203
Rosen, 17
rotation
active, 280
effect on spin state, 281
group, 188
of wave function, 279
passive, 280
standard representation on $L^2(\mathbb{R}^3)$, 280
rotation representation, 439
Rydberg, 218
constant, 235
formula, 235

Schrödinger equation, 5, 26
derivation, 426
time-independent, 6
Schrödinger evolution, 422
Schwartz space, 136
self-adjoint operator, 29, 31
commuting, 118
in terms of resolutions of the identity, 121
semiclassical state, 162
series of spectral lines for hydrogen, 235
shell, 237
subshell, 238

signature
  of a quadratic form, 302
simple harmonic oscillator
  classical treatment, 3
  Hamilton–Jacobi approach, 106
  quantum treatment, 4
skew symmetry, 390
SO(4) symmetry of hydrogen, 253
special unitary group, 182
spectral decomposition in the Dirac calculus, 75
spectral theorem, 36, 53
spectroscopy
  nomenclature, 236
  notation, 237
spectrum, 44, 58
  complete, 75
  continuous, 59
  discrete, 59
  multiplicity, 59
  point, 59
  Weyl sequence, 62
spherical coordinates, n-dimensional, 206
spherical harmonics, 200, 201
  as solutions for Kepler problem, 256
  in n dimensions, 205
    integral equation for, 255
spin, 276
  coupled, 344
  one-half system, 275
  operator triad, 276
  particle of spin j, 290
  Pauli matrices, 181
  representation, 181, 320
  topology of spin states, 372
spin group, 301, 314
  accidental isomorphisms, 318
  as double cover of special orthogonal group, 317
  general definition, 315
  representation, 320
  spin structure, 320
spin–orbit
  coupling, 287
  interaction, 299
  term, 299
spinor, 283, 320
  Dirac spinor, 323
  harmonics, 299
  in Pauli's theory, 291
  Pauli spinor, 321
  wave equation, 299
spring constant, 3
Stark effect, 237
state
  coherent, 162
  maximally mixed, 374
  mixed, 361, 367
  normalized, 25
  pure, 361, 367
  semiclassical, 162
  state space, 25
    physical state space, 66
  state vector, 9, 25
    nonnormalizable, 51, 71
stationary
  path, 85
  trajectory, 87
  value, 85
statistical mechanics, 408
  canonical ensemble, 417
  fundamental assumption or principle of equal a priori probabilities, 409
  fundamental law, 417
  maximum-term method, 412
  partition function, 414
statistical operator, 377
statistics
  Bose–Einstein, 380, 416
  Fermi–Dirac, 380, 416
  Maxwell–Boltzmann, 415
  spin–statistics theorem, 380
stereographic projection, 248
Stone, 29, 31, 36
Stone's theorem, 428
summability
  Abel summable series, 260
  Cesàro summable series, 260
symmetric
  algebra, 388
  operator, 383, 387
  product, 383, 387
    in the symmetric algebra, 388
symmetry
  of the Hamiltonian, 441

tempered distribution, 137
tensor algebra, 334
tensor product
  algebraic, 328
  analytic, 328
  antisymmetric, 388

tensor product, cont.
as state space of multi-component systems, 329
of representations, 332
symmetric, 386
to incorporate intrinsic observables, 288
test function, 137
time evolution, 26, 424
trace, 365
trace class operator, 376
transition amplitude, 512
translation representation, 437
Trotter product formula, 491

uncertainty principle
Fourier, 132
general, 129
Heisenberg, 125
unitary evolution, 424

vacuum state, 400
variance, 42
variational derivative, 84
von Neumann, 17, 29, 31, 36
impossibility of hidden variables, 17, 341

wave function, 9
wave mechanics, 25, 26
wedge product, 384, 389
Weyl
algebra, 123
relations, 121, 123
sequence, 62

Yukawa potential, 246

Zeeman effect, 237

Printed in the United States
by Baker & Taylor Publisher Services